- iron out misconceptions or problems that arose during the lesson
- recap on new vocabulary introduced
- start an investigation or problem-solving activity so that pupils can continue it
- set the homework for the lesson
- use a combination of these.

Group work

There are many opportunities for you to set up group work using the Pupil Book. What you choose should reflect the abilities and level of independence of your class. What may work well with one class may not work well with another.

Here are some ideas of the choices you could make.

You could:

- Use some of the questions in the exercise for group or pair work. Choose questions that are nearer the end of the exercise as a general rule, as the first few are usually intended for skill development and consolidation of key points.
- Use a Discussion or Practical and ask pupils to report their results back as part of the plenary.
- Use an Investigation or Puzzle and ask pupils to explain what strategies they used as part of the plenary.
- Ask a group to make up their own question or puzzle similar to one in the Pupil Book and give it to the rest of the class to do.
- Adapt or use the **Starter Activity** as group work, especially if there are several ideas given or you used a starter activity of your own.
- Use the extension ideas given in each section as group work.

Planning charts

This section also includes comprehensive planning charts for each chapter. They show the links of the Programme of Study and the *Framework for Teaching Mathematics* to the book. National Curriculum levels are also provided. Pupil Book page references are provided for each objective.

Homework

This is a comprehensive resource providing Homework support for every exercise in the Pupil Book. Answers are included. Review questions, Investigations and or Puzzles and Practicals from the pupil books can also be used.

Key Objectives

Year 8

◆ **Add, subtract, multiply and divide integers.**
Chapter 2 Integers, Powers and Roots
Adding and subtracting integers, page 36
Multiplying and dividing integers, page 39
Framework examples pages 48–51

◆ **Use the equivalence of fractions, decimals and percentages to compare proportions; calculate percentages and find the outcome of a given percentage increase or decrease.**
Chapter 5 Fractions
Fractions and decimals, page 102
Ordering fractions, page 105
Chapter 6 Percentages, Fractions, Decimals
Converting fractions, decimals and percentages, page 119
Percentage of — mentally, page 123
Percentage of — written and calculator methods, page 125
Percentage increase and decrease, page 127
Mixed percentage problems, page 131
Framework examples pages 70–77

◆ **Divide a quantity into two or more parts in a given ratio; use the unitary method to solve simple word problems involving ratio and direct proportion.**
Chapter 7 Ratio and Proportion
Direct proportion, page 137
Simplifying ratios, page 139
Ratio and proportion, page 141
Solving ratio and proportion problems, page 143
Dividing in a given ratio, page 145
Framework examples pages 78–81

◆ **Use standard column procedures for multiplication and division of integers and decimals, including by decimals such as 0·6 or 0·06; understand where to position the decimal point by considering equivalent calculations.**
Chapter 4 Written and Calculator Calculation
Multiplying, page 83
Dividing, page 85
Dividing by a decimal, page 88
Framework examples pages 104–107

◆ **Simplify or transform linear expressions by collecting like terms; multiply a single term over a bracket.**
Chapter 8 Expressions, Formulae and Equations
Brackets, page 166
Collecting like terms, page 168
Framework examples pages 116–119

◆ **Substitute integers into simple formulae.**
Chapter 8 Expressions, Formulae and Equations
Substituting into expressions, page 172
Substituting into formulae, page 175
Framework examples pages 138–143

◆ **Plot the graphs of linear functions, where y is given explicitly in terms of x; recognise that equations of the form $y = mx + c$ correspond to straight-line graphs.**
Chapter 10 Graphs
Graphing functions, page 233
Equations of straight-line graphs, page 236
Framework examples pages 164–167

◆ **Identify alternate and corresponding angles; understand a proof that the sum of the angles of a triangle is 180° and of a quadrilateral is 360°.**
   Chapter 11 Lines and Angles
   Angles made with intersecting and parallel lines, page 272
   Using geometrical reasoning to find angles, page 274
   Angles in triangles, page 277
   Angles in quadrilaterals, page 281
   Framework examples pages 178–183

◆ **Enlarge 2-D shapes, given a centre of enlargement and a positive whole-number scale factor.**
   Chapter 13 Coordinates and Transformations
   Enlargement, page 320
   Framework examples pages 212–215

◆ **Use straight edge and compasses to do standard constructions.**
   Chapter 12 Shape, Construction and Loci
   Construction, page 301
   Constructing triangles and quadrilaterals, page 304
   Framework examples pages 220–223

◆ **Deduce and use formulae for the area of a triangle and parallelogram, and the volume of a cuboid; calculate volumes and surface areas of cuboids.**
   Chapter 14 Measures, Perimeter, Area and Volume
   Perimeter and area, page 348
   Surface area and volume, page 355
   Framework examples pages 234–241

◆ **Construct, on paper and using ICT, a range of graphs and charts; identify which are most useful in the context of a problem.**
   Chapter 15 Data Collection
   Two-way tables, page 374
   Framework examples page 255
   Chapter 16 Analysing Data. Drawing and Interpreting Graphs
   Finding the median, range and mode from a stem-and-leaf diagram, page 388
   Framework examples page 259
   Compound bar charts and line graphs, page 391
   Frequency diagrams, page 395
   Drawing pie charts, page 397
   Scatter graphs, page 400
   Surveys, page 407
   Framework examples pages 262–267

◆ **Find and record all possible mutually exclusive outcomes for single events and two successive events in a systematic way.**
   Chapter 17 Probability
   Calculating probability, page 416
   Calculating probability by listing outcomes, page 418
   Framework examples pages 278–281

◆ **Identify the necessary information to solve a problem; represent problems and interpret solutions in algebraic, geometric or graphical form.**
   Chapter 14 Measures, Perimeter, Area and Volume
   Estimating, page 343
   Practical question 1, page 344
   Chapter 9 Sequences and Functions
   Sequences in practical situations, page 213
   Exercise 8 question 5, page 214
   The specific material listed above covers the examples given in the Framework examples page 27.
   Many other questions in New National Framework Mathematics also cover this objective.
   Framework examples pages 26–27

◆ **Use logical argument to establish the truth of a statement.**

Using and applying mathematics to solve problems

2–35 Applying mathematics and solving problems

2–25 • Solve more demanding problems and investigate in a range of contexts: number, algebra, shape, space and measures, and handling data; compare and evaluate solutions.
All of the examples listed in the Framework examples pages 2–25 are covered in the Pupil Book. They are integrated throughout the relevant sections.

26–7 • **Identify the necessary information to solve a problem; represent problems and interpret solutions in algebraic, geometric or graphical form,** using correct notation and appropriate diagrams.
Chapter 14 Measures, Perimeter, Area and Volume
Estimating   page 343
Practical question 1   page 344
Chapter 9 Sequences and Functions
Sequences in practical situations   page 213
Exercise 8 question 5   page 214

28–9 • Solve more complex problems by breaking them into smaller steps or tasks, choosing and using efficient techniques for calculation, algebraic manipulation and graphical representation, and resources, including ICT.
Chapter 2 Integers, Powers and Roots
Squares and square roots   page 51
Puzzle question 3   page 54
Cubes and cube roots   page 55
Puzzle   page 57
Chapter 14 Measures, Perimeter, Area and Volume
Surface area and volume   page 355
Exercise 10 Review 5   page 356
Chapter 6 Percentages, Fractions, Decimals
Percentage increase and decrease   page 127
Discussion   page 127

30–1 • **Use logical argument to establish the truth of a statement;** give solutions to an appropriate degree of accuracy in the context of the problem.
Chapter 2 Integers, Powers and Roots
Divisibility   page 44
Investigation: Consecutive Numbers   page 46
Chapter 14 Measures, Perimeter, Area and Volume
Surface area and volume   page 355
Investigation: Making boxes is similar   page 358
Chapter 7 Ratio and Proportion
Solving ratio and proportion problems   page 143
Exercise 4 questions 9 and Review 4   page 144

32–5 • Suggest extensions to problems, conjecture and generalise; identify exceptional cases or counter-examples.
Chapter 2 Integers, Powers and Roots
Prime factor decomposition   page 47
Investigation: Factors question 4   page 49
Chapter 14 Measures, Perimeter, Area and Volume
Perimeter and area   page 348
Investigation: Triangles in Cubes   page 353
Exercise 7 question 15   page 350
Chapter 9 Sequences and Functions
Sequences in practical situations   page 213
Exercise 8 question 7   page 214
Chapter 12 Shape, Construction and Loci
Properties of triangles, quadrilaterals and polygons   page 287
Exercise 2 question 6 and review 3   page 289

Numbers and the number system

36–47 Place value, ordering and rounding

36–9 • Read and write positive integer powers of 10; multiply and divide integers and decimals by 0·1, 0·01.
Chapter 1 Place Value, Ordering and Rounding
Powers of ten   page 16
Adding and subtracting multiples of 0·1, 0·01 and 0·001   page 17
× and ÷ by multiples of 10, 100 and 1000   page 19
Multiplying and dividing by 0·1 and 0·01   page 20

40–1 • Order decimals.
Chapter 1 Place Value, Ordering and Rounding
Putting decimals in order   page 24

42–5 • Round positive numbers to any given power of 10; round decimals to the nearest whole number or to one or two decimal places.
Chapter 1 Place Value, Ordering and Rounding
Rounding to powers of ten   page 28
Rounding to decimal places   page 30
Rounding is used in many other sections of the Pupil Book.

48–59 Integers, powers and roots

48–51 • **Add, subtract, multiply and divide integers.**
Chapter 2 Integers, Powers and Roots
Adding and subtracting integers   page 36
Multiplying and dividing integers   page 39

52–5 • Recognise and use multiples, factors (divisors), common factor, highest common factor, lowest common multiple and

primes; find the prime factor decomposition of a number (e.g. $8000 = 2^6 \times 5^3$).
Chapter 2 Integers, Powers and Roots
Divisibility   page 44
Prime factor decomposition   page 47
Highest common factor and lowest common multiple   page 50

56–9 • Use squares, positive and negative square roots, cubes and cube roots, and index notation for small positive integer powers.
Chapter 1 Place Value, Ordering and Rounding
Powers of 10   page 16
Chapter 2 Integers, Powers and Roots
Squares and square roots   page 51
Cubes and cube roots   page 55

60–81 Fractions, decimals, percentages, ratio and proportion

60–5 • Know that a recurring decimal is a fraction; use division to convert a fraction to a decimal; order fractions by writing them with a common denominator or by converting them to decimals.
Chapter 5 Fractions
Fractions of shapes   page 98
One number as a fraction of another   page 99
Fractions and decimals   page 102
Ordering fractions   page 105

66–9 • Add and subtract fractions by writing them with a common denominator; calculate fractions of quantities (fraction answers); multiply and divide an integer by a fraction.
Chapter 5 Fractions
Adding and subtracting fractions   page 108
Fraction of — multiplying an integer by a fraction   page 111
Dividing an integer by a fraction   page 113

70–7 • Interpret percentage as the operator 'so many hundredths of' and express one given number as a percentage of another; **use the equivalence of fractions, decimals and percentages to compare proportions; calculate percentage and find the outcome of a given percentage increase or decrease.**
Chapter 6 Percentages, Fractions, Decimals
Converting fractions, decimals and percentages   page 119
Percentage of — mentally   page 123
Percentage of — written and calculator   page 125
Percentage increase and decrease   page 127
Mixed percentage problems   page 131

78–81 • Consolidate understanding of the relationship between ratio and proportion; reduce a ratio to its simplest form, including a ratio expressed in different units, recognising links with fraction notation; **divide a quantity into two or more parts in a given ratio; use the unitary method to solve simple word problems involving ratio and direct proportion.**
Chapter 7 Ratio and Proportion
Direct proportion   page 137
Simplifying ratios   page 139
Ratio and proportion   page 141
Solving ratio and proportion problems   page 143
Dividing in a given ratio   page 145

Calculations

82–7 Number operations and the relationships between them

82–5 • Understand addition and subtraction of fractions and integers, and multiplication and division of integers; use the laws of arithmetic and inverse operations.
These principles are developed throughout Chapters 1 to 5 and in algebra in Chapter 8.

86–7 • Use the order of operations, including brackets, with more complex calculations.
Chapter 2 Integers, Powers and Roots
Order of operations with integers   page 43
Chapter 4 Written and Calculator Calculation
Using brackets on a calculator   page 92

88–103 Mental methods and rapid recall of number facts

88–91 • Recall known facts, including fraction to decimal conversions; use known facts to derive unknown facts, including products involving numbers such as 0·7 and 6, and 0·03 and 8.
Chapter 1 Place Value, Ordering and Rounding
Multiplying and dividing by 0.1 and 0.01   page 20
Chapter 3 Mental Calculation
Adding and subtracting   page 62
Multiplying and dividing   page 66
Chapter 5 Fractions
Fractions and decimals   page 102

92–101 • Consolidate and extend mental methods of calculation, working with decimals, fractions and percentages, squares and square roots, cubes and cube roots; solve word problems mentally.
Chapter 2 Integers, Powers and Roots
Square and square roots   page 51
Cubes and cube roots   page 55

190–1 • Know that if two 2-D shapes are congruent, corresponding sides and angles are equal.
Chapter 12 Shape, Construction and Loci
Congruence   page 293

198–201 • Know and use geometric properties of cuboids and shapes made from cuboids; begin to use plans and elevations.
Chapter 12 Shape, Construction and Loci
Describing and sketching 3-D shapes. Constructing nets   page 295
Plans and elevations   page 297

202–17 Transformations

202–11 • Transform 2-D shapes by simple combinations of rotations, reflections and translations, on paper and using ICT; identify all the symmetries of 2-D shapes.
Chapter 13 Coordinates and Transformations
Combinations of transformations   page 314
Symmetry   page 317

212–15 • Understand and use the language and notation associated with enlargement; **enlarge 2-D shapes, given a centre of enlargement and a positive whole-number scale factor;** explore enlargement using ICT.
Chapter 13 Coordinates and Transformations
Enlargement   page 320

216–17 • Make simple scale drawings.
Chapter 13 Coordinates and Transformations
Scale drawing   page 324

218–19 Coordinates

218–19 • Given the coordinates of points A and B, find the mid-point of the line segment AB.
Chapter 13 Coordinates and Transformations
Coordinates and transformations   page 313
Finding the mid-point of a line   page 328

220–7 Construction and loci

220–3 • **Use straight edge and compasses to construct:**
– the mid-point and perpendicular bisector of a line segment;
– the bisector of an angle;
– the perpendicular from a point to a line;
– the perpendicular from a point on a line;
construct a triangle, given three sides (SSS); use ICT to explore these constructions.
Chapter 12 Shape, Construction and Loci
Construction   page 301
Constructing triangles and quadrilaterals   page 304

224–7 • Find simple loci, both by reasoning and by using ICT, to produce shapes and paths, e.g. an equilateral triangle.
Chapter 12 Shape, Construction and Loci
Loci   page 307

228–41 Measures and mensuration

228–31 • Use units of measurement to estimate, calculate and solve problems in everyday contexts involving length, area, volume, capacity, mass, time, angle and bearings; know rough metric equivalents of imperial measures in daily use (feet, miles, pounds, pints, gallons).
Chapter 14 Measures, Perimeter, Area and Volume
Metric conversions   page 335
Metric and imperial equivalents   page 339
Units, measuring instruments and accuracy   page 340
Estimating   page 343

232–3 • Use bearings to specify direction.
Chapter 14 Measures, Perimeter, Area and Volume
Bearings   page 344

234–7 • **Deduce and use formulae for the area of a triangle, parallelogram** and trapezium; calculate areas of compound shapes made from rectangles and triangles.
Chapter 14 Measures, Perimeter, Area and Volume
Perimeter and area   page 348

238–41 • **Know and use the formula for the volume of a cuboid; calculate volumes and surface areas of cuboids** and shapes made from cuboids.
Chapter 14 Measures, Perimeter, Area and Volume
Surface area and volume   page 355

Handling data

248–55 Specifying a problem, planning and collecting data

248–9 • Discuss a problem that can be addressed by statistical methods and identify related questions to explore.
Chapter 15 Data Collection
Surveys — the problem or question   page 376

250–1 • Decide which data to collect to answer a question, and the degree of accuracy needed; identify possible sources.
Chapter 15 Data Collection
Collecting the data   page 377

252–5 • Plan how to collect the data, including sample size; construct frequency tables with given equal class intervals for sets of continuous data; design and use two-way tables for discrete data.
Chapter 15 Data Collection
Discrete and continuous data   page 372
Grouping continuous data   page 373

Two-way tables   page 374
Planning a survey   page 378

254–5 • Collect data using a suitable method, such as observation, controlled experiment, including data logging using ICT, or questionnaire.
Chapter 15 Data Collection
Planning a survey   page 378

256–67 Processing and representing data, using ICT as appropriate

256–61 • Calculate statistics, including with a calculator; recognise when it is appropriate to use the range, mean, median and mode and, for grouped data, the modal class; calculate a mean using an assumed mean; construct and use stem-and-leaf diagrams.
Chapter 16 Analysing Data. Drawing and Interpreting Graphs
Mode and range   page 382
Mean   page 383
Median, mean, mode, range   page 385
Finding the median, range and mode from a stem-and-leaf diagram   page 388

262–7 • **Construct, on paper and using ICT:**
– pie charts for categorical data;
– bar charts and frequency diagrams for discrete and continuous data;
– simple line graphs for time series;
– simple scatter graphs;
identify which are most useful in the context of the problem.
Chapter 16 Analysing Data. Drawing and Interpreting Graphs
Compound bar charts and line graphs   page 391
Frequency diagrams   page 395
Drawing pie charts   page 397
Scatter graphs   page 400

268–75 Interpreting and discussing results

268–71 • Interpret tables, graphs and diagrams for both discrete and continuous data, and draw inferences that relate to the problem being discussed; relate summarised data to the questions being explored.
Chapter 16 Analysing Data. Drawing and Interpreting Graphs
Interpreting graphs and tables   page 402

272–3 • Compare two distributions using the range and one or more of the mode, median and mean.
Chapter 16 Analysing Data. Drawing and Interpreting Graphs
Comparing data   page 389

272–3 • Communicate orally and on paper the results of a statistical enquiry and the methods used, using ICT as appropriate; justify the choice of what is presented.
Chapter 16 Analysing Data. Drawing and Interpreting Graphs
Surveys   page 407

276–85 Probability

276–7 • Use the vocabulary of probability when interpreting the results of an experiment; appreciate that random processes are unpredictable.
Chapter 17 Probability
Language of probability   page 414

278–81 • Know that if the probability of an event occurring is p, then the probability of it not occurring is $1 - p$; **find and record all possible mutually exclusive outcomes for single events and two successive events in a systematic way,** using diagrams and tables.
Chapter 17 Probability
Calculating probability   page 416
Calculating probability by listing outcomes   page 418

282–3 • Estimate probabilities from experimental data; understand that:
– if an experiment is repeated there may be, and usually will be, different outcomes;
– increasing the number of times an experiment is repeated generally leads to better estimates of probability.
Chapter 17 Probability
Estimating probability from experiments   page 420

284–5 • Compare experimental and theoretical probabilities in different contexts.
Chapter 17 Probability
Comparing calculated probability with experimental probability   page 423

NOTES • Key objectives are highlighted in **bold type**.
• Page references are to the supplement of examples for Years 7, 8 and 9.

Mapping to the sample medium-term plans

YEAR 8: AUTUMN TERM

Teaching objectives for the oral and mental activities

These objectives are covered by using the Mental Starter Ideas given under that section heading in the Teacher's Notes section of this Teacher Planning Pack.

- Order, add, subtract, multiply and divide integers.
- Multiply and divide decimals by 10, 100, 1000.
- Count on and back in steps of 0·4, 0·75, $\frac{3}{4}$ ...
- Round numbers, including to one or two decimal places.
- Know and use squares, positive and negative square roots, cubes of numbers 1 to 5 and corresponding roots.
- Convert between fractions, decimals and percentages.
- Find fractions and percentages of quantities.

- Know or derive complements of 0·1, 1, 10, 50, 100, 1000.
- Add and subtract several small numbers or several multiples of 10, e.g. 250 + 120 − 190.
- Use jottings to support addition and subtraction of whole numbers and decimals.
- Calculate using knowledge of multiplication and division facts and place value, e.g. 432 × 0·01, 37 ÷ 0·01.
- Recall multiplication and division facts to 10 × 10.
- Use factors to multiply and divide mentally, e.g. 22 × 0·02, 420 ÷ 15.

- Multiply and divide a two-digit number by a one-digit number.
- Use partitioning to multiply, e.g. 13 × 1·4.
- Use approximations to estimate the answers to calculations, e.g. 39 × 2·8.

- Solve equations, e.g. 3*a* − 2 = 31.

- Visualise, describe and sketch 2-D shapes.
- Estimate and order acute, obtuse and reflex angles.

- Use metric units (length, mass, capacity) and units of time for calculations.
- Use metric units for estimation (length, mass, capacity).
- Convert between m, cm and mm, km and m, kg and g, litres and ml, cm² and mm².

- Discuss and interpret graphs.

- Apply mental skills to solve simple problems.

Notes

1   Under each objective, the relevant sections from *New National Framework Mathematics* are listed. It is not intended that all the material in each section is covered by all pupils. You will need to choose work appropriate for your pupils, teaching programme and teaching style. There is plenty of choice.

2   Some of the Year 8 teaching programme objectives are revisited more than once in this medium-term plan. Each time the objective is given, **all** of the sections from *New National Framework Mathematics* that cover this objective are listed. You will need to split the material to suit your class and teaching programme.

Teaching objectives for the main activities

	SUPPORT	CORE	EXTENSION
	From the Y7 teaching programme. *This is covered by the 8* book, the support section in the 8 Core book and the worksheets from the Teacher Resource Pack, as listed.*	From the Y8 teaching programme.	From the Y9 teaching programme. *This is covered by the 8 Plus book.*
Number/algebra 1 (6 hours) Integers, powers and roots (48–59)	● Understand negative numbers as positions on a number line; order, add and subtract positive and negative integers in context. *Worksheet 6*	● **Add, subtract, multiply and divide integers.** ***Chapter 2 Integers, Powers and Roots*** *Adding and subtracting integers page 36* *Multiplying and dividing integers page 39*	

- Use the prime factor decomposition of a number.

- Use ICT to estimate square roots and cube roots.
- Use index notation for integer powers and simple instances of the index laws.

Sequences and functions

(144–157)

● Use simple tests of divisibility. *Worksheet 5*	● Recognise and use multiples, factors (divisors), common factor, highest common factor, lowest common multiple and primes; find the prime factor decomposition of a number (e.g. $8000 = 2^6 \times 5^3$). **Chapter 2 Integers, Powers and Roots** *Divisibility page 44* *Prime factor decomposition page 47* *Highest common factor and lowest common multiple page 50*
● Recognise the first few triangular numbers, squares of numbers to at least 12×12 and the corresponding roots. *Worksheet 7*	● Use squares, positive and negative square roots, cubes and cube roots, and index notation for small positive integer powers. **Chapter 1 Place Value, Ordering and Rounding** *Powers of ten page 16* **Chapter 2 Integers, Powers and Roots** *Squares and square roots page 51* *Cubes and cube roots page 55* ● Generate and describe integer sequences. **Chapter 9 Sequences and Functions** *Writing sequences from flow charts page 200* *Generating sequences by multiplying and dividing page 201* *Counting forwards or backwards in increasing or decreasing steps page 203* *Showing sequences with geometric patterns page 204* *Arithmetic sequences page 205* *Predicting the next few terms of a sequence page 207*
● Generate terms of a simple sequence given a rule. *Worksheet 19*	● Generate terms of a linear sequence using term-to-term and position-to-term definitions of the sequence, on paper and using a spreadsheet or graphical calculator. **Chapter 9 Sequences and Functions** *Writing sequences from term-to-term rules and rules for the nth term page 208* *Describing linear sequences page 211*
● Generate sequences from practical contexts and describe the general term in simple cases. *Worksheet 19*	● Begin to use linear expressions to describe the nth term of an arithmetic sequence, justifying its form by referring to the activity or practical context from which it was generated. **Chapter 9 Sequences and Functions** *Sequences in practical situations page 213* *Finding the rule for the nth term page 217*

Mapping to the sample medium-term plans

		SUPPORT From the Y7 teaching programme. *This is covered by the 8* book, the support section in the 8 Core book and the worksheets from the Teacher Resource Pack, as listed.*	CORE From the Y8 teaching programme.	EXTENSION From the Y9 teaching programme. *This is covered by the 8 Plus book.*
Shape, space and measures 1 (6 hours) Geometrical reasoning: lines, angles and shapes (178–189)		● Use correctly the vocabulary, notation and labelling conventions for lines, angles and shapes. *Worksheets 23, 24 and 26* ● **Identify parallel and perpendicular lines; know the sum of angles at a point, on a straight line and in a triangle,** and recognise vertically opposite angles. *Worksheets 23 and 25* ● Use angle measure; distinguish between and estimate the size of acute, obtuse and reflex angles. *Worksheet 24*	● **Identify alternate angles and corresponding angles;** *Chapter 11 Lines and Angles* *Angles made with intersecting and parallel lines page 272* *Using geometrical reasoning to find angles page 274* **understand a proof that:** – **the sum of the angles of a triangle is 180° and of a quadrilateral is 360°;** *Chapter 11 Lines and Angles* *Angles in triangles page 277* *Angles in quadrilaterals page 281* – the exterior angle of a triangle is equal to the sum of the two interior opposite angles. *Chapter 11 Lines and Angles* *Angles in triangles page 277* ● Solve geometrical problems using side and angle properties of equilateral, isosceles and right-angled triangles and special quadrilaterals, explaining reasoning with diagrams and text; classify quadrilaterals by their geometric properties. *Chapter 11 Lines and angles* *Angles in triangles page 277* *Chapter 12 Shape, Construction and Loci* *Visualising and sketching 2-D shapes page 286* *Properties of triangles, quadrilaterals and polygons page 287* *Tessellations page 291*	● Explain how to find, calculate and use: – the sums of the interior and exterior angles of quadrilaterals, pentagons and hexagons; – the interior and exterior angles of regular polygons. ● **Solve problems using properties of angles, of parallel and intersecting lines, and of triangles and other polygons.**

Mapping to the sample medium-term plans

	SUPPORT	CORE	EXTENSION

Construction (220–223)	● Use a ruler and protractor to:
	  – measure and draw lines to the nearest millimetre and angles, including reflex angles, to the nearest degree;
	  Worksheet 24
	  – construct a triangle given two sides and the included angle (SAS) or two angles and the included side (ASA).
	  Worksheet 27
		● **Use straight edge and compasses to construct:**
		  – **the mid-point and perpendicular bisector of a line segment;**
		  – **the bisector of an angle;**
		  – **the perpendicular from a point to a line;**
		  – **the perpendicular from a point on a line.**
		  Chapter 12 Shape, Construction and Loci
		  Construction page 301
		
		● Investigate in a range of contexts: shape and space.
		  Throughout the above sections
			● Know the definition of a circle and the names of its parts.
			● Use straight edge and compasses to construct a triangle, given right angle, hypotenuse and side (RHS).

Solving problems (14–17)			

Handling data 1 (6 hours)
Probability
(276–283)
	SUPPORT
	From the Y7 teaching programme.
	This is covered by the 8* book, the support section in the 8 Core book and the worksheets from the Teacher Resource Pack, as listed.
	
	● **Understand and use the probability scale from 0 to 1; find and justify probabilities based on equally likely outcomes in simple contexts.**
	Worksheet 39
	
	● Collect data from a simple experiment and record in a frequency table; estimate probabilities based on this data.
	Worksheet 35
		CORE
		From the Y8 teaching programme.
		
		● Use the vocabulary of probability when interpreting the results of an experiment; appreciate that random processes are unpredictable.
		Chapter 17 Probability
		  Language of probability page 414
		● Know that if the probability of an event occurring is p, then the probability of it not occurring is $1 - p$; **find and record all possible mutually exclusive outcomes for single events and two successive events in a systematic way,** using diagrams and tables.
		Chapter 17 Probability
		  Calculating probability page 416
		  Calculating probability by listing outcomes page 418
		● Estimate probabilities from experimental data; understand that:
		  – if an experiment is repeated there may be, and usually will be, different outcomes;
		  – increasing the number of times an experiment is repeated generally leads to better estimates of probability.
		Chapter 17 Probability
		  Estimating probability from experiments page 420
			EXTENSION
			From the Y9 teaching programme.
			This is covered by the 8 Plus book.
			
			● Identify all the mutually exclusive outcomes of an experiment; **know that the sum of probabilities of all mutually exclusive outcomes is 1 and use this when solving problems.**
			
			● Compare experimental and theoretical probabilities in a range of contexts; appreciate the difference between mathematical explanation and experimental evidence.

	SUPPORT	CORE	EXTENSION
Number 2 (6 hours) Fractions, decimals, percentages (60–77)	From the Y7 teaching programme. *This is covered by the 8* book, the support section in the 8 Core book and the worksheets from the Teacher Resource Pack, as listed.*	From the Y8 teaching programme.	From the Y9 teaching programme. *This is covered by the 8 Plus book.*
	● Use fraction notation to express a smaller whole number as a fraction of a larger one; **simplify fractions by cancelling all common factors and identify equivalent fractions;** convert terminating decimals to fractions. *Worksheet 12*	● Know that a recurring decimal is a fraction; use division to convert a fraction to a decimal; order fractions by writing them with a common denominator or by converting them to decimals. **Chapter 5 Fractions** *Fractions of shapes page 98* *One number as a fraction of another page 99* *Fraction and decimals page 102* *Ordering fractions page 106*	
	● Add and subtract fractions with common denominators; calculate fractions of quantities (whole-number answers); multiply a fraction by an integer. *Worksheet 12*	● Add and subtract fractions by writing them with a common denominator; calculate fractions of quantities (fraction answers); multiply and divide an integer by a fraction. **Chapter 5 Fractions** *Adding and subtracting fractions page 108* *Fraction of — multiplying an integer by a fraction page 111* *Dividing an integer by a fraction page 113*	● Use efficient methods to **add, subtract, multiply and divide fractions,** interpreting division as a multiplicative inverse; cancel common factors before multiplying or dividing.
	● Understand percentage as the 'number of parts per 100'; calculate simple percentages. *Worksheet 13*	● Interpret percentage as the 'number of hundredths of' and express one given number as a percentage of another; **use the equivalence of fractions, decimals and percentages to compare proportions; calculate percentages and find the outcome of a given percentage increase or decrease.** **Chapter 6 Percentages, Fractions, Decimals** *Converting fractions, decimals and percentages page 119* *Percentage of — mentally page 123* *Percentage of — written and calculator methods page 125* *Percentage increase and decrease page 127* *Mixed percentage problems page 131*	● Solve problems involving percentage changes.
Calculations (82–85, 88–101)		● Understand addition and subtraction of fractions; use the laws of arithmetic and inverse operations. **Chapter 5 Fractions** *Adding and subtracting fractions page 108*	

Mapping to the sample medium-term plans

- Consolidate the rapid recall of number facts, including positive integer complements to 100 and multiplication facts to 10 × 10, and quickly derive associated division facts.
  Worksheet 8

- Recall known facts, including fraction to decimal conversions; use known facts to derive unknown facts, including products such as 0·7 and 6, and 0·03 and 8.
  Chapter 1 Place Value, Ordering and Rounding
  Multiplying and dividing by 0·1 and 0·01   page 20
  Chapter 3 Mental Calculation
  Multiplying and dividing   page 66
  Chapter 5 Fractions
  Fractions and decimals   page 102

- Consolidate and extend mental methods of calculation, working with decimals, fractions and percentages; solve word problems mentally.
  Chapter 3 Mental Calculation
  Chapter 5 Fractions
  Chapter 6 Percentages, Fractions, Decimals
  Note: All of these chapters have sections which cover this objective.

Algebra 2 (6 hours)
Equations and formulae
(112–119, 138–143)

SUPPORT
From the Y7 teaching programme.
This is covered by the 8* book, the support section in the 8 Core book and the worksheets from the Teacher Resource Pack, as listed.

- **Use letter symbols to represent unknown numbers or variables;** know the meanings of the words *term, expression* and *equation.*
  Worksheet 15

- Simplify linear algebraic expressions by collecting like terms.
  Worksheet 16

CORE
From the Y8 teaching programme.

- Begin to distinguish the different roles played by letter symbols in equations, formulae and functions; know the meanings of the words *formula* and *function.*
  Chapter 8 Expressions, Formulae and Equations
  Understanding algebra   page 158
- Know that algebraic operations follow the same conventions and order as arithmetic operations; use index notation for small positive integer powers.
  Chapter 8 Expressions, Formulae and Equations
  Understanding algebra   page 158
  More understanding algebra   page 160
  Simplifying expressions — multiplying and dividing   page 164
- **Simplify or transform linear expressions by collecting like terms; multiply a single term over a bracket.**

EXTENSION
From the Y9 teaching programme.
This is covered by the 8 Plus book.

- Use index notation for integer powers and simple instances of the index laws.

- Simplify or transform algebraic expressions by taking out single term common factors.

Mapping to the sample medium-term plans

		Chapter 8 Expressions, Formulae and Equations
		  Brackets page 166
		  Collecting like terms page 168
		  Writing expressions page 178
		  More writing and simplifying expressions page 179
		• Use formulae from mathematics and other subjects; **substitute integers into simple formulae**, and positive integers into expressions involving small powers (e.g. $3x^2 + 4$ or $2x^3$); derive simple formulae.
		Chapter 8 Expressions, Formulae and Equations
		  Substituting into expressions page 172
		  Substituting into formulae page 175
		  Writing and finding formulae page 182	

	SUPPORT From the Y7 teaching programme. *This is covered by the 8* book, the support section in the 8 Core book and the worksheets from the Teacher Resource Pack, as listed.*	**CORE** From the Y8 teaching programme.	**EXTENSION** From the Y9 teaching programme. *This is covered by the 8 Plus book.*

Shape, space and measures 2
(6 hours)
Measures and mensuration
(228–231, 234–241)
	• **Convert one metric unit to another** (e.g. grams to kilograms); **read and interpret scales on a range of measuring instruments.**
	Worksheet 32
		• Use units of measurement to estimate, calculate and solve problems in everyday contexts involving length, area, volume, capacity, mass, time and angle; know rough metric equivalents of imperial measures in daily use (feet, miles, pounds, pints, gallons).
		Chapter 14 Measures, Perimeter, Area and Volume
		  Metric conversions page 335
		  Metric and imperial equivalents page 339
		  Units, measuring instruments and accuracy page 340
		  Estimating page 343
			• Convert between area measures (mm^2 to cm^2, cm^2 to m^2, and vice versa) and between volume measures (mm^3 to cm^3, cm^3 to m^3, and vice versa).

	• Know and use the formula for the area of a rectangle; calculate the perimeter and area of shapes made from rectangles. *Worksheet 33*	• **Deduce and use formulae for the area of a triangle, parallelogram** and trapezium; calculate areas of compound shapes made from rectangles and triangles. ***Chapter 14 Measures, Perimeter, Area and Volume*** *Perimeter and area page 348*	• **Know and use the formulae for the circumference and area of a circle.**
	• Calculate the surface area of cubes and cuboids. *Worksheet 33*	• **Know and use the formula for the volume of a cuboid; calculate volumes and surface areas of cuboids** and shapes made from cuboids. ***Chapter 14 Measures, Perimeter, Area and Volume*** *Surface area and volume page 355*	• Calculate the surface area and volume of right prisms.
Solving problems (18–21)		• Investigate in a range of contexts: measures. *Throughout the above sections*	

YEAR 8: SPRING TERM

Teaching objectives for the oral and mental activities

These objectives are covered by using the Mental Starter Ideas given under that section heading in the Teachers' Notes section of this Teacher Planning Pack.

- Order, add, subtract, multiply and divide integers.
- Round numbers, including to one or two decimal places.
- Know and use squares, positive and negative square roots, cubes of numbers 1 to 5 and corresponding roots.
- Know or derive quickly prime numbers less than 30.
- Convert between improper fractions and mixed numbers.
- Find the outcome of a given percentage increase or decrease.

- Know complements of 0·1, 1, 10, 50, 100, 1000.
- Add and subtract several small numbers or several multiples of 10, e.g. $250 + 120 - 190$.
- Calculate using knowledge of multiplication and division facts and place value, e.g. $432 \times 0·01$, $37 \div 0·01$, $0·04 \times 8$, $0·03 \div 5$.
- Recall multiplication and division facts to 10×10.
- Use factors to multiply and divide mentally, e.g. $22 \times 0·02$, $420 \div 15$.
- Multiply and divide a two-digit number by a one-digit number.
- Multiply by near 10s, e.g. 75×29, $8 \times {}^{-}19$.
- Use partitioning to multiply, e.g. $13 \times 1·4$.

- Use approximations to estimate the answers to calculations, e.g. $39 \times 2·8$.
- Solve equations, e.g. $n(n - 1) = 56$.
- Visualise, describe and sketch 2-D shapes, 3-D shapes and simple loci.
- Estimate and order acute, obtuse and reflex angles.
- Use metric units (length, area and volume) and units of time for calculations.
- Use metric units for estimation (length, area and volume).
- Recall and use the formula for perimeter of rectangles and calculate areas of rectangles and triangles.
- Calculate volumes of cuboids.
- Discuss and interpret graphs.
- Apply mental skills to solve simple problems.

Notes

1 Under each objective, the relevant sections from *New National Framework Mathematics* are listed. It is not intended that all the material in each section is covered by all pupils. You will need to choose work appropriate for your pupils, teaching programme and teaching style. There is plenty of choice.

2 Some of the Year 8 teaching programme objectives are revisited more than once in this medium-term plan. Each time the objective is given, **all** of the sections from *New National Framework Mathematics* that cover this objective are listed. You will need to split the material to suit your class and teaching programme.

Teaching objectives for the main activities

	SUPPORT	CORE	EXTENSION
	From the Y7 teaching programme. *This is covered by the 8* book, the support section in the 8 Core book and the worksheets from the Teacher Resource Pack, as listed.*	From the Y8 teaching programme.	From the Y9 teaching programme. *This is covered by the 8 Plus book.*
Algebra 3 (6 hours) Sequences, functions, graphs	• Express simple functions in words. *Worksheet 20*	• Express simple functions in symbols; represent mappings expressed algebraically. **Chapter 9 Sequences and Functions** Functions *page 218* Finding the function given the input and output *page 221* Properties of Functions *page 224*	• Find the inverse of a linear function.

(160–177)	• Plot graphs of linear functions (y given implicitly in terms of x), e.g. $ay + bx = 0$, $y + bx + c = 0$, on paper and using ICT; **given values for m and c, find the gradient of lines given by equations of the form $y = mx + c$.**
	
	• Discuss and interpret distance–time graphs.
		• Generate points in all four quadrants and **plot the graphs of linear functions, where y is given explicitly in terms of x**, on paper and using ICT; **recognise that equations of the form $y = mx + c$ correspond to straight-line graphs.**
		Chapter 10 Graphs
		  Graphing functions page 233
		  Equations of straight-line graphs page 236
		• Construct linear functions arising from real-life problems and plot their corresponding graphs; discuss and interpret graphs arising from real situations.
		Chapter 10 Graphs
		  Reading and plotting real-life graphs page 240
		  Distance/time graphs page 246
		  Interpreting and sketching real-life graphs page 248
			• Generate coordinate pairs that satisfy a simple linear rule; recognise straight-line graphs parallel to the x-axis or y-axis.
			Worksheet 21

Number 3 (9 hours)
Place value
(36–47)
	EXTENSION
	From the Y9 teaching programme.
	This is covered by the 8 Plus book.
	• Extend knowledge of integer powers of 10; multiply and divide by any integer power of 10.
		CORE
		From the Y8 teaching programme.
		• Read and write positive integer powers of 10; multiply and divide integers and decimals by 0·1, 0·01.
		Chapter 1 Place Value, Ordering and Rounding
		  Powers of ten page 16
		  Adding and subtracting multiples of 0·1, 0·01 and 0·001 page 17
		  × and ÷ by multiples of 10, 100 and 1000 page 19
		  Multiplying and dividing by 0·1 and 0·01 page 20
		• Order decimals.
		Chapter 1 Place Value, Ordering and Rounding
		  Putting decimals in order page 24
		• Round positive numbers to any given power of 10: round decimals to the nearest whole number or to one or two decimal places.
		Chapter 1 Place Value, Ordering and Rounding
		  Rounding to powers of ten page 28
		  Rounding to decimal places page 30
			SUPPORT
			From the Y7 teaching programme.
			This is covered by the 8* book, the support section in the 8 Core book and the worksheets from the Teacher Resource Pack, as listed.
			• Understand and use decimal notation and place value; multiply and divide integers and decimals by 10, 100 and 1000, and explain the effect.
			Worksheets 1 and 2
			• Round positive whole numbers to the nearest 10, 100 or 1000 and decimals to the nearest whole number or one decimal place.
			Worksheet 4

SUPPORT
From the Y7 teaching programme.
This is covered by the 8* book, the support section in the 8 Core book and the worksheets from the Teacher Resource Pack, as listed.

Calculations (92–107, 110–111)	• Consolidate and **extend mental methods of calculation to include decimals, fractions and percentages**, accompanied where appropriate by suitable jottings. *Worksheets 8, 12 and 13*	• Consolidate and extend mental methods of calculation, working with decimals, squares and square roots, cubes and cube roots; solve word problems mentally. ***Chapter 2 Integers, Powers and Roots*** *Squares and square roots page 51* *Cubes and cube roots page 55* ***Chapter 3 Mental Calculation*** *Adding and subtracting page 62* *Multiplying and dividing page 66* *Solving problems mentally page 70* • Make and justify estimates and approximations of calculations. ***Chapter 3 Mental Calculation*** *Making estimates page 73* *Estimating answers to calculations page 74*	• Extend mental methods of calculation, working with decimals, fractions, percentages, factors, powers and roots.
		• Consolidate standard column procedures for addition and subtraction of integers and decimals with up to two places. ***Chapter 4 Written and Calculator Calculation*** *Adding and subtracting page 81*	• Use standard column procedures to add and subtract integers and decimals of any size, including a mixture of large and small numbers with differing numbers of decimal places.
	• **Multiply and divide three-digit by two-digit whole numbers; extend to multiplying and dividing decimals with one or two places by single-digit whole numbers.** *Worksheet 11*	• **Use standard column procedures for multiplication and division of integers and decimals, including by decimals such as 0·6 or 0·06; understand where to position the decimal point by considering equivalent calculations.** ***Chapter 4 Written and Calculator Calculation*** *Multiplying page 83* *Dividing page 85* *Dividing by a decimal page 88* • Check a result by considering whether it is of the right order of magnitude and by working the problem backwards. ***Chapter 4 Written and Calculator Calculation*** *Checking answers page 89*	• Multiply and divide by decimals, dividing by transforming to division by an integer.
Calculator methods (108–109)	• Carry out calculations with more than one step using brackets and the memory. *Worksheet 10*	• Carry out more difficult calculations effectively and efficiently using the function keys of a calculator for sign change, powers, roots and fractions; use brackets and the memory.	• Use a calculator efficiently and appropriately to perform complex calculations with numbers of any size, knowing not to round during intermediate steps of a calculation.

		Chapter 2 Integers, Powers and Roots *Integers on a calculator page 42* ***Chapter 4 Written and Calculator Calculation*** *Using brackets on a calculator page 92* *Using the calculator memory page 93* ***Chapter 5 Fractions*** *Fractions and decimals page 102* *Adding and subtracting fractions page 108* ● Enter numbers and interpret the display of a calculator in different contexts (negative numbers, fractions, decimals, percentages, money, metric measures, time). ***Chapter 2 Integers, Powers and Roots*** *Integers on a calculator page 42* ***Chapter 4 Written and Calculator Calculation*** *Using brackets on a calculator page 92* *Using the calculator memory page 93* ***Chapter 5 Fractions*** *Fractions and decimals page 102* *Adding and subtracting fractions page 108* *Practice at interpreting the calculator display is given throughout many sections.*	

	SUPPORT From the Y7 teaching programme. *This is covered by the 8* book, the support section in the 8 Core book and the worksheets from the Teacher Resource Pack, as listed.*	**CORE** From the Y8 teaching programme.	**EXTENSION** From the Y9 teaching programme. *This is covered by the 8 Plus book.*
Shape, space and measures 3 (6 hours) Geometrical reasoning: lines, angles and shapes (190–191) Transformations (202–215	● Recognise and visualise the transformation and symmetry of a 2-D shape: – reflection in given mirror lines, and line symmetry; – rotation about a given point, and rotation symmetry; – translation; explore these transformations and symmetries using ICT. *Worksheets 30 and 31*	● Know that if two 2-D shapes are congruent, corresponding sides and angles are equal. ***Chapter 12 Shape, Construction and Loci*** *Congruence page 293* ● Transform 2-D shapes by simple combinations of rotations, reflections and translations, on paper and using ICT; identify all the symmetries of 2-D shapes. ***Chapter 13 Coordinates and Transformations*** *Coordinates and transformations page 313* *Combinations of transformations page 314* *Symmetry page 317*	● **Know that translations, rotations and reflections preserve length and angle and map objects on to congruent images;** identify reflection symmetry in 3-D shapes.

Mapping to the sample medium-term plans

- Understand and use the language and notation associated with enlargement; **enlarge 2-D shapes, given a centre of enlargement and a positive whole-number scale factor;** explore enlargement using ICT.
  Chapter 13 Coordinates and Transformations
  Enlargement   page 320

- Enlarge 2-D shapes, given a centre of enlargement and a negative whole-number scale factor, on paper; identify the scale factor of an enlargement as the ratio of the lengths of any two corresponding line segments; recognise that enlargements preserve angle but not length, and understand the implications of enlargement for perimeter.
- **Use proportional reasoning to solve a problem:** interpret and use ratio in a range of contexts.

Ratio and proportion (78–81)

- Understand the relationship between ratio and proportion; solve simple problems about ratio and proportion using informal strategies.
  Worksheet 14

- Consolidate understanding of the relationship between ratio and proportion; reduce a ratio to its simplest form, including a ratio expressed in different units, recognising links with fraction notation.
  Chapter 7 Ratio and Proportion
  Direct proportion   page 137
  Simplifying ratios   page 139
  Ratio and proportion   page 141
  Solving ratio and proportion problems   page 143

Algebra 4 (6 hours)
Equations and formulae
(112–113, 122–125, 138–143)

SUPPORT
From the Y7 teaching programme.
This is covered by the 8* book, the support section in the 8 Core book and the worksheets from the Teacher Resource Pack, as listed.

- **Use letter symbols to represent unknown numbers or variables;** know the meanings of the words *term, expression* and *equation.*
  Worksheet 15
- Construct and solve simple linear equations with integer coefficients (unknown on one side only) using an appropriate method (e.g. inverse operations).
  Worksheet 18

CORE
From the Y8 teaching programme.

- Begin to distinguish the different roles played by letter symbols in equations, formulae and functions; know the meanings of the words *formula* and *function.*
  Chapter 8 Expressions, Formulae and Equations
  Understanding algebra   page 158
- Construct and solve linear equations with integer coefficients (unknown on either or both sides, without and with brackets) using appropriate methods (e.g. inverse operations, transforming both sides in the same way).
  Chapter 8 Expressions, Formulae and Equations
  Writing equations   page 181
  Writing and solving equations   page 184
  Solving equations by transforming both sides   page 188

EXTENSION
From the Y9 teaching programme.
This is covered by the 8 Plus book.

- **Construct and solve linear equations with integer coefficients** (with and without brackets, negative signs anywhere in the equation, positive or negative solution), **using an appropriate method.**
- Use formulae from mathematics and other subjects; substitute numbers into expressions and formulae; derive a formula and, in simple cases, change its subject.

Mapping to the sample medium-term plans

	SUPPORT	CORE	EXTENSION
			EXTENSION From the Y9 teaching programme. *This is covered by the 8 Plus book.*
• Use formulae from mathematics and other subjects; **substitute integers into simple formulae**, including examples that lead to an equation to solve; derive simple formulae. **Chapter 8 Expressions, Formulae and Equations** *Substituting into formulae page 175* *Writing and finding formulae page 182*	**SUPPORT** From the Y7 teaching programme. *This is covered by the 8* book, the support section in the 8 Core book and the worksheets from the Teacher Resource Pack, as listed.*	**CORE** From the Y8 teaching programme.	
Handling data 2 (6 hours) Handling data (248–273)	• Given a problem that can be addressed by statistical methods, suggest possible answers. *Worksheet 35* • Design a data collection sheet or questionnaire to use in a simple survey; construct frequency tables for discrete data. *Worksheets 35 and 37*	• Discuss a problem that can be addressed by statistical methods and identify related questions to explore. **Chapter 15 Data Collection** *Surveys — the problem or question page 376* • Decide which data to collect to answer a question, and the degree of accuracy needed; identify possible sources. **Chapter 15 Data Collection** *Collecting the data page 377* • Plan how to collect the data, including sample size; design and use two-way tables for discrete data. **Chapter 15 Data Collection** *Two-way tables page 374* *Planning a survey page 378* • Collect data using a suitable method, such as observation, controlled experiment using ICT, or questionnaire. **Chapter 15 Data Collection** *Planning a survey page 378* **Chapter 16 Analysing Data. Drawing and Interpreting Graphs** *Surveys page 407*	• Discuss how data relate to a problem; identify possible sources, including primary and secondary sources. • Gather data from specified secondary sources, including printed tables and lists from ICT-based sources.

Mapping to the sample medium-term plans

Solving problems
(28–29)

- Calculate statistics for small sets of discrete data:
  - find the mode, median and range;
  - calculate the mean, including from a simple frequency table, using a calculator for a larger number of items.
  Worksheet 38

- Construct, on paper and using ICT, graphs and diagrams to represent data, including:
  - bar-line graphs;
  use ICT to generate pie charts.
  Worksheet 36

- Write a short report of a statistical enquiry and illustrate with appropriate diagrams, graphs and charts, using ICT as appropriate; justify choice of what is presented.

- Calculate statistics, including with a calculator; recognise when it is appropriate to use the range, mean, median and mode; construct and use stem-and-leaf diagrams.
  Chapter 16 Analysing Data. Drawing and Interpreting Graphs
  Mode and range   page 382
  Mean   page 383
  Median, mean, mode range   page 385
  Finding the median, range and mode from a stem-and-leaf diagram   page 388

- **Construct, on paper and using ICT:**
  - **pie charts for categorical data;**
  - **bar charts and frequency diagrams for discrete data;**
  - **simple scatter graphs;**
  identify which are most useful in the context of the problem.
  Chapter 16 Analysing Data. Drawing and Interpreting Graphs
  Compound bar charts and line graphs   page 391
  Drawing pie charts   page 397
  Scatter graphs   page 400

- Interpret tables, graphs and diagrams for discrete data and draw inferences that relate to the problem being discussed; relate summarised data to the questions being explored.
  Chapter 16 Analysing Data. Drawing and Interpreting Graphs
  Interpreting graphs and tables   page 402

- Communicate orally and on paper the results of a statistical enquiry and the methods used, using ICT as appropriate; justify the choice of what is presented.
  Chapter 16 Analysing Data. Drawing and Interpreting Graphs
  Surveys   page 407

- Solve more complex problems by breaking them into smaller steps or tasks, choosing and using resources, including ICT.
  Throughout the above sections.

- Interpret graphs and diagrams and draw inferences to support or cast doubt on initial conjectures; have a basic understanding of correlation.

YEAR 8: SUMMER TERM

Teaching objectives for the oral and mental activities

These objectives are covered by using the Mental Starter Ideas given under that section heading in the Teachers' Notes section of this Teacher Planning Pack.

- Order, add, subtract, multiply and divide integers.
- Multiply and divide decimals by 10, 100, 1000, 0·1, 0·01.
- Round numbers, including to one or two decimal places.
- Know and use squares, cubes, roots and index notation.
- Know or derive prime factorisation of numbers to 30.
- Convert between fractions, decimals and percentages.
- Find the outcome of a given percentage increase or decrease.

- Know complements of 0·1, 1, 10, 50, 100.
- Add and subtract several small numbers or several multiples of 10, e.g. 250 + 120 − 190.
- Use jottings to support addition and subtraction of whole numbers and decimals.
- Calculate using knowledge of multiplication and division facts and place value, e.g. $432 \times 0·01$, $37 \div 0·01$, $0·04 \times 8$, $0·03 \div 5$.
- Recall multiplication and division facts to 10×10.
- Use factors to multiply and divide mentally, e.g. $22 \times 0·02$, $420 \div 15$.
- Multiply by near 10s, e.g. 75×29, $8 \times {}^{-}19$.

- Use partitioning to multiply, e.g. $13 \times 1·4$.
- Use approximations to estimate the answers to calculations, e.g. $39 \times 2·8$.
- Solve equations, e.g. $n(n-1) = 56$, __ + __ = $^{-}46$.
- Visualise, describe and sketch 2-D shapes, 3-D shapes and simple loci.
- Estimate and order acute, obtuse and reflex angles.
- Use metric units (length, mass, capacity, area and volume) and units of time for calculations.
- Use metric units for estimation (length, mass, capacity, area and volume).
- Convert between m, cm and mm, km and m, kg and g, litres and ml, cm^2 and mm^2.
- Discuss and interpret graphs.
- Calculate a mean using an assumed mean.
- Apply mental skills to solve simple problems.

Notes

1   *Under each objective, the relevant sections from New National Framework Mathematics are listed. It is not intended that all the material in each section is covered by all pupils. You will need to choose work appropriate for your pupils, teaching programme and teaching style. There is plenty of choice.*

2   *Some of the Year 8 teaching programme objectives are revisited more than once in this medium-term plan. Each time the objective is given, all of the sections from New National Framework Mathematics that cover this objective are listed. You will need to split the material to suit your class and teaching programme.*

Teaching objectives for the main activities

	SUPPORT	CORE	EXTENSION
	From the Y7 teaching programme. *This is covered by the 8* book, the support section in the 8 Core book and the worksheets from the Teacher Resource Pack, as listed.*	From the Y8 teaching programme.	From the Y9 teaching programme. *This is covered by the 8 Plus book.*
Number 4 (6 hours) Calculations (82–87, 92–107, 110–111)		● Understand addition and subtraction of fractions and integers, and multiplication and division of integers; use the laws of arithmetic and inverse operations. *Chapter 2 Integers, Powers and Roots* *Adding and subtracting integers page 36* *Multiplying and dividing integers page 39* *Order of operations with integers page 43*	● **Understand the effects of multiplying and dividing by numbers between 0 and 1.**

Mapping to the sample medium-term plans

• Consolidate and **extend mental methods of calculation to include decimals, fractions and percentages,** accompanied where appropriate by suitable jottings. *Worksheets 8, 12 and 13*	***Chapter 5 Fractions*** *Adding and subtracting fractions page 108* • Use the order of operations, including brackets, with more complex calculations. ***Chapter 2 Integers, Powers and Roots*** *Order of operations with integers page 43* ***Chapter 4 Written and Calculator Calculation*** *Using brackets on a calculator page 92* • Consolidate and extend mental methods of calculation, working with decimals, fractions and percentages, squares and square roots, cubes and cube roots; solve word problems mentally. ***Chapter 2 Integers, Powers and Roots*** *Squares and square roots page 51* *Cubes and cube roots page 55* ***Chapter 3 Mental Calculation*** *Adding and subtracting page 62* *Multiplying and dividing page 66* *Solving problems mentally page 70* ***Chapter 5 Fractions*** *Much of this chapter is done mentally* ***Chapter 6 Percentages, Fractions, Decimals*** *Converting fractions, decimals and percentages page 119* *Percentage of — mentally page 123* • Make and justify estimates and approximations of calculations. ***Chapter 3 Mental Calculation*** *Making estimates page 73* *Estimating answers to calculations page 74* • Consolidate standard column procedures for addition and subtraction of integers and decimals with up to two places. ***Chapter 4 Written and Calculator Calculation*** *Adding and subtracting page 81*	• Understand the order of precedence and effect of powers. • Extend mental methods of calculation, working with decimals, fractions, percentages, factors, powers and roots. • Use standard column procedures to add and subtract integers and decimals of any size.
• **Multiply and divide three-digit by two-digit whole numbers; extend to multiplying and dividing decimals with one or two places by single-digit whole numbers.** *Worksheet 11*	• **Use standard column procedures for multiplication and division of integers and decimals, including by decimals such as 0·6 or 0·06; understand where to position the decimal point by considering equivalent calculations.**	• Multiply and divide by decimals, dividing by transforming to division by an integer.

22

Mapping to the sample medium-term plans

Chapter 4 Written and Calculator Calculation
Multiplying   *page 83*
Dividing   *page 85*
Dividing by a decimal   *page 88*
● Check a result by considering whether it is of the right order of magnitude and by working the problem backwards.
Chapter 4 Written and Calculator Calculation
Checking answers   *page 89*
● Use units of measurement to estimate, calculate and solve problems in everyday contexts.
Chapter 14 Measures, Perimeter, Area and Volume
Metric conversions   *page 335*
Metric and imperial equivalents   *page 339*
Units, measuring instruments and accuracy   *page 340*
Estimating   *page 342*

Measures
(228–231)

● **Convert one metric unit to another** (e.g. grams to kilograms).
Worksheet 32

SUPPORT
From the Y7 teaching programme.
This is covered by the 8* book, the support section in the 8 Core book and the worksheets from the Teacher Resource Pack, as listed.

CORE
From the Y8 teaching programme.

EXTENSION
From the Y9 teaching programme.
This is covered by the 8 Plus book.

Algebra 5 (8 hours)
Equations and formulae
(116–137)

● Simplify linear algebraic expressions by collecting like terms.
Worksheet 16

● Construct and solve simple linear equations with integer coefficients (unknown on one side only) using an appropriate method (e.g. inverse operations).
Worksheet 18

● **Simplify or transform linear expressions by collecting like terms; multiply a single term over a bracket.**
Chapter 8 Expressions, Formulae and Equations
Brackets   *page 166*
Collecting like terms   *page 168*
More writing and simplifying expressions   *page 179*
● Construct and solve linear equations with integer coefficients (unknown on either or both sides, without and with brackets) using appropriate methods (e.g. inverse operations, transforming both sides in the same way).
Chapter 8 Expressions, Formulae and Equations
Writing equations   *page 181*
Writing and solving equations   *page 184*
Solving equations by transforming both sides   *page 188*

● Simplify or transform algebraic expressions by taking out single term common factors.

● **Construct and solve linear equations with integer coefficients** (with and without brackets, negative signs anywhere in the equation, positive or negative solution), **using an appropriate method.**

- Use systematic trial and improvement methods and ICT tools to find approximate solutions of equations such as $x^3 + x = 20$.
- Solve problems involving direct proportion using algebraic methods, relating algebraic solutions to graphical representations of the equations; use ICT as appropriate.
- Plot graphs of linear functions (y given implicitly in terms of x), e.g. $ay + bx = 0$, $y + bx + c = 0$, on paper and using ICT.

- Use trial and improvement methods where a more efficient method is not obvious.

EXTENSION
From the Y9 teaching programme.
This is covered by the 8 Plus book.

Sequences, functions and graphs (164–177)

- Generate coordinate pairs that satisfy a simple linear rule; recognise straight-line graphs parallel to the x-axis or y-axis.
Worksheet 21 and 22

- Begin to use graphs and set up equations to solve simple problems involving direct proportion.
Chapter 8 Expressions, Formulae and Equations
Solving equations using a graph   page 192

- **Plot the graphs of linear functions, where y is given explicitly in terms of x,** on paper and using ICT.
Chapter 10 Graphs
Graphing functions   page 233
- Construct linear functions arising from real-life problems and plot their corresponding graphs; discuss and interpret graphs arising from real situations.
Chapter 10 Graphs
Reading and plotting real-life graphs   page 240
Distance/time graphs   page 246
Interpreting and sketching real-life graphs   page 248
- Solve more demanding problems and investigate in a range of contexts: algebra.
Throughout the book

Solving problems (6–13, 28–29)

- **Break a complex calculation into simpler steps, choosing and using appropriate and efficient operations, methods** and resources, including ICT.

- Solve more complex problems by breaking them into smaller steps or tasks, choosing and using efficient techniques for algebraic manipulation.
Throughout the section/book

SUPPORT
From the Y7 teaching programme.
This is covered by the 8* book, the support section in the 8 Core book and the worksheets from the Teacher Resource Pack, as listed.

CORE
From the Y8 teaching programme.

- Represent problems mathematically, making correct use of symbols, words, diagrams, tables and graphs.
Throughout support worksheets

- Solve more demanding problems and investigate in a range of contexts: number and measures.
Throughout the section/book
- **Identify the necessary information to solve a problem; represent problems and interpret solutions in algebraic or graphical form,** using correct notation.
Throughout the section/book

Solving problems (6 hours)
Solving problems

(2–35)

24

Mapping to the sample medium-term plans

	Support	Core	Extension

Ratio and proportion (78–81)

- **Break a complex calculation into simpler steps, choosing and using appropriate and efficient operations, methods** and resources, including ICT.
  Throughout support worksheets

- Understand the significance of a counter-example.
  Throughout support worksheets

- Understand the relationship between ratio and proportion; solve simple problems about ratio and proportion using informal strategies.
  Worksheet 14

- Solve more complex problems by breaking them into smaller steps or tasks, choosing and using efficient techniques for calculation.
  Throughout the section/book

- **Use logical argument to establish the truth of a statement;** give solutions to an appropriate degree of accuracy in the context of the problem.
  Throughout the section/book

- Suggest extensions to problems, conjecture and generalise; identify exceptional cases or counter-examples.
  Throughout the section/book

- Consolidate understanding of the relationship between ratio and proportion; reduce a ratio to its simplest form, including a ratio expressed in different units, recognising links with fraction notation; **divide a quantity into two or more parts in a given ratio; use the unitary method to solve simple word problems involving ratio and direct proportion.**
  Chapter 7 Ratio and Proportion
  Direct proportion   page 137
  Simplifying ratios   page 139
  Ratio and proportion   page 141
  Solving ratio and proportion problems   page 143
  Dividing in a given ratio   page 145

- Solve increasingly demanding problems and evaluate solutions; explore connections in mathematics across a range of contexts.

- **Present a concise, reasoned argument, using symbols, diagrams and graphs and related explanatory text.**

- **Use proportional reasoning to solve a problem, choosing the correct numbers to take as 100%, or as a whole;** compare two ratios; interpret and use ratio in a range of contexts, including solving word problems.

Shape, space and measures 4
(9 hours)
Geometrical reasoning: lines, angles and shapes

(198–201)

SUPPORT
From the Y7 teaching programme.
This is covered by the 8* book, the support section in the 8 Core book and the worksheets from the Teacher Resource Pack, as listed.

- Use 2-D representations to visualise 3-D shapes and deduce some of their properties.
  Worksheet 29

- Use ruler and protractor to construct simple nets of 3-D shapes, e.g. cuboid, regular tetrahedron, square-based pyramid, triangular prism.
  Worksheet 29

CORE
From the Y8 teaching programme.

- Know and use geometric properties of cuboids and shapes made from cuboids; begin to use plans and elevations.

EXTENSION
From the Y9 teaching programme.
This is covered by the 8 Plus book.

- Visualise and use 2-D representations of 3-D objects; analyse 3-D shapes through 2-D projections, including plans and elevations.

		• Use and interpret maps, scale drawings.
		• Use straight edge and compasses to construct a triangle, given right angle, hypotenuse and side (RHS).
		• Calculate the surface area and volume of right prisms.

Transformations (216–217)	*Chapter 12 Shape, Construction and Loci* Describing and sketching 3-D shapes. *Constructing nets page 295* *Plans and elevations page 297*	
	• Make simple scale drawings. *Chapter 13 Coordinates and Transformations* *Scale drawing page 324*	
Coordinates (218–219)	• Use conventions and notation for 2-D coordinates in all four quadrants; find coordinates of points determined by geometric information. *Worksheet 28*	• Given the coordinates of points A and B, find the mid-point of the line segment AB. *Chapter 13 Coordinates and Transformations* *Finding the mid-point of a line page 328*
Construction and loci (220–227)	• Use a ruler and protractor to: – measure and draw lines to the nearest millimetre and angles, including reflex angles, to the nearest degree; – construct a triangle given two sides and the included angle (SAS) or two angles and the included side (ASA); explore these constructions using ICT. *Worksheets 23, 24 and 27*	• **Use straight edge and compasses to construct:** – a triangle, given three sides (SSS); use ICT to explore this construction. *Chapter 12 Shape, Construction and Loci* *Constructing triangles and quadrilaterals page 304*
		• Find simple loci, both by reasoning and by using ICT, to produce shapes and paths, e.g. an equilateral triangle. *Chapter 12 Shape, Construction and Loci* *Loci page 307*
		• Use bearings to specify direction. *Chapter 14 Measures, Perimeter, Area and Volume* *Bearings page 344*
Mensuration	• Calculate the surface area of cubes and cuboids. *Worksheet 33*	• **Know and use the formula for the volume of a cuboid; calculate volumes and surface areas of cuboids** and shapes made from cuboids. *Chapter 14 Measures, Perimeter, Area and Volume* *Surface area and volume page 355*
(232–233, 238–241)		

Mapping to the sample medium-term plans

	SUPPORT	CORE	EXTENSION
Handling data 3 (7 hours) Handling data (248–275)	From the Y7 teaching programme. *This is covered by the 8* book, the support section in the 8 Core book and the worksheets from the Teacher Resource Pack, as listed.*	From the Y8 teaching programme.	From the Y9 teaching programme. *This is covered by the 8 Plus book.*
	● Given a problem that can be addressed by statistical methods, suggest possible answers. *Worksheet 35*	● Discuss a problem that can be addressed by statistical methods and identify related questions to explore. **Chapter 15 Data Collection** *Surveys — the problem or question page 376* ● Decide which data to collect to answer a question, and the degree of accuracy needed: identify possible sources. **Chapter 15 Data Collection** *Collecting the data page 377*	● Discuss how data relate to a problem; identify possible sources, including primary and secondary sources.
	● Design a data collection sheet or questionnaire to use in a simple survey; construct frequency tables for discrete data, grouped where appropriate in equal class intervals. *Worksheets 35 and 37*	● Plan how to collect the data, including sample size; construct frequency tables with given equal class intervals for sets of continuous data. **Chapter 15 Data Collection** *Grouping continuous data page 373* *Planning a survey page 378*	● **Design a survey or experiment to capture the necessary data from one or more sources; determine the sample size and degree of accuracy needed; design, trial and if necessary refine data collection sheets;** construct tables for large discrete and continuous sets of raw data, choosing suitable class intervals.
	● Calculate statistics for small sets of discrete data: – find the mode, median and range, and the modal class for grouped data; – calculate the mean, including from a simple frequency table, using a calculator for a larger number of items. *Worksheet 38*	● Collect data using a suitable method, such as observation, controlled experiment, including data logging using ICT, or questionnaire. **Chapter 15 Data Collection** *Planning a survey page 378* ● Calculate statistics, including with a calculator; calculate a mean using an assumed mean; know when it is appropriate to use the modal class for grouped data. **Chapter 16 Analysing Data. Drawing and Interpreting Graphs** *Mode and range page 382* *Mean page 383*	

- Construct, on paper and using ICT, graphs and diagrams to represent data, including:
  – frequency diagrams for grouped discrete data;
  use ICT to generate pie charts.
  Worksheet 36

- **Construct, on paper and using ICT:**
  – **bar charts and frequency diagrams for continuous data;**
  – **simple line graphs for time series;**
  identify which are most useful in the context of the problem.
  Chapter 16 Analysing Data. Drawing and Interpreting Graphs
    Compound bar charts and line graphs   page 391
    Frequency diagrams   page 395
- Interpret tables, graphs and diagrams for continuous data and draw inferences that relate to the problem being discussed; relate summarised data to the questions being explored.
  Chapter 16 Analysing Data. Drawing and Interpreting Graphs
    Interpreting graphs and tables   page 402
- Compare two distributions using the range and one or more of the mode, median and mean.
  Chapter 16 Analysing Data. Drawing and Interpreting Graphs
    Comparing data   page 389
- Communicate orally and on paper the results of a statistical enquiry and the methods used, using ICT as appropriate; justify the choice of what is presented.

- Compare two or more distributions and make inferences, using the shape of the distributions, the range of data and appropriate statistics.

- Write a short report of a statistical enquiry and illustrate with appropriate diagrams, graphs and charts, using ICT as appropriate; justify choice of what is presented.

  Chapter 16 Analysing Data. Drawing and Interpreting Graphs
    Surveys   page 407
- Compare experimental and theoretical probabilities in different contexts.
  Chapter 17 Probability
    Comparing calculated probability with experimental probability   page 423
- Solve more complex problems by breaking them into smaller steps or tasks, choosing and using graphical representation, and also resources, including ICT.
  Throughout above sections

- Appreciate the difference between mathematical explanation and experimental evidence.

Probability
(284–285)

Solving problems
(28–29)

1

Teachers' Notes

1 Place Value, Ordering and Rounding

Topic in Chapter	Links to Programme of Study
	Ma2 1a–l are integrated throughout. Only obvious ones have been listed as part of the links.
Powers of ten *page 16*	**Ma2 2b** ... use the notation for small integer powers ... **Ma2 2d** use decimal notation ...
Adding and subtracting multiples of 0·1, 0·01 and 0·001 *page 17*	
× and ÷ by multiples of 10, 100 and 1000 *page 19*	**Ma2 3a** ... multiply or divide any number by powers of 10, and any positive number by a number between 0 and 1 ...
Multiplying and dividing by 0·1 and 0·01 *page 20*	
Putting decimals in order *page 24*	**Ma2 2d** ... order decimals **Ma3 4a** ... convert measurements from one unit to another ...
Rounding to powers of ten *page 28*	**Ma2 2a** use their previous understanding of integers and place value to deal with arbitrarily large positive numbers and round them to a given power of 10 ...
Rounding to decimal places *page 30*	**Ma2 3h** round to the nearest integer ... **Ma2 4d** give solutions in the context of a problem to an appropriate degree of accuracy ...

Links to Framework for Teaching Mathematics: 8 Teaching Programme
Applying mathematics and solving problems are integrated throughout.
Place value, ordering and rounding ● Read and write positive integer powers of 10; multiply and divide integers and decimals by 0·1, 0·01. **Integers, powers and roots** ● Use ... index notation for small positive integer powers.
Place value, ordering and rounding ● Order decimals.
Place value, ordering and rounding ● Round positive numbers to any given power of 10; round decimals to the nearest whole number or to one or two decimal places.

Links to Attainment Targets

Level 4 ... and order decimals to three decimal places
Level 5 Pupils use their understanding of place value to multiply and divide whole numbers and decimals by 10, 100 and 1000.
Level 6 Pupils order and approximate decimals when solving numerical problems ...
Level 7 ... They understand the effects of multiplying and dividing by numbers between 0 and 1 (begin)

Topic area	Powers of ten	pupil book page 16

Framework Objectives

✓ Read and write positive integer powers of ten, **Framework examples page 37**
✓ Use ... index notation for small positive integer powers, **Framework examples page 57**

Sheets Worksheet 1 for Support   Homework Sheet 1.1   Starter Resource Sheet 1

Resources You could use decimal place value cards, place value charts, number fans and number/digit cards including ones with a decimal point. Blu-Tack or removable sticky tape are needed.

 ## Starter ideas and activities

Mental starter ideas
- Ask some place value questions such as questions 1 and 2 in the Number Support chapter on page 7 of the Pupil Book.
- Ask for the answers to some powers, such as 4^2, 10^2, 2^3, 10^3, ...
- Ask students to say powers in the form '10^2 is 10 to the power of 2' etc.

Starter activity
- **Make a place value chart.**
  1 Use Starter Resource Sheet 1 to make cards. Give one card to each of 20 pupils.
  2 Ask the pupils with 1, 10, 100, 1000, ... to line up in order.
  3 Ask the pupils with 1, 10^2, 10^3, 10^4, ... to line up facing their matching card.
  4 On the whiteboard attach the card using Blu-Tack or removable sticky tape to make a place value chart as shown below.
  5 Ask other pupils to write the word names such as 'units', 'tens', ... above the chart.

Millions	Hundreds of thousands	etc.				
1 000 000	100 000	10 000	1000	100	10	1
10^6	10^5	10^4	10^3	10^2	10^1	1

- **Discuss how very large numbers are written using powers of 10.**

 ## Teaching points and activities

- It is important for pupils to realise that powers of 10 underpin our number system. Make sure pupils understand this by emphasising the place value chart given on page 16 of the Pupil Book.
- Pupils often read the decimal part of a number as a whole number. For example, 1·496 is read as one point four hundred and ninety-six instead of one point four nine six. This can lead pupils to mistakenly think that 1·496 is larger than 1·58, for example, because 496 is greater than 58.
- When finding the number that is, say, 10 000 less than 1 800 000 encourage pupils to read the number as one million, eight hundred thousand because then it is easy to do the subtraction mentally as 'eight hundred thousand minus ten thousand is seven hundred and ninety thousand'.

Support
- See **Place value. Reading and writing numbers** on page 1 of the Pupil Book.
- If pupils have difficulty with Practice questions 1 and 2 use Worksheet 1 questions 1, 2, 3, 5, 6, 8 (page 2 in the Teacher Resource Pack)
- You could use the Puzzle on page 17 of the Pupil Book to revise decimal place value.

Extension
- See **Sticks and stones** (Pupil Book page 15). Challenge pupils to add and subtract bigger numbers using this number system and other number systems.
- See **Puzzle** (Pupil Book page 17). Pupils could make up their own puzzles similar to these and give them to someone else to solve.

Plenary ideas
- Mix up or remove some of the cards you attached to the whiteboard in the starter. Ask pupils to put them back in the right place.
- Consolidate the fact that powers of 10 underpin our number system.

Homework Review questions at the end of Exercise 1 (Pupil Book page 16) and/or Homework Sheet 1.1 (page 184 of this pack).

Links Link this topic to Index Notation.
Link to Science: Exercise 1 questions 3, 4, Review 2; screen on page 16.
Link to Geography: Exercise 1 question 3.
Link to Design and Technology: Exercise 1 question 3.
Link to Powers: Exercise 1 questions 1, 2, 4 and Reviews 1 and 2.
Link to Mental Calculation: Exercise 1 question 5 and Review 4.

Topic area	Adding and subtracting multiples of 0·1, 0·01 and 0·001	pupil book page 17

Framework Objectives ✓ Read and write positive integer powers of ten, **Framework examples page 37**

Sheets Worksheet 1 for Support   Homework Sheet 1.2   Resource Sheet pages 190 and 191
Teacher Resource Pack

Resources You could use a counting stick and sticky labels for the starter activity.

Starter ideas and activities

Mental starter ideas
- Ask pupils to count on and back in steps of 0·1, 0·01 and progress to steps of 0·001. You could use a number line to help prompt less able pupils. You could progress to multiples of 0·1 and 0·01, such as 0·4 and 0·05 etc.
- Ask some questions such as practice questions 37 and 55 on pages 11 and 12 in the Number Support chapter of the Pupil Book.

Starter activity
Adding and subtracting 0·1, 0·01 and 0·001 using a scale
1 Put scales such as one of these across the full width of the board or OHP (or use a counting stick and sticky labels).

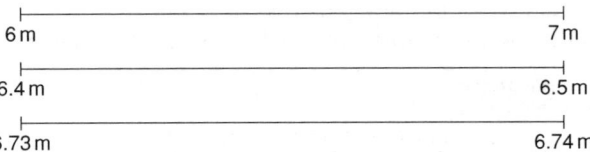

2 Fill in the tenths between 6 m and 7 m first and ask pupils questions such as:
   What is 0·1 or 0·04 or ... more than 6·4?
   What is 0·1 or 0·05 or ... less than 6·4?
   What do I have to subtract to get from 6·8 to 6·3?
3 Now fill in the hundredths between 6·4 and 6·5.
   Ask questions, such as:
   What is 0·01 or 0·07 or ... more than 6·42?
4 Ask how you could divide the line between 6 and 7 into thousandths.
   Ask questions, such as:
   What is 0·001 or 0·006 or ... more or less than 6·73?

Teaching points and activities

- Check that pupils can count on and back in steps of 0·1 and 0·01 (Practice questions 37 and 55 of the Number Support chapter.)
- Recap on the place value of 1 in 0·1, 0·01 and 0·001.
- A number line or place value chart is helpful for pupils who have difficulty adding or subtracting multiples of 0·1, 0·01 and 0·001.
- Pupils have difficulty subtracting a larger digit from a smaller digit, such as subtracting 0·07 from 8·82. Encourage them to name 8 tenths and 2 hundredths as 82 hundredths and then subtract 7 hundredths from this.

Support
- See **Place value. Reading and writing numbers** on page 1 of the Pupil Book. If pupils have difficulty with Practice questions 37 and 55 use Worksheet 1 questions 4 and 7 (page 2 in the Teacher Resource Pack). Pupils should have done the previous section, *Powers of ten*.

Extension
- See **Starred questions** in Exercise 2 (Pupil Book page 18).

Plenary ideas
- Use Exercise 2 questions 6, 7 and 8. Ask pupils to demonstrate how to do these and promote discussion of place value. These questions illustrate the importance of this topic in measurement. In question 8 discuss conversion between metric units.

Homework Review questions at the end of Exercise 2 (Pupil Book page 18) and/or Homework Sheet 1.2 (page 184 of this pack).

Links Link this topic to Science: Exercise 2 questions 3 and 6.
Link to Measures: Exercise 2 questions 3 and 7.
Link to Geography: Exercise 2 Review 2.
Link to Design and Technology: Exercise 2 question 8.

Topic areas	**× and ÷ by multiples of 10, 100 and 1000**	pupil book page 19
	Multiplying and dividing by 0·1 and 0·01	pupil book page 20

Framework Objectives ✓ ... multiply and divide integers and decimals by 0·1, 0·01, **Framework examples pages 39, 89 and 97**

Sheets Worksheet 2 for Support   Homework Sheets 1.3 and 1.4   Resource Sheet pages 191 and 192 Teacher Resource Pack   ICT Worksheet Bigger or Smaller on page 117 of the ICT Support section of the Teacher Resource Pack.

Resources You could use a place value board. You could use mini-whiteboards and digit cards. **Investigation: Multiplying and Dividing by 0·1 and 0·01** (Pupil Book page 20), calculator or spreadsheet package and resource sheet page 191 (Teacher Resource Pack).

 ## Starter ideas and activities

Mental starter ideas Choose a number, such as 6, and multiply it by 10 then 100 then 1000.
Progress to numbers such as 1·7 and 0·78 etc.
Do the same for division, first choosing a number such as 550 and then progressing to numbers such as 5·6 and 0·78 etc.

Starter activity
● **Multiplying and dividing by 10, 100 and 1000**
Use number cards 1 to 10 and a decimal point card and some zero cards.
1 Ask pupils to make the number 2.3. Now ask these pupils to make the number that is 2 × 2.3, now 20 × 2.3, now 200 × 2.3, now 2000 × 2.3.
2 Repeat with other numbers and multiples.
3 Emphasise that we can multiply by multiples in two steps. Point out the pattern of the answers.
4 Repeat for some simple divisions such as 9.6 ÷ 3, 9.6 ÷ 30, 9.6 ÷ 300, 9.6 ÷ 3000.
● **Investigating multiplying and dividing by 0·1 and 0·01**
Use the Investigation on page 20 of the Pupil Book and work with the whole class to fill in the chart. You could use a place value board instead of a calculator or spreadsheet.

 ## Teaching points and activities

● Use patterns, such as the one below, to help with understanding. See the screen on page 19 for more examples.
    6·3 × 3 = 18·9
    6·3 × 30 = 189
    6·3 × 300 = 1890
    6·3 × 3000 = 18 900
This can be extended to include patterns that include multiplying by 0·1 and 0·01, or dividing by 0·1 and 0·01.
● Emphasise that it is the digits that move and not the decimal point. Use place value charts to illustrate this if necessary.
● Pupils often have difficulty remembering which way the digits move for multiplication and division. Encourage pupils to look at the answer to check it is sensible. When a positive number has been multiplied by a multiple of 10, 100 or 1000 the answer should be bigger.
● It is important to emphasise that 0·1 equals $\frac{1}{10}$ and that multiplying by $\frac{1}{10}$ is the same as dividing by 10. Likewise emphasise that 0·01 equals $\frac{1}{100}$ and so multiplying by 0·01 is the same as multiplying by $\frac{1}{100}$, which is the same as dividing by 100. It is helpful to make the link to 'finding a fraction of' and that 'of' means multiply.
● The second-to-last dot in the Discussion on page 21 will need very careful explanation and quick revision on how to find the area of a rectangle.

Support
● See **Multiplying and dividing by 10, 100 and 1000** on page 1 of the Pupil Book (Number Support chapter). If pupils have difficulty with Practice questions 14 and 26 use Worksheet 2 (page 2 in the Teacher Resource Pack).

Extension
● See **starred questions** in Exercises 3 and 4 (Pupil Book pages 19 and 23).
● See **Investigation: × and ÷ by 0·1 and 0·01** (Pupil Book page 20). Challenge pupils to come up with a rule for multiplying and dividing by 0·1 and 0·01 and show it is true for both whole numbers and decimals.
● See **Discussion** (Pupil Book page 21). Challenge pupils to write the answers to the questions in this discussion before discussing them.

Plenary ideas
● Use Exercise 3 questions 5 and 6 and ask pupils to explain how they got their answers.

- Challenge pupils to come up with some way of remembering which way to move the digits when multiplying or dividing.
- Use Exercise 4 question 9. Ask pupils to explain the answers.

Homework Review questions at the end of Exercises 3 and 4 (Pupil Book pages 19 and 23) and/or Homework Sheets 1.3 and 1.4 (page 185 of this pack)

ICT
- See Investigation: × and ÷ by Multiples of 10, 100 and 1000 (Pupil Book page 20). Use Smile for Windows: Numeracy: Tenners to reinforce understanding.
- See Investigation: × and ÷ by 0·1 and 0·01 (Pupil Book page 20). You will need a spreadsheet package. Use ICT Worksheet 'Bigger or Smaller' in the ICT support section of the Teacher Resource Pack (page 117).

Links Link to Science: Exercise 4 question 5.
Link to Geography: Exercise 4 question 8.
Link to PE: Exercise 3 question 4.
Link to Measurement: Exercise 4 question 8, Review 3.

Topic area	**Putting decimals in order**	pupil book page 24

Framework Objectives ✓ Order decimals, **Framework examples page 41**

Sheets Worksheet 3 for Support   Homework Sheet 1.5   Resource Sheet pages 192 and 193 Teacher Resource Pack

Resources Number/digit cards for the Starter Activity. You could use a decimal number line. You could use mini-whiteboards, cards with decimals on and small acetates with decimals on. You could use a calculator in the game Ladders. A graphical calculator for the Practical (Pupil Book page 28). Smile for Windows: Sense of Number Box D and Smile for Windows: Numeracy: Number lines. OHP for Plenary.

 ## Starter ideas and activities

Mental starter ideas
- Ask questions such as Practice questions 6, 41 and 46 on pages 8 and 11 in the Number Support chapter of the Pupil Book.
- Ask some quick questions, such as 'Which is greater 1·45 or 1·453?'

Starter activity
- **Game: Guess my number**
  Play a quick game of guess my number by choosing a number between either 50 and 100 or 1 and 2 with 1, 2 or 3 decimal places and asking pupils to ask questions that require a yes/no answer.
  For example: *Is the number greater than 1·5?*
                Is the number of hundredths greater than the number of tenths?
- **Putting numbers in order**
  1 Put 5 or 6 decimal numbers with 2 or 3 decimal places on cards and ask for volunteers to come to the front and hold them.
  2 Ask another pupil to come and arrange them in order explaining their method as they do so. Make the point that you look at the most significant digit (the one with the highest value) first.
  3 You could put the numbers on a number line to help. If you are going to do this make sure the numbers you choose are suitable. If you choose numbers too far apart in value you will need a very long number line for them.

 ## Teaching points and activities

- Some revision on conversion between metric units may be needed before pupils can do Exercise 5 question 5.
- Make sure pupils can put whole numbers and decimals with 2 decimal places in order and that they remember what > and < mean.
- Encourage pupils who have difficulty to use a number line.
- Remind pupils that 5·897 is read as 'five point eight nine seven' and not as 'five point eight hundred and ninety-seven'. Otherwise pupils may be misled into thinking that 5·897 is greater than 5·98.

Support
- See **Putting numbers in order** on page 1 of the Pupil Book (Number Support chapter). If pupils have difficulty with Practice questions 6, 11, 18, 41, 46 and 67 use Worksheet 3 (page 3 in the Teacher Resource Pack).

Extension
- See **Puzzle** (Pupil Book page 27). Encourage pupils to find at least 13 paths. See **Game: Ladders** (Pupil Book page 27). Challenge pupils to choose numbers between 1 and 2 or between 0·5 and 0·6 and so on.

Plenary ideas
- Have an enlarged copy of the diagram for the Puzzle on page 27 of the Pupil Book on an OHP. Ask pupils to come up and add paths to the diagram using different coloured pens, while the rest of the class act as watchdog for mistakes.
- Play the game on page 27 of the pupil book as a whole class.

Homework Review questions at the end of Exercise 5 (Pupil Book page 27) and/or Homework Sheet 1.5 (page 186 of this pack).

Graphical calculator For command sequences to use in doing the Practical (Pupil Book page 28), see the Graphical Calculator Support section (page 181 of the Teacher Resource Pack).

ICT   Smile for Windows: Sense of Number Box D — use with Putting Decimals in Order on page 24 of the Pupil Book.

Smile for Windows: Numeracy: Number lines — use with Putting Decimals in Order on page 24 of the Pupil Book.

See pages 114 and 115 of the ICT section of the Teacher Resource Pack.

Links   Link this topic to ordering fractions (Pupil Book page 105) and graphs (Pupil Book page 232).

Link to Metric Conversions: Exercise 5 questions 5, 6, Review 4.

Link to Measures: Exercise 5 questions 3, 11 and Review 2.

Link to Handling Data: Exercise 5 question 11.

Link to Science: Exercise 5 Review 2.

Topic areas	Rounding to powers of ten	pupil book page 28
	Rounding to decimal places	pupil book page 30

Framework Objectives ✓ Round positive numbers to any given power of 10; round decimals to the nearest whole number or to one or two decimal places, **Framework examples pages 43 and 45**

Sheets Worksheet 4 for Support   Homework Sheets 1.6 and 1.7

Resources A number line for the Starter Activity. You could use a decimal number line. A calculator.

 ## Starter ideas and activities

Mental starter ideas
- Ask some questions on rounding, such as:
  round 345 to the nearest 10/100
  round 86·7 to the nearest 10/100
  round 56 789 to the nearest 1000 and so on.

- Ask some quick questions about rounding to the nearest whole number, such as:
  What is 6·7 to the nearest whole number?
  What is 8·72 ℓ to the nearest litre?
  What is £5·87 to the nearest pound?

Starter activity
- **Game: Rounding**
  1 Ask pupils to write down three multiples of 10 between 100 and 200.
  2 Call out some numbers between 100 and 200 and ask pupils to round them to the nearest 10. If the answer is one of their three numbers, they cross it out. The first to have their three numbers crossed out is the winner.
  3 You could play this with multiples of 100 or 1000 also.
- **Rounding to the nearest power of ten**
  1 Tell pupils that Canada has the longest coastline in the world. It is 90 908 km long.
  2 Ask for a volunteer to draw a number line and put this number on it. The end points of the number line must be decided. They could be 10 000 and 100 000, or 80 000 and 100 000, or 89 000 and 91 000.
  3 Keep asking for volunteers until all of these number lines have been drawn.
  4 Ask 'What is 90 908 rounded to the nearest 100 000, 10 000, 1000, 100, 10?'
  5 Repeat with another fact which is in the hundred thousands.
- **Rounding to decimal places using a number line to help**
  1 Put some number with 3 decimal places on the board.
  2 Ask for volunteers to put each number on a number line. (You may have to have these pre-drawn or you could ask a more able group to draw their own.)
  3 Ask what each number is to 2 decimal places.
  4 Ask how you could round these numbers to 1 decimal place.
  5 Give some more numbers with 3 or more decimal places and ask how to round each to one decimal place and then 2 decimal places without using a number line.

 ## Teaching points and activities

- Raise the question of how accurate numbers need to be in different situations, especially in measurement situations.
- Check that pupils can round to the nearest 10, 100 and 1000.
  Use questions 9, 44 and 70 in the Number Support chapter pages 8, 11 and 13.
  Some of Exercise 6 question 1 (Pupil Book page 28) also recaps this.
- Remind pupils that halfway values are rounded up and that this is a convention.
- Encourage pupils to use or visualise a number line when rounding.
- Pupils sometimes find it difficult to find the greatest and least values. Encourage them to use a number line and also to work backwards from the chosen answer to see if it would have in fact been rounded to the amount given. Also encourage checking other values to see if there is a greater or smaller value possible.
- Make sure pupils can round to the nearest whole number and that they understand that to do this they look at the digit in the next place (tenths).
- Emphasise that to round to a number we look at the next significant digit only and ignore all the rest.
- Sometimes pupils will round numbers such as 11.02345 as 11 to 1 d.p. Point out that it should be 11.0 as the zero tells us that it was rounded to 1 d.p. and not to the nearest whole number.

- When rounding the answer to a calculation encourage pupils to estimate the answer first. Consider the context to help decide what to round the answer to. Exercise 7 question 8 gives practice at this.
- Discuss situations when we would give the answer as a recurring decimal rather than round the decimal.

Support
- See **Rounding** on page 2 of the Pupil Book (Number Support chapter).
  If pupils have difficulty with Practice questions 9 or 70 use Worksheet 4 (page 4 in the Teacher Resource Pack).

Extension
- See **starred questions** in Exercises 6 and 7 (Pupil Book pages 28 and 31).
- See **Discussion** (Pupil Book page 30). Encourage pupils to come up with other similar situations.

Plenary ideas
- Use Exercise 6 question 2 as a class discussion.
- Write some measurements on the board. Say what the measurements are being used for. Ask pupils to round the measurements sensibly. For example:
Length of foot	22·8 cm
Mass of apple	149·92 g
Capacity of mug	248 mℓ
- Use Exercise 7 question 8 as a class discussion. Ask pupils to suggest an answer and explain why they chose it.
- Use the Discussion on page 30 of the Pupil Book if you didn't use it for the Starter Activity.

Homework
Review questions at the end of Exercises 6 and 7 (Pupil Book pages 28 and 31) and/or Homework Sheets 1.6 and 1.7 (pages 187 and 188 of this Pack).

Links
Link this topic to graphs: Exercise 6 question 4.
Link to Measures: Discussion page 30; Exercise 7 question 6, Review 3.
Link to Geography: Exercise 7 Review 6.
Link to Design and Technology: Exercise 7 question 3.
Link to finding the mean: Exercise 7 questions 6, 8a, 8b.
Link to Area: Exercise 7 question 7.

2 Integers, Powers and Roots

Topic in Chapter	Links to Programme of Study *Ma2 a–l are integrated throughout.*		Links to Framework for Teaching Mathematics: 8 Teaching Programme *Applying mathematics and solving problems are integrated throughout.*
Adding and subtracting integers *page 36*	**Ma2 2a**	... understand and use negative numbers, both as positions and translations on a number line; order integers ...	**Integers, powers and roots** ● **Add, subtract, multiply and divide integers.**
Multiplying and dividing integers *page 39*	**Ma2 3a**	add, subtract, multiply and divide integers ...	**Calculator methods** ● Carry out more difficult calculations effectively and efficiently using the function keys for sign change ...
	Ma2 3i	develop a range of strategies for mental calculation; derive unknown facts from those they know	● Enter numbers and interpret the display in different contexts (negative numbers ...).
Integers on a calculator *page 42*	**Ma2 3o**	use calculators effectively and efficiently; know how to enter complex calculations using brackets (for example, for negative numbers, ...)	
Order of operations with integers *page 43*	**Ma2 3b**	use brackets and the hierarchy of operations; know how to use the ... distributive laws to do mental calculations more efficiently.	**Number operations and the relationships between them** ● Use the order of operations, including brackets, with more complex calculations.
	Ma2 3i	develop a range of strategies for mental calculation; derive unknown facts from those they know	
Divisibility *page 44*	**Ma2 2a**	... use the concepts and vocabulary of factor (divisor), multiple, common factor, highest common factor, least common multiple, prime number and prime factor decomposition.	**Integers, powers and roots** ● Recognise and use multiples, factors (divisors), common factor, highest common factor, lowest common multiple and primes; find the prime factor decomposition of a number (e.g. $8000 = 2^6 \times 5^3$).
Prime factor decomposition *page 47*	**Ma2 3a**	... find the prime factor decomposition of positive integers (for example, $8000 = 2^6 \times 5^3$)	**Applying mathematics and solving problems** ● Suggest extensions to problems, conjecture and generalise; identify exceptional cases or counter-examples.
Highest common factor and lowest common multiple *page 50*	**Ma2 3g**	... recall all multiplication facts to 10×10, and use them to derive quickly the corresponding division facts ...	● **Use logical argument to establish the truth of a statement ...**
	Ma2 3i	develop a range of strategies for mental calculation; derive unknown facts from those they know	

Topics in Chapter	Links to Programme of Study		Links to Framework for Teaching Mathematics: 8 Teaching Programme
Squares and square roots *page 51*	**Ma2 2b**	use the terms square, positive and negative square root (knowing the square root sign denotes the positive square root), cube, cube root ...	**Integers, powers and roots** ● Use squares, positive and negative square roots, cubes and cube roots, and index notation for small positive integer powers.
Cubes and cube roots *page 55*	**Ma2 3g**	... recall the cubes of 2, 3, 4, 5 and 10 ...	**Mental methods and rapid recall of number facts** ● Consolidate and extend mental methods of calculation, ... working with ... squares and square roots, cubes and cube roots ...
	Ma2 3i	develop a range of strategies for mental calculation; derive unknown facts from those they know	**Calculator methods** ● Carry out more difficult calculations effectively and efficiently, using the function keys for ... powers, roots ...
	Ma2 3p	use the function keys for ... squares, square roots ...	**Applying mathematics and solving problems** ● Solve more demanding problems and investigate in a range of contexts ...
	Ma2 4a	draw on their knowledge of ... simple integer powers and their corresponding roots ...	

Links to Attainment Targets

Level 4 ... pupils use a range of mental methods of computation with the four operations, including mental recall of multiplication facts up to 10 × 10 and quick derivation of corresponding division facts ... Pupils recognise and describe ... relationships including multiple, factor and square.

Level 5 ... They order, add and subtract negative numbers in context.

Topic area	**Adding and subtracting integers**	pupil book page 36

Framework Objectives
✓ Add, subtract, ... integers, Framework examples page 49
✓ Carry out more difficult calculations effectively and efficiently using the function keys for sign change ...
✓ Enter numbers and interpret the display in different contexts (negative numbers ...)

Sheets
Worksheet 5 for Support   Homework Sheet 2.1   Resource Sheet pages 193 and 194
(Teacher Resource Pack)   Starter Resource Sheet 2

Resources
A cube with the numbers 1, 2, 3, ⁻1, ⁻2, ⁻3 on the faces for Game: To the Dungeon. You could use number lines or number cards. Thermometer for mental starter and number cards for plenary.

Starter ideas and activities

Mental starter ideas
- Ask questions such as Practice questions 12, 28, 29 and 32 on pages 9 and 10 in the Number Support chapter of the Pupil Book.
- Put the temperature 10 °C on the board. Ask pupils what the temperature would be if it rose 5 °C, dropped 15 °C and so on.

Starter activity
- **Game: To the Dungeon**
  Play the game on page 35 of the Pupil Book.
- **Adding and subtracting integers**
  Use the Discussion on page 36 of the Pupil Book. This gets fairly complicated in parts so you may want to only use parts of it for a less able class. Before doing the Discussion you may want to go through some examples using first a thermometer and then a number line. (Use Starter Resource Sheet 2.)

Teaching points and activities

- Some pupils will find the Discussion points on page 36 difficult. It is not essential that this is covered as Exercise 1 can be completed without having done so.
- Encourage the use of a number line to help if pupils are having difficulty.
- Encourage pupils to see that a negative number and a subtraction sign may indicate either a position or a translation on the number line.
- Pupils may know the rule that two negatives make a positive and use it to work out ⁻7 – 8 as 15. Encourage the use of a number line to show that ⁻7 gives the position on the line and the subtraction sign gives a translation to the left to ⁻15.

Support
- See **Integers** on page 2 of the Pupil Book (Number Support chapter).
- If pupils have difficulty with Practice questions 3, 12, 23, 28, 29, 32, 52, 54, 75 and 76 use Worksheet 6 (page 5 in the Teacher Resource Pack).
- See **Game: To the Dungeon** (Pupil Book page 35).

Extension
- See **starred questions** in Exercise 1 (Pupil Book page 37)
- See **Discussion** (Pupil Book page 36). Challenge pupils to write the answers to the questions before discussing them.
- See **Puzzle** (Pupil Book page 39). Challenge pupils to make up their own puzzle like this.

Plenary ideas
- Go over adding and subtracting integers using a number line. Also revise the character boxes in the teaching screen on page 36 of the Pupil Book.
- Use Exercise 1 question 3, 7 or 8. Ask pupils to write possible answers on the board and explain how they got them.
- Use number cards and ask pupils to, for example, find two cards that add to ⁻5 or have a difference of 3. Discuss the answers, using a number line to verify.

Homework
Review questions at the end of Exercise 1 (Pupil Book page 37) and/or Homework Sheet 2.1 (page 189 of this pack).

Graphical calculator
See page 181 of the Teacher Resource Pack for command sequences to use to add and subtract integers.

Links
Link this topic to understanding the operations addition and subtraction, and substituting positive and negative numbers in expressions and formulae (Pupil Book page 172).

| Topic areas | **Multiplying and dividing integers** | pupil book page 39 |
| | **Integers on a calculator** | pupil book page 42 |

Framework Objectives

✓ ... multiply and divide integers, Framework examples page 51
✓ Carry out more difficult calculations effectively and efficiently using the function keys for sign change ..., **Framework examples pages 51 and 109**
✓ Enter numbers and interpret the display in different contexts (negative numbers ...), **Framework examples pages 51 and 109**

Sheets Homework Sheets 2.2 and 2.3 Resource Sheet pages 194, 195 and 196 Teacher Resource Pack

Resources Mini-whiteboards for the Starter Activity. A calculator.

Starter ideas and activities

Mental starter ideas
Revise tables with quick questions.
Give some sums and differences, including negatives, and ask pupils to estimate the answers. For example, $^-486 + 228 \approx \, ^-500 + 200 = \, ^-300$.

Starter activity
● **Multiplying and dividing with patterns**
  1 Put some patterns on the board like the ones in Exercise 2 question 1.
  2 Ask for volunteers to come and write the next lines of the pattern on the board or on their mini-whiteboards or similar.
● **Rules of multiplying and dividing integers**
  1 Draw out with questions what the rules are for multiplying (see top of screen on page 39 of the Pupil Book)
  2 *What is the inverse of multiplying?*
  3 Draw out with questions the rules for dividing. (See the character in the screen on page 39 and 40 of the Pupil Book.)
  4 Write some sums, such as $^-3 \times 4 = \, ^-12$, and ask pupils to use inverse operations to write two divisions from this. ($^-12 \div 4 = \, ^-3$ and $^-12 \div \, ^-3 = 4$)
● **Sign change key and using the calculator**
  1 Ask pupils to find the sign change key on their calculators.
  2 Give some simple sums to do on the calculator, e.g. $^-4 + 3 = \, ^-1$, $7 + \, ^-10 = \, ^-3$ and so on. This is so pupils can check mentally that the answer is correct.

Teaching points and activities

● Encourage pupils to use their multiplication facts to work out divisions.
● Use patterns to help pupils discover the rules for multiplying and dividing integers.
● Be aware that different calculators may have different keys to give a sign change.
● Spend some time going over inverse operations, i.e. that dividing by a negative number is the inverse of multiplying by a negative number. You could extend this to give a group of facts, for example:
  if $\quad ^-2 \times \, ^-4 = \, ^-8$
  then $^-4 \times \, ^-2 = \, ^-8$
  and $\quad ^-8 \div \, ^-4 = \, ^-2$
  and $\quad ^-8 \div \, ^-2 = \, ^-4$.
● In this book a **raised** negative sign is used to show a negative number. Point out to pupils that this is not always the case in all written material.
● Encourage pupils to estimate the answers to calculations.

Support
● Pupils should have done multiplying and dividing whole numbers and using a calculator. Pupils should have covered the previous section in this chapter.

Extension
● See **starred questions** in Exercise 2 (Pupil Book page 40).

Plenary ideas
● Put some patterns on the board for pupils to continue. Point out that the resulting patterns can be used to deduce the rules for multiplying and dividing integers.
● Use Exercise 2 question 10 and ask for volunteers to fill in numbers. Then ask the class if the number filled in is possible and what other possibilities there are. You could have two charts on the board at the same time with the alternative solutions showing.

Homework
Review questions at the end of Exercises 2 and 3 (Pupil Book pages 40 and 42) and/or Homework Sheets 2.2 and 2.3 (page 190 of this pack).

Links
Link this topic to substituting positive and negative numbers into expressions and formulae on (Pupil Book page 172).
Link to sequences: Discussion page 39; Exercise 2 question 1.

© New National Framework Mathematics 8 Nelson Thornes Ltd

Topic area	Order of operations with integers	pupil book page 43

Framework Objectives ✓ Use the order of operations, including brackets, with more complex calculations, **Framework examples pages 51 and 87**

Sheets Worksheet 8   Homework Sheet 2.4   Resource Sheet page 196 Teacher Resource Pack

Starter ideas and activities

Mental starter ideas Ask questions such as Practice question 7 on page 8 in the Number Support chapter of the Pupil Book. You could divide the class into groups for parts c and d and have a race to see who can find an answer first, second, etc.

Starter activity
- **Revising order of operations with whole numbers**
  1 *What does BIDMAS stand for?*
  2 Give a string of numbers such as 5, 6, 8, 2, 4 and see who can come up with the greatest answer using brackets and the four operations. You could also give a specific answer that you have already worked out and see who can get this answer first.
- **Order of operations with integers**
  Now give a string of numbers which includes a negative number, such as ⁻3, 5, 8, ⁻7 and ask who can come up with the largest answer or give a specific answer and have a race to see who can get this answer first.

Teaching points and activities

- Go over the use of the word BIDMAS to aid remembering the order of operations.
- Remind pupils that brackets are not needed for a calculation such as $8 + (3 \times 7)$ because if the brackets weren't there, the multiplication would still be done first.
- Emphasise that we always work from left to right doing all the brackets, then go back to the beginning and work from left to right doing all the powers, then go back to the beginning and work from left to right doing all the multiplication **and** division, and then go back to the beginning and work from left to right doing all the addition **and** subtraction.
- Point out that when multiplying out a bracket we are using the distributive law. Make the link to algebra.
- Emphasise that when there is a division line the whole numerator must be divided by the whole denominator and so sometimes brackets are needed to ensure the numerator and denominator are calculated first. Show that $\frac{8+3}{7-2}$ is the same as $(8 + 3) \div (7 - 2)$.

Support
- See **Order of operations** on page 4 of the Pupil Book (Number Support chapter). If pupils have difficulty with Practice questions 7 and 30 use Worksheet 8 question 7 and Worksheet 10 questions 1 to 3 (pages 7 and 8 Teacher Resource Pack).

Extension
- See **Starred questions** in Exercise 4 (Pupil Book page 44).

Plenary ideas
- Go over Exercise 4 question 2 on page 44 of the Pupil Book, but instead of finding the largest number, find as many different answers as possible. You could ask pupils to come and write their different answers on the board.
- Give some questions which highlight the problem areas, such as:
  $16 \div 4 \times ⁻2$ or $5 \times 6 \div ⁻2$ (we must work from left to right),

  $\dfrac{5 \times ⁻2}{3 + 2}$ (brackets must be inserted).
- Point out the links, especially the link to algebra topics.

Homework Review question at the end of Exercise 4 (Pupil Book page 44) and/or Homework Sheet 2.4 (page 191 in this pack).

Links Link this topic to Order of Operations using the calculator (page 92 pupil book) and order of operations in algebra (page 172 pupil book).

Topic area	**Divisibility**	pupil book page 44

Framework Objectives
✓ Recognise and use multiples, factors (divisors), common factor, **Framework examples page 53**
✓ Problem solving, **Framework examples page 31**

Sheets
Worksheet 5 for Support   Homework Sheet 2.5   Resource Sheet page 190 Teacher Resource Pack

Resources
A graphical calculator. Mini-whiteboards for starter activity.

Starter ideas and activities

Mental starter ideas
Ask some quick questions about the division facts to 10×10.

Starter activity
● **Revision of divisibility rules**
  1 Put Practice question 33 on page 11 in the Number Support chapter of the Pupil Book — see Resource Sheet (page 190 Teacher Resource Pack).
  2 Revise the divisibility rules first (Pupil Book page 2).
  3 Ask for volunteers to shade the answers.
  4 Before the shading is done, ask everyone else to put the answer of their mini-whiteboard or similar.
● **Divisibility of larger numbers**
  Use the Discussion on page 45 of the Pupil Book. This should bring out the point that the divisor must be written as the product of two numbers which have no common factors.

Teaching points and activities

● A recap of factors may be needed before looking at divisibility by larger numbers.
● The Discussion on page 45 of the Pupil Book is important. Pupils need to understand the answers to both parts to be able to answer some of the questions in the exercise. They must understand that the factors chosen to check divisibility must have no factors in common other than 1. If you do not do the Discussion on page 45, you will need to bring out this point somehow, e.g. when finding if a number is divisible by, say, 18 we can write 18 as the product of either 6×3 or 9×2. To check if a number is divisible by 18, checking if it is divisible by 9 and 2 will work, but checking if it is divisible by 6 and 3 will not because 6 and 3 have a common factor of 3.
● Exercise 5 question 6 could be used as a discussion. This is a good use of a graphical calculator because the display stays on the screen.

Support
● See **Divisibility** on page 2 of the Pupil Book (Number Support chapter).
● If pupils have difficulty with Practice questions 33 and 42 use Worksheet 5 (page 4 of the Teacher Resource Pack).

Extension
● See **starred questions** in Exercise 5 (Pupil Book page 45).
● See **Investigation: Consecutive Numbers** (Pupil Book page 46). Encourage pupils to use algebra to prove that the sum of any five consecutive numbers is divisible by 5.
● See **Investigation: Divisibility by 7 and 11** (Pupil Book page 46). Ask pupils to write their working for part 2 as a report, including an explanation of what they did.

Plenary ideas
● Remind the class of the divisibility rules they should already know.
● Choose and go over an example of divisibility by larger numbers in which it is essential that pupils realise the factors of the larger number must have no common factors other than 1.

Homework
Review questions at the end of Exercise 5 (Pupil Book page 45) and/or Homework Sheet 2.5 (page 191 of this pack).

Graphical calculator
Exercise 5 question 6. Check if a number is prime. See page 182 of the Teacher Resource Pack.

Links
Link this topic to Dividing (Pupil Book page 85) and factors (Pupil Book pages 47–50).

Topic areas	Prime factor decomposition	pupil book page 47
	Highest common factor and lowest common multiple	pupil book page 50

Key Framework Objectives

✓ Recognise and use multiples, factors (divisors), common factor, highest common factor, lowest common multiple and primes; find the prime factor decomposition of a number (e.g. $8000 = 2^6 \times 5^3$), **Framework examples pages 53 and 55**

✓ Problem solving, **Framework examples pages 7 and 33**

Sheets Worksheet 7 for Support   Homework Sheets 2.6 and 2.7   Resource sheet page 197 Teacher Resource Pack

Resources You could use a factor-finder and/or a prime number chart.

Starter ideas and activities

Mental starter ideas

● Ask some questions such as Practice questions 4, 8, 20 and 74 on pages 8 and 14 in the Number Support chapter of the Pupil Book.

● Ask pupils to chant the multiples of simple numbers including simple fractions and decimals.

Starter activity

● **Finding highest common factors and lowest common multiples**
  1 Ask for the factors of 24. Write each factor as it is given on a card or piece of paper.
  2 Ask for volunteers to come and hold one of the cards.
  3 Do the same for the factors of 20.
  4 *Which is the highest common factor?*
  5 Now ask for the multiples of 8. Write each multiple up to 64 on a card. Ask for volunteers to come and hold one of the cards.
  6 Do the same for the first 8 multiples of 12.
  7 *Which is the lowest common multiple?*

● **Prime factor decomposition**
  Demonstrate how to write a number as the product of prime numbers using a table and then using a factor tree. Ask pupils to demonstrate others.

Teaching points and activities

● Make sure pupils know the prime numbers up to 30 off by heart and understand what a prime number and prime factor are.

● Pupils often find **a** common factor rather than the **highest** common factor. Encourage pupils to check if the common factor they have found is the highest. They can do this by dividing both numbers by this factor and seeing if the answers to these two divisions have any common factors. If they have then the number they divided by is not the highest common factor.

Support

● See **Multiples, factors, primes, squares, triangular numbers** on page 3 of the Pupil Book (Number Support chapter). If pupils have difficulty with Practice questions 4, 8, 20, 24, 56, 74 and 77 use Worksheet 7 questions 1, 2, 3, 4a, 4b, 4c, 5, 9, 12 (page 6 of the Teacher Resource Pack).

Extension

● See **Starred questions** in Exercise 7 (Pupil Book page 50).

● See **Investigation: Factors** (Pupil Book page 49). In part 1 challenge pupils to find some 3-digit numbers. In part 4, challenge pupils to use algebra.

● See **Puzzle** (Pupil Book page 49), question 4.

Plenary ideas

● Choose two or three parts of Exercise 7 question 7 on page 50 of the Pupil Book to go over in detail. Also go over prime factor decomposition of the numbers in one of the parts of question 7.

● Start the Puzzles or Investigation on page 49 of the Pupil Book and ask pupils to finish it for homework. Discuss strategies for doing these.

● Discuss the answers to Exercise 7 questions 4 and 5 and ask pupils to explain how they found their answers.

Homework Review questions at the end of Exercises 6 and 7 (Pupil Book pages 48 and 50) and/or Homework Sheets 2.6 and 2.7 (page 192 of this pack).

Links Link this topic to Adding and Subtracting Fractions (Pupil Book page 108) and Prime Numbers (Pupil Book pages 3 and 47).

Link to cancelling fractions: Exercise 7 question 6.

| Topic area | **Squares and square roots** | pupil book page 51 |

Framework Objectives

✓ Use squares, positive and negative square roots, ..., **Framework examples pages 57 and 59**
✓ Carry out more difficult calculations effectively and efficiently, using the function keys for ... powers, roots ..., **Framework examples page 109**
✓ Problem solving, **Framework examples pages 7 and 29**

Sheets Worksheets 7 and 8   Homework Sheet 2.8

Resources A mini-whiteboard for the Starter Activity. You could use squares, to show how square numbers are formed, or a counting stick. A calculator with square and square root keys.

 Starter ideas and activities

Mental starter ideas
- Ask questions such as Practice questions 22, 47 and 62 on pages 10, 11 and 13 in the Number Support chapter of the Pupil Book.
- *What is a square number?*
Ask pupils to give you the first 12 square numbers.
Give some square numbers and ask for the square roots.

Starter activity
- **Positive and negative square roots**
  1 Revise the meaning of the word square root.
  2 Ask pupils to write on their mini-whiteboards the answer to 4×4 and $^-4 \times ^-4$.
    What is the square root of 16?
  3 Pupils will probably say 4, so ask 'Is there another number that is also the square root of 16?'
  4 *What is the inverse of squaring a number?*
    What does $(^-6)^2$ equal?
    What is the square root of 36? (Make sure pupils give both 6 and $^-6$ as the answer.)
  5 Give some more square numbers and ask for the both the positive and negative square roots.
- **Finding square roots using factors**
  1 Ask pupils to write 144 as the product of two square numbers.
    How could you use this to find the square root of 144? (9×16)
    Does $\sqrt{144} = \sqrt{9 \times 16}$? Does $\sqrt{144}$ equal $\sqrt{9} \times \sqrt{16}$?
  2 Give some more examples and ask pupils to make a general statement about $\sqrt{mn}$ and $\sqrt{m} \times \sqrt{n}$.
- **Square roots on the calculator**
  Ask pupils to find some squares and square roots on their calculators. A calculator for an OHP is useful. You could divide the class in two and play a game of 'Beat the Calculator' by putting some square roots on the board that can be done mentally and ask one group to find them mentally and the other using the calculator.

 Teaching points and activities and activities

- Recap on inverse operations, estimating, triangular numbers and consecutive numbers.
- Make sure pupils know the first 12 square numbers and their square roots.
- Emphasise that squaring and square rooting are inverse operations.
- Encourage pupils to estimate answers when squaring and square rooting. Use known square numbers and their square roots as upper and lower bounds.
- Revise order of operations before beginning the exercise. It is needed in Exercise 8 questions 1 and 5.
- Pupils may try and do questions such as $\sqrt{118 - 47}$ as $\sqrt{118} - \sqrt{47}$. Point out that the $\sqrt{}$ sign acts like a bracket and the answer to the calculation under the $\sqrt{}$ sign must be worked out first. If using the calculator, a bracket must be put around the calculation under the $\sqrt{}$ sign.
- Pupils get confused about the difference between $^-6^2$ and $(^-6)^2$. Point out that in the first one we square 6 first to get 36 and then make the answer negative to get $^-36$. In the second one we are squaring $^-6$ (because of the bracket) which is $^-6 \times ^-6 = 36$.
- Exercise 8 question 11 is good for revising or introducing trial and improvement.

Support
- See **Multiples, factors, primes, squares, triangular numbers** on page 3 of the Pupil Book (Number Support chapter).
- If pupils have difficulty with Practice questions 22, 47, 62, 63, 78 use Worksheet 7 questions 4d, 6, 7, 8, 10, 11, 13 (page 6 of the Teacher Resource Pack) and Worksheet 8 question 13 (page 7 of the Teacher Resource Pack).

Extension
- See **Starred questions** in Exercise 8 (Pupil Book page 52).
- See **Puzzle**, (Pupil Book page 54) question 4.
- See **Investigation: Sum and difference of two squares** (Pupil Book page 55). Challenge pupils to also find the first five 3-digit numbers that can be written as the sum of two squares in part **a**. They are 100, 101, 104, 106 and 109. In **b** challenge them to find numbers between 30 and 50 which can be written as the difference of two square numbers. They are 32, 35, 36, 39, 40, 44, 45 and 48.

Plenary ideas
- Give some simple square root questions such as $\sqrt{81}$ and then extend by giving an example where the square root is found by factorising.
- Recap on the meaning of the terms factorise, square, square number, positive and negative square root. Ask pupils to explain the meaning of these terms using an example to illustrate.
- Begin the investigation on page 55 of the pupil book and ask pupils to finish it for homework.

Homework
Review questions at the end of Exercise 8 (Pupil Book page 52) and/or Homework Sheet 2.8 (page 193 of this pack).

Graphical calculator
Use a calculator to estimate square roots. Exercise 8 question 11. See page 184 of the Teacher Resource Pack.

Links
Link this topic to finding factors (Pupil Book page 47) and Estimating (Pupil Book page 74).

Topic area	**Cubes and cube roots**	pupil book page 55

Framework Objectives
✓ Use ... cubes and cube roots ..., **Framework examples pages 57 and 59**
✓ Carry out more difficult calculations effectively and efficiently, using the function keys for ... powers, roots ..., **Framework examples page 109**
✓ Problem solving, **Framework examples pages 7 and 29**

Sheets Homework Sheet 2.9

Resources Mini-whiteboards for Mental Starter. An OHP for the Starter activities. A calculator, preferably with cube and cube root keys. Multilink cubes. A graphical calculator for Investigation: End Digits (Pupil Book page 57).

Starter ideas and activities

Mental starter ideas Revise rounding to 1 and 2 decimal places and give some numbers for pupils to round. You could ask them to write their answers on their mini-whiteboards.
You could do this as multi-choice at first if pupils are having difficulty.

You could revise tables, squares and square roots and writing numbers using indices.

Starter activity
● **Cubes** Use the Discussion on page 55 of the Pupil Book. You could use small cubes to model the cubes in the discussion or ask for volunteers to build them.
● **Cube roots**
  1 Draw a function machine on the board or on an OHP.
  2 Show cubing 5 using the function machine. Ask 'How can we go back the other way?'
  3 Draw the inverse function machine.
  4 Cube some other numbers using the function machine and find the cube root using the inverse function machine. Ask 'What can you say about cubing and cube rooting?'
  5 Ask pupils to find how to cube and cube root using a calculator.
  6 Give some numbers to cube and cube root and discuss the answers and the rounding of them. Encourage pupils to estimate first using some known cube numbers.

$5 \rightarrow$ cubed $\rightarrow 5 \times 5 \times 5 = 125$

$5 \leftarrow$ cube root $\leftarrow 125$

Teaching points and activities

● Some calculators may not have a cube root button. Pupils would then need to use the ⬚ button.
● Emphasise that cubing and finding the cube root are inverse operations.
● Make sure pupils can round to 1 and 2 decimal places before doing Exercise 9 questions 4 and 5.
● Emphasise that pupils need to learn the cubes of 1, 2, 3, 4, 5 and 10.
● Pupils often cube something like $(^-5)^3$ as $^-5^3$. Show the need for the brackets when using a calculator or give a similar explanation to the one given for squaring negative numbers in the previous section.

Support ● Pupils should have covered the previous section, *Squares and square roots*.

Extension
● See **Starred question** in Exercise 9 (Pupil Book page 56)
● See **Puzzle** (Pupil Book page 57).
● See **Investigation: End Digits** (Pupil Book page 57). Ask pupils to write a report.

Plenary ideas
● Use the Puzzle on page 57 of the Pupil Book as a class discussion. Revise the cube numbers they know first.
● Play a game of Bingo. Ask pupils to write down nine of these numbers 1, 4, 8, 9, 16, 25, 27, 36, 49, 64, 81, 100, 121, 125, 144, 1000.
  Ask questions, such as:
    What is the cube of 3?
    What is the square root of 121?
    What is the cube root of 64?
  If a pupil has the answer they cross it out.
  The first person to cross out all nine numbers is the winner.
● You could start the Investigation on page 57 and get pupils to finish it for homework. Pupils could use an ordinary calculator to do the first part at home if they do not have a graphical one. They would need to write down the answers as they get them.

Homework   Review questions at the end of Exercise 9 (Pupil Book page 56) and/or Homework Sheet 2.9 (page 193 of this pack).

Graphical calculator   Investigation: End Digits. You could use a graphical calculator.

Links   Link to sequences: Investigation: End Digits, page 57, and Exercise 9 question 6.
Link to volume: Discussion page 55

3 Mental Calculation

Topic in Chapter	Links to Programme of Study		Links to Framework for Teaching Mathematics: 8 Teaching Programme
	Ma2 a–l are integrated throughout.		*Applying mathematics and solving problems are integrated throughout.*
Adding and subtracting *page 62*	**Ma2 3a**	Add, subtract ... integers and then any number ...	**Integers, powers and roots** ● Add, subtract, multiply and divide integers.
	Ma2 3b	... know how to use the commutative, associative ... laws to do mental ... calculations more efficiently	**Number operations and the relationships between them** ● Understand addition and subtraction of integers, and multiplication and division of integers; use the laws of arithmetic and inverse operations.
	Ma2 3g	recall all positive integer complements to 100 (for example 37 + 63 = 100)	**Mental methods and rapid recall of number facts** ● Recall known facts ...; use known facts to derive unknown facts, including products involving numbers such as 0·7 and
	Ma2 3i	develop a range of strategies for mental calculation ... add and subtract mentally numbers with up to two decimal places (for example, 13·76 − 5·21, 20·08 + 12·4)	6 and 0·03 and 8. ● Consolidate and extend mental methods of calculation, working with decimals ...
Multiplying and dividing *page 66*	**Ma2 3a**	... multiply and divide integers and then any number ...	
	Ma2 3b	... know how to use the commutative, associative and distributive laws to do mental ... calculations more efficiently	
	Ma2 3g	... recall all multiplication facts to 10 × 10, and use them to derive quickly the corresponding division facts ...	
	Ma2 3i	develop a range of strategies for mental calculation ... multiply and divide numbers with no more than one decimal digit (for example, 14·3 × 4, 56·7 ÷ 7), using factorisation when possible	

Topic in Chapter	Links to Programme of Study	Links to Framework for Teaching Mathematics: 8 Teaching Programme
Solving problems mentally *page 70*	**Ma2 1a** ... select appropriate strategies to use for numerical ... problems **Ma2 1d** select efficient techniques for numerical calculation ... **Ma2 3i** develop a range of strategies for mental calculation ... **Ma2 4a** draw on their knowledge of the operations and the relationships between them ... **Ma2 4b** select appropriate operations, methods and strategies to solve number problems ...	**Mental methods and rapid recall of number facts** ● Recall known facts ... use known facts to derive unknown facts, including products involving numbers such as 0·7 and 6, and 0·03 and 8. ● Consolidate and extend mental methods of calculation, working with decimals ...; solve word problems mentally. **Applying mathematics and solving problems** ● Solve more demanding problems and investigate in a range of contexts: number, algebra, shape, space and measures and handling data.
Making estimates *page 73* Estimating answers to calculations *page 74*	**Ma2 1e** make mental estimates of the answers to calculations ... **Ma2 3h** ... estimate answers to problems involving decimals **Ma2 4c** use a variety of checking procedures, including ... considering whether the result is of the right order of magnitude	**Mental methods and rapid recall of number facts** ● Make and justify estimates and approximations of calculations. **Checking results** ● Check a result by considering whether it is of the right order of magnitude ... .

Links to Attainment Targets

Level 4 In solving number problems, pupils use a range of mental methods of computation with the four operations, including mental recall of multiplication facts up to 10 × 10 and quick derivation of corresponding division facts.

In solving problems ... without a calculator, pupils check the reasonableness of their results by reference to ... the size of the numbers.

Level 5 ... They use all four operations with decimals to two places. They check their solutions by ... estimating using approximations.

Level 6 Pupils ... approximate decimals when solving numerical problems ...

Level 7 In making estimates, pupils round ... and multiply and divide mentally (begin)

Topic area   **Adding and subtracting**   pupil book page 62

Framework Objectives
- ✓ Add, subtract ... integers, **Framework examples pages 89 and 93**
- ✓ Understand addition and subtraction of integers; ... use the laws of arithmetic ..., **Framework examples pages 89 and 93**
- ✓ Recall known facts ...; use known facts to derive unknown facts ..., **Framework examples pages 89 and 93**
- ✓ Consolidate and extend mental methods of calculation, working with decimals ..., **Framework examples pages 89 and 93**
- ✓ Solve word problems mentally, **Framework examples page 101**

Sheets   Worksheet 8 for Support   Homework Sheet 3.1   Resource Sheet page 198 Teacher Resource Pack

Resources   You could use a number line.

Starter ideas and activities

Mental starter ideas   Ask pupils questions which revise complements.
What do you have to add to 57 to get 100?   What is 1000 – 456?
What must I add to 4·7 to get 10?   What does 0·64 + 0·36 equal?
Ask some questions such as Practice questions 15, 31 and 36 on pages 10 and 11 in the Number Support chapter of the Pupil Book.

Starter activity   **Strategies for adding and subtracting mentally**
Put some additions and subtractions on the board and have pupils come up and explain how they would find the answers. Encourage them to show their method by drawing or using a number line.
Examples   5·7 + 3·8   5700 – 1632   6·9 + 2·1   2·39 + 3·9   0·68 + 0·54
Discuss other ways of doing each and which way is the most efficient.

Teaching points and activities

- Encourage the use of all the strategies given in the screen on page 62 of the Pupil Book.
- Pupils often have trouble with subtraction.
  Encourage the use of a number line and methods such as partitioning and compensation.
- Pupils sometimes try to use one method to work out every example. Encourage pupils to develop a range of strategies and discuss how some methods are more efficient than others sometimes. You can do this by asking pupils to do the same sum in several different ways and then discussing the methods. With practice they will learn to recognise an efficient method for the calculation given.
- This topic needs constant reinforcement and it is a good idea to give a couple of mental calculations frequently.

Support
- See **Mental calculation** on page 3 of the Pupil Book (Number Support chapter).
  If pupils have difficulty with Practice questions 13a, 15, 17, 31 and 36 use Worksheet 8 questions 3, 4, 5, 8 and 10 (page 7 of the Teacher Resource Pack).

Extension
- See **Starred questions** in Exercise 1 (Pupil Book page 63).
- See **Puzzles** (Pupil Book page 66). Pupils could be asked to make a puzzle similar to part 2 and give this to someone else to solve.
- See **Investigation: Magic Squares** question 5 (Pupil Book page 65). Making up a magic square of their own is not easy unless the formula is known.

Plenary ideas
- Put a calculation on the board as you did at the start and ask for suggestions of ways to do it, e.g. 18·3 + 6·9 or 8·03 – 4·64.
- Start the Puzzle on page 66 of the Pupil Book and ask pupils to finish it for homework.
- Start the Investigation on page 65 of the Pupil Book and ask pupils to finish it for homework.

Homework   Review questions at the end of Exercise 1 (Pupil Book page 63) and/or Homework Sheet 3.1 (page 194 of this pack).

Links   Link this topic to understanding the operations addition and subtraction.
Link to Adding and subtracting integers: Exercise 1 question 1, Review 1.

Topic area	**Multiplying and dividing**	pupil book page 66

Framework Objectives
- ✓ ... **multiply and divide integers, Framework examples pages 85, 89 and 97**
- ✓ Understand ... multiplication and division of integers; use the laws of arithmetic ..., **Framework examples pages 85, 89 and 97**
- ✓ Recall known facts ...; use known facts to derive unknown facts, including products involving numbers such as 0·7 and 6, and 0·03 and 8, **Framework examples pages 85, 89 and 97**
- ✓ Consolidate and extend mental methods of calculation, working with decimals ..., **Framework examples pages 85 and 97**
- ✓ Solving word problems mentally, **Framework examples page 101**

Sheet Worksheet 8 for Support   Homework Sheet 3.2   Resource Sheet pages 198 and 199 Teacher Resource Pack

Resources You could use number lines.

Starter ideas and activities

Mental starter ideas Revise tables and quick multiplications and divisions such as $4 \times 5 \times 7$ and $63 \div 3$ and so on.

Starter activity
- **Game to revise multiplying and dividing**
  Choose a target number, say 144 or 240 or 300 and start at 2. Divide the class into two groups which take turns. Ask pupils to multiply or divide the answer to the last calculation by a one-digit number. The winner is the group who reaches the target number, e.g. $2 \times 4 = 8$ and $8 \times 9 = 72$ and $72 \times 2 = 144$.
- **Strategies for multiplying and dividing mentally**
  Discuss methods to do 12×2.5 and have pupils come up and demonstrate these using a number line if appropriate. Use two or three different ways.
- **Relating multiplying to finding areas**
  Put some rectangles on the board or on an OHP and ask pupils to find the area. Discuss and ask pupils to demonstrate methods.

Teaching points and activities

- A quick reminder of the rules for multiplying and dividing integers may be needed.
- Encourage the use of all the strategies given on pages 66 and 67 of the Pupil Book.
- Pupils sometimes try to use one method to work out every example. Encourage pupils to develop a range of strategies and discuss how some methods are more efficient than others sometimes. You can do this by asking pupils to do the same sum in several different ways and then discussing the methods. With practice they will learn to recognise an efficient method for the calculation given.
- Care needs to be taken when going over the method for questions such as $1600 \div 800$. $\frac{1600}{800} = \frac{1600}{8 \times 100}$ which is the same as $1600 \div 8 \div 100$. Pupils may be inclined to say it equals $1600 \div 8 \times 100$.
- This topic needs constant reinforcement and it is a good idea to give couple of mental calculations frequently.

Support
- See **Mental calculation** on page 3 of the Pupil Book (Number Support chapter).
- If pupils have difficulty with Practice questions 5, 13b, 27, 35, 40, 43 and 69 use Worksheet 8 questions 1, 2, 11, 12 and 14 (page 7 of the Teacher Resource Pack).

Extension
- See **starred questions** in Exercise 2 (Pupil Book page 68).

Plenary ideas
- As a summary of the different methods covered in this lesson, go over Exercise 2 question 9 on page 68 of the Pupil Book.
- Put these three calculations on the board and ask pupils to come up with the most efficient way of doing each.
  14×2.5 (probably $10 \times 2.5 + 4 \times 2.5$)
  2.3×16 (probably $2.3 \times 2 \times 2 \times 2 \times 2$)
  56×21 (probably $56 \times 20 + 56$)

Homework Review questions at the end of Exercise 2 (Pupil Book page 68) and/or Homework Sheet 3.2 (page 195 of this pack).

Links Link this topic to understanding the operations multiplication and division, multiplying and dividing integers (Pupil Book page 39) and place value (Pupil Book page 16).
Link to integers: Exercise 2 questions 2, 3, 4, 7b, 8 and Reviews 2, 3 and 4.

| **Topic area** | **Solving problems mentally** | pupil book page 70 |

Framework Objectives ✓ Consolidate and extend mental methods of calculation working with decimals ...; solve word problems mentally, **Framework examples page 101**

Sheets Worksheet 8 for Support   Homework Sheet 3.3

Resources You could use number lines.

Starter ideas and activities

Mental starter ideas Ask questions such as Practice questions 13 and 27 on pages 9 and 10 in the Number Support chapter of the Pupil Book.

Starter activity **Vocabulary**
Discuss vocabulary associated with each of the four operations.
  What words tell us to subtract? (difference, how much more, etc.)
  What words tell us to multiply? and so on.
You could use Exercise 3 question 1 and discuss methods of doing each.

Teaching points and activities

- Recap the meaning of mean, assumed mean, percentage of, scale, BIDMAS.
- Pupils often have difficulty interpreting word problems. Encourage them to re-read the question and work out what they know and what they are asked to find out. Knowing the vocabulary that is associated with each operation is helpful.

Support
- Pupils should have covered the previous two sections, *Adding and subtracting* and *Multiplying and dividing*.
- See **A star is born**, the Starter on page 61 of the Pupil Book.
- Use Worksheet 8 questions 6 and 9. (Teacher Resource Pack page 7)

Extension
- See **starred questions** in Exercise 3 (Pupil Book page 70).
- See **Puzzles** (Pupil Book page 73). Challenge pupils to make up more puzzles such as this to give to a partner to solve.

Plenary ideas
- Use the Puzzles on page 73 of the Pupil Book.
- Use one of the questions from Exercise 3, such as question 10, 15 or 17. Discuss the strategies pupils could use to work these out.

Homework Review questions at the end of Exercise 3 (Pupil Book page 70) and/or Homework Sheet 3.3 (page 196 of this pack).

Links Link this topic to understanding the operations addition, subtraction, multiplication and division, adding, subtracting, multiplying and dividing mentally (Pupil Book pages 62–69).
Link to Scale drawing: Exercise 3 question 9.
Link to Measurement: Exercise 3 questions 1b, c, d, e, l, 2, 3, 4, 7.
Link to Angles: Exercise 3 question 1f, 1p, Review 5.
Link to Handling Data: Exercise 3 questions 1k, 1q, 1r, 13, 16, Review 5.
Link to Percentages: Exercise 3 question 8, Review 2.
Link to Order of operations: Exercise 3 question 18.
Link to Perimeter and Area: Exercise 3 question 1g, i, n, 12.
Link to Science: Exercise 3 questions 10, 13, 16.

| Topic areas | **Making estimates** | pupil book page 73 |
| | **Estimating answers to calculations** | pupil book page 74 |

Framework Objectives

✓ Make and justify estimates and approximations of calculations, **Framework examples pages 103 and 111**

✓ Check a result by considering whether it is of the right order of magnitude, **Framework examples pages 103 and 111**

Sheets   Worksheet 9 for Support   Homework Sheet 3.4

Resources   Mini-whiteboards for the Starter Activity. You could use number lines. A calculator.

 Starter ideas and activities

Mental starter ideas

Ask for situations where an estimate is good enough and some where an exact answer is needed. Prompt pupils to think about science, newspaper reports, school newsletters, technology, etc.

Ask questions such as Practice questions 10 and 21 on pages 8 and 10 in the Number support chapter of the Pupil Book.

Starter activity

● **When is an estimate good enough?**
  You could use Exercise 4 on page 74 as a class discussion.

● **Estimating answers in different ways**

  1 Ask pupils to estimate the answer to $\dfrac{128 \times 29}{21}$.

  2 Ask pupils to put their answers on their mini-whiteboards. Then ask some pupils to demonstrate how they got their estimates.

● **Estimating amounts — a practical exercise**
  For homework the night before you could ask pupils to do one of the Practical questions on page 74 of the Pupil Book and then discuss the answers pupils got.

 Teaching points and activities

● Emphasise that different situations call for different degrees of accuracy when making an estimate. For example, an estimate for the time taken to walk to school does not need to be as accurate as the time taken to run a marathon.

● Emphasise that there is often more than one possible estimate. The examples on the screen on pages 74 and 75 of the Pupil Book illustrate this.

● Pupils will often always round to the nearest power of 10 rather than looking for 'nice numbers'. Encourage pupils to look for nice numbers first.

Support

● See **Estimating** on page 4 of the Pupil Book (Number Support chapter).
  If pupils have difficulty with Practice questions 10, 21, 34, 35c and 38 use Worksheet 9 (page 8 of the Teacher Resource Pack).

Extension

● See **starred questions** in Exercise 5 (Pupil Book page 75).

Plenary ideas

● Talk about general situations where it is good enough to just give estimates of answers. You could use some of the ideas in the Practical on page 74 of the Pupil Book.

● Ask pupils to find an estimate for 467 × 24 and 677 ÷ 48 in two different ways. Discuss the answers.

● You could discuss some of the answers to Exercise 5 question 6.

Homework

Review questions at the end of Exercises 4 and 5 (Pupil Book pages 74 and 75) and/or Homework Sheet 3.4 (page 197 of this pack).

Links

Link this topic to understanding the operations addition, subtraction, multiplication and division, adding, subtracting, multiplying and dividing mentally (Pupil Book pages 62–69), rounding (Pupil Book pages 28–30) and checking answers (Pupil Book page 89).
Link to Science: Exercise 5, questions 6k, l.
Link to Geography: Exercise 5 question 6j.
Link to Perimeter and area: Exercise 5 question 7.
Link to Measures: Exercise 5 questions 6g, 6h and Review 5b.
Link to Order of operations; Exercise 5 question 4 and Review 4.
Link to Squares and square roots: Exercise 5 question 5.

4 Written and Calculator Calculation

Topic in Chapter	Links to Programme of Study	Links to Framework for Teaching Mathematics: 8 Teaching Programme
	Ma2 a–l are integrated throughout.	*Applying mathematics and solving problems are integrated throughout.*
Adding and subtracting *page 81*	**Ma2 1a** ... select appropriate strategies to use for numerical ... problems **Ma2 1d** select efficient techniques for numerical calculation ... **Ma2 3a** add, subtract ... integers and then any number **Ma2 3j** use standard column procedures for addition and subtraction of integers and decimals **Ma2 4b** select appropriate operations, methods and strategies to solve number problems ...	**Number operations and the relationships between them** ● Understand addition and subtraction of ... integers, and multiplication and division of integers; use the laws of arithmetic and inverse operations. **Written methods** ● Consolidate standard column procedures for addition and subtraction of integers and decimals with up to two places. ● **Use standard column procedures for multiplication and division of integers and decimals, including by decimals such as 0·6 or 0·06; understand where to position the decimal point by considering equivalent calculations.** **Applying mathematics and solving problems** ● Solve more demanding problems and investigate in a range of contexts: number ... ● Solve more complex problems ... choosing and using efficient techniques for calculation ...
Multiplying *page 83*	**Ma2 1a** ... select appropriate strategies to use for numerical ... problems **Ma2 1d** select efficient techniques for numerical calculation ... **Ma2 1e** make mental estimates of the answers to calculations; use checking procedures to monitor the accuracy of their results. **Ma2 3a** ... multiply ... integers and then any number; multiply ... any positive number by a number between 0 and 1 ... **Ma2 3h** round to the nearest integer **Ma2 3k** use standard column procedures for multiplication of integers and decimals, understanding where to position the decimal point ... **Ma2 4b** select appropriate operations, methods and strategies to solve number problems	
Dividing *page 85*	**Ma2 1a** ... select appropriate strategies to use for numerical ... problems **Ma2 1d** select efficient techniques for numerical calculation ...	
Dividing by a decimal *page 88*	**Ma2 3a** ... divide integers and then any number; ... divide ... any positive number by a number between 0 and 1; ... **Ma2 3k** ... solve a problem involving division by a decimal by transforming it to a problem involving division by an integer **Ma2 4b** select appropriate operations, methods and strategies to solve number problems ...	

Topic in Chapter	Links to Programme of Study	Links to Framework for Teaching Mathematics: 8 Teaching Programme
Checking answers *page 89*	**Ma2 1e** ... use checking procedures to monitor the accuracy of their results **Ma2 4a** draw on their knowledge of the operations and the relationships between them ... **Ma2 4c** use a variety of checking procedures, including working the problem backwards, and considering whether a result is of the right order of magnitude	**Checking results** ● Check a result by considering whether it is of the right order of magnitude and by working the problem backwards. **Number operations and the relationships between them** ● ... use inverse operations.
Using brackets on a calculator *page 92* Using the calculator memory *page 93*	**Ma2 3b** use brackets and the hierarchy of operations ... **Ma2 3o** use calculators effectively and efficiently: know how to enter complex calculations using brackets (for example ... for the division of more than one term), know how to enter a range of calculations including those involving measures ... **Ma2 3q** understand the calculator display, interpreting it correctly (for example in money calculations, and when the display has been rounded by the calculator), ...	**Number operations and the relationships between them** ● Use the order of operations, including brackets, with more complex calculations. **Calculator methods** ● Carry out more difficult calculations effectively and efficiently ...: ... use brackets and the memory. ● Enter numbers and interpret the display in different contexts ...

Links to Attainment Targets

Level 4 They use efficient written methods of addition and subtraction and of short multiplication and division. They add and subtract decimals to two places ... In solving problems with or without a calculator, pupils check their results by reference to their knowledge of the context or the size of the numbers.

Level 5 They use all four operations with decimals to two places ... Pupils understand and use an appropriate non-calculator method for solving problems that involve multiplying and dividing any three-digit number by any two-digit number. They check their solutions by applying inverse operations or estimating using approximations.

Level 6 Pupils ... approximate decimals when solving numerical problems ...

Level 7 In making estimates, pupils round ... and multiply and divide mentally (begin) ... Pupils solve numerical problems involving multiplication and division with numbers of any size, using a calculator efficiently and appropriately (begin).

Topic area	Adding and subtracting	pupil book page 81

Framework Objectives
✓ Understand addition and subtraction of ... integers, ... use the laws of arithmetic and inverse operations, **Framework examples page 105**
✓ **Consolidate standard column procedures for addition and subtraction of integers and decimals with up to two places, Framework examples page 105**
✓ Problem solving, **Framework examples page 7**

Sheets Worksheet 11 for Support   Homework Sheet 4.1   Resource Sheet page 200 Teacher Resource Pack

Resources You could use a number line to help with estimates.

Starter ideas and activities

Mental starter ideas
- Ask questions that revise complements to 1.
- Give a decimal such as 5·2 or 4·65 and ask pupils to
  find the complement to 10,
  double it,
  multiply it by 5,
  divide it by 5 and so on.

Starter activity
- **Revising standard written method**
  Give questions such as Practice question 16 on page 9 of the Number Support chapter and revise standard written methods for doing these. Ask pupils to come and demonstrate on the board how to do each. Discuss possible different ways of doing it.
- **Game: Make a sum**
  Divide the class into two groups
  Give a target number, such as 156 or 49·8 or 0·879, and then ask for pupils to come up with one addition and one subtraction that gives this target number. Give just 3 or 4 minutes for this.
  Ask pupils to swap calculations with someone in the other group to mark it. Give one point for each correct calculation. You could give an extra point for any calculations that have used three numbers or have used two numbers with different numbers of decimal places.

Teaching points and activities

- Point out that units must be the same when adding or subtracting measurements.
- Make sure pupils know their addition facts and complements to 1, 10, 50, 100 and 1000.
- Pupils sometimes have difficulty if there are different numbers of digits in numbers being added or subtracted. Point out that a zero can be put in as a place-filler.
  For example: 5·34 – 3·672
  The number 5·34 has no thousandths but we can put a zero there to show this and write the sum as

      5·340
    −3·672

- When there are different numbers of digits in numbers being added or subtracted pupils sometimes left- or right-justify all the numbers. Remind pupils we must line up numbers of the same place value. In decimals, lining up the decimal point ensures this happens.

Support
- See **Written calculation** on page 4 of the Pupil Book (Number Support chapter).
- If pupils have difficulty with Practice question 16 use Worksheet 11 questions 1, 2, 8a, 8b (page 9 of the Teacher Resource Pack).
- See **Back to Front**, the Starter on page 80 of the Pupil Book.

Extension
- Some pupils will not need to cover this section at all.
- See **starred questions** in Exercise 1 (Pupil Book page 81).
- See **Investigation: Ten pounds and eighty-nine pence** (Pupil Book page 83).
  Pupils could be asked to think about why this works, but the algebra to prove it is beyond most.

Plenary ideas
- Talk about the Investigation on page 83 of the Pupil book and ask what pupils found out when they started with different amounts of money.
- Use Exercise 1 question 7 or 8. Discuss strategies for doing these.
- Put these calculations on the board and ask pupils what is wrong with each.

$$
\begin{array}{r} 5.578 \\ +\ 45.67 \\ \hline \end{array}
\qquad
\begin{array}{r} 5.578 \\ +\ 45.67 \\ \hline \end{array}
\qquad
\begin{array}{r} 7.896 \\ -\ 3.428 \\ \hline 4.472 \end{array}
$$

- Point out the links, especially the link to estimating.

Homework   Review questions at the end of Exercise 1 (Pupil Book page 81) and/or Homework Sheet 4.1 (page 198 of this pack).

Links   Link this topic to understanding the operations addition and subtraction, adding and subtracting mentally (Pupil Book page 62), rounding (Pupil Book pages 28–32), checking answers (Pupil Book page 89) and estimating (Pupil Book page 74).
Link to Geography: Exercise 1 question 3.
Link to Science: Exercise 1 question 4, Review 3.
Link to Design and Technology: Exercise 1 question 6, Review 2.

| Topic area | **Multiplying** | pupil book page 83 |

Framework Objectives
✓ Understand ... multiplication and division of integers; use the laws of arithmetic and inverse operations, **Framework examples page 105**
✓ **Use standard column procedures for multiplication of integers and decimals, including by decimals such as 0·6 or 0·06; understand where to position the decimal point by considering equivalent calculations, Framework examples page 105**

Sheets Worksheet 11 for Support   Homework Sheet 4.2   Resource Sheet page 200 Teacher Resource Pack

Resources You could use a number line to help with estimates.

Starter ideas and activities

Mental starter ideas Revise tables. You could play a tables game such as bingo or memory.
Revise multiplying by multiples of whole numbers, e.g. 21×30.
Give some practice at estimating, such as $3·72 \times 21·7 \approx 4 \times 20 = 80$.

Starter activity **Strategies for multiplying**
1 Give a multiplication, such as $31·6 \times 3·6$, and discuss ways of working it out.
2 Ask pupils to estimate the answer first and then show the grid method and the long multiplication method of finding the answer. Ask 'Which method do you think is best? Why?'

Teaching points and activities

- Some pupils will be able to work out the answers to some of the questions in the exercise mentally. Encourage pupils to always do as much as possible mentally before using a formal written method.
- Encourage pupils to always estimate the answer first and then check their answers against this estimate to check it is the right order of magnitude. Encourage pupils to check the answer in other ways, such as looking at the last digits.
- When using the grid method some pupils have trouble multiplying by multiples of 0·1 and 0·01, for example $0·02 \times 8$.
  Remind them that 0·02 is 2 hundredths and 8×2 hundredths is 16 hundredths which is 0·16. Or use patterns to help: $2 \times 8 = 16$
  $0·2 \times 8 = 1·6$
  $0·02 \times 8 = 0·16$
- Pupils sometimes forget when multiplying by, say, 65 to multiply by 60 and then 5 rather than 6 and then 5. Encourage them to multiply by the 60 first or even write 65 as $60 + 5$.
- Link the grid method and the long multiplication methods.

Support
- See **Written calculation** on page 4 of the Pupil Book (Number Support chapter).
  If pupils have difficulty with Practice questions 38 and 66 use Worksheet 11 questions 3, 4, 7, 8c, 9 (page 9 of the Teacher Resource Pack).

Extension
- See **Starred** questions in Exercise 2 (Pupil Book page 84).
- See **Puzzle** (Pupil Book page 85) question 2. Challenge pupils to make up one of their own and give it to someone else to solve.

Plenary ideas
- Put a multiplication on the board, such as $5·76 \times 3·9$, and discuss how to do it. Ask individual pupils to show how to do it or get each pupil to just do the next step and then ask another to take over. Emphasise the teaching points given above and in particular how to determine the position of the decimal point in the answer.
- Discuss the puzzles on page 85 of the pupil book and strategies for doing these.

Homework Review questions at the end of Exercise 2 (Pupil Book page 84) and/or Homework Sheet 4.2 (page 199 of this pack).

Links Link this topic to understanding the operation multiplication, multiplying mentally (Pupil Book page 66), rounding (Pupil Book pages 28–32), estimating (Pupil Book page 74), multiplying by powers of 10 (Pupil Book page 19) and checking answers (Pupil Book page 89).
Link to measures: Exercise 2 questions 6, 7, Review 2.
Link to Design and Technology: Exercise 2 question 7.

Topic area	**Dividing**	pupil book page 85

Framework Objectives
✓ Understand ... multiplication and division of integers; use the laws of arithmetic and inverse operations, **Framework examples page 107**
✓ **Use standard column procedures for multiplication and division of integers and decimals, including by decimals such as 0·6 or 0·06; understand where to position the decimal point by considering equivalent calculations, Framework examples page 107**
✓ Problem solving, **Framework examples page 3**

Sheets
Worksheet 11 for Support   Homework Sheet 4.3

Resources
A set of multiplication dominoes for the Starter Activity. You could use a number line to help with estimates.

 ## Starter ideas and activities

Mental starter ideas
- Ask lots of divisions to 100. You could do this as a game, for example Bingo.
- Practise estimating the answers to divisions.
- Ask for the remainders to simple divisions. For example, the remainder when 54 is divided by 7.

Starter activity
- **Revising dividing**
  You could use Practice question 65 on page 13 of the Number Support chapter as revision.
- **Game: Dominoes**
  Play dominoes with a set of multiplication dominoes that have multiplications such as 0·5 × 8 and 9 × 0·6, etc.
- **Remainders**
  1 Discuss ways of giving a remainder. For example, the answer to 568 ÷ 12 could be given as 47 R 4 or as $47\frac{1}{3}$ or as 47·3 to 1 decimal place.
  2 Discuss remainders in context. For example, a school needs to transport 89 children to a sporting event by minibuses. Each minibus holds 14 pupils. How many will be needed? How many seats will be left over?
- **Strategies for dividing**
  1 Put a division such as 67·5 ÷ 18 on the board.
  2 Ask for an estimate of the answer.
  3 Discuss ways to find the answer
  4 Now do one like 72·3 ÷ 14. Find the answer to 1 d.p. *Can we stop now?*
  5 Emphasise the need to go to one more d.p. than we want the answer to.

 ## Teaching points and activities

- Make sure pupils know their tables and are familiar with finding remainders. Ensure they can interpret the remainder in word problems.
- Pupils often find division hard and it may be necessary to revise some simpler division calculations first, such as 91 ÷ 20. It is especially important that pupils estimate the answer to all divisions to confirm the position of the decimal point in the answer.
- Emphasise the need to keep decimal points underneath one another.
- Pupils find it hard to work out that, for example, 3·5 is a bit more than 14 × 0·2 (see the division in the screen on page 86). Encourage them to think of 35 and decide what the closest multiple of 14 is to this. Likewise, in the next example given in the screen for 6·7 think of 67 and use estimation or jottings to work out what is the closest multiple of 16 to 67. 16 × 4 = 64 and so 16 × 0·4 is 6·4.

Support
- See **Written calculation** on page 4 of the Pupil Book (Number Support chapter).
  If pupils have difficulty with Practice questions 65 and 79 use Worksheet 11 questions 5, 6 (page 9 of the Teacher Resource Pack).

Extension
- See **starred questions** in Exercise 3 (Pupil Book page 92).

Plenary ideas
- Discuss Exercise 3 question 10 or 11 and strategies for doing them. For question 11 ask pupils to check whether there is more than one answer.
- Put a division with a remainder on the board, appropriate to the level that the class has reached, and ask individuals to come and show each step. Discuss how to estimate to confirm the position of the decimal point.

Homework Review questions at the end of Exercise 3 (Pupil Book page 86) and/or Homework Sheet 4.3 (page 200 of this pack).

Links Link this topic to understanding the operation division, dividing mentally (Pupil Book page 66), rounding (Pupil Book pages 28–32), estimating (Pupil Book page 74), multiplying and dividing by powers of 10 (Pupil Book page 19) and checking answers (Pupil Book page 89).
Link to measures: Exercise 3 question 9.
Link to Geography: Exercise 3 question 7.

| Topic area | **Dividing by a decimal** | pupil book page 88 |

Framework Objectives

✓ Understand ... multiplication and division of integers; use the laws of arithmetic and inverse operations, **Framework examples page 107**

✓ **Use standard column procedures for ... division of integers and decimals, including by decimals such as 0·6 or 0·06; understand where to position the decimal point by considering equivalent calculations, Framework examples page 107**

Sheets Homework Sheet 4.4   Starter Resource Sheet 3

Resources Set of dominoes for the Starter Activity. You could use a number line to help with estimates.

Starter ideas and activities

Mental starter ideas Revise tables.

Starter activity
- **Dividing by a decimal**
  Use the Discussion on page 88 of the Pupil Book.
- **Estimating and equivalent calculations**
  Put an example similar to the one in the screen on page 88 on the board.
  How could we work out an estimate of the answer?
  Is there another calculation we could do that would be equivalent to this one?
- **Game: Dominoes**
  Give some more practice at both working out an estimate and finding an equivalent calculation. You could do this using a set of dominoes or by having a preprepared OHP with multi-choice questions on it, such as Starter Resource Sheet 3.

Teaching points and activities

- Pupils have trouble grasping the idea of doing an equivalent calculation. You could show using a calculator that it is in fact equivalent. This may convince some, but to aid understanding of this concept you could use a balance machine and show that you have balanced both sides by doing the same thing to both sides.

Example

$$\frac{20}{0.04} = 500$$

$$\frac{20 \times \mathbf{100}}{0.04} = 500 \times \mathbf{100}$$

$$\frac{20 \times 100}{0.04 \times \mathbf{100}} = 500 \times \frac{100}{100}$$

$$\frac{2000}{4} = 500$$

Support
- Pupils should have covered the previous section, *Dividing*.

Extension
- See **starred question** in Exercise 4 (Pupil Book page 88).

Plenary ideas
- Discuss Exercise 4 question 2 as a class but go on and do the calculation, asking a different pupil to do each step one by one. You could use another question that pupils had difficulty with.

Homework Review questions at the end of Exercise 4 (Pupil Book page 88) and/or Homework Sheet 4.4 (page 200 of this pack).

Links Link this topic to understanding the operation division, dividing mentally (Pupil Book page 66), dividing by 10, 100 and 1000 (Pupil Book page 19), rounding (Pupil Book pages 28–32), estimating (Pupil Book page 74) and checking answers (Pupil Book page 89).

| Topic area | Checking answers | pupil book page 89 |

Framework Objectives
✓ Check a result by considering whether it is of the right order of magnitude and by working the problem backwards, **Framework examples pages 103 and 111**
✓ ... use inverse operations, **Framework examples page 111**

Sheets Homework Sheet 4.5

Resources A calculator.

 Starter ideas and activities

Mental starter ideas Give some calculations and ask pupils to write down three other calculations that they know are true. For example, if you give $38·6 \times 3·4 = 131·24$ then we know that $3·4 \times 38·6 = 131·24$, and $131·24 \div 3·4 = 38·6$, and $131·24 \div 38·6 = 3·4$. Begin with some that use each of the four operations.
Ask pupils to write some calculations as an equivalent calculation using partitioning.
Examples $23·4 \times 12 = 23·4 \times 10 + 23·4 \times 2$ or $23·4 \times 12 = 20 \times 12 + 3 \times 12 + 0·4 \times 12$

Starter activity **Strategies for checking answers**
Give some calculations such as Exercise 5 question 1 (you could use this actual question).
Is the answer right or wrong? Why?
What mistake might have been made to get the answer given?

 Teaching points and activities

- Emphasise the need to check all answers, especially those done on a calculator. In particular talk about the fact that sometimes different calculations give different answers.
- Pupils often use only estimating to check answers. Estimating only checks if the answer is in the right order of magnitude and will not pick up all errors.
- Encourage pupils to develop a range of strategies like the ones given in the screen on pages 89 and 90 for checking answers.

Support
- See **Estimating** on page 4 of the Pupil Book (Number Support chapter).
- Try Practice questions 80, 81, 82 and 84 and ask pupils to check their answers. You could ask pupils what 'sensible' might mean in the mathematical context.

Extension
- Ask pupils to make up some examples themselves, like the ones in the exercise, and give them to someone else to decide if the answer is sensible.

Plenary ideas
- Choose a calculation such as $48 \times 12 = 576$ and go through the different ways this answer can be checked. (Seeing if it is sensible, using inverse operations, using an equivalent calculation and checking the last digits.)
- Discuss the answers to Exercise 5 questions 8 and/or 9. Emphasise the different ways of checking answers.

Homework Review questions at end of Exercise 5 (Pupil Book page 90) and/or Homework Sheet 4.5 (page 200 of this pack).

Links Link this topic to rounding (Pupil Book pages 28–32), estimating (Pupil Book page 74), checking the solution to an equation by substituting (Pupil Book page 184).
Link to Measures: Exercise 5 questions 1a, 1d.
Link to Science: Exercise 5 Review 1.

| Topic areas | Using brackets on a calculator | pupil book page 92 |
| | Using the calculator memory | pupil book page 93 |

Framework Objectives

✓ Use the order of operations, including brackets, with more complex calculations, **Framework examples pages 87 and 109**

✓ Carry out more difficult calculations effectively and efficiently ... use brackets and the memory, **Framework examples pages 87 and 109**

✓ Enter numbers and interpret the display in different contexts ..., **Framework examples page 109**

Sheets   Worksheet 10 for Support   Homework Sheet 4.6   Resource Sheet (page 201 Teacher Resource Pack)

Resources   A calculator. OHP for plenary.

 ## Starter ideas and activities

Mental starter ideas

Give a simple calculation with brackets and ask pupils to work it out without a calculator.
Example $5(8 + 3) + 4 =$
Now try it on the calculator.
Do you get the same answer?

Starter activity

● **Keying sequences**
  1 Put up three different keying sequences for this calculation: $4(8 + 7) \div (12 - 7)$.
      key $4(8 + 7) \div 12 - 7$
      key $4 \times (8 + 7 \div (12 - 7))$
      key $4(8 + 7) \div (12 - 7)$
  2 Divide the class into four groups. Give three of the groups one keying sequence each and ask the fourth group to work it out mentally. *Which keying sequence is correct*?
  3 You could repeat this for other calculations.

● **Calculator memory**
  1 Demonstrate how to use the calculator memory.
  2 Put up the calculation $15 + 25 \div 15 + 5$.
      How could you do this calculation using brackets?
      How could you do this calculation using the calculator memory?

 ## Teaching points and activities

● Discuss when it would be useful to use the memory keys on a calculator.

● Encourage pupils to estimate the answers to calculations. Point out that it is very easy to key a wrong button.

● Point out that different calculators may have different memory keys.

● Emphasise that as long as an equals is not keyed part way through a calculation, the calculator will work out the operations in the correct order.

● It should be stressed that the horizontal line in a calculation such as $\frac{13 + 29}{11 - 4}$ acts as a bracket on both the numerator and denominator, i.e. the calculation becomes $(13 + 29) \div (11 - 4)$.

● In Exercise 6 question 4 the memory can be used to store the value 175. It can then be recalled for each calculation. Point out to pupils that this reduces the likelihood of keying the number in wrongly as well as saving time.

Support

● Pupils should have covered *Using a calculator* and *Checking answers* in this chapter and *Order of operations* in the Mental Calculation chapter.

● See *Using a calculator* on page 5 of the Pupil Book.

● If pupils have difficulty with Practice question 80 use Worksheet 10 question 4 (page 8 of the Teacher Resource Pack).

Extension   ● See **starred question** in Exercise 6 (Pupil Book page 92).

Plenary ideas

● Discuss Exercise 6 question 5 (or 6 to extend thinking). See if pupils got different answers.

● Discuss Exercise 7 question 5.

Homework   Review questions at the end of Exercises 6 and 7 (Pupil Book page 92 and 94) and/or Homework Sheet 4.6 Exercises 1 and 2 (page 201 of this pack).

Links   Link this topic to understanding the operations addition, subtraction, multiplication and division, order of operations (Pupil Book page 43) and rounding (Pupil Book pages 28–32).
Link to Design and Technology: Exercise 7 question 1.
Link to PE: Exercise 7 question 2.

5 Fractions

Topic in Chapter	Links to Programme of Study *Ma2 a–l are integrated throughout.*	Links to Framework for Teaching Mathematics: **8 Teaching Programme** *Applying mathematics and solving problems are integrated throughout.*
Fractions of shapes *page 98* One number as a fraction of another *page 99*	**Ma2 2c** use fraction notation; understand equivalent fractions, simplifying a fraction by cancelling all common factors **Ma2 3c** ...express a given number as a fraction of another	**Fractions, decimals, percentages, ratio and proportion** ● Know that a recurring decimal is a fraction; use division to convert a fraction to a decimal; order fractions by writing them with a common denominator or by converting them to decimals.
Fractions and decimals *page 102* Ordering fractions *page 105*	**Ma2 2c** ...order fractions by rewriting them with a common denominator **Ma2 2d** use decimal notation and recognise that each terminating decimal is a fraction (for example, $0 \cdot 137 = \frac{137}{1000}$) **Ma2 3c** ...; perform short division to convert a simple fraction to a decimal **Ma2 3g** recall ... the fraction-to-decimal conversion of familiar simple fractions (for example, $\frac{1}{2}, \frac{1}{4}, \frac{1}{5}, \frac{1}{10}, \frac{1}{100}, \frac{1}{3}, \frac{2}{3}, \frac{1}{8}$) **Ma2 3p** (and how to enter a fraction as a decimal *on the calculator*)	● ... use the equivalence of fractions, decimals ... to compare proportions. **Mental methods and rapid recall of number facts** ● Recall known facts, including fraction to decimal conversions ... ● Consolidate and extend mental methods of calculation, working with decimals, fractions ... **Calculator methods** ● Carry out more difficult calculations effectively and efficiently, using the function keys for ... fractions ... ● Enter numbers and interpret the display in different contexts (... fractions, decimals ...)
Adding and subtracting fractions *page 108*	**Ma2 3c** ...add and subtract fractions by writing them with a common denominator **Ma2 3l** use efficient methods to calculate with fractions **Ma2 3p** use the function keys for ... fractions ...	**Fractions, decimals, percentages, ratio and proportion** ● Add and subtract fractions by writing them with a common denominator. **Mental methods and rapid recall of number facts** ● Consolidate and extend mental methods of calculation, working with ... fractions ... **Calculator methods** ● Carry out more difficult calculations effectively and efficiently, using the function keys for ... fractions ... ● Enter numbers and interpret the display in different contexts.

Topic in Chapter	Links to Programme of Study		Links to Framework for Teaching Mathematics: 8 Teaching Programme
Fraction of — multiplying an integer by a fraction *page 111*	**Ma2 3c**	calculate a given fraction of a given quantity, expressing the answer as a fraction; ...	**Fractions, decimals, percentages, ratio and proportion** ● ... calculate fractions of quantities (fraction answers); multiply and divide an integer by a fraction. **Mental methods and rapid recall of number facts** ● Consolidate and extend mental methods of calculation, working with ... fractions ...
Dividing an integer by a fraction *page 113*	**Ma2 3d**	understand and use unit fractions as multiplicative inverses for example, by thinking of multiplication by $\frac{1}{5}$ as division by 5, or multiplication by $\frac{6}{7}$ as multiplication by 6 followed by division by 7 (or vice versa); multiply ... a given fraction by an integer ...	
	Ma2 3l	use efficient methods to calculate with fractions including cancelling common factors before carrying out the calculation.	

Links to Attainment Targets

Level 4 They recognise ... proportions of a whole and use simple fractions ... to describe these.

Level 5 They reduce a fraction to its simplest form by cancelling common factors ...
They calculate fractional ... parts of quantities and measurements ...

Level 6 Pupils are aware of which number to consider as ... a whole, in problems involving comparisons, and use this to evaluate one number as a fraction ... of another.
They understand and use the equivalences between fractions, decimals ...
They add and subtract fractions by writing them with a common denominator.

Topic areas	**Fractions of shapes**	pupil book page 98
	One number as a fraction of another	pupil book page 99

Framework Objectives ✓ Consolidate and extend mental methods of calculation, working with ... fractions ...,
Framework examples page 61

Sheets Worksheet 12 for Support   Homework Sheet 5.1   Starter Resource Sheet 4

Resources A counting stick and cards, some with fraction questions and some with the answers on for the Starter Activity. You could use fraction pieces, fraction number lines and a counting stick.

Starter ideas and activities

Mental starter ideas

- Use Starter Resource Sheet 4.
  Ask pupils to estimate the fraction that is shaded.
- Ask some questions such as Practice questions 19, 25, 39, 58 and 59 of the Number Support chapter.

Starter activity

- **One number as fraction of another**
  1 Ask questions such as:
   What fraction of 4 is 1? (You may need to draw a diagram with four parts and one shaded to help.)
   How did you work this out?
   What fraction of 12 is 5?
   What do notice about the number that comes after 'of' and the denominator?
  2 You could use a counting stick or a scale on an OHP to make this more visual by putting one hand on 4 and one hand on 1 or one hand on 5 and the other on 12.
- **Game — Match up**
  1 Make a set of cards with questions such as 'What fraction of 24 is 8?'
  2 Make another set with the answers on them.
  3 Have the same number of cards as there are pupils in your class.
  4 Give a card to each pupil in your class and then ask them to find their matching card and pair up.
- **Fraction of the day**
  You could go over the day's school timetable and talk about what fraction of the day is spent on each subject or activity.

Teaching points and activities

- Recap cancelling fractions to their simplest form.
- Emphasise that fractional parts of a shape must be of equal size.
- Remind pupils that when finding one number as a fraction of another the numbers must have the same units.
- Pupils sometimes forget to add numbers together to find the total in questions such as question 8 of Exercise 2. Encourage pupils to ask themselves what they are finding the fraction **of**. In question 8 they are finding the fraction of the **total** number of pupils in Nazir's class. So 14 and 18 must first be added to find the total number of pupils in Nazir's class.
- Point out that when the sectors of a pie chart are expressed as a fraction, the fractions should add to 1.

Support

- See **Fractions** on page 5 of the Pupil Book (Number Support chapter).
- If pupils have difficulty with Practice questions 19, 25, 39, 51, 58 and 59 use Worksheet 12 questions 1, 3, 5, 6, 7, 8, 10a and 11 (page 10 of the Teacher Resource Pack).

Extension

- See **Investigation: Unit Fractions** (Pupil Book page 98) part 3.
- See **starred questions** in Exercise 2 (Pupil Book page 100).
- See **Practical** (Pupil Book page 102). Ask pupils to display their answers using Power Point or in some other interesting way.

Plenary ideas
- You could give a simple pie chart to the class and ask individual pupils to express each sector as a fraction of the whole, e.g.

Members of a Golf Club

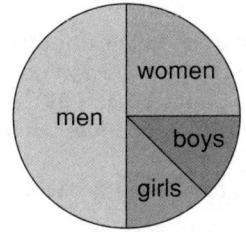

$\frac{1}{2}$ of the members are men.

$\frac{1}{4}$ of the members are women.

$\frac{1}{8}$ of the members are boys.

$\frac{1}{8}$ of the members are girls.

- You could start the investigation on page 98 of the pupil book and ask pupils to finish it for homework.
- Discuss Exercise 2 questions 10 or 11. Both of these illustrate the need to find a total first.
- You could start the practical on page 102 of the pupil book.
- You could do some examples as part of the plenary such as finding the fraction of your class that own a pet.

Homework
Review questions at the end of Exercises 1 and 2 (Pupil Book pages 99 and 100) and/or Homework Sheet 5.1 Exercises 1 and 2 (page 202 of this pack).

Links
Link to pie charts: Exercise 1 questions 2, 3, Review 2.
Link to proportion: Exercise 2 question 7.
Link to measures: Exercise 2 questions 1, 2, 3.
Link to Science: Exercise 2 question 10.
Link to angles: Exercise 2 Review 1.
Link to enlargement and scale factor: Exercise 2 question 12.
Link to Handling Data: Exercise 2 questions 7, 11, Review 4.

| **Topic area** | **Fractions and decimals** | pupil book page 102 |

Framework Objectives
- ✓ Know that a recurring decimal is a fraction; use division to convert a fraction to a decimal; order fractions by writing them with a common denominator or by converting them to decimals, **Framework examples pages 65 and 99**
- ✓ Recall known facts, including fraction to decimal conversions ..., **Framework examples page 99**
- ✓ Consolidate and extend mental methods of calculation, working with ... fractions ..., **Framework examples page 99**

Sheets   Worksheet 13 for Support   Homework Sheet 5.2   Resource Sheet page 201 Teacher Resource Pack   ICT Worksheet 'Investigating Fractions' page 119 of the Teacher Resource Pack.

Resources   For the Starter Activity you may need a square divided into tenths and one divided into hundredths, mini-whiteboards and some SNAP fraction/decimal cards. A calculator. You could use fraction/decimal dominoes and decimal number lines. Computer and spreadsheet package.

Starter ideas and activities

Mental starter ideas
- Ask questions such as Practice questions 19, 39, 51 and 59 of the Number Support chapter. You may have used some of these in the previous section.
- Ask for the factors of 10 and 100.

Starter activity
- **Decimals as fractions**
  Revise the meaning of 0·1 and 0·01 and multiples of these. You could use a square divided into tenths and one divided into hundredths to help.
    What is four tenths as a decimal?
    What is 0·79 as a fraction?
  Ask pupils to put the answers on their mini-whiteboards.
  To start with make sure the answers won't cancel and then move on to ones that do and use this to revise cancelling.
- **Game: Fraction/Decimal match up**
  Make a set of SNAP fraction/decimal cards to revise the fraction and decimal equivalents pupils should know.
- **Equivalent Fractions**
    Use the numbers 2, 4, 3 and 6 or 15, 16, 20 and 24 to make two equivalent fractions.
  Use statements like this to revise cancelling fractions to their lowest terms.

Teaching points and activities

- Make sure pupils can find equivalent fractions and cancel fractions to their lowest terms.
- Make sure pupils are aware that fractions and division are linked. Pupils often forget that $\frac{2}{3}$ means $2 \div 3$. Ensure pupils can do divisions such as $5 \div 8$. You may need to revise this.
- It is important for pupils to understand the relationship between fractions and decimals rather than just learn a method of converting from one to the other. Use diagrams (squares divided into tenths and hundredths) to illustrate the equivalences.
- Once understanding has been established, encourage pupils to learn the equivalences such as
  $0.125 = \frac{1}{8}$ and $0.3 = \frac{1}{3}$, as well as $0.1 = \frac{1}{10}$ and $0.01 = \frac{1}{100}$ and $0.25 = \frac{1}{4}$, etc.
  These can then be used to work out other conversions.
- Emphasise that all fractions can be made into decimals and all terminating and recurring decimals can be converted to fractions.
- It is helpful for pupils to discover the answers to the Investigation on page 104. i.e. that fractions with denominators that have prime factors other than 2 or 5 will give a recurring decimal.

Support
- See **Decimals, fractions and percentages** on page 6 of the Pupil Book (Number Support chapter). If pupils have difficulty with Practice questions 48a–l and 49 use Worksheet 13 questions 2a–f and 3 (page 11 of the Teacher Resource Pack).

Extension
- See **Investigation: Prime Factors and Recurring Decimals** (Pupil Book page 104).

Plenary ideas
- Ask pupils for an example of a terminating and a recurring decimal.
- Discuss the answers to Exercise 4 question 7 and 9.
- Start the Investigation on page 104 of the Pupil Book and ask pupils to finish it for homework.

Homework Review questions at the end of Exercises 3 and 4 (Pupil Book pages 102 and 104) and/or Homework Sheet 5.2 Exercises 1 and 2 (page 203 of this pack).

ICT See ICT Worksheet 'Investigating Fractions' on page 119 of the ICT Support section in the Teacher Resource Pack.

Links Link this topic to mental calculation (Pupil Book page 61), place value (Pupil Book page 16), rounding (Pupil Book pages 28–32), equivalent fractions, cancelling fractions to their lowest terms (Pupil Book page 5), percentages (Pupil Book page 119) and prime factors (Pupil Book page 47).
Link to Science: Exercise 4 question 7a, 7c, 7d.
Link to PE: Exercise 4 question 7b.

| | **Topic area** | **Ordering fractions** | pupil book page 105 |

Framework Objectives
✓ ... order fractions by writing them with a common denominator or by converting them to decimals, **Framework examples page 65**
✓ Recall known facts, including fraction to decimal conversions ..., **Framework examples page 65**
✓ Consolidate and extend mental methods of calculation, working with ... fractions ..., **Framework examples page 65**

Sheets   Homework Sheet 5.3   Resource Sheet pages 201 and 202 Teacher Resource Pack

Resources   A scale divided into appropriate fractions for the Mental Starter. You could use fraction pieces and fraction number lines. A calculator.

 Starter ideas and activities

Mental starter ideas
● Ask simple questions such as
    Which is bigger $\frac{1}{2}$ or $\frac{1}{3}$?
    You may need to have some scales divided into the appropriate amounts, one under the other, so that pupils can compare fractions.
● Ask questions such as
    Write $\frac{2}{3}$ with a denominator of 6.

Starter activity   **Which is bigger?**
    Which is larger, $\frac{2}{5}$ or $\frac{3}{8}$? How could you work it out?
    Show all the methods pupils may come up with (or ask them to come up and explain them) including changing both to decimals and writing both with the same denominator.
    Which way do you think is best?
    Will this always be the best way?

 Teaching points and activities

● Make sure pupils can change fractions to decimals and find common denominators.
● Make sure pupils can convert between mixed numbers and improper fractions.
● Make the link between finding the common denominator and the LCM.

Support   ● Pupils should have covered the previous sections in this chapter.

Extension   ● See **starred questions** in Exercise 5 (Pupil Book page 106).
● See **Puzzle** (Pupil Book page 108). Challenge students to make up other puzzles such as this to give to classmates to solve.

Plenary ideas   ● Ask what fraction of the day pupils spend sleeping, eating, at school and talk about ordering these fractions.
● Discuss the answers to Exercise 5 question 8.
● Discuss the Puzzle on page 108 of the Pupil Book.

Homework   Review questions at the end of Exercise 5 (Pupil Book page 106) and/or Homework Sheet 5.3 (page 204 of this pack).

Links   Link this topic to equivalent fractions, mixed numbers and improper fractions (Pupil Book page 5).

Topic area	Adding and subtracting fractions	pupil book page 115

Framework Objectives

✓ Add and subtract fractions by writing them with a common denominator, **Framework examples page 67**

✓ Consolidate and extend mental methods of calculation, working with ... fractions ..., **Framework examples page 67**

✓ Carry out more difficult calculations effectively and efficiently, using the function keys for ... fractions ..., **Framework examples page 109**

✓ Problem solving, **Framework examples page 7**

Sheets Worksheet 12 for support   Homework Sheet 5.4

Resources You could use fraction pieces for the Starter Activity. You could use fraction number lines. A calculator.

Starter ideas and activities

Mental starter ideas Use fraction pieces and make some fraction sums, e.g. use the fraction pieces for $\frac{1}{2}$ and $\frac{1}{4}$ and ask pupils for the sum of $\frac{1}{2}$ and $\frac{1}{4}$.

Starter activity **Using diagrams to add and subtract fractions**

1 Put this on the board or on an OHP.

$\frac{1}{3} + \frac{3}{4}$

2 Draw a diagram divided into thirds and shade $\frac{1}{3}$ of it.

3 Draw a diagram the same size as the first and divide it into quarters and shade $\frac{3}{4}$ of it.

How could I divide up both of these diagrams so they have the same number of parts?

How did you work this out?

How could I use these diagrams to find the answer to $\frac{1}{3} + \frac{3}{4}$?

4 Draw diagrams to answer these.

$\frac{1}{4} + \frac{5}{12}$ $\frac{7}{8} - \frac{3}{4}$

How else could you find the answer? (By finding a common denominator.)

Teaching points and activities

- Recap LCM, simplest form and equivalent fractions.
  Some pupils prefer to use the diagrams to add and subtract initially. Make the link between the diagrams and finding a common denominator.
- Pupils sometimes multiply the denominators together to find the common denominator and then cross multiply to find the numbers to add together for the numerator. Discourage this as it does not aid understanding and can lead to pupils working with very large numbers that are more difficult to calculate mentally.
- Once pupils have seen how easy adding and subtracting is using the fraction button on the calculator it is hard to stop them using it. Point out that they may be asked to do a fraction calculation without the calculator and therefore must know how to do this.

Support
- Pupils should have covered the previous sections in this chapter.
- See **Fractions** on page 5 (Number Support chapter in the Pupil Book).

Extension
- See **The whole sound of music**, the Starter on page 97 of the Pupil Book.
- See **starred questions** in Exercise 6 (Pupil Book page 109).
- See **Investigation: Egyptian Fractions** (Pupil Book page 110) Question 4: challenge pupils to see who can find the most fractions that can be written as the sum of unit fractions.

Plenary ideas
- Take a fraction addition such as $\frac{2}{5} + \frac{3}{4}$ and ask pupils to demonstrate how this can be done by changing to fractions with the same denominator, but also by using the calculator.
- Use the Puzzle on page 110 of the Pupil Book. See how many different ways the class can answer each part.
- Start the Investigation on page 110 and ask pupils to finish it for homework. This Investigation is quite challenging but is a very good way of getting pupils to practice.

Homework Review questions at the end of Exercise 6 (Pupil Book page 109) and/or Homework Sheet 5.4 (page 204 of this pack).

Links Link this topic to mental calculation (Pupil Book page 61) and equivalent fractions.

| Topic area | Fraction of — multiplying an integer by a fraction | pupil book page 111 |

Framework Objectives
✓ ... calculate fractions of quantities (fraction answers); multiply and divide an integer by a fraction, **Framework examples pages 67 and 69**
✓ Consolidate and extend mental methods of calculation, working with ... fractions ..., **Framework examples pages 83 and 99**

Sheets Worksheet 12   Homework Sheet 5.5

Resources Cards with equivalent fraction statements on for the Starter Activity.

Starter ideas and activities

Mental starter ideas
- Ask questions such as Practice questions 57 and 73 of the Number Support chapter.
- Make a spider web with a number, such as 12, in the middle.

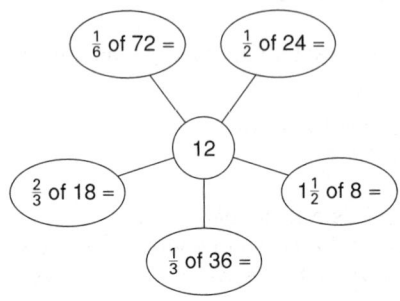

Starter activity
Equivalent statements and multiplying an integer and a fraction
1 Remind pupils that in maths 'of' means multiply and we can multiply in any order.
$\frac{3}{4} \times 8$ is the same as $\frac{3}{4}$ of 8, and $8 \times \frac{3}{4}$, and $8 \div 4 \times 3$, and $8 \times 3 \div 4$, and $3 \times \frac{1}{4} \times 8$.
2 Make cards with some equivalent statements, such as the ones above, and ask pupils to make matching sets of six cards.
3 *How could you find the answer to $7 \times \frac{3}{5}$?*
We are allowed to have a fraction in the answer.
4 Ask for a volunteer to show how to do it on the board or on an OHP.
(7 lots of $\frac{3}{5}$ is $7 \times 3 \times \frac{1}{5}$, which is $21 \times \frac{1}{5}$, which is $\frac{21}{5}$, which is $4\frac{1}{5}$).

Teaching points and activities

- Stress 'of' means multiply.
- Make sure pupils can find 'fraction of' mentally. Exercise 7 question 1 gives practice at this. Remind them that it is easier when finding $\frac{5}{8}$, for example, to find $\frac{1}{8}$ first by dividing by 8 and then multiplying the answer by 5.
- Pupils sometimes multiply the denominator **and** the numerator by the integer, e.g. $7 \times \frac{3}{5} = \frac{21}{35}$. Encourage pupils to see $7 \times \frac{3}{5}$ as 7 lots of $\frac{3}{5}$ which is $\frac{21}{5}$. You could use a number line or scale divided into fifths to show this.
- Spend some time going over the cancelling of fractions in examples such as $\frac{3}{4} \times \frac{36}{1}$. Pupils get confused when the cancelling involves more than one fraction.

Support
- See **Fractions, decimals and percentages** on page 5 of the Pupil Book (Number Support). If pupils have difficulty with Practice questions 13a, 57, 73a–g use Worksheet 12 questions 2, 9, 12 (page 10 of the Teacher Resource Pack).

Extension
- See **starred questions** in Exercise 7 (Pupil Book page 112).

Plenary ideas
- Tell the class you had £40 but spent $\frac{2}{5}$ of it on food. Ask them how they would work out how much you had spent.
- Discuss Exercise 7 questions 6a, 6m and 7.

Homework Review questions at the end of Exercise 7 (Pupil Book page 112) and/or Homework Sheet 5.5 (page 205 of this pack).

Links Link this topic to mental calculation (page 61 of the Pupil Book.)
Link to measures: Exercise 7 question 4, Review 2.

5 Fractions

| Topic area | Dividing an integer by a fraction | pupil book page 113 |

Framework Objectives
✓ multiply and divide an integer by a fraction, **Framework examples pages 69**
✓ Consolidate and extend mental methods of calculation, working with ... fractions ..., **Framework examples page 69**

Sheets Homework Sheet 5.6

Resources You could use circles divided into parts and a whiteboard marker to colour them in.

Starter ideas and activities

Mental starter ideas Practise tables.

Starter activity **Introducing dividing an integer by a fraction**
Use the Discussion on page 113 of the Pupil Book.

Teaching points and activities

- Encourage pupils to use diagrams and number lines to help as this is a difficult topic for pupils to grasp initially.
- Remind pupils that multiplying and dividing are inverse operations.
- Some pupils mistakenly think that when we divide, the answer **always** gets smaller. Emphasise that this is only the case when the divisor is greater than 1. When we divide by a number less than one, the answer is bigger.

Support
- Pupils should have covered the previous sections in this chapter.

Extension
- This whole section could be considered as extension.

Plenary ideas
- Discuss a real-life example, such as 'How many quarters would you get if you cut up 3 apples?' Ask pupils to write this down as a calculation and discuss.
- Discuss Exercise 8 question 7, doing it using diagrams and also just with numbers.
- Ask pupils to work out $6 \times \frac{5}{2}$ and $6 \div \frac{2}{5}$. Ask what they notice about the two answers.

Homework Review questions at the end of Exercise 8 (Pupil Book page 114) and/or Homework Sheet 5.6 (page 205 of this pack).

Links Link this topic to mental calculation (Pupil Book page 61).

6 Percentages, Fractions, Decimals

Topic in Chapter	Links to Programme of Study	Links to Framework for Teaching Mathematics: 8 Teaching Programme
	Ma2 a–l are integrated throughout.	*Applying mathematics and solving problems are integrated throughout.*
Converting fractions, decimals and percentages *page 119*	**Ma2 2e** understand that 'percentage' means 'number of parts per 100' and use this to compare proportions (for example, 10% means 10 parts per 100 ...) **Ma2 3e** convert simple fractions of a whole to percentages of the whole and vice versa ... **Ma2 3o** use calculators effectively and efficiently ...	**Fractions, decimals, percentages, ratio and proportion** ● Interpret percentage as the operator 'so many hundredths of' and express one given number as a percentage of another; **use the equivalence of fractions, decimals and percentages to compare proportions ...** **Mental methods and rapid recall of number facts** ● Consolidate and extend mental methods of calculation, working with decimals, fractions and percentages ...
Percentage of — mentally *page 123*	**Ma2 2e** ...interpret percentage as the operator 'so many hundredths of' (... 15% of means $\frac{15}{100} \times y$) **Ma2 3e** ... understand the multiplicative nature of percentages as operators ...	**Fractions, decimals, percentages, ratio and proportion** ● ... calculate percentages ... **Mental methods and rapid recall of number facts** ● Consolidate and extend mental methods of calculation, working with ... percentages ...
Percentage of — written and calculator methods *page 125* Percentage increase and decrease *page 127* Mixed percentage problems *page 131*	**Ma2 2e** ...interpret percentage as the operator 'so many hundredths of' ... **Ma2 3e** ... understand the multiplicative nature of percentages as operators ... (for example, 20% discount on £150 gives a total calculated as £(0·8 × 150)) **Ma2 3m** solve simple percentage problems, including increase and decrease (for example, simple interest, VAT, discount, ... income tax ...) **Ma2 3o** use calculators effectively and efficiently ...	**Fractions, decimals, percentages, ratio and proportion** ● ... calculate percentages and find the outcome of a given percentage increase or decrease ... **Calculator methods** ● Carry out more difficult calculations effectively and efficiently ... ● Enter numbers and interpret displays in different contexts (... percentages ...) ... **Applying mathematics and solving problems** ● Solve more demanding problems in a range of contexts: number ...

Links to Attainment Targets

Level 4   They recognise ... proportions of a whole and use ... percentages to describe these.

Level 5   They calculate ... percentage parts of quantities and measurements, using a calculator where appropriate.

Level 6   Pupils are aware of which number to consider as 100 per cent or a whole, in problems involving comparisons, and use this to evaluate one number as a fraction or percentage of another.
They understand and use the equivalence between fractions, decimals and percentages.

Level 7   They understand and use proportional changes, calculating the result of any proportional change using only multiplicative methods. (begin)

| Topic area | **Converting fractions, decimals and percentages** pupil book page 119 |

Framework Objectives
✓ Interpret percentage as the operator 'so many hundredths of' and express one given number as a percentage of another; **use the equivalence of fractions, decimals and percentages to compare proportions ..., Framework examples pages 71, 75 and 111**
✓ Consolidate and extend mental methods of calculation, working with decimals, fractions and percentages ..., **Framework examples pages 111**

Sheets
Worksheet 13 for Support   Homework Sheet 6.1   Resource Sheet page 202 Teacher Resource Pack

Resources
Dominoes and a calculator for the Starter Activity. You could use percentage dominoes, a counting stick, place value chart and hundreds square. A number line.

 ## Starter ideas and activities

Mental starter ideas
- Ask questions such as Practice questions 48, 49, 60 and 68 on pages 12 and 13 in the Number Support chapter of the Pupil Book.
- You could use a number line with sticky labels to show that $\frac{1}{10} = 0 \cdot 1 = 10\%$. Remind pupils of place value, 0·56 is $\frac{56}{100} = 56\%$, etc.

Starter activity
- **Percentages and everyday life**
  1 Ask pupils for some examples of when percentages are used.
  2 Use the **How much?** Starter on page 118 of the Pupil Book.
- **Game: Dominoes or Odd one out**
  1 If pupils need practice at converting between simple decimals, fractions and percentages you could use a set of dominoes (Taskmaster or a set of your own) or you could play 'Odd one out' by giving out sets of four, such as 78%, 0·78, $\frac{39}{50}$, 7·8 or 0·07, $\frac{7}{10}$, $\frac{7}{100}$, 7%.
  2 Ask for reasons why it is the odd one out.
- **Strategies for converting fractions and decimals to percentages**
  1 *We know that 0·87 × 87%. What have we multiplied 0·87 by to get 87%?* (100%)
  2 Discuss the fact that this is the same as $0 \cdot 87 = \frac{87}{100} = 87\%$. Multiplying by 100% just allows us to miss out writing the decimal as a fraction with denominator 100.
  3 It is important to bring out the point that we have not just multiplied by 100 but 100% because it is **not true** to say that 0·87 = 0·87 × 100 = 87%. The % sign must be in the second step to make a true statement.

 ## Teaching points and activities

- Make sure pupils understand that decimals, fractions and percentages are all ways of expressing proportions of something. Make sure pupils understand place value, i.e. $\frac{1}{10} = 0 \cdot 1$ and $\frac{1}{100} = 0 \cdot 01$, etc. You may need to use diagrams or a counting stick to remind some.
- Remind pupils to always cancel fractions if they can.
- When changing a fraction or decimal into a percentage by multiplying by 100%, emphasise that it is important to include the % sign to make the statement correct, i.e. $\frac{5}{7} = 0 \cdot 714$ (3 d.p.) = 0·714 × 100% = 71·4%.
- Point out that we usually round a percentage to the nearest per cent depending on the context. Sometimes we round to 1 or 2 d.p.
- Emphasise that some common fractions are written as mixed number percentages, e.g. $\frac{1}{3} = 33\frac{1}{3}\%$ and $\frac{2}{3} = 66\frac{2}{3}\%$ and $\frac{1}{8} = 12\frac{1}{2}\%$.
- When comparing proportions we need to convert them all to fractions, or all to decimals, or all to percentages. It is often easiest to compare if they are all percentages.
- Remind pupils to always try a mental method first, then a written method and only use a calculator when they have to.
- Some pupils have difficulty understanding percentages greater than 100%, e.g. 113%, is $1\frac{37}{100}$ or 1·37. Emphasise that 100% is one whole. Give some examples in everyday life of percentages greater than 100%, e.g. lambing percentages or growth.

Support
- See **Decimals, fractions and percentages** on page 6 of the Pupil Book (Number Support chapter).
- If pupils have difficultly with Practice questions 48, 49, 58, 60, 68 and 85 then use Worksheet 12 question 10b and Worksheet 13 questions 1, 2g–l and 6 (pages 10 and 11 of the Teacher Resource Pack).

Extension
- See **starred questions** in Exercise 1 (Pupil Book page 120)
- See **Practical** (Pupil Book page 123). This Practical can be made into an extension exercise by encouraging pupils to display their results in a table or pie chart. This way the link can be made to Handling Data.

Plenary ideas
- Discuss the answers to Exercise 1 questions 11 and 12 and ask pupils to demonstrate their methods. You could also do the same for Exercise 1 questions 13 and 14.
- Use the discussion on page 122 of the pupil book to emphasise that rounding will sometimes give us a total that does not add to 100%.
- Discuss ways of doing the practical on page 123 of the pupil book.

Homework
Review questions at the end of Exercise 1 (Pupil Book page 120) and/or Homework Sheet 6.1 (page 206 of this pack)

Links
Link this topic to mental calculation (Pupil Book page 61), cancelling fractions to their lowest terms (Pupil Book page 5), probability (Pupil Book page 414), pie charts and bar charts (Pupil Book pages 397 and 391).
Link to measures: Exercise 1 questions 6, 7, 8, Review 2.
Link to Science: Exercise 1 question 10.
Link to Geography: Exercise 1 question 9.

Topic area	**Percentage of — mentally**	pupil book page 123

Framework Objectives
✓ ... calculate percentages ..., **Framework examples pages 73, 99 and 101**
✓ Consolidate and extend mental methods of calculations, working with ... percentages ..., **Framework examples pages 73, 99 and 101**

Sheets Worksheet 13 for Support   Homework Sheet 6.2

Resources You could use percentage dominoes.

 ## Starter ideas and activities

Mental starter ideas Ask questions such as Practice question 64 on page 13 of the Number Support chapter.

Starter activity **Strategies for finding 'percentage of' mentally**
Give this problem:
The top speed of a car is given in the manual as 160 km/h. How fast would the car be going if it went 40% of this speed? What about 60%? What about 5%, 29%, 72%, ... ?
Discuss ways of getting the answer mentally by finding 10% first.
What if the manual was wrong and the top speed was 125% of the speed given in the manual?

 ## Teaching points and activities

- Make sure pupils can find fractions of amounts.
- Emphasise the use of 100%, 25% and 10% to work out other percentages such as 5%, 1%, 30%, etc. Stress that 10% can be found by dividing by 10 or multiplying by 0·1, and that 25% can be found by dividing by 4.
- Encourage pupils to check that the units in the answer are correct, e.g. if the calculation is 40% of 225 **m** then the answer must in **metres**. Some pupils may put % as the units.
- VAT calculations are a good example of an everyday use of this topic.
  $17\frac{1}{2}\%$ can be found by finding $10\% + 5\% + 2\frac{1}{2}\%$.

Support
- See **Decimals, fractions and percentages** on page 6 of the Pupil Book (Number Support chapter). If pupils have difficulty with Practice questions 61 and 64 use Worksheet 13 questions 4 and 5 (page 11 of the Teacher Resource Pack).

Extension
- See **starred questions** in Exercise 2 (page 124 Pupil Book)
- You could extend this topic by doing some of the questions given in the next section mentally.

Plenary ideas
- Ask pupils when they may need to find percentages of quantities. Choose a topic like banking or schools or hospitals and ask for examples in these specific areas.
- Discuss Exercise 2 question 7.

Homework Review questions at the end of Exercise 2 (Pupil Book page 124) and/or Homework Sheet 6.2 Exercise 1 (page 207 of this pack).

Links Link this topic to mental calculation (Pupil Book page 61) and converting fractions, decimals and percentages (Pupil Book page 119).
Link to Measurement: Exercise 2 question 1e–w, Review 1b to j.

> **Topic area** **Percentage of — written and calculator methods** pupil book page 125

Framework Objectives
✓ **Calculate percentages ..., Framework examples page 73**
✓ Enter numbers and displays in different contexts (... percentages ...), **Framework examples page 73**

Sheets Worksheet 13 for support   Homework Sheet 6.2

Resources A calculator.

 ## Starter ideas and activities

Mental starter ideas Revise × and ÷ by 10 and 100. You could start with a number and draw a spider diagram for ×10, ×100, ×1000, ÷10, ÷100, ÷1000

Starter activity
● **Finding 'percentage of' using a written or calculator method**
  1 Give pupils a similar problem to this one:
     Yesterday I looked at a stereo for £456.80. The sales person told me I would have to pay an 18% deposit. How much deposit would I pay? (You could make up one of you own that would be of interest to your class.)
     How could we work this out? (Using a written method or a calculator.) Ask pupils to come and demonstrate how to do this.
  2 Encourage the use of the fraction, decimal and finding 1% first methods given on page 125 of the Pupil Book.
  3 Discuss which method is best.
● **Estimating the answers**
  Ask how you could estimate the answer to 18% of 456·80. (20% of 450 or 20% of 500)

 ## Teaching points and activities

● Encourage pupils to use mental methods if possible and to always estimate the answer to the calculation.
● Recap how to find the position of the decimal point in a decimal multiplication.
  Some pupils learn a method for doing these calculations and then get it wrong, e.g. 58% of 64·8 = 58/64·8 × 100. It helps if pupils read the calculation from left to right as 58% of 64·8 and then write down the parts one by one. 58% is $\frac{58}{100}$, 'of' means ×, the number we are finding 58% of is 64·8.

Support
● Pupils should have covered the previous sections in this chapter. Also see **percentage of a quantity using a calculator** on page 6 (Number Support chapter). If pupils have difficulty with Practice question 83 use Worksheet 13 question 7 (page 11 of the Teacher Resource Pack).

Extension
● See **starred questions** in Exercise 3 (Pupil Book page 125)

Plenary ideas
● Discuss where in everyday life you find 'percentage of' problems and put a real-life example on an OHP. One that relates to something topical would be good. Ask pupils to demonstrate how to find the answer. Emphasise the teaching points given above.
● Discuss the answers to Exercise 3 questions 9, 11, 13 or 14. Discuss the different methods that can be used.

Homework Review questions at the end of Exercise 3 (Pupil Book page 125) and/or Homework Sheet 6.2 Exercise 2 (page 207 of this pack).

Links Link this topic to using the calculator (Pupil Book page 6) and converting fractions, decimals and percentages (Pupil Book page 119).
Link to Geography: Exercise 3 question 9b, Review 3a.
Link to measures: Exercise 3 questions 1, 3, 9a, Reviews 1 and 2.
Link to Handling Data: Exercise 3 question 9b, 12, 13, Review 4.
Link to Area: Exercise 3 question 10.

| **Topic area** | **Percentage increase and decrease** | pupil book page 127 |

Framework Objectives

✓ **Calculate percentages and find the outcome of a given percentage increase or decrease ..., Framework examples page 77**

✓ Enter numbers and displays in different contexts (... percentages ...), **Framework examples page 77**

✓ Problem solving, **Framework examples pages 3, 11, 29**

Sheets   Homework Sheet 6.3   ICT Worksheets 'Discount Deals' and 'Population Increase' Teacher Resource Pack pages 122 and 123

Resources   A calculator. A graphical calculator. Computer and spreadsheet package.

 Starter ideas and activities

Mental starter ideas   Practise some mental 'percentage of' calculations such as 45% of 80 m, $17\frac{1}{2}$% of £120, ...

Starter activity
- **Introducing percentage increase and decrease**
  Use the Discussion on page 127 of the Pupil Book.
- **Using a multiplier to find percentage increase and decrease**
  Is taking away 20% of 200 m the same as 80% of 200 m. Why?
  Ask pupils to explain their answer using a diagram if they want to.
  Is adding on 30% of 200 the same as 130% of 200? Why?

 Teaching points and activities

- Pupils may have difficulty with the idea that a 20% increase is the same as finding 120% and that a 15% decrease is the same as finding 85%. Encourage pupils to use the one-step method of doing these calculations. It is important that they understand why it can be done this way. You may need to use a diagram to illustrate this.

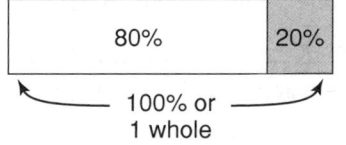
20% taken off a whole is the same as 80% of the whole.

- Encourage pupils to estimate the answer.
- Recap that a percentage greater than 100% is greater than 1 whole, e.g. 130% = $1\frac{3}{10}$.

Support
- Students should have covered all the earlier sections in this chapter.

Extension
- See **starred questions** in Exercise 4 (Pupil Book page 129).
- See the last bullet point of the **Discussion** on page 127 of the Pupil Book.
- See **Investigation: Population increase** on page 131 of the Pupil Book. Ask pupils to write a report on their answers showing how they found them and ask more 'what if' questions to explore.

Plenary ideas
- If you did not use the Discussion on page 127, the first five bullets could be discussed as part of the plenary.
- Discuss the answers to Exercise 4 question 1 or 12 or 13 to extend thinking.
- Discuss the Puzzle on page 131 of the Pupil Book.

Homework   Review questions at the end of Exercise 4 (Pupil Book page 129) and/or Homework Sheet 6.3 (page 208 in this pack).

Graphical calculator   Investigation: Population increase (Pupil Book page 131).

ICT   Practical (Pupil Book page 131). See ICT Worksheet 'Discount Deals' on page 122 of the ICT Support Section in the Teacher Resource Pack.
Investigation: Population increase (Pupil Book page 131). See ICT Worksheet 'Population Increase' on page 123 of the ICT Support section in the Teacher Resource Pack.

Links   Link this topic to percentage of (Pupil Book pages 123 and 125), enlargement (Pupil Book page 320), scale drawing (Pupil Book page 324) and area and volume (Pupil Book pages 348 and 355).
Link to Science: Investigation: Population increase (Pupil Book page 131).
Link to measures: Puzzle (Pupil Book page 131).

| Topic area | Mixed percentage problems | pupil book page 131 |

Framework Objectives

✓ Express one given number as a percentage of another, **Framework examples page 75**
✓ Use the equivalence of fractions, decimals and percentage to compare proportions. Calculate percentages and find the outcome of a given percentage increase or decrease ..., **Framework examples page 77**
✓ Enter numbers and displays in different contexts (... percentages ...), **Framework examples page 77**
✓ Problem solving, **Framework examples pages 3 and 29**

Sheets Homework Sheet 6.4

Resources A calculator.

 ## Starter ideas and activities

Starter activity You could remind pupils of the points given in the previous section in this chapter before they begin the exercise.

 ## Teaching points and activities

- This section gives practice at all types of percentage problems and could be used as a revision exercise or homework.
- Recap finding VAT.

Support
- Pupils should have covered all the previous sections in this chapter.

Extension
- See **starred questions** in Exercise 5 (Pupil Book page 131)

Plenary ideas
- Ask pupils how many different real-life situations they can think of in which percentages are used.
- Discuss Exercise 5 questions 7 and 11. Ask pupils to demonstrate strategies for doing these.

Homework Review questions at the end of Exercise 5 (Pupil Book page 131) and/or Homework Sheet 6.4 (page 208 in this pack).

Links Link to Design and Technology: Exercise 5 questions 1, 10.
Link to measures: Exercise 5 question 2.
Link to Handling Data: Exercise 5 question 10, 11, Review 4.

7 Ratio and Proportion

Topic in Chapter	Links to Programme of Study *Ma2 a–l are integrated throughout.*		Links to Framework for Teaching Mathematics: 8 Teaching Programme *Applying mathematics and solving problems are integrated throughout.*
Direct proportion *page 137*	**Ma2 3n**	solve word problems about ratio and proportion, including using informal strategies and the unitary method of solution (for example, given that m identical items cost £y, then one item costs £y/m and n items cost £$(n \times y/m)$, the number of items that can be bought for £z is $z \times m/y$)	**Fractions, decimals, percentages, ratio and proportion** ● ... reduce a ratio to its simplest form, including a ratio expressed in different units, recognising links with fraction notation ... **use the unitary method to solve simple word problems involving** ratio and **direct proportion**
	Ma2 4a	draw on their knowledge of the operations and the relationships between them, ... to solve problems involving ratio and proportion	
Simplifying ratios *page 139*	**Ma2 2f**	use ratio notation, including reduction to its simplest form and its various links to fraction notation	**Fractions, decimals, percentages, ratio and proportion** ● Consolidate understanding of the relationship between ratio and proportion; reduce a ratio to its simplest form; ... **use the unitary method to solve simple word problems involving ratio and direct proportion.**
Ratio and proportion *page 141*	**Ma2 2e**	understand that 'percentage' means number of parts per 100 ... use this to compare proportions	**Applying Mathematics and Solving Problems** ● Solve more demanding problems and investigate in a range of contexts: number ... ● **Use logical argument to establish the truth of a statement**
Solving ratio and proportion problems *page 143*	**Ma2 2f** **Ma2 2g**	use ratio notation recognise where fractions or percentages are needed to compare proportion; identify problems that call for proportional reasoning ...	
	Ma2 3n	solve word problems about ratio and proportion, including using informal strategies and the unitary method of solution ...	
	Ma2 4a	draw on their knowledge of the operations and the relationships between them ... to solve problems involving ratio and proportion ...	
Dividing in a given ratio *page 145*	**Ma2 3f**	divide a quantity in a given ratio (for example, share £15 in the ratio 1 : 2)	**Fractions, decimals, percentages, ratio and proportion** ● ... **divide a quantity into two or more parts in a given ratio**

Links to Attainment Targets

Level 5   ... solve simple problems involving ratio and direct proportion.
Level 6   ... and calculate using ratios in appropriate situations ...
Level 7   They understand and use proportional changes, calculating the result of any proportional change using only multiplicative methods. (begin)

Topic area	Direct proportion	pupil book page 137

Framework Objectives

✓ ... use the unitary method to solve simple word problems involving ... **direct proportion, Framework examples pages 79**
✓ Problem solving, **Framework examples page 5**

Sheets

Worksheet 14   Homework Sheet 7.1
ICT Worksheets 'Constant Multiplier' and 'Currency Converter' Teacher Resource Pack pages 125 and 126

Resources

Red and blue counters for the Starter Activity. A computer and a spreadsheet package.

 ## Starter ideas and activities

Mental starter ideas

● Ask questions such as Practice question 45 on page 11 of the Number Support chapter.
● Revise percentage, decimal and fraction equivalences.
    What is 0·2 as a fraction? percentage?
    What is $\frac{13}{20}$ as a decimal? percentage?
    What is $12\frac{1}{2}$% as a decimal? fraction?
    What is 150% as a fraction? decimal?

Starter activity

Direct proportion
1 Demonstrate the example given in the teaching screen on page 137 of the Pupil Book with some red and blue counters.
2 Ask pupils to come and show you how much red paint you would need for, say, 5 ℓ, 10 ℓ, ... (shown as counters).
3 Ask *What if we had 8 ℓ of blue paint?* (*This can't be shown without chopping the counters up so how else could we do it?*)
4 Try and draw out from the pupils the method of finding how much red paint to use with 1 ℓ first and then multiply this by 8.

 ## Teaching points and activities

● Emphasise the point that the amounts stay in the same proportion. That is, if we find the ratio, say, of blue paint to red paint in the example on page 137, this ratio is always the same. Also emphasise that if we plotted the amounts of red paint and blue paint as coordinate pairs on a grid they would form a straight line.
● Try and make clear the distinction between direct proportion and proportion.

Support

● See **Ratio and proportion** on page 7 of the Pupil Book (Number Support chapter). If pupils have difficulty with Practice questions 45 and 72 then use worksheet 14 questions 1 and 2 (page 12 of the Teacher Resource Pack).

Extension

● See **starred questions** in Exercise 1 (Pupil Book page 138). You could ask pupils to try doing some of the questions in the exercise mentally.

Plenary ideas

● Show the class a real-life straight line graph on an OHP, for example a kilometre to mile conversion graph. Discuss the fact that as kilometres double, so do miles. Thus talk and ask questions about direct proportion.
● Discuss the answers to Exercise 1 question 2, asking pupils to demonstrate how they did these. Encourage pupils to find the cost of one unit first.
● Discuss Exercise 1 question 8, 9 or 10 to extend thinking a little.

Homework

Review questions at the end of Exercise 1 (Pupil Book page 138) and/or Homework Sheet 7.1 (page 209 of this pack).

ICT

Practical (Pupil Book page 139). **Question 1**: See ICT Worksheet 'Constant Multiplier' on page 125 in the ICT Support Section of the Teacher Resource Pack. **Question 2**: See ICT Worksheet 'Currency Converter' on page 126 in the ICT Support Section of the Teacher Resource Pack.

Links

Link this topic to conversion graphs (Pupil Book page 240), graphs of linear relationships (Pupil Book page 233) and problems involving ratio and proportion (Pupil Book page 143).
Link to Design and Technology: Exercise 1 questions 1, 6, 7.
Link to measures: Exercise 1 question 10.

| Topic area | Simplifying ratios | Pupil Book page 139 |

Framework Objectives ✓ ... reduce a ratio to its simplest form, including a ratio expressed in different units, recognising links with fraction notation;..., **Framework examples page 81**

Sheets Worksheet 14 for support Homework Sheet 7.2

Resources Mini-whiteboards or digit cards for the Mental Starter. You could use Multilink or counters.

 ## Starter ideas and activities

Mental starter ideas
- Revise cancelling fractions to their simplest form. You could ask pupils to hold up the answers on mini-whiteboards or using digit cards.
- Ask questions such as Practice Question 53 on page 12 of the Number Support chapter.

Starter activity **Equivalent ratios on a line**
1 Demonstrate equivalent ratios using a line, i.e. show the ratio 2 : 3 as 2 parts to 3 or as 4 parts to 6 or as 6 parts to 9. Just keep making the parts smaller.
2 Ask pupils to show you a line divided in the ratio 3 : 4 in three different ways.
3 *What is the ratio 6: 8 in its simplest form?*
   What is the ratio 8 : 12 in its simplest form?

 ## Teaching points and activities

- Ask pupils to come up with some real-life situations in which ratios are used. Discuss these.
- Emphasise the similarities with cancelling fractions to their simplest form. (You must divide each part of the ratio by the highest common factor.)
- Emphasise that ratios in their simplest form do not contain a decimal or fraction.
- Recap HCF. Point out that when we cancel ratios, we do so by dividing each part by the HCF.
- Pupils find simplifying ratios which contain fractions difficult. Encourage them to try multiplying by the denominator of the part that has the fraction or, if both parts have a fraction, by the common denominator (LCM of the denominators). When there are decimals it is often easiest to multiply all parts by 10, 100, ... (whatever is needed to give a whole number in all parts of the ratio) and then simplify the resulting ratio.

Support
- Pupils should have covered **Fractions** in the support chapter. See **Ratio and Proportion** on page 7 (Number Support chapter).
- If pupils have difficulty with Practice questions 50a and 53 use Worksheet 14 question 4 (page 12 of the Teacher Resource Pack).

Extension
- See **starred** questions in Exercise 2 (Pupil Book page 140).
- You could give some three-part ratios with fractions and decimals for pupils to try.

Plenary ideas
- Discuss Exercise 2 question 5 (or 11 and 12 to extend thinking).
- Give these four ratios to simplify
   $4 : 16 : 24$, $2\frac{1}{2} : 3$, $2 \cdot 4 : 3$, $420 \text{ g} : 2 \text{ kg}$
Ask pupils to demonstrate how to do each.

Homework Review questions at the end of Exercise 2 (Pupil Book page 140) and/or Homework Sheet 7.2 (page 209 of this pack)

Links Link this topic to fraction notation (Pupil Book page 98) and converting between measures (Pupil Book page 335).
Link to Design and Technology: Exercise 2 question 9.
Link to measures: Exercise 2 questions 6, 7, 10, 11, Review 4.
Link to Science: Exercise 2 question 4.

Topic areas	Ratio and proportion	pupil book page 141
	Solving ratio and proportion problems	pupil book page 143

Framework Objectives
✓ Consolidate understanding of the relationship between ratio and proportion; ... **use the unitary method to solve simple word problems involving ratio and direct proportion, Framework examples pages 5, 31 and 81**
✓ Problem solving, **Framework examples pages 5 and 31**

Sheets Worksheet 14 for Support   Homework Sheets 7.3 and 7.4

Resources Ten sets of cards with related ratios and proportions for the Starter Activity. A calculator.

Starter ideas and activities

Mental starter ideas
Revise the difference between ratio and proportion and ask some questions such as Practice questions 50 and 72 of the Number Support chapter.

Starter activity
The difference between ratio and proportion
1 Make 10 sets of cards with related ratios and proportions like this set. (50 cards in total)
   Black counters 2   Ratio of black to white is 2 : 3   Proportion of black is $\frac{2}{5}$
   White counters 3   Proportion of white is 60%
2 Give out the cards to pupils and ask them to put them into groups. You could divide the class into two groups and give one set of 50 cards to each group and have a race to see who can put them together into 10 correct sets first.
Or If you don't want to make cards you could ask for five volunteers to come to the front of the class. Put hats (or similar) on two of them.
   What is the ratio of those with hats to those without?
   What proportion have hats? Don't have hats? What is this as a fraction/percentage/decimal?

Teaching points and activities

● Emphasise that ratio expresses part-to-part whereas proportion expresses part-to-whole.
● Emphasise that a proportion can be given as a fraction, decimal or percentage.
● Pupils sometimes forget to find the total when finding a proportion, e.g. if there are 2 white birds and 8 blue birds sometimes pupils, when asked for the proportion of white birds, simply put the smaller number given over the larger to get $\frac{2}{8}$ or $\frac{1}{4}$.
   Encourage pupils to ask themselves what **total** they are finding the proportion **of**.
   Ask them to check if they have this total or if they need to find it first.
● Recap scale drawings and pie charts for question 8 of the exercise.

Support
● Pupils should have covered the previous sections in this chapter.
● See **Ratio and Proportion** page 7 (Number Support chapter).
● If pupils have difficulty with Practice questions 50 and 72 use Worksheet 14 questions 3 and 6 (page 12 of the Teacher Resource Pack).

Extension
● See **starred questions** in Exercise 4 (Pupil Book page 143).
● See **Puzzle** (Pupil Book page 145). Challenge pupils to write an explanation for their answer to this Puzzle.

Plenary ideas
● Discuss the answers to Exercise 3 question 2 as a class.
● Discuss Exercise 4 questions 2 and 9.

Homework
Review questions at the end of Exercises 3 and 4 (Pupil Book pages 142 and 143) and/or Homework Sheets 7.3 and 7.4 (page 210 of this pack)

Links
Link this topic to fraction notation and cancelling fractions to their lowest form (Pupil Book pages 5 and 199) and the equivalence between fractions, decimals and percentages (Pupil Book page 119), enlargement (Pupil Book page 320) and scales (Pupil Book page 324).
Link to Design and Technology: Exercise 3 Review; Exercise 4 questions 3, 6, 9, Review 4; Puzzle.
Link to measures: Exercise 3 question 4; Exercise 4 questions 1, 4, 7, 10.
Link to Geography: Exercise 4 question 2, 11.
Link to Handling Data: Exercise 4 question 8.
Link to area: Exercise 4 question 13.
Link to PE: Exercise 4 Review 2.

| Topic area | **Dividing in a given ratio** | Pupil Book page 145 |

Framework Objectives
- ✓ ... divide a quantity into two or more parts in a given ratio, **Framework examples page 81**
- ✓ Solving problems, **Framework examples page 5**

Sheets Worksheet 14 for Support   Homework Sheet 7.5   Resource Sheet page 203 Teacher Resource Pack

Resources Counters on an OHP for the starter activity. You could use a line divided into parts or a counting stick.

 ## Starter ideas and activities

Mental starter ideas Revise division facts such as $45 \div 3$, $188 \div 4$, $2400 \div 5$, $128 \div 8$, ...

Starter activity **Use concrete materials to demonstrate dividing in a given ratio**
1. Have 36 counters on an OHP or give them out to pupils.
2. *I have £36 which I want to divide between my two sisters Lizzy and Sis in the ratio 2 : 1.*
   How many shares will Lizzy get? What about Sis?
   How many shares will there be altogether?
   How many pounds (counters) are there in one share? (36 ÷ 3)
   How many will there be in 2 shares?
   How much would each sister get? How can you check that you are right?
   If I divide the counters in the ratio 4 : 5 how many shares will there be?
   How many counters will there be in one share?
   How many will there be in 4 shares? 5 shares?

   If three people bought a raffle ticket in the ratio 2 : 1 : 3 and it won a prize of £6600, how much should each get?
   How could you work this out? (By adding up the total number of shares to get 6 and then finding out how much one share is: £6600 ÷ 6 = £1100.)
3. You could now do the first few from Exercise 5 question 1 with the class.

 ## Teaching points and activities

- Pupils sometimes have difficulty understanding why we need to work out the total number of shares. It helps to work with money (use money or counters), which they are familiar with, and actually divide it up in the required ratio as in the starter activity. Some pupils may need to continue working with concrete materials to help with understanding.
- Encourage pupils to check their answers by checking that the total of all the shares is the same as the total that was to be divided up.

Support
- Pupils should have covered the previous sections in this chapter.
- **See Ratio and proportion** page 7 (Number Support chapter in the Pupil Book). If pupils have difficulty with Practice question 71, use Worksheet 14 question 5 (page 12 of the Teacher Resource Pack).

Extension
- See **starred questions** in Exercise 5 (Pupil Book page 146).
- See **Investigation: Triangle Paths** (Pupil Book page 147). Challenge pupils to divide the sides of the triangles into harder ratios. Try the investigation with different shapes.

Plenary ideas
- Discuss the answers to Exercise 5 question 1, asking pupils to demonstrate how they did these.
- Discuss the answer to Exercise 5 question 7 as a class (the amount does not come out to be an exact amount and so lends itself to discussing how you would round the amount sensibly).
- Start the Investigation: Triangle Paths on page 147 and ask pupils to finish it for homework.

Homework Review questions at the end of Exercise 5 (Pupil Book page 146) and/or Homework Sheet 7.5 (page 210 of this pack).

Links Link to Design and Technology: Exercise 5 question 3.
Link to shapes: Exercise 5 question 5; Investigation: Triangle Paths.
Link to measures: Exercise 5 questions 3, 4, Reviews 2, 3.

8 Expressions, Formulae and Equations

Topic in Chapter	Links to Programme of Study *Ma2 a–l are integrated throughout.*		Links to Framework for Teaching Mathematics: 8 Teaching Programme *Applying mathematics and solving problems are integrated throughout.*
Understanding algebra *page 158* More understanding algebra *page 160*	**Ma2 1d** **Ma2 1g** **Ma2 5a** **Ma2 5b**	select efficient techniques for ... algebraic manipulation develop correct and consistent use of notation, symbols ... distinguish the different roles played by letter symbols in algebra, knowing that letter symbols represent definite unknown numbers in equations defined quantities or variables in formulae (for example $V = IR$), ... and in functions they define new expressions or quantities by referring to known quantities ... (for example, $y = 2 - 7x$) ... understand that the transformation of algebraic expressions obeys and generalises the rules of arithmetic; ... distinguish in meaning between the words 'equation', 'formula', ... and 'expression'.	**Equations, formulae and identities** • Begin to distinguish the different roles played by letter symbols in equations, formulae and functions; know the meanings of the words *formula* and *function*. • Know that algebraic operations follow the same conventions and order as arithmetic operations; use index notation for small positive integer powers.
Simplifying expressions — multiplying and dividing *page 164* Brackets *page 166* Collecting like terms *page 168*	**Ma2 1d** **Ma2 5b**	select efficient techniques for ... algebraic manipulation ... simplify or transform algebraic expressions by collecting like terms (for example, $x^2 + 3x + 5 - 4x + 2x^2 = 3x^2 - x + 5$(begin)), by multiplying a single term over a bracket	• **Simplify or transform linear expressions by collecting like terms; multiply a single term over a bracket.**
Substituting into expressions *page 172* Substituting into formulae *page 175*	**Ma2 5c** **Ma2 5f**	... substitute positive and negative numbers into expressions such as $3x^2 + 4$ and $2x^3$ use formulae from mathematics and other subjects (for example, formulae for the area of a triangle ...); substitute numbers into a formula; ... change its subject (for example ... find the perimeter of a rectangle given its area A and length l of one side)	**Equations, formulae and identities** • **Simplify or transform linear expressions by collecting like terms; multiply a single term over a bracket.** • Use formulae from mathematics and other subjects; **substitute integers into simple formulae**, including examples that lead to an equation to solve, and positive integers into expressions involving small powers (e.g. $3x^2 + 4$ or $2x^3$); derive simple formulae.

Topic in Chapter	Links to Programme of Study	Links to Framework for Teaching Mathematics: 8 Teaching Programme
Writing expressions *page 178*	**Ma2 1f** represent problems and solutions in algebraic forms ...	● Construct and solve linear equations with integer coefficients (unknown on either or both sides, without and with brackets) using appropriate methods (e.g. inverse operations, transforming both sides in same way).
More writing and simplifying expressions *page 179*	**Ma2 1g** develop correct and consistent use of notation, symbols ... when solving problems	● Begin to use graphs and set up equations to solve simple problems involving direct proportion.
Writing equations *page 181*	**Ma2 1d** select efficient techniques for ... algebraic manipulation	**Applying mathematics and solving problems**
Writing and finding formulae *page 182*	**Ma2 1f** ... present and interpret solutions in the context of the original problem ...	● solve more demanding problems and investigate in a range of contexts: ... algebra ...
Writing and solving equations *page 184*	**Ma2 1g** develop correct and consistent use of notation, symbols ... when solving problems	● **... represent problems and interpret solutions in algebraic ... or graphical form,** using correct notation and appropriate diagrams.
Solving equations by transforming both sides *page 188*	**Ma2 5d** set up simple equations (for example, find the angle a in a triangle with angles a, $a + 10$, $a + 20$), solve simple equations (for example, $5x = 7$, $3(2x + 1) = 8$, ...) by using inverse operations or by transforming both sides in the same way	● Solve more complex problems ... choosing and using efficient techniques for ... algebraic manipulation and graphical representation and resources, including ICT.
Solving equations using a graph *page 192*	**Ma2 5e** solve linear equations, with integer coefficients, in which the unknown appears on either side or both sides of the equation; solve linear equations that require prior simplification of brackets, including those that have negative signs occurring ... in the equation ...	

Links to Attainment Targets

Level 5 They construct, express in symbolic form, and use formulae involving one or two operations. They use brackets appropriately.

Level 6 They formulate and solve linear equations with whole-number coefficients.

Level 7 They evaluate algebraic formulae, substituting ... decimals and negative numbers. (begin) Pupils manipulate algebraic formulae, equations and expressions. (begin)

Topic area	Understanding algebra	pupil book page 158

Framework Objectives
✓ Begin to distinguish the different roles played by letter symbols in equations, formulae and functions; know the meaning of the words *formula* and *function*, Framework examples page 113

Sheets Worksheet 15 for support   Homework Sheet 8.1   Starter Resource Sheet 5

Resources You could use missing-number dominoes.

 ## Starter ideas and activities

Mental starter ideas Revise adding and subtracting 2- and 3-digit numbers mentally.
Give some examples such as 56 + 37, and 450 – 327, etc.

Starter activity
● **Expressions and equations and formulae**
  1 Remind pupils what an expression and an equation are (see screen on page 158 of the Pupil Book).
  2 On an OHP or the board write some expressions and equations (you could use the ones in Exercise 1). Ask pupils to come and put a tick by the expressions and a square beside the equations. Ask them to explain how they know.
  3 Do the same for some expressions, equations and some formulae.
● **Introducing functions**
  Put a function machine on the board or on an OHP.
    How could we write the rule for this function machine?
    The rule is called a function. What do you notice about it? (It has an x and a y and it gives the rule for the relationship between them or it defines y in terms of x.)
● **Writing expressions**
  1 Put the class into groups of 3 or 4.
  2 Make a set of pairs of cards with an expression written with × and ÷ signs and the simplified version.
    Example $3 \times n$ and $3n$, $a \times b$ and ab, $4 \times (s + t)$ and $4(s + t)$, $(h + 4) \div 6$ and $\dfrac{(h + 4)}{6}$
  You could use Starter Resource Sheet 5.
  3 Give a set of about 20 cards to each group of 3 or 4 pupils and ask them to match them up.

 ## Teaching points and activities

● Some pupils have difficulty understanding the difference between equations, formulae and functions. They all have equals signs. Emphasise that in a formula, the letters always stand for something specific and there is always more than one variable (letter). Stress that a function is a special sort of equation used especially to describe the relationship between the input and output of a function machine, or the x- and y-coordinates of a graph. (Make the link between these.)
● Some pupils have difficulty as soon as letters are introduced. Emphasise often that a letter always stands for a number. Give as many concrete examples as possible, such as:
    A function machine always adds 10 to every number that goes into it so we can write a general rule using letters for the number put in and the number coming out.
    y (the number that comes out) = x (the number that goes in) + 10.
● Explain that there are some conventions we use when writing expressions:
    we write the number before the letter (because this is the way we read it),
    we usually write the letters in alphabetical order if there is more than one, i.e. we write pq rather than qp,
    we usually write $1n$ as n and $1p$ as p, etc.
    Encourage pupils to use these conventions as this avoids problems later on.

Support ● See **Writing expressions** and **Working with expressions — Algebra and arithmetic** on page 150 (Algebra Support chapter of Pupil Book). If pupils have difficulty with Practice questions 3, 4 and 8 use Worksheet 15 questions 1, 2 and 3 (page 13 of the Teacher Resource Pack).

Extension ● See **starred question** in Exercise 3 (Pupil Book page 159).

Plenary ideas ● Give an example of each of the following as a summary of the section: an expression (one with brackets); an equation; a formula; a function.
● Discuss the answers to Exercise 3 question 1 or Exercise 4 question 1.

Homework Review questions at the end of Exercises 1, 2, 3 and 4 (Pupil Book pages 158, 159 and 160) and/or Homework Sheet 8.1 Exercises 1, 2, 3 and 4 (page 211 of this pack).

Links Link this topic to writing and solving equations (Pupil Book pages 181 and 184), writing expressions (Pupil Book page 178) and understanding the operations addition, subtraction, multiplication and division.
Link to area: Exercise 4 question 4.
Link to sequences: Exercise 2 question 2.
Link to Science: Exercise 3 Review 1b.

Topic area	More understanding algebra	pupil book page 160

Framework Objectives
✓ Begin to distinguish the different roles played by letter symbols in equations, formulae and functions; know the meaning of the words *formula* and *function*, **Framework examples pages 113 and 115**
✓ Know that algebraic operations follow the same conventions and order as arithmetic operations; ..., **Framework examples page 115**

Sheets
Worksheet 16 for Support   Homework Sheet 8.2   Starter Resource Sheet 6
ICT Worksheets 'Inverses' and 'True or False' (Teacher Resource Pack pages 129 and 130)

Resources
True/false cards for the Starter Activity. You could use missing-number dominoes.
Computer and a spreadsheet package.

 Starter ideas and activities

Mental starter ideas
- Revise order of operations by asking questions similar to Practice questions 7 and 30 of the Number Support chapter.
- Revise inverse operations by putting pupils into groups of four. Ask pupils to take turns at making up a true number sentence, such as $56 + 23 = 79$. Ask each of the other pupils in the group to make up a calculation that must also be true to make a set of four facts, such as:
  $56 + 23 = 79$ $23 + 56 = 79$ $79 - 56 = 23$ $79 - 23 = 56$
  You could give out cards for pupils to make into sets instead.
- Ask questions such as Practice question 7 on page 154 in the Algebra Support chapter.

Starter activity
- **What does equals mean?**
  Use the Discussion on page 160 of the Pupil Book.
- **Order of operations and algebra**
  Put some expressions on the OHP or board such as $3n + 4$ and $3(n + 4)$ and $9 - n^2$.
  Which part of this expression would we work out first if we knew the value of n?
- **True/false**
  Make a set of true/false cards (Starter Resource Sheet 6).
  Give one to each pupil and ask him or her to hold it up and say 'true' or 'false' and explain why it is true or false.

 Teaching points and activities

- Recap order of operations, the commutative rule and inverse operations.
- Pupils sometimes have difficulty with algebra, but not with arithmetic. It is important to continually point out that algebra follows the same rules as arithmetic and if difficulties arise give an example from arithmetic to illustrate.

Support
- See **Writing expressions** and **Working with expressions — Algebra and arithmetic** on page 150 (Algebra Support chapter of the Pupil Book).
- If pupils have difficulty with Practice question 7 then use Worksheet 16 question 1 (page 14 of the Teacher Resource Pack).

Extension
- See **starred question** in Exercise 6 (Pupil Book page 162).

Plenary ideas
- Use some coloured rods to make up questions similar to question 2 of the exercise. Ask pupils to come and make true equations.
- Discuss Exercise 6 question 1 (or 4 for extension). Ask pupils to explain how they got their answers.

Homework
Review questions at the end of Exercises 5 and 6 (Pupil Book pages 161 and 162) and/or Homework Sheet 8.2 Exercises 1 and 2 (page 212 of this pack).

ICT
For the Practical (Pupil Book page 163) use ICT Worksheet 'Inverses' on page 129 of the ICT Support section of the Teacher Resource Pack.
For Exercise 5 question 3 use ICT Worksheet 'True or False' on page 130 of the ICT Support section of the Teacher Resource Pack.

Links
Link this topic to equations (Pupil Book pages 181 and 184), writing expressions (Pupil Book page 178) and understanding the operations addition, subtraction, multiplication and division.

| Topic area | **Substituting into formulae** | pupil book page 175 |

Framework Objectives ✓ Use formulae from mathematics and other subjects; **substitute integers into simple formulae,** including examples that lead to an equation to solve ..., **Framework examples page 141**

Sheets Worksheet 17 for support Homework Sheet 8.7 Starter Resource Sheet 8

Resources Cards for the Starter Activity—Starter Resource Sheet 8 plus extras.

 ## Starter ideas and activities

Mental starter ideas Ask questions such as Practice questions 1 and 5 on pages 153 and 154 of the Algebra Support chapter in the Pupil Book.

Starter activity **Substituting into formulae**
Use a similar activity to the one for substituting into expressions using cards and volunteers. Choose a formula such as $A = \frac{b \times h}{2}$ where A is the area of a triangle and b is the base of the triangle and h is the perpendicular height.

 ## Teaching points and activities

- Emphasise the need for care with units. Encourage pupils to check that all the units given are the ones that the formula requires, e.g. if the formula says that s is the distance in metres then s must be in metres when substituting for it. The answer should have units.
- Some pupils may find Exercise 13 difficult. Recap inverse operations and solving equations using inverse operations to help.

Support
- You should have covered the previous sections in this chapter.
- See **Formulae** page 151 (Algebra Support chapter in the Pupil Book). If pupils have difficulty with Practice questions 1, 2 and 5 use Worksheet 17 (page 15 of the Teacher Resource Pack).

Extension
- See **starred questions** in Exercises 12 and 13 (Pupil Book pages 175 and 177). See **Puzzle** (Pupil Book page 176). Ask pupils to make up a puzzle like these of their own to give to someone else to work out.

Plenary ideas
- Discuss the answers to Exercise 12 question 5.
- Discuss Exercise 12 question 8 or 9 to extend thinking.
- Start the Puzzle on page 176 of the Pupil Book. Ask pupils for strategies to do this. You may finish this but if not ask pupils to finish it for homework.

Homework Review questions at the end of Exercises 12 and 13 (Pupil Book pages 175 and 177) and/or Homework Sheet 8.7 Exercises 1 and 2 (page 216 of this pack).

Links Link this topic to substituting into expressions (Pupil Book page 172).
Link to area and perimeter: Exercise 12 questions 1, 5, Review 1; Exercise 13 question 4.
Link to Science: Exercise 12 questions 2, 3; Exercise 13 questions 1, 2, Review.
Link to measures: Exercise 12 questions 6, 8, 9, Reviews 2 and 3.

| Topic area | Writing expressions | pupil book page 178 |
| | More writing and simplifying expressions | pupil book page 179 |

Framework Objectives

✓ Know that algebraic operations follow the same conventions as arithmetic operations; ..., **Framework examples pages 117 and 143**

✓ **Simplify or transform linear expressions by collecting like terms; multiply a single term over a bracket, Framework examples pages 117 and 143**

✓ Problem solving, **Framework examples page 7**

Sheets   Homework Sheets 8.8 and 8.9

Starter ideas and activities

Mental starter ideas
- Ask quick questions such as Practice question 11 on page 154 of the Algebra Support chapter.
- Ask some quick questions on multiplication and division, such as double 34, 12 × 8 and 33 × 0·3.

Starter activity
- **Writing expressions**
  1 Put the class into a circle or several smaller circles.
  2 Ask someone to start and say 'I have n marbles'. The next person says 'I have ... more/less than that' or 'I have twice/three times/half ... as many, how many marbles do I have?' This is said to the next person in the circle. The next person answers, for example, $n + 3$ or $2n$ or whatever the answer is. The next person says 'I have ..., how many marbles do I have?'
  3 Have someone write the expressions on the board as each person gives their answer. Check if it is correct. Sometimes the expressions can be simplified. Prompt to get the expression simplified. Stop when the expressions get too complicated and start again with n marbles. Having several smaller circles may avoid this.
  4 If pupils have difficulty you could use an object such as a box to represent n marbles.
- **Equivalent expressions**
  Use the Discussion on page 179 of the Pupil Book.
  You may need to discuss some of the questions from exercise 15 as well.

Teaching points and activities

- Point out that sometimes expressions can be written in more than one way and sometimes the expression written can be simplified.
- Recap simplifying expressions, cancelling fractions, collecting like terms. Some pupils will find the section 'more writing and simplifying expressions' difficult and need lots of prompts. You could put the questions from the exercise on an OHP and go over them one by one getting pupils to do more and more on their own each time.

Support
- Pupils should have covered the previous sections in this chapter.

Extension
- See **starred questions** in Exercises 14 and 15 (Pupil Book pages 178 and 180).
- See **Puzzle** (Pupil Book page 181). Challenge pupils to find facts about the sum of different numbers of consecutive numbers, of consecutive odd numbers and consecutive even numbers.

Plenary ideas
- Discuss Exercise 14 question 5 (or question 8 to extend thinking and ask for strategies).
- Discuss some of the questions in Exercise 15.
- Discuss the Puzzle on page 181 of the Pupil Book and ask pupils to finish it for homework.

Homework   Review questions at the end of Exercises 14 and 15 (Pupil Book pages 178 and 180) and/or Homework Sheet 8.8 and 8.9 (page 217 of this pack)

Links   Link this topic to understanding algebra (Pupil Book page 158).
Link to measures: Exercise 14 questions 4, 5.
Link to perimeter and area: Exercise 15 questions 1, 3, 4, 5, Review.

Topic area	**Writing equations**	pupil book page 181

Framework Objectives ✓ Construct ... linear equations with integer coefficients ..., **Framework examples page 123**

Sheets Worksheet 18 for support   Homework Sheet 8.10   Starter Resource Sheet 9

Resources None required

 ## Starter ideas and activities

Mental starter ideas Ask questions such as Practice question 25 on page 156 of the Algebra Support chapter in Pupil Book. Some of these may be too hard as mental questions.

Starter activity **Writing equations**
Use Starter Resource Sheet 9 and put the class into groups of four or five to write an equation for the scenario given on the card. Swap the cards around the groups and then compare and discuss the answers.

 ## Teaching points and activities

- Make sure pupils know what an equation is and ask for some examples.
- Point out that pupils can use any letter to represent the unknown in an equation.
- Remind pupils what 'equals' means.
- Some pupils have difficulty knowing where to start. Encourage them to write down expressions for all the things they are told or can work out. Tell them to ask themselves if any of these expressions are equal or are equal to other information given in the question. For example, for Exercise 16 question 1 they could write $f + s$ people went to the play altogether. For part a they could ask 'What is 1500 equal to?'
  For part b they might write 'Twice as many people went on the second night so $2f$ people went on the second night. What else is this equal to?'
- Encourage pupils to check that the equation they have written is true.

Support • See **Equations** on page 151 (Algebra Support chapter of the Pupil Book). If pupils have difficulty with Practice questions 19 and 25 then use Worksheet 18 questions 2a, 3 (page 15 of the Teacher Resource Pack).

Extension • See **starred questions** in Exercise 16 (Pupil Book page 181)

Plenary ideas • Give a question such as:
  I am n years old. My mother is m years old. My mother is twice as old as I am. Write an equation for this. (Try and make up a true statement like this one, of interest to the class.)
- Discuss the answers to Exercise 16 question 2.

Homework Review questions at the end of Exercise 16 (Pupil Book page 181) and/or Homework Sheet 8.10 (page 218 of this pack).

Links Link this topic to understanding algebra (Pupil Book page 158).

Topic area	Writing and finding formulae	pupil book page 182

Framework Objectives
✓ ... derive simple formulae, **Framework examples page 143**
✓ Problem solving, **Framework examples page 35**

Sheets
Homework Sheet 8.11   ICT Worksheet 'Formulae' (Teacher Resource Pack page 133)
Starter Resource Sheet 10

Resources
A computer and a spreadsheet package.

Starter ideas and activities

Mental starter ideas
Practise some of the number objectives that your class needs to practice.

Starter activity
Writing formulae
Use Starter Resource Sheet 10.
Ask for volunteers to come and match up the correct statement with its formula.

Teaching points and activities

- Make sure pupils know what a formula is and ask for some examples from other subject areas.
- Remind pupils that in most formulae, units are needed.
- You could use the Investigation on page 183 of the Pupil Book as group work.

Support
- Pupils should have already covered *Substituting into Formulae*.

Extension
- See **Investigation: Deriving Formulae** question 4 (Pupil Book page 183).

Plenary ideas
- Start the Investigation on page 183 and ask pupils to finish it for homework.
- Talk about the relevance of writing formulae to sequences in practical situations.

Homework
Review question at the end of Exercise 17 (Pupil Book page 182) and/or Homework Sheet 8.11 (page 218 of this pack).

ICT
See Practical (Pupil Book page 184). See ICT Worksheet 'Formulae' on page 133 of the ICT Support section in the Teacher Resource Pack.

Links
Link this topic to sequences in practical situations (Pupil Book page 213) and finding the rule for the nth term (Pupil Book page 217).
Link to Science: Exercise 17 question 1b.
Link to measures: Exercise 17 question 1a, 1c.
Link to area: Exercise 17 question 1e.
Link to shapes: Investigation: Deriving Formulae questions 2, 3, 4.
Link to sequences: Investigation: Deriving Formulae questions 1, 2, 3.

| Topic area | **Writing and solving equations** | pupil book page 184 |

Framework Objectives
✓ Construct and solve linear equations with integer coefficients (... without and with brackets) using appropriate methods (e.g. inverse operations ...), **Framework examples pages 123 and 125**
✓ Problem solving, **Framework examples pages 7, 9 and 35**

Sheets   Worksheet 18 for support   Homework Sheet 8.12   Starter Resource Sheet 11

Resources   You could use Algebra Domino Packs (Taskmaster). OHP for plenary.

 ## Starter ideas and activities

Mental starter ideas   Ask some questions such as Practice question 28a–f on page 156 of the Algebra Support chapter in the Pupil Book.

Starter activity
- **Solving equations**
  1 Put this equation on the board or OHP: $8m + 7 = 27$.
    How could you solve this equation?
    You may need to show one first using a function and inverse function machine.

    Solve $6n + 4 = 34$

    $$n \rightarrow \boxed{\times 6} \xrightarrow{6n} \boxed{+4} \rightarrow 6n + 4$$

    $$5 \leftarrow \boxed{\div 6} \xleftarrow{30} \boxed{-4} \leftarrow 34$$

    $n = 5$

  2 Tell them that we don't usually draw function and inverse function machines when solving an equation.
    How do we usually write down the steps to solve the equation?
  3 Give some more equations to solve. (You could use some from Exercise 18 question 1.)
- **Solving equations with brackets**
  Use the Discussion on page 185 of the Pupil Book.
- **Collecting like terms**
  Use Starter Resource Sheet 11
  1 Before doing Exercise 19 give an example such as the one in the screen on page 186 of the Pupil Book. Ask for volunteers to come and write the expressions for the amounts Paula, Rosa and Ralph have on a card (or use Starter Resource Sheet 11). Have these pupils stand out the front of the class holding their cards.
    What equation could we write from this?
  2 Ask other pupils to come and hold cards with + on them, one with = and one with 144 to make the equation.
    How could we solve this equation? (Collect the like terms first.)
  3 Now put the equation on the board or OHP and ask for volunteers to come and show how to solve it step by step. You could do this with cards but it gets a bit complicated to follow.

 ## Teaching points and activities

- Recap inverse operations, substitution, multiplying out brackets and collecting like terms.
- Pupils often have difficulty writing down the steps to solve an equation correctly. Emphasise that each step must be written down and that the equals must mean equals. Pupils often write things such as:
  $$3c + 7 = 19$$
  $$= 19 - 7$$
  $$= 12 \div 3$$
  $$= 4$$

It is important to correct such mistakes and make sure that pupils understand why they are incorrect. A discussion about this is useful.

- Emphasise that when solving equations all the like terms must be collected first. It is helpful to ask pupils to create a flowchart of how to solve an equation which includes decision boxes such as:

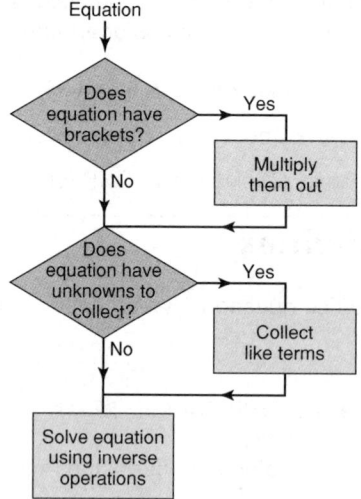

Support
- Pupils should have covered the previous section *Writing equations*.
- See **Equations** on page 151 (Algebra Support chapter in the Pupil Book).
  If pupils have difficulty with Practice questions 19b, 28 and 30 use Worksheet 18 questions 1, 2b, 4 and 5 (page 16 Teacher Resource Pack).

Extension
- See **starred** questions in Exercises 18 and 19 (Pupil Book pages 185 and 186).

Plenary ideas
- Discuss the answers to Exercise 19 question 6 or 7 (or 8 to extend thinking). If you use question 6 you could put the pyramids onto an OHP. Check the answer by substituting.

Homework
Review questions at the end of Exercises 18 and 19 (Pupil Book pages 185 and 186) and/or Homework Sheet 8.12 Exercises 1 and 2 (page 219 of this pack)

Links
Link this topic to understanding algebra (Pupil Book page 158).
Link to measures: Exercise 19 questions 7a, Review 3a, 3c.
Link to area and perimeter: Exercise 19 questions 2, Review 2.

Topic area	Solving equations by transforming both sides	pupil book page 188

Framework Objectives ✓ Construct and solve linear equations with integer coefficients (unknown on either or both sides, (without brackets ...) using appropriate methods (e.g. ... transforming both sides in same way), **Framework examples page 125**

Sheets Homework Sheet 8.13   Resource Sheet page 209

Resources None needed.

 ## Starter ideas and activities

Mental starter ideas
- Practise multiplying and dividing a 2-digit number by a 1-digit number.
- Practise solving simple equations such as $3a - 2 = 31$.

Starter activity **Introducing transforming both sides of an equation**
Use the Discussion on page 188 of the Pupil Book.
Alternatively you could do a practical example with counters on scales, then add or subtract similar numbers of counters to each pan to show how they still balance. Then double or halve the number of counters on each side.

 ## Teaching points and activities

- Stress that an equation will still be true as long as you do the same thing to both sides.
- Pupils sometimes forget to transform **both** sides in the same way. Emphasise this by encouraging them to write down what they do to both sides using a different coloured pen such as red. This highlights the fact that what has been done to each side is exactly the same.
- When there are unknowns on both sides, pupils sometimes automatically subtract (or add) the amount on the left-hand side of the equation. Point out that this sometimes leads to a negative number in front of the unknown, which makes the equation harder to solve. It is easier to add or subtract the smaller amount of unknown.
- Point out that while you can often choose to use either inverse operations or transforming both sides to solve an equation, inverse operations cannot be used until the unknowns are collected together on one side of the equation.

Support
- Pupils should have covered the previous section *Writing equations*.
- See **Equations** on page 151 (Algebra Support chapter in the Pupil Book). Try Practice questions 19, 28 and 30 if you haven't already done them in a previous section.

Extension
- See **starred** questions in Exercises 21 (Pupil Book page 190).
- See **Discussion** (Pupil Book page 188). The third bullet point of the discussion can be extended to solve harder equations.

Plenary ideas
- Ask some individual pupils to explain the transforming both sides method using whatever they want to. This could be a diagram or a set of scales or a step-by-step mathematical explanation. You could give a particular equation for pupils to work on, such as $8x + 4 = 5x + 40$.
- Discuss Exercise 21 question 4 or 5 and ask pupils to demonstrate how to do it.
- Discuss strategies for solving the Puzzle on page 191 of the Pupil Book

Homework Review questions at the end of Exercises 20 and 21 (Pupil Book pages 189 and 190) and/or Homework Sheet 8.13 (page 220 of this pack)

Links Link this topic to understanding algebra (Pupil Book page 158).
Link to angles: Exercise 21 question 2.

Topic area	**Solving equations using a graph**	pupil book page 192

Framework Objectives

✓ Construct and solve linear equations with integer coefficients ... using appropriate methods ..., **Framework examples page 137**

Sheets   Homework Sheet 8.14   Resource Sheet pages 210 and 211 Teacher Resource Pack

Resources   You could use Algebra Domino Packs (Taskmaster).

Starter ideas and activities

Mental starter ideas   Practise adding, subtracting, multiplying and dividing integers and decimals.

Starter activity   **Using a graph**
Use the Discussion on page 192 of the Pupil Book.

Teaching points and activities

- Emphasise the link to direct proportion. When the ratio of corresponding pairs is always the same, then the plotted coordinate pairs will form a straight line. From this we can write a formula.
- It is helpful to link this to function machines (draw one) and show the coordinate pairs as input and output. Pupils can sometimes write down the formula more easily from this. The function machine for the example given in the discussion is:

number of bags → [× cost] → cost of peanuts

x → [× £3] → y

Function is $y = 3x$

Support
- Pupils should have covered the previous section *Writing and solving equations*.

Extension
- See **Discussion** (Pupil Book page 192). The second bullet point of the Discussion can be used to extend thinking to cases when the graph may not be a straight line and so the variables are not in direct proportion. The equation will therefore not be linear. What does this mean?

Plenary ideas
- Discuss the answers to Exercise 22 question 2, asking pupils all the things they can tell about graphs of straight lines: direct proportion, writing a formula from the straight line and so on. You could use a real-life example of interest to the pupils instead and tell them something such as 'The school wants to buy some cricket bats. They are £12 each but we can get them at a special price of £10.' Then go on and ask similar questions to those asked in question 2.

Homework   Review question at the end of Exercise 22 (Pupil Book page 192) and/or Homework Sheet 8.14 (page 221 of this pack).

Links   Link this topic to understanding algebra (Pupil Book page 158).
Link to Graphs: all of Exercise 22.
Link to measures: Exercise 22 question 3, Review.
Link to PE: Exercise 22 question 3.

9 Sequences and Functions

Topic in Chapter	Links to Programme of Study *Ma2 a–l are integrated throughout.*		Links to Framework for Teaching Mathematics: 8 Teaching Programme *Applying mathematics and solving problems are integrated throughout.*
Writing sequences from flow charts *page 200*	**Ma2 1i**	explore, identify, and use pattern and symmetry in algebraic contexts, investigating whether particular cases can be generalised further	**Sequences functions and graphs** ● Generate and describe linear sequences. ● Generate terms of a linear sequence using term-to-term and position-to-term definitions of the sequence, on paper and using a spreadsheet or graphical calculator. ● Begin to use linear expressions to describe the nth term of an arithmetic sequence, justifying its form by referring to the activity or practical context from which it was generated.
Generating sequences by multiplying and dividing *page 201*	**Ma2 6a**	generate common integer sequences (including sequences of odd or even integers, powers of 2, power of 10, triangular numbers)	
Counting forwards or backwards in increasing or decreasing steps *page 203*			
Showing sequences with geometric patterns *page 204*			
Arithmetic sequences *page 205*			
Predicting the next few terms of a sequence *page 207*			**Applying mathematics and solving problems** ● Identify the necessary information to solve a problem ... ● Suggest extensions to problems, conjecture and generalise; identify exceptional cases or counter-examples.
Writing sequences from term-to-term rules and rules for the nth term *page 208*	**Ma2 1i**	explore, identify and use pattern and symmetry in algebraic contexts, investigating whether particular cases can be generalised further	
Describing linear sequences *page 211*	**Ma2 6a**	generate common integer sequences ...	
Sequences in practical situations *page 213*	**Ma2 6b**	find the first terms of a sequence given a rule arising naturally from a context (for example ... from a regularly increasing spatial pattern); find the rule (and express it in words) for the nth term of a sequence	
Finding the rule for the nth term *page 217*	**Ma2 6c**	generate terms of a sequence using term-to-term and position-to-term definitions of the sequence; use linear expressions to describe the nth term of an arithmetic sequence, justifying its form by referring to the activity or context from which it was generated	
Functions *page 218*	**Ma2 5a**	distinguish the different roles played by letter symbols in algebra, knowing that letter symbols ... in functions ... define new expressions or quantities by referring to known quantities (for example, $y = 2 - 7x$)	**Sequences, functions and graphs** ● Express simple functions in symbols; represent mappings expressed algebraically. **Equations, formulae and identities** ● Begin to distinguish the different roles played by letter symbols in equations, formulae and functions; know the meanings of the words *formula* and *function*.
Finding the function given the input and output *page 221*			
Properties of functions *page 224*	**Ma2 6d**	express simple functions ... in symbols ...	

Links to Attainment Targets

Level 6   When exploring number sequences, pupils find and describe in words the rule for the next term or nth term of a sequence where the rule is linear. ... They represent mappings expressed algebraically, ...

Level 7   Pupils find and describe in symbols the next term or nth term of a sequence ... (begin)

Topic area	**Writing sequences from flow charts**	pupil book page 200

Framework Objectives	✓ Generate and describe linear sequences, **Framework examples page 145**
Sheets	Homework Sheet 9.1 Starter Resource Sheet 12
Resources	Mini-whiteboards for the Starter Activity.

Starter ideas and activities

Mental starter ideas
Practise counting on and back in steps of 0·4, 0·75, $\frac{3}{4}$, ... Use a variety of starting numbers including negative numbers.

Starter activity
- **Flow charts for everyday activities**
  1 Use Starter Resource Sheet 12 to demonstrate a **flow chart** for making a cup of coffee.
  2 Ask for volunteers to create a flow chart for boiling an egg (or any other simple task).
  3 Ask pupils to draw the boxes for the flow chart one by one on pieces of paper and then stand out at the front in a line to create the flow chart.
  4 Check to see that everything is in the correct order.

- **Flow charts to generate a sequence**
  1 Now give a mathematical example such as the one in the screen on page 200. Draw the chart on the OHP or board.
  2 Choose some numbers one by one (or you could ask pupils to choose).
  3 Ask pupils to write the output on mini-whiteboards (or paper) and hold them up.
  4 You could do this with volunteers out at the front holding up the steps of the flow chart and have other volunteers as the numbers going through the chart explaining the steps.
  Example I am the number 4. Now I go through the 'add on 3' box and come out as the number 7. Now I must write my number down. I am less than 12 so I go back to the 'add on 3' box.

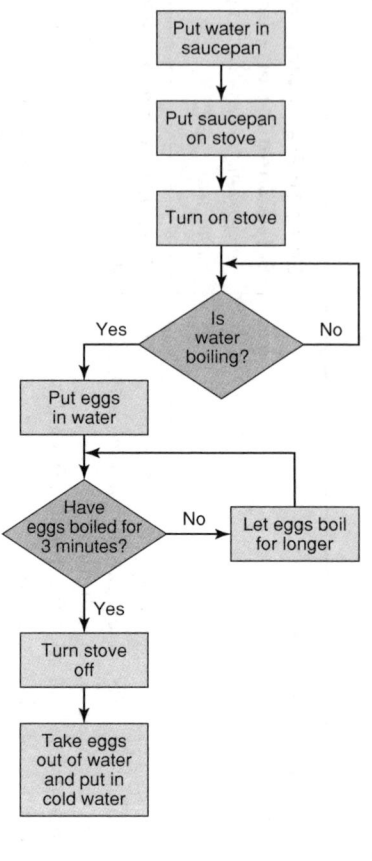

Teaching points and activities

- Emphasise the need to check the answer and check that the decision box has been satisfied exactly.

Support
- See **Sequences** on page 151 (Algebra Support chapter in the pupil book).

Extension
- You could challenge pupils to make up their own flow chart and sequence to give to a partner to work out.

Plenary ideas
- Ask a volunteer or individual to make up a flow chart for the rest of the class.
- Discuss Exercise 1 question 1e or 1f.
- Consolidate vocabulary used during the lesson.

Homework
Review question at the end of Exercise 1 (Pupil Book page 200) and/or Homework sheet 9.1 (page 222 of this pack).

Links
Link this topic to mental calculation (Pupil Book page 61).

| Topic area | Generating sequences by multiplying and dividing | pupil book page 201 |

Framework Objectives
✓ Generate ... linear sequences, **Framework examples page 145**
✓ Generate terms of a linear sequences using term-to-term ... definitions of the sequence, on paper and using a ... graphical calculator, **Framework examples page 145**
✓ Problem solving, **Framework examples page 11**

Sheets Homework Sheet 9.2

Resources A counting stick and mini-whiteboards for the Starter Activity. A graphical calculator.

 Starter ideas and activities

Mental starter ideas
● Ask questions such as Practice question 6 on page 154 of the Algebra Support chapter in the pupil book.
You could use a counting stick to help if needed.
● Practise multiplying and dividing integers.

Starter activity
● **Revise vocabulary**
Revise the meaning of the words 'term', 'infinite' and 'finite' and 'consecutive'.
Give the example in the screen on page 201 of the Pupil Book.
● **Examples of sequences made by × and ÷**
1 Ask for examples of sequences made by multiplying and dividing.
Give me a starting number and a number to multiply by.
2 You could ask for volunteers to come and write these on the board or OHP:
Write down the first five terms of the sequence.
You could do this with pupils standing out the front holding mini-whiteboards or similar.
3 Do some division ones as well but you will need to direct these or the numbers will get messy.
What if we started with a negative number/fraction or multiplied/divided by a negative number?
Pupils may need to use a counting stick to help.

 Teaching points and activities

● Make sure pupils can multiply and divide integers.
● This is a good (and easy) place to use a graphical calculator.
● Emphasise that the differences between consecutive terms are not the same when multiplying and dividing by a constant number. The sequence ascends or descends in **unequal** steps. You may need to explain what a constant is.

Support
● See **Sequences** on page 151 (Algebra Support chapter in the Pupil Book).

Extension
● See **starred questions** in Exercise 2 (Pupil Book page 202).
● See **Practical** (Pupil Book page 202) 1c, 1d and 4.
● See **Reductions** on page 199 of the Pupil Book. Pupils could explore the multiples of other numbers or sequences such as in part b but with a number other than 2.

Plenary ideas
● Discuss the answers to Exercise 2 question 2 and the strategies for doing these.
● Discuss the Practical on page 202. You will need a graphical calculator on an OHP for this or a graphical calculator for each pupil or pair of pupils.
● Use the Discussion on page 203 of the Pupil Book.

Homework Review question at the end of Exercise 2 (Pupil Book page 202) and/or Homework sheet 9.2 (page 222 of this pack).

Graphical Calculator Practical (Pupil Book page 202). See page 185 of the Teacher Resource Pack.
Discussion (Pupil Book page 203). A graphical calculator can be used for this.

Links Link this topic to mental calculation (Pupil Book page 61).
Link to PE: teaching screen.

| Topic areas | Counting forwards or backwards in increasing or decreasing steps | pupil book page 203 |
| | Showing sequences with geometric patterns | pupil book page 204 |

Framework Objectives
✓ Generate ... linear sequences, **Framework examples pages 145 and 147**
✓ Generate terms of a linear sequence using term-to-term ... definitions of the sequences, on paper, **Framework examples pages 145 and 147**

Sheets Worksheet 19 for support   Homework Sheet 9.3   ICT Worksheet 'Pascal's or the Chinese Triangle' (Teacher Resource Pack page 135)

Resources Digit cards and paper for the Starter Activity. You could use a counting stick. Square and triangular dot paper is useful. A computer and a spreadsheet.

 Starter ideas and activities

Mental starter ideas
- Practise counting on or back in constant steps.
- Ask pupils to write down the first 12 square numbers and the first six powers of 2.

Starter activity
- **Sequences that increase or decrease in unequal steps**
  You will need digit cards and paper.
  1 Have five cards with +1, +2, +3, +4, +5 and give these to five pupils.
  2 Ask a pupil to come and write a very large 1 on a piece of paper and hold it up.
  3 Ask the pupil with card +1 to stand beside, but just behind, this person.
    What would the next term be if we add 1 on?
  4 Ask the pupil who answered to write the answer, 2, in large print on a piece of paper and come and stand next to, but just in front of, the person with the card +1.
  Carry on until you have the pupils lined up holding the cards as follows:

  5   *What do you notice about this sequence?*
     What would the first five terms of the sequence that starts at 100 and goes down by 1, 2, 3, 4, 5, ... be?
  6 Ask for some other sequences that ascend or descend by unequal steps. Give starting numbers and the steps.
- **Showing sequences with geometric patterns.**
  Use the Discussion on page 204 of the Pupil Book.

 Teaching points and activities

- Some pupils have difficulty keeping track of the numbers being added and subtracted to get the next term. Encourage them to write these in a different colour or small print between the terms.

$$100 \overset{+1}{,} 99 \overset{+2}{,} 97 \overset{+3}{,} 94 \overset{+4}{,} 90 \overset{+5}{,} 85 , ...$$

- When describing these sequences pupils sometimes say that they are, for example, 'adding one more each time'. This is not correct. Encourage them to describe the sequence more precisely, e.g. 'the sequence starts at 1 and the difference between terms is 1 more each time 'or' the sequence starts at 1 and ascends by increasing steps of 1, then 2, then 3, then 4, and so on.'

Support
- See **Sequences** on page 151 (Algebra Support chapter in the Pupil Book).
  If pupils have difficulty with Practice Question 6 then use Worksheet 19 questions 1, and 3 (page 17 of the Teacher Resource Pack).
- Pupils should have covered the previous sections in this chapter.

Extension
- See **Starred question** in Exercise 3 (Pupil Book page 203).
- See **Discussion** (Pupil Book page 204). Pupils could be challenged to draw diagrams for harder sequences such as 1 + 3, 3 + 5, 5 + 7, ...
- See **Investigation** (Pupil Book page 205). Pupils could be extended by having a competition to see who can find the most patterns in Pascal's Triangle.

© New National Framework Mathematics 8 Nelson Thornes Ltd

Plenary ideas
- Ask pupils to give you an example of a sequence which
  Increases by steps of 1, 2, 3, 4, 5, ...
  Decreases by steps of 2, 4, 6, 8, 10, ... and so on.
  Pupils will need to choose their own starting number. You could put the different answers on the board or OHP and discuss why they are different. (Start at different numbers.)
- Discuss Exercise 3 question 4.
- Ask pupils if they can think of other real-life situations where a sequence made by counting up or down in increasing or decreasing steps is used.

Homework
Review questions at the end of Exercise 3 (Pupil Book page 203) and/or Homework Sheet 9.3 (page 222 of this pack).
Part or all of the Investigation on page 222 in the Pupil Book could be set for homework.

ICT
ICT could be used for the Investigation: Pascal's Triangle on page 205 of the Pupil Book. See ICT Worksheet 'Pascal's or the Chinese Triangle' on page 135 of the ICT Support section in the Teacher Resource Pack.

Links
Link this topic to indices and mental calculation (Pupil Book page 61).
Link to PE: Exercise 3 questions 3, 4, Review 2.
Link to measures: Exercise 3 question 2.

Topic area	Arithmetic sequences	pupil book page 205

Framework Objectives
✓ Generate and describe linear sequences. **Framework examples page 147**
✓ Generate terms of a linear sequence using term-to-term ... definitions of the sequence, on paper ..., **Framework examples page 147**

Sheets   Homework Sheet 9.4   Resource Sheet page 212 Teacher Resource Pack

Resources   A counting stick for the Starter Activity.

Starter ideas and activities

Mental starter ideas   Ask question such as Practice question 10 on page 154 of the Algebra Support chapter.

Starter activity
● **Introducing arithmetic sequences**
Define an arithmetic sequence and give some examples. Ask pupils for some more examples, preferably from real life (house numbers, the date of every Saturday in a month, the cost of 1, 2, 3, 4, 5, 6, ... items, etc.).
Use a counting stick to help if needed.
● **Play true/false**
Divide the class into groups of three or four and ask these questions. They are allowed to discuss them and provide a group answer. Each group that gets one correct gets 1 point.
True of false?
The sequence of odd numbers starting at 1 is an arithmetic sequence.
The multiples of 3 starting at 6 is not an arithmetic sequence.
The sequence of square numbers starting at 1 is an arithmetic sequence.
The powers of 2 starting at 2 is not an arithmetic sequence.
The multiples of any number is an arithmetic sequence.
A sequence which ascends in constant steps is not an arithmetic sequence.
A sequence with the rule 'multiply the previous term by 2' is not an arithmetic sequence.
You may need to use a counting stick to show whether or not the differences between consecutive terms is constant.

Teaching points and activities

● It is important for the future that pupils understand the terms 'constant' and 'linear' as applied to arithmetic sequences. Emphasise that arithmetic sequences are called linear sequences because they ascend or descend in equal steps.
● Some pupils have difficulty remembering which is a and which is d. It is helpful to think that a is the **start** of the alphabet so it is the **starting** number and d stands for **difference** between the terms, i.e. the number ad**d**ed.
● When a sequence descends such as 20, 18, 16, 14, ... some pupils will say that $d = 2$ instead of $^-2$. Emphasise that d is the number ad**d**ed and that sometimes this is a negative number.

Support   ● Pupils should have covered the previous sections in this chapter.

Extension   ● See **starred questions** in Exercise 4 (Pupil Book page 205).

Plenary ideas
● Ask some pupils to give you an a and d for an arithmetic sequence. Ask the rest of the class to give the sequence. Suggest that whole numbers be used initially and then progress to negatives, decimals and possibly fractions.
● Discuss the answers to Exercise 4 question 4.
● Discuss Exercise 4 question 6, 7, 8 or 9. (Questions 8 and 9 are good ones to do with more able classes.)

Homework   Review questions at the end of Exercise 4 (Pupil Book page 205) and/or Homework Sheet 9.4 (page 223 in this pack).

Links   Link this topic to mental calculation (Pupil Book page 61) and factors and multiples (Pupil Book pages 3, 47 and 50).

| Topic area | **Predicting the next few terms of a sequence** | pupil book page 207 |

Framework Objectives
✓ Generate ... linear sequences, **Framework examples page 145**
✓ Generate terms of a linear sequence using term-to-term ... definitions of the sequence, on paper ..., **Framework examples page 145**

Sheets   Homework Sheet 9.5

 Starter ideas and activities

Mental starter ideas   Give a variety of mental questions. You could include some mental problem solving such as the questions on page 70 of the Pupil Book.

Starter activity   Use the Discussion on page 207 of the Pupil Book.
What do you need to know to be able to continue a sequence with absolute certainty?
(The rule.)

 Teaching points and activities

- Emphasise that unless the rule is given, any continuation of a sequence is only a prediction and not a certainty.
- Encourage pupils to come up with as many predictions as possible. It is helpful to share these to emphasise the point that the rule is needed to be sure how a sequence continues.
- Encourage pupils to look for the difference between terms and if they are equal or unequal, to see whether the sequence is ascending or descending, and to look for other patterns.

Support   • Pupils should have covered the previous sections in this chapter.

Extension
- See **starred questions** in Exercise 5 (Pupil Book page 207).
- See **Discussion** (Pupil Book page 208). Challenge pupils to see who can come up with the most answers.

Plenary ideas
- Discuss the answers to Exercise 5 question 2.
- Discuss some of the Exercise 5 review questions.
- Use the Discussion on page 208 of the Pupil Book and ask for volunteers to fill in the missing numbers and explain why they chose these numbers.
- Write the first three numbers of a sequence (e.g. 1, 2, 3, ... or 2, 4, 8, ...) on the board and challenge the class to come up with as many different sequences as possible. Ensure they can explain how they got their answers. You could begin with just the first two numbers of a sequence as this will give more scope for a variety of answers.

Homework   Review questions at the end of Exercise 5 (Pupil Book page 207) and/or Homework Sheet 9.5 (page 223 in this pack).

Links   Link this topic to mental calculation (Pupil Book page 61).

Topic area	**Writing sequences from term-to-term rules and rules for the *n*th term**	pupil book page 208

Framework Objectives
✓ Generate and describe linear sequences. **Framework examples pages 149 and 151**
✓ Generate terms of a linear sequence using term-to-term rules and position-to-term definitions of the sequence on paper and using a spreadsheet or graphical calculator, **Framework examples pages 149 and 151**
✓ Problem solving, **Framework examples page 9**

Sheets
Worksheet 19   Homework Sheet 9.6   Starter resource Sheet 8   ICT Worksheets 'Generating Sequences' and 'Sequence Differences' (Teacher Resource Pack pages 136 and 137)

Resources
Mini-whiteboards, digit cards, *n* cards, operation cards for the Starter Activity. A calculator. A graphical calculator. A computer and a spreadsheet package.

Starter ideas and activities

Mental starter ideas
Ask questions such as Practice question 23 on page 156 of the Algebra Support chapter. Practise substitution.

Starter activity
- **Revision of vocabulary**
  Revise the vocabulary and '*n*th term' and 'general term' and the notation '*T(n)*'.
- **Rules for sequences**
  1   *How might the rule for a sequence be given?*
  2   Ask for volunteers to give some examples such as 'it starts at 4 and each term is 4 more than the previous term' or 'first term 8', rule 'multiply by 3'.
      You could have volunteers at the front holding paper, cards or mini-whiteboards to illustrate the answers.
  3   Point out that these sorts of rules are called term-to-term rules. Ask why this might be.
  4   *If we wanted to find the 60th term how would the rule need to be given for us to find this without writing out 60 terms?* (As a rule for the *n*th term or as a general rule.)
  5   Give some examples of a rule for the *n*th term, e.g. $2n + 3$ or $100 - n$ or $0{\cdot}1n + 10$ or $n^2 + 3$.
      Can you suggest any others?
- **Show how to substitute into the rule for the *n*th term**
  You will need digit cards and an *n* card, and operation cards (Starter Resource Sheet 8).
  1   Put the rule $T(n) = 60 - 2n$ on the board.
  2   Ask pupils to come and show this using cards as $60 - 2 \times n$.
  3   Give the card 18 to a pupil and ask 'What do we have to do to find the 18th term? (Substitute 18 for *n*.)
  4   Ask the pupil holding the card to change places with the person holding the *n* card.
  5   Ask the other pupils to work out the answer on their whiteboards.
  6   Repeat for the 25th term and the 60th term.

Teaching points and activities

- Make sure pupils can add and subtract, including negative numbers.
- Some pupils have difficulty with the terminology $T(n)$. It helps if each time you use this you point out that it is the *n*th term or the general term. It is also helpful to often use $T(1)$, $T(2)$, ... when working with sequences so that pupils get used to seeing it.

Support
- Pupils should have covered the previous sections in this chapter.
- See **Sequences** on page 151 (Algebra Support chapter in the Pupil Book).
  If pupils have difficulty with Practice questions 10, 23 and 27 use Worksheet 19 questions 4 and 7 (page 17 of the Teacher Resource Pack).

Extension
- See **starred questions** in Exercise 6 (Pupil Book page 209).
- See **Investigation: Fibonacci sequence** (Pupil Book page 210). Challenge pupils to research other patterns within a Fibonacci sequence, as in question 2. The Internet has a lot of information about Fibonacci sequences.

Plenary ideas
- Discuss the answers to Exercise 6 question 1 parts d, h and I and/or question 3 parts c, e and g. Ask pupils to demonstrate how they got their answers.
- Discuss Exercise 6 question 6. Pupils could use graphical calculators or you could use one connected to a screen.

Homework   Review questions at the end of Exercise 6 (Pupil Book page 209) and/or Homework Sheet 9.6 (page 223 in this pack).

Graphical calculator   See Practical (Pupil Book page 211). See the Graphical Calculator Support section for notes on this (page 186 of the Teacher Resource Pack).
See Exercise 6 question 6 and Review 5. See Graphical Calculator Support section (page 186 of Teacher Resource Pack).

ICT   For Practical (Pupil Book page 211) question 2 see ICT Worksheet 'Generating Sequences' and ICT Worksheet 'Sequence Differences' on pages 136 and 137 of the ICT Support section of the Teacher Resource Pack.

Links   Link this topic to mental calculation (Pupil Book page 61), formulae (Pupil Book page 175) and substituting (Pupil Book pages 172 and 175).
Link to Science: Investigation: Fibonacci sequence (Pupil Book page 210).

Topic area	**Describing linear sequences**	pupil book page 211

Framework Objectives   ✓ Generate and describe linear sequences, **Framework examples page 151**

Sheets   Worksheet 19 for support   Homework Sheet 9.7   Starter Resource Sheet 13

 Starter ideas and activities

Mental starter ideas
- Ask questions such as Practice question 15 on page 155 of the Algebra Support chapter. You will need to put the descriptions on the board or OHP. You could use Starter Resource Sheet 13.
- Practise writing down multiples. You could also ask pupils to write down the sequence of numbers that are all one less than the multiples of 3 or 5 or 10 ...

Starter activity   **Describing linear sequences**
Use the Discussion on page 211 of the Pupil Book.

 Teaching points and activities

- Remind pupils what a linear sequence is (one which ascends or descends in equal steps).
- Encourage pupils to find the constant difference. Emphasise that the constant difference is the number which multiplies n.
- Encourage pupils to see the patterns, for example:
  $T(n) = 3n$ is the multiples of 3 starting at 3.
  $T(n) = 3n + 1$ is the sequence of numbers, one more than the multiples of 3, starting at 4.
  $T(n) = 3n + 2$ is the sequence of numbers, 2 more than the multiples of 3, starting at 5.
- Sometimes pupils miss out the starting number in the description. Emphasise that we need to know where the sequence starts or several different sequences could match the description.

Support
- Pupils should have covered the previous sections in this chapter.
  If pupils have difficulty with question 15 on page 155 use Worksheet 19 question 5 (page 17 Teacher Resource Pack).

Extension
- See **starred questions** in Exercise 7 (Pupil Book page 212).

Plenary ideas
- Discuss the differences between these sequences.
  $T(n) = 2n$
  $T(n) = 2n + 1$
  $T(n) = 2n + 4$
  $T(n) = 6 - 2n$
- Discuss the answers to Exercise 7 question 7.
- Discuss Exercise 7 question 8.

Homework   Review questions at the end of Exercise 6 (Pupil Book page 212) and/or Homework Sheet 9.7 (page 224 of this pack).

Links   Link this topic to expressions and formulae (Pupil Book page 175).

| Topic area | Sequences in practical situations | pupil book page 213 |

Framework Objectives

✓ Begin to use linear expressions to describe the nth term of an arithmetic sequence, justifying its form by referring to the activity or practical context from which it was generated, **Framework examples pages 155 and 157**

✓ Problem solving, **Framework examples pages 7 and 27**

Sheets   Worksheet 19 for support   Homework Sheet 9.8

Starter ideas and activities

Mental starter ideas   Practise multiplying and dividing decimals by 10, 100 and 1000.

Starter activity   **Finding an expression for the nth term**
Use the Discussion on page 213 of the Pupil Book.
You may need to do another example such as the one given in the screen on page 214 of the Pupil Book and ask some questions similar to those in the discussion.

Teaching points and activities

- Recap writing expressions.
- Emphasise that the rule for the nth term should be worked out by looking at the pattern of the diagrams. Encourage pupils to check, by substituting, that the rule they come up with does in fact give the terms of the sequence.

Support
- Pupils should have covered the previous sections in this chapter.
- See **Sequences** on page 151 (Algebra Support chapter in the Pupil Book).
  If pupils have difficulty with Practice question 20 use Worksheet 19 question 6 (page 17 of the Teacher Resource Pack).

Extension
- See **starred questions** in Exercise 8 (Pupil Book page 214).

Plenary ideas
- Discuss the answers to Exercise 8 question 3 or 6 (question 6 for extension).

Homework   Review questions at the end of Exercise 8 (Pupil Book page 214) and/or Homework Sheet 9.8 (page 225 of this pack).

Links   Link this topic to integers (Pupil Book page 36), understanding algebra (Pupil Book page 158) and writing expressions and simplifying (Pupil Book page 166–179).
Link to Art: Exercise 7 questions 1, 3, Reviews 1 and 2.
Link to shapes: Exercise 7 questions 4, 7.

Topic area	Finding the rule for the nth term	pupil book page 217

Framework Objectives

✓ Begin to use linear expressions to describe the nth term of an arithmetic sequence ..., **Framework examples page 157**

Sheets Homework Sheet 9.9

Resources Digit cards for the Starter Activity. A graphical calculator.

 Starter ideas and activities

Mental starter ideas Practise using factors to multiply and divide mentally.

Starter activity **Finding the rule for the nth term**

1 Put this sequence on the board or have pupils hold it up pieces of paper at the front of the class: 3, 5, 7, 9, 11, ...

2 *How could you describe the rule for the sequence?*
(Add two to the previous terms will probably be the most likely answer.)

3 *If I wanted to know what the 120th term was, how could I find this?*
(You could write down all the terms up to 120 or you could find a rule for the general or nth term.)

4 *How could I find the rule for the general or nth term?*
(Draw a table and find the differences between terms.)
What does the rule generally look like?

5 Make a difference table either on the board or by asking pupils with the digit card 2 to stand in between the pupils holding up the numbers of the sequence.
What does this constant difference of 2 tell us?
(That the number multiplying n in the rule for the nth term is 2.)

6 *How can we find the rest of the rule?*
(By finding what it has to be for the first term and then checking for other terms. For this example the rule is $2n + 1$.)

 Teaching points and activities

● Emphasise that the constant difference gives the number multiplying n in the rule for the nth term. Explain why this is by referring back to the section on sequences in practical situations.

● Sometimes pupils find the constant difference as, say, 4 and then write the rule as $n + 4$. Encourage pupils to check by substituting that the rule they come up with does give the correct sequence.

Support ● Pupils should have covered the previous sections in this chapter.
● See **Sequences** on page 151 (Algebra Support chapter in the Pupil Book).

Extension ● This whole section could be regarded as extension work.

Plenary ideas ● Give one of the sequences from question 1 or make up a new one and ask pupils to come and demonstrate how to find the answer. Encourage the use of a difference table.

Homework Review question at the end of Exercise 9 (Pupil Book page 218) and/or Homework Sheet 9.9 (page 226 of this pack).

Graphical calculator This is to be used for Exercise 9 question 4 (Pupil Book page 218). See the Graphical Calculator Support section page 187 in the Teacher Resource Pack.

Links Link this topic to understanding algebra (Pupil Book page 158) and writing expressions (Pupil Book pages 178 and 179).

| Topic area | Functions | pupil book page 218 |

Framework Objectives ✓ Express simple functions in symbols; represent mappings expressed algebraically, **Framework examples page 161**

Sheets Worksheet 20 for support   Homework Sheet 9.10   Resource Sheet pages 212–215 Teacher Resource Pack   ICT Worksheet 'Functions' (Teacher Resource Pack page 138).

Resources You could use a box or boxes with a mini-whiteboard stuck on the front to act as a function machine. The operations can then be changed as required. A computer and a spreadsheet package.

Starter ideas and activities

Mental starter ideas Ask questions such as Practice question 16 on page 155 of the Algebra Support chapter.

Starter activity 1 Put a function machine on the board or OHP, e.g.

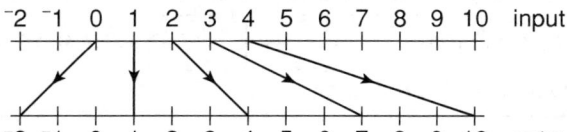

2 *How could you show the output from this function machine?* (As arrows from the right-hand side of it. Draw out that you could use a table and ask for volunteers to draw one.)

3 Ask for some input numbers and fill in the table. Progress to input numbers that are decimals and fractions.

4 Draw a mapping diagram such as this on the board or OHP.

```
⁻2 ⁻1  0  1  2  3  4  5  6  7  8  9  10   input
 ├──┼──┼──┼──┼──┼──┼──┼──┼──┼──┼──┼──┼──┤

                                         y = 3x − 2

 ├──┼──┼──┼──┼──┼──┼──┼──┼──┼──┼──┼──┼──┤
⁻2 ⁻1  0  1  2  3  4  5  6  7  8  9  10   output
```

Put 'input' beside the top line and 'output' beside the bottom line. Write the function $y = 3x − 2$ beside it.

5 Ask for volunteers to draw a mapping diagram for $y = 2x − 3$.

Teaching points and activities

- Recap simple function machines.
- Recap integer addition, subtraction, multiplication and division.
- Recap the meaning of 'function'.

Support ● See **Functions** on page 152 (Algebra Support chapter in the Pupil Book). If pupils have difficulty with Practice questions 16 and 26 use Worksheet 20 questions 1 and 2 (page 17 of the Teacher Resource Pack).

Extension ● See **starred questions** in Exercise 10 (Pupil Book page 219).

Plenary ideas ● Discuss Exercise 10 question 7 as a class.

Homework Review questions at the end of Exercise 10 (Pupil Book page 219) and/or Homework Sheet 9.10 (page 226 of this pack).

ICT See Practical (Pupil Book page 221). See ICT Worksheet 'Functions' on page 138 of the ICT Support section in the Teacher Resource Pack.

Links Link this topic to writing sequences from rules and writing sequences given the nth term (Pupil Book page 208), understanding algebra Pupil Book page 158), writing expressions (Pupil Book page 178) and mental calculation (Pupil Book page 61).
Link to enlargement: Exercise 9 question 7.

| Topic area | **Finding the function given the input and output** | pupil book page 221 |

Framework Objectives ✓ Express simple functions in symbols; represent mappings expressed algebraically, **Framework examples page 163**

Sheets Homework Sheet 9.11

Resources You could use a box or boxes with a mini-whiteboard stuck on the front to act as a function machine. The operations can then be changed as required.

 ## Starter ideas and activities

Mental starter ideas Practise all the operations with integers, e.g. $^-4 + 3$, $^-7 \times 3$, $3 + {}^-4 - {}^-4$.

Starter activity
- **Finding the function**
  Use the Discussion on page 221 of the Pupil Book.
- **Game: Guess my rule**
  You could play a game of guess my rule such as the one suggested in the Discussion.
- **Finding the rule for a two-step function machine**
  1 Give a function machine such as this one and ask for the rule.

  $$1, 2, 3, 4 \rightarrow \boxed{} \rightarrow \boxed{} \rightarrow 2, 5, 8, 11$$

  What do you notice about the output? (It forms a sequence.)
  How could you find the rule for this function machine? (By finding the rule for the sequence.)
  2 Demonstrate this using a difference table.
  3 To extend thinking repeat with a different function machine and the inputs 3, 5, 7, 11, 15.

  $$3, 5, 7, 11, 15 \rightarrow \boxed{} \rightarrow \boxed{} \rightarrow 3, 9, 15, 27, 39$$

  How could you find the rule?

 ## Teaching points and activities

- Some pupils try and combine the two operations of a two-stage function machine into one operation or mix up the order. If this is the case encourage pupils to find the answer in two stages with the output from the first stage written down.
- When required to find two missing operations, stress the importance of writing the input and output numbers in order first. Then relate finding the operations to finding the rule for the nth term.

Support • Pupils should have covered the previous section *Functions*.

Extension • See **starred questions** in Exercise 11 (Pupil Book page 223).

Plenary ideas • Discuss any of the questions from Exercise 11 that pupils had difficulty with.

Homework Review questions at the end of Exercise 11 (Pupil Book page 223) and/or Homework Sheet 9.11 (page 227 of this pack).

Links Link this topic to understanding algebra (Pupil Book page 158), finding the rule for the nth term (Pupil Book page 217).

Topic area	Properties of functions	pupil book page 224

Framework Objectives ✓ Express simple functions in symbols; represent mappings expressed algebraically, **Framework examples page 163**

Sheets Worksheet 20   Homework Sheet 9.12   Resource Sheet page 215 Teacher Resource Pack Starter Resource Sheet 14 (optional)

Resources You could use a box or boxes with a mini-whiteboard stuck on the front to act as a function machine. The operations can then be changed as required.

 ## Starter ideas and activities

Mental starter ideas Practise multiplication and division by 0·1 and 0·01.

Starter activity **Properties of functions when combining operations**
Use the Discussion on page 224 of the Pupil Book.
Alteratively use Starter Resource Sheet 14.
Ask the following questions.
For part 1   *What single operation could replace the two given?*
For part 2   *Both of these function machines give the same output.*
            The rule for one is ... what is the rule for the other?
            Are these the same rule expressed two different ways?
For part 3   *What happens if we change the order of the operations?*
For part 4   *What must the missing operations be for the output to be the same as the input?*
            Does it matter in what order we have the operations?
There is a summary of the properties on page 224 of the Pupil Book.

 ## Teaching points and activities

- Most of these properties have been seen before by pupils. This section pulls them together as a summary.
- Recap multiplying out brackets and relate this to two ways of expressing the same function.
- Recap inverse operations.
- Some pupils have difficulty seeing that the function is usually different if we change the order of operations in a function machine. You may need to give lots of examples to convince them.
- Some pupils have difficulty finding the inverse of a function. It is helpful to relate this to solving an equation using inverse operations.
- Recap BIDMAS. Emphasise that the inverse operations must be in the reverse order to the operations. This is important for understanding solving equations using inverse operations.

Support
- Pupils should have covered the previous section *Functions*.
- See **Functions** on page 152 (Algebra Support chapter in the Pupil Book).
  If pupils have difficulty with Practice question 29 use Worksheet 20 question 3 (page 18 of the Teacher Resource Pack)

Plenary ideas
- Discuss the properties given on page 224 of the Pupil Book and ask for an example of each.

Homework Review questions at the end of Exercise 12 (Pupil Book page 225) and/or Homework Sheet 9.12 (page 227 of this pack).

Links Link this topic to inverse operations solving equations (Pupil Book page 184), inverse operations (Pupil Book page 162), formulae (Pupil Book page 175), sequences (Pupil Book page 200) and mental calculation (Pupil Book page 61).

10 Graphs

Topic in Chapter	Links to Programme of Study *Ma2 a–l are integrated throughout.*	Links to Framework for Teaching Mathematics: 8 Teaching Programme *Applying mathematics and solving problems are integrated throughout.*
Graphing functions *page 233*	**Ma2 6e** use the conventions for coordinates in the plane; plot points in all four quadrants ... plot graphs of functions in which y is given explicitly in terms of x (for example, $y = 2x + 3$), ...	**Sequences, functions and graphs** ● Generate points in all four quadrants and **plot the graphs of linear functions where y is given explicitly in terms of x**, on paper and using ICT; **recognise that equations of the form $y = mx + c$ correspond to straight-line graphs.**
Equations of straight-line graphs *page 236*	**Ma2 6e** ... recognise (when values are given for m and c) that equations of the form $y = mx + c$ correspond to straight-line graphs in the coordinate plane **Ma2 6h** find the gradient of lines given by equations of the form $y = mx + c$ (when values are given for m and c); investigate the gradients of parallel lines ... (for example, knowing that $y = 5x$ and $y = 5x - 4$ represent parallel lines) (begin)	
Reading and plotting real-life graphs *page 240* Distance/time graphs *page 246* Interpreting and sketching real-life graphs *page 248*	**Ma2 1f** represent problems and solutions in ... graphical forms; move from one form of representation to another to get different perspectives on the problem; present and interpret solutions in the context of the original problem **Ma2 5g** ... relate algebraic solutions to graphical representations of the equations **Ma2 6f** Construct linear functions arising from real-life problems and plot their corresponding graphs; discuss and interpret graphs arising from real situations (for example, distance/time graph for an object moving with constant speed)	**Sequences, functions and graphs** ● Construct linear functions arising from real-life problems and plot their corresponding graphs; discuss and interpret graphs arising from real situations. **Applying mathematics and solving problems** ● **Represent problems and interpret solutions in ... graphical form**, using correct notation and appropriate diagrams. ● Solve more complex problems by ... choosing and using efficient techniques for ... graphical representation, and resources including ICT.

Links to Attainment Targets

Level 5 Pupils use and interpret coordinates in all four quadrants.
Level 6 They represent mappings expressed algebraically, and use Cartesian coordinates for graphical representation interpreting general features.
Level 8 Pupils sketch and interpret graphs of linear ... functions, and graphs that model real situations (begin).

Topic area	Graphing functions	pupil book page 233

Framework Objectives

✓ Generate points in all four quadrants and **plot the graphs of linear functions where y is given explicitly in terms of x,** on paper and using ICT; **recognise that equations of the form $y = mx + c$ correspond to straight line graphs, Framework examples pages 165 and 167**

Sheets

Worksheet 21 for support   Homework Sheet 10.1   Starter Resource Sheet 15   Resource Sheet page 216

Resources

Mini-whiteboards for Starter Activities. Graph paper.

Starter ideas and activities

Mental starter ideas

- Practise all four operations with integers, e.g. $4 \times {}^-2$, $3 \times {}^-1 + 3$ and $2 \times {}^-3 - 2$.
- Give a function, such as $y = 2x$, and ask for the value of y for particular x values.
- Give a function, such as $y = x + 2$, and draw up a table such as this one.

x	$^-1$	0	2
y			

What are the y-values? Write them on your mini-whiteboards.

Starter activity

- **Drawing straight-line graphs**
  1 Draw a set of axes on the OHP or board, or use Starter Resource Sheet 15. You could also draw it on the floor or pavement.
  2 Put the equation $y = 3x + 2$ on the board or OHP.
  3 Draw up a table such as this one and ask for volunteers to fill it in, demonstrating what they are doing. You may have to do one entry first.

x	$^-1$	0	2
y			

  4 Ask for volunteers to come and plot the points and draw a straight line through them. If you have drawn the axes on the floor or pavement you could ask pupils to stand at the points.
    If I choose another point on the line what can you say about the coordinates of the point? (They will satisfy $y = 3x - 2$, i.e. if we substitute them in for x and y in $y = 3x - 2$ the number on each side of the equals sign will be the same.)
    What can you say about all of the points that lie on the straight line? (If a linear sequence is plotted on a graph then the points will be in an imagined straight line.)
- **Plotting sequences**
  Use the Discussion on page 234 of the Pupil Book.

Teaching points and activities

- Make sure pupils can plot coordinates in all four quadrants.
- Go over the conventions for drawing graphs — titles, labels, scales, how to plot points, how to join points. Sometimes pupils are not careful about these.
- Emphasise that functions of the form $y = ax + b$ are linear and the graph will always be a straight line. All of the points that lie on the line will satisfy the equation of the line and any coordinate pair that satisfies the equation of the line will lie on the line.
- Pupils sometimes get confused between the terms function and equation of the line. Emphasise that a function is a special sort of equation e.g. $y = 2x + 3$ is a function and it is also the equation of the straight line we get if we draw the graph of the function by plotting points.
- Pupils sometimes have trouble substituting negative values of x into the equation. Encourage them to substitute the positive ones first and then look for a pattern.

Support

- See **Graphs** on page 152 (Algebra Support chapter of the Pupil Book). If pupils have difficulty with Practice questions 12 and 13 then use Worksheet 21 question 1 and 2 (page 18 of the Teacher Resource Pack).
- See **It's a puzzle** (Pupil Book page 232).

Extension
- See **Starred questions** in Exercise 1 (Pupil Book page 234).

Plenary ideas
- Recap on the words 'linear function' and 'coordinate pair' and 'straight line graph'.
- Discuss the answers to Exercise 1 questions 5, 6 and 7 and discuss the difference between the graph of a function and the graph of a sequence. Link linear function with linear sequences.

Homework

Review questions at the end of Exercise 1 (Pupil Book page 234) and/or Homework Sheet 10.1 (page 228 of this pack).

Links

Link this topic to coordinates (Pupil Book page 313).
Link to Sequences: Discussion page 234; Exercise 1 questions 7, 8, 9, Review 3.
Link to Functions: especially Exercise 1 question 6.

Topic area	Equations of straight line graphs	pupil book page 236

Framework Objectives
✓ Generate points in all four quadrants and **plot the graphs of linear functions where y is given explicitly in terms of x,** on paper and using ICT; **recognise that equations of the form $y = mx + c$ correspond to straight line graphs,** Framework examples pages 165 and 167
✓ Problem solving, **Framework examples pages 9 and 3**

Sheets
Worksheet 21 for support Homework Sheet 10.2 Starter Resource Sheets 16 and 17
ICT Worksheets 'Drawing a Graph on a Spreadsheet' and 'Drawing Graphs with a Graph Plotter' (Teacher Resource Pack pages 140 and 141)

Resources
Graph paper. A computer, a graph plotter and a spreadsheet package. A graphical calculator.

Starter ideas and activities

Mental starter ideas
Ask questions such as Practice questions 18 and 21 on pages 155 and 156 of the Algebra Support chapter in the Pupil Book.

Starter activity
Features of graphs
Use Starter Resource Sheets 16 and 17.
Tell pupils that the graph of an equation of the form $y = mx + c$ is always a straight line.
From looking at these graphs what do you think the m represents?
What do you think the c represents?

Teaching points and activities

- Sometimes pupils get m and c mixed up. You could ask them to devise a way to remember that m represents the slope and c represents the y-intercept.
- Emphasise that in equations such as $y = x - 3$, the negative goes with the 3.
- Emphasise that the greater the value of m the steeper the slope. Each time you go along one unit horizontally a steeper slope rises more than a shallower slope.
- Point out that if m is negative the slope of the graph is the same way as the middle stroke of an upper-case N.

Support
- Pupils should have covered the previous section in this chapter.
- See **Graphs** on page 152 (Algebra Support chapter of the Pupil Book). If pupils have difficulty with Practice questions 18 and 21 use Worksheet 21 questions 3 and 4 (pages 18 and 19 of the Teacher Resource Pack).

Extension
- See **starred questions** in Exercise 2 (Pupil Book page 237).
- See **Investigation: Features of graphs** A2 and C (Pupil Book page 236).

Plenary ideas
- Discuss the answers to Exercise 2 question 7 (or question 8 to extend thinking).
- You could summarise the topic by giving the class a list of straight-line equations and asking them:
  Which have the same gradient?
  Which have a negative/positive gradient?
  Which has the steepest slope?
  Which is parallel to the x-axis? y-axis?
  Which have the same y-intercept?

Homework
Review questions at the end of Exercise 2 (Pupil Book page 237) and/or Homework Sheets 10.2 (page 229 of this pack).

Graphical calculator
See Investigation: Features of graphs (Pupil Book page 236). B and C can be done using a graphical calculator. A2 can be checked using a graphical calculator.
See Practical (Pupil Book page 239). Use a graphical calculator. Notes are given in the Graphical Calculator Support section (page 187 of the Teacher Resource Pack).

ICT
See Investigation: Features of graphs (Pupil Book page 236).
See ICT Worksheet 'Drawing a Graph using a Spreadsheet' and 'Drawing Graphs with a Graph Plotter' on pages 140 and 141 of the ICT Support section of the Teacher Resource Pack.

Links
Link this topic to coordinates (Pupil Book page 313).

Topic area	**Reading and plotting real-life graphs**	pupil book page 240

Framework Objectives   ✓ Construct linear functions arising from real-life problems and plot their corresponding graphs; discuss and interpret graphs arising from real situations, **Framework examples page 173**

Sheets   Worksheet 22 for Support   Homework Sheet 10.3   Resource Sheet pages 217–219 Teacher Resource Pack   Starter Resource Sheet 18

Resources   None needed.

 Starter ideas and activities

Mental starter ideas   Give some simple scales to read such as these.

Starter activity   **Plotting real-life graphs**
Use Starter Resource Sheet 18.
This equation gives the distance a car travels.
1   *How could we plot the graph of this car's journey?*
       How many points do we need to plot?
       What values would be good to choose for t ?
2   Draw up the table for *s* and *t* on the OHP.
3   Ask for volunteers to come and plot the points.
       Is it sensible to draw a line through these points?
       What else do we have to do to finish the graph? (title, label axes)
       How far has the car travelled after three and a half seconds? How can you work this out using the graph? Is it an exact answer or just an estimate?

 Teaching points and activities

- Emphasise the points made in the screen on page 240 of the Pupil Book about how to plot a real-life graph.
- Encourage pupils to plot at least three points and probably four to get an accurate graph. Otherwise it is very easy for pupils to get an inaccurate line.
- Choosing suitable scales for the axes is difficult for some pupils. Encourage them work out what the maximum and minimum values are for each axis and then divide each axis into sensible equal divisions. Emphasise the equal divisions.
- Emphasise that a graph must have a title and the axes must be labelled.
- Encourage pupils to always draw as big a graph as possible as it is then easier to plot points accurately.

Support
- Pupils should have covered the previous section in this chapter.
- See **Graphs** on page 152 (Algebra Support chapter of the Pupil Book). If pupils have difficulty with Practice questions 22 use Worksheet 22 (page 19 of the Teacher Resource Pack).

Extension
- See **Investigation: Collision course** (Pupil Book page 245). Ask pupils to write a report on their findings.

Plenary ideas
- Discuss the answers to Exercise 3 question 5. This question includes most of the important points you need to emphasise.
- Start the Investigation on page 245 and ask pupils to finish it for homework. Discuss strategies for doing this.

Homework   Review questions at the end of Exercise 3 (Pupil Book page 241) and/or Homework Sheet 10.3 (page 230 of this pack).
You could ask pupils to find examples of real-life graphs in newspapers or magazines and write a few sentences about them.

Graphical calculator   See **Practical** (Pupil Book page 245).

Links   Link this topic to coordinates (Pupil Book page 313) and graphs of functions (Pupil Book pages 233–239).
Link to converting between measures: teaching notes, Exercise 3 Review 3.
Link to measures: teaching notes; Exercise 3 questions 1, 3, 5.
Link to Handling Data: Discussion page 241; Exercise 3 question 8.
Link to Science: Exercise 3 question 7, Review 1.

© **New National Framework Mathematics 8 Nelson Thornes Ltd**

| Topic area | Distance/time graphs | pupil book page 246 |

Framework Objectives
✓ Construct linear functions arising from real-life problems and plot their corresponding graphs; discuss and interpret graphs arising from real situations, **Framework examples page 173**

Sheets
Homework Sheet 10.4   Resource Sheet pages 220 and 221 Teacher Resource Pack Starter Resource Sheet 19

Starter ideas and activities

Mental starter ideas
Practise approximations to estimate the answers to calculations.

Starter activity
- **Interpreting distance/time graphs**
  Use Starter Resource Sheet 19.
  What does the line labelled A tell us about Anna's journey?
  Write next to the line 'Anna is walking at a constant speed'.
  Repeat this for B, C and D.
- **Drawing distance/time graphs**
  Use the Discussion on page 246 of the Pupil Book.

Teaching points and activities

- Emphasise the points made in the screen on page 246 of the Pupil Book.
  It is helpful if pupils understand these. Use explanations such as if the distance increases by the same amount for each minute that Anna walks she must be walking at the same speed each minute. Show this using a pointer and the graph (Starter Resource Sheet 19).
  If the distance is not increasing at all while time is going by then Anna must be stopped.
  If the distance she is from home is getting less as each minute passes then she must be walking towards home.

Support
- Pupils should have covered the previous sections in this chapter.

Extension
- Pupils could make up a simple journey of their own and draw a distance/time graph for it.

Plenary ideas
- Ask a pupil to describe a simple journey, such as walking to the park or school or some other place, and ask the rest of the class to sketch the graph of it. (This will work better with pupils who have a good grasp of this topic.)
- Discuss the answers to Exercise 4 questions 2 and 4.

Homework
Review questions at the end of Exercise 4 (Pupil Book page 246) and/or Homework Sheet 10.4 (page 231 of this pack).
You could also ask pupils to sketch a graph of an activity they do.

Links
Link this topic to drawing straight-line graphs (Pupil Book pages 233–239).
Link to measures: Exercise 4 questions 1, 2, 3, 4.
Link to PE: Exercise 4 Review.

Topic area | **Interpreting and sketching real-life graphs** | pupil book page 248

Framework Objectives

✓ Construct linear functions arising from real-life problems and plot their corresponding graphs; discuss and interpret graphs arising from real situations, **Framework examples pages 173, 175 and 177**

Sheets Homework Sheet 10.5 Resource Sheet page 221 Teacher Resource Pack

 ## Starter ideas and activities

Mental starter ideas Give a range of mental activities that your class needs to practise.

Starter activity **Interpreting graphs**
Use the Discussion on page 248 of the Pupil Book. You could put the graphs onto OHP.

 ## Teaching points and activities

- Encourage pupils to think about the real-life situation being shown by the graph by looking at the labels on the axes.
- Emphasise that a straight positive slope means that as one variable is increasing at a constant rate the other is also increasing at a constant rate. A straight negative slope means that as the x-axis variable is increasing the y-axis variable is decreasing.
- Encourage pupils to look at general trends.
- Point out that this section is particularly relevant to science and geography.

Support
- Pupils should have covered the previous section in this chapter.
- See **Graphs** on page 152 (Algebra Support chapter of the Pupil Book).

Extension
- See **starred questions** in Exercise 5 (Pupil Book page 249).
- See **Discussion** (Pupil Book page 253).

Plenary ideas
- Discuss the answers to Exercise 5 question 5 or 8.
- Discuss some of the Exercise 5 review questions.
- Use the Discussion on page 248 of the Pupil Book to extend thinking (you could use this at another point in the lesson). Then you could ask pupils to sketch their own graphs, on the same grid, of a bucket and a bath being filled with water from the same flow of water from a tap. Discuss the answers.
- Point out the links, especially the link to line graphs.

Homework Review questions at the end of Exercise 5 (Pupil Book page 249) and/or Homework Sheet 10.5 (page 232 of this pack).

Links Link this topic to interpreting graphs (Pupil Book page 402) and line graphs (Pupil Book page 391).
Link to Science: Exercise 5 questions 8, 9, Review 4.
Link to measures: Exercise 5 questions 3, 4, 5, Review 1b, 1c, 2, 3.
Link to PE: Exercise 5 questions 2c, 6, 7.
Link to Design and Technology: Exercise 5 question 2d.

11 Lines and Angles

Topic in Chapter	Links to Programme of Study *Ma2 a–l are integrated throughout.*		Links to Framework for Teaching Mathematics: 8 Teaching Programme *Applying mathematics and solving problems are integrated throughout.*
Angles made with intersecting and parallel lines *page 272* Using geometrical reasoning to find angles *page 274*	Ma3 1a	select problem-solving strategies and resources, including ICT, to use in geometrical work, and monitor their effectiveness	**Geometrical reasoning: lines, angles and shapes** ● **Identify alternate angles and corresponding angles.** **Applying mathematics and solving problems** ● Solve more demanding problems and investigate in a range of contexts: ... shape, space and measures ... ● Identify the necessary information to solve a problem. Represent problems and interpret solutions in ... geometric ... form. ● Solve more complex problems by breaking them into smaller steps. ● **Use logical argument to establish the truth of a statement.**
	Ma3 1d	interpret, discuss and synthesise geometrical information presented in a variety of forms	
	Ma3 1e	communicate mathematically, making use of geometrical diagrams and related explanatory text	
	Ma3 1f	use precise language and exact methods to analyse geometrical configurations	
	Ma3 1k	show step-by-step deduction in solving a geometric problem	
	Ma3 2a	recall and use properties of angles at a point, angles on a straight line (including right angles), perpendicular lines, and opposite angles at a vertex	
	Ma3 2c	use parallel lines, alternate angles and corresponding angles ...	

Angles in triangles *page 277*	Ma3 1h	distinguish between practical demonstration, proof, conventions, facts, definitions and derived properties	**Geometrical reasoning: lines, angles and shapes**
			● **... understand a proof that**
			   – **the sum of the angles of a triangle is 180°
			   and of a quadrilateral is 360°**
			   – the exterior angle of a triangle is equal to the sum of the two interior opposite angles.
			● Solve geometrical problems using side and angle properties of equilateral, isosceles and right-angled triangles ... explaining reasoning with diagrams and text.
			
			Applying mathematics and solving problems
			● **Use logical argument to establish the truth of a statement.**

	Ma3 1k	show step-by-step deduction in solving a geometric problem	
	Ma3 2c	... understand ... a proof that the angle sum of a triangle is 180 degrees; understand a proof that the exterior angle of a triangle is equal to the sum of the interior angles at the other two vertices	
	Ma3 2d	use angle properties of equilateral, isosceles and right-angled triangles; ...	
Angles in quadrilaterals *page 281*	Ma3 1k	show step-by-step deduction in solving a geometrical problem	
	Ma3 2d	... explain why the angle sum of any quadrilateral is 360 degrees	

Links to Attainment Targets

Level 5 ... Pupils know the angle sum of a triangle and that of angles at a point.

Level 6 ... They solve problems using ... angle properties of intersecting and parallel lines, and explain these properties.

| Topic area | **Angles made with intersecting and parallel lines** | pupil book page 272 |

Framework Objectives
✓ **Identify alternate and corresponding angles, Framework examples pages 181 and 183**

Sheets
Worksheet 23 for support   Homework Sheet 11.1   ICT Worksheets 'Three intersecting Lines', 'Parallel Lines' and 'A Quadrilateral between Parallel Lines' (Teacher Resource Pack pages 144, 145 and 146) Starter Resource Sheet 20

Resources
An OHP, protractor, tracing paper and mini-whiteboards for the Starter Activity. A computer and Geometer's Sketchpad for Practical (Pupil Book page 272). Sheets of acetate or tracing paper.

 Starter ideas and activities

Mental starter ideas
● Use Starter Resource Sheet 20.
Ask pupils to estimate the size of each angle and name it as either acute, right, obtuse, straight or reflex. Ask pupils to put the angles in order of size.
Use a protractor to measure the angles or ask pupils to come and demonstrate this.
● Ask pupils questions such as Practice questions 3 and 4 on page 264 of the Shape, Space and Measures Support chapter.

Starter activity
● **Introducing the relationships**
Use the Practical on page 272 of the Pupil Book.
(You could use see-through tracing paper instead of acetate or at worst make an OHP of the diagrams in the Practical and ask the same questions as in the Practical.)
Show pupils that alternate angles can be rotated to fit exactly on top of one another and that one corresponding angle can be translated to fit exactly on top of the other.
Why do you think the angles have been given these names?
● **Complementary and supplementary angles**
Explain what these are and give some mental practice at finding them. Pupils could write the answers on their whiteboards. You could divide the class into two groups and have a competition to see who can write down the most complementary (or supplementary) pairs in two minutes (or similar time).

 Teaching points and activities

● Some pupils have difficulty remembering the names of and identifying alternate and corresponding angles. It may help to point out that alternate angles are on alternate sides of the same line. Corresponding angles are in corresponding positions.
● Some pupils may try to measure the angles to find their size.
Emphasise that most of the diagrams given are not drawn to scale and so it is not possible to find the size of angles by measuring them.

Support
● See **Lines and angles** on page 258 (Shape, Space and Measures Support chapter in the Pupil Book). If pupils have difficulty with Practice questions 3, 4, 10 and 15 use Worksheet 24 questions 1–7 (page 20 of the Teacher Resource Pack).
● See **Picture perfect** on page 271 of the Pupil Book.

Extension
● See **Practical** (Pupil Book page 272). Pupils could be asked to write a report on their findings, especially if this exercise is done using ICT.

Plenary ideas
● Recap on the words complementary, supplementary, corresponding angles and alternate angles. Ask pupils to come and give examples of each.
● Revise angles on a straight line, angles at a point and vertically opposite angles and tell pupils they will be needing these for the next lesson.
● Discuss the answers to Exercise 1 question 5.

Homework
Review questions at the end of Exercise 1 (Pupil Book page 273) and/or Homework Sheet 11.1 (page 233 of this pack).

ICT
For the Practical (Pupil Book page 272) use Geometer's SketchPad, a dynamic geometry software package. See ICT Worksheets 'Three intersecting Lines', 'Parallel Lines' and 'A Quadrilateral between Parallel Lines' on pages 144, 145 and 146 of the ICT Support section in the Teacher Resource Pack.

Links
Link this topic to transformations (Pupil Book page 314).
Link to Design and Technology: Exercise 1 question 4.

| Topic area | **Use geometrical reasoning to find angles** | pupil book page 274 |

Framework Objectives
✓ **Identify alternate and corresponding angles, Framework examples pages 181 and 183**
✓ Solve geometrical problems ... explaining reasoning with diagrams and text, **Framework examples page 183**

Sheets Worksheets 24 and 25 for Support   Homework Sheet 11.2   Starter Resource Sheet 21

Resources Mini-whiteboards for the Starter Activity. A calculator.

 Starter ideas and activities

Mental starter ideas Ask questions such as Practice question 20a–c, e and i on page 267 in the Shape, Space and Measures Support chapter of the Pupil Book.

Starter activity Use Starter Resource Sheet 21.
1 *What do you think prove means?* (That it must be true always with no exceptions.)
2 *Write down what you think the first step to prove that e = 38 is. And the next step.*
3 Discuss the fact that sometimes we have to find the size of an angle that is not named. We must name it somehow so we can refer to it in our proof.
  Is there another way we could prove that e = 38? (This may have come out above.)
4 Go through a similar process for the next examples asking pupils to write down the steps one by one on their whiteboards or paper, or ask for pupils to come and show them on the board or OHP.

 Teaching points and activities

● Make sure pupils can solve equations where they have to collect like terms first and equations with unknowns on both sides. If you have not covered this topic yet make a note to include questions 6 and Review 3 when you do that topic. It is good link between algebra and shape, space and measures.
● This is a very important section. It is worthwhile spending time ensuring pupils understand that writing down the steps one by one and the reasons is very important. Geometrical reasoning is an important skill to have for future mathematics. It is best to make sure that pupils write each step of their process and reasoning on a new line. Make sure equals signs are used correctly.
● Sometimes pupils find it hard to know where to start. Encourage them to follow the steps given in the screen on page 274 of the Pupil Book. Sometimes having a checklist of all the angle properties can help.
● Pupils sometimes try to take short cuts or write mathematically incorrect statements such as $b + 36 = 180$
  $$= 180 - 36$$
  $$= 144$$
  Encourage them to write down every step and check that it is a true statement.

Support ● See **Calculating angles** on page 259 (Shape, Space and Measures Support chapter of the Pupil Book). If pupils have difficulty with Practice questions 20a–c, e and i and 30b use Worksheet 25 question 1a–e (page 21 of the Teacher Resource Pack).

Extension ● See **starred questions** in Exercise 2 (Pupil Book page 275).

Plenary ideas ● Discuss the meaning of the words geometrical reasoning and prove.
● Discuss the answer to Exercise 2 question 1.
● Put some examples of wrong reasoning or mathematical statements on the board or OHP and ask pupils to explain what is wrong with each. You could use some that you have seen in pupils' work.

Homework Review questions at the end of Exercise 2 (Pupil Book page 275) and/or Homework Sheet 11.2 (page 234 in this pack).

Links Link this topic to angle properties (Pupil Book page 258) and parallel and intersecting lines (Pupil Book page 272).
Link to solving equations: Exercise 2 question 6, Review 3.

<table>
<tr><td>**Topic area**</td><td>**Angles in triangles**</td><td>pupil book page 277</td></tr>
</table>

Framework Objectives
- ✓ Use logical argument to establish the truth of a statement, **Framework examples page 183**
- ✓ **... understand a proof that**
  - **the sum of the angles of a triangle is 180° ...**
  - the exterior angle of a triangle is equal to the sum of the two interior opposite angles, **Framework examples page 183**
- ✓ Solve geometrical problems using side and angle properties of equilateral, isosceles and right-angled triangles ... explaining reasoning with diagrams and texts, **Framework examples page 183**
- ✓ Problem solving, **Framework examples page 17**

Sheets
Worksheet 25 for support   Homework Sheet 11.3   Resource Sheet page 225 Teacher Resource Pack   ICT Worksheet 'Angles in a Triangle and on a Straight Line' (Teacher Resource Pack page 147).

Resources
A computer and Geometer's SketchPad. A calculator.

Starter ideas and activities

Mental starter ideas
Ask questions such as Practice question 20d, f, g and h on page 267 in the Shape, Space and Measures Support chapter in the Pupil Book.

Starter activity
Proving the angles of a triangle add to 180° and the exterior angle equals the sum of the two opposite interior angles
Use the Investigation on page 277 of the Pupil Book.
You will need to prompt more with lower ability classes.

Teaching points and activities

- Remind pupils that it is important to have just one statement and reason on each line in a proof.
- Remind pupils that the diagrams are not drawn to scale so measuring the angles will not give the correct answer.
- Encourage pupils to write down all the steps and reasons, especially if they are asked to prove something.
- If pupils are having difficulty encourage them to write down what they know. Rotating a diagram can sometimes help. A checklist of all the angle properties can sometimes be helpful too.
- Discuss situations in real life where angles in triangles may be used.

Support
- See **Angles** on page 258 (Shape, Space and Measures Support chapter of the Pupil Book). If pupils have difficulty with Practice questions 20d, f, g, h, use Worksheet 25 questions 1f–i, 2b (page 21 of the Teacher Resource Pack).
- Pupils should have covered the previous section in this chapter.

Extension
- See **starred questions** in Exercise 3 (Pupil Book page 279).

Plenary ideas
- Recap on the findings of the Investigation on page 277.
- Discuss the answers to Exercise 3 question 7 or 8.
- Discuss Exercise 3 question 10 or 11 for extension.
- Use the Puzzle on page 281 of the Pupil Book.

Homework
Review questions at the end of Exercise 3 (Pupil Book page 279) and/or Homework Sheet 11.3 (page 235 of this pack).

ICT
See Practical (Pupil Book page 277). Use ICT Worksheet 'Angles in a Triangle and on a Straight Line' on page 147 of the ICT Support section of the Teacher Resource Pack.

Links
Link this topic to triangles (Pupil Book page 259).
Link to bearings: Exercise 3 question 10.
Link to shapes: Exercise 3 questions 4, 11.
Link to solving equations: Exercise 3 question 7.

| Topic area | Angles in quadrilaterals | pupil book page 281 |

Framework Objectives

✓ Use logical argument to establish the truth of a statement, **Framework examples pages 183 and 187**

✓ **... understand a proof that**
  – the sum of the angles of a ... quadrilateral is 360° ..., **Framework examples page 183**

✓ Solve geometrical problems using side and angle properties of equilateral, isosceles and right-angled triangles ... explaining reasoning with diagrams and texts, **Framework examples pages 183 and 187**

Sheets   Worksheet 25 for support   Homework Sheet 11.4

Resources   Mini-whiteboards for the Starter Activity. A calculator.

Starter ideas and activities

Mental starter ideas   Tell pupils the target number is 360. Go around the class giving pupils a number and asking them what number must be added to it to make 360. Then give two numbers and ask what the third must be to make 360. Then give three numbers and ask what the fourth must be to make it 360.

Starter activity   **Sum of angles in a quadrilateral**
Use the Investigation on page 281 of the Pupil Book.
Put an example, such as the one in the screen, on the OHP or board and ask pupils to write down each step on their whiteboards or ask for volunteers to put the steps on the board.

Teaching points and activities

● Pupils sometimes have problems starting more complex problems. Encourage them to write down what they know and what they can work out and then see how this can be used to work out what they have been asked to find.

● Pupils sometimes just write down the answer and a few jottings. Stress the importance of writing everything down step by step and checking that it is mathematically correct. Again, remind them that there should be just one statement per line.
You may need to demonstrate this many times.

Support   ● See **Properties of quadrilaterals** on page 259 (Shape Space and Measures Support chapter of the Pupil Book). If pupils have difficulty with Practice questions 7 and 30 use Worksheet 25 question 2a (page 21 of the Teacher Resource Pack).

● Pupils should have covered previous sections in this chapter.

Extension   ● See **starred questions** in Exercise 4 (Pupil Book page 282).

Plenary ideas   ● Discuss Exercise 4 question 4 (question 5 to extend thinking).

Homework   Review questions at the end of Exercise 4 (Pupil Book page 282) and/or Homework Sheet 11.4 (page 236 of the Teacher Resource Pack).

Links   Link this topic to angles in a triangle (Pupil Book page 277), geometrical reasoning (Pupil Book page 274) and shapes (Pupil Book page 286–290).

12 Shape, Construction and loci

Topic in Chapter	Links to Programme of Study *Ma2 a–l are integrated throughout.*		Links to Framework for Teaching Mathematics: 8 Teaching Programme *Applying mathematics and solving problems are integrated throughout.*
Visualising and sketching 2-D shapes *page 286*	**Ma3 1a**	select problem-solving strategies and resources, including ICT, to use in geometrical work, and monitor their effectiveness	**Geometrical reasoning: lines, angles and shapes** ● Solve geometrical problems using side and angle properties of equilateral, isosceles and right-angled triangles and special quadrilaterals, explaining reasoning with diagrams and text; classify quadrilaterals by their geometric properties.
Properties of triangles, quadrilaterals and polygons *page 287*	**Ma3 1b**	select and combine known facts and problem-solving strategies to solve complex problems	
Tessellations *page 291*	**Ma3 1c**	identify what further information is needed to solve a problem; break complex problems down into a series of tasks	● Know that if 2-D shapes are congruent, corresponding sides and angles are equal. **Transformations** ● ... identify all the symmetries of 2-D shapes.
Congruence *page 293*	**Ma3 1e**	communicate mathematically, making use of geometrical diagrams and related explanatory text	**Applying mathematics and solving problems** ● Suggest extensions to problems, conjecture and generalise; identify exceptional cases or counter-examples.
	Ma3 1f	use precise language and exact methods to analyse geometrical configurations	
	Ma3 1j	explore connections in geometry; ...	
	Ma3 1k	show step-by-step deduction in solving a geometrical problem	
	Ma3 1n	identify exceptional cases when solving geometrical problems	
	Ma3 2c	... understand the properties of parallelograms ...	
	Ma3 2d	use angle properties of equilateral, isosceles and right-angled triangles	
	Ma3 2f	recall the essential properties of special types of quadrilateral, including square, rectangle, parallelogram, trapezium and rhombus; classify quadrilaterals by their geometric properties	

Topic in Chapter	Links to Programme of Study		Links to Framework for Teaching Mathematics: 8 Teaching Programme
Describing and sketching 3-D shapes. Constructing nets *page 295*	**Ma3 1e**	communicate mathematically, making use of geometrical diagrams and related explanatory text	**Geometrical reasoning: lines, angles and shapes** ● Know and use geometric properties of cuboids and shapes made from cuboids, begin to use plans and elevations.
	Ma3 1f	use precise language and exact methods to analyse geometrical configurations	
	Ma3 2j	explore the geometry of cuboids (including cubes), and shapes made from cuboids	
	Ma3 2k	use 2-D representations of 3-D shapes and analyse 3-D shapes through 2-D projections …	
	Ma3 4d	… construct cubes, … and other 3-D shapes from given information	
Plans and elevations *page 297*	**Ma3 2j**	explore the geometry of cuboids (including cubes) and shapes made from cuboids	
	Ma3 2k	use 2-D representations of 3-D shapes and analyse 3-D shapes through 2-D projections and cross-sections, including plans and elevations.	
Construction *page 301* Constructing triangles and quadrilaterals *page 304*	**Ma3 4c**	use straight edge and compasses to do standard constructions, including … the mid-point and perpendicular bisector of a line segment, the perpendicular from point to a line, the perpendicular from a point on a line, and the bisector of an angle	**Construction and loci** ● Use straight edge and compasses to construct: – the mid-point and perpendicular bisector at a line segment; – the bisector of an angle; – the perpendicular from a point to a line. – the perpendicular from a point on a line.
Loci *page 307*	**Ma3 4j**	find loci, both by reasoning and by using ICT to produce shapes and paths …	**Construction and loci** ● Find simple loci, both by reasoning and by using ICT, to produce shapes and paths …

Links to Attainment Targets

Level 4 Pupils make 3-D mathematical models by linking given faces or edges, draw common 2-D shapes in different orientations on grids.

Level 5 They identify all the symmetries of 2-D shapes.

Level 6 Pupils recognise and use common 2-D representations of 3-D objects. They know and use the properties of quadrilaterals in classifying different types of quadrilateral. They solve problems using angle and symmetry properties of polygons … They devise instructions for a computer to generate … paths.

Level 7 They determine the locus of an object moving according to a rule (begin).

Level 8 Pupils understand and use congruence (begin).

New National Framework

<table>
<tr><td>**Topic area**</td><td>**Visualising and sketching 2-D shapes**</td><td>pupil book page 286</td></tr>
</table>

Framework Objectives

✓ Solve geometrical problems using side and angle properties of equilateral, isosceles and right-angled triangles and special quadrilaterals, explaining reasoning with diagrams and text; classify quadrilaterals by their geometric properties, **Framework examples pages 185 and 187**

✓ Problem solving, **Framework examples page 15**

Sheets Homework Sheet 12.1

Resources You could have copies of the shapes available for checking answers. Pictures or posters with geometrical patterns. Squared paper. A computer.

Starter ideas and activities

Mental starter ideas
- Ask questions such as Practice question 7 on page 265 of the Shape Space Measures Support chapter.
- Ask pupils questions such as:
  Visualise a square with one corner cut off. Describe the shape left.
  What if it was a rectangle with one corner cut off?
  What if two corners were cut off the square?

Starter activity **Imagining shapes**
Use the Discussion on page 286 of the Pupil Book.
You could also use some of the other questions in Exercise 1 as a discussion.

Teaching points and activities

- Make sure pupils know the names of all the common 2-D shapes.
- You could do this whole section as a discussion or whole class exercise asking pupils to come and demonstrate their answers on the board or OHP. Or encourage pupils to check their answers by making copies of the shapes.
- Encourage pupils to try and find more than one answer to most of the questions. This exercise is a good place to revise the properties of 2-D shapes that pupils should know.

Support
- See **2-D shapes** on page 259 (Shape, Space and Measures Support chapter of the Pupil Book). If pupils have difficulty with question 7 use Worksheet 26 questions 1 and 2 (page 22 Teacher Resource Pack).

Extension
- See **Investigation: Rectangles and Squares** on page 287 of the Pupil Book. In questions 1 and 2 challenge pupils to use larger rectangles and write a report on their findings.

Plenary ideas
- Discuss the answers to as many of the Exercise 1 questions as possible and have pupils demonstrate their answers. Encourage as many different answers as possible.
- Discuss ideas for the Practical on page 286 and ask pupils to finish it for homework. (It may take a few days and could be set as a longer-term project.)
- You could start the Investigation on page 287 and ask pupils to finish it for homework. Discuss strategies for making sure all of the ways are drawn.

Homework Review questions at the end of Exercise 1 (Pupil Book page 286) and/or Homework Sheet 12.1 (page 237 of this pack).

ICT You could use a computer to make a poster or booklet for the Practical on page 286.

Links Link this topic to triangles, quadrilaterals and polygons (Pupil Book page 287).

Topic area	**Properties of triangles, quadrilaterals and polygons**	pupil book page 287

Framework Objectives

✓ Solve geometrical problems using side and angle properties of equilateral, isosceles and right-angled triangles and special quadrilaterals, explaining reasoning with diagrams and text; classify quadrilaterals by their geometric properties. **Framework examples pages 185 and 187**

✓ ... identify all the symmetries of 2-D shapes, **Framework examples pages 185 and 187**

✓ Problem solving, **Framework examples pages 15 and 35**

Sheets

Worksheet 26 for support   Homework Sheet 12.2   Resource Sheets pages 227 to 232 Teacher Resource Pack   ICT Worksheets 'Different Parallelograms' and 'Making Shapes with Logo' (Teacher Resource Pack pages 155 and 156)

Resources

Investigation: Properties. A computer and Geometer's SketchPad, a dynamic geometry software package, for Investigation: Properties of Shapes. Large copies of an equilateral triangle, isosceles triangle, square, rectangle, parallelogram, rhombus, kite, isosceles trapezium, trapezium and arrowhead, a 3 × 3 pinboard and square dotty paper. A computer and MswLogo.

Starter ideas and activities

Mental starter ideas

You could play true/false with your class. Divide the class into groups and give them these statements to decide if each is true or false. They must come up with a group answer. Give 1 point for each correct answer and take 1 point off for each wrong answer.

> *A quadrilateral has 4 sides.*
> *A kite is symmetrical.*
> *A parallelogram has just one pair of parallel sides.*
> *A trapezium has just one pair of parallel sides.*
> *A square has 4 lines of symmetry.*
> *An equilateral triangle has exactly 2 lines of symmetry.*
> *A rhombus has 4 equal angles.*
> *An arrowhead (or delta) has exactly 2 equal angles.*
> *The diagonals of a rhombus are both lines of symmetry.*
> *A triangle can have a reflex angle.*
> *A quadrilateral with 4 equal sides must be a square.*

Starter activity

● **Properties of 2-D shapes**
  Use part A of the Investigation: Properties of Shapes (Pupil Book page 287). The results of this are needed for pupils to be able to do the exercise. You could also do parts B and C if there is time or set these for homework.

Teaching points and activities

● Many pupils have trouble remembering all of the properties of quadrilaterals and triangles. Emphasise that these can be worked out from the symmetry properties. It is therefore important that they can tell you the symmetry properties of each shape with ease. This may mean revising this on a regular basis.

● It may be helpful for some pupils to have a copy of each of the shapes made out of a fairly durable material such as a sheet of plastic. This way they can fold or rotate the shapes to work out the properties.

● Order of rotation symmetry may need revising, especially the fact that if the order of rotation symmetry of a shape is 1, it has no rotation symmetry.

Support

● See **2-D Shapes** on page 259 Shape, Space and Measures Support chapter. If pupils have difficulty with Practice questions 7, 17, 30a use Worksheet 26 question 1 (page 22 of the Teacher Resource Pack).

● See **Piecing it together** (Pupil Book page 285).

Extension

● See **starred questions** in Exercise 2 (Pupil Book page 289).

● See **Investigation: Properties of Shapes** (Pupil Book page 287) part B.
  Challenge pupils to find as many different groups as possible. This is a good opportunity to make links to area and perimeter and to transformations.

● See **Investigation: Polygons and right angles** (Pupil Book page 291). Challenge pupils to try and find an expression for the greatest number of right angles in an n-sided polygon. It might be necessary to give them the hint that the expression might be different for odd and even values of n.

- **Practical** (Pupil Book page 291) question 6. Encourage pupils to explore many different sorts of designs using polygons and quadrilaterals and, if possible, print them off for display.

Plenary ideas
- Recap on the symmetry properties of each shape and discuss how these can be used to find the other properties.
- Discuss the answers to Exercise 2 question 2 or 6.
- Discuss Exercise 2 question 7 or 8 to extend thinking.

Homework   Review questions at the end of Exercise 2 (Pupil Book page 289) and/or Homework Sheet 12.2 (page 237 of this pack).

ICT   For Investigation: Properties (Pupil Book page 287) use Geometer's SketchPad, a dynamic geometry software package. See ICT Worksheet 'Different parallelograms' on page 151 of the ICT Support section in the Teacher Resource Pack.
For Practical (Pupil Book page 291) use MswLogo. See ICT Worksheet 'Making Shapes with Logo' on page 155 of the ICT Support section in the Teacher Resource Pack.

Links   Link this topic to angles (Pupil Book pages 258 and 272), area, perimeter, transformations (Investigation: Properties of Shapes).
Link to symmetry: Investigation: Properties of Shapes (Pupil Book page 287) and all of Exercise 2.

| Topic area | Tessellations | pupil book page 291 |

Framework Objectives

✓ Solve geometrical problems using side and angle properties of equilateral, isosceles and right-angled triangles and special quadrilaterals, explaining reasoning with diagrams and text; classify quadrilaterals by their geometric properties, **Framework examples page 189**

Sheets Starter Resource Sheet 22   ICT Worksheet 'Tessellations' (Teacher Resource Pack page 152).

Resources A computer and Geometer's SketchPad, a dynamic geometry software package.

 ## Starter ideas and activities

Mental starter ideas
- Revise the angle properties of 2-D shapes. You could do this by putting a shape on the board. Ask everyone to stand up and go round the class asking pupils to tell you a property of this shape. If a pupil can't tell you a property but the next pupil can then the first pupil sits down.
- Revise angles at a point add to 360°.

Starter activity
- **Tessellations — what are they?**
  Use Starter Resource Sheet 22.
  1 Make an OHP of the shapes on the starter resource sheet and then cut them out.
  2 Remind pupils what a tessellation is.
  3 Show the hexagon.
      Do you think this shape will tessellate?
  4 Ask someone to come and demonstrate.
  5 Repeat for the other shapes.
- **How do we make a tessellation?**
  Use the Discussion on page 292 of the Pupil Book.

 ## Teaching points and activities

- Some pupils find visualising tessellations very difficult. Encourage them to use cut-outs of the shapes they are working with.
- Emphasise that the shape can be rotated, translated, reflected or a combination of these to make the tessellation. You may need to recap reflection, rotation and translation.
- You may need to give a clue to pupils that they need to think about angles when trying to explain why shapes will or will not tessellate. Pupils may find these explanations difficult.

Support
- Pupils should have covered the previous section, *Properties of triangles*, *quadrilaterals and polygons*.

Extension
- See **Investigation: Tessellations** (Pupil Book page 292) questions 3 and 4.

Plenary ideas
- Summarise what a tessellation is by displaying some.
- Ask pupils to come up and show the results of the Investigation on page 292.

Homework Pupils could be given part of the Investigation to do at home.

ICT See Discussion (Pupil Book page 292). A computer tiling software package may be used here.
See Investigation: Tessellations question 3 on page 292. See ICT Worksheet 'Tessellations' on page 152 of the ICT Support section of the Teacher Resource Pack.

Links Link this topic to transformations (Pupil Book pages 261 and 313), symmetry (Pupil Book page 317), angles (Pupil Book page 258) and angles made with parallel lines (Pupil Book page 272).
Link to Art: whole topic.

| Topic area | Congruence | pupil book page 293 |

Framework Objectives
✓ Know that if 2-D shapes are congruent, corresponding sides and angles are equal,
Framework examples page 191

Sheets   Homework Sheet 12.3   Resource Sheet (page 232 Teacher Resource Pack)

Resources   A copy of the shapes in the Practical on page 293 of the Pupil Book. Protractor and ruler.
Tracing paper, a 3 × 3 pinboard or 3 × 3 dotty grid, a 4 × 4 pinboard or 4 × 4 dotty grid, a
5 × 5 pinboard or 5 × 5 dotty grid for the Practical.

Starter ideas and activities

Mental starter ideas   Ask what congruent means.
Ask pupils to give you examples in the classroom of shapes that are congruent. Ask 'If we
reflect/rotate/translate a shape is the image congruent to the original?'

Starter activity   **Congruent?**
Use the Practical on page 293 of the Pupil Book.

Teaching points and activities

- Some pupils find it difficult to see that if we reflect or rotate a shape it is still the same
  shape. Demonstrate this with a cut out shape and show that we can turn it and rotate it in
  many different ways and it is still the same shape.
- Some pupils may need to use tracing paper to do the exercise. Encourage pupils to try
  without it first.

Support   ● Pupils should have covered the previous sections in this chapter.

Extension   ● See **starred questions** in Exercise 3 (Pupil Book page 293).

Plenary ideas   ● Discuss the answers to Exercise 3 question 3 (or question 4 to extend thinking).
- Start the Practical on page 294 of the Pupil Book and ask pupils to finish it for homework.
  Discuss strategies for finding all of the possible ways.
- Ask pupils to name congruent shapes they can see around them.

Homework   Review questions at the end of Exercise 3 (Pupil Book page 293) and/or Homework
Sheet 12.3 (page 240 of this pack).

Links   Link this topic to transformations (Pupil Book page 261).

© **New National Framework Mathematics 8 Nelson Thornes Ltd**

Topic area	Describing and sketching 3-D shapes. Constructing nets	pupil book page 295

Framework Objectives
✓ Know and use geometric properties of cuboids and shapes made from cuboids, **Framework examples pages 199**
✓ Problem solving, **Framework examples page 15**

Sheets
Worksheet 28 for support   Homework Sheet 12.4   Resource Sheet page 233 Teacher Resource Pack

Resources
You will need a frameworks kit or Multilink/centimetre cubes and isometric paper (Resource Sheet pages 261 to 263 Teacher Resource Pack) for the Starter Activity. Multilink/centimetre cubes and a photograph or poster for the Practical (Pupil Book page 295).

Starter ideas and activities

Mental starter ideas
Ask questions such as Practice questions 21, 28 and 29 on pages 267 and 268 of the Shape, Space and Measures Support chapter.

Starter activity
- **Describing 3-D shapes**
  Use the Practical (Pupil Book page 295).
  Put a photograph on an OHP and ask round the class for a description of shapes.
- **Isometric drawing**
  1 Make an OHP of some isometric (triangle dotty) paper (Resource Sheet pages 261 to 263 Teacher Resource Pack).
  2 *How do we draw a cube on this paper?*
  3 Ask for volunteers to show this.
     What about two cubes?
  4 Make Multilink models and ask for volunteers to draw them.
  5 You could make this into a group exercise and award points for correctly drawn shapes. Each group would be given the same shape to draw and some isometric paper.

Teaching points and activities

- Make sure pupils know what a net is and can sketch nets for simple shapes.
- The isometric paper must be the right way round. Encourage pupils to check this before they attempt to draw on it.
- Encourage pupils to use models of the shapes in many of the questions in Exercise 4 to help them answer the questions.
- Nets can be difficult for some pupils. Initially pupils could build a model of the shape and cut out the net they have drawn to see if it makes the correct 3-D shape. Encourage them to visualise as much as possible by 'seeing' the model in their head as a 3-D shape.
- Make pupils aware of the fact that isometric drawings can be made without using isometric paper.

Support
- See **3-D Shapes** on page 260 (Shape, Space and Measures Support chapter of the Pupil Book). If pupils have difficulty with Practice questions 21, 28 and 29 use Worksheet 29 (page 24 of the Teacher Resource Pack).

Extension
- See **starred questions** in Exercise 4 (Pupil Book page 296).

Plenary ideas
- Use the Puzzle on page 297. You could have a model of this cube ready to help them solve the puzzle. You could suggest using a table to help solve it.
- Ask pupils to describe some 3-D shapes they can see in the room. Encourage them to use words such as face, edge, vertex, intersect, point, parallel and perpendicular.

Homework
Review questions at the end of Exercise 4 (Pupil Book page 296) and/or Homework Sheet 12.4 (page 239 of this pack).

Links
Link this topic to 3-D shapes (Pupil Book page 260).

| Topic area | Plans and elevations | pupil book page 297 |

Framework Objectives ✓ ... begin to use plans and elevations, **Framework examples page 201**

Sheets Homework Sheet 12.5   Resource Sheet page 234 Teacher Resource Pack

Resources Models made from blocks for the Starter Activity. You could use a set of solid shapes or a frameworks kit. Multilink or centimetre cubes and isometric paper for the Practical (Pupil Book page 300).

 ## Starter ideas and activities

Mental starter ideas Ask questions such as:
Visualise what the top of a can of soup looks like if you are a fly flying directly above it. What shape would you see?

Starter activity
- **Views from the top**
  Use the Discussion on page 297 of the Pupil Book.
- **Other views**
  1 Divide the class into groups and give each group a model made from four blocks, similar to the one in the screen on page 298. Label the front, the top and the side in some way.
  2 Ask each group to draw the view from the side, the view from above and the view from the front. You may have to demonstrate one first.
  3 Ask them then to pass the drawings on to the next group without the model that was given. This group must try and make the model from the drawings.
  If a group cannot make the model because they think the drawing they were given is wrong, they can go back to the group that drew the views and ask to see the model.
  4 If the drawings are wrong the group trying to make the model gets 10 points. A correctly drawn set of views gets 10 points and a correctly made model gets 10 points.

 ## Teaching points and activities

- Encourage pupils to imagine they are very small creatures standing looking at the side or front and then flying over the top. Tell them to imagine they can only see the face they are drawing.
- If pupils have difficulty then real models made with Multilink cubes may help.

Support
- See **Nets** on page 261 (Shape, Space and Measures Support chapter of the Pupil Book). Try Practice questions 28 and 41.

Extension
- See **Practical** (Pupil Book page 300). Challenge pupils to complete this whole practical and write up a report which shows all the drawings, and answers all the questions.

Plenary ideas
- Have some models of 3-D shapes made with Multilink or centimetre cubes. Ask pupils to sketch the front and side elevations and plan views of each of them.
- Discuss the answers to Exercise 5 question 3.
- Use the Practical on page 300 of the Pupil Book. This ties together all the main points while extending thinking as well.

Homework Review questions at the end of Exercise 5 (Pupil Book page 298) and/or Homework Sheet 12.5 (page 239 of this pack).

Links Link this topic to 3-D shapes (Pupil Book page 295).

Topic area **Construction**   pupil book page 301

Framework Objectives
✓ Use straight edge and compasses to construct:
 – the mid-point and perpendicular bisector of a line segment
 – the bisector of an angle
 – the perpendicular from a point to a line
 – the perpendicular from a point on a line, Framework examples page 221

Sheets Homework Sheet 12.6   Resource Sheet pages 235 and 236 Teacher Resource Pack
Starter Resource Sheet 23

Resources Compasses and ruler.

 ## Starter ideas and activities

Mental starter ideas Are these true or false? If it is false then change it so that it is true.
 A rhombus has 4 equal sides.
 A rhombus has 2 equal diagonals.
 A rhombus has 4 equal angles.
 The diagonals of a rhombus bisect the angles.
 A rhombus has 2 lines of symmetry.
 A rhombus has rotation symmetry of order 2.

Starter activity ● **Constructions**
Use Starter Resource Sheet 23.
Go through each step one by one and ask pupils to volunteer to show you what to do.
Now do the Discussion on page 302 of the Pupil Book.

 ## Teaching points and activities

● Some pupils do not measure and draw carefully enough to get an accurate drawing. Emphasise the need to measure and join lines and draw through points very very carefully.
● Go over the meaning of 'mid-point', 'bisector' and 'perpendicular'.
● Pupils are more likely to remember these constructions if they understand them a little. The Discussion on page 302 will help some pupils understand them, but for others it may be a little too complicated.

Extension ● See **starred questions** in Exercise 6 (Pupil Book page 303).

Plenary ideas ● Discuss Exercise 6 question 2 and ask pupils to demonstrate how they did it.
● Go over the first six steps of the Puzzle on page 304 of the Pupil Book.

Homework Review questions at the end of Exercise 6 (Pupil Book page 303) and/or Homework Sheet 12.6 (page 240 of this pack).

Links Link this topic to properties of a rhombus (Pupil Book pages 259 and 287), angles (Pupil Book page 272) and loci (Pupil Book page 309).

| Topic area | **Constructing triangles and quadrilaterals** | pupil book page 304 |

Framework Objectives
✓ Construct a triangle given 3 sides (SSS); use ICT to explore these constructions, **Framework examples page 223**
Problem solving, **Framework examples page 17.**

Sheets
Worksheet 27 for Support   Homework Sheet 12.7   Starter Resource Sheet 24
ICT Worksheets 'Constructions on a Line' and 'Constructing a Triangle with 3 Known Lengths' (Teacher Resource Pack pages 153 and 154).

Resources
Protractor, ruler and compasses. A computer and Geometer's SketchPad, a dynamic geometry software package.

Starter ideas and activities

Mental starter ideas
Revise all the different sorts of triangles and their properties. You could do this by putting a triangle on the board and asking pupils who know the name of it to stand up. Ask those who can tell you its symmetry properties to stay standing, those who can tell you the angle properties to stay standing and then the side properties to stay standing. Ask one of the pupils standing to come and tell all of the answers.

Starter activity
- **Revising constructing triangles using a ruler and protractor**
  Revise drawing a triangle using a protractor and ruler. You could use Practice questions 37 and 41 on pages 269 and 270 of the Shape, Space and Measures Support chapter in the Pupil Book.
- **Constructing a triangle given three sides**
  Use Starter Resource Sheet 24.
  1 Discuss how we might use compasses to construct a triangle.
  2 Ask pupils to come up and draw in each step.
- **Constructing quadrilaterals**
  Use the Discussion on page 305 of the Pupil Book.

Teaching points and activities

- Emphasise the need to draw and measure accurately.

Support
- See **'Constructing Triangles'** on page 260 (Shape, Space and Measures Support chapter of the Pupil Book). If pupils have difficulty with Practice questions 37 and 41 use Worksheet 27 (page 23 of the Teacher Resource Pack).

Plenary ideas
- Discuss Exercise 7 question 4 as a class or draw it on the OHP step by step asking pupils to come and demonstrate how.
- Point out the links, especially the link to scale drawings.

Homework
Review questions at the end of Exercise 7 (the Pupil Book page 305) and/or Homework Sheet 12.7 (page 241 of this pack)

ICT
See Practical (Pupil Book page 306). You will need ICT Worksheets 'Constructions on a Line' and 'Constructing a Triangle with three Known Lengths' on pages 153 and 154 of the ICT Support section of the Teacher Resource Pack.

Links
Link this topic to 2-D shapes (Pupil Book pages 259 and 287).
Link to scale drawings: Exercise 7 question 4, Review 3.
Link to nets: Exercise 7 question 5, Review 4.

| Topic area | Loci | pupil book page 307 |

Framework Objectives   ✓ Find simple loci, both by reasoning and by using ICT, to produce shapes and paths ..., **Framework examples pages 225 and 227**

Sheets   Homework Sheet 12.8   Starter Resource Sheet 25   ICT Worksheet 'Investigating Loci' (Teacher Resource Pack page 156).

Resources   A computer and MswLogo.

 ## Starter ideas and activities

Mental starter ideas   Use the Discussion on page 307. If you use the first bullet point you could use Starter Resource Sheet 25.

Starter activity
- **Discovering some common loci**
  Use the Practical on page 308 of the Pupil Book. You could change the practical and use non-perpendicular sides instead of wall A and wall B in part 3. This would illustrate the case for an angle that is not a right angle.

 ## Teaching points and activities

- Pupils sometimes just draw one point as the locus. Emphasise that the locus is the set of all the points that satisfy the conditions given. Encourage pupils to draw the position of several points that do satisfy the conditions before deciding what the locus is.
- Encourage pupils to become familiar with some of the common loci such as:
  The set of points equidistant from a fixed point, P, is a circle.
  The set of points equidistant from two intersecting lines is the bisector of the angle between the lines.
  The set of points that are equidistant from two fixed points, P and Q, is the perpendicular bisector of the line joining the points.

Support
- Pupils should have covered the previous two sections in this chapter.

Plenary ideas
- Discuss the answers to all the Exercise 8 questions.
- Ask pupils to sketch on the board the paths of some common objects such as:
  A car going along the road
  The tip of a clock hand
  The head of someone walking parallel to a straight hedge.
- Point out the links, especially the link to construction.

Homework   Review questions at the end of Exercise 8 (Pupil Book page 308) and/or Homework Sheet 12.8 (page 242 of this pack).

ICT   See **Practical** (Pupil Book page 309). You will need ICT Worksheet 'Investigating Loci' on page 156 of the ICT Support section in the Teacher Resource Pack.

Links   Link to construction (Pupil Book page 301).

13 Coordinates and Transformations

Topic in Chapter	Links to Programme of Study *Ma2 a–l are integrated throughout.*		Links to Framework for Teaching Mathematics: 8 Teaching Programme *Applying mathematics and solving problems are integrated throughout.*
Coordinates and transformations *page 313*	**Ma3 1a**	select problem-solving strategies and resources, including ICT, to use in geometrical work, and monitor their effectiveness	
Combinations of transformations *page 314*	**Ma3 1j**	explore connections in geometry; …	**Transformations** ● Transform 2-D shapes by simple combinations of rotations, reflections and translations, on paper and using ICT; identify all the symmetries of 2-D shapes.
	Ma3 3a	understand that rotations are specified by a centre and an (anticlockwise) angle; use right angles, fractions of a turn or degrees to measure the angle of rotation; understand that reflections are specified by a mirror line, translations by a distance and direction, …	
	Ma3 3b	recognise and visualise rotations, reflections and translations …; transform 2-D shapes by translation, rotation and reflection …	
	Ma3 3e	… use axes and coordinates to specify points in all four quadrants; locate points with given coordinates; find the coordinates of points identified by geometrical information (for example, find the coordinates of the fourth vertex of a parallelogram with vertices at (2, 1)(⁻7, 3) and (5, 6)); …	
Symmetry *page 317*	**Ma3 3b**	recognise and visualise rotations, reflections … including reflection symmetry of 2-D shapes, and rotation symmetry of 2-D shapes …	
Enlargement *page 320*	**Ma3 3a**	understand that … enlargements are specified by a centre and positive scale factor	**Transformations** ● Understand and use the language and notation associated with enlargement; **enlarge 2-D shapes, given a centre of enlargement and a positive whole-number scale factor;** explore enlargement using ICT.
	Ma3 3c	recognise, visualise and construct enlargements of objects using positive integer scale factors greater than one …	
	Ma3 3d	recognise that enlargements preserve angle but not length; identify the scale factor of an enlargement as the ratio of the lengths of any two corresponding line segments and apply this to triangles (begin)	

Topic in Chapter	Links to Programme of Study	Links to Framework for Teaching Mathematics: 8 Teaching Programme
Scale drawing *page 324*	**Ma3 3d** ... use and interpret maps and scale drawings	**Transformations** ● Make simple scale drawings.
Finding the mid-point of a line *page 328*	**Ma3 3e** ... find the coordinates of the mid-point of the line segment AB, given points A and B ...	**Coordinates** ● Given the coordinates of points A and B, find the mid-point of the line segment AB.

Links to Attainment Targets

Level 4   They reflect simple shapes in a mirror line.

Level 5   They identify all the symmetries of 2-D shapes.

Level 6   They devise instructions for a computer to generate and transform shapes ...
They enlarge shapes by a positive whole-number scale factor.

Level 7   Pupils enlarge shapes ... and appreciate the similarity of the resulting shapes (begin).

Topic areas	Coordinates and transformations	pupil book page 313
	Combinations of transformations	pupil book page 314

Framework Objectives
✓ Transform 2-D shapes by simple combinations of rotations, reflections, and translations, on paper and using ICT; ..., **Framework examples pages 203, 205 and 209**
✓ Problem solving, **Framework examples page 17**

Sheets
Worksheets 28 and 30 for support   Homework Sheets 13.1 and 13.2   Resource Sheet pages 237–239 Teacher Resource Pack   Starter Resource Sheet 26

Resources
You could use a grid drawn on acetate for the Starter 'Flag it up'. Mini-whiteboards for the Starter Activity. Graph paper. Pin and mirror. Square dotty paper.

 ## Starter ideas and activities

Mental starter ideas
You could use Practice questions 9 and 18 on pages 265 and 267 of the Shape, Space and Measures Support chapter to revise coordinates. Put these on an OHP or use Starter Resource Sheet 26 and ask for volunteers to do them.

Starter activity
● **Revise rotating, translating and reflecting**
You could do **Flag it up** on page 312 of the Pupil Book.
or **1** Ask pupils to copy this set of axes on to their whiteboards.

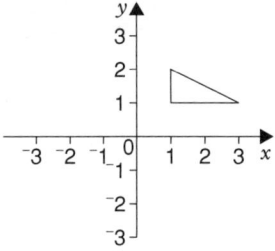

2 Ask pupils to reflect the shape in the y-axis. Ask for the coordinates of the new triangle.
3 Now ask pupils to rotate the new triangle 270° (discuss the fact that if no direction is given we assume it is anticlockwise). Ask for the coordinates of the new triangle (after rotation).
4 Ask pupils to translate the latest image up 4 squares and left 1 square. Ask for the coordinates of the triangle.

● **Combinations of transformations**
Use the Discussion on page 314 of the Pupil Book to introduce this topic. You could also do Exercise 1 question 1 instead and then move on to questions 2, 3, 4 and 5 to do as a class. Pupils could then go on and do questions 6 and 7 individually.

 ## Teaching points and activities

● Recap line graphs for $y = a$ and $x = b$ where a and b represent numbers.
● Some pupils have trouble visualising a rotation. Encourage them to attempt to find the answer by visualising and then check it using tracing paper. This way pupils get immediate feedback and can correct what they have done if necessary. Practice does help visualisation skills to develop.
● Reflections could be checked using a mirror or tracing paper. Emphasise that for reflection, corresponding points must be equidistant from the mirror line.
● Emphasise that the image after a translation, reflection or rotation is congruent to the object.
● The inverse of a rotation sounds complicated when written down. Encourage pupils to explain it in their own words so that they realise it is not as complicated as it sounds. Emphasise that there are two inverse rotations but they both have the same outcome.

Support
● See **Coordinates** on page 261 (Shape, Space and Measures Support chapter of the Pupil Book).
● If pupils have difficulty with Practice questions 9, 18, 19, 22, 23, 24, 33 and 43 use Worksheets 28 and 30 (pages 23 and 25 of the Teacher Resource Pack).

Extension
● See **starred questions** in Exercise 2 (Pupil Book page 314).
● See **Practical** (Pupil Book page 317). Encourage pupils to colour their designs and try out different designs.

Plenary ideas
● Discuss the answers to Exercise 2 questions 6 and 7, asking pupils to demonstrate strategies for finding the answers to question 7.
● Start the Practical on page 317 and ask pupils to finish it for homework.

Homework Review questions at the end of Exercises 1 and 2 (Pupil Book pages 313 and 314) and/or Homework Sheets 13.1 and 13.2 (page 243 of this pack).

ICT See **Practical** (Pupil Book page 317) part B. Use ICT Worksheet 'Tessellations' on page 153 of the ICT Support section of the Teacher Resource Pack. You may have used this in Chapter 12.

Links Link this topic to coordinate pairs (Pupil Book page 261), properties of triangles, quadrilaterals and polygons (Pupil Book pages 259 and 287).
Link to Design and Technology: Practical (Pupil Book page 317).

> **Topic area**   **Symmetry**   pupil book page 317

Framework Objectives   ✓ ... identify all the symmetries of 2-D shapes, **Framework examples page 211**

Sheets   Worksheet 31 for support   Homework Sheet 13.3   Starter Resource Sheet 27   Resource Sheets (pages 239 and 227 to 231 Teacher Resource Pack)

Resources   Pinboard or square dotty paper for the Starter Activity. Tracing paper. A 3×3 pinboard or square dotty paper. A 5×5 pinboard or square dotty paper. Squared paper.

 Starter ideas and activities

Mental starter ideas
- Ask questions such as Practice Questions 5, 6, 13, and 39 on pages 264, 266 and 270 in Shape, Space and Measures Support chapter in the Pupil Book. Use Starter Resource Sheet 27.
- Revise the names of the special triangles and quadrilaterals. You could do this by playing 'What is my shape?'. Give pupils a description of a shape such as 'it has 3 sides, all equal' or 'it has 3 sides and 2 equal angles', etc.
  What is my shape?

Starter activity
- **Reflection and rotation symmetry**
  Use Practical (Pupil Book part A page 318). You could divide the class into groups with one pinboard per group or work with the whole class and use square dotty paper.
  Ask each group to show the shape with the most symmetry properties.

 Teaching points and activities

- Encourage pupils to visualise rotating the shape or visualise folding it to see how many times one half fits exactly on top of the other.
- Pupils may need copies of the shapes for Exercise 3 question 2. You could use resource sheets on pages 227 to 231 of the Teacher Resource Pack. The special quadrilateral that causes most confusion is usually the parallelogram. Encourage pupils to use a copy of the shape so they can turn and fold it. Or they could use a mirror and tracing paper.
- It can be confusing for pupils that shapes with order of rotation symmetry 1 do not have rotation symmetry. This will need to be stressed. Also, when finding the order of rotation symmetry, stress that the starting and ending position must NOT both be counted.

Support
- See **Symmetry** on page 262 (Shape, Space and Measures Support chapter of the Pupil Book).
- If pupils have difficulty with Practice questions 5, 6, 13, 17 and 39 use Worksheet 31 (page 27 of the Teacher Resource Pack).

Extension
- See **Practical** (Pupil Book page 318). Challenge pupils to see who can find the most answers.
- See **Investigation: Pentominoes** (Pupil Book page 319) starred section.

Plenary ideas
- You could ask pupils to draw the following shapes on the board:
  a shape with 1 or 2 or 3 lines of symmetry
  a shape with order of rotation symmetry 1 or 2 or 3 or 4
- Discuss the answer to Exercise 3 question 2 and/or 3. Question 3 would be a good questions to ask pupils to demonstrate how they got their answers.
- Start the Investigation on page 319 and ask pupils to finish it for homework. Discuss strategies for checking that all twelve pentominoes have been drawn and there are none the same.
- Point out the links, especially the link to properties of 2-D shapes.

Homework   Review questions at the end of Exercise 3 (Pupil Book page 318) and/or Homework Sheet 13.3 (page 244 of this pack).

Links   Link this topic to properties of 2-D shapes (Pupil Book pages 259 and 287), transformations (Pupil Book pages 261 and 314), Design and Technology, and Art.

┌───┐
│ **Topic area**   **Enlargement**   pupil book page 320 │
└───┘

Framework Objectives   ✓ Understand and use the language and notation associated with enlargement; **enlarge 2-D shapes, given a centre of enlargement and a positive whole number scale factor;** explore enlargement using ICT, **Framework examples pages 213 and 215**

Sheets   Homework Sheet 13.4   Starter Resource Sheet 28   Resource Sheet pages 240–242 Teacher Resource Pack   ICT Worksheet 'Enlargement' (Teacher Resource Pack page 158).

Resources   Grid paper on an OHP. Squared paper. 1 cm square dotty paper. A dynamic geometry software package.

 ## Starter ideas and activities

Mental starter ideas   Revise some number mental strategies appropriate for your class.

Starter activity   ● **Introducing enlargement**
Use the Discussion on page 320 of the Pupil Book.
● **Scale factor of enlargement**
Use Starter Resource Sheet 28.
How many times longer is this length on the larger tree than on the smaller tree? (2)
What about this length? This one?
Is it the same for all the lengths?
Discuss the words scale factor.
How do we draw an enlargement?
Use Starter Resource Sheet 28.
Demonstrate how to draw A′ and then ask pupils how to find B′ and so on.
What is $\frac{OA'}{OA}$? *(2) What about* $\frac{OB'}{OB}$? *What about ...*
What would these ratios be if the scale factor was 3?

 ## Teaching points and activities

● When drawing an enlargement on a grid it is easier to count squares than to measure. Show pupils how to do this but emphasise that the ratio of distances from the centre of enlargement of corresponding points is always the same as the scale factor.
● You could point out that we can have a scale factor of 1. Ask what the enlargement would look like.
● Emphasise that the centre of enlargement does not change when we enlarge a shape.

Support   ● Exercise 4 on page 320 of the Pupil Book should be done easily by most pupils.

Extension   ● See **starred questions** in Exercise 5 (Pupil Book page 323).
● See **Practical** (Pupil Book page 322 parts B and C). See the ICT Support section of the Teacher Resource Pack (page 158).

Plenary ideas   ● Recap on what an enlargement is and ask questions such as:
What is the scale factor?
What is the ratio of corresponding lengths on the enlargement and the original equal to? (The scale factor.)
Which of these remains the same when we enlarge a shape? Lengths. Angles. Area.
What happens to the centre of enlargement when we enlarge a shape?
● Ask for some examples of enlargement in everyday life.
● Discuss the answers to Exercise 5 question 4 or the results of the practical on page 322.

Homework   Review questions at the end of Exercises 4 and 5 (Pupil Book pages 320 and 323) and/or Homework Sheet 13.4 Exercises 1 and 2 (pages 244 and 245 of this pack).

ICT   See Practical (Pupil Book page 322). Use ICT Worksheet 'Enlargement' on page 158 of the Teacher Resource Pack.

Links   Link this topic to coordinates (Pupil Book page 261), ratio and proportion (Pupil Book page 137) and making scale drawings (Pupil Book page 324).
Link to Art and Photography: Discussion (Pupil Book page 320).
Link to 2-D shapes (Pupil Book page 287).

| Topic area | Scale drawing | pupil book page 324 |

Framework Objectives ✓ Make simple scale drawings, **Framework examples page 217**

Sheets Homework Sheet 13.5   Starter Resource Sheet 29

Resources Ruler, metre ruler, tape measure and scissors.

Starter ideas and activities

Mental starter ideas
Ask some questions such as:
A drawing is made so that each centimetre on it represents 4 m in real life. On the drawing a bus is 3 cm long. How long is the bus in real life?
What about a tree 5 cm high on the drawing?
What about a house 2·5 cm high on the drawing?
A map has a scale of 1 cm represents 1000 cm. On the map a road is 5 cm long. How long is this in real life, in metres? in kilometres?

Starter activity
● **Estimating from known distances**
Use Starter Resource Sheet 29.
  In real life a man is about 1·8 m high. How could we use this to work out the height of the child/tree/garage/light?/

● **Scale drawing**
1 Ask for a volunteer to come to the front of the class.
2 Measure his or her height in cm and round the answer to the nearest 10 cm (for ease of mental calculation for the scaling).
3 Put this height on the board.
4 Now draw a scale drawing of this person using the scale 1 cm represents 10 cm.
So if the pupil you measured was 150 cm tall then draw a stick person 15 cm tall. Tell the class how high you have drawn the person and ask 'What scale did I use?'
5 Ask for a volunteer to come and draw the person using one of these scales.
  1 cm represents 5 cm   1 cm represents 4 cm   1 cm represents 20 cm
  What scale has ... used?
6 You could repeat this a few times.

● **Making a scale drawing**
Ask pupils to measure the classroom in metres. Make a sketch of the room with the real life measurements on it. Say:
  I want to make a scale drawing of this room. What scale should I use?
Discuss suitable scales and then make a scale drawing of the room.

Teaching points and activities

● Relate scale drawings to scale factors in enlargement and also to ratios. Emphasise that when expressing scales as a ratio, the units must be the same.
● Sometimes pupils have trouble with the units, for example, if the scale on a drawing is 1 cm represents 4 m, then some pupils will multiply by 4 and then put cm in the answer.
● Pupils sometimes find word problems, such as Exercise 6 questions 3 to 9, hard to interpret. Encourage pupils to make a sketch or write sentences such as this for question 3:
  1 cm represents 400 cm
so 2 cm represents 2 × 400 = 800 cm or 8 m
  3 cm represents 3 × 400 = 1200 cm or 12 m
● Finding a suitable scale challenges some pupils. Encourage them to keep dividing the largest real life measurement by whole number amounts such as 2, 5, 10, 100, etc. until they find an answer that is about how large they want the length to be on the drawing.

Support
● Pupils should have covered the section on enlargement.

Extension
● See **starred questions** in Exercise 6 (Pupil Book page 325).
● See **Practical** on page 327 of the Pupil Book. Challenge pupils to write up their work and include accurate scale drawings.

Plenary ideas
● Discuss the ways of representing scales on maps. Especially discuss the meaning of these scales:
  1 cm represents 2 m   1 : 200
● Ask pupils to share their scale drawing from the practical on page 327 and explain why they chose the scale they did. Ask other pupils to comment.

Homework   Review questions at the end of Exercises 6 and 7 (Pupil Book pages 325 and 328) and/or Homework Sheet 13.5 Exercise 1 and 2 (page 246 of this pack).

Links   Link this topic to ratio and proportion (Pupil Book page 136), enlargement (Pupil Book page 320) and measurements (Pupil Book page 334).
Link to Geography: Exercise 6 questions 5, 7.
Link to Science: Exercise 6 Review 1.
Link to Design and Technology: Exercise 6 questions 3, 6, 8, 9, Review 3; Practical 2 on page 327; most other parts of the whole section.

| Topic area | Finding the mid-point of a line | pupil book page 328 |

Framework Objectives ✓ Given the coordinates of points A and B, find the mid-point of the line segment AB, **Framework examples page 219**

Sheets Homework Sheet 13.6

Resources A prepared OHP for the Starter Activity.

 ## Starter ideas and activities

Mental starter ideas Revise some number objectives that are appropriate for your class.

Starter activity
- **Finding the mid-point of a horizontal or vertical line**
  Use the Discussion on page 328 of the Pupil Book. You may like to put the diagrams on an OHP.
- **Finding the mid-point of a line**
  After Exercise 8 use the Discussion on page 329 of the Pupil Book to introduce the formula for finding the mid-point of a line. You may want to put the diagram on an OHP.

 ## Teaching points and activities

- Some pupils find understanding this a challenge. Use diagrams and lots of examples until understanding takes place. Encourage understanding before learning the formula off by heart. Some pupils may need considerable help in understanding the notation that is used.
- Encourage pupils to see the formula as finding the mean of the coordinates. Recap finding the mean of two values.

Support
- Pupils should have covered the *Coordinates and transformations* section.

Extension
- This whole section could be regarded as extension.

Plenary ideas
- Ask pupils to explain how to find the mid-point of a line without using the formula.
- Put this diagram on the board.

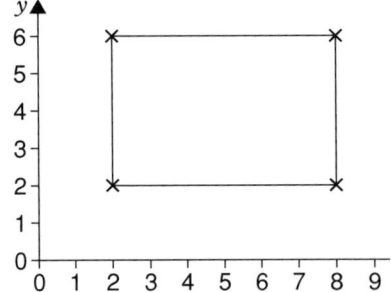

Ask 'How could we find the coordinates of the exact centre of the rectangle?' (By finding the mid-point of one of the diagonals.)
- Discuss the answers to Exercise 9 question 2.

Homework Review questions at the end of Exercises 8 and 9 (Pupil Book pages 329 and 330) and/or Homework Sheet 13.6 (page 247 of this pack).

Links Link this topic to coordinates (Pupil Book page 313), mean (Pupil Book page 383).

14 Mesures, Perimeter, Area and Volume

Topic in Chapter	Links to Programme of Study *Ma2 a–l are integrated throughout.*	Links to Framework for Teaching Mathematics: 8 Teaching Programme *Applying mathematics and solving problems are integrated throughout.*
Metric conversions *page 335* Metric and imperial equivalents *page 339* Units, measuring instruments and accuracy *page 340* Estimating *page 343*	**Ma3 4a** ... convert measurements from one unit to another; know rough metric equivalents of pounds, feet, miles, pints and gallons; make sensible estimates of a range of measures in everyday settings **Ma3 4i** convert between area measures ...	**Measures and mensuration** ● Use units of measurement to estimate, calculate and solve problems in everyday contexts involving length, area, ... capacity, mass, ...; know rough metric equivalents of imperial measures in daily use (feet, miles, pounds, pints, gallons). **Applying mathematics and solving problems** ● **Identify the necessary information to solve a problem.**
Bearings *page 344*	**Ma3 4b** understand angle measure, using the associated language (for example, use bearings to specify direction)	**Measures and mensuration** ● Use units of measurement to ... calculate and solve problems in everyday contexts involving ... angles and bearings. ● Use bearings to specify direction.
Perimeter and area *page 348*	**Ma3 4f** find areas of rectangles, recalling the formula, understanding the connection to counting squares and how it extends this approach; recall and use the formulae for the area of a parallelogram and a triangle ... calculate perimeters and areas of shapes made from triangles and rectangles	**Measures and mensuration** ● Use units of measurement to ... calculate and solve problems in everyday contexts involving length, area ... ● **Deduce and use formulae for the area of a triangle, parallelogram** and trapezium; calculate areas of compound shapes made from rectangles and triangles. **Applying mathematics and solving problems** ● Suggest extensions to problems, conjecture and generalise; identify exceptional cases or counter-examples.

Topic in Chapter	Links to Programme of Study	Links to Framework for Teaching Mathematics: 8 Teaching Programme
Surface area and volume *page 355*	**Ma3 3g** find volumes of cuboids, recalling the formula and understanding the connection to counting cubes and how it extends this approach; calculate volumes of ... shapes made from cubes and cuboids	**Measures and mensuration** ● Use units of measurements to ... calculate and solve problems in everyday contexts involving ... area, volume ... ● **Know and use the formula for the volume of a cuboid; calculate volumes and surface areas of cuboids and shapes made from cuboids.** **Applying mathematics and solving problems** ● Solve more demanding problems and investigate in a range of contexts ... ● **Use logical argument to establish the truth of a statement.**

Links to Attainment Targets

Level 4   They choose and use appropriate units and instruments, interpreting with appropriate accuracy, numbers on a range of measuring instruments.

Level 5   They know the rough metric equivalents of imperial units still in daily use and convert one metric unit to another. They make sensible estimates of a range of measures in relation to everyday situations. Pupils understand and use the formula for the area of a rectangle.

Level 6   They understand and use appropriate formulae for finding ... areas of rectilinear figures and volumes of cuboids when solving problems.

Level 7   They calculate ... areas and volumes in plane shapes and ... Pupils appreciate the imprecision of measurement ...(begin)

| Topic area | Metric conversions | Pupil Book page 335 |

Framework Objectives

✓ Use units of measurement to estimate, calculate and solve problems in everyday contexts involving length, area, ... capacity, mass ...; know rough metric equivalents of imperial measures in everyday use (feet, miles, pounds, pints, gallons), **Framework examples page 229**

✓ Problem solving, **Framework examples page 21**

Sheets Worksheet 32   Homework Sheet 14.1   Resource Sheet page 243 Teacher Resource Pack

Resources A calculator. You could use or make a measures dominoes pack.

Starter ideas and activities

Mental starter ideas
- Ask questions such as Practice questions 2, 14 and 27 on pages 264 and 266 of the Shape, Space and Measures Support chapter in the Pupil Book.
- Revise multiplication and division by 10, 100 and 1000.

Starter activity
- **Revise metric conversions**
  1 Divide the class in to groups of five or six.
  2 Ask each group to present a way of remembering the conversions given. They could use a mind map, a mnemonic, a riddle or whatever they like.
  3 Ask some groups to present their ideas to the rest of the class.
- **Introducing other metric units**
  1 Discuss tonnes and what might be measured in tonnes. Give 1 tonne = 1000 kg.
  2 Repeat for 1 hectare = 10 000 m^2
  3 Ask for volunteers to ask questions such as:
    An elephant has a mass of 7 tonnes, how many kg is this?
    A park is 4·6 ha, how many m^2 is this?
    The pupil who answers could ask the next question.
  4 Give the conversions: 1 ℓ = 1000 cm^3 1 mℓ = 1 cm^3 1000 ℓ = 1 m^3
    Ask some questions round the class. You could use Exercise 2 question 1.
    For what sorts of things would we give the volume in cm^3 rather than mℓ or litres?
    For what sorts of things would we give the volume in mℓ or litres rather than cm^3.

Teaching points and activities

- Some pupils get confused whether to multiply or divide when converting between units. Encourage them to look at the answer to see if it is sensible, e.g. if a pupil converts 45 m to cm by dividing and gets 0·45 cm, ask them if 45 m could possibly be the same as 0·45 cm. It may also help to point out that you will always get more of the smaller unit so to change a larger unit, such as litres, to a smaller unit such as millilitres we must multiply to get more millilitres than litres.
- Some pupils find the new unit conversions for capacity hard to remember. You may need to keep them on the board or OHP to begin with. Ask some volunteers to come and explain how they might remember them. Encourage pupils to think of the sizes of the cube containers with sides 1 cm, 10 cm and 1 m and then think how much water each could hold.
- If pupils are finding word problems difficult, it may help to sketch a diagram.

Support
- See **Measures** on page 262 (Shape, Space and Measures Support chapter of the Pupil Book). If pupils have difficulty with Practice questions 2, 14, 27, 38a and 38b use Worksheet 32 questions 2, 3, 5 and 9a, b, and c (page 28 of the Teacher Resource Pack).

Extension
- See the Starter **Makeover** (Pupil Book page 334). Challenge pupils to write a brief report on this.
- See **starred questions** in Exercise 1 (Pupil Book page 335).
- See **Puzzle** (Pupil Book page 338).

Plenary ideas
- Discuss the answers to Exercise 2 questions 4, 5 and 6.

Homework Review questions at the end of Exercises 1 and 2 (Pupil Book pages 335 and 337) and/or Homework Sheet 14.1 Exercises 1 and 2 (page 248 of this pack).

Links Link this topic to multiplying and dividing by 10, 100 and 1000 (Pupil Book page 19).
Link to area: Exercise 1 questions 3, 8, 9, 10, 13, 14, 19, Review 3.

Topic area **Metric and imperial equivalents**   pupil book page 339

Framework Objectives ✓ Use units of measurement to estimate, calculate and solve problems in everyday contexts involving length, area, ... capacity, mass ...; know rough metric equivalents of imperial measures in everyday use (feet, miles, pounds, pints, gallons), **Framework examples page 229**

Sheets Worksheet 32 for support   Homework Sheet 14.2

Resources A calculator.

 ## Starter ideas and activities

Mental starter ideas Revise some of the metric/imperial equivalents that pupils know.
Ask questions such as Practice question 34 on page 269 of the Shape, Space and Measures Support chapter in the Pupil Book.

Starter activity
- **Imperial units still in use**
  Discuss where imperial units are still used in Britain.
- **Introducing metric and imperial equivalents**
  Discuss the rough metric equivalents given on page 339 of the Pupil Book:
  1 pound is a bit less than $\frac{1}{2}$ kg. If we use this to convert 6 lbs to kg as 'a bit less than 3 kg and then convert 3 kg into lbs we get 6·6 lbs. Why didn't we get back to 6 lb?
  Explain that we could say that 1 lb is about 0·45 kg and this would give us a closer estimate, but we try and choose a rough equivalent that is easy to work with.
  Use the metric/imperial equivalents to convert 5 pints to mℓ and then convert this back to pints.
  Do you get back to 5 pints? Why not? (Similar reason to above.)
  You could do some of the questions from the exercise as a whole class; say, one part of each of Exercise 3 questions 1 to 7.

 ## Teaching points and activities

- Revise what ≈ means.
- Emphasise that these are only rough equivalents. The answers will be approximate.

Support
- See **Measures** on page 263 (Shape, Space and Measures Support chapter of the Pupil Book). If pupils have difficulty with Practice questions 34, 38c and 38d use Worksheet 32 question 8 and 9d (page 28 of the Teacher Resource Pack).

Plenary ideas
- Discuss situations where imperial units are still commonly used.
  Discuss where you might need to do a conversion between metric and imperial.
- Discuss the answers to Exercise 3 questions 9 to 13.
- Discuss the Exercise 3 review question. You could put it onto an OHP and ask pupils to come and give the answers explaining how they did it.

Homework Review questions at the end of Exercise 3 (Pupil Book page 339) and/or Homework Sheet 14.2 (page 249 of this pack).

Links Link to Geography: Exercise 3 question 7.
Link to Design and Technololgy: Exercise 3 questions 11 and 13.

| Topic area | Units, measuring instruments and accuracy | pupil book page 340 |
| | Estimating | pupil book page 343 |

Framework Objectives
✓ Use units of measurement to estimate, calculate and solve problems in everyday contexts involving length, area, ... capacity, mass ...; know rough metric equivalents of imperial measures in everyday use (feet, miles, pounds, pints, gallons), **Framework examples page 231**
✓ Problem solving, **Framework examples page 21**

Sheets Worksheet 32   Homework Sheets 14.3 and 14.4

Resources Equipment needed to make a clock. Tape measure and chalk.

 Starter ideas and activities

Starter activity
● **Discuss units, measuring instruments and accuracy**
Use the Discussion on page 340 of the Pupil Book.

● **Using benchmarks to help when estimating**
Use the Discussion on page 343 of the Pupil Book.
Alternatively, take a walk with the class around the school grounds and ask pupils to estimate lengths, masses, capacities, areas and volumes of some things that you already know the answers to.

 Teaching points and activities

● Stress to pupils that in everyday life it is sometimes very important to use accurate measurements, but other times an estimate will do.
● You could use Exercise 4 as a discussion or group exercise.
● Some pupils have more difficulty estimating measurements of some things than others. Encourage them to use 'benchmarking', e.g. if they don't know the mass of a bicycle they could think about how many 10 kg bags of potatoes would be about the same mass. Encourage them to develop as many benchmarks as possible.
● Giving a range is not an exact science. There are a lot of correct answers. Ask pupils to explain how they arrived at the answer they gave. Encourage pupils to think about a particular example in each part of question 1. For question 1 part 1 they could think about the step on a particular staircase rather than give the range for all steps on staircases. The range for all would be very large because it might include things such as doll's house steps right up to steps on a staircase with very large steps.

Support
● See **Estimating** under **Measures** on page 263 (Shape, Space and Measures Support chapter of the Pupil Book). If pupils have difficulty with Practice questions 1, 16 and 32 use Worksheet 32 questions 4 and 7 (page 28 of the Teacher Resource Pack).

Extension
● See **Practical** (Pupil Book page 342). Challenge pupils to write a report giving the method, diagrams, any problems encountered and how they overcame them and any alternative methods that could make a more accurate clock.

Plenary ideas
● Discuss the answers to as many Exercise 4 questions as possible.
● Ask pupils to explain how they did Practical part 1 on page 342 of the Pupil Book if you used this as part of the lesson.
● Discuss what pupils used as benchmarks to answer Exercise 5 question 1. Discuss the range given.
● Discuss and do Exercise 5 question 3.
● Ask pupils to explain how they did one of the parts of the Practical on page 344 of the Pupil Book if you used this as part of a lesson.

Homework Review questions at the end of Exercises 4 and 5 (Pupil Book pages 341 and 343) and/or Homework Sheet 14.3 and 14.4 (page 249 of this pack).

Links Link this topic to Science and other subject areas such as History (the water clock, practical Pupil Book page 342) and rounding (Pupil Book pages 28–30).
Link to Design and Technology: Discussion (Pupil Book page 340).
Link to Science, Geography, Design and Technology, Art, PE.

| Topic area | Bearings | pupil book page 344 |

Framework Objectives
✓ Use units of measurement to ... calculate and solve problems in everyday contexts involving ... angles and bearings, **Framework examples page 233**
✓ Use bearings to specify direction, **Framework examples page 233**.
✓ Problem solving, **Framework examples page 21**

Sheets Worksheet 34 for support   Homework Sheet 14.5   Resource Sheets (pages 243 and 244 Teacher Resource Pack)

Resources A protractor for the Starter Activity. A ruler.

Starter ideas and activities

Mental starter ideas Revise angles. Ask questions such as:
How many degrees are there in a full turn? a straight line?
If I turn through 143°, how many more degrees must I turn through to turn a full turn?
What do angles on a straight line add to?
What do you know about corresponding angles on parallel lines?

Starter activity
● **Introducing bearings**
1 Ask for two volunteers to come and stand a distance apart and not lined up.
2 Draw a North line on the board next to the pupil closest to it.
  How could I describe where pupil A is standing with reference to pupil B? (You may need to prompt to get that an angle from North could be used.)
3 *We call this a bearing. What sorts of people use bearings?*
4 Because it is hard to measure the angle here draw a sketch on the OHP.
  How could I measure the angle from North at pupil A to pupil B?
  Use a protractor to measure the angle, pointing out that we always measure in a clockwise direction from North.
5 *What if I wanted the bearing of pupil A from Pupil B?*
  Draw on the sketch the North line at pupil B. Ask someone to demonstrate how to find the bearing.
6 Tell pupils that if you know the bearing of B from A you can find the bearing of A from B without measuring.
● **Finding the bearing of B from A when you know the bearing of A from B**
Use the Discussion on page 345.

Teaching points and activities

● It is important to explain that when the bearing of A **from B** is wanted, that the word **from** indicates where the North line is drawn.
● In Exercise 6 question 2 you may need to warn pupils to note the direction the planes are flying.

Support
● Pupils should have covered **Angles** on page 258 of the Pupil Book.
● Also see **Compass directions** on page 264 (Shape, Space and Measures Support chapter of the Pupil Book). If pupils have difficulty with Practice question 35 use Worksheet 33 (page 29 of the Teacher Resource Pack).

Extension
● See **starred questions** in Exercise 6 (Pupil Book page 346).
● See **Practical** (Pupil Book page 348).

Plenary ideas
● Discuss the answers to Exercise 6 question 2 or 3.
● Discuss Exercise 6 question 7 or 8 to extend thinking.
● Make sure you have gone over examples in which the bearing is an acute angle, an obtuse angle and a reflex angle.

Homework Review questions at the end of Exercise 6 (Pupil Book page 346) and/or Homework Sheet 14.5 (page 250 of this pack).

Links Link this topic to angles and lines (Pupil Book page 271) and scale drawings (Pupil Book page 324).
Link to Geography throughout the section.

Topic area	Perimeter and area	pupil book page 348

Framework Objectives
✓ Use units of measurement to ... calculate and solve problems in everyday contexts involving length, area ..., **Framework examples pages 235 and 237**
✓ **Deduce and use formulae for the area of a triangle, parallelogram** and trapezium; calculate areas of compound shapes made from rectangles and triangles, **Framework examples pages 235 and 237**
✓ Problem solving, **Framework examples page 19**

Sheets Worksheet 33 Homework Sheet 14.6

Resources Copies of the shapes in the Discussion. 1 cm squared paper. A pinboard or square dotty paper for the Investigation: Moving vertices. A map for the Practical.

 ## Starter ideas and activities

Mental starter ideas
- Revise the meanings of the words area and perimeter.
  Ask questions such as Practice questions 25 and 26 on pages 68 in the Shape, Space and Measure Support chapter of the Pupil Book. You will need to draw the diagrams on the board or OHP.
- Revise the units for perimeter and area by asking questions such as:
  Which of these units would be used for the perimeter of a netball court?
  cm m m^2 m^3 km
  Which of these units would be used for the area of a netball court?
  cm^3 m^2 km^2 cm m

Starter activity
- **Introducing perimeter of a non-right-angled triangle, parallelogram and trapezium**
  Use the Discussion on page 348 of the Pupil Book.

 ## Teaching points and activities

- Encourage pupils to understand how the formulae have been derived. This helps them to remember them. (You could use the Discussion on page 348.) It could be useful if you had copies of the shapes.
- Emphasise the need to be careful with units. Every answer should have a unit.
- Emphasise for triangles, parallelograms and trapeziums that h must be the perpendicular height. Refer back to the discussion about deriving the formulae.
- Pupils may need reminding of the order of operations when using the formula for the area of a trapezium.
- For some questions pupils may have to solve an equation. (When the unknown is not the subject — Exercise 7 questions 8, 9 and 10.) You may need to revise this.

Support
- See **Perimeter, area and surface area** on page 263 (Shape, Space and Measures Support chapter of the Pupil Book). If pupils have difficulty with Practice questions 25, 26 and 44 use Worksheet 34 questions 1, 2, 3, 4, 5, 7 and 8 (page 30 of the Teacher Resource Pack).

Extension
- See **starred questions** in Exercise 7 (Pupil Book page 350).
- See **Investigations: Moving vertices, question 3; Hexagons; and Triangles in Cubes** (Pupil Book page 353).

Plenary ideas
- Ask pupils to come and show how the formula for either the triangle, parallelogram or trapezium is derived. You may need to have the diagrams from the Discussion available.
- Discuss the answer to Exercise 7 question 5.
- Discuss the answer to Exercise 7 questions 8, 9 and 10 to extend thinking.
- Start either part of the Investigation on page 353 and ask pupils to finish it for homework.

Homework Review questions at the end of Exercises 7 and 8 (Pupil Book pages 350 and 354) and/or Homework Sheet 14.6 Exercises 1 and 2 (page 251 of this pack).

Links Link this topic to formulae (Pupil Book page 175).
Link to Art and Design and Technology: Exercise 7 question 8, Review 1.
Link to Properties of 2-D Shapes; Investigations: Moving vertices, Hexagons, Triangles in cubes (page 353).

| Topic area | Surface area and volume | pupil book page 355 |

Framework Objectives

✓ Use units of measurement to ... calculate and solve problems in everyday contexts involving ... area, volume ..., **Framework examples pages 239 and 241**

✓ **Know and use the formula for the volume of a cuboid; calculate volumes and surface areas of cuboids** and shapes made from cuboids, **Framework examples pages 239 and 241**

✓ Problem solving, **Framework examples pages 19, 29 and 31**

Sheets Worksheet 33 for Support   Homework Sheet 14.7

Resources Centicubes, measuring cylinder and cuboid-shaped objects for the Practical (Pupil Book page 355). 1 cm squared paper and thin card for the Investigation. A calculator. A box that can be opened out.

Starter ideas and activities

Mental starter ideas Revise what 'surface area' means and relate it to the net of a 3-D shape.
Ask questions such as Practice question 42 on page 270 in the Shape, Space and Measure Support chapter of the Pupil Book.

Starter activity
● **Introducing the volume of a cuboid**
Use the Discussion on page 355 of the Pupil Book.
You could use centimetre cubes to make models.
You could ask for volunteers to come and make models (cuboids) and ask the rest of the class to find the volume until they can generalise the formula.

Teaching points and activities

● Illustrate the idea of surface area by taking a box and opening it out. Thus show them that the total area is the sum of the areas of the faces.

● Some pupils get mixed up between the units for area and the units for volume. Point out that for volume they are multiplying three measurements together and so the units are units^3.

● You could extend pupils in Exercise 9 question 6 to see that the volume could be found by using 'volume of a prism = area of the end face × the length of prism.'

Support
● See **Perimeter, Area and Surface Area** in the Shape, Space and Measures Support chapter (Pupil Book page 263). If pupils have difficulty with Practice questions 42 and 45 use Worksheet 34 question 6 (page 30 of the Teacher Resource Pack).

Extension
● See **starred questions** in Exercise 9 (Pupil Book page 356).
● See **Investigation: Making boxes** (Pupil Book page 358). Challenge pupils to investigate nets for boxes with other dimensions.

Plenary ideas
● Give pupils the volume of a cuboid and ask them to find possible dimensions, e.g. 'The volume of a cuboid is 48 cm^3. What might the length, width and height be? What if the cuboid had volume 72 m^3?'
Ask whether it is possible to have a cube with whole-number lengths for the sides and volume 48 m^3, 64 m^3, 121 m^3, 125 m^3?

● Discuss the answers to Exercise 9 question 4.

● Start the Investigation on page 358 of the Pupil Book and ask pupils to finish it for homework. You could give pupils some 1 cm squared paper to help.

Homework Review questions at the end of Exercise 9 (Pupil Book page 356) and/or Homework Sheet 14.7 (page 252 of this pack).

Links Link this topic to formulae (Pupil Book page 175), finding perimeter and area (Pupil Book page 348), lines and angles (Pupil Book page 271), shapes (Pupil Book page 287).
Link to nets: Investigation: Making boxes (Pupil Book page 358).

15 Data Collection

Topic in Chapter	Links to Programme of Study *Ma2 a–1 are integrated throughout.*	Links to Framework for Teaching Mathematics: 8 Teaching Programme *Applying mathematics and solving problems are integrated throughout.*
Discrete and continuous data *page 372* Grouping continuous data *page 373*	**Ma4 3a** design and use data-collection sheets for grouped discrete and continuous data …	**Specifying a problem, planning and collecting data** ● Plan how to collect the data, …; construct frequency tables with given equal class intervals for sets of continuous data; design and use two-way tables for discrete data.
Two-way tables *page 374*	**Ma4 3c** design and use two-way tables for discrete and grouped data	
Surveys — the problem or question *page 376*	**Ma4 1a** (i) specify the problem and plan: formulate questions in terms of the data needed, and consider what inferences can be drawn from the data; decide what data to collect … **Ma4 1a** (ii) collect data from a variety of suitable sources, including experiments and surveys, and primary and secondary sources **Ma4 1b** identify what further information is required to pursue a particular line of enquiry **Ma4 1e** interpret, discuss and synthesise information … **Ma4 1f** communicate mathematically, making use of diagrams and related explanatory text **Ma4 1g** examine critically, and justify, their choice of mathematical presentation of problems involving data **Ma4 2b** identify questions that can be addressed by statistical methods **Ma4 2d** identify which primary data they need to collect and in what form …	**Specifying a problem, planning and collecting data** ● Discuss a problem that can be addressed by statistical methods and identify related questions to explore. ● Decide which data to collect to answer a question …; identify possible sources. ● Plan how to collect the data, including sample size; … ● Collect data using a suitable method, such as observation, controlled experiment, including data logging using ICT or questionnaire.
Collecting the data *page 377* Planning a survey *page 378*	**Ma4 3a** design and use data-collection sheets for grouped discrete and continuous data; collect data using various methods including observation, controlled experiment, data logging, questionnaires and surveys **Ma4 3b** gather data from secondary sources, including printed tables and lists from ICT-based sources	

Topic area	Discrete and continuous data	pupil book page 372
	Grouping continuous data	pupil book page 373

Framework Objectives ✓ ... construct frequency tables with given equal class intervals for sets of continuous data; design and use two-way tables for discrete data ... **Framework examples page 253**

Sheets Worksheet 37   Homework Sheet 15.1 and 15.2   Starter Resource Sheet 30   Resource Sheet pages 245 and 246 Teacher Resource Pack.

 Starter ideas and activities

Mental starter ideas Revise tally marks by giving some examples and asking what they stand for.
Revise class intervals by putting some unequal ones on the board and asking what is wrong with them.

Starter activity
- **What is the difference?**
  Use the Discussion on page 372 of the Pupil Book.
- **The difference between grouping discrete and grouping continuous data**
  Give the examples in the screen on page 373 of the Pupil Book (Starter Resource Sheet 30). Then use the Discussion on page 373 of the Pupil Book.
  Give pupils some other cross-country times and ask them in which class interval they should be put.
- **Grouping some continuous data from the class**
  Ask everyone in the class to measure their hand span.
    Is this discrete or continuous data?
    How could we group this data?
  Draw a table up on the board and then ask each person to come and call out their hand span and put it on the table. Ask the rest of the class to watch for mistakes.

 Teaching points and activities

- Emphasise that discrete data is usually counted and continuous data is usually measured.
- Some pupils get confused by data that has fractions but is discrete, e.g. shoe sizes can have halves but it is discrete data because it can only have particular values.
- Some pupils have trouble understanding $6 \leqslant x < 7$ and $6–$.
  Emphasise that class intervals must not overlap and so the values must go from 6 right up to as close as you can get to 7 without including 7 so that the next class interval can start at 7. Emphasise also that both $6–$ and $6 \leqslant \cdots < 7$ are the same.
- There is only one question in the exercise where pupils have to choose suitable class intervals for data but when they do surveys in a later chapter they will need to be able to do this. Some pupils may need guidance with this question and you could do it as part of the plenary at the end. Tell pupils that we usually have between 5 and 10 class intervals. Draw out that this will depend on how much data there is and how spread it is. Too many or too little makes it difficult to see the trends in how the data is distributed.
- The points at the ends of class intervals cause some pupils trouble. Encourage them to interpret what the class intervals mean and where the end points lie before starting to fill in the tally chart. For example pupils must decide that 160 is in the class interval $150 < m \leqslant 160$ and not the next class interval because we read this as m is greater than 150 but less than **or equal** to 160.

Support
- See **Grouped data** on page 363 (Handling Data Support chapter in the Pupil Book).
- If pupils have difficulty with Practice question 22 use Worksheet 37 (page 34 of the Teachers Resource Pack).

Plenary ideas
- Ask pupils to give you three examples of continuous data and three examples of discrete data. Discuss the examples given.
- Discuss the answers to Exercise 1 question 1.
- Discuss the answers to Exercise 2 question 6, in particular the class intervals to choose.

Homework Review questions at the end of Exercises 1 and 2 (Pupil Book page 372) and/or Homework Sheet 15.1 and 15.2 (page 253 of this pack).

Links Link to measurement: Discussion (Pupil Book page 372); parts of Exercise 1.
Link to measurement: Exercise 2 question 1, 4, Review.
Link to PE: Discussion (Pupil Book page 373); Exercise 2 questions 2, 3.

| Topic area | **Two-way tables** | pupil book page 374 |

Framework Objectives ✓ design and use two-way tables for discrete data, **Framework examples page 255**

Sheets Worksheet 36 for support   Homework Sheet 15.3

Starter ideas and activities

Mental starter ideas Ask questions such as Practice question 12 on page 367 of the Handling Data Support chapter.

Starter activity **Drawing a two-way table**
1 Give the class these instructions:
  If you are a boy stand over here and if you are a girl stand over there.
  Boys divide yourselves into left-handed and right-handed and girls do the same.
2 You should have four groups of people now.
  How could we show the numbers in these groups in a table? (You could get them to stand in four sections like a two-way table.)
  From the table how could we find the total number of boys/girls/left-handed people/right-handed people?

Teaching points and activities

- Emphasise that the name next to each row or column tells you what the **total** of that row stands for. The individual boxes tell you the number of ... with **both** attributes that relate to that box.
- Point out that two-way tables are a very easy way to compare data.

Support
- If pupils have difficulty with question 12 on page 367 of the Handling Data Support chapter use Worksheet 36 question 6 (pages 31 to 34 of the Teacher Resource Pack).

Extension
- See **starred questions** in Exercise 3 (Pupil Book page 375).

Plenary ideas
- Discuss Exercise 3 question 5 to extend thinking.

Homework Review questions at the end of Exercise 3 (Pupil Book page 375) and/or Homework Sheet 15.3 (page 254 of this pack).

Links Link to measurement: Exercise 3 question 5.
Link to PE: Exercise 3 Review 1.

Topic areas	Surveys — the problem or question	pupil book page 376
	Collecting the data	pupil book page 377
	Planning a survey	pupil book page 378

Framework Objectives

✓ Discuss a problem that can be addressed by statistical methods and identify related questions to explore, **Framework examples page 249**

✓ Decide which data to collect to answer a question ...; identify possible sources, **Framework examples page 251**

✓ Plan how to collect the data, including sample size; ..., **Framework examples page 253**

✓ Collect data using a suitable method, such as observation, controlled experiment, including data logging using ICT or questionnaire, **Framework examples pages 253 and 255**

Sheets   Worksheet 35 for support   Homework Sheet 15.4

Resources   Reference material for collecting data.

Starter ideas and activities

Mental starter ideas   Ask some questions such as Practice question 16 on page 368 of the Handling Data Support chapter.

Starter activity
- **Discuss the use of survey results**
  Use **At the Polls** on page 371 of the Pupil Book.
- **Collecting the data**
  Use Discussions on pages 376 and 377 (both of them).

Teaching points and activities

- Emphasise that a sample must be representative of the whole population that a survey relates to.
- A lot of the topics in this chapter are heading towards teaching pupils to prepare their own survey. It may be more useful to ask pupils to choose a survey topic and then cover the relevant sections as required. It is important that pupils get an overall view of the process of planning and carrying out a survey. The cycle given in the Framework, section 1 page 18, is useful. This is also given as a note in the Practical on page 408 of the Pupil Book.
  When pupils do the practical on pupil book page 378, make sure they keep this plan somewhere safe so that they could use it as the basis of the survey they do in the next chapter.

Support
- See **Planning and collecting data** on page 362 (Handling Data Support chapter of the Pupil Book). If pupils have difficulty with Practice question 16 try Worksheet 35 question 3 (page 31 of the Teacher Resource Pack).

Extension
- There is nothing specific but this section will naturally lend itself to extension through the Discussions and Exercise 4.

Plenary ideas
- Ask some pupils to come and show the survey plans they made. Discuss these.

Homework   Review questions at the end of Exercise 4 (Pupil Book page 378) and/or Homework Sheet 15.4 (page 254 of this pack).

ICT   See ICT Support section page 159 in the Teacher Resource Pack for some suggestions of Internet sources for planning a survey.

Links   Link to Science: Discussion (Pupil Book page 376) numbers 2 and 3; parts of Exercise 4 question 1; Practical (Pupil Book page 378).
Link to PE: Discussion (Pupil Book page 376) number 4; parts of Exercise 4 question 1.
Link to Geography: Practical (Pupil Book page 378).

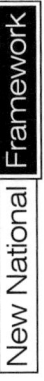

16 Analysing Data. Drawing and Interpreting Graphs

Topic in Chapter	Links to Programme of Study *Ma2 a–l are integrated throughout.*	Links to Framework for Teaching Mathematics: 8 Teaching Programme *Applying mathematics and solving problems are integrated throughout.*
Mode and range *page 382*	**Ma4 4a** draw and produce, using paper and ICT, ... stem-and-leaf diagrams	**Processing and representing data, using ICT as appropriate**
Mean *page 383*	**Ma4 4b** calculate mean, range and median of small data sets with discrete then continuous data; identify the modal class for grouped data	• Calculate statistics, including with a calculator; recognise when it is appropriate to use the range, mean, median and mode and, for grouped data, the modal class; calculate a mean using an assumed mean; construct and use stem-and-leaf diagrams.
Median, mean, mode, range *page 385*		
Finding the median, range and mode from a stem-and-leaf diagram *page 388*	**Ma4 5d** Compare distribution and make inferences using ... measures of average and range	**Interpreting and discussing results** • Compare two distributions using the range and one or more of the mode, median and range.
Comparing data *page 389*		
Compound bar charts and line graphs *page 391*	**Ma4 4a** draw and produce, using paper and ICT, pie charts for categorical data and diagrams for continuous data, including line graphs for time series, scatter graphs, frequency diagrams ...	**Processing and representing data, using ICT as appropriate** • **Construct, on paper and using ICT:** – pie charts for categorical data – bar charts and frequency diagrams for discrete and continuous data; – simple line graphs for time series; – simple scatter graphs; **identify which are most useful in the context of the problem:**
Frequency diagrams *page 395*		
Drawing pie charts *page 397*		
Scatter graphs *page 400*	**Ma4 5f** have a basic understanding of correlation (begin)	
Interpreting graphs and tables *page 402*	**Ma4 1h** apply mathematical reasoning, explaining and justifying inferences and deductions	**Applying mathematics and solving problems** • Solve more demanding problems and investigate in a range of contexts: ... handling data; compare and evaluate solutions.
	Ma4 1i explore connections in mathematics and look for cause and effect when analysing data	• **Use logical argument to establish the truth of a statement;** give solutions to an appropriate degree of accuracy in the context of the problem.
	Ma4 5a relate summarised data to the initial questions	• Suggest extensions to problems, conjecture and generalise, identify exceptional cases or counter-examples.
	Ma4 5b interpret a wide range of graphs and diagrams and draw conclusions	**Interpreting and discussing results**
	Ma4 5c look at data to find patterns and exceptions	• Interpret tables, graphs and diagrams for both discrete and continuous data and draw inferences that relate to the problem being discussed; relate summarised data to the questions being explored.
	Ma4 5d compare distributions and make inferences ...	

Specifying a problem, planning and collecting data

- Discuss a problem that can be addressed by statistical methods and identify related questions to explore.
- Decide which data to collect to answer a question and the degree of accuracy needed; identify possible sources.
- Plan how to collect the data including sample size …
- Collect data using a suitable method, such as observation, controlled experiment, including data logging using ICT, or questionnaire.

Processing and representing data, using ICT as appropriate

- **Construct, on paper and using ICT:**
  - **pie charts for categorical data**
  - **bar charts and frequency diagrams for discrete and continuous data;**
  - **simple line graphs for time series;**
  - **simple scatter graphs;**
  - **Identify which are most useful in the context of the problem.**

Interpreting and discussing results

- Communicate orally and on paper the results of a statistical enquiry and the methods used, using ICT as appropriate; justify the choice of what is presented.

Ma4 1a	carry out each of the four aspects of the handling data cycle to solve problems:
	(i) specify the problem and plan: formulate questions in terms of the data needed, and consider what inferences can be drawn from the data; decide what data to collect (including sample size and data format) and what statistical analysis is needed
	(ii) collect data from a variety of suitable sources, including experiments and surveys, and primary and secondary sources
	(iii) process and represent the data: turn the raw data into usable information that gives insight into the problem
	(iv) interpret and discuss the data: answer the initial question by drawing conclusions from the data
Ma4 1b	identify what further information is required to pursue a particular line of enquiry
Ma4 1c	select and organise the appropriate mathematics and resources to use for a task
Ma4 1d	review progress as they work; check and evaluate solutions
Ma4 1e	interpret, discuss and synthesise information presented in a variety of forms
Ma4 1f	communicate mathematically, making use of diagrams and related explanatory text
Ma4 1g	examine critically, and justify, their choice of mathematical presentation of problems involving data
Ma4 2b	identify questions that can be addressed by statistical methods
Ma4 2d	identify which primary data they need to collect and in what format
Ma4 2e	design an experiment or survey; decide what secondary data to use
Ma4 3a	… collect data using various methods including observation, controlled experiment, data logging, questionnaires and surveys
Ma4 3b	gather data from secondary sources, including printed tables and lists from ICT-based sources

Links to Attainment Targets

Level 4   They understand and use the mode and range to describe sets of data. They group data, where appropriate, in equal class intervals, represent collected data in frequency diagrams and interpret such diagrams. They construct and interpret simple line graphs.

Level 5   Pupils understand and use the mean of discrete data. They compare two simple distributions, using the range and one of the mode, median or mean. They interpret graphs and diagrams, including pie charts, and draw conclusions.

Level 6   They construct and interpret frequency diagrams. They construct pie charts. Pupils draw conclusions from scatter diagrams, and have a basic understanding of correlation.

Level 7   They determine the modal class and estimate the … range of sets of grouped data … (begin)

| Topic area | **Mode and range** | pupil book page 382 |
| | **Mean** | pupil book page 383 |

Framework Objectives

✓ Calculate statistics, including with a calculator; recognise when it is appropriate to use the range, mean median and mode and, for grouped data, the modal class; calculate a mean using an assumed mean, **Framework examples pages 257, 259 and 261**

✓ Problem solving, **Framework examples page 25**

Sheets Worksheet 38 for support   Homework Sheets 16.1 and16·2   ICT Worksheet 'Finding Means with a Spreadsheet' (Teacher Resource Pack page 163)

Resources A calculator. A computer and a spreadsheet. Access to the internet.

 ## Starter ideas and activities

Mental starter ideas
- Give pupils some numbers including decimal numbers and ask them to put them in order.
- Ask pupils some questions such as Practice questions 4, 6, 8, 9a and 10 on pages 366 and 367 in the Handling Data Support chapter.
- Revise adding and subtracting positive and negative numbers, including decimals.

Starter activity
- **Mode and range**
  Recap what these words mean.
    Why might it be useful to know the mode? Why might it be useful to know the range?
- **Mean**
  Recap what the formula for the mean is and how to find it from a frequency table.
  (Practice question 17 on page 368 of the Handling Data Support chapter).
    How could we use a spreadsheet to find the mean from a table of values?
  You could use a computer connected to a screen to demonstrate.
- **Assumed mean**
  Ask 10 pupils to give you their height in cm. Put these on the board or OHP.
    What would you guess was the mean of these heights?
  Go on and demonstrate how to use the assumed mean to find the mean of the heights.

 ## Teaching points and activities

- Encourage pupils to check that they have inserted the right formula into the cells in the spreadsheet by checking that the answer is correct using a calculator.
- The assumed mean is best used if it can be worked out mentally or with jottings. Otherwise it is no quicker than finding the mean using the formula.

Support
- See **Mode, median, mean and range** on page 364 (Handling Data Support chapter). If pupils have difficulty with Practice questions 4, 6, 9a, b, 10, 17 and 29 use Worksheet 38 questions 1, 2, 5 and 6 (page 34 of Teacher Resource Pack).

Extension
- See **starred questions** in Exercise 1 (Pupil Book page 382).

Plenary ideas
- Consolidate vocabulary used during the lesson: mean, mode, range, frequency distribution, assumed mean.

Homework Review questions at the end of Exercises 1, 2 and 3 (Pupil Book pages 382, 383 and 384) and/or Homework Sheets 16.1 and 16·2 Exercises 1 and 2 (page 255 of this pack).

ICT See the ICT Support section of the Teacher Resource Pack to find a possible website.
Exercise 1 question 4 on page 382 of the Pupil Book.
See Exercise 2 on page 383 of the Pupil Book.
Use ICT Worksheet 'Finding Means with a spreadsheet' on page 163 of the ICT Support section in the Teacher Resource Pack.

Links Link this topic to rounding to 2 decimal places (Pupil Book page 30).
Link to Geography: Exercise 1 questions 3, 4.
Link to PE: Exercise 1 questions 1, 2.
Link to Science: Exercise 3 Review.
Link to adding and subtracting integers: using the assumed mean method in Exercise 3.

| **Topic area** | **Median, mean, mode, range** | pupil book page 385 |

Framework Objectives
✓ Calculate statistics, including with a calculator; recognise when it is appropriate to use the range, mean, median and mode and, for grouped data, the modal class; calculate a mean using an assumed mean..., **Framework examples pages 257, 259 and 261**

Sheets   Worksheet 38 for Support   Homework Sheet 16.3

Resources   Internet or CD-ROM or database.

 ## Starter ideas and activities

Mental starter ideas   Give six numbers including decimals and ask pupils to find the mean using the assumed mean method.

Starter activity
● **When is the median, mean, mode or range useful?**
Discuss when the mode, median and mean are useful (see screen on page 385 of the Pupil Book).
Use the Discussion on page 385 of the Pupil Book to further extend pupils' understanding of when each of these is used.

 ## Teaching points and activities

● Emphasise that each of the mean, median and mode summarises the data into one single value. Each tells us something a little different.
● Some pupils get the mean and median mixed up. It may help to point out that median and middle both have a d in the middle. Often when people talk about the average they are referring to the mean.

Support
● Pupils should have covered the previous sections in this chapter, *Mode and Range* and *Mean*.
● See **Mode, median, mean and range** on page 364 (Handling Data Support chapter of the Pupil Book). If pupils have difficulty with Practice questions 8, 9c and 29 use Worksheet 38 questions 3 and 4 (page 35 of the Teacher Resource Pack). Pupils may have already done some of these questions.

Extension
● See **starred questions** in Exercise 4 (Pupil Book page 386).
● See **Puzzle** (Pupil Book page 387) question 2d.

Plenary ideas
● You could ask pupils for some suggestions of when each of the mode, mean, median and range would be useful. Ask them to give some actual examples.
● Discuss the Exercise 4 question 2.
● Use the Puzzle on page 387 to extend thinking.

Homework   Review questions at the end of Exercise 4 (Pupil Book page 386) and/or Homework Sheet 16.3 (page 256 of this pack)

ICT   See Practical (Pupil Book page 387). The Internet, or a CD-ROM or database could be used. See ICT Support section (page 159 of the Teacher Resource Pack).

Links   Link to rounding to 2 decimal places (Pupil Book page 30).
Link to two-way tables: Exercise 4 question 6.
Link to PE: Exercise 4 question 2, 7, Review 2.
Link to Geography: Exercise 4 question 4, Practical.
Link to Science: Puzzle question 1.

| Topic area | Finding the median, range and mode from a stem-and-leaf diagram | pupil book page 388 |

Framework Objectives ✓ ... construct and use stem-and-leaf diagrams **Framework examples page 259**

Sheets Homework Sheet 16.4

 ## Starter ideas and activities

Mental starter ideas
- Add and subtract numbers using partitioning.
- Revise finding the median by giving an even number of values and asking pupils to find the median.

Starter activity

Making a stem-and-leaf diagram of ages

1 Draw this empty stem and leaf diagram on the board or OHP.
2 Explain what a stem-and-leaf diagram is by asking a student the age of the second eldest person in his or her household. Then put this age on the diagram. Do this a few times and then ask for some volunteers to come and put an age on the diagram.
3 Now put the data in order for each leaf.
  How could we find the median of these ages? (By working out which will be the middle value and counting along.)
  How could you find the range? The mode?

Note: You could use a ready made stem-and-leaf diagram here and then make one as suggested here for the plenary.

```
1 ┤
2 ┤
3 ┤
4 ┤
5 ┤
6 ┤
7 ┤
8 ┤
```

 ## Teaching points and activities

- Emphasise that the data values must be in order for the final version. If the stem-and-leaf is used to collect the data then it must be ordered once it has been collected.
- Some pupils count from the wrong end of the line of leaves when finding the median. Emphasise that the counting must be done **in order**.
- Point out that a stem-and-leaf diagram has all of the actual data values on it but they are put into a diagram that gives a useful picture of the trends.
- Emphasise that the stem can have different values, i.e. tens, hundreds, tenths, etc.

Support
- Pupils should have covered the previous sections in this chapter.

Plenary ideas
- Go over the structure of a stem-and-leaf diagram and ask pupils when it would be useful to use one.
- You could make a stem and leaf diagram such as the one suggested in the Starter Activity if you didn't make one there.
- Discuss exercise 5 Question 1

Homework Review questions at the end of Exercise 5 (Pupil Book page 388) and/or Homework Sheet 16.4 (page 257 of this pack)

Links Link to Geography: teaching notes (Pupil Book page 388); Exercise 5 question 1b, Review.

| | **Topic area** | **Comparing data** | pupil book page 389 |

Framework Objectives
- ✓ Compare two distributions using the range and one or more of the mode, median and range, **Framework examples page 273**
- ✓ Calculate statistics, including with a calculator; recognise when it is appropriate to use the range, mean, median and mode and, for grouped data, the modal class;.., **Framework examples pages 257, 259 and 261**

Sheets Worksheet 38 for support   Homework Sheet 16.5   Starter Resource Sheet 31   ICT Worksheets 'Interrogating a database 1' and 'Interrogating a database 2' (Teacher Resource Pack pages 164 and 165)

Resources A computer and a spreadsheet package. The Internet or a database.

 ## Starter ideas and activities

Mental starter ideas Revise finding the mean, median, mode and range.
You could do this by putting up this data set and asking pupils to find the values mentally.

56   59   63   68   75   80

- Comparing data

Starter activity Use Starter Resource Sheet 31.
Which one of these brands of electric jug would you buy? Why?
Who does not agree with (the person who just gave the answer to the last question)? Why not?
Draw out the reasons given in the screen on page 389.

 ## Teaching points and activities

- Emphasise that sometimes when you are comparing data there is no absolute correct answer. It may depend on what you consider to be most important when making your choice.
- Emphasise the importance of using the range when comparing data, as this tells you how spread out the data is.

Support
- Pupils should have covered the previous sections in this chapter.
- If pupils have difficulty with question 29 on page 370 of pupil book, use Worksheet 38 question 7 (page 35 Teacher Resource pack).

Extension
- See **starred questions** in Exercise 6 (Pupil Book page 389).

Plenary ideas
- Discuss the answers to Exercise 6 questions 2 or 4.
- Discuss the Practical on page 391 and how pupils could collect the data and where from.

Homework Review questions at the end of Exercise 6 (Pupil Book page 389) and/or Homework Sheet 16.5 (page 257 of this pack).

ICT For Practical (Pupil Book page 391) see ICT Worksheets 'Interrogating a database 1' and 'Interrogating a database 2' on pages 164 and 165 of the Teacher Resource Pack for the notes on how to interrogate a database and find the mean, median, mode and range of two sets of data to compare.

Links Link to PE: Exercise 6 question 2; Practical (page 391).
Link to Science: Exercise 6 question 3, Review; Practical (page 391).
Link to Geography: Exercise 6 question 4; Practical (page 391).
Link to Two-way Tables: Exercise 6 question 1.
Link to Stem-and-leaf diagrams: Exercise 6 question 5.

Topic area	**Compound bar charts and line graphs**	pupil book page 391

Framework Objectives
✓ Construct, on paper and using ICT;
- bar charts and frequency diagrams for discrete and continuous data;
- simple line graphs for time series;
identify which are most useful in the context of the problem, Framework examples pages 263 and 265

Sheets
Worksheet 36 for support   Homework Sheet 16.6   Resource Sheet pages 246–249 Teacher Resource Pack   ICT Worksheets 'Bar Charts' and 'Drawing a Line Graph' (Teacher Resource Pack pages 166 and 167) Starter Resource Sheet 32

Resources
Graph paper. Temperature probe and graphical calculator. Internet access. A computer and a spreadsheet package.

 ## Starter ideas and activities

Mental starter ideas
Ask questions such as:
What sorts of data do we use a bar chart for? What about a line graph?

Starter activity
- Look at graphs
Use Starter Resource Sheet 32.
   What must all graphs have? (A title and axes labelled.)
   What else does the compound bar chart have?
Look at the graph labelled 'Sources of vitamin A in the British diet'.
   Do you think this is a good way to show this data? What does it tell us?
Look at the line graph.
   What sort of data do we put on a line graph?
   What does the little squiggly symbol on the vertical axis tell us?
   What does the graph show?

 ## Teaching points and activities

- Some pupils have difficulty drawing graphs if the grid is not given (questions 4 and 5 in the exercise). Encourage them to find the highest and lowest values for each axis first. For a compound bar chart this means adding amounts together to find a total for each category.
- Encourage looking at the values to see how accurate the data is when choosing what graph paper to use. You may need to give some pupils help to get started.
- Emphasise the need to give the graph a title, label the axes and have a key if there is more than one set of data on the same set of axes. The axes must be divided into equal intervals.
- Emphasise that line graphs are used for showing changes over time.

Support
- See **Displaying data** on page 362 (Handling Data Support chapter of the Pupil Book). If pupils have difficulty with Practice questions 1 and 22 use Worksheet 36 question 1 (pages 31 and 32 of the Teacher Resource Pack).

Extension
- See **starred questions** in Exercise 7 (Pupil Book page 392).
- See **Practical** (Pupil Book page 394). Challenge pupils to do this as a survey using the steps given for planning and collecting data on page 378 of the Pupil Book.

Plenary ideas
- Recap on what a graph needs including a key for more than one data set on the axes.
- Discuss Exercise 7 question 4 (or 5 to extend thinking).

Homework
Review questions at the end of Exercise 7 (Pupil Book page 392) and/or Homework Sheet 16.6 (page 258 of this pack)

Graphical calculator
A graphical calculator is needed for the Practical on page 394.

ICT
See ICT Worksheets 'Bar charts' and 'Drawing a Line Graph' on pages 166 and 167 in the ICT Support section of the Teacher Resource Pack.

Links
Link this topic to graphs (Pupil Book page 232).
Link to Science: Exercise 7 questions 4, 5; Practical (Pupil Book page 394).
Link to Geography: Exercise 7 Review.

Topic area	Frequency diagrams	pupil book page 395

Framework Objectives   ✓ Construct, on paper and using ICT:
  – bar charts and frequency diagrams for discrete and continuous data;
  identify which are most useful in the context of the problem, Framework examples page 263

Sheets   Homework Sheet 16.7   Resource Sheet pages 249 and 250 Teacher Resource Pack Starter resource Sheet 33

Resources   Computer for drawing graphs. Internet access for the Practical.

Starter ideas and activities

Mental starter ideas   Give some practice at reading scales.

Starter activity   **Drawing a frequency graph**
1 Discuss how we graph grouped discrete data.
2 Use Starter Resource Sheet 33.
   You could collect some data on thumb lengths from your class to make this more interesting.
3 *How could we graph this data on these axes?* (Pupils will probably want to do the same as for discrete data and have the class intervals written underneath.)
4 *Could we have the bars joined like this?* (Draw the first two bars in.)
5 Draw the rest of the chart and add the labels between the bars on it.
   Why do we have the labels between the bars and not under them? (Draw out an answer such as 'Because there is no gap between the class intervals. One goes right up to the next one.')

Teaching points and activities

- Some pupils get mixed up between bar charts for grouped data and frequency diagrams. Emphasise the differences: the bar chart is for discrete data, which is counted, so there is a whole number space between bars; frequency diagrams have the bars touching with the labels between the bars. One class interval goes right up to the next with no gaps in between because the data is measured.
- Pupils get confused by the fact that the graph looks the same whether the original class intervals were $30 \leqslant m < 35$ or $30 < m \leqslant 35$.
  Point out that if they were just given the graph, they would need to be told how the class intervals had been set up to know which class interval the end point values such as 30 and 35 were in.

Support
- Pupils should have covered the previous section in this chapter, *Compound bar charts and line graphs*.

Extension
- See **Practical** (Pupil Book page 396). Challenge pupils to do this as a survey using the steps given for planning and collecting data on page 378 of the Pupil Book.

Plenary ideas
- Discuss the Practical on page 396. Discuss any suggestions pupils have and summarise the main points, such as the number of class intervals, the class intervals being of equal width, etc. Remind pupils of the main points when drawing a frequency diagram, e.g. the bars must touch and the labels are put between the bars.
- Discuss the answers to Exercise 8 question 3 or 4.

Homework   Review questions at the end of Exercise 8 (Pupil Book page 395) and/or Homework Sheet 16.7 (page 259 of this pack).

ICT   **Practical** (Pupil Book page 396): you could use the Internet to collect data.

Links   Link to measurement (Pupil Book page 334).
Link to PE: Exercise 8 question 3, Review: Practical (page 296).
Link to Science: Exercise 8 questions 1, 4.

| **Topic area** | **Drawing pie charts** | pupil book page 397 |

**Framework
Objectives**
✓ **Construct, on paper and using ICT:**
  – pie charts for categorical data;
  **identify which are most useful in the context of the problem, Framework examples
  page 263**
  Problem solving, **Framework examples page 25**

Sheets
Worksheet 36 for Support   Homework Sheet 16.8   Resource Sheet page 251 Teachers
Resource Pack   Starter Resource Sheet 34   ICT Worksheet 'Pie Charts' (Teacher Resource
Pack page 168)

Resources
OHP protractor for the Starter Activity. Protractor, compasses and ruler. A computer and a
spreadsheet package for the Practical.

Starter ideas and activities

Mental starter ideas
● Revise 'fraction and percentage of'. Ask questions such as:
$\frac{1}{5}$ of 360 $\frac{3}{8}$ of 360   30% of 360

Starter activity
● **Revising pie charts**
Revise what a pie chart is. Emphasise that a pie chart tells us the proportion for each
category and that the angle for each sector represents this proportion. The bigger the
angle the bigger the proportion.
What sort of data do we usually display on a pie chart? (categorical data)
● **Drawing a pie chart**
Use Starter Resource Sheet 34.
1 Ask 20 or 30 pupils (or a number that divides into 360 without remainder) how many
  videos they have watched in the last fortnight .
2 Fill in the frequency table.
  What fraction haven't watched any? One? Two? ...
  Fill in fractions.
3 *How could I use these fractions to draw a pie chart?*
  Fill in the calculations for finding the angle for each sector. Ask for volunteers to do this.
4 *How do I draw the pie chart?*
  Draw the pie chart using an OHP protractor.
  You could ask for volunteers to draw in each angle.

Teaching points and activities

● Some pupils try to learn the steps for drawing a pie chart as an algorithm. Encourage
understanding of what each step is achieving. Emphasise that a pie chart shows us
proportions and so we have to draw it using proportions.
● Emphasise that the angle of each sector is the proportion of 360° for this category.
● Encourage pupils to draw the angles starting with the largest. This cuts down the
percentage error for the smallest sector.
● Point out the use of pie charts in many other subject areas, especially Geography and
Science.

Support
● See **Pie chart** in the Handling Data Support chapter (Pupil Book page 363).
If pupils difficulty with Practice questions 13, 26, 27 try Worksheet 36 questions 5 and 10
(pages 33 and 34 of the Teacher Resource Pack).

Extension
● See **starred questions** in Exercise 9 (Pupil Book page 398).
● See Practical (Pupil Book page 399). Notes are given for this in the ICT Support section
(page 168 of Teacher Resource Pack).

Plenary ideas
● Discuss the answers to Exercise 9 question 4. The angles don't add up to 360° when
rounded for this example. Discuss the reason for this.

Homework
Review questions at the end of Exercise 9 (Pupil Book page 398) and/or Homework
Sheet 16.8 (page 259 of this pack).

ICT
For practical (Pupil Book page 399) see ICT Worksheet 'Pie charts' on page 168 of the ICT
Support section in the Teacher Resource Pack.

Links
Link this topic to percentages (Pupil Book page 118), proportion (Pupil Book page 136) and
angles (Pupil Book page 271).
Link to PE: Exercise 9 question 1c. Link to Science: Exercise 9 question 1a.

Topic area Scatter graphs

pupil book page 400

Framework Objectives
✓ Construct, on paper and using ICT:
  – simple scatter graphs;
  identify which are most useful in the context of the problem, Framework examples page 267

Sheets
Homework Sheet 16.9 for support   Resource Sheet page 252 Teacher Resource pack
Starter Resource Sheet 35   ICT Worksheet 'Scatter Graphs' (Teacher Resource Pack page 169)

Resources
A computer and a spreadsheet package for the Practical. Graph paper.

Starter ideas and activities

Mental starter ideas
- Revise coordinates by putting a grid on board or OHP and asking for the coordinates of points.
- Use Starter Resource Sheet 35.
  Discuss the relationship between the variables on the two axes, e.g. as time is increasing the temperature is increasing/decreasing/staying the same/etc.

Starter activity
- **Introducing scatter graphs**
  Make an OHP of the screen on Pupil Book page 400.
  You could ask 10 or so pupils for the amount of time they watched TV last night and the amount of homework they did, and use this data to draw a graph.
  Ask for volunteers to come and plot the points on a prepared grid.
- **What might we use a scatter graph for?**
  Use the Discussion on page 400.

Teaching points and activities

- Emphasise the need to plot points carefully.
- Some pupils have difficulty interpreting the trend from a scatter graph. Encourage them to ask themselves 'As the variable along the horizontal axis gets bigger what happens to the variable on the vertical axis?' If it also generally gets bigger then we can say 'As ... increases so does ... '
  In Year 9 pupils will progress to drawing a line of best fit. The focus in Year 8 is on seeing the general trend.

Support
- Pupils should have covered the previous sections in this chapter.

Extension
- See Practical (Pupil Book page 402). Use a spreadsheet to draw a scatter graph.

Plenary ideas
- Discuss the answers to Exercise 10 questions 3, 4 or 5.

Homework
Review questions at the end of Exercise 10 (Pupil Book page 400) and/or Homework Sheet 16.9 (page 260 of this pack)

ICT
See Practical (Pupil Book page 402). Use ICT Worksheet 'Scatter Graphs' on page 169 of the ICT Support section in the Teacher Resource Pack.

Links
Link this topic to two-way tables (Pupil Book page 374).
Link to PE: Exercise 10 question 3.
Link to measurement: Exercise 10 questions 2, 5, Review.

Topic area	Interpreting graphs and tables	pupil book page 402

Framework Objectives ✓ Interpret tables, graphs and diagrams for both discrete and continuous data, and draw inferences that relate to the problem being discussed: relate summarised data to the questions being explored, **Framework examples pages 269 and 271**
✓ Problem solving, **Framework examples page 25**

Sheets Worksheet 36 for Support   Homework Sheet 16.10   Starter Resource Sheet 36   Resource Sheet (page 253 Teacher Resource Pack).

Resources Graph paper. Published graphs for the Practical.

Starter ideas and activities

Mental starter ideas Ask questions such as Practice questions 1d, 7, 11, 13, 20, 25, 26 and 27 on pages 366, 367, 368, 369 and 370 in the Handling Data Support chapter. You may have covered some of these in previous sections.

Starter activity
● **Revise finding percentages**
Revise finding percentages. Ask questions such as:
In a village there are 5000 people. 2200 of them are female. What percentage are female?
● **Interpreting graphs**
Use **'Watch it!'** on page 381 of the Pupil Book.
Use Starter Resource Sheet 36 which has the graphs from the Discussion on page 402.

Teaching points and activities

● Define 'compare' and 'contrast'.
● You may need to recap on the different sorts of graphs that might be given and the types of data they are used for e.g. a bar chart usually displays categorical data.
● Sometimes pupils have difficulty working out what a graph is about. Encourage them to look first at the title and then the labels on the axes. If there is a key, then look at this as well. Encourage them also to look at the type of graph it is.
● Point out that a graph may look complicated when it isn't.
● Pupils sometimes have difficulty reading the scale on the population pyramids. Emphasise that the reading from these is only an estimate. Pupils may also need to be prompted on how to find the percentages asked for. Remind pupils that a percentage is a proportion.
● You could do some of the questions from Exercise 11 as a discussion or group exercise. Pupils may get more out of it this way.
● Recap what is meant by a 'trend'.

Support
● Pupils should have covered previous sections in this chapter.
● See **Displaying data** on page 362 (Handling Data Support chapter of the Pupil Book). If pupils have difficulty with Practice questions 5, 7, 19, 20, 25, 26 and 28 use Worksheet 36 (pages 31 to 34 of the Teacher Resource Pack). Some of these questions may already have been tried.

Extension
● See **starred questions** in Exercise 11 (Pupil Book page 403).

Plenary ideas
● Show Starter Resource Sheet 36 again and either finish it if there was not time to do so for the Starter or recap on the main points of interpreting graphs and tables.
● Discuss any of the answers to the Exercise 11 questions.
● You could take a graph from a newspaper or magazine and ask what sort of information could be gained from it.

Homework Review questions at the end of Exercise 11 (Pupil Book page 403) and/or Homework Sheet 16.10 (page 261 of this pack).

Links Link to Geography: Exercise 11 questions 3, 6, 7, Review 2.
Link to bar charts and line graphs and pie charts: Discussion (Pupil Book page 402); Exercise 11 all questions.
Link to percentages: Exercise 11 questions 6, 7, 8, Review 1.

| Topic area | **Surveys** | pupil book page 407 |

Framework Objectives

✓ Discuss a problem that can be addressed by statistical methods and identify related questions to explore, **Framework examples page 249**

✓ Decide which data to collect to answer a question, and the degree of accuracy needed; identify possible sources, **Framework examples page 251**

✓ Plan how to collect the data, including sample size.., **Framework examples page 253**

✓ Collect data using a suitable method, such as observation, controlled experiment, including data logging using ICT or questionnaire, **Framework examples page 255**

✓ Communicate orally and on paper the results of a statistical enquiry and the methods used, using ICT as appropriate; justify the choice of what is presented, **Framework examples page 273**

Surveys

● This section enables pupils to draw together much of what they have learnt in chapters 15 and 16.

● This section is best given as a longer term project for pupils to do. It could be discussed as part of the plenary at the end of one of the other sections in this chapter to get pupils started on it. It could be done later in the year as a summary of what pupils have learnt. It could be done alongside the learning of the other sections in Chapters 15 and 16.

● Emphasise the cycle given in the practical on page 408. Pupils may like to make a copy of this in their books. (see Resource Sheet page 254 in the Teacher Resource Pack).

● Pupils should probably check their topic with you before they start their survey.

17 Probability

New National Framework

Topic in Chapter	Links to Programme of Study *Ma2 a–l are integrated throughout*	Links to Framework for Teaching Mathematics: 8 Teaching Programme *Applying mathematics and solving problems are integrated throughout.*
Language of probability *page 414*	**Ma4 5h** use the vocabulary of probability in interpreting results involving uncertainty and prediction	**Probability** ● Use the vocabulary of probability when interpreting the results of an experiment; ...
Calculating probability *page 416*	**Ma4 4c** understand and use the probability scale **Ma4 4d** understand and use estimates or measures of probability from theoretical models, including equally likely outcomes ...	● Know that if the probability of an event occurring is p, then the probability of it not occurring is $1 - p$; **find and record all possible mutually exclusive outcomes for single events and two successive events in a systematic way,** using diagrams and tables.
Calculating probability by listing outcomes *page 418*	**Ma4 4e** list all outcomes for single events, and for two successive events, in a systematic way **Ma4 4f** identify different mutually exclusive outcomes and know that the sum of the probabilities of all these outcomes is 1	
Estimating probability from experiments *page 420*	**Ma4 1a** recognise the limitations of any assumptions, and the effects that varying the assumptions could have on conclusions drawn from the data analysis	**Probability** ● Estimate probabilities from experimental data; understand that: – if an experiment is repeated there may be, and usually will be, different outcomes; – increasing the number of times an experiment is repeated generally leads to better estimates of probability. ● Compare experimental and theoretical probabilities in different contexts.
Comparing calculated probability with experimental probability *page 423*	**Ma4 4d** understand and use estimates or measures of probability from theoretical models, including equally likely outcomes, or from relative frequency **Ma4 5i** compare experimental data and theoretical probabilities **Ma4 5j** understand that if they repeat an experiment, they may — and usually will — get different outcomes, and that increasing sample size generally leads to better estimates of probability and population characteristics	

Links to Attainment Targets

Level 5 They understand and use the probability scale from 0 to 1. Pupils find and justify probabilities and approximations to these, by selecting and using methods based on equally likely outcomes and experimental evidence, as appropriate. They understand that different outcomes may result from repeating an experiment.

Level 6 When dealing with a combination of two experiments, pupils identify all outcomes, using diagrammatic, tabular or other forms of communication. In solving problems, they use their knowledge that the total probability of all mutually exclusive outcomes of an experiment is 1.

Level 7 Pupils understand relative frequency as an estimate of probability and use this to compare outcomes of experiments (begin).

© **New National Framework Mathematics 8** Nelson Thornes Ltd

177

| Topic area | Language of probability | pupil book page 414 |

Framework Objectives
✓ Use the vocabulary of probability when interpreting the results of an experiment; ...,
Framework examples page 277
✓ Problem solving, **Framework examples page 23**

Sheets   Worksheet 39 for support   Homework Sheet 17.1

Resources   A number line or counting stick and mini-whiteboards for Starter for the Starter Activity.

Starter ideas and activities

Mental starter ideas
- Ask questions such as Practice questions 2 and 3 on page 366 of the Handling Data Support chapter.
- Practise converting between fractions, decimals and percentages. You could do this using a number line or counting stick.
- Practise finding the bigger fraction from a choice of two by either converting them both to decimals or by finding a common denominator. *Example* Which is bigger, $\frac{5}{8}$ or $\frac{7}{12}$?

Starter activity
- **Recap on the language**
  Use the Discussion on page 414 of the Pupil Book.
- **Random and unpredictable**
  1 Discuss the words 'random' and 'unpredictable'.
  2 Ask pupils to use their mini-whiteboards to show true (T) or false (F) for these:
     *If I take a sweet from a bag without looking, this is a **random** event.*
     *The outcome of tossing a dice is **predictable**.*
     *The outcome of taking a piece of fruit without looking from a bag of mixed fruit is **unpredictable**.*
- **More likely**
  1 Ask for 16 volunteers. Divide them in to a group of 6 and a group of 10.
  2 Give 4 of the group of 6 an object (you could put a hat on their heads or give them a pen to hold or ...). Give 6 of the group of 10 the same thing.
     I am going to choose one person from each of these groups at random (pull the name out of a hat maybe). From which group am I more likely to get someone who (has a hat on or is holding a pen or ...)? How can you be sure?
  3 Encourage answers that talk about greater proportion and ask them to show it is true mathematically.
     Pupils will need to show that $\frac{4}{6} > \frac{6}{10}$ or $\frac{2}{3} > \frac{3}{5}$.

Teaching points and activities

- Make sure pupils can find the larger of two fractions by changing them both to decimals or by finding a common denominator.
- Emphasise that probability is linked to proportion. If there is a greater proportion of red counters in bag A than in bag B then the probability of taking a red counter from bag A is greater than from bag B.
- Pupils often intuitively know that the probability of one thing is greater than another but cannot explain how they know this. Encourage pupils to explain that if there is more of one item than another in a group, then the probability of choosing this item at random is greater.

Support
- See **Probability** on page 365 (Handling Data Support chapter of the Pupil Book).
  If pupils have difficulty with Practice questions 2, 3, 14 use Worksheet 39 questions 1, 2, 3 and 5 (page 36 of the Teacher Resource Pack).

Plenary ideas
- Recap on the vocabulary used and ask for examples of each.
  | Likely | Unlikely | Random | Unpredictable |
  | Equally likely | Proportion | Most likely | |
- Discuss the answers to Exercise 1 questions 5 and 6.
- Repeat the Starter Activity 'more likely' with different numbers in each group.

Homework   Review questions at the end of Exercise 1 (Pupil Book page 414) and/or Homework Sheet 17.1 (page 262 of this pack).

Links   Link this topic to equivalent fractions and fraction to decimal conversions (Pupil Book page 102).
Link to bar graphs: Exercise 1 question 6.

Topic area **Calculating probability**

pupil book page 416

Framework Objectives
✓ Know that if the probability of an event occurring is p, then the probability of it not occurring is $1 - p$; **Framework examples page 279**
Problem solving, **Framework examples page 23**.

Sheets Worksheet 39 for support Homework Sheet 17.2 Starter Resource Sheet 37

Starter ideas and activities

Mental starter ideas
Ask questions such as Practice questions 15, 18 and 21 on pages 368 and 369 in the Handling Data Support chapter of the Pupil Book.
You could use an OHP of these to help (Starter Resource Sheet 37).

Starter activity
● **Revise probability**
  1 *If the letters of the alphabet are put into a bag and one is taken at random, what is the probability that the letter will be a vowel? A consonant? An E?*
  2 Encourage pupils to think about proportion. What proportion of the letters in the bag are vowels? consonants? Es?
    Note: You could actually put the letters into a bag so that it is more realistic.
  3 *Does each letter have an equally likely chance of being chosen?*
    What if I put the letters of the word MATHEMATICS in a bag?
    What is the probability if a letter is taken at random of the letter being an M?
    Does each letter have an equally likely chance of being chosen? (Yes)
● **Probability of an event not happening**
  1 *What is the probability of an event that is certain to happen?* (1)
    Is it certain that I will either come to school tomorrow or not come to school tomorrow?
    If the probability that I will come to school is 0·98, what is the probability that I won't come to school? (0·02) *How do you know this?* ($1 - 0·98 = 0·02$)
  2 Ask some similar questions using percentages and fractions instead.

Teaching points and activities

● You may need to revise decimal to percentage to fraction conversions and revise the probability scale.
● Some pupils have trouble with calculating probability. Emphasise that probability tell us the likelihood of an event happening or the proportion of times we would expect it to happen.
● Emphasise that if an event is impossible it will happen 0 times out of the total number of possible outcomes or 0% of the time so the probability is 0. If an event is certain to happen then it happens 100% of the time which is 1. All other probabilities lie between 0 and 1.
● Emphasise that an event is certain to either happen or not happen and so the probability of it happening plus the probability of it not happening must add to 1.
● Emphasise that we can only use the formula for calculating probability if the events are equally likely.

Support
● Pupils should have covered the previous section in this chapter.
● See **Probability** on page 365 (Handling Data Support Chapter of the Pupil Book)
  If pupils have difficulty with Practice questions 18 and 21 use Worksheet 39 questions 6 and 7 (page 36 of the Teacher Resource Pack).
● See **In a spin** on page 413 of the Pupil Book.

Extension
● See **starred questions** in Exercise 2 (Pupil Book page 416).

Plenary ideas
● Recap on the vocabulary: Event Sample Sample space
● Discuss the answers to Exercise 2 question 6 or 7 (or 10, 11 and 12 for extension).
● Use the Discussion on page 418 of the Pupil Book. This introduces the word 'biased'.

Homework
Review questions at the end of Exercise 2 (Pupil Book page 416) and/or Homework Sheet 17.2 (page 262 of this pack).

Links
Link this topic to decimals, fractions and percentages (Pupil Book page 119).
Link to PE: Exercise 2 question 3.

| Topic area | Calculating probability by listing outcomes | pupil book page 418 |

Framework Objectives

✓ Know that if the probability of an event occurring is p, then the probability of it not occurring is $1 - p$; **find and record all possible mutually exclusive outcomes for single events and two successive events in a systematic way, using diagrams and tables, Framework examples page 281**
Problem solving, **Framework examples page 23**

Sheets Worksheet 39 for support   Homework Sheet 17.3

Resources A coin and a dice for the Starter Activity.

 ## Starter ideas and activities

Mental starter ideas Ask questions like 'What are the possible outcomes when I toss a die? spin this spinner?'

Starter activity
● **Finding outcomes**
   1 Use the Discussion on page 418.
   2 Have a coin and a dice.
      I am about to toss these two together. What are all the possible outcomes I could get?
      List them on the board. Tell pupils we call the list of all possible outcomes the sample space.
      What is the probability I will get a head and an odd number?
      List the ways to get a head and an odd number (H1, H3, H5). Ask someone to work out the probability. Ask for this as a fraction, decimal and percentage.

 ## Teaching points and activities

● Pupils sometimes try to calculate the probability without listing all the outcomes. This can lead to some being missed. Encourage pupils to use tables or lists to display the outcomes. This way the number of favourable outcomes can be seen more clearly.

Support
● Pupils should have covered the previous section in this chapter.
● See **Probability** on page 365 (Handling Data Support chapter of the Pupil Book)
If pupils have difficulty with Practice question 15 use Worksheet 39 question 4 (page 35 of the Teacher Resource Pack).

Extension ● See **starred questions** in Exercise 3 (Pupil Book page 419).

Plenary ideas
● Recap on the term 'sample space'. Ask for an example.
● Discuss the answers to Exercise 3 question 5. You could put the table on an OHP.

Homework Review questions at the end of Exercise 3 (Pupil Book page 419) and/or Homework Sheet 17.3 (page 263 of this pack).

Links Link this topic to fractions, decimals and percentages (Pupil Book page 119) (Pupil Book page 108).

Topic areas	**Estimating probability from experiments**	pupil book page 420
	Comparing calculated probability with experimental probability	pupil book page 423

Framework Objectives

✓ Estimate probabilities from experimental data; understand that:
  – if an experiment is repeated there may be, and usually will be, different outcomes;
  – increasing the number of times an experiment is repeated generally leads to better estimates of probability, **Framework examples page 283**
✓ Compare experimental and theoretical probabilities in different contexts, **Framework examples page 285**

Sheets Worksheet 39   Homework Sheet 17.4   Starter Resource Sheet 8   ICT Worksheets 'Throwing Dice 1' and 'Throwing Dice 2' (Teacher Resource Pack pages 172 and 173).

Resources Number cards for the Starter Activity. Two fair dice for the Practical on page 421. A computer and spreadsheet package, card, a dice, a 10-sided fair dice and Internet access for the Practical on page 423. Five cards with a different design on each for the Investigation on page 424.

Starter ideas and activities

Mental starter ideas

Ask questions such as Practice question 23 on page 369 of the Handling Data Support chapter.

Starter activity

● **Estimating probability from experiments**
  1   *Is it always possible to calculate a probability?*
      How could you find the probability of the next car that passes the school having just one occupant?
      Encourage pupils to talk about doing experiments and how many would need to be done to get an accurate estimate.
  2   Use the Discussion on page 421 of the Pupil Book.
or 1   Put some number cards (five of the numbers 1 to 9 with two of the same number) into a hat or bag (you could use Starter Resource Sheet 8). Pull out a card at a time and ask a pupil to record the results on a tally chart, then replace the card.
  2   Do this 10 times.
  3   Ask pupils to predict what the five cards you put in the hat were.
  4   Discuss how they made this prediction.
      Take the cards out. The prediction will probably be wrong.
  5   Repeat the experiment 50 times and ask for a prediction.
      What can you say about the number of times an experiment is repeated?

Teaching points and activities

● Recap frequency diagrams.
● Talk about what is meant by *calculated* probability, *theoretical* probability and *experimental* probability.
● Emphasise that the probability of some events cannot be calculated. The only way we can estimate the probability of these is by doing an experiment.
● Emphasise that the more times an experiment is repeated the more accurate the estimate of the probability is. Back this up with the evidence that when we do an experiment where we can find the theoretical probability, the greater the number of times the experiment is repeated the closer it gets to the theoretical probability. Pupils will get a greater understanding of this by simulating some experiments on the computer. See the ICT section below.
● To save time you may need to get lots of groups to do the same experiment and then combine results to get a large enough number of repeats. Emphasise that when an experiment is repeated or done by different groups the results are very likely to be different. The greater the number of trials the more similar the results will be.

Support ● Pupils should have covered the previous sections in this chapter. If pupils have difficulty with question 23 on page 369 of the Handling Data Support Chapter use Worksheet 39 question 8 (page 36 of the Teacher Resource Pack).

Extension ● See the second part of **Practical** (Pupil Book page 423) part A. Challenge pupils to calculate the theoretical probability and compare their results with these.

Pupils could draw a bar chart with the theoretical and experimental probabilities next to each other.

Part D: There are notes on this in the ICT Support section (page 170 of the Teacher Resource Pack).

Plenary ideas
- Ask pupils to present the conclusion they got from the Practical on page 421 or the one on page 423 of the Pupil Book. You could do part B of the one on page 423 using a computer attached to a display screen.
- Discuss situations when we can't calculate the theoretical probability and we may need to do an experiment (fault testing in manufacturing is a good example).
- Discuss the answers to Exercise 4 question 6.
- You could do the Investigation on page 424 with the whole class.

Homework Review questions at the end of Exercise 4 (Pupil Book page 421) and/or Homework Sheet 17.4 (page 263 of this pack).

ICT See Practical (Pupil Book page 423) part B. See ICT Worksheets 'Throwing Dice 1' and 'Throwing Dice 2' on pages 172 and 173 in the ICT Support section of the Teacher Resource Pack.
For Practical (page 423) part E see the lottery website at www.lottery.co.uk

Links Link to calculating probability (Pupil Book page 416) and the equivalence of fractions, decimals and percentages (Pupil Book page 119).

2

Homework Sheets

Powers of ten

1   Write these as powers of ten.

   a a hundred thousand   **b** ten million   ***c** ten thousand million

2   Write the numbers in brackets in words.

   a Earth came into being $4 \cdot 6 \times 10^9$ years ago.

   b The first vertebrate appeared 5×10^8 years ago.

   c The last dinosaurs existed $6 \cdot 5 \times 10^6$ years ago.

   d The last Ice Age began $1 \cdot 1 \times 10^5$ years ago.

3   Write in figures the number that is

   a 9 less than two and a half thousand

   b 4 more than three and a quarter million.

Adding and subtracting multiples of $0 \cdot 1$, 0.01 and $0 \cdot 001$

1

$\overline{3 \cdot 05}$	$\overline{7 \cdot 319}$	$\overline{0 \cdot 393}$		$\overline{6 \cdot 347}$	$\overline{7 \cdot 112}$	$\overline{1 \cdot 93}$	$\overline{1 \cdot 352}$		$\overline{7 \cdot 6}$	$\overline{3 \cdot 05}$	$\overline{4 \cdot 625}$	$\overline{6 \cdot 347}$	
$\overline{6 \cdot 347}$	$\overline{7 \cdot 112}$	$\overline{2 \cdot 72}$		$\overline{7 \cdot 6}$	$\overline{3 \cdot 05}$	$\overline{0 \cdot 393}$	$\overline{1 \cdot 88}$	$\overline{4 \cdot 625}$	$\overline{6 \cdot 347}$	$\overline{3 \cdot 05}$		$\overline{7 \cdot 319}$	$\overline{4 \cdot 625}$	$\overline{7 \cdot 6}$
				N				**N**					
$\overline{3 \cdot 05}$	$\overline{7 \cdot 319}$	$\overline{0 \cdot 393}$		$\overline{2 \cdot 73}$	$\overline{1 \cdot 115}$	$\overline{1 \cdot 93}$	$\overline{1 \cdot 352}$	$\overline{2 \cdot 73}$	$\overline{4 \cdot 625}$	$\overline{4 \cdot 813}$	$\overline{0 \cdot 393}$	
$\overline{7 \cdot 54}$	$\overline{7 \cdot 319}$	$\overline{0 \cdot 804}$	$\overline{3 \cdot 91}$	$\overline{3 \cdot 91}$	$\overline{1 \cdot 115}$	$\overline{7 \cdot 6}$						

Write the letter beside each calculation above its answer in the box.

N $2 \cdot 63 + 0 \cdot 1 = 2 \cdot 73$ **A** $4 \cdot 635 - 0 \cdot 01$ **H** $7 \cdot 318 + 0 \cdot 001$

Y $0 \cdot 814 - 0 \cdot 01$ **O** $7 \cdot 312 - 0 \cdot 2$ **C** $1 \cdot 931 - 0 \cdot 001$

M $4 \cdot 773 + 0 \cdot 04$ **E** $0 \cdot 453 - 0 \cdot 06$ **W** $2 \cdot 18 - 0 \cdot 3$

P $7 \cdot 49 + 0 \cdot 05$ **I** $1 \cdot 035 + 0 \cdot 08$ **T** $2 \cdot 98 + 0 \cdot 07$

K $1 \cdot 442 - 0 \cdot 09$ **R** $6 \cdot 339 + 0 \cdot 008$ **S** $8 - 0 \cdot 4$

D $2 \cdot 81 - 0 \cdot 09$ **L** $4 - 0 \cdot 09$

*2 Paula estimated the weight of a parcel to be $2 \cdot 5$ kilograms. The actual weight of the parcel was 40 grams less than this. What was the weight of the parcel?

Homework Sheet 1.3　8 Core

× and ÷ by multiples of 10, 100 and 1000

1　**a**　40 × 70　　**b**　20 × 300　　**c**　800 ÷ 40　　**d**　2000 ÷ 50

　　e　30 × 60 000　**f**　50 000 ÷ 50　**g**　24 ÷ 600　　**h**　43 × 30 000

　　i　3200 ÷ 80　　**j**　3300 ÷ 3000　**k**　16 ÷ 40

2　A trout hatchery has 40 ponds, each one containing 8000 trout. How many trout are there altogether?

3　George put his mother's photographs into albums. There were 8000 photographs and each album held 200. How many albums did George need?

*4　Make up four different multiplications that have an answer of 12 000.

*5　Make up four different divisions that have an answer of 0·03.

Homework Sheet 1.4　8 Core

Multiplying and dividing by 0·1 and 0·01

1　**a**　Which of these is the same as 6·72 ÷ 0·1?

　　　A 6·72 ÷ 10　　**B** 6·72 × 100　　**C** 6·72 × 10　　**D** 6·72 ÷ 100

　　b　Which of these is equivalent to 0·23 × 0·01?

　　　A 0·23 ÷ 0·01　　**B** 0·23 × 10　　**C** 0·23 ÷ 100　　**D** 0·23 × 100

　　c　Which of these is the same as 1·94 ÷ 0·01?

　　　A 1·94 × 100　　**B** 1·94 × 10　　**C** 1·94 ÷ 100　　**D** 1·94 ÷ 10

2

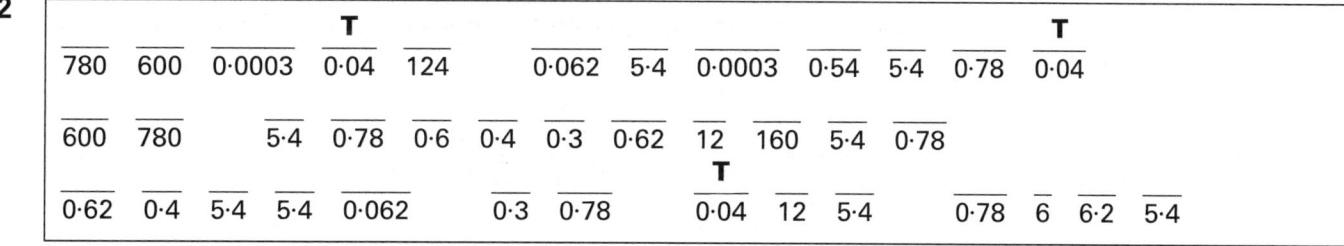

Put the letter beside each question above its answer in the box.

T 0·4 × 0·1 = 0·04　　**P** 6·2 × 0·01　　**L** 0·04 ÷ 0·1　　**F** 7·8 ÷ 0·01

D 0·62 ÷ 0·1　　　　**R** 0·3 × 0·001　　**I** 30 × 0·01　　**N** 78 × 0·01

U 0·6 ÷ 0·1　　　　　**M** 1·6 ÷ 0·01　　**O** 6 ÷ 0·01　　**G** 3 × 0·2

H 1·2 ÷ 0·1　　　　　**Y** 1·24 ÷ 0·01　　**E** 9 × 0·6　　**C** 0·9 × 0·6

S 3·1 × 0·2

3　Tina had six pieces of ribbon, each 0·4 m long. What total length of ribbon did she have?

Putting decimals in order

1   Are these true or false?

    a $6·12 < 6·124$ **b** $2·708 < 2·79$ **c** $4·68 > 4·672$

    d $3·407 > 3·41$ **e** $0·20035 < 0·2016$ **f** $^-6·3 < \,^-6·05$

    g $^-0·803 > \,^-0·82$

2   Six swimmers were timed in a medley race. Their times, in seconds, were

        18·67, 17·9, 17·86, 18·9, 18·09, 18·59

    Put these times in order, fastest first.

3   ___ > ___ > ___

    Put these numbers in the gaps.

    a 0·67, 0·607, 0·76   **b** 8·42, 8·24, 8·4

    c 1·3072, 1·372, 1·327   **d** 0·0043, 0·034, 0·00403

4   Put these in ascending order.

    a 637 m, 0·67 km, 0·607 km, 673 m

    b 1827 g, 1·72 kg, 1072 g, 1·087 kg

    c 13 400 mℓ, 13·41 ℓ, 13 104 mℓ, 13·041 ℓ, 13·14 ℓ

5   Which number is halfway between these?

    a 0·2 and 0·3   **b** 0·05 and 0·08   **c** $^-0·6$ and $^-0·9$ **d** $^-3·6$ and $^-0·4$

6   Rewrite the decimals in order, smallest first, and their attached balloons should spell out the name of a shape. What is it?

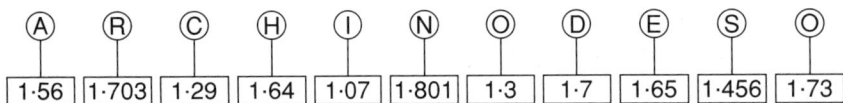

    Ⓐ   Ⓡ   Ⓒ   Ⓗ   Ⓘ   Ⓝ   Ⓞ   Ⓓ   Ⓔ   Ⓢ   Ⓞ

    | 1·56 | 1·703 | 1·29 | 1·64 | 1·07 | 1·801 | 1·3 | 1·7 | 1·65 | 1·456 | 1·73 |

7 $4·93 \leqslant m \leqslant 5·03$

    Give 10 values m could be.

Rounding to powers of ten

1 Round

    a 647 to the nearest 10

    b 2369 to the nearest 100

    c 25 248 to the nearest 1000

    d 7299 to the nearest 1000

    e 8426 to the nearest 100

    f 6245 to the nearest 10

    g 72 850 to the nearest 100

    h 87 236 to the nearest 10 000

    i 784 231 to the nearest 100 000

    j 4 684 298 to the nearest million.

2 16 483 pizza slices were sold at Wimbledon one year. The television newscaster reported 16 500 pizza slices sold. Was this figure rounded to the nearest ten or nearest hundred?

3 2 784 000 people were reported to have visited Alton Towers one year. The number was given to the nearest hundred.

    a What was the smallest number of people that could have visited?

    b What was the largest number that could have visited?

*4 It was reported that 9·7 million people died in World War I.

    a This figure has been rounded to the nearest hundred thousand. What is the greatest and smallest number of people that could have died?

    b One report said that nearly ten million died. Another said 'Just over nine and a half million died'. Which report is more accurate? Explain.

    c A historian wrote that during the Battle of the Somme, the British, French and Germans lost 1 265 173 casualties altogether. Round this number to the nearest hundred thousand.

Rounding to decimal places

1 Paul rounds these calculator displays to 1 decimal place. What answers should he get?

a `1.672`   **b** `18.3184`

c `26.056`   **d** `6.031`

d `11.2948`   **f** `17.998`

2 If Paul rounded the calculator displays in **question 1** to the nearest hundredth, what answers should he get?

3 **a** Round 2846 g to the nearest tenth of a kilogram.

   b Round 6283 m to the nearest tenth of a kilometre.

   c Round 238 cm to the nearest tenth of a metre.

4 Approximate these answers to 2 decimal places.

   a 46 ÷ 7   **b** 19 ÷ 3   **c** 17 ÷ 9

5 Three friends spent £17·87 on a present for a colleague. They shared the cost between them. How much should each pay, to the nearest penny?

6 Round the answers to these sensibly and say what you have rounded them to.
On a sponsored cycle ride Niki covered a distance of 3850 km in 29 days.

   a What distance did she ride, on average, each day?

   b If she cycled for 8 hours each day, what distance did she ride, on average, each minute?

Adding and subtracting integers

1 Add two numbers to get the number in the circle above. What goes in the top circle of each?

a 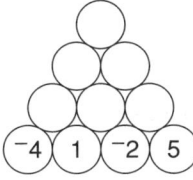 **b** 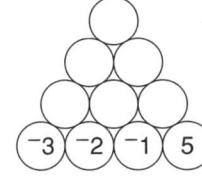 **c**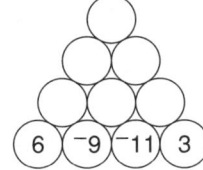

2 Copy the circles in **question 1** again. Subtract the number on the right from the number on the left to get the number above.

What number goes in the top circle of each?

3 Find the answers to these mentally.

a ⁻4 + ⁻11   **b** ⁻13 + 29   **c** 43 + ⁻18   **d** ⁻17 − ⁻14

e ⁻31 + 28 − 17 + 23   **f** ⁻16 − 27 + 32 − 14 + 29

4 Look at the integers in the box.

⁻1	3	0	⁻4	5	⁻2	1	⁻3

Rewrite the following, replacing each * with an integer from the box.

The integers may be used more than once.

a * + * = ⁻6

b * + * = 3

c * − * = ⁻1

d * − * = 7

e * − * = 7   (a different answer from part **d**)

5 Copy the square below and use the integers ⁻4, ⁻3, ⁻2, ⁻1, 0, 1, 2, 3 and 4 once each to complete a magic square.

Every row, column and diagonal should add up to 0.

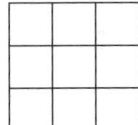

Multiplying and dividing integers

1 **a** $^-5 \times {}^-3$ **b** $4 \times {}^-6$ **c** $^-2 \times 7$ **d** $^-8 \times {}^-4$

    e $6 \times {}^-9$ **f** $^-7 \times 7$ **g** $^-9 \times {}^-7$ **h** $^-36 \div {}^-9$

    i $^-48 \div 6$ **j** $72 \div {}^-9$ **k** $\dfrac{^-42}{7}$ **l** $\dfrac{16}{^-8}$

2 **a** How many $^-3$s are in $^-33$?

    b $^-6 \times {}^-5 = 30$. Use the numbers $^-6$, $^-5$ and 30 to write down three more facts.

3 Start with the 4 in the snake's tail and work along its length, multiplying or dividing each answer by the number in the next segment.

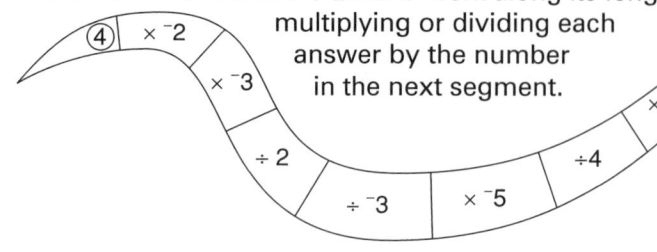

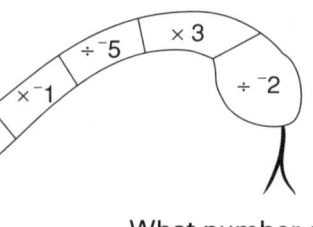

What number do you end up with in the snake's head?

Integers on a calculator

												C		
$\overline{21}$ $\overline{428}$ $\overline{^-44}$		$\overline{77}$ $\overline{^-18}$ $\overline{323}$ $\overline{105}$ $\overline{^-44}$ $\overline{^-525}$ $\overline{21}$		$\overline{^-19}$ $\overline{7}$ $\overline{^-525}$ $\overline{^-162}$ $\overline{77}$ $\overline{^-44}$

| $\overline{^-204}$ $\overline{4067}$ | $\overline{21}$ $\overline{428}$ $\overline{^-44}$ | $\overline{^-1\cdot2}$ $\overline{65}$ $\overline{^-13\cdot8}$ $\overline{424}$ | $\overline{^-204}$ $\overline{^-525}$ |

| $\overline{^-204}$ $\overline{4067}$ | $\overline{21}$ $\overline{428}$ $\overline{^-44}$ | $\overline{^-1\cdot2}$ $\overline{7}$ $\overline{21}$ $\overline{21}$ $\overline{65}$ **C** $\overline{^-162}$ $\overline{^-19\cdot2}$ $\overline{^-525}$ |

Write the letter beside each question above its answer.

C $^-73 + {}^-89 = {}^-162$ **L** $173 + {}^-96$ **Y** $278 - {}^-146$ **N** $^-49 \times {}^-83$

H $12\,412 \div {}^-29$ **S** $84 \times {}^-6\cdot25$ **G** $17\cdot5 - {}^-87\cdot5$ **O** $^-416 \div {}^-6\cdot4$

M $1102 \div {}^-58$ **A** $^-61 - {}^-43$ **E** $^-17\cdot5 + {}^-26\cdot5$ **I** $^-2\cdot4 \times 85$

R $^-8\cdot5 \times {}^-38$ **K** $^-24 \times {}^-32 \div {}^-40$ **D** $4\cdot6 \times {}^-2\cdot4 \div 0\cdot8$ **T** $168 \div {}^-24 \times {}^-3$

U $18\cdot7 + {}^-24\cdot2 - {}^-12\cdot5$ **B** $^-1\cdot8 \times {}^-2\cdot4 \div {}^-3\cdot6$

CHAPTER 2

Order of operations with integers

30.	31.	32.	33.	34.	35.	36.	37.	38.	39.
20.	21.	22.	23.	24.	25.	26.	27.	28.	29.
10.	11.	12.	13.	14.	15.	16.	17.	18.	19.
0.	1.	2.	3.	4.	5.	6.	7.	8.	9.
⁻10.	⁻9.	⁻8.	⁻7.	⁻6.	⁻5.	⁻4.	⁻3.	⁻2.	⁻1.
⁻20.	⁻19.	⁻18.	⁻17.	⁻16.	⁻15.	⁻14.	⁻13.	⁻12.	⁻11.
⁻30.	⁻29.	⁻28.	⁻27.	⁻26.	⁻25.	⁻24.	⁻23.	⁻22.	⁻21.

Do the calculations in order from **a** to **v**. Join the answers on the grid.

a, **b** and **c** are already done. What picture do you get?

a $^-4 \times {}^-5 + {}^-2 = 18$

b $^-3(^-8 + {}^-1) = 27$

c $^-4 \times {}^-6 - 1 = 23$

d $^-9 \times {}^-4 + (8 \times {}^-\frac{1}{2})$

e $30 - {}^-2 \times {}^-4$

f $^-4 \times {}^-2 - {}^-3$

g $\dfrac{^-8 \times 5}{^-4}$

h $2(7 - {}^-5) - ({}^-8 \times {}^-3)$

i $2 \times {}^-3 - {}^-7$

j $^-6 \times 2 + \dfrac{^-8}{^-2}$

k $\dfrac{^-10 \times {}^-4 - {}^-2 \times {}^-2}{^-2}$

l $3(^-5 + {}^-3) - {}^-7$

m $\dfrac{^-8 \times {}^-5}{^-4} - \dfrac{^-15}{5}$

n $4 \times {}^-2 - 2(^-5 + 3)$

o $\dfrac{^-6 \times {}^-5 + {}^-2}{^-7 + 5}$

p $(^-3 \times 6) - ({}^-2 + 1 + {}^-4)$

q $\dfrac{^-17 - {}^-5}{^-2 \times {}^-2}$

r $^-3 \times {}^-7 - ({}^-3 \times {}^-5 - 2)$

s $^-4(2 \times {}^-4 + 1)$

t $^-5 \times {}^-6 + (1 + {}^-2)$

u $\dfrac{^-8 \times 5 - {}^-2}{^-2}$

v $\dfrac{^-7 \times 5 + {}^-2}{^-2}$

Divisibility

1

13 794	3456	9255	2592	1590

Which of these numbers are divisible by

a 6   **b** 15   **c** 18   **d** 24?

2 A farmer harvests more than 1000 cabbages from his field. He can pack them into boxes of 5, 6 or 9 without any gaps in the boxes.

What is the smallest number of cabbages he could have harvested?

Prime factor decomposition

1 Use a table to write these numbers as a product of prime factors in index notation.

    a 120   **b** 168   **c** 210   **d** 264   **e** 6300

2 What are the missing numbers, A, B, C and D, on these factor trees?

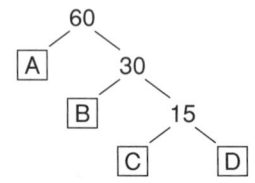
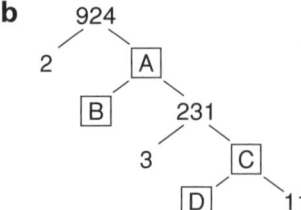
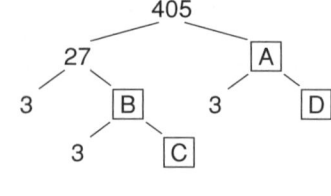

3 Use a table or a factor tree to write each of these as a product of prime factors in index notation.

    a 468   **b** 700   *****c** 12 474   *****d** 7497

*****4** Find the two lowest odd numbers that each have exactly four factors.

Highest common factor and lowest common multiple

1 What is the **i** HCF **ii** LCM of these pairs of numbers?

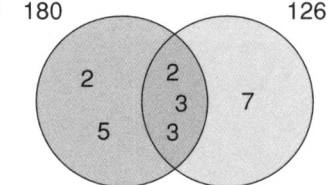
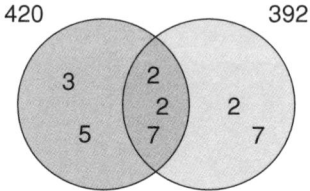

2 Find the LCM of the pairs of numbers in **question 1**.

3 Fill in this crossnumber puzzle.

Across

  1 LCM of 168 and 1260

  4 HCF of 84 and 315

  7 LCM of 9 and 15

  8 LCM of 84 and 172

10 HCF of 528 and 672

11 HCF of 375 and 500

12 LCM of 49 and 91

14 LCM of 9 and 21

17 HCF of 168 and 392

18 LCM of 32 and 44

20 HCF of 147 and 196

21 HCF of 336 and 616

22 LCM of 88 and 605

24 LCM of 18 and 24

26 LCM of 100 and 125

27 HCF of 288 and 480

Down

  2 HCF of 1638 and 2730

  3 LCM of 81 and 93

  5 LCM of 42 and 99

  6 HCF of 308 and 476

  8 HCF of 1188 and 2772

  9 LCM of 32 and 56

13 LCM of 36 and 88

15 LCM of 168 and 441

16 HCF of 60 and 105

18 HCF of 273 and 390

19 LCM of 28 and 126

20 LCM of 16 and 50

22 HCF of 588 and 630

23 HCF of 495 and 630

25 LCM of 12 and 18

CHAPTER 2

Powers and roots   Except for questions 5, 6, and 7.

1   Find the answers to these mentally.

   a $(14 - 9)^2$ **b** $(8 + 3)^2$ **c** $(^-7)^2$ **d** $\sqrt{52 - 27}$ **e** $\sqrt{5^2 - 4^2}$ **f** $(4 \times 9 - 28)^2$

2   Give the positive and negative square roots of these.   **a** 36   **b** 121

3   Find these by factorising.   **a** $\sqrt{256}$ **b** $\sqrt{576}$ **c** $\sqrt{900}$

4 $\sqrt{30}$ lies between 5 and 6. Explain how you know this without doing the calculation.

5   Use your calculator to find these. Give the answer to 1 decimal place.

   a $6 \cdot 1^2$ **b** $(^-9 \cdot 4)^2$ **c** $\sqrt{17 - 6}$ **d** $\sqrt{4^2 + 7^2}$ **e** $\dfrac{3 \cdot 9^2 - 2 \cdot 3^2}{1 \cdot 7}$

6   The sum of the squares of four consecutive even numbers is 504.

   What are the four even numbers?

7   Find the missing digits.

   a $(*5)^2 = 6*5$ **b** $(1*)^2 = 3*1$ **c** $(*3)^2 = 5**9$

Cubes and cube roots   Except for questions 3 and 4.

1   Find the missing numbers mentally.

   a 4 cubed equals ....   **b** 8 is the cube of ....

   c .... is the cube of 5   **d** 3 cubed equals ....

   e 1 is the cube root of ....   **f** The cube root of 1000 is ....

   *__g__ 6 is the cube root of ....

2   Find these mentally.

   a 5^3 **b** $\sqrt[3]{1}$ **c** 3^3 **d** $\sqrt[3]{216}$ **e** 10^3

3   Use your calculator to find these. Round **f** onwards to 2 decimal places.

   a $(^-9)^3$ **b** 13^3 **c** $1 \cdot 7^3$ **d** $\sqrt[3]{6859}$ **e** $\sqrt[3]{4 \cdot 096}$

   f $(3 \cdot 54)^3$ **g** $\sqrt[3]{69}$ **h** $\sqrt[3]{138}$ **i** $\sqrt[3]{666}$

$$10 = 1 + 1 + 8 \qquad\qquad 62 = 8 + 27 + 27$$
$$= 1^3 + 1^3 + 2^3 \qquad\qquad = 2^3 + 3^3 + 3^3$$

4

3	10	17	36	55	62	99	129	136

   All of the numbers in the box can be written as the sum of three cubes. Write all of them in this way.

Adding and subtracting

1 a $4 + {}^-7 + 5 + {}^-10 + 2$　　　　**b** ${}^-6 - 4 - {}^-3 + 7 + {}^-5$

　c $0·4 + {}^-0·9 - {}^-0·3 + 0·7 + {}^-0·2$　　**d** $1·3 - {}^-0·2 - 0·8 + {}^-0·4 - 0·3$

2 a $63 + 19$　　　**b** $280 + 170$　　**c** $310 - 180$　　**d** $840 - 260$

　e $2500 + 1700$　**f** $5400 - 2700$　**g** $1·8 + 3·4$　　**h** $5·3 - 1·8$

　i $0·73 - 0·28$　　**j** $0·603 - 0·018$　**k** $47 + 135 + 68$　**l** $7·6 - 3·9 + 1·4$

3 Fill in this magic square.

6			
19	7	14	10
		9	
12	16		17

4

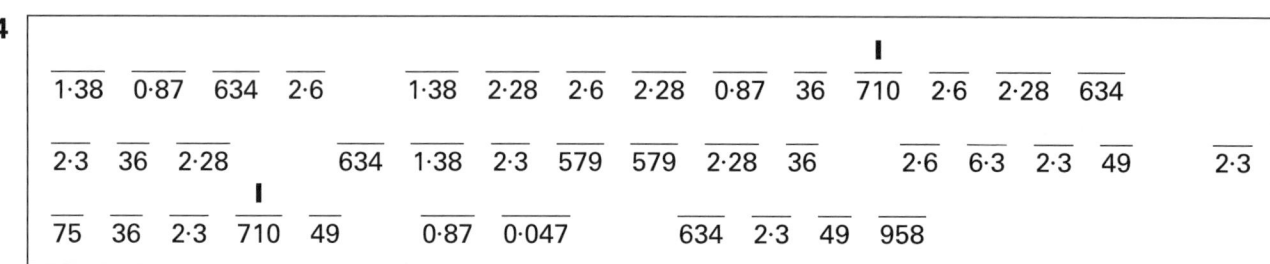

Write the letter beside each calculation above its answer in the box.

I $230 + 480 = 710$　**R** $83 - 47$　　**G** $170 - 95$　　**T** $1·7 + 0·9$

H $2·8 + 3·5$　　　**D** $1760 - 802$　**O** $2·49 - 1·62$　**A** $8·2 - 5·9$

F $0·071 - 0·024$　**S** $276 + 358$　**L** $728 - 149$　　**N** $76 - 27$

E $1·82 + 0·46$　　**M** $5·1 - 3·72$

*5 Find the values of the digits A, B, C and D which make this true.

$$
\begin{array}{r}
AC·BA \\
-\ \ B·AC \\
\hline
AD·AB \\
\end{array}
$$

Multiplying and dividing

1 Fill in the missing numbers in this number chain.

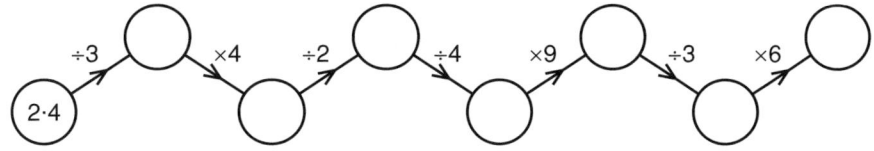

2 a 3×0.6 **b** 0.4×0.7 **c** $0.7 \div 2$

3 a 3.6×2.5 **b** $^-1.5 \times {}^-2.7$ ***c** $^-2.4 \times 1.75$

4

					D							
$^-36$	56	25	25	4	3210	$^-36$	2800	25	102	8·4	3210	0·23

D												
280	$^-36$	47	1800	102	720	$^-36$	720	4·8	102	25	720	0·23

8·4	$^-300$	25	25	4·8	25

Write the letter beside each calculation above its answer in the box.

D $40 \times 70 = 2800$ **S** 30×60 **K** $800 \div 200$ **N** 240×3

H 642×5 **R** $^-4 \times 75$ **A** $1.5 \times {}^-24$ **T** $1.15 \div 5$

Y $235 \div 5$ **I** $714 \div 7$ **W** $672 \div 12$ **E** $350 \div 14$

G 2.4×3.5 **C** $^-1.5 \times {}^-3.2$

5 a Find the product of 270 and 4. Add 13 to it.

b If $18 \times 23 = 414$, what is 1.8×2.3?

c Find the quotient when 765 is divided by 15.

Remember the quotient is the answer when two numbers are divided.

d Add the difference between 270 and 145 to the sum of 18 and 33.

6 The product of two numbers is 108. One number is three times the other. What are the two numbers?

*7 Which two numbers have a product of 832 and a quotient of 13?

Solving problems mentally

1 **a** A baby is 18 weeks old. How many days old is it?

   b What is the perimeter of a square of side 6·2 cm?

   c Eight tickets to a concert cost £13·20 altogether. How much is each ticket?

   d How many minutes and seconds in 384 seconds?

2 Paul bought a camera on hire purchase. He paid a deposit of £25 and six monthly payments of £12·75.

   a How much did the camera cost him altogether?

   b How much could he have saved by paying the cash price of £86·50?

3 Find x and y if:

   a The sum of x and y is 15.   ***b** The sum of x and y is 3·7.

      The product of x and y is 54.   The product of x and y is 3.

4 A firm produces 20 000 golf balls each day. They are packed in boxes of 50. How many boxes are needed each day?

5 These are the heights of six friends, measured in centimetres.

     148, 162, 151, 169, 172, 158

   What is their mean height?

6 Farmer Jones keeps cows and hens. He has 41 creatures altogether, which have 118 legs between them. How many cows and how many hens does he have?

7 Find the sum of the numbers from 1 to 20.

   Is there a quick way of doing this?

8 I am a three-digit number less than 200.

   I am divisible by 21.

   The sum of my digits is an odd number greater than 10.

   What number am I?

9 I am a three-digit number. The sum of my digits is 6.

   I am divisible by both 12 and 13.

   What number am I?

Exercise 1   Making estimates

1   Would an EXACT number or an ESTIMATED number be needed for these.

   a A doctor prescribes a number of tablets to be taken by a patient each day.

   b A film critic wants to know how many people watched a film during its first week of release.

   c A solicitor wants to know how much deposit a client needs to pay for a house.

   d A mother wants to know how many children will be coming to her son's birthday party.

   e A farmer wants to know how many kilograms of potatoes are grown in one field in a year.

Exercise 2   Estimating answers to calculations

1   Choose the best approximation for each calculation.

   a 272×24 **A** 200×20 **B** 300×20 **C** 300×30

   b $524 \div 48$ **A** $500 \div 50$ **B** $500 \div 40$ **C** $600 \div 50$

   c $8 \cdot 9 \div 1 \cdot 8$ **A** $8 \div 1$ **B** $9 \div 1$ **C** $9 \div 2$

   d $13 \cdot 9 \times 27 \cdot 4$ **A** 10×30 **B** 20×20 **C** 20×30

2   Estimate the answer to these calculations. Show how you found your estimate.

   a 217×138 **b** 873×27 **c** $274 \div 32$ **d** $928 \div 18 \cdot 6$

   e $8 \cdot 32 \times 1 \cdot 77$ **f** $0 \cdot 63 \times 4 \cdot 29$ **g** $11 \cdot 29 \div 0 \cdot 483$ **h** $19 \div 3 \cdot 86$

3   Write down a calculation you could do to find an estimate for these. Find the estimate.

   a $\dfrac{183 + 216}{38}$ **b** $\dfrac{18 \cdot 6 \times 29 \cdot 4}{6 \cdot 38}$ **c** $\dfrac{41 \cdot 8 - 18 \cdot 29}{7 \cdot 82 \times 5 \cdot 19}$

4   For each of these, write down a calculation you could do to estimate the answer.

   a A school bought 218 textbooks costing £2·75 each. What was the total cost?

   b 2852 football fans returned home on 42 seater coaches. How many coaches were needed?

   c An office received 83 phonecalls per hour. How many phonecalls were received during a 42-hour week?

CHAPTER 3

Adding and subtracting

1

						E			E
4·43	0·7	5·57	8·14	12·59	9·17		4·43	5·68	2·64	9·38	5·9		23·6	8·64	9·38

	E								
8·64	9·38	23·6	0·7	0·7	5·57		0·29	23·6	9·61	5·9

Write the letter beside each calculation above its answer in the box.

E 6·2 + 4·9 − 1·72 = 9·38

Y 2 + 6·5 + 1·83 − 4·76

T 6·92 + 7·38 − 4·69

O 1·34 + 9·07 + 3·62 − 8·35

X 3 + 3·3 − 0·33 − 3·33

G 6·8 + 1·9 − 3·73 + 4·2

R 0·8 + 8·8 − 0·88 − 0·08

S 8 − 3·6 + 2·9 − 1·4

A 2·8 + 13·7 + 28·4 − 21·3

I 11·29 − 4·88 + 1·73

L 4·4 + 0·44 − 4·14

N 18 − 7·63 + 1·98 + 0·24

B 1 − 0·74 + 0·03

F 7·3 − 0·07 + 1·92 − 4·73

2 A lorry can carry a maximum load of 2·8 tonnes. If 0·64 tonnes of coal are loaded into it, followed by another 0·9 tonnes, how many more tonnes will it hold?

3 A jar full of jam weighs 0·35 kg. The empty jar weighs 86 g. What is the weight of jam in the jar?

4 Jodie decided to save some of her pocket money every week. She decided to put 1p in her money box the first week, 2p in the second week, 4p in the third week, 8p in the fourth week and so on. She gave up after 11 weeks. Write down how much Jodie put in her money box each week and add it to find the total she had saved after 11 weeks.

5 **a** Write down all the numbers that can be made using all three of the digits 7, 3 and 1 and a decimal point. There must be at least one digit both after and before the decimal point.

b Find the difference between the largest and smallest numbers.

c Find the sum of the two middle numbers.

d Find the total of the difference between the lowest two numbers and the difference between the highest two numbers.

Multiplying

1

					H					
17·019	7·912	0·9984	1·36	5·824	17·019	218·4	5·824	0·265	4·576	7·912

		H								
4·576	94·6	1·36	218·4	66·3	2·276	66·3	7·912	1·16	5·824	2·276

					H				
66·3	1·53	66·3	1·16	7·68	1·36	218·4	4·576	0·9984	0·332

Write the letter beside each calculation above its answer in the box.

H $2·6 \times 84 = 218·4$ **Y** $1·6 \times 4·8$ **R** $0·8 \times 1·45$ **T** $2·72 \times 0·5$

S $86 \times 1·1$ **I** $2·08 \times 2·2$ **G** $8·3 \times 0·04$ **O** $7·28 \times 0·8$

E $39 \times 1·7$ **V** $1·7 \times 0·9$ **B** $1·06 \times 0·25$ **A** $4·6 \times 1·72$

F $11·38 \times 0·2$ **N** $1·56 \times 0·64$ **P** $18·3 \times 0·93$

2 Find the cost of these. You may have to round your answer to the nearest penny.

a Kelly bought 1·25 kg of treacle toffee.

b Colin bought 0·12 kg of chocolate limes.

c Claire bought 0·24 kg of pear drops.

d Craig bought 160 g of mint imperials.

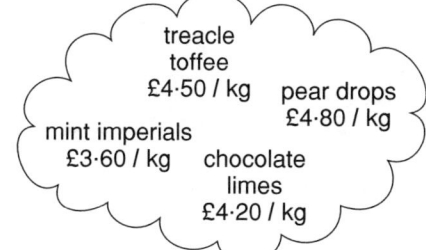

treacle toffee
£4·50 / kg   pear drops
£4·80 / kg
mint imperials
£3·60 / kg   chocolate limes
£4·20 / kg

3 Find two numbers with a product of 459 and a difference of 10.

4 The currency exchange rate between British pounds and continental euro is £1 = 1·64 euro. How many euro would you get for £75?

5 What values might A, B and C have?

$$
\begin{array}{r}
\text{A·BA} \\
\times \quad \text{AA} \\
\hline
\text{AC·CA}
\end{array}
$$

Is this the only answer?

Can you find another set of values for A, B and C?

CHAPTER 4

Dividing

1   Chocolate bars are packed into boxes of 24.

   a How many boxes can be filled using 394 chocolate bars?

   b How many chocolate bars are left over?

2   Give the answer to these as a decimal.

   a 635 ÷ 25   **b** 372 ÷ 16   **c** 88·4 ÷ 13

3   Give the answers to these to 1 decimal place.

   a 538 ÷ 14   **b** 857 ÷ 18   **c** 73·4 ÷ 17

4   A brand of coffee comes in three different sized jars.

Small 250 g £2.05

Medium 500 g £3.96

Large 700 g £5.56

   Which size gives the best value for money?

Dividing by a decimal

1   Write an equivalent division you could do to find the answer to these.

   a 48 ÷ 0·2   **b** 761 ÷ 0·03   **c** 8 ÷ 0·05   **d** 11 ÷ 1·6

2   Calculate these.

   a 24 ÷ 0·6   **b** 125 ÷ 0·05   **c** 78 ÷ 1·2   **d** 70 ÷ 2·5

3   Give the answers to these to 1 decimal place.

   a 248 ÷ 0·9   **b** 22 ÷ 0·06   **c** 91 ÷ 2·4   **d** 642 ÷ 1·4

Checking answers   Use a calculator for question 1 but not question 2

1   Check the answers to these using a method of your choice. Show how you did it.

   a 8·2 × 6·18 = 47·376   **b** 792 ÷ 1·2 = 660   **c** 6·8 × 4·3 = 29·42

2   Without using a calculator, choose a possible answer for each calculation. Explain your choice.

   a 4·8 × 6·9   **A** 35·42   **B** 34·61   **C** 33·12

   b 4·3 × 1·2   **A** 5·61   **B** 5·16   **C** 6·51

Exercise 1   Using brackets on a calculator

1

									S							
‾3·5	‾2·5	‾4	‾4·2		‾1·4	‾2·8	‾4	‾4	‾3		‾2	‾2·5	‾5·3	‾23·8	‾2·8	‾1·2	‾5·3

	S								
‾2	‾2·8	‾3	‾23·8	‾2·5	‾6·25		‾2·5	‾1·2	‾4

Write the letter beside each calculation above its answer in the box.

S $4·8 \div (3·2 - 1·6) = 3$

I $(1·6 \times 2·4) \div (1·53 + 1·67)$

R $\dfrac{(8·6 + 3·4)^2}{(6 - 1·2)^2}$

F $(8·5 - 5·7) \times (6 - 4·5)$

A $17·64 \div (4·8 + 1·5)$

C $\dfrac{(8·4 \div 0·42)}{(5·76 + 4·24)}$

T $(5·2 + 1·6) \times 3·5$

O $8 \div (2 + 1·2)$

N $(11·6 + 13·8) - (12·4 + 7·7)$

B $8·316 \div \{6 - (0·2 \times 0·3)\}$

G $1·4 \times (0·8 + 1·7)$

L $60 \div \{3 \times (4 + 1)\}$

2 Put brackets in the following, where necessary, to make it true.

$2 + 5 \times 4 - 1 + 6 + 8 \div 7 = 19$

Exercise 2   Using the calculator memory

1 Bruce bought 5 golf balls at £1·74 each, 3 packets of tees at 58p each and 2 sets of golf gloves at £6·29 each. How much change would he get from £25?

2 1 pint ≈ 0·57 litres

How many litres are equivalent to

a 5 pints   **b** 11 pints   **c** 20 pints?

CHAPTER 4

Exercise 1   Fractions of shapes

1   What fraction of this shape is shaded?

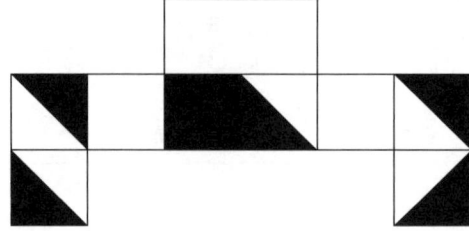

2   The pie chart shows the visitors to a new museum during its first day of opening.

Estimate the fraction of visitors that were

a men   **b** women.

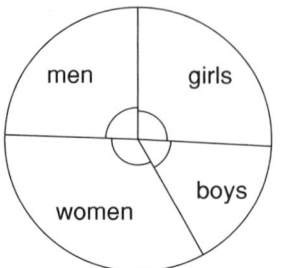

Exercise 2   One number as a fraction of another

1   Write these as fractions of a minute.

   a 20 seconds   **b** 10 seconds   **c** 50 seconds   **d** 45 seconds

2 a What fraction of 20 is 14?

   b What fraction of 75 is 60?

   c Give 85 as a fraction of 120.

3 a What fraction of £1 is 40p?

   b What fraction of 1 metre is 75 cm?

   c What fraction of 1 gallon is 2 pints?

4   An adult set of teeth contains 8 Incisors, 4 Canines, 12 Premolars and 8 Molars. What fraction are

   a Incisors   **b** Canines   **c** Premolars?

*__*5__   Exactly one half of this rectangle has been shaded.

   Draw three rectangles and find interesting ways of shading exactly one half of each.

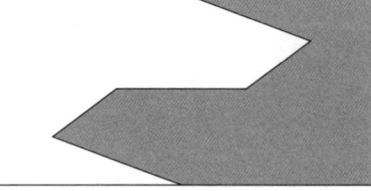

Exercise 1   Fractions and decimals

1

$4\frac{3}{50}$ $\frac{2}{5}$ $\frac{6}{125}$ $\frac{1}{200}$ $\frac{7}{20}$ **A** $\frac{6}{125}$ $\frac{221}{500}$ $\frac{7}{20}$ $4\frac{33}{50}$ $4\frac{3}{50}$ $4\frac{131}{200}$ $\frac{2}{5}$ $\frac{5}{8}$ $4\frac{33}{50}$

$\frac{2}{5}$ $4\frac{33}{50}$ **A** $4\frac{3}{50}$ $\frac{6}{25}$ $\frac{7}{20}$ $4\frac{3}{5}$ $\frac{2}{5}$ $4\frac{131}{200}$ $\frac{2}{5}$

Write the letter beside each calculation above its answer in the box.

Write these as fractions in their simplest form.

A $0\cdot35 = \frac{7}{20}$ **W** $0\cdot24$ **B** $0\cdot005$ **O** $0\cdot625$

I $0\cdot4$ **E** $0\cdot442$ **M** $0\cdot048$ **H** $4\cdot6$

S $4\cdot06$ **N** $4\cdot66$ **L** $4\cdot655$

Exercise 2   Fractions and decimals   Except for questions 2 and 3.

1   Write these as decimals.

   a $\frac{3}{5}$ **b** $\frac{17}{20}$ **c** $\frac{21}{25}$ **d** $\frac{3}{20}$ **e** $\frac{19}{50}$

   f $1\frac{11}{20}$ **g** $3\frac{5}{8}$ **h** $\frac{85}{50}$ **i** $\frac{18}{5}$ **j** $\frac{11}{4}$

2   Use your calculator to write these as decimals. Round to 2 d.p.

   a $\frac{7}{13}$ **b** $\frac{4}{9}$ **c** $\frac{5}{7}$ **d** $\frac{19}{31}$

3   **a** Write these as decimals.

     $\frac{1}{11},$ $\frac{2}{11},$ $\frac{3}{11},$ $\frac{4}{11}$

   b Describe the pattern.

   c Use the pattern to write down the decimals for these.

     $\frac{5}{11},$ $\frac{6}{11},$ $\frac{7}{11},$ $\frac{8}{11}$

Ordering fractions

1 Which of < or > goes in the box?

   a $\frac{7}{8} \square \frac{3}{4}$ **b** $\frac{2}{3} \square \frac{11}{15}$ **c** $\frac{2}{7} \square \frac{3}{8}$ **d** $\frac{3}{11} \square 0·3$

   e $0·8 \square \frac{15}{17}$ **f** $\frac{7}{9} \square \frac{3}{4}$ **g** $\frac{8}{19} \square \frac{13}{29}$

2 In the Burns family, 5 of the 8 children are girls. In the Jones family, 4 of the 7 children are girls. Which family had the greatest fraction of girls?

3 Put these fractions in order from smallest to largest.

   a $\frac{3}{4}, \quad \frac{7}{9}, \quad \frac{2}{3}, \quad \frac{5}{6}, \quad \frac{7}{12}, \quad \frac{1}{2}$ **b** $\frac{2}{5}, \quad \frac{1}{4}, \quad \frac{4}{15}, \quad \frac{1}{6}, \quad \frac{5}{12}, \quad \frac{3}{10}, \quad \frac{7}{20}$

4   **a** Show these fractions with an arrow. $\frac{4}{20}, \quad \frac{20}{25}, \quad \frac{9}{15}, \quad \frac{9}{5}, \quad \frac{26}{20}$

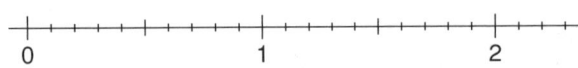

   b Show these fractions with an arrow. $\frac{2}{5}, \quad \frac{7}{28}, \quad \frac{9}{25}, \quad \frac{11}{20}, \quad \frac{6}{50}$

Adding and subtracting fractions

1 **a** $\frac{2}{3} - \frac{1}{6}$ **b** $\frac{3}{4} + \frac{1}{12}$ **c** $\frac{7}{9} - \frac{2}{3}$ **d** $\frac{5}{16} + \frac{3}{8}$

   e $\frac{2}{7} + \frac{1}{3}$ **f** $\frac{17}{20} - \frac{3}{8}$ **g** $\frac{2}{5} + \frac{3}{4} - \frac{7}{10}$

2 Tapforth School collected some money for charity. They gave $\frac{2}{5}$ of it to Guide Dogs for the Blind, $\frac{3}{7}$ of it to Cancer Research and the rest to Childline. What fraction of the total amount collected went to Childline?

3 What digits could go in the △ to make these true?

   a $\frac{\triangle}{8} + \frac{\triangle}{\triangle} = \frac{7}{\triangle}$ **b** $\frac{\triangle}{6} + \frac{\triangle}{\triangle\triangle} = \frac{11}{\triangle\triangle}$

   Find as many ways as you can for each.

*4 A bath was $\frac{7}{10}$ full of water. 12 litres was run out of the bath. The bath was then $\frac{1}{2}$ full. How many litres does the bath hold altogether?

Fraction of. Multiplying an integer by a fraction

1 Give these as mixed numbers.

a $\frac{7}{8}$ of 5   **b** $\frac{2}{5}$ of 11   **c** $\frac{5}{9}$ of 8

2 Paul threw the javelin 65 metres. Jack only threw it $\frac{9}{10}$ as far as Paul. How far did Jack throw the javelin? Give your answer as a mixed number.

3 Use cancelling to find the answers to these.

a $\frac{2}{9}$ of 12   **b** $\frac{3}{4}$ of 26   **c** $\frac{7}{20} \times 15$ **d** $2\frac{3}{4} \times 6$

e $1\frac{3}{8} \times 22$ **f** $72 \times \frac{11}{27}$ **g** $3\frac{1}{12} \times 4$

Dividing an integer by a fraction

1 Write a division for each of these.

a How many fifths in 2?   **b** How many quarters in 5?

2 **a** $4 \div \frac{1}{7}$ **b** $3 \div \frac{1}{9}$ **c** $5 \div \frac{1}{3}$ **d** $6 \div \frac{1}{4}$

3 What goes in the box?

a $\frac{3}{4}$ of 24 = 18,   so $18 \div \frac{3}{4} = \square$

b $\frac{2}{9}$ of 36 = 8,   so $8 \div \frac{2}{9} = \square$

4 Use the diagrams to help find the answers.

a $6 \div \frac{3}{5} =$

b $10 \div \frac{2}{3} =$

5 **a** $14 \div \frac{7}{10}$ **b** $8 \div \frac{2}{5}$ **c** $9 \div \frac{3}{4}$

6 Maria has 10 cakes. She gives $\frac{2}{5}$ of a cake to each of her friends. How many friends receive a piece of cake?

CHAPTER 5

Converting fractions, decimals and percentages Except for questions 3 and 4.

1 a Convert these to percentages　　**i** 0·71　　**ii** 0·395　　**iii** 1·613

  b Convert these to fractions in their lowest terms.

    i 72%　　**ii** 80%　　**iii** 125%

2 Jessica received the following marks in her summer tests:

Geography	History	Maths	Science	English
35 out of 50	17 out of 20	54 out of 75	8 out of 10	19 out of 25

In which subject did Jessica get the highest mark? What percentage did she get?

 3 Put these in order from smallest to largest. Only use a calculator if you need to.

  a $\frac{3}{8}$, 0·38, $\frac{1}{2}$, 45%　　　　**b** 0·68, $\frac{3}{5}$, 64%, $\frac{13}{20}$

  c $1\frac{2}{5}$, 1·49, 138%, $1\frac{11}{25}$　　**d** 0·83, $\frac{17}{20}$, 71%, $\frac{3}{4}$

 4

					N			**N**		
56·7%	82·2%	56·7%	85·0%	51·1%	91·7%	29·3%	19·4%	91·7%	64·7%	46·3%
			N							
42·9%	46·3%	12·1%	51·1%	91·7%	30·8%	19·4%	55·4%			
		N								
62·1%	61·6%	56·7%	91·7%	42·9%	62·1%	46·3%	85·0%	19·4%	56·7%	55·4%

Write the letter beside each calculation above its answer in the box.

Write these as percentages to 1 d.p.

N $\frac{11}{12} = 91·7$　　**G** $\frac{4}{13}$　　**I** $\frac{3}{7}$　　**T** $\frac{18}{29}$　　**E** $\frac{6}{31}$

A $\frac{38}{67}$　　**L** $\frac{17}{141}$　　**D** $\frac{83}{101}$　　**R** $\frac{46}{83}$　　**H** $\frac{517}{839}$

V $\frac{213}{727}$　　**O** $\frac{23}{45}$　　**U** $\frac{11}{17}$　　**Y** $\frac{91}{107}$　　**S** $\frac{19}{41}$

5 Are these true or false?

  a $\frac{3}{5} < 0·62$　　　**b** $38\% < \frac{3}{8}$　　　**c** $26\% > 0·2$　　　**d** $\frac{17}{20} > 0·8$

  e $\frac{17}{25} < 71\%$　　**f** $17\% > \frac{7}{40}$　　**g** $0·09 > 10\%$　　**h** $\frac{7}{5} < 1·4$

Exercise 1　Percentage of — mentally

1　**a** 25% of 300 litres　**b** 60% of 40 m　　**c** 15% of £80

　　d 95% of 1500 g　　**e** 110% of 120 cm　**f** 2% of £5500

　　g 11% of 600 g　　**h** 47% of 400 mℓ　**i** 23% of £20

2　Your body is 70% water. What weight of water is in a person weighing 90 kg?

3　Graham paid a 15% deposit on a television costing £840. How much deposit did he pay?

Exercise 2　Percentage of — written and calculator

Except for question 2.

1　Use a written method to find these.

　　a 35% of £70　　**b** 12% of 600 m　　**c** 43% of £18

　　d 84% of 150 km　**e** $12\frac{1}{2}$% of 216 cm　**f** 26% of 480 litres

2　**a** 13% of 270 g　　**b** 61% of £260　　**c** 3% of 855 km

　　d 48% of 650 m　**e** $17\frac{1}{2}$% of £36　　**f** 29% of 345 mℓ

3　18·5% of the world's population died from Black Death in the 14th Century. If the world's population was 405 million, how many died as a result of Black Death?

4　A lawnmower is in the sale at two garden centres. Jack's Garden Centre is offering £45 off and Greenthumb Garden Centre is giving 24% off. Which Garden Centre is cheaper?

Jack's Garden Centre

Lawn-mower HG70 £180 £45 off

Greenthumb Garden Centre

24% off the price of £180 Lawn-mower HG70

5　A survey of 8500 people revealed that 65% of them owned a home computer and 80% of those who did had Internet access.

　　a How many owned a home computer?

　　b How many had Internet access?

6　Jackie bought a washing machine for £349 + VAT ($17\frac{1}{2}$%). How much VAT did she pay? Round your answer sensibly.

Percentage increase and decrease

1 a Increase 448 by 35%.   **b** Decrease 2850 by 38%.

2 A packet of soap powder was on offer with 15% extra free. If the packet originally contained 1400 g, how much did it actually contain in the offer?

3 The entry fee to the zoo in October is reduced by 25% on the Summer price of £8·40. Joe is a pensioner, so gets a further 10% reduction on the October price. How much does Joe pay to get into the zoo in October?

*4 Four children estimate the number of sweets in a jar. Sophie was 5% out and Henry was 30% out. The estimates of the four children are:

   750, 806, 589, 642

How many sweets were in the jar?

Mixed percentage problems

1 Homes were found for 112 of the 175 dogs in a rescue centre within a month. What percentage was this?

2 Jacks earns £6·50 an hour. On a Sunday, he earns an extra 48%.

   a How much per hour does he earn on a Sunday?

   Jack works a 40-hour week at the standard rate and 5 hours overtime on a Sunday.

   b How much is Jack's weekly wage?

   c How much tax does Jack pay if he is taxed at 22% on his total earnings? Give your answer to the nearest penny.

3 20 badgers were counted in one small area of an English county. This was $\frac{1}{12}$ of the badgers believed to exist in the county. If 84 of the badgers in the county were less than twelve months old, what percentage of the county's badger population were twelve months old or more?

Direct proportion

1 **a** 5 metres of copper pipe cost £1·85. What would 7 metres cost?

b 9 pencils cost £3·06. What would 4 pencils cost?

2 350 mℓ of a yoghurt drink costs 56p. What would 100 mℓ cost?

3 Bill and Bob are decorators. Bill is younger than Bob and works faster. For every 4 window frames Bill paints, Bob only manages to paint 3. If Bob paints 198 window frames in a month, how many does Bill paint?

4 A bead necklace has 4 purple beads for every 7 black beads. If the necklace contains 96 purple beads, how many black beads does it contain?

5 This recipe makes 20 biscuits.

a How many grams of butter would be needed to make 15 biscuits?

b How much sugar would be needed to make 36 biscuits?

c How much flour would be needed for 64 biscuits?

100 g	Butter
30 g	Sugar
1	Lemon
1	Egg yolk
170 g	Flour

Simplifying ratios

1 Write these ratios in their simplest form.

a 3 : 21 : 9   **b** 60 : 75 : 165   **c** 14 : 63 : 112

2 The three angles of a triangle are 84°, 60° and 36°. Write the ratio of the smallest to the largest angle in its simplest form.

3 Write these ratios in their simplest form.

a $2\frac{2}{3} : 4$ **b** $10 : 6\frac{1}{4}$ **c** 3·5 : 15

4 Write these as ratios in their simplest form.

a £2·50 : 85p   **b** 450 m : 2·5 km   **c** 40 min : $1\frac{1}{2}$ hours

5 Paula takes 1 min 15 seconds to complete a race. Her friend Julie takes 1 min 20 seconds. Write Paula's time to Julie's time as a ratio in its simplest form.

CHAPTER 7

Ratio and proportion

1 A box of chocolates contains 20 dark chocolates, 25 milk chocolates and 15 white chocolates.

   a What is the ratio of dark to white chocolates?

   b What is the ratio of dark to milk to white chocolates?

   c What fraction of the box is milk chocolates?

   d What percentage of the box is white chocolates?

2 A playgroup had 8 two-year-olds, 12 three-year-olds and 20 four-year-olds.

   a Write the ratio of two-year-olds to four-year-olds.

   b What proportion of the playgroup were three-year-olds?

   c What percentage of the playgroup were older than two?

Solving ratio and proportion problems

1 Bornham United won, drew and lost matches in the ratio 5 : 4 : 7.

   If they lost 21 matches, how many did they win?

2 A rum punch is made from lemonade, orange juice and rum in the ratio 7 : 4 : 1. If a jug of rum punch contains 280 ml of orange juice, how much lemonade and how much rum does it contain?

3 Kim, Gail and Colin inherited money from their uncle in the ratio 5 : 4 : 3 respectively.

   Colin got the smallest share of £450.

   How much did their uncle leave them altogether?

4 A model of a railway engine is made on a scale of 1 : 40.

   a If the height of the model is 6·5 cm, how high is the real engine in metres?

   b If the real engine is 8·5 m long, how long is the model in centimetres?

Dividing in a given ratio

1 **a** Share £160 in the ratio 1 : 4 : 5.

   b Share £405 in the ratio 2 : 5 : 8.

   c Share 384 g in the ratio 3 : 5 : 8.

2 A metal alloy uses 1 part tin, 7 parts copper and 3 parts zinc.

   How much of each is in 462 g of the alloy?

3 Shaun, Grant and Christina shared the profits of their business in the ratio of their investments, which were £5000, £8000 and £7000 respectively. If the business made a profit of £58 000 in the first year, how much did each receive?

CHAPTER 7

Exercise 1   Understanding algebra — expressions and equations

1   Which of these are expressions and which are equations? Explain how you can tell.

$4x + 7 = 15$ $12m - 5$ $\dfrac{6y}{5}$ $2(a + 6)$ $x - 3 = 8$

2   **a**   Is $5c + 2$ an expression or an equation? Can c have any value?

   b   Is $t + 5 = 7$ an expression or an equation? Can t have any value or does it have a particular value?

Exercise 2   Understanding algebra — formulae

1   The relationship between temperatures in °Fahrenheit, F, and °Celsius, C, is $F = \frac{9}{5}C + 32$. Is $F = \frac{9}{5}C + 32$ a formula or an equation? Explain.

2   One of these is an equation, one is an expression and one is a formula. Which is which?

   $p + 2q$ $2p + 3 = 8$ $p = 2q + 4$ where p is the total cost and q is the quantity of pastry in kg.

Exercise 3   Functions

1   In each of the following, there is one equation, one expression, one formula and one function. Which do you think is which?

   a $L = 4M + 50$, $5q + 3$, $6m + 3 = 18$, $y = 5x - 4$

   b $2(3m - p)$, $4(x + 2) = 24$, $y = 4 - \frac{1}{2}x$, $P = 4 - 3q$

   c $v^2 - u^2 = 2as$, $y = \dfrac{x}{3} + 2$, $\dfrac{7x - 4}{5}$, $3 - 2m = 5$

   d $y = \dfrac{4 - 3x}{7}$, $P = 4s$, $\dfrac{2a + 4}{3} = 5$, $\dfrac{p^2 - 4}{3p}$

Exercise 4   Rules for writing expressions

1   Write these without multiplication or division signs.

   a $2 \times m$ **b** $q \times 8$ **c** $x \div 8$ **d** $4 \times (x + 7)$

   e $(4 + p) \div 5$ **f** $2 \times a \times b$ **g** $p \times q \times 7$ **h** $6m \div 5$

   i $a \times a \times a \times 2$ **j** $2x \times 2x$

2   Write an expression for each of these.

   a m multiplied by n **b**   half of y

   c   four times m plus three   **d**   a third of the sum of a and b

   e   a quarter of the product of c and d **f**   five times the sum of s and t

Exercise 1 More understanding algebra — algebraic operations

1 Which part of these expressions would you work out first?

a $6 + 2x$ **b** $8 - m^2$ **c** $2(x + 7)$ **d** $4 - 3(7 - t)$

e $3y^2 + 4$ **f** $\dfrac{2(a + 4)}{5}$

2 Write true or false for each of these.

a $abc = cab$ **b** $p + q + r = pqr$ **c** $x + y + z = z + x + y$

3 Martina wrote this

Explain what is wrong with it.

Write it out correctly.

$$a + b = 26 + 29$$
$$= 26 + 20$$
$$= 46 + 9$$
$$= 55$$

Exercise 2 Inverse operations

1 Find two matching equations from the box for each of these. You could check by choosing numbers to substitute for the unknowns.

a $p - q = 3$

b $\dfrac{a}{2} = b$

c $4x = y$

d $6m = n$

e $t + r = 5$

$p = 3 - q$	$p = q + 3$	$pq = 3$	$q = p - 3$	$3 - p = q$
$b = 2a$	$ab = 2$	$2b = a$	$a + b = 2$	$\dfrac{a}{b} = 2$
$x = \dfrac{y}{4}$	$xy = 4$	$x = 4y$	$4 = \dfrac{y}{x}$	$4 - y = x$
$n = 6 \times m$	$6mn = 1$	$\dfrac{m}{n} = 6$	$mn = 6$	$6 = \dfrac{n}{m}$
$rt = 5$	$t = \dfrac{5}{r}$	$t = r - 5$	$5 - r = t$	$5 - t = r$

2 What is the inverse of each of these?

a subtracting 9

b multiplying by 4

c adding 3, then dividing by 2

d dividing by 6, then subtracting 10

e subtracting 4, then multiplying by 5

3 Choose the correct answer.

a If $2x + 3 = y$, then **A** $\dfrac{y}{2} - 3 = x$ **B** $\dfrac{y + 3}{2} = x$ **C** $\dfrac{y - 3}{2} = x$.

b If $4(m + 3) = n$, then **A** $m = 4n - 3$ **B** $m = \dfrac{n}{4} - 3$ **C** $m = \dfrac{n}{4} + 3$.

Simplifying expressions — multiplying and dividing

1 Simplify these.

    a $3 \times 6m$ **b** $4a \times 5$ **c** $2x - 6q$ **d** $^-6 \times 3t$

2 Write these using indices.

    a $c \times c \times c \times c \times c$ **b** $x \times x \times x \times x$ **c** $e \times e \times e \times e \times e \times e \times e \times e$

3 Simplify these.

    a $x^2 \times x$ **b** $p^2 \times p^2$ **c** $3y \times 2y$ **d** $2m \times 7m$

4 Simplify these.

    a $\dfrac{p^2}{p}$ **b** $\dfrac{y^3}{y}$ **c** $r^5 \div r^3$ **d** $m^4 \div m^4$

    e $\dfrac{8q^3}{2q}$ **f** $\dfrac{12b^4}{4b}$ **g** $\dfrac{6m^2}{3m}$

5 What is the width of these rectangles?

    a $36x^2$ $\leftarrow 9x \rightarrow$ **b** $24y^2$ $\leftarrow 8y \rightarrow$ **c** $54n^2$ $\leftarrow 9n \rightarrow$

6

| $\overline{2x}$ | $\overline{\dfrac{2b^2}{3}}$ | $\overline{20}$ | $\overline{r^2}$ | $\overline{r^2}$ | $\overline{1}$ | $\overline{r^2}$ | $\overset{\textbf{I}}{\overline{p^4}}$ | $\overline{\dfrac{2b^2}{3}}$ | | $\overline{2x}$ | $\overline{\dfrac{2b^2}{3}}$ | | $\overline{\dfrac{b^2}{3}}$ | $\overline{2x}$ | $\overline{\dfrac{2b^2}{3}}$ | $\overline{2}$ |
| $\overline{2x}$ | $\overline{\dfrac{2b^2}{3}}$ | | $\overline{2x}$ | $\overline{20}$ | | $\overline{r^2}$ | $\overline{m^3}$ | $\overline{4a}$ | $\overline{6y^2}$ | $\overline{r^2}$ | $\overline{\dfrac{2b^2}{3}}$ | $\overline{\dfrac{2b^2}{3}}$ | $\overline{2}$ | $\overline{6y^2}$ | $\overline{2x}$ | $\overset{\textbf{I}}{\overline{p^4}}$ | $\overline{20}$ |

Simplify these expressions.

Write the letter beside each expression above its answer in the box.

    I $p^3 \times p = p^4$ **P** $\dfrac{12a}{3}$ **N** $\dfrac{20x}{x}$ **Z** $q^2 \div q^2$

    E $r^4 \div r^2$ **X** $\dfrac{m^5}{m^2}$ **S** $\dfrac{10b^2}{15}$ **R** $2y \times 3y$

    A $\dfrac{40x}{20}$ **T** $6q^2 \div 3q^2$ **F** $\dfrac{2b^3}{6b}$

Brackets

1 Multiply out the brackets.

a $2(a + 7)$ **b** $11(x - 2)$ **c** $4(3 - q)$ **d** $5(y - 6)$

e $^-3(m + 8)$ **f** $^-4(j - 3)$ **g** $^-6(4 - f)$ **h** $7(2m + 5)$

i $4(6 - 4m)$ **j** $3(2a - 4b)$ **k** $^-7(5t - 3r)$ ***l** $14(24a - 21b)$

Homework Sheet 8.5   8 Core

New National Framework

Exercise 1   Collecting like terms

1

		o						
$\overline{4a - 3}$ $\overline{5a + 3b}$ $\overline{5b^2}$		$\overline{9b - 2}$ $\overline{6a + 9b}$ $\overline{4b}$ $\overline{8a^2}$ $\overline{4a + 2}$ $\overline{5b^2}$ $\overline{a + b}$ $\overline{4a - 3}$						

o							
$\overline{5a^2}$ $\overline{6a + 9b}$ $\overline{5a^2}$ $\overline{5b^2}$	$\overline{4a + 10b}$ $\overline{3a + 1}$ $\overline{a + b}$	$\overline{5b^2}$ $\overline{2a + b}$ $\overline{5b^2}$ $\overline{6a + 2b}$ $\overline{5b^2}$ $\overline{8a^2}$					

Simplify these. Write the letter beside each above its answer in the box.

O $2a + 3b + 4a + 6b$ **W** $5a + 3b - a + 7b$ **H** $8a + b - 3a + 2b$ **L** $6a + 2b - 4a - b$

$\quad = 6a + 9b$

G $6a - 4 - 2a + 6$ **Y** $8b + 5 + b - 7$ **S** $7a + 2b - 6a - b$ **T** $5a + 1 - a - 4$

U $2a + 3b - 2a + b$ **V** $a + b + 5a + b$ **A** $a - 4 + 2a + 5$ **N** $5a^2 + 3a^2$

E $6b^2 - b^2$ **P** $2a^2 + 4a^2 - a^2$

2 This diagram shows a pentagon. Write an expression for the perimeter of the pentagon. Simplify the expression.

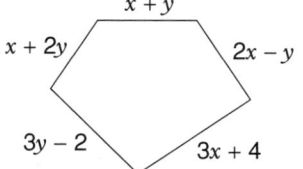

3 The number in each box is found by adding the numbers in the two boxes below it.

Find the missing expressions. Simplify them.

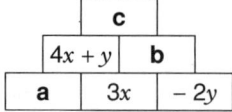

Exercise 2   Multiplying out brackets, then simplifying

1 Multiply out the brackets and then simplify.

a $2(x + 3) + 5(x + 1)$ **b** $6(a + 4) + 3(a - 2)$

c $4(m - 7) + 5(m + 4)$ **d** $9(2y + 4) + 2y$

e $6 + 4(k + 7)$ **f** $5 - (x + 2)$

g $11 - 2(2m + 3)$ **h** $6(4p - q) - (3q - 2p)$

CHAPTER 8

Substituting into expressions

1

$^-60$	$^-4$	$^-60$	30		$^-60$	$^-4$	$^-0{\cdot}5$	30		28	30	$2{\cdot}5$	
	R											**R**	
$^-4$	12	11	2	11	$^-2{\cdot}5$	30	$^-0{\cdot}5$	$^-0{\cdot}5$	10	2	12	6	6	$^-2{\cdot}5$

If $a = 2$, $b = 1{\cdot}5$ and $c = {}^-4$, evaluate these. Write the letter that is beside each above its answer in the box.

R $3a^2 = 12$ **A** $7a^2 + 2$ **Y** $2a^3 - 6$ **I** $10b - 2a$

S $a^2 - b$ **W** $2(a + 8b)$ **L** $a + b + c$ **E** $5a + c$

G $\dfrac{^-c}{a}$ **N** $\dfrac{2b + 2c}{a}$ **O** $\dfrac{4(c - a)}{a - c}$ **C** $5(c - 2a^2)$

2 If $m = 5$, $n = 3$, and $p = 8$, find the value of these.

a $\dfrac{21}{n}$ **b** $\dfrac{p}{2} + m$ **c** $2n^2$ **d** $p^2 - m^2$

e $\dfrac{2m + p}{n}$ **f** $\sqrt{m^2 - n^2}$ **g** $(m + n)^2 - p^2$ **h** $m(n^2 - p)$

3 The formula for the nth triangular number is $\dfrac{n(n + 1)}{2}$.

a The fourth triangular number is given by $\dfrac{n(n + 1)}{2}$ when $n = 4$.

Find the fourth triangular number.

b Find the tenth triangular number.

Exercise 1　Substituting into formulae

1 The formula for the time taken to cook a turkey is $T = 20w + 30$ where T is the time in minutes and w is the weight of the turkey in pounds.

Find the time needed to cook a turkey weighing

a 9 pounds　　**b** 12 pounds　　**c** 16 pounds.

2 The cost of taking a youth club on an outing is given by the formula

$$c = 6n + 30m$$

where c is the cost in pounds, n is the number of people going and m is the number of minibuses hired.

a Find c when $n = 40$ and $m = 4$.

b Find c when $n = 65$ and $m = 7$.

3 A room is l metres long, w metres wide and h metres high. The formula for the areas of the walls, A m^2, is

$$A = 2h(l + w).$$

Find A when

a $l = 7$ m, $w = 5$ m, $h = 3$ m

b $l = 11$ m, $w = 8$ m, $h = 2.8$ m.

4 The diagram shows a square picture frame of side L cm, for a square picture of side m cm. The formula for calculating the area of the surround is $A = L^2 - m^2$. Find the area of the surround when

a $L = 20$ cm, $m = 10$ cm　　**b** $L = 25$ cm, $m = 12$ cm

c $L = 24$ cm, $m = 16$ cm.

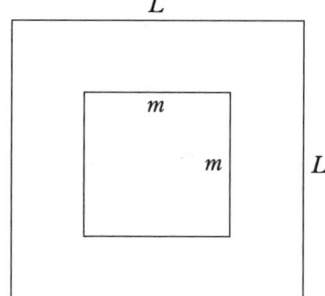

Exercise 2　Substituting into a formula

1 The formula for the area of a triangle, A, is $A = \frac{1}{2}bh$ where b is the base and h is the height.

a Find A when $b = 7$ cm and $h = 8$ cm.　　**b** Find h when $b = 4$ cm and $A = 20$ cm^2.

2 If a stone is dropped from a cliff top, the distance, s m, it falls in time, t, sec is given by the formula $s = 5t^2$.

a Find s when $t = 6$ seconds.　　***b** Find t when $s = 405$ m.

CHAPTER 8

Writing expressions, formulae and equations

1 Kirk gets m pounds pocket money each month.

   a Write an expression for the total amount of pocket money Kirk got in a year.

   b Lenny gets twice as much as Kirk, less £1. Write an expression for the amount Lenny gets.

   c Max gets half as much as Lenny. Write an expression for the amount Max gets.

2 Apples cost x pence each and bananas cost y pence each. Write an expression for the total cost, in pence, of 6 apples and 4 bananas.

3 Donna scores t points in a school quiz.

   a Sunita scores 3 points less than Donna. How many points does Sunita score?

   b Lauren scores twice as many points as Sunita. How many points does Lauren score?

More writing and simplifying algebraic expressions

1 Mark made a stair gate with wooden strips.

   a Write an expression for the total length of wooden strips used.

   b Write a different expression for the total length of wooden strips used.

   c Show that the expressions you wrote in **a** and **b** are equivalent.

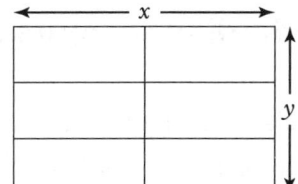

2 Write three different expressions for the area of the room shown in the diagram.

   Simplify your expressions to show that they are equivalent.

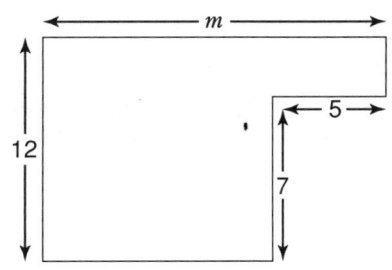

*3 Try to prove, using algebra, that the sum of any seven consecutive numbers is divisible by 7.

Writing equations

1 Pooch Parlour grooms k dogs in one week and Groom Room grooms t dogs in the same week.

Write an equation for each of these.

   a Altogether Pooch Parlour and Groom Room groom 840 dogs that week.

   b Groom Room groomed three times as many dogs as Pooch Parlour.

  ***c** If Pooch Parlour had groomed 160 more dogs, they would have groomed half the number of dogs groomed by Groom Room.

2 Kate is n years old and her daughter Tara is h years old. Write an equation for each of these.

   a Tara is 19 years younger than Kate.

   b The sum of Kate's and Tara's ages is 32.

   c If Kate were 3 years younger, she would be 4 times as old as Tara.

   d Three times the difference between Kate's and Tara's ages is 52.

Writing and finding formulae

1 Write a formula for each of these.

   a The number of yards, y, is found by dividing the number of feet, f, by 3.

   b The cost, c, of hiring a hedgecutter is found by multiplying the number of days hired, d, by 4 and adding 7.

   c The number of sandwiches needed, s, is found by multiplying the number of people, p, by 6 and adding 16.

   d The number of bags of cement, b, needed to lay a concrete garage base is found by multiplying the length of the garage, L, by the width, W and dividing by 5.

Exercise 1   Writing and solving equations

1   Solve these equations.

   a $4x = 14$ **b** $7x + 4 = 46$ **c** $4 = \dfrac{20}{y}$

   d $37 = 8m - 3$ **e** $2 \cdot 3a + 3 \cdot 6 = 22$ **f** $5 \cdot 2m - 6 \cdot 8 = 14$

2   Write equations for each of these. Solve the equations.

   a Jack is p years old. Rupert is 7 years older. Three times the sum of Jack and Rupert's ages is 57. How old is Jack?

   b I think of a number, y. I multiply it by 9, then subtract 17. The answer is 163. What number did I think of?

   c If I buy 7 books at c pounds each and 8 magazines at £2 each, the total cost is £61·50. How much is each book?

3   Solve these equations.

   a $5(x + 11) = 70$ **b** $6(2m - 8) = 84$ **c** $4(3b + 6) = 96$

   d $3(5m - 6) = 42$ ***e** $5(3x + 2) = 37$ ***f** $10(7y + 8) = 164$

Exercise 2   More solving equations

1   Solve these equations.

   a $2x + 10 + 3x - 3 = 22$ **b** $3n + 2 + 4n - 6 = 52$

   c $10x + 4 - 4x - 5 = 11$ **d** $4(k + 2) + 2(k - 1) = 30$

   e $3(p - 3) + 4(2p + 1) = 50$ **f** $6(2c + 3) - 4(c - 1) = 38$

2   The perimeter of this rectangle is 162 cm. Write and solve an equation to find the width of the rectangle, h.

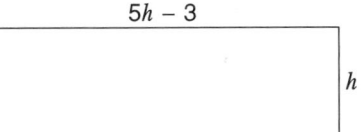

3   Write and solve an equation for each of these.

   a The mass of eight packets of biscuits and the box they are stored in is 2500 g. If the box weighs 340 g, what is the mass of each packet of biscuits?

   b Ruth buys two boxes of chocolates. There are n chocolates in the first box and six fewer chocolates in the second box. If there are 50 chocolates altogether in the two boxes, how many chocolates were in the first box?

   c Adding 5 to 4 times a number, then dividing by 3, gives the answer 11. Find the number.

CHAPTER 8

Exercise 1   Solving equations by transforming both sides

1   Solve these equations by doing the same to both sides.

a $4x - 8 = 20$

b $7 + 5a = 22$

c $\dfrac{k}{4} = 8$

d $11y = 44$

e $2m - 5 = 11$

f $6t + 3 = 27$

2   Solve these equations to fill in this crossnumber.

Across

1 $n - 6 = 8$

2 $\dfrac{n+4}{3} = 5$

4 $5n + 10 = 120$

5 $3n + 2 = 20$

8 $\dfrac{n}{3} = 16$

9 $3n - 20 = 25$

10 $\dfrac{n}{2} + 3 = 8$

12 $3 + 5n = 48$

Down

1 $2n + 4 = 28$

3 $n + 3 = 20$

4 $\dfrac{n}{5} + 2 = 7$

6 $\dfrac{n}{6} = 3$

7 $2n + 8 = 50$

8 $2n + 13 = 93$

10 $\dfrac{n}{4} = 4$

11 $2n - 20 = 50$

Exercise 2   Solving equations

1   Solve these equations.

a $6x = 4x + 8$

b $2p + 3 = 7p - 17$

c $3y + 14 = y + 20$

d $4m - 5 = 2m + 9$

2   I think of a number. If I multiply it by 6 and add 4, I get the same answer as when I multiply it by 9 and subtract 14. Find the number.

3   Lisa and Glen have the same number of CDs. They store their CDs in equal-sized CD holders. Lisa has 5 holders, each containing n CDs, and 6 extra CDs. Glen has 4 holders, each containing n CDs, and 18 extra CDs. How many CDs fit into one CD holder?

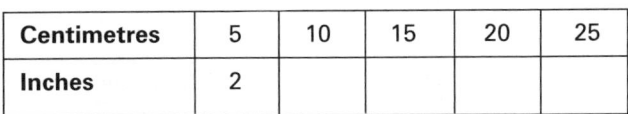
Solving equations using a graph

1 **a** There are 5 centimetres to every 2 inches.

Copy and fill in this table.

b Work out the ratio   centimetres : inches
for each pair of values on the table.
What do you notice?

Centimetres	5	10	15	20	25
Inches	2				

c Is the number of inches directly proportional to
the number of centimetres? Explain.

Inches versus centimetres

d Plot the graph of centimetres against inches.

Do the points lie in a straight line?

e Write a formula for the relationship between
inches (y) and centimetres (x).

$y =$

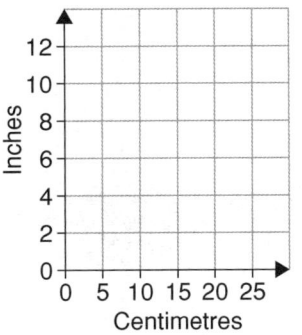

f Find the number of inches in these numbers of
centimetres.

   i 65 cm   **ii** 80 cm

2 William was sponsored £20 for every 3 laps of the athletics track he ran.

Number of laps	3	6	9	12	15
Sponsor money in £	20				

a Copy and fill in this table.

b Draw a graph to illustrate this relationship, putting the number of laps on the horizontal axis. Do the
points lie in a straight line?

c Write a formula for the relationship between sponsor money (y) and number of laps (x).

$y =$

d Use this relationship to find the sponsor money William would get if he completed

   i 21 laps   **ii** 30 laps.

CHAPTER 8

Writing sequences from flow charts

1 Write down the sequences given by these flow charts.

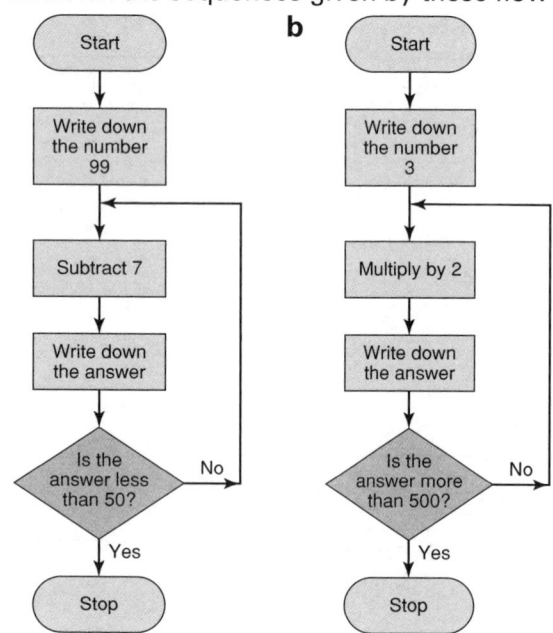

a

Start

Write down the number 99

Subtract 7

Write down the answer

Is the answer less than 50? — No

Yes

Stop

b

Start

Write down the number 3

Multiply by 2

Write down the answer

Is the answer more than 500? — No

Yes

Stop

Generating sequences by multiplying and dividing

1 Write down the first six terms of these sequences.

   a *1st term* 5 *rule* multiply by 2   **b** *1st term* 1 *rule* multiply by ⁻4

   c *1st term* 120 *rule* divide by 2   **d** *1st term* ⁻32 *rule* divide by ⁻2

2 Each of these sequences continues in the same way. Write down the next two terms.

   a 2, 10, 50, ...   **b** 9, 3, 1, ...   **c** 100, 20, 4, ...

Counting forwards or backwards in increasing or decreasing steps

1 Write down the first six terms of these sequences.

   a Start at 3 and count forwards by 2, 3, 4, 5, ...

   b Start at 0 and count forwards by 2, 4, 6, 8, ...

   c Start at 70 and count backwards by 1, 3, 5, 7, ...

2 Valentina decorates porcelain plates in a factory. She completes 20 plates during her first week. Each subsequent week she completes 1, 2, 3, 4, ... more plates than the previous week.

   a Write down how many plates she completed in each of the first 6 weeks.

   b Write down how many plates she completed in the tenth week.

Arithmetic sequences

1   Write down the first six terms of these arithmetic sequences.

   a $a = 4, d = 6$ **b** $a = 5, d = 9$ **c** $a = 14, d = {}^-5$

2   Write down the values of a and d for these arithmetic sequences.

   a 9, 13, 17, 21, ...   **b** 16, 9, 2, $^-5$, $^-12$, ...   **c** $^-8$, $^-5$, $^-2$, 1, 4, ...   **d** 0·8, 0·3, $^-0·2$, $^-0·7$, ...

Predicting the next few terms of a sequence

1   Predict the next three terms of these sequences.

   For each sequence decide if it

   A ascends by equal steps   **B** ascends by unequal steps

   C descends by equal steps   **D** descends by unequal steps

   a 80, 68, 56, 44, 32, ...   **b** 4, 11, 18, 25, 32, ...   **c** 1, 3, 6, 10, 15, ...   **d** 75, 73, 69, 63, 55, ...

2   Shahid was asked how the sequence 3, 6, 12, ... might continue.

   Give two possible answers Shahid could give. Explain your answers.

Writing sequences from term-to-term rules and rules for the nth term

1   Write down the first five terms of these sequences.

	1st term	term-to-term rule
a	10 000	Divide by 5
b	2	Multiply by $^-3$
c	7	Add consecutive even numbers starting with 2
d	1, 4	Add the two previous terms
e	3	Double the number, then subtract 2

2   The nth term of a sequence is given. Write the first five terms.

   a $T(n) = 2n + 7$ **b** $T(n) = 106 - 5n$ **c** $T(n) = 3n + 0·5$ **d** $T(n) = 0·3n$

3   For each of the sequences in **question 2**, find $T(15)$.

4   Describe each of the sequences in **question 2**, using a term-to-term rule.

5   The term-to-term rule for a sequence is   'add 8'. What might the first term and the rule be if

   a all the terms are multiples of 8   **b** all the terms are 3 more than the multiples of 8

   c all the terms are even   **d** at least 4 terms are negative.

Describing linear sequences

1 Paul said that $T(n) = 8n + b$ would generate multiples of 8.

   a For what values of b is Paul correct?

   b Describe the sequence generated by $T(n) = 8n - 2$.

2 $T(n) = 5n + b$

   Describe the sequence that is generated when

   a $b = 0$ **b** $b = 5$ **c** $b = 20$ **d** $b = 1$ **e** $b = {}^-1$.

3 **a** Describe the sequence generated by $T(n) = 99 - 9n$

   b Write a rule for the nth term of a sequence that

      i ascends by fours   **ii** descends by threes.

Homework Sheet 9.7   8 Core

New National Framework

Describing linear sequences

1 Paul said that $T(n) = 8n + b$ would generate multiples of 8.

   a For what values of b is Paul correct?

   b Describe the sequence generated by $T(n) = 8n - 2$.

2 $T(n) = 5n + b$

   Describe the sequence that is generated when

   a $b = 0$ **b** $b = 5$ **c** $b = 20$ **d** $b = 1$ **e** $b = {}^-1$.

3 **a** Describe the sequence generated by $T(n) = 99 - 9n$

   b Write a rule for the nth term of a sequence that

      i ascends by fours   **ii** descends by threes.

Arithmetic sequences

1   Write down the first six terms of these arithmetic sequences.

   a $a = 4, d = 6$ **b** $a = 5, d = 9$ **c** $a = 14, d = {}^-5$

2   Write down the values of a and d for these arithmetic sequences.

   a 9, 13, 17, 21, ...   **b** 16, 9, 2, ${}^-5$, ${}^-12$, ...   **c** ${}^-8$, ${}^-5$, ${}^-2$, 1, 4, ...   **d** 0·8, 0·3, ${}^-0·2$, ${}^-0·7$, ...

Predicting the next few terms of a sequence

1   Predict the next three terms of these sequences.

   For each sequence decide if it

   A ascends by equal steps   **B** ascends by unequal steps

   C descends by equal steps   **D** descends by unequal steps

   a 80, 68, 56, 44, 32, ...   **b** 4, 11, 18, 25, 32, ...   **c** 1, 3, 6, 10, 15, ...   **d** 75, 73, 69, 63, 55, ...

2   Shahid was asked how the sequence 3, 6, 12, ... might continue.

   Give two possible answers Shahid could give. Explain your answers.

Writing sequences from term-to-term rules and rules for the nth term

1   Write down the first five terms of these sequences.

	1st term	term-to-term rule
a	10 000	Divide by 5
b	2	Multiply by ${}^-3$
c	7	Add consecutive even numbers starting with 2
d	1, 4	Add the two previous terms
e	3	Double the number, then subtract 2

2   The nth term of a sequence is given. Write the first five terms.

   a $T(n) = 2n + 7$ **b** $T(n) = 106 - 5n$ **c** $T(n) = 3n + 0·5$ **d** $T(n) = 0·3n$

3   For each of the sequences in **question 2**, find T(15).

4   Describe each of the sequences in **question 2**, using a term-to-term rule.

5   The term-to-term rule for a sequence is   'add 8'. What might the first term and the rule be if

   a all the terms are multiples of 8   **b** all the terms are 3 more than the multiples of 8

   c all the terms are even   **d** at least 4 terms are negative.

Describing linear sequences

1 Paul said that $T(n) = 8n + b$ would generate multiples of 8.

   a For what values of b is Paul correct?

   b Describe the sequence generated by $T(n) = 8n - 2$.

2 $T(n) = 5n + b$

   Describe the sequence that is generated when

   a $b = 0$ **b** $b = 5$ **c** $b = 20$ **d** $b = 1$ **e** $b = {}^-1$.

3 **a** Describe the sequence generated by $T(n) = 99 - 9n$

   b Write a rule for the nth term of a sequence that

   i ascends by fours   **ii** descends by threes.

Homework Sheet 9.7 8 Core

New National Framework

Describing linear sequences

1 Paul said that $T(n) = 8n + b$ would generate multiples of 8.

   a For what values of b is Paul correct?

   b Describe the sequence generated by $T(n) = 8n - 2$.

2 $T(n) = 5n + b$

   Describe the sequence that is generated when

   a $b = 0$ **b** $b = 5$ **c** $b = 20$ **d** $b = 1$ **e** $b = {}^-1$.

3 **a** Describe the sequence generated by $T(n) = 99 - 9n$

   b Write a rule for the nth term of a sequence that

   i ascends by fours   **ii** descends by threes.

Sequences in practical situations

1   Jane is making a cross-stitch sample with a sequence of growing patterns.

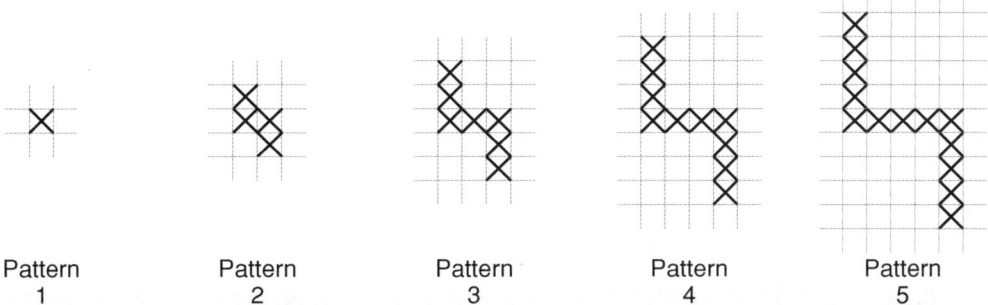

| Pattern | Pattern | Pattern | Pattern | Pattern |
| 1 | 2 | 3 | 4 | 5 |

a What sequence is generated by the number of crosses in Pattern 1, Pattern 2, Pattern 3, ...?
Predict the next few terms.
Explain how you got this sequence, referring to the diagrams.

b Write an expression for the number of crosses needed for Pattern n.
Justify your expression by referring to the diagrams.

2   Martin is making square photo frames, edged with small square tiles.

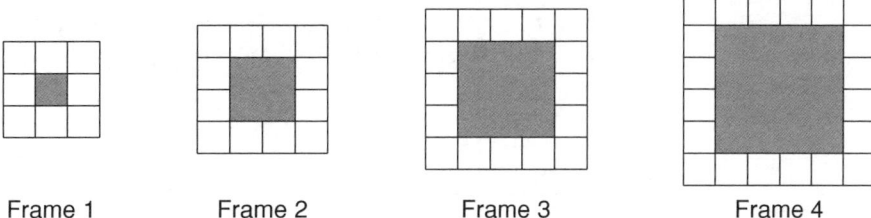

Frame 1   Frame 2   Frame 3   Frame 4

a What sequence is generated by the number of small square edging tiles in Frame 1, Frame 2, Frame 3, ...?
Predict the next few terms.
Explain how you got this sequence, referring to the diagrams.

b Write an expression for the number of square tiles needed for Frame n.
Justify your expression by referring to the diagrams.

3   Julie is stacking triangular boxes for a display.

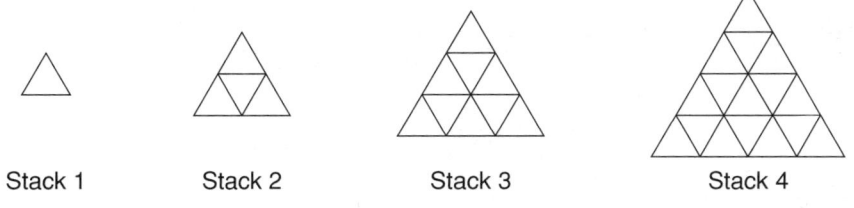

Stack 1   Stack 2   Stack 3   Stack 4

a What sequence is generated by the 'point up' triangular boxes?

b What sequence is generated by the total number of boxes in each stack?

CHAPTER 9

Finding the rule for the *n*th term.

1 Write an expression for the *n*th term.

    a 2, 5, 8, 11, 14, ...   **b** 7, 11, 15, 19, 23, ...   **c** 63, 57, 51, 45, 39, ...   **d** 2, ⁻1, ⁻4, ⁻7, ⁻10, ...

2 8, 17, 26, 35, 44, ...

    The sequence continues in the same way.

    a Write the next three terms in the sequence.   **b** Explain the rule for finding the next term.

    c What is the 15th term of the sequence?   **d** What is the *n*th term of the sequence?

Functions

1 Complete the output column for each function machine.

    a $x \rightarrow$ [multiply by 5] $\rightarrow$ [subtract 3] $\rightarrow y$

Input	Output
7	
20	
1·5	
$4\frac{1}{2}$	

    b $x \rightarrow$ [add 4] $\rightarrow$ [divide by 3] $\rightarrow y$

Input	Output
8	
29	
5·6	
$2\frac{3}{8}$	

2 Fill these mapping diagrams in for each function machine and input given.

    a 0, 1, 2, 3, 4 $\rightarrow$ [add 1] $\rightarrow$ [multiply by 2] $\rightarrow y$

```
 ⁻1  0  1  2  3  4  5  6  7  8  9 10 11

 ⁻1  0  1  2  3  4  5  6  7  8  9 10 11
```

    b ⁻2, ⁻1, 0, 1, 2 $\rightarrow$ [multiply by 3] $\rightarrow$ [subtract 4] $\rightarrow y$

```
 ⁻10 ⁻9 ⁻8 ⁻7 ⁻6 ⁻5 ⁻4 ⁻3 ⁻2 ⁻1  0  1  2  3  4  5  6

 ⁻10 ⁻9 ⁻8 ⁻7 ⁻6 ⁻5 ⁻4 ⁻3 ⁻2 ⁻1  0  1  2  3  4  5  6
```

3 Draw an input/output table and a mapping diagram for this function machine and given input.

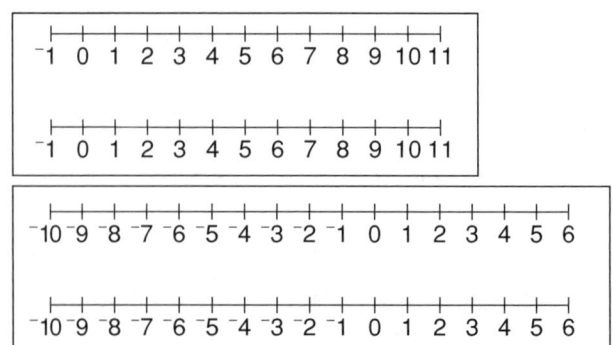

⁻2, ⁻1, 0, 1, 2 $\rightarrow$ [add 3] $\rightarrow$ [divide by 2] $\rightarrow y$

4 Write true or false for these.

    a The lines on the mapping diagram for $x \rightarrow x + 3$ will be parallel.

    b The lines on the mapping diagram for $y = 3x$, if extended backwards will all meet at a point on the zero line.

CHAPTER 9

Sequences in practical situations

1 Jane is making a cross-stitch sample with a sequence of growing patterns.

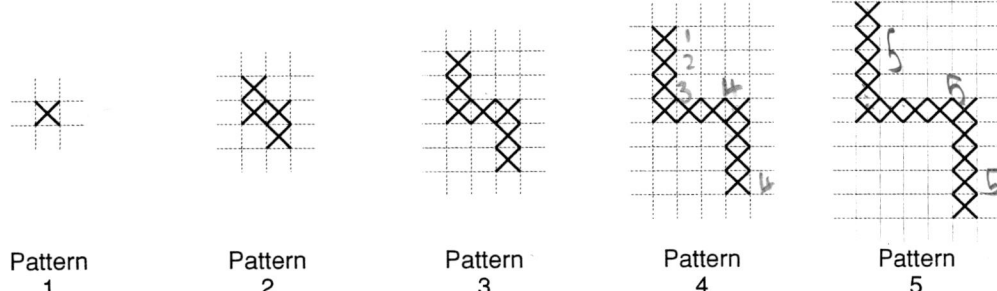

Pattern 1　　Pattern 2　　Pattern 3　　Pattern 4　　Pattern 5

a What sequence is generated by the number of crosses in Pattern 1, Pattern 2, Pattern 3, ...?
Predict the next few terms.
Explain how you got this sequence, referring to the diagrams.

b Write an expression for the number of crosses needed for Pattern n.
Justify your expression by referring to the diagrams.

2 Martin is making square photo frames, edged with small square tiles.

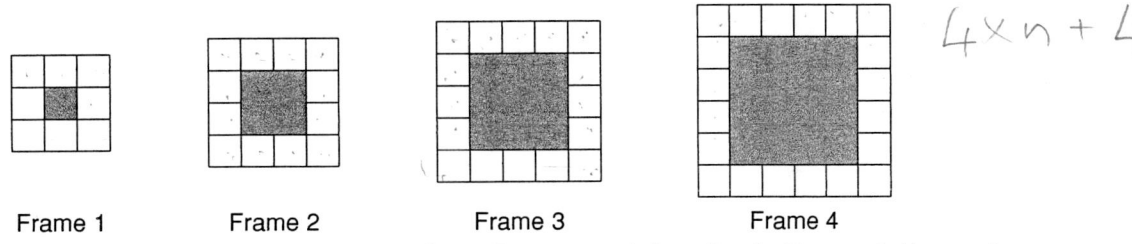

Frame 1　　Frame 2　　Frame 3　　Frame 4

$4 \times n + 4$

a What sequence is generated by the number of small square edging tiles in Frame 1, Frame 2, Frame 3, ...?
Predict the next few terms.
Explain how you got this sequence, referring to the diagrams.

b Write an expression for the number of square tiles needed for Frame n.
Justify your expression by referring to the diagrams.

3 Julie is stacking triangular boxes for a display.

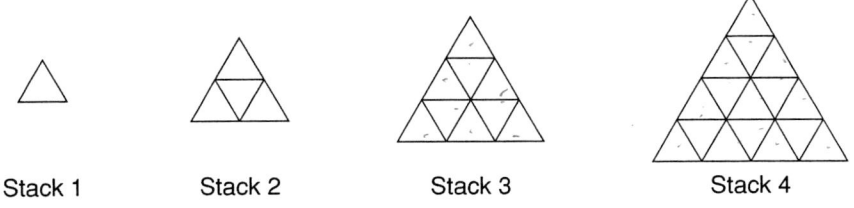

Stack 1　　Stack 2　　Stack 3　　Stack 4

a What sequence is generated by the 'point up' triangular boxes?

b What sequence is generated by the total number of boxes in each stack?

Finding the rule for the *n*th term.

1 Write an expression for the *n*th term.

    a 2, 5, 8, 11, 14, ...   **b** 7, 11, 15, 19, 23, ...   **c** 63, 57, 51, 45, 39, ...   **d** 2, ⁻1, ⁻4, ⁻7, ⁻10, ...

2 8, 17, 26, 35, 44, ...

    The sequence continues in the same way.

    a Write the next three terms in the sequence.   **b** Explain the rule for finding the next term.

    c What is the 15th term of the sequence?   **d** What is the *n*th term of the sequence?

Functions

1 Complete the output column for each function machine.

a $x \rightarrow$ [multiply by 5] $\rightarrow$ [subtract 3] $\rightarrow y$

Input	Output
7	
20	
1·5	
$4\frac{1}{2}$	

b $x \rightarrow$ [add 4] $\rightarrow$ [divide by 3] $\rightarrow y$

Input	Output
8	
29	
5·6	
$2\frac{3}{8}$	

2 Fill these mapping diagrams in for each function machine and input given.

a 0, 1, 2, 3, 4 $\rightarrow$ [add 1] $\rightarrow$ [multiply by 2] $\rightarrow y$

```
⁻1  0  1  2  3  4  5  6  7  8  9  10 11

⁻1  0  1  2  3  4  5  6  7  8  9  10 11
```

b ⁻2, ⁻1, 0, 1, 2 $\rightarrow$ [multiply by 3] $\rightarrow$ [subtract 4] $\rightarrow y$

```
⁻10 ⁻9 ⁻8 ⁻7 ⁻6 ⁻5 ⁻4 ⁻3 ⁻2 ⁻1  0  1  2  3  4  5  6

⁻10 ⁻9 ⁻8 ⁻7 ⁻6 ⁻5 ⁻4 ⁻3 ⁻2 ⁻1  0  1  2  3  4  5  6
```

3 Draw an input/output table and a mapping diagram for this function machine and given input.

 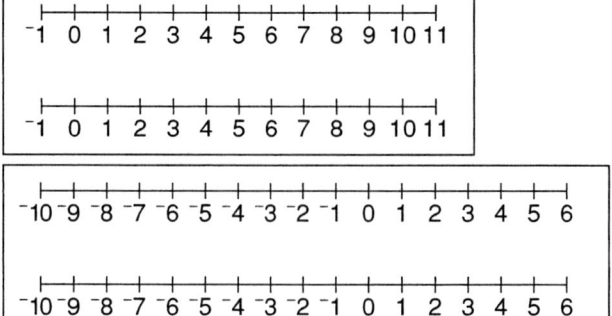

⁻2, ⁻1, 0, 1, 2 $\rightarrow$ [add 3] $\rightarrow$ [divide by 2] $\rightarrow y$

4 Write true or false for these.

    a The lines on the mapping diagram for $x \rightarrow x + 3$ will be parallel.

    b The lines on the mapping diagram for $y = 3x$, if extended backwards will all meet at a point on the zero line.

Finding the function given the input and output

1 Find the function for these.

a   1, 4, 2, 5, 3 → ☐ → ☐ → 5, 14, 8, 17, 11

b   4, 1, 3, 2, 5 → ☐ → ☐ → 14, ⁻1, 9, 4, 19

c   8, 4, 10, 6, 2 → ☐ → ☐ → 12, 4, 16, 8, 0

d   1, 5, 9, 4, 8 → ☐ → ☐ → 5, 21, 37, 17, 33

Properties of functions

1   What single operation could replace these?

a → add 4 → add 7 →

b → multiply 2 → multiply 6 →

c → add 2 → subtract 5 →

d → multiply 8 → divide 2 →

2   Find two different ways of writing the rule for this function.

1, 2, 3, 4 → ☐ → ☐ → ⁻3, 0, 3, 6

3   What are the missing operations?

a   1, 7, 4, 0 → divide by 2 → add 5 → ? → ? → 1, 7, 4, 0

b   6, 10, 4, 7 → multiply by 6 → subtract 8 → ? → ? → 6, 10, 4, 7

c   8, 11, 2, 9 → add 4 → divide by 10 → ? → ? → 8, 11, 2, 9

4   Find the input. Draw inverse function machines to help.

a ___ → add 4 → divide by 3 → 5

b ___ → multiply by 2 → add 5 → 17

c ___, ___, ___ → divide by 2 → subtract 6 → 1, ⁻2, 8

d ___, ___, ___ → subtract 3 → multiply by 5 → 30, 10, ⁻5

5   Find the inverse function for these.

a $x \rightarrow 6x - 4$ **b** $x \rightarrow \frac{x}{3} - 2$ **c** $x \rightarrow 3(x + 4)$ **d** $x \rightarrow \frac{x + 7}{2}$

CHAPTER 9

Graphing functions

1 Which of these coordinate pairs lie on the line $y = 3x + 2$?

   a (3, 2)   **b** (2, 3)   **c** (1, 5)   **d** (3, 11)   **e** (2, 7)

2 Which of these coordinate pairs satisfy the rule $y = 2x - 5$?

   a (3, 1)   **b** (2, 5)   **c** (2, ⁻1)   **d** (⁻5, 2)   **e** ($-\frac{1}{2}$, ⁻6)

3　**a** Copy and complete the table for $y = 3x - 4$.

   b Write down the coordinate pairs for the points.

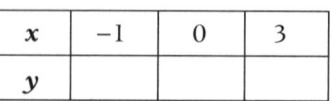

x	−1	0	3
y			

   c On the grid, plot the three points.

     Draw and label the line with the equation $y = 3x - 4$.

   d Write down the coordinate pairs for two other points that lie on the line.

     Do the coordinate pairs satisfy the rule $y = 3x - 4$?

   e Does $y = 3x - 4$ go through the point (8, 20)? Explain.

   f Will the point (⁻11, ⁻29) lie on the line? Explain.

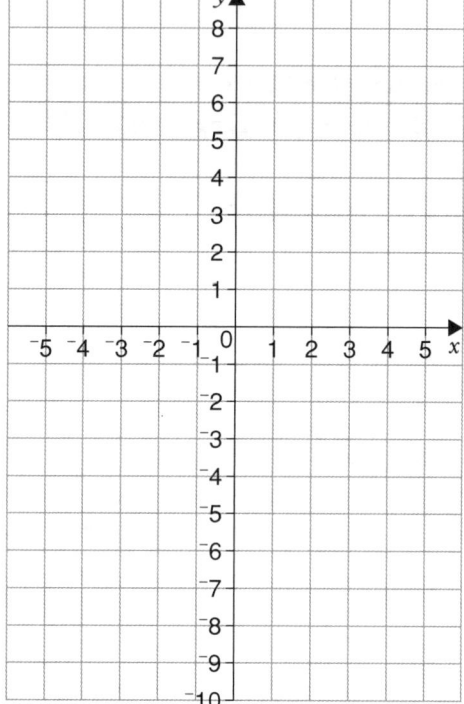

4 This table shows the sequence for the number of tiles in a tiling pattern.

Pattern number	1	2	3	4	5
Number of tiles	1	4	7	10	13

   a Plot the sequence on a grid. Do the points lie in a straight line?

   b Is 25 a term of the sequence? Explain.

   c Would it be sensible to join the points with a straight line? Explain.

Equations of straight-line graphs

1 Look at the equations of straight-line graphs given in the box.

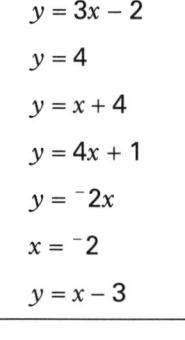

$$y = 3x - 2$$
$$y = 4$$
$$y = x + 4$$
$$y = 4x + 1$$
$$y = {}^-2x$$
$$x = {}^-2$$
$$y = x - 3$$

  a Which one will

    i   be the steepest (but not vertical)

    ii   be horizontal

    iii have a negative slope

    iv be parallel to the y-axis?

  b Which two are parallel to each other?

2 Match these equations with the lines on the grid.

  A $y = {}^-x + 3$

  B $x = 5$

  C $y = 3x - 6$

  D $y = 3x$

  E $y = {}^-5$

  F $y = {}^-x$

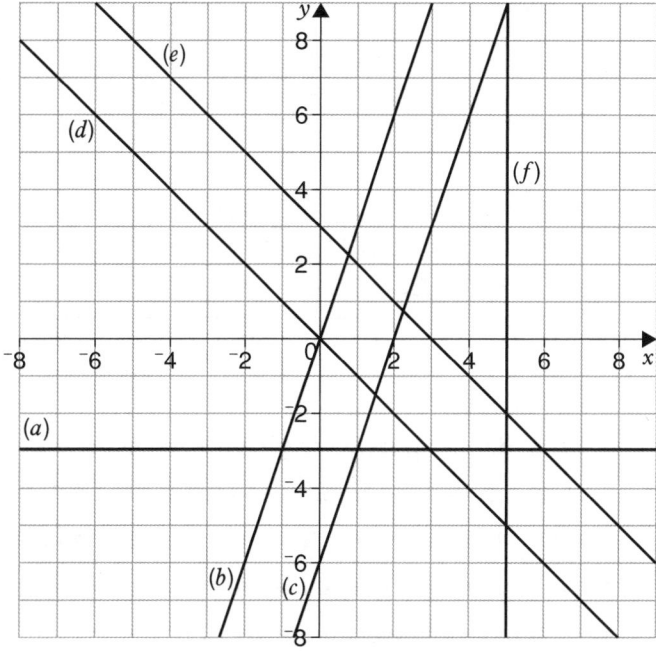

*3 Write down the equation of a line which

  a is parallel to $y = {}^-5x + 7$ and passes through the origin

  b crosses the y-axis at (0, 4) and is parallel to $y = x + 9$.

*4 **a** A line with the equation $y = mx + 4$ goes through the point (3, 10).
    What is the value of m?

  b A line with the equation $y = mx - 5$ goes through the point (2, 1).
    What is the value of m?

  c A line with the equation $y = 4x + c$ goes through the point (${}^-1$, ${}^-2$).
    What is the value of c?

CHAPTER 10

Reading and plotting real-life graphs

1 This graph shows the mass of a kitten and a puppy at different ages.

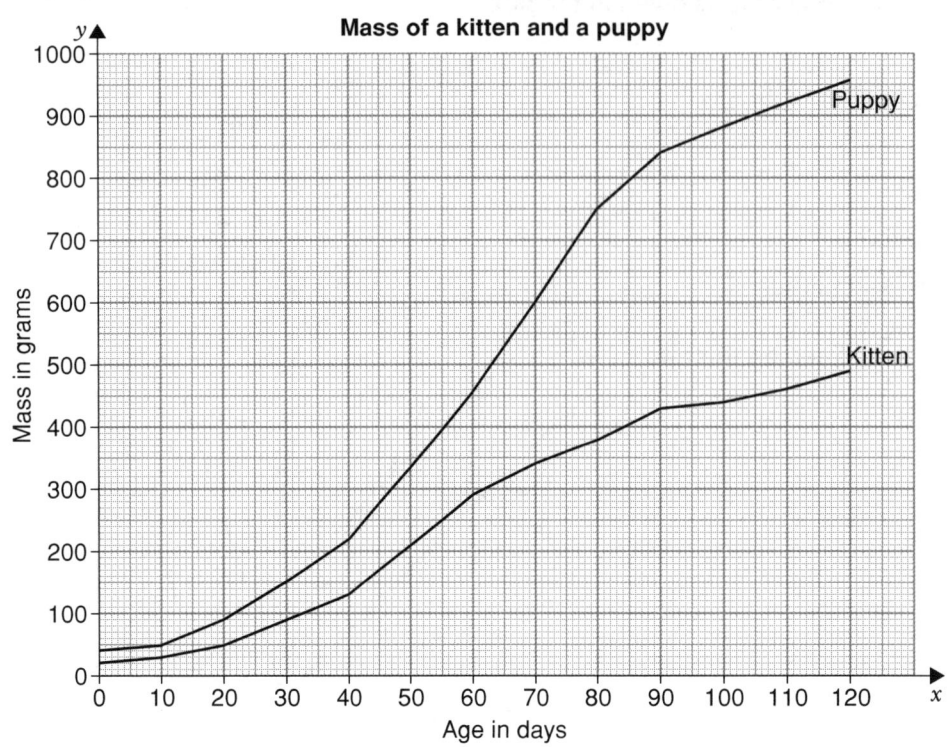

Mass of a kitten and a puppy

a What did the puppy weigh when it was 40 days old?

b How old was the kitten when it weighed 430 g?

c How much more did the puppy weigh than the kitten at 100 days old?

d What do you estimate the kitten weighed at 75 days old?

e How old do you estimate the puppy was when it weighed 810 g?

f What does the comparison of the graphs tell you about the growth of the kitten and puppy?

2 Luc is taking part in a sponsored swim. For each length of the pool he completes, he raises £4·50.

a Complete this table.

b On this grid, show the relationship 'money raised' versus 'number of lengths completed'.

Number of lengths completed	1	2	3	4	5
Money raised in £					

c Is the relationship linear?

d Should the points be joined with a straight line? Explain.

3 The cost of hiring a car is given by the relationship $C = 0·5m + 16$ when C is the cost in pounds and m is the number of miles driven.

a Use the relationship to complete these coordinates

(10, 21) (20,) (30,) (40,) (100,)

b Plot these five points on a grid and draw the line that goes through them.

c Use your graph to estimate the cost when the car is driven 68 miles.

d Use your graph to estimate the number of miles driven when the cost of hiring is £58.

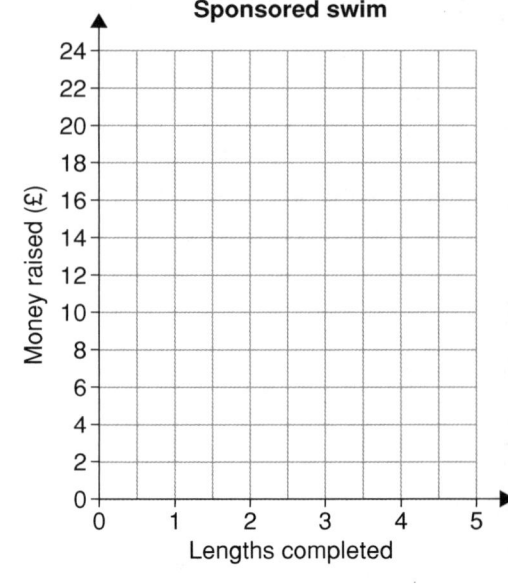

Sponsored swim

Distance/time graphs

1 Paula walked 600 metres to her friend's house. It took 5 minutes. She stayed there for 10 minutes, then she walked on with her friend to the park, which is 1600 m from Paula's house. This took 15 minutes. The girls stayed at the park for 20 minutes, then Paula left her friend and walked home in 10 minutes. Finish the graph for Paula's walk.

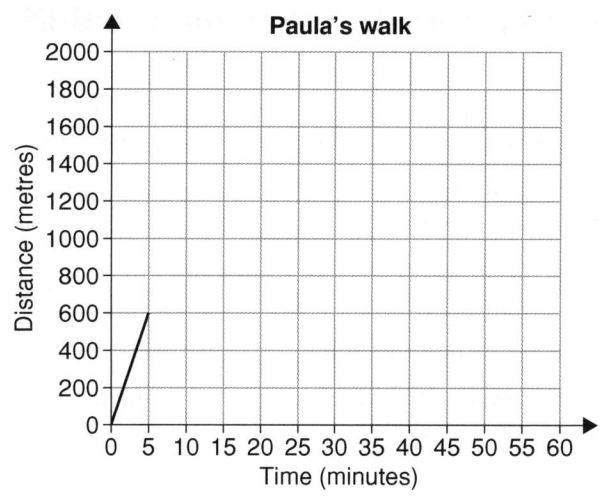

Paula's walk

2 This graph shows Helen's car journey from home to the golf club and back home again.

  a How far away is the golf club?

  b How long did Helen take to get there?

  c For how long did Helen stay at the golf club?

  d Helen stopped on the way home to do some shopping. How far away from home are the shops?

  e How long did it take Helen to get home from the shops?

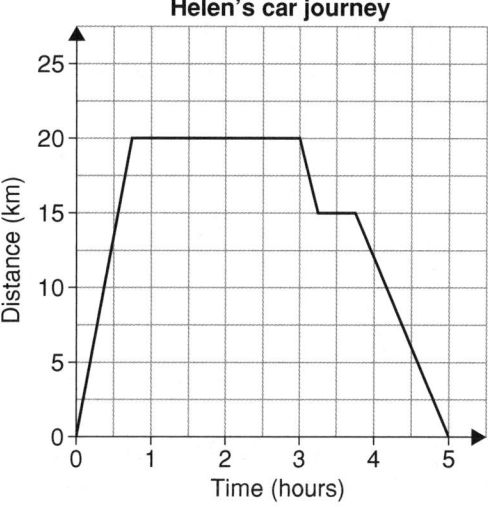

Helen's car journey

CHAPTER 10

Interpreting and sketching real-life graphs

1 The graph shows the number of drinks in a factory drinks machine at various times in the day.

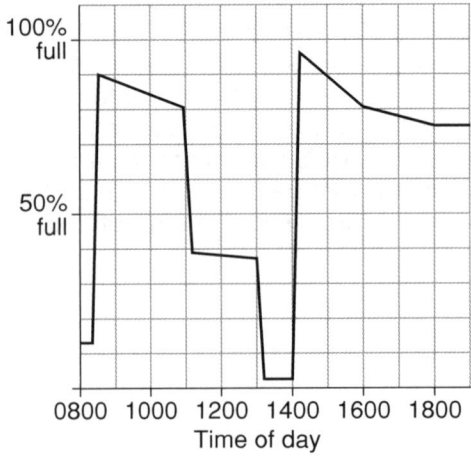

a The machine was filled up twice during the day. At what times did this happen?

b At what time do you estimate the factory has a morning tea break?

c At what time is the factory lunchtime?

d Explain why the first and last sections of the graph are horizontal.

2 Explain these graphs.

Say how the variables on the axes are related. Use words like 'increases', 'decreases' and 'stays the same'.

a

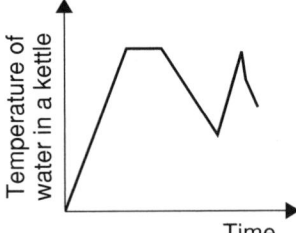

b

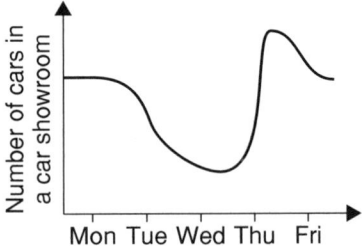

3 The diagram shows part of a track down which a marble runs. The marble is held at A and let go.

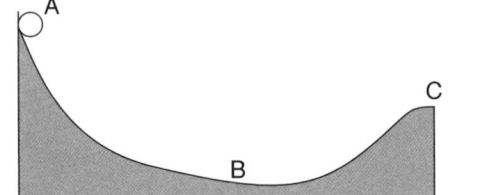

Draw the graph of the speed of the marble as it travels along the track from A to B to C. Mark A, B and C on your graph.

CHAPTER 11

Angles made with intersecting and parallel lines

1   Look at this diagram and write down

    a   a pair of equal angles

    b   a pair of supplementary angles

    c   a pair of corresponding angles

    d   a pair of alternate angles.

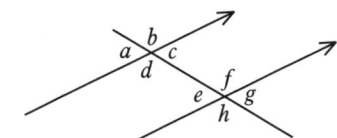

2   Find the size of the marked angle. Give a reason.

    a 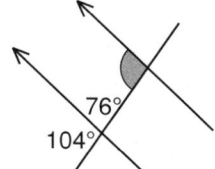   **b**    **c**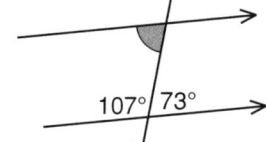

3   For each of the following diagrams, write down the letters of all the angles that are equal to angle a.

    a 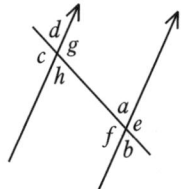   **b** 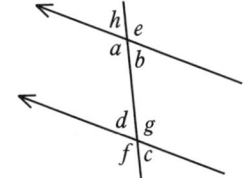   **c**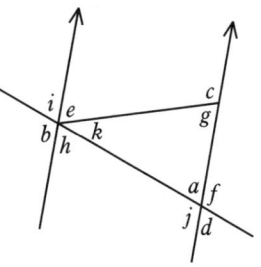

4   Which of these pairs of angles are complementary angles?

    A 76° and 14°   **B** 123° and 57°   **C** 38° and 52°

5   Copy the diagram and write in the sizes of all the missing angles.

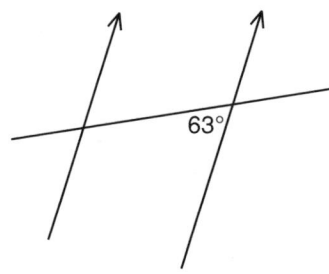

Using geometrical reasoning to find angles

1   Find the size of the angles marked m and n. Write down all your working and reasons.

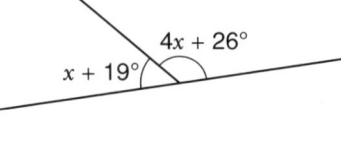

2   Write and solve equations to find the values of x.

a

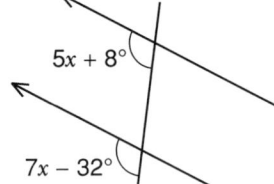

b

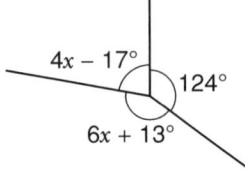

c

3   Are the lines AB and CD parallel? Explain.

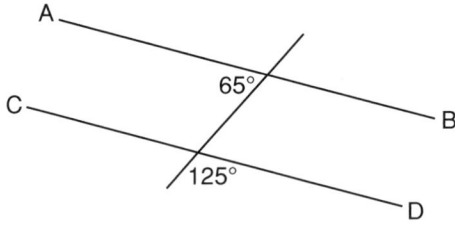

4   In the diagram, x has the value 69°. Prove that this is true, showing your working clearly and giving reasons.

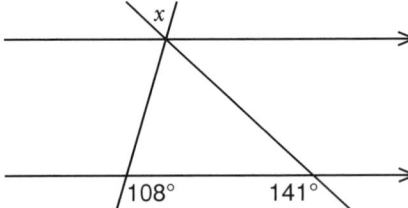

Angles in triangles

1 Calculate the value of each angle marked with letters. Using these values, fill in the crossnumber.

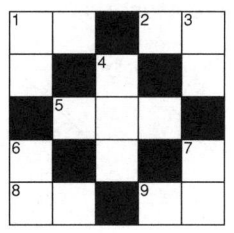

Across	Down
1 a	**1** b
2 j	**3** e
5 g	**4** i
8 d	**6** h
9 f	**7** c

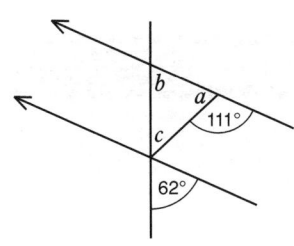
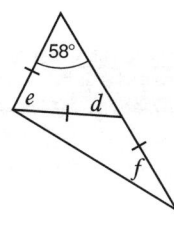
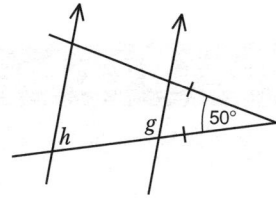
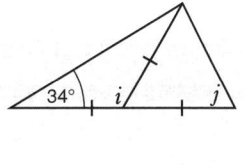

2 Find the size of the angles marked with letters. Write down all your working and give reasons.

a

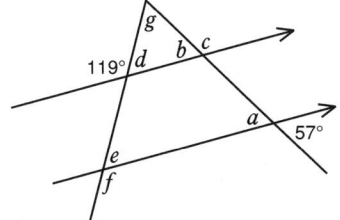

b

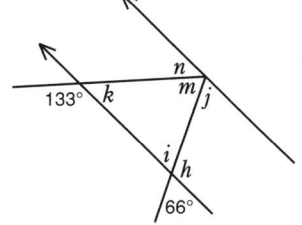

3 Prove that angles x = angle y. Write down all your working and give reasons.

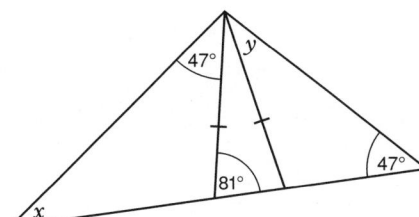

4 Write an equation for each of these and solve it to find x.

a

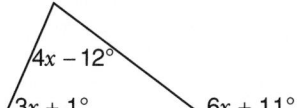

b

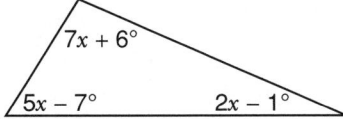

CHAPTER 11

Angles in quadrilaterals

1 Find the value of x. Give reasons.

a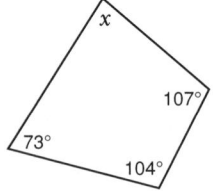

b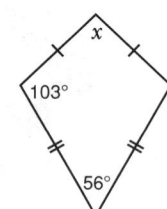

c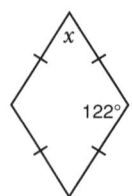

Homework Sheet 11.4 8 Core

New National Framework

Angles in quadrilaterals

1 Find the value of x. Give reasons.

a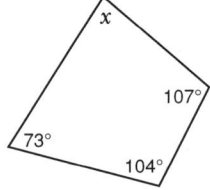

b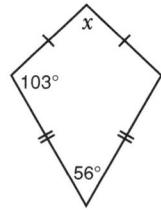

c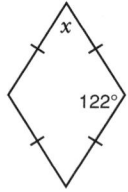

Visualising and sketching 2-D shapes

1 Imagine two identical kites put together along sides of equal length. What shape is formed? Is this the only possible shape? Explain your answer.

Properties of triangles, quadrilaterals and polygons

1 Name all the special quadrilaterals which have

   a 2 axes of symmetry

   b rotation symmetry of order 2

   c diagonals which are perpendicular.

2 Which special quadrilateral could I be?

   a My diagonals bisect each other.

     I have two pairs of equal sides and no right angles.

   b My diagonals intersect at right angles, but are not equal.

     I have one axis of symmetry.

3 Are these statements true or false? Explain your answer.

   a Some rectangles are squares.

   b All rhombuses are squares.

   c Some parallelograms are rhombuses.

4 Join each shape to all the properties that belong to it. The first one has been done for you.

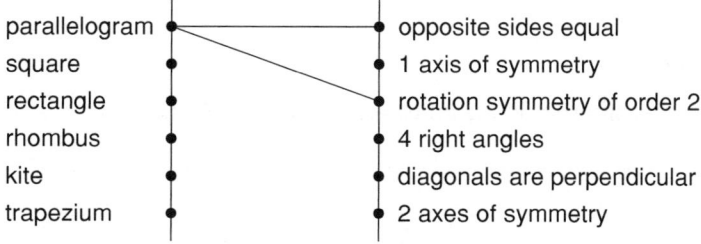

5 **a** I am a triangle with one axis of symmetry and no rotation symmetry.

     What kind of triangle am I?

   b I am a special quadrilateral with no lines of symmetry and no rotation symmetry.

     What could I be?

CHAPTER 12

Congruence

1 In the diagram, name all the shapes that are congruent to

a A
b J
c P
d H
e F

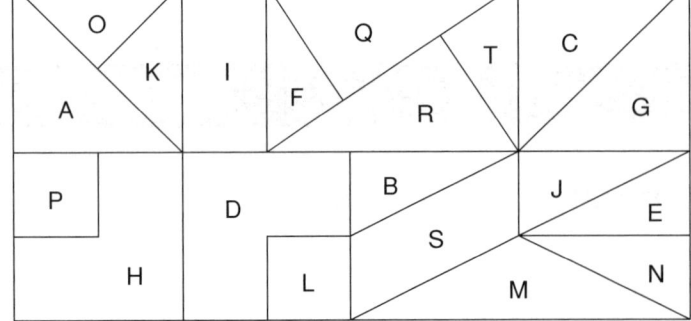

2 Name all the shapes that are congruent to A.

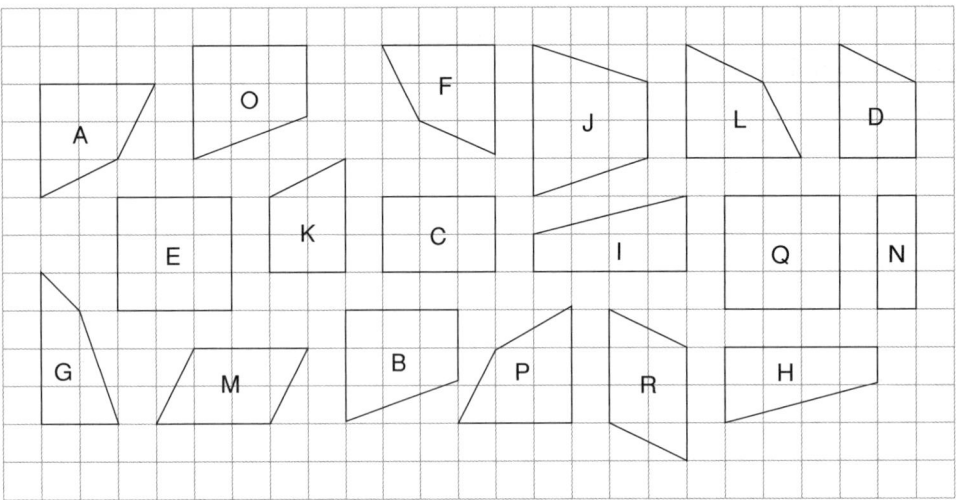

3 These triangles are all congruent.

a Name the side in each triangle that is equal to AC in triangle ABC.

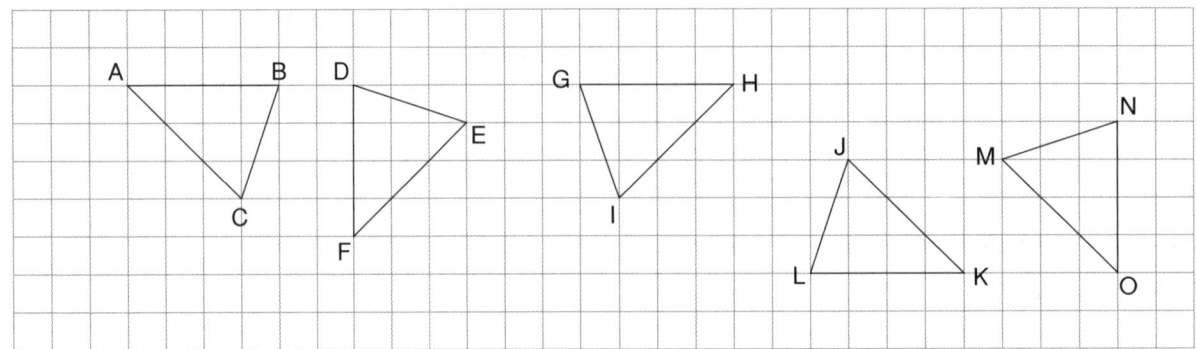

b Name the angle in each triangle that is equal to angle B in triangle ABC.

c Name the shortest side in all the triangles.

Describing and sketching 3-D shapes. Constructing nets.

1   Imagine two cubes. Paint one face on each red. Paint the other five faces on each cube yellow.

  a Put the two cubes together with the red faces at opposite ends.

    i   At how many edges do a red and yellow face meet?

    ii At how many edges do two yellow faces meet?

    iii How many yellow faces does the new shape have?

  b Put the two cubes together with both red faces facing in the same direction.

    i   How many yellow faces does the shape have?

    ii At how many edges do two yellow faces meet now?

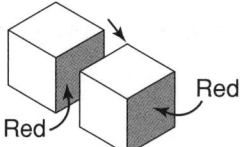

2   Draw these shapes on isometric dot paper.

a 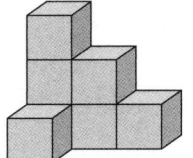 **b** 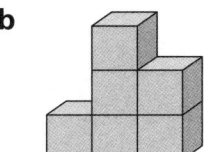 **c**

3   On the drawings you made in **question 2**, use dashed lines to show the positions of the hidden edges.

Plans and elevations

1   Draw the front elevation, side elevation and plan view of these shapes.

a 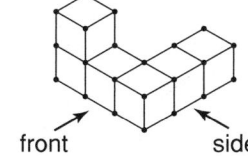 **b** 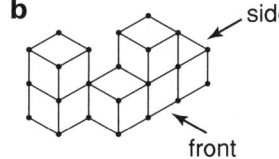 **c**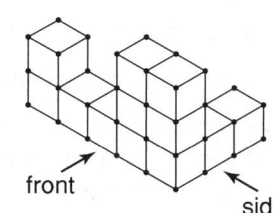

2   On isometric paper, draw the shapes that these represent. Draw each from the view shown by the arrow.

a 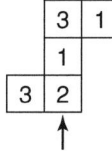 **b**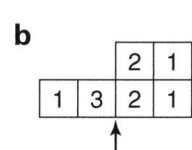

Construction

1 **a** ABCD is a rectangle 7 cm long and 4 cm wide. E is a point on CD, 3 cm from C.

b Construct the line through E that is perpendicular to CD.

c Construct the bisector of angle B.

d Label the point where the perpendicular to CD and the bisector of angle B meet as X.

Measure and write down the length of AX.

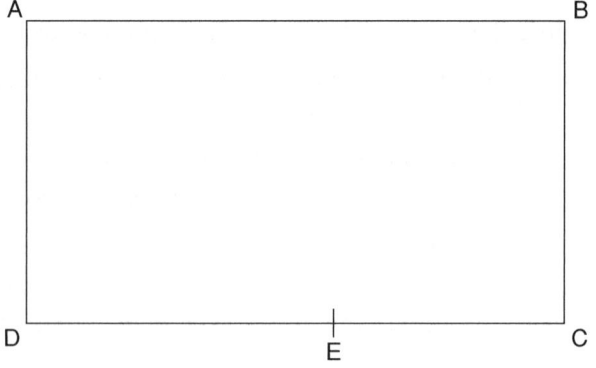

2 **a** The diagram shows a plan of a field PQRS.

b The farmer wants to put a footpath across the field from Q to SR and perpendicular to SR.

Use your compasses to construct the line of the footpath.

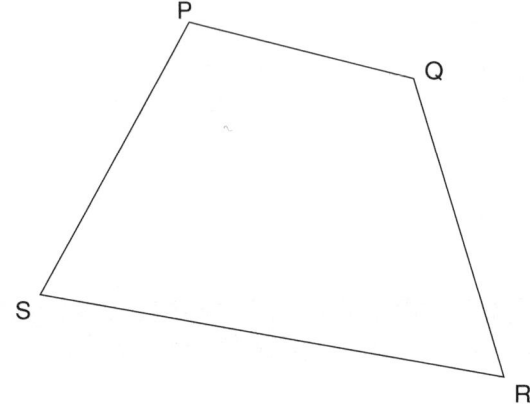

3 **a** Construct the perpendicular bisector of LN.

b Construct the perpendicular bisector of MN.

c Put your compass point on the point where the two bisectors meet and the pencil point on L. Draw a circle. Do you notice anything?

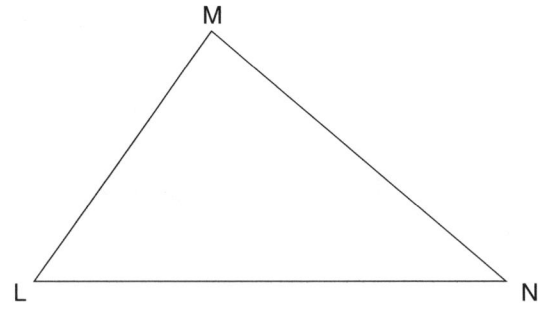

4 **a** Construct the bisector of angle Z.

b Construct the bisector of angle Y.

c With your compass point on the point where the two bisectors meet, try to draw a circle inside the triangle that touches each side of the triangle. This is called the inscribed circle of a triangle.

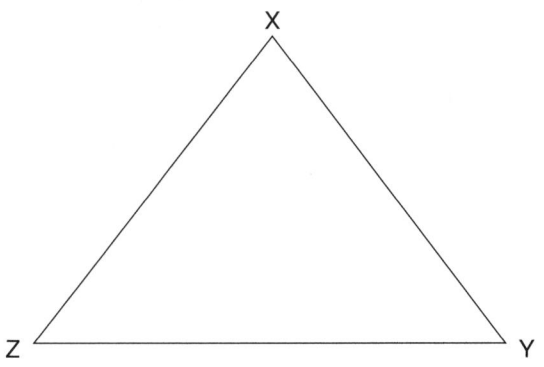

New National Framework

Constructing triangles and quadrilaterals

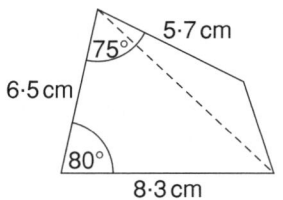

1　Use your ruler and protractor to construct this quadrilateral accurately.

Measure and write down the length of the dotted line.

2　Construct a parallelogram with sides of 6·2 cm and 3·8 cm and an angle of 52°.

Measure and write down the length of the longest diagonal.

3　Jade is making a wooden frame to mount a clock on. A plan of her frame is shown.

Use your ruler, protractor and compasses to make an accurate drawing of the frame.

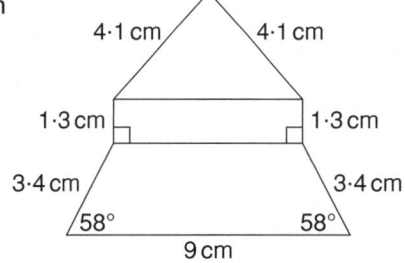

4　Use ruler, compasses and set square to construct a net for each of these.

a
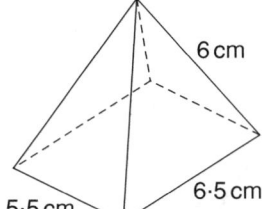

b
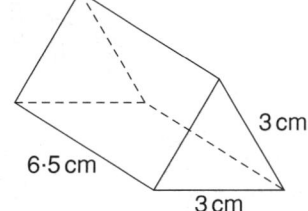

c
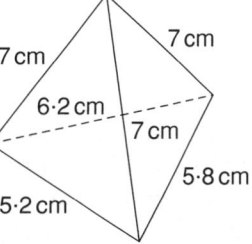

© **New National Framework Mathematics 8** Nelson Thornes Ltd

Locus

1 Sophie takes her dog, Flash, to agility classes.

a First, Flash has to walk up and over a ramp, like the one shown in the diagram.

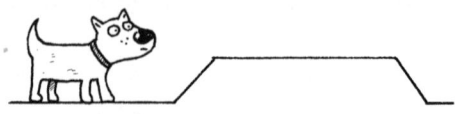

Copy the diagram and draw the locus of Flash's head as he goes over the ramp.

Describe the locus.

b Then Flash has to pick up a ball attached to one end of a rope. The other end of the rope is attached to a stake in the ground. Flash has to run for 1 minute, keeping hold of the ball and keeping the rope stretched.

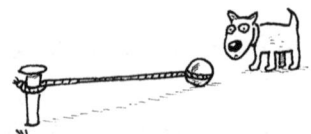

Describe and sketch the locus of Flash's path.

c Next, Flash has to run down a track between the two rails, keeping the same distance away from both rails all the time.

Describe and sketch the locus of the path Flash runs on.

d Finally, Flash has to pick up a dumbell from a bucket and run to put it in another bucket. The buckets are at the opposite corners of a square fenced arena and Flash must keep the same distance away from the fences on either side of him as he runs. Describe and sketch the locus of Flash's run.

Coordinates and transformations

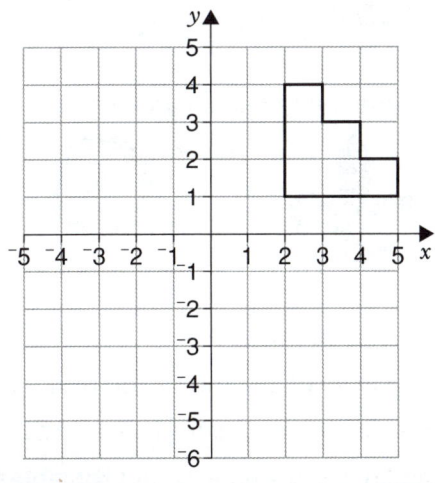

1 **a** Write down the coordinates of all the vertices of the shape.

  b Write down the coordinates of the vertices when the original shape is

  i reflected in the y-axis

  ii rotated 90° clockwise about the origin

  iii translated 6 units to the left and 5 units down

  iv reflected in the line $y = {}^-1$.

Combinations of transformations

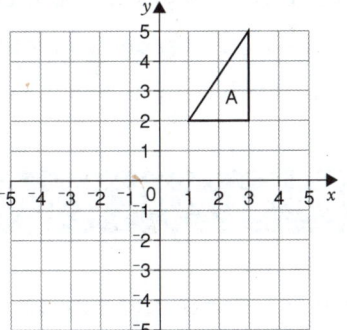

1 **a** Reflect triangle A in the x-axis. Label the image triangle B.

  b Rotate triangle B through 180° about the origin. Label the image C.

  c What single transformation is equivalent to these two transformations, mapping A onto C?

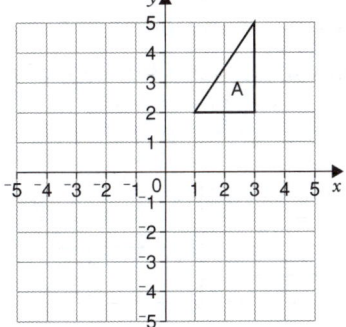

2 **a** Reflect triangle A in the y-axis. Label the image M.

  b Draw line $y = x$, which passes through $({}^-2, {}^-2)$, $({}^-1, {}^-1)$, $(0, 0)$, $(1, 1)$, ...

  c Draw the reflection of triangle M in the line $y = x$. Label the image N.

  d What single transformation would move triangle N back to triangle A?

3 Which of the transformations below will map

  a P to Q   **b** P to R   **c** P to S   **d** Q to S?

  A reflection in the x-axis

  B reflection in the line $y = x$

  C rotation of 90° about the origin

  D translation 7 units left and 6 units down

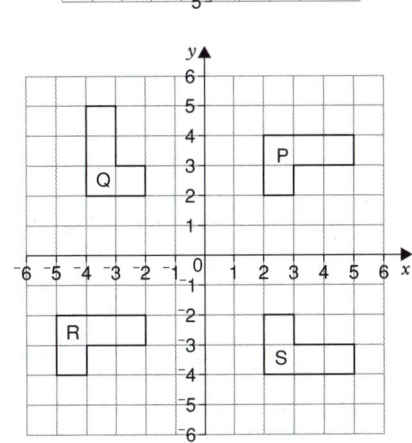

4 Use the diagram in **question 3**.

  Which combination of transformations below will map

  a P to S   **b** Q to R   **c** Q to S?

  A reflection in the y-axis, followed by rotation 90° clockwise about the origin

  B translation 2 units left, 3 units down, followed by rotation of 90° clockwise about the point $({}^-6, {}^-2)$

  C reflection in the y-axis, followed by a rotation of 180° about the origin

CHAPTER 13

Symmetry

1 Describe the reflection and rotation symmetry of these.

a    b    c    d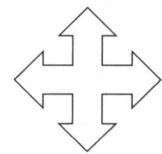

2 Draw a shape which has

   a two lines of symmetry and rotation symmetry of order 2

   b one line of symmetry and no rotation symmetry

   c no lines of symmetry, but rotation symmetry of order 4.

3 Show how you could put these two shapes together to make

   a a shape with reflection symmetry, but no rotation symmetry

   b a shape with rotation symmetry, but no reflection symmetry

   c a shape with both reflection symmetry and rotation symmetry.

Exercise 1   Enlargement

1 Each shape on the left has been enlarged to the shape on the right.

Give the scale factor for each of these enlargements.

a 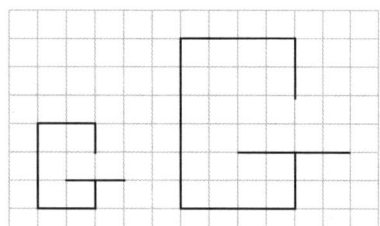   b 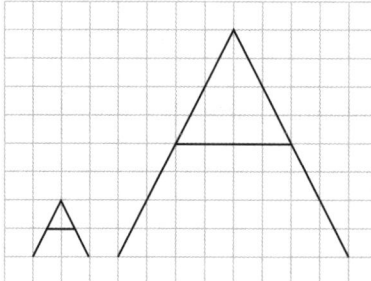   c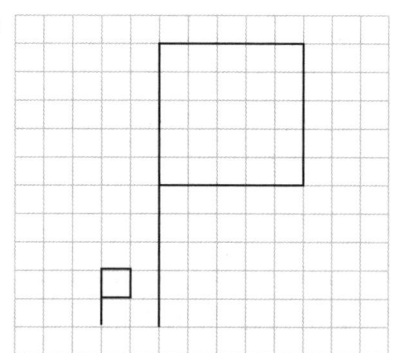

2 LMN has been enlarged to L'M'N'. What is the scale factor of the enlargement?

a 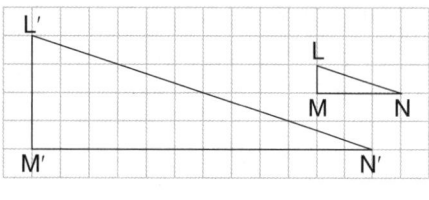   b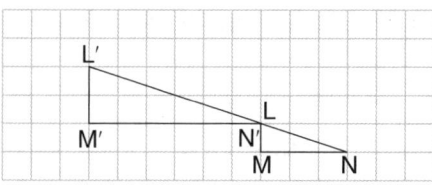

c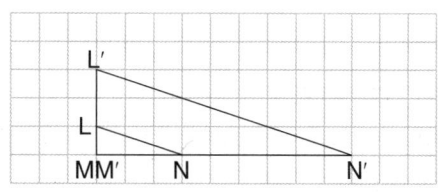

Exercise 2 More enlargement

1 Draw three sets of axes on squared paper with x- and y-values from 0 to 16. Copy each of these shapes onto one of the sets of axes. Enlarge each by the scale factor given. Use the origin as your centre of enlargement for each. Write down the coordinates of P′.

a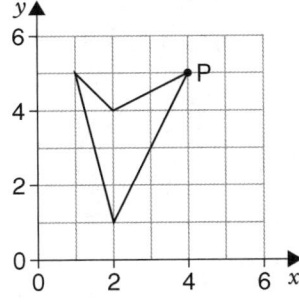

b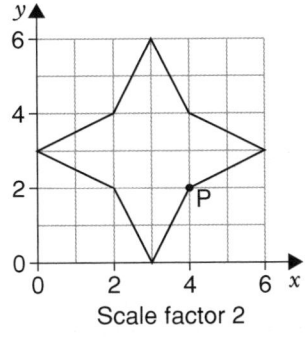

Scale factor 2

c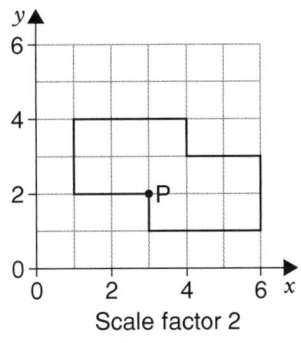

Scale factor 2

2 Draw a set of axes with x- and y-values from 0 to 20.

Copy this diagram onto the axes.

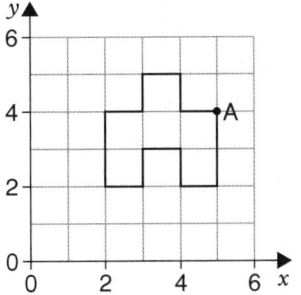

a With centre of enlargement (0, 0), enlarge this shape by a scale factor of 4. Write down the coordinates of A′.

b With centre of enlargement (3, 3), enlarge the shape by a scale factor of 2. Write down the coordinates of A′ for this enlargement.

3 Draw a set of axes with x- and y-values from ⁻10 to +10 on both axes. Plot the points P(1, 4), Q(3, ⁻2) and R(⁻4, ⁻3) and join up to make triangle PQR. With centre of enlargement (2, ⁻1), enlarge triangle PQR by a scale factor of 2. Label P′, Q′ and R′ and write down the coordinates for each.

CHAPTER 13

Exercise 1   Scale drawing

1   The picture shows a giraffe. A man of average height is standing beside it. Estimate the height of the giraffe.

2   Isaac has a scale drawing of his house and garden. The scale used is 1 cm represents 5 m.

   a On the scale drawing, the width of the front of the house is 2·4 cm. How wide is Isaac's house?

   b On the scale drawing, the length of the driveway is 4·5 cm. How long is Isaac's driveway?

   c On the scale drawing, the length of the garage is 1·5 cm. How long is the real garage?

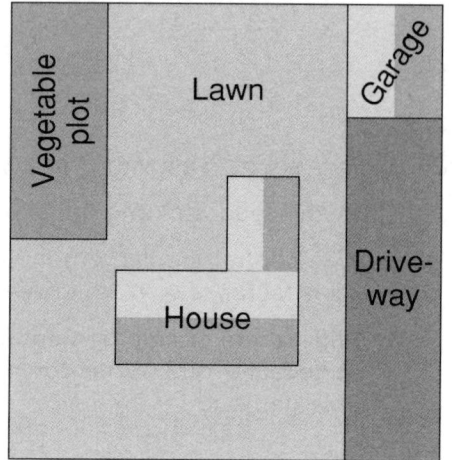

3   Kirsten drew a scale drawing of her classroom. In real life, the room is 8 m by 6 m. What is the scale of the drawing?

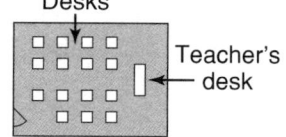

4   On a map a road is 10 cm long. In real-life the road is 20 km long. What is the scale of the map?

5   In a scale drawing a playground is 4 cm long. In real-life the playground is 20 m long. What is the scale used?

6   A map uses a scale of 1 cm to represent 50 m. What actual distance do these measurements on the map represent?

   a 3 cm   **b** 8·5 cm   **c** 1·2 cm

Exercise 2   Making a scale drawing

1   Make an accurate scale drawing of this room. Use the scale

   1 cm represents 0·5 m.

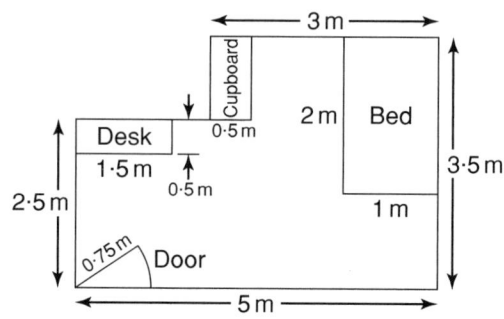

Exercise 1   Finding the mid-point of a vertical or horizontal line

1   Find the coordinates of the mid-points of these horizontal and vertical lines, MN.

   a M(2, 6)   N(2, 10)   **b** M(1, 4)   N(5, 4)   **c** M($^-$4, 3)   N(6, 3)

   d M($^-$2, $^-$7)   N($^-$2, $^-$4)   **e** M(1, $^-$3)   N(1, 8)

Exercise 2   Finding the mid-point of a line segment

1   Find the coordinates of the mid-point of the line PQ.

   a   P(2, 5),   Q(4, 2)   **b** P($^-$3, $^-$4),   Q(5, 6)   **c** P($^-$8, 2),   Q($^-$3, 7)   **d** P(3, $^-$5),   Q($^-$6, 11)

2   C(3, $^-$2) is the mid-point of a line AB. A is the point (7, 4). Plot the points A and C on a grid and find the coordinates of B.

Exercise 1   Metric conversions

1   A video about evolution shown in the Zoo Education Centre lasts 23 minutes. If the video is played 12 times in a day, for how many hours and minutes does it run altogether?

2   Each malaria tablet Paula takes gives her protection for 5 days. She takes 11 tablets during her holiday. How many weeks and days protection did Paula have?

3   The drive belt on a factory machine lasts for 4 months. If the machine has had 14 drive belts since it was installed in the factory, how many years and months has it been in the factory?

4   **a** Joe has a farm of 64 hectares in area. How many square metres is this?

   b The area of the school playing fields is 46 000 m^2. How many hectares is this?

5   The furthest away anyone has ever spat a cherry stone is 29·3 m. How many centimetres is this?

6   Nellie the elephant weighed 6185 kg. How many tonnes is this?

7   You produce about 40 mℓ of saliva every hour. How many litres will you produce in a week?

8   A team of firemen started a long distance charity walk at 12 noon on Sunday 12th May. They finished 6188 minutes later. What day, date and time did they finish?

Exercise 2   Capacity

1

		T													
360	800	8	9·4	0·8	2760	9400	2·76	2·76	800	0·8	94		36	800	3·6	9·4

T									
8	80	9·4	2·76	3·6	9·4		9·4	0·276	9·4	94

Write the letter beside each measurement above its answer in the box.

T 8000 cm^3 in ℓ = 8   **V** 3600 ℓ in m^3 **I** 9·4 ℓ in cm^3 **P** 2·76 m^3 in ℓ

H 36 mℓ in cm^3 **R** 800 ℓ in m^3 **Y** 276 cm^3 in ℓ **C** 0·36 m^3 in ℓ

A 0·8 ℓ in cm^3 **S** 94 cm^3 in mℓ **W** 0·08 m^3 in ℓ **E** 9400 ℓ in m^3

L 2760 m^3 in ℓ

2   A bottle of medicine holds 400 mℓ.

   a How many cℓ are in the bottle?

   b The bottle provides 16 doses of medicine. How many cm^3 are in each dose?

Metric and imperial equivalents

1   Kerryn lives about 60 miles away from her grandparents. About how far away is this in kilometres?

2   The longest beard grown by a man was 215 inches long. How long is that in metres?

3   Your body contains about 36 litres of water. How many pints of water does your body contain?

4   Anne's recipe for a cake needed 7 ounces of sugar. What mass is that in grams?

5   Paul weighed his dog, Jasper. Jasper weighed 15 kg. How many pounds is this?

6   Colin fills up the trough in the garden using a one-gallon watering can. He needed 7 cans of water to fill the trough. How many litres of water does the trough hold?

Units, measuring instruments and accuracy

1   Choose a sensible degree of accuracy for these.

Choose from the box.

a The length of the river Thames.

b The thickness of a ruler.

c The mass of an aeroplane.

d The time to walk to the local shops.

e The capacity of a water jug.

f The mass of an earthworm.

g Your waist measurement.

nearest g
nearest 100 g
nearest kg
nearest tonne
nearest mℓ
nearest 100 mℓ
nearest ℓ
nearest mm
nearest cm
nearest m
nearest 10 m
nearest 100 m
nearest km
nearest second
nearest minute
nearest hour

CHAPTER 14

2   Do you think these measurements are given to the

A nearest centimetre   **B** nearest metre
C nearest 10 m   **D** nearest 100 m?

a Vikram estimates his road to be 800 m long.

b Polly says her school tie is 1·25 metres long.

c Ryan says he lives 3500 m away from the swimming pool.

Estimating

1   Using metric units, write down an approximate measurement for each of these. Give a range for each.

a height of a door

b mass of a banana (with skin)

c circumference of your head

d capacity of a teacup

e length of a television remote control

f capacity of a bucket.

Bearings

1 **a** South is the same as a bearing of   **A** 000°   **B** 090°   **C** 180°   **D** 270°.

    b The bearing 225° is the same as   **A** NE   **B** SE   **C** NW   **D** SW.

2 **a** What is the bearing of J from K?

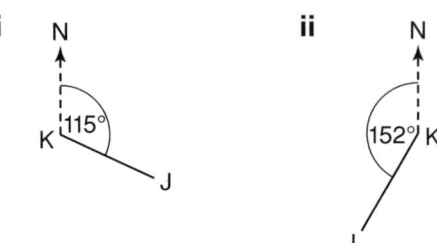

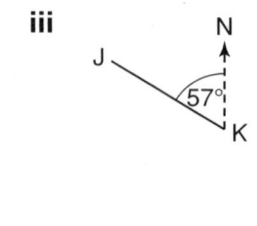

    b What is the bearing of C from D?

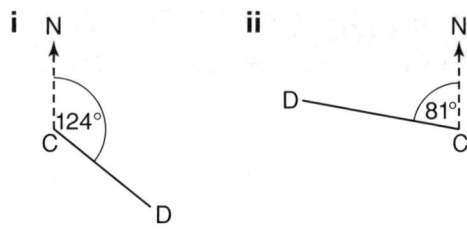

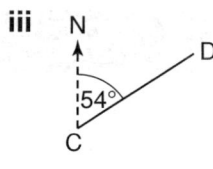

3 The map shows the relative positions of four villages in Africa.

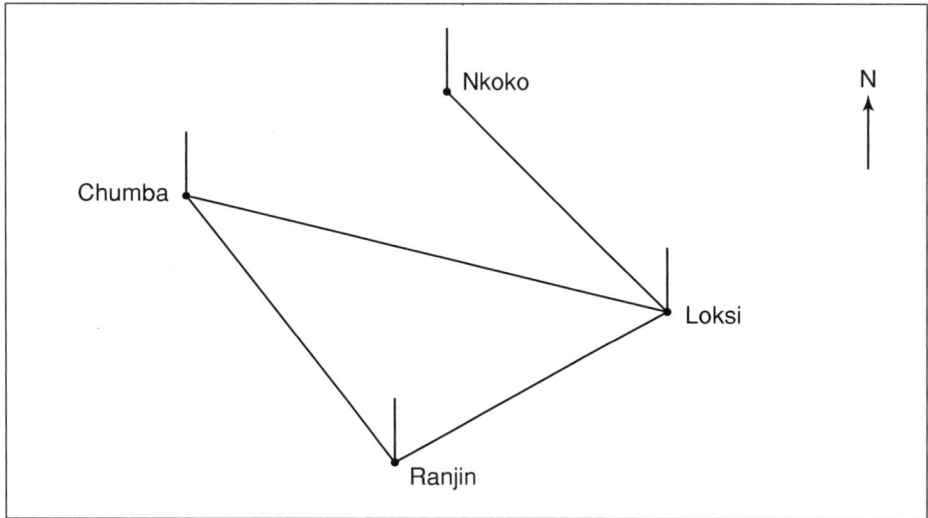

    a Which village is on a bearing of about 070° from Ranjin?

    b What is the bearing of Loksi from Nkoko?

    c What is the bearing of Chumba from Loksi?

    d What is the bearing of Ranjin from Chumba?

    e Without measuring, write down the bearing of Chumba from Ranjin.

Exercise 1 Perimeter and area

1 Calculate the areas of these triangles.

a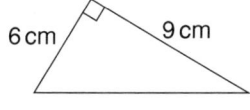
6 cm   9 cm

b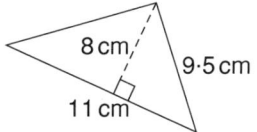
8 cm   9·5 cm
11 cm

c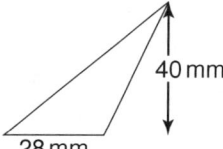
40 mm
28 mm

2 Find the area of these shapes.

a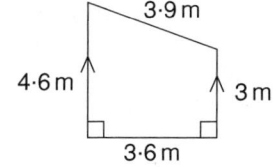
3·9 m
4·6 m   3 m
3·6 m

b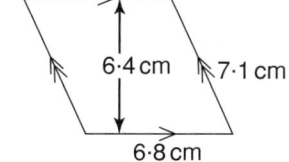
6·4 cm   7·1 cm
6·8 cm

c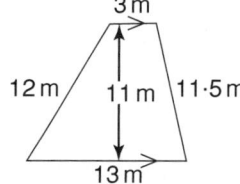
3 m
12 m   11 m   11·5 m
13 m

3 Find the perimeter of each of the shapes in **question 2**.

4 This diagram shows the end face of a triangular tent.

The area of this triangular face is 1·26 m^2.

If the base is 1·8 m, what is the height of the tent?

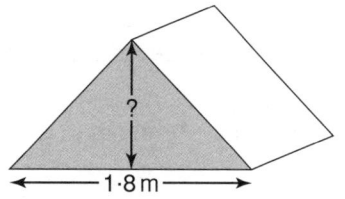
?
1·8 m

*5 Dave uses rectangular paving slabs to pave a patio. Part of the patio is shown in the diagram.

What is the area of each paving slab?

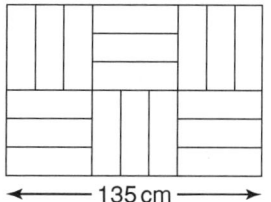
135 cm

Exercise 2 More area

1 Use a method of your choice to find the area of each of these shapes.

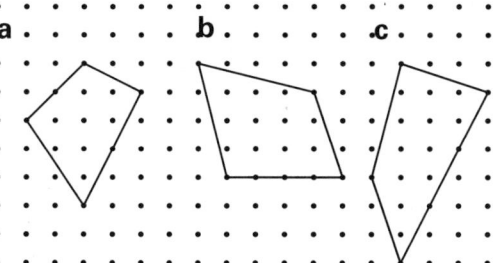
a   **b**   **c**

CHAPTER 15

Surface area and volume   Only use a calculator if you need to

1 Find the volume and surface area of each of these.

a

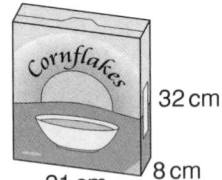

32 cm
8 cm
21 cm

b

Doggy Chocs
7 cm
14 cm
12 cm

c

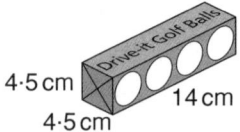

Drive-it Golf Balls
4·5 cm
4·5 cm
14 cm

2 Lucy's paddling pool is a cuboid in shape.

What volume of water will it hold when it is three-quarters full?

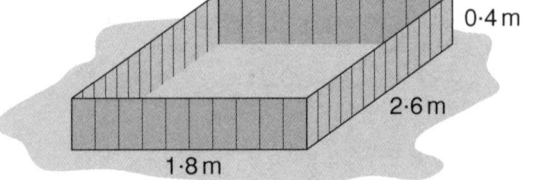

0·4 m
2·6 m
1·8 m

3 Philip is packing dominoes into a box. Each domino is 5·5 cm by 3 cm by 0·5 cm.

How many dominoes can Philip fit into the box?

1·5 cm
6 cm
33 cm

4 a Find the volume of this set of steps by splitting it up into cuboids.

b Find the surface area of the set of steps.

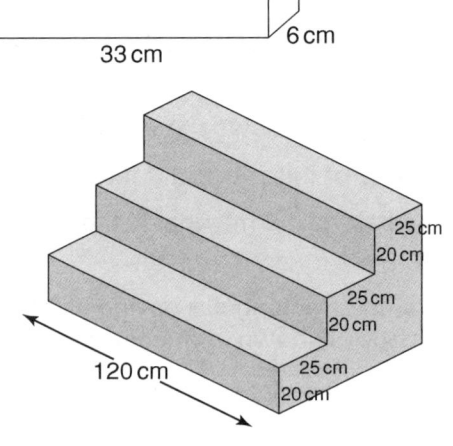

25 cm
20 cm
25 cm
20 cm
120 cm
25 cm
20 cm

***5 a** Write down the dimensions of all the cuboids that can be made, such that length + width + height = 12 cm.

If all the dimensions are whole numbers of centimetres,

b work out the volumes of all the cuboids in part **a**.
Which cuboid has the greatest volume?

Discrete and continuous data

1   Which of the following are discrete data and which are continuous data?

  a the number of cars in a car-park

  b the price of bus fares

  c the temperature in a classroom

  d the number of pupils in a chess club

  e the masses of new-born puppies

Grouping continuous data

1   The time taken in seconds for each of 20 students to sort a shuffled pack of playing cards into suits is given below.

47 68 53 72 45 66 82 59 62 77

50 71 70 64 49 56 63 61 79 83

  a Complete this frequency table

  b How many students took more than 60 seconds but less than or equal to 70 seconds?

  c How many students took more than 70 seconds?

  d How many students took less than or equal to 60 seconds?

Time taken (secs)	Tally	Frequency
$40 < t \leqslant 50$		
$50 < t \leqslant 60$		
$60 < t \leqslant 70$		
$70 < t \leqslant 80$		
$80 < t \leqslant 90$		

  e Explain why the class intervals given on the table are more useful than the intervals $25 < t \leqslant 50$, $50 < t \leqslant 100$, $100 < t \leqslant 150$.

CHAPTER 15

Two-way tables

1   This table shows the numbers of boys and girls who chose different languages to study. The pupils could only choose one language.

	French	German	Spanish
Boys	27	36	24
Girls	34	22	26

a How many boys chose to study Spanish?

b How many pupils chose to study German altogether?

c How many girls were there altogether?

d Compare the choices made by the boys and the girls.

2   An electrical goods shop did a survey on how customers paid for televisions, video players or DVD players. They recorded how many customers paid cash, by cheque or by credit card. Design a two-way table the shop manager could use to collect the data.

Homework Sheet 15.4   8 Core

New National Framework

Surveys

1   Design a survey to find out what pupils in your year group intend to do with their lives and careers after Year 11.

a Write down four relevant questions you could include in the survey.

b Write down some possible results.

c Write down what data you would need to collect.

d Design a collection sheet or questionnaire to collect the data.

e Suggest a suitable sample size and describe how you would choose this sample.

f Say whether your data source is a primary or secondary source.

Mode and range

1 This table shows the time taken, in minutes, for 30 pupils to run backwards down a corridor.

Time taken (t) in seconds	Frequency
$8 < t \leqslant 10$	4
$10 < t \leqslant 12$	11
$12 < t \leqslant 14$	8
$14 < t \leqslant 16$	5
$16 < t \leqslant 18$	2

   a What is the modal class for the time taken?

   b If the shortest time was 8·3 seconds and the longest was 16·9 seconds, what is the range?

Homework Sheet 16.2   8 Core

New National Framework

Exercise 1   Mean

1 Evan recorded how many cars drove past his house in five-minute intervals one day. These are his results.

Number of cars per five-minute interval	0	1	2	3	4	5	6	7	8	9
Frequency	6	8	12	16	17	28	31	20	11	4

Use a spreadsheet or calculator to find the mean number of cars passing Evan's house per five-minute interval. Give your answer to 1 d.p.

Exercise 2   Assumed means

1 Jenna measured the lengths of earthworms in a biology experiment.

Use an assumed mean to find the mean length of an earthworm from Jenna's results.

12·8 cm, 10·4 cm, 11·2 cm, 13·1 cm, 12·5 cm, 10·9 cm, 11·8 cm, 12·1 cm

2 Massimo counted the number of drawing pins in 10 boxes.

    68, 63, 65, 66, 69, 61, 59, 64, 58, 60

Use an assumed mean to find the mean number of drawing pins in each box.

CHAPTER 16

Median, mean, mode, range

1 Twenty-four pupils in Kathleen's class did a sponsored charity walk. This list gives the amounts they raised in pounds

17 24 58 36   4 19 27 30 85 16 25 15

11   9 72 64 18 38 22 47 62 12 16 31

a Find the median amount raised.

b What goes in the gap?

About half of Kathleen's class each raised more than .... pounds.

c Find the mean amount raised, to the nearest penny.

d Find the range of the amounts raised.

2 Sixteen pupils in Jed's class did a sponsored cycle ride and raised these amounts, in pounds.

56   38   27   18   26   32   17   21

29   35   41   44   19   36   14   28

a Find the mean amount raised, to the nearest penny.

b Find the range of the amounts raised.

c Which class, Jed's or Kathleen's in **question 1**, raised the most, on average, per pupil?

d Which class, Jed's or Kathleen's in **question 1**, had the greatest range of amounts raised.

3 Marko and Lorna were playing a game with two dice. Each time they threw the dice, they added the two numbers together. Marko threw the dice four times and scored 2, 5, 7 and 8. Lorna also threw the dice four times and scored 3 and 6 on two of her throws. What could she have scored on the other two throws, if the mean and range of her scores was the same as Marko's?

4 Ten shoppers were asked to give a mark, out of five, rating the taste of a new ice cream.

Nine of the marks are 4,   2,   3,   5,   3,   2,   1,   3,   2.

What is the tenth mark if

a the mode is 2

b the mean is 3

c the median and the mode are the same

d the mean is the same as the median?

Finding the median, range and mode from a stem-and-leaf diagram

1 Lawrence wanted to compare the number of phonecalls per hour received in two different mail-order offices, Brentons and Fashion Extra.

This list shows his results for Brentons.

36, 44, 21, 37, 39, 42, 28, 19, 33, 26, 48, 31, 40, 25, 39

38, 17, 30, 47, 32, 28, 25, 20, 39, 19, 34, 46, 35, 39

a Construct a stem-and-leaf diagram to show Lawrence's results.

b How many times did Brentons receive between 40 and 49 phonecalls per hour?

c Find the median, mode and range of the number of phonecalls per hour.

This stem-and-leaf diagram shows Lawrence's results for Fashion Extra.

d Find the median, mode and range for Fashion extra.

e Compare the number of phonecalls per hour received by Brentons and Fashion Extra.

Number of phonecalls per hour at Fashion Extra	
1	1 2 4 7 7 8
2	0 1 3 3 3 4 5 7 8 8 9 9
3	0 0 1 1 2 3 4 4 6 7 9 9
4	0 1 1 3 4 6 6

Comparing data

1 Hester is trying two different kinds of fertilisers, one organic and one chemical, on her raspberry canes to see how they increase the crop.

This data gives the mass in grams of raspberries produced by each cane for the two different fertilisers.

Organic Fertiliser (18 canes)

306	254	418	72	126	385	112	218	164
381	412	103	87	422	333	186	294	188

Chemical Fertiliser (19 canes)

326	315	198	218	344	238	129	213	169	246
159	286	302	188	261	147	319	243	172	

a Find the mean, median and range for the canes grown using organic fertiliser.

b Find the mean, median and range for the canes grown using chemical fertiliser.

c Write a brief report comparing the effect that each type of fertiliser has on the crop.

d Which fertiliser would Hester recommend? Explain why.

CHAPTER 16

Compound bar charts and line graphs

1 The following table gives the number, in millions, of males and females in different age ranges in England. The data was collected in the 2001 census.

Age range	0–9	10–19	20–29	30–39	40–49	50–59	60–69	70–79	80–89	90 and over
Number of males (millions)	3·10	3·20	3·07	3·77	3·26	3·05	2·21	1·57	0·60	0·07
Number of females (millions)	2·95	3·06	3·15	3·89	3·31	3·11	2·34	2·02	1·14	0·24

a Complete this compound bar chart.

b What conclusions can you draw about the distribution of males and females in the different age ranges?

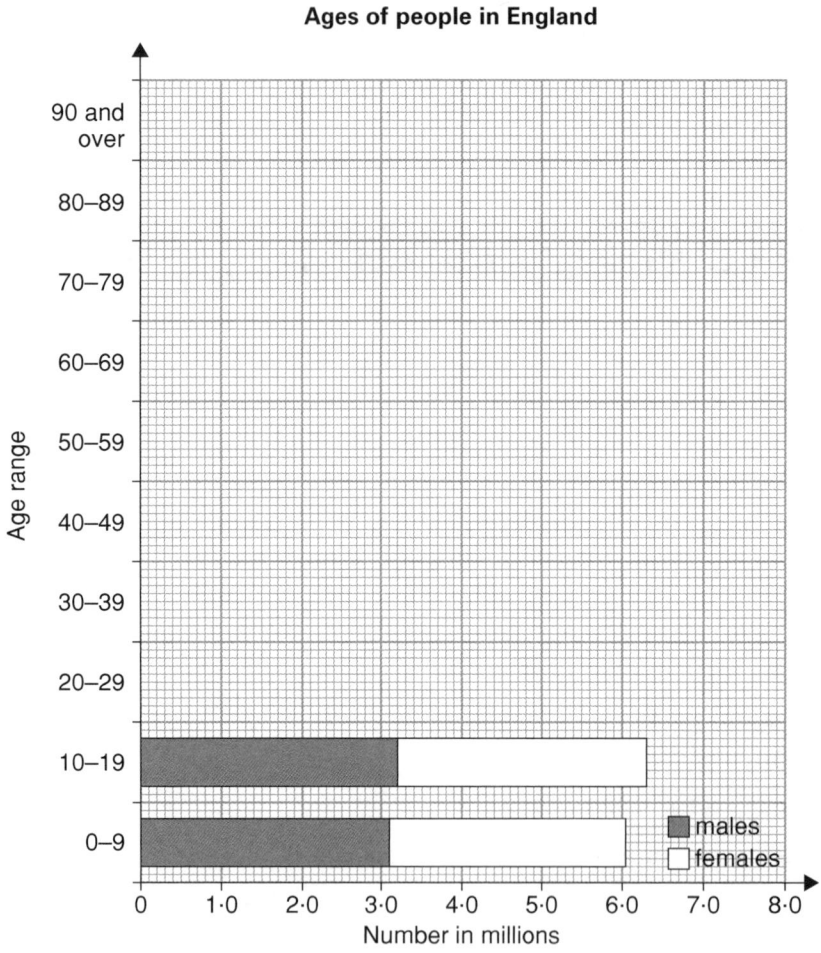

2 The table shows the English and Maths Key Stage 3 Test results for Little Moreton School.

Level	3	4	5	6	7	8
Number of pupils in English	42	37	58	68	19	10
Number of pupils in Maths	36	30	42	67	34	25

a Draw a line graph for this data. Put the levels on the horizontal axis. Put both sets of data on the same graph. Use a continuous line for English and a broken line for Maths.

b Compare the results in the two subjects and comment on your findings.

Frequency diagrams

1   Kane was doing a survey about the time taken to travel to school by the pupils in his year. These are his results.

Time taken (minutes)	0–	5–	10–	15–	20–	25–	30–	35–	40–	45–
Number of pupils	23	38	27	36	21	17	13	8	4	2

a Draw a frequency diagram for this data.

b How many pupils took less than 15 minutes?

c How many pupils took at least 30 minutes?

d Can you tell how many pupils took exactly 15 minutes? Explain.

Drawing pie charts

1   40 people used a Sports Centre one afternoon.

This table shows the activities they did in the Sports Centre.

Activity	Gym	Swimming	Badminton	Aerobics
Number of people	8	5	12	15

Fill in the missing numbers to calculate the angles of the sectors.

Gym $\frac{8}{40} \times 360° = $ _____ °

Swimming $\frac{5}{40} \times 360° = $ _____ °

Badminton $\frac{12}{} \times = $ _____ °

Aerobics   _____ × _____ = _____ °

Draw a circle with radius 5 cm.

Show the data on a pie chart.

2   This table shows the numbers of different types of animals treated by a veterinary practice in one day.

Type of animal	Dog	Cat	Hamster	Sheep	Cow
Number treated	17	14	2	22	5

Draw a pie chart for this data.

CHAPTER 16

Scatter graphs

1   Vikram collected data about the heights and shoe sizes of 16 people.

The table shows his results.

Height (cm)	183	167	169	172	175	174	180	178	181	173	168	175	174	185	167	171
Shoe size	11	6	$6\frac{1}{2}$	$8\frac{1}{2}$	8	7	10	9	$10\frac{1}{2}$	$7\frac{1}{2}$	7	$7\frac{1}{2}$	$8\frac{1}{2}$	$10\frac{1}{2}$	$5\frac{1}{2}$	$7\frac{1}{2}$

a Copy the axes onto graph paper and draw a scatter diagram to show the data.

b Does someone who is tall generally have larger feet?

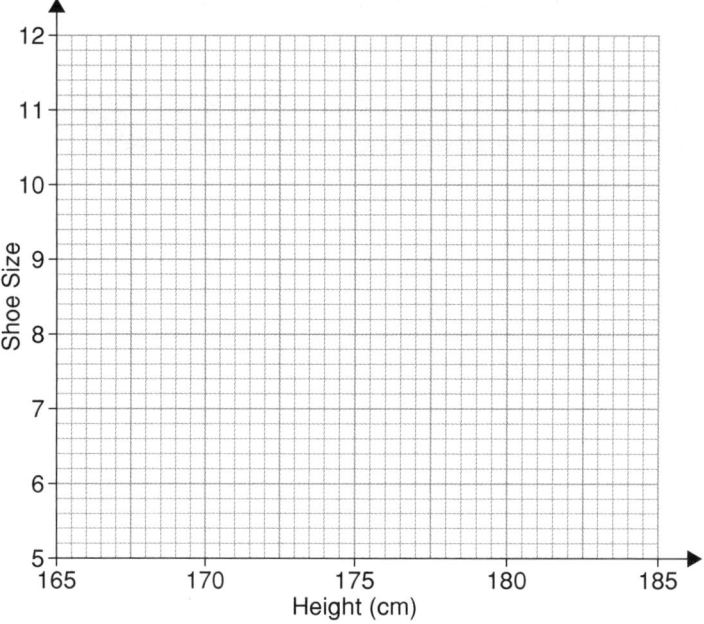

2   The table shows the test results of 12 students in a Science and a French test.

Science result	24	8	4	24	12	20	40	32	36	48	17	35
French result	28	38	41	34	31	31	22	25	23	22	36	21

a Draw a scatter graph for the data.

b Does the graph tell you that someone who is good at Science tends to be not so good at French or tends to also be good at French?

Interpreting graphs and tables

1 The graph shows how the average weekly household expenditure on fuel and power, alcoholic drinks, tobacco and fares and other travel costs has changed since 1995.

  a How has the expenditure on fuel and power changed?

  b How has the expenditure on fares and other travel costs changed?

  c What might affect the trends shown in expenditure on tobacco in the future?

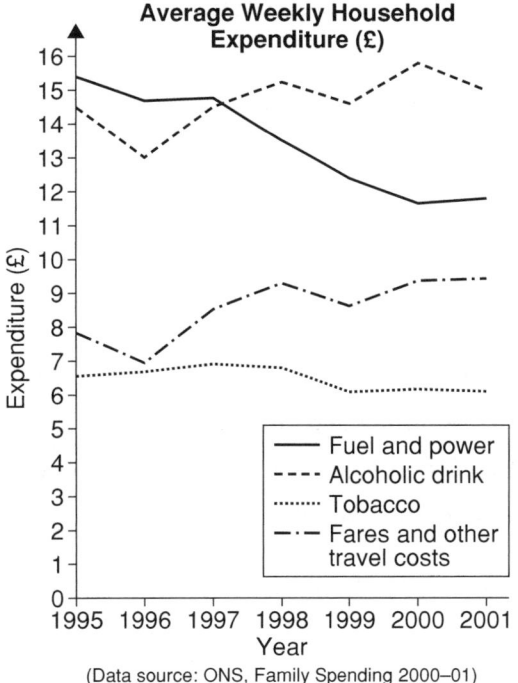

Average Weekly Household Expenditure (£)

(Data source: ONS, Family Spending 2000–01)

2 The graph shows the percentage of overweight children in the United States from the 1960s to the year 2000.

  a Comment on the general trend of overweight children.

  b Is the age group 6–11 more prone to being overweight than the age group 12–19, or the other way round, or does it vary?

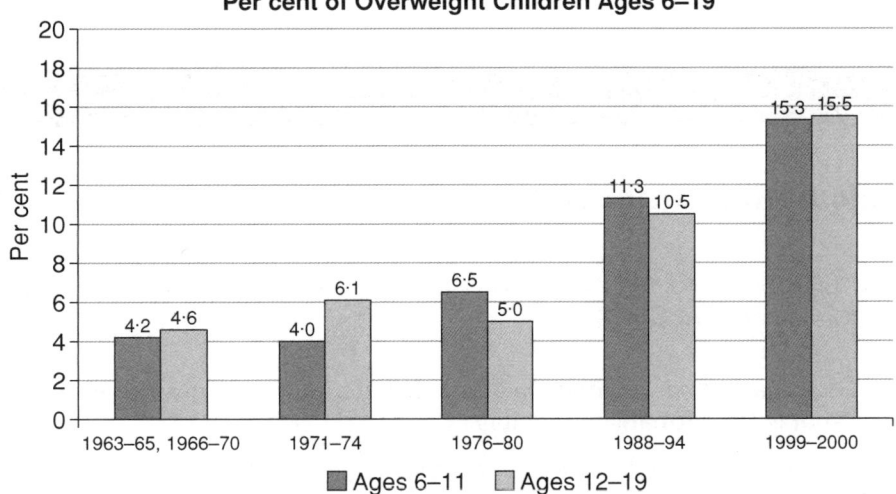

Per cent of Overweight Children Ages 6–19

(Data source: National Centre for Health Statistics)

3 The pie charts show the proportions of money spent on housing, food, clothes, motoring and household services in 1974 and 2001.

  a Which categories did households spend a lower proportion on in 2001 than they did in 1974?

  b Which category has doubled in proportion from 1974 to 2001?

  c Households in 2001 spent a higher proportion of their budget on motoring than they did in 1974. Why do you think this is?

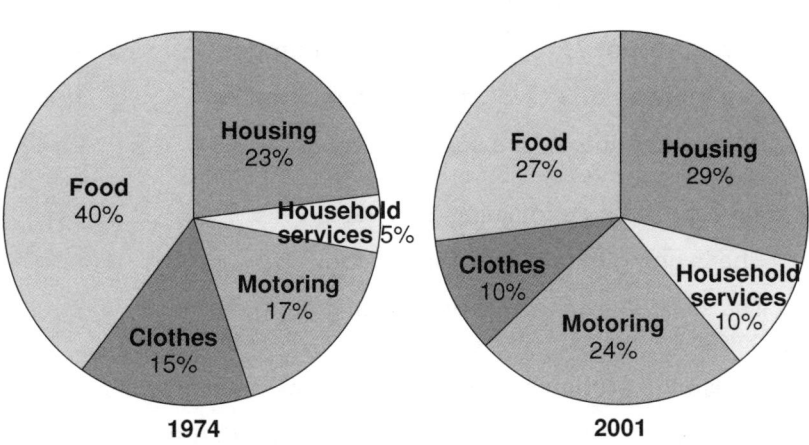

CHAPTER 16

Language of probability

1 Nathan is playing a game with cards with numbers on.

2 3 4 1 2 5 2 4 1 3 2 2

He shuffles the cards and picks one out without looking.

Which number is he most likely to pick out? Why?

2 Millie likes dark chocolates. She is offered a chocolate from a choice of two boxes.

Box A contains 8 dark and 13 milk chocolates.

Box B contains 13 dark and 24 milk chocolates.

From which box should Millie choose? Why?

3 Gelsomino is firing darts at this target. Which area is he most likely to hit — the bullseye, the circle round the bullseye or the square back board?

Explain why.

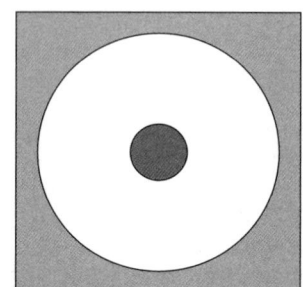

Calculating probability

1 The probability that Keeley will see her friend Sam on the way to school is $\frac{4}{9}$. What is the probability that Keeley will not see Sam?

2 Hazel has 12 baby rabbits in a hutch. 7 are black, 4 are white and 1 is grey. She lifts one out at random. What is the probability that it is

a white **b** not white **c** black **d** not black **e** black or white?

3 Gareth is throwing a 12-sided dice with the numbers 1 to 12 on it.

What is the probability that it lands on

a 7 **b** not 7 **c** an even number **d** a prime number

e a multiple of 3 **f** a square number **g** not a square number

h an odd number or a square number?

4 Faye has a bag containing 30 sweets.

She is going to take one without looking.

The scale shows the probability of getting each colour.

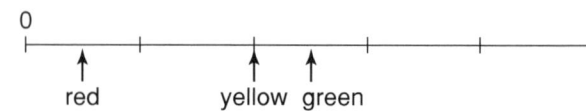

a Use the probabilities to work out how many sweets of each colour are in Faye's bag.

b What is the probability of getting **i** a red, green or yellow sweet **ii** a purple sweet?

Calculating probabilities by listing outcomes

1   Three boys, Aidan, Brian and Colin and three girls, Daphne, Ellen and Fiona are all in a competition to win a holiday.

Only one boy and one girl can win.

a Complete the table to show who could win the holiday.

Aidan	Aidan						
Daphne	Ellen						

b What is the probability that Brian and Fiona will both win the holiday?

2   A dice is tossed and this spinner is spun.

The numbers on the dice and the spinner are multiplied.

a Complete the table to show the sample space.

b Find the probability that the dice and spinner will give an answer of

   i   12   **ii** an odd number

   iii a number less than 10

   iv at least 15.

spinner

dice

×	1	2	3	4	5	6
1	1	2				
2		4				
3						
4						

Estimating probability from experiments

1   Ben did a survey of 100 pupils in his school to find out how many wore glasses. He found that 37 did.

Estimate the probability that the next pupil Ben meets will wear glasses.

2   One day the school tuck shop sold the items shown in the table.

a Use the table to estimate the probability that the first item sold in the tuck shop the next day was

   i a chocolate bar   **ii** an apple

b If data was collected on a different day would the results be the same? Explain.

c If the data was collected from the tuck shop for a month rather than one day, would the estimated probabilities be more accurate? Explain.

Crisps	27
Drinks	49
Apples	13
Chocolate bars	31

CHAPTER 17

Homework Sheet 1.1

Powers of ten
1 a 10^5 b 10^7 c 10^{10}
2 a Four point six times ten to the power of nine
   b Five times ten to the power of eight
   c Six point five times ten to the power of six
   d One point one times ten to the power of five
3 a 2491 b 3 250 004

Homework Sheet 1.2

Adding and subtracting multiples of 0·1, 0·01 and 0·001
1 THE ROCK STAR ROD STEWART HAS THE NICKNAME PHYLLIS
2 2·46 kg

Homework Sheet 1.3

× and ÷ by multiples of 10, 100 and 1000
1 a 2800   b 6000   c 20   d 40
   e 1 800 000 f 1000   g 0·04   h 1 290 000
   i 40   j 1·1   k 0·4
2 320 000 trout
3 40 albums
4 There are many possible answers. One is 20 × 600.
5 There are many possible answers. One is 15 ÷ 500.

Homework Sheet 1.4

Multiplying and dividing by 0·1 and 0·01
1 a C 6·72 × 10   b C 0·23 ÷ 100   c A 1·94 × 100
2 FORTY PERCENT OF ENGLISHMEN SLEEP IN THE NUDE
3 2·4 m

Homework Sheet 1.5

Putting decimals in order
1 a True   b True   c True   d False
   e True   f True   g True
2 17·86, 17·9, 18·09, 18·59, 18·67, 18·9
3 a 0·76 > 0·67 > 0·607   b 8·42 > 8·4 > 8·24
   c 1·372 > 1·327 > 1·3072   d 0·034 > 0·0043 > 0·00403
4 a 0·607 km, 637 m, 0·67 km, 673 m
   b 1072 g, 1·087 kg, 1·72 kg, 1827 g
   c 13·041 ℓ, 13 104 mℓ, 13·14 ℓ, 13 400 mℓ, 13·41 ℓ
5 a 0·25 b 0·065 c ⁻0·75 d ⁻2·0
6 ICOSAHEDRON
7 There are many different answers.

Homework Sheet 1.6

Rounding to powers of ten
1 a 650   b 2400   c 25 000   d 7000
   e 8400   f 6250   g 72 900   h 90 000
   i 800 000   j 5 000 000
2 The nearest hundred
3 a 2 783 500   b 2 784 499
4 a Greatest: 9 749 999   Smallest: 9 650 000
   b Just over nine and a half million is more accurate because 9 700 000 is nearer to 9 500 000 than to 10 000 000.
   c 1 300 000 or 1·3 million

Homework Sheet 1.7

Rounding to decimal places
1 a 1·7 b 18·3 c 26·1 d 6·0 e 11·3 f 18·0
2 a 1·67 b 18·32 c 26·06 d 6·03 e 11·29 f 18·00
3 a 2800 g b 6300 m c 240 cm
4 a 6·57 b 6·33 c 1·89
5 £5·96
6 a 133 km each day to the nearest kilometre
   b 280 m each minute to the nearest 10 metres

Homework Sheet 2.1

Adding and subtracting integers
1 a ⁻2 b ⁻7 c ⁻51
2 a ⁻18 b ⁻3 c ⁻5
3 a ⁻15 b 16 c 25 d ⁻3 e 3 f 4
4 a ⁻4 + ⁻2 = ⁻6   OR   ⁻3 + ⁻3 = ⁻6
   b ⁻2 + 5 = 3   OR   3 + 0 = 3
   c ⁻3 − ⁻2 = ⁻1   OR   ⁻4 − ⁻3 = ⁻1
   d 5 − ⁻2 = 7 ⎫ either way round
   e 3 − ⁻4 = 7 ⎭

5

⁻3	2	1
4	0	⁻4
⁻1	⁻2	3

Homework Sheet 2.2

Multiplying and dividing integers
1 a 15   b ⁻24   c ⁻14   d 32
   e ⁻54   f ⁻49   g 63   h 4
   i ⁻8   j ⁻8   k ⁻6   l ⁻2
2 a 11   b ⁻5 × ⁻6 = 30
                 30 ÷ ⁻5 = ⁻6
                 30 ÷ ⁻6 = ⁻5
3 3

Homework Sheet 2.3

Integers on the calculator
THE LARGEST MUSCLE IN THE BODY IS IN THE
BUTTOCKS

Homework Sheet 2.4

Order of operations with integers
1   A pig

Homework Sheet 2.5

Divisibility
1   a   13 794, 3456, 2592, 1590   **b** 9255, 1590
   c   3456, 2592   **d** 3456, 2592
2   1080

Homework Sheet 2.6

Prime factor decomposition
1   a $2^3 \times 3 \times 5$ **b** $2^3 \times 3 \times 7$ **c** $2 \times 3 \times 5 \times 7$
   d $2^3 \times 3 \times 11$ **e** $2^2 \times 3^2 \times 5^2 \times 7$
2   a   A = 2, B = 2, C = 3, D = 5 [or C = 5, D = 3]
   b   A = 462, B = 2, C = 77, D = 7
   c   A = 15, B = 9, C = 3, D = 5
3   a $2^2 \times 3^2 \times 13$ **b** $2^2 \times 5^2 \times 7$ **c** $2 \times 3^4 \times 7 \times 11$
   d $3^2 \times 7^2 \times 17$
4   15 and 21

Homework Sheet 2.7

Highest common factor and lowest common multiple
1   a   18   **b** 28
2   a   1260   **b** 5880
3
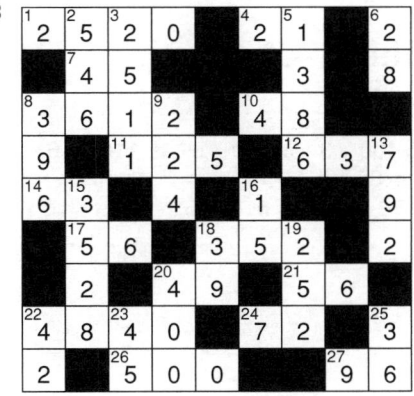

Homework Sheet 2.8

Powers and roots
1   a 25   **b** 121   **c** 49   **d** 5   **e** 3   **f** 64
2   a +6, ⁻6   **b** +11, ⁻11
3   a 16   **b** 24   **c** 30
4   $5^2 = 25$ and $6^2 = 36$. 30 lies between 25 and 36, so $\sqrt{30}$
   lies between 5 and 6.
5   a 37·2   **b** 88·4   **c** 3·3   **d** 8·1   **e** 5·8
6   8, 10, 12, 14
7   a 2, 2   **b** 9, 6   **c** 7, 3, 2

Homework Sheet 2.9

Cubes and cube roots
1   a 64   **b** 2   **c** 125   **d** 27
   e 1   **f** 10   **g** 216
2   a 125   **b** 1   **c** 27   **d** 6   **e** 1000
3   a ⁻729   **b** 2197   **c** 4·913   **d** 19
   e 1·6   **f** 44·36   **g** 4·10   **h** 5·17
   i 8·73
4   $3 = 1^3 + 1^3 + 1^3$ $17 = 1^3 + 2^3 + 2^3$
   $36 = 1^3 + 2^3 + 3^3$ $55 = 1^3 + 3^3 + 3^3$
   $99 = 2^3 + 3^3 + 4^3$ $129 = 1^3 + 4^3 + 4^3$
   $136 = 2^3 + 4^3 + 4^3$

Homework Sheet 3.1

Adding and subtracting
1   a ⁻6   **b** ⁻5   **c** 0·3   **d** 0
2   a 82   **b** 450   **c** 130   **d** 580
   e 4200   **f** 2700   **g** 5·2   **h** 3·5
   i 0·45   **j** 0·585   **k** 250   **l** 5·1
3

6	18	11	15
19	7	14	10
13	9	20	8
12	16	5	17

4   MOST METEORITES ARE SMALLER THAN A GRAIN
   OF SAND
5   A possible answer is A = 1, B = 3, C = 8, D = 5
   18·31
   −3·18
   ‾‾‾‾‾
   15·13

Homework Sheet 3.2

Multiplying and dividing

2 a 1·8 **b** 0·28 **c** 0·35

1

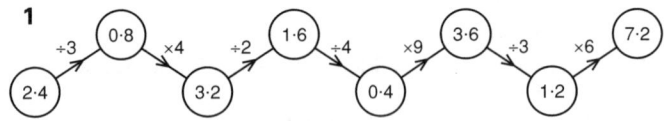

3 a 9 **b** 4·05 **c** ⁻4·2
4 A WEEK HAD EIGHT DAYS IN ANCIENT GREECE
5 a 1093 **b** 4·14 **c** 51 **d** 176
6 18 and 6
7 104 and 8

Homework Sheet 3.3

Solving problems mentally

1 a 126 days old **b** 24·8 cm **c** £1·65
  d 6 minutes 24 seconds
2 a £101·50 **b** £15
3 a 6 and 9 **b** 1·2 and 2·5
4 400 boxes
5 160 cm
6 18 cows and 23 hens
7

$$20 + \frac{1}{19} + \frac{2}{18} + \frac{3}{17} + \frac{4}{16} + \frac{5}{15} + \frac{6}{14} + \frac{7}{13} + \frac{8}{12} + \frac{9}{11} + 10$$

$$= 20 \times 10 + 10$$

$$= 210$$

8 168
9 312

Homework Sheet 3.4

Ex 1 Making estimates

1 a Exact **b** Estimate **c** Exact **d** Exact
  e Estimate
2 a C **b** C **c** A **d** B

Ex 2 Estimating answers to calculations

1 a B 300 × 20
  b A 500 ÷ 50
  c C 9 ÷ 2
  d A 10 × 30
2 Possible answers are:
  a 200 × 100 = 20 000   **b** 900 × 30 = 27 000
  c 300 ÷ 30 = 10   **d** 900 ÷ 20 = 45
  e 8 × 2 = 16   **f** 0·6 × 4 = 2·4
  g 10 ÷ 0·5 = 20   **h** 20 ÷ 4 = 5

3 a $\frac{200 + 200}{40} = 10$ **b** $\frac{20 \times 30}{6} = 100$

  c $\frac{40 - 20}{8 \times 5} = \frac{1}{2}$ or 0·5

4 a 200 × £3 = £600
  b 3000 ÷ 40 = 75 coaches
  c 80 × 40 = 3200 phonecalls

Homework Sheet 4.1

Adding and subtracting

1 FLYING FOXES ARE REALLY BATS
2 1·26 tonnes
3 264 g
4

1st	2nd	3rd	4th	5th	6th
1p	2p	4p	8p	16p	32p

7th	8th	9th	10th	11th
64p	£1·28	£2·56	£5·12	£10·24

  Total = £20·47
5 a 1·37, 1·73, 3·17, 3·71, 7·13, 7·31, 13·7, 17·3, 31·7,
    37·1, 71·3, 73·1
  b 71·73
  c 21·01
  d 2·16

Homework Sheet 4.2

Multiplying

1 PANTOPHOBIA IS THE FEAR OF EVERYTHING
2 a £5·63 **b** £0·51 **c** £1·15 **d** £0·58
3 17 and 27
4 123 Euros
5 A = 1, B = 2, C = 3
  No, you could have
  A = 1, B = 3, C = 4
  or A = 1, B = 4, C = 5
  or A = 1, B = 5, C = 6 etc.

Homework Sheet 4.3

Dividing

1 a 16 **b** 10
2 a 25·4 **b** 23·25 **c** 6·8
3 a 38·4 **b** 47·6 **c** 4·3
4 The Medium 500 g jar.

Homework Sheet 4.4

Dividing by a decimal

1 a 480 ÷ 2 **b** 76 100 ÷ 3 **c** 800 ÷ 5 **d** 110 ÷ 16
2 a 40 **b** 2500 **c** 65 **d** 28
3 a 275·6 **b** 366·7 **c** 37·9 **d** 458·6

Homework Sheet 4.5

Checking answers
Possible answers are:
1 **a** $8 \cdot 2 \times 6 \cdot 18$ is approximately $8 \times 6 = 48$. But it should be more than 48, because $8 \cdot 2$ is greater than 8 and $6 \cdot 18$ is greater than 6. So $47 \cdot 376$ must be wrong.
   b $792 \div 1 \cdot 2 = 7920 \div 12$ (Equivalent calculation)
      $= 660$ Correct
   c $6 \cdot 8 \times 4 \cdot 3$ must end in 4 because $8 \times 3 = 24$, so $29 \cdot 42$ cannot be correct.
2 **a** C $33 \cdot 12$ $8 \times 9 = 72$, so the answer must end in 2. It must be less than 5×7 which is 35.
   b B $5 \cdot 16$ $3 \times 2 = 6$, so the answer must end in 6.

Homework Sheet 4.6

Ex 1 Using brackets on the calculator
1 GOLF BALLS CONTAIN CASTOR OIL
2 $2 + 5 \times (4 - 1) + (6 + 8) \div 7 = 19$

Ex 2 Using the calculator memory
1 £1·98
2 **a** 2·85 litres   **b** 6·27 litres   **c** 11·4 litres

Homework Sheet 5.1

Ex 1 Fractions of shapes
1 $\frac{7}{20}$
2 **a** About $\frac{1}{4}$ **b** About $\frac{1}{3}$

Ex 2 One number as a fraction of another
1 **a** $\frac{1}{3}$ **b** $\frac{1}{6}$ **c** $\frac{5}{6}$ **d** $\frac{3}{4}$
2 **a** $\frac{7}{10}$ **b** $\frac{4}{5}$ **c** $\frac{17}{24}$
3 **a** $\frac{2}{5}$ **b** $\frac{3}{4}$ **c** $\frac{1}{4}$
4 **a** $\frac{1}{4}$ **b** $\frac{1}{8}$ **c** $\frac{3}{8}$
5 There are many possible answers.

Homework Sheet 5.2

Ex 1 Fractions and decimals
1 SIMBA MEANS LION IN SWAHILI

Ex 2 Fractions and decimals
1 **a** 0·6   **b** 0·85   **c** 0·84   **d** 0·16
   e 0·38   **f** 1·55   **g** 3·625   **h** 1·7
   i 3·6   **j** 2·75
2 **a** 0·54 **b** 0·44 **c** 0·71 **d** 0·61
3 **a** 0·090909 ...   0·181818 ...   0·272727 ...   0·363636 ...
   b The answers repeat the multiples of 9.
   c 0·454545 ..., 0·545454 ..., 0·636363 ..., 0·727272 ...

Homework Sheet 5.3

Ordering fractions
1 **a** > **b** < **c** < **d** < **e** < **f** > **g** <
2 The Burns family
3 **a** $\frac{1}{2}, \frac{7}{12}, \frac{2}{3}, \frac{3}{4}, \frac{7}{9}, \frac{5}{6}$ **b** $\frac{1}{6}, \frac{1}{4}, \frac{4}{15}, \frac{3}{10}, \frac{7}{20}, \frac{2}{5}, \frac{5}{12}$

4 **a**
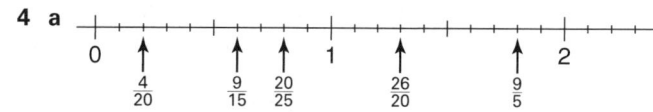

   b
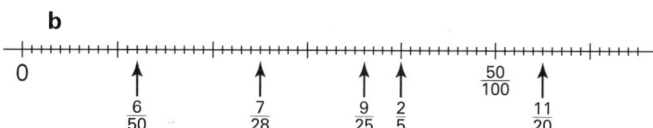

Homework Sheet 5.4

Adding and subtracting fractions
1 **a** $\frac{3}{6} = \frac{1}{2}$ **b** $\frac{10}{12} = \frac{5}{6}$ **c** $\frac{1}{9}$ **d** $\frac{11}{16}$ **e** $\frac{13}{21}$ **f** $\frac{19}{40}$ **g** $\frac{9}{20}$
2 $\frac{6}{35}$ to Childline.
3 Possible answers are:

   a $\frac{1}{8} + \frac{3}{8} = \frac{7}{8}$ **b** $\frac{1}{6} + \frac{9}{12} = \frac{11}{12}$
   $\frac{2}{8} + \frac{5}{8} = \frac{7}{8}$ $\frac{2}{6} + \frac{7}{12} = \frac{11}{12}$
   $\frac{3}{8} + \frac{1}{2} = \frac{7}{8}$ $\frac{3}{6} + \frac{5}{12} = \frac{11}{12}$
   $\frac{4}{8} + \frac{3}{8} = \frac{7}{8}$ $\frac{4}{6} + \frac{3}{12} = \frac{11}{12}$
   $\frac{5}{8} + \frac{2}{8} = \frac{7}{8}$ $\frac{5}{6} + \frac{1}{12} = \frac{11}{12}$
   $\frac{6}{8} + \frac{1}{8} = \frac{7}{8}$

4 60 litres

Homework Sheet 5.5

Fraction of. Multiplying an integer by a fraction
1 **a** $4\frac{3}{8}$ **b** $4\frac{2}{5}$ **c** $4\frac{4}{9}$
2 $58\frac{1}{2}$ metres
3 **a** $2\frac{2}{3}$ **b** $19\frac{1}{2}$ **c** $5\frac{1}{4}$ **d** $16\frac{1}{2}$
   e $30\frac{1}{4}$ **f** $29\frac{1}{3}$ **g** $12\frac{1}{3}$

Homework Sheet 5.6

Dividing an integer by a fraction
1 **a** $2 \div \frac{1}{5}$ **b** $5 \div \frac{1}{4}$
2 **a** 28 **b** 27 **c** 15 **d** 24
3 **a** 24 **b** 36
4 **a** 10 **b** 15
5 **a** 20 **b** 20 **c** 12
6 25 friends

Homework Sheet 6.1

Fractions, decimals, percentages
1 **a i** 71% **ii** 39·5% **iii** 161·3%
   b i $\frac{18}{25}$ **ii** $\frac{4}{5}$ **iii** $1\frac{1}{4}$
2 History 85%
3 **a** $\frac{3}{8}$, 0·38, 45%, $\frac{1}{2}$ **b** $\frac{3}{5}$, 64%, $\frac{13}{20}$, 0·68
   c 138%, $1\frac{2}{5}$, $1\frac{11}{25}$, 1·49 **d** 71%, $\frac{3}{4}$, 0·83, $\frac{17}{20}$
4 A DAY ON VENUS IS LONGER THAN ITS YEAR
5 **a** True **b** False **c** True **d** True
   e True **f** False **g** False **h** False

Homework Sheet 6.2

Ex 1   Percentage of — mentally
1 **a** 75 litres **b** 24 m **c** £12 **d** 1425 g
   e 132 cm **f** £110 **g** 66 g **h** 188 ml
   i £4·60
2 63 kg
3 £126

Ex 2   Percentage of — written and calculator
1 **a** £24·50 **b** 72 m **c** £7·74 **d** 126 km
   e 27 cm **f** 124·8 litres
2 **a** 35·1 g **b** £158·60 **c** 25·65 km **d** 312 m
   e £6·30 **f** 100·05 ml
3 74 925 000 or 74·925 million
4 Jack's Garden Centre
5 **a** 5525 **b** 4420
6 £61·08

Homework Sheet 6.3

Percentage increase and decrease
1 **a** 604·8 **b** 1767
2 1610 g
3 £5·67
4 620 Sweets

Homework Sheet 6.4

Mixed percentage problems
1 64%
2 **a** £9·62 **b** £308·10 **c** £67·78
3 65%

Homework Sheet 7.1

Direct proportion
1 **a** £2·59 **b** £1·36
2 16p
3 264 window frames
4 168 black beads
5 **a** 75 g butter **b** 54 g sugar **c** 544 g flour

Homework Sheet 7.2

Simplifying ratios
1 **a** 1 : 7 : 3 **b** 4 : 5 : 11 **c** 2 : 9 : 16
2 3 : 7
3 **a** 2 : 3 **b** 8 : 5 **c** 7 : 30
4 **a** 50 : 17 **b** 9 : 50 **c** 4 : 9
5 15 : 16

Homework Sheet 7.3

Ratio and proportion
1 **a** 4 : 3 **b** 4 : 5 : 3 **c** $\frac{5}{12}$ **d** 25%
2 **a** 2 : 5 **b** $\frac{3}{10}$ or 0·3 or 30% **c** 80%

Homework Sheet 7.4

Solving ratio and proportion problems
1 15 matches
2 490 ml lemonade, 70 ml rum
3 £1800
4 **a** 2·6 m **b** 21·25 cm

Homework Sheet 7.5

Dividing in a given ratio
1 **a** £16 : £64 : £80 **b** £54 : £135 : £216
   c 72 g : 120 g : 192 g
2 42 g tin, 294 g copper, 126 g zinc
3 Shaun received £14 500, Grant received £23 200, Christina received £20 300

Homework Sheet 8.1

Ex 1   Understanding algebra — expressions and equations
1 Equations: $4x + 7 = 15$, $x - 3 = 8$. They contain the = sign.

   Expressions: $12m - 5$, $\frac{6y}{5}$, $2(a + 6)$

2 **a** An expression. c can have any value.
   b An equation. t has only one particular value.

Ex 2   Understanding algebra — formulae
1 It is a formula because it is a rule which gives a relationship between real quantities.
2 Equation: $2p + 3 = 8$, expression: $p + 2q$, formula: $p = 2q + 4$

1

Worksheets

Place value. Reading and writing numbers

1 What is the value of the digit 4 in these?

    a 468   **b** 47 809   **c** 68·42   **d** 11·674   **e** 8·04

2 Write these as decimal numbers.

    a Two thousand five hundred and eight point three six

    b Eight hundred and fourteen thousand and eleven point six one five

    c Thirteen hundredths

    d $83\frac{21}{1000}$

3 What is

    a 10 more than 9768   **b** 1000 more than 24 785   **c** 10 less than 907?

4 Paul has weighed out 21·35 g of copper sulphate crystals. How much must he add or subtract to get

    a 22·35 g   **b** 21·25 g   **c** 21·38 g   **d** 21·34 g   **e** 20·85 g?

5 Write these in words.   **a** 36 802   **b** 605 004·37   **c** 2 041 689·048

6 Which number is halfway between these?

    a 2940 and 2950   **b** 0·4 and 0·5   **c** 2·3 and 2·7   **d** 4·8 and 5·1

7 What is

    a 0·01 more than 3·27   **b** 0·1 less than 18·56   **c** 0·1 more than 6·95   **d** 0·01 less than 3·205?

8 Make the largest and smallest number you can using

    a the digits 8, 4, 1 and 2

    b the digits 3, 7, 9 and 5 and a decimal point. The decimal point must have at least one figure before and after it.

Multiplying and dividing by 10, 100 and 1000

1 What is

    a 43 × 10   **b** 16 × 100   **c** 12 × 1000   **d** 670 ÷ 10   **e** 83 000 ÷ 100

    f 1·4 × 10   **g** 12·89 ÷ 10   **h** 0·2 ÷ 10   **i** 307·4 ÷ 1000   **j** 8·6 × 100?

2 A 10p coin measures 2·4 cm across.

    a How long is a line of ten 10p coins?   **b** How long is a line of one hundred 10p coins?

    c How much is a line of one thousand 10p coins worth?

3 Tins of baked beans cost 37p. Ten tins are put in a box. Ten boxes are put in a crate.

    a How many tins of baked beans are there in a crate?   **b** How much do 100 crates cost?

4 Which number goes in the square?

    a 400 ÷ ☐ = 4   **b** 3·7 × ☐ = 3700   **c** 8 ÷ ☐ = 0·8   **d** 6·2 × ☐ = 62   **e** 0·3 ÷ ☐ = 0·03

© New National Framework Mathematics 8 Nelson Thornes Ltd

Putting numbers in order

1 Which of these statements are true?

    a 279 > 297   **b** 20 409 < 20 490   **c** 4000 > 3999

2 Write these numbers in order from smallest to largest.

    a 2687, 6278, 6782, 7286, 2867, 8276   **b** 4·6, 0·46, 6·04, 46, 40·6, 6·004, 4·06

3 Which is greater, 284 762 miles or 284 726 miles?

4 Which of <, > or = makes these sentences correct?

    a 4 + 9 ... 6 + 7   **b** 13 − 7 ... 2 + 5   **c** ⁻3 ... ⁻7   **d** ⁻1 ... 0   **e** 8 − 4 ... 4 − 8

5 Here are some number cards.

    You can use each card once to make a number like this:

    a What is the smallest number you can make using all four cards?

    b Use some of these four number cards to make numbers that are as close as
        possible to the number written below.

    Examples   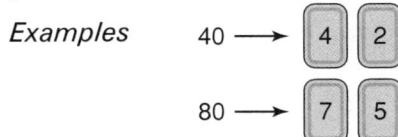

    You must not use the same card more than once in each answer.

    i 50 ⟶ ☐☐   **ii** 30 ⟶ ☐☐

    iii 500 ⟶ ☐☐☐   **iv** 3000 ⟶ ☐☐☐☐

6 Ten competitors completed a race in the following times:

    Darren 37·4 sec, Nigel 41·62 sec, Paul 38·04 sec, Malcolm 39·6 sec

    John 37·09 sec, Daniel 38·2 sec, Hassan 39·55 sec, Kirk 40·8 sec

    Philip 37·43 sec, Silvio 40·68 sec.

    Write the competitors in the order in which they finished, fastest first.

7 Kathleen works on the cheese stall at the market. She cuts up a large block of Cheddar
    into pieces for display and weighs each one accurately. These are their masses.

    0·43 kg,   0·527 kg,   0·418 kg,   0·602 kg,   0·5 kg,   0·48 kg,   0·56 kg,   0·407 kg,   0·509 kg

    Put the masses in order from lightest to heaviest.

8 Which of <, > or = goes in the box?

    a 4500 mm ☐ 4·505 m   **b** 2·885 kg ☐ 299 g   **c** 3·6 km ☐ 3597 m

9 **a** Make the largest number you can with these cards. There must be at least
        one digit after the decimal point.   

    b Make the second largest number, with at least one digit after the decimal point.

    c Make the second smallest number, with at least one digit before the decimal point.

NUMBER

Rounding

1 Jack did a survey of the number of cars entering a supermarket car park per hour at four different times of the day. This is his result.

early morning 782   late morning 1247   early afternoon 998   evening 345

a Round each of these numbers to the nearest 10.

b Round each of them to the nearest 100.

2 Tiffany's hens have produced 290 eggs in a year, rounded to the nearest 10. What is the smallest number of eggs the hens could have produced?

3 You have 96 528 km of tubing in your arteries and veins. Round this number

a to the nearest hundred   **b** to the nearest thousand.

4 Round the answers to these sensibly.

a Five friends shared the cost of a meal which came to £38·69 in total. How much did each one have to pay?

b The total weight of Paul's seven baby guinea pigs is 1185 g. How heavy, on average, is each guinea pig?

5 Australia covers an area of 7 620 000 km² to the nearest ten thousand km².

a What is the smallest area it could cover?

b What is the largest area it could cover to the nearest km²?

Divisibility

1 Shade triangles with numbers that are

a divisible by 10   **b** divisible by 5

c divisible by 4   **d** divisible by 3.

What shape does the shading make?

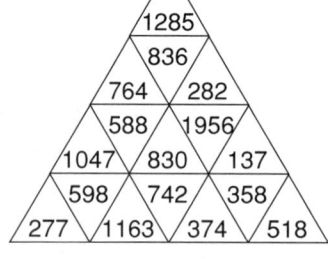

2 **a** Which of these is divisible by both 3 and 5?

645,   204,   1335,   365,   935

b Which of these is divisible by both 4 and 9?

224,   2516,   1044,   338,   612

c Which of these is divisible by 8?

216,   274,   548,   656,   380

Integers

1

```
   +----+----+----+----+----+----+----+----+----+----+----+----+----+----+----+  °C
  ⁻80  ⁻70  ⁻60  ⁻50  ⁻40  ⁻30  ⁻20  ⁻10   0   10   20   30   40   50   60   70
```

Mark these temperatures on the number line.

A 37 °C   Normal body temperature
B ⁻39 °C   Mercury freezes solid
C ⁻78 °C   Carbon dioxide snow changes to gas

2 a Put these temperatures in order, coldest first.

   3 °C,   ⁻8 °C,   0 °C,   ⁻2 °C,   4 °C,   ⁻5 °C

b Put these numbers in descending order.

   ⁻3·6,   2·7,   ⁻0·4,   1·9,   0·75,   ⁻4·1

3 Paula has £35 in her bank account. Write her new bank balance as a positive
or negative number, if

   a she withdraws £20   **b** she then withdraws a further £30.

4 The temperature at midnight was ⁻11 °C. How much has it risen if the temperature
is now   **a** 3 °C   **b** ⁻7 °C   **c** 0 °C   **d** 14 °C?

5 At a certain time one day, the temperature in London is 14 °C and the temperature
in Moscow is ⁻5 °C. What is the difference between the temperatures in
London and Moscow?

6 A submarine lies at ⁻185 m. It then rises 110 m towards the surface.
How deep is it now?

7 A diver dives down to ⁻23 m, swims up 8 m and then dives down another 19 m to
photograph a fish. How deep is the diver now?

8 [2] [⁻3] [⁻2] [5] [0] [⁻1] [4]

   a Which number card makes this true?   [⁻2] + [ ] + [5] = [0]

   b Which two number cards make this true?   [0] + [⁻1] + [ ] + [ ] = [⁻2]

9 Put these integers on a number line: ⁻4,   3,   6,   ⁻5,   ⁻7

   Which number is halfway between these?

   a ⁻5, ⁻7   **b** 6, ⁻4   **c** ⁻7, 3   **d** ⁻5, 6

10 a Write down the next three lines of this pattern.

$$2 + {}^-1 = 1$$
$$2 + {}^-2 = 0$$
$$2 + {}^-3 = {}^-1$$
$$2 + {}^-4 = {}^-2$$

   b Use your pattern to write down the answer to 2 + ⁻9.

11 Place the numbers from ⁻6 to 2 in the circles, so that
each line has a total of ⁻3.

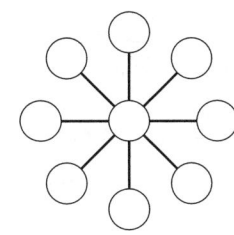

NUMBER

Multiples, factors, primes, squares, triangular numbers

1 Write down the first five multiples of **a** 4   **b** 7.

2 Write down the factor pairs of **a** 12   **b** 36   **c** 84.

3 Write down the first six prime numbers.

4 From the box write down

   a the multiples of 4

   b the prime numbers

   c the factors of 30

   d the square numbers.

2	1	21	9
4	11	16	19
7	15	35	30

5 a Write down the factors of 36.

   b Write down the factors of 60.

   c Write down all the common factors of 36 and 60.

   d What is the highest common factor of 36 and 60?

6 a Write down all the triangular numbers less than 60.

   b Write down all the square numbers between 30 and 60.

7 Shade

   a the multiples of 6

   b the prime numbers

   c the square numbers

   d the factors of 30.

18	8	610	62	1
21	6	80	2	88
28	39	900	26	14
32	4	56	15	46
42	11	27	33	11

   What letter of the alphabet does the shading make?

8 Find the answers to these.

   a $\sqrt{64}$ **b** $\sqrt{25}$ **c** $\sqrt{81}$ **d** $\sqrt{1}$ **e** $\sqrt{121}$

9 a The numbers 357 and 753 can be made using the digits 3, 5 and 7.
     What other numbers can be made with these three digits?
     Each digit can only be used once.

   b What is the largest multiple of 3 that can be made from the digits 3, 5 and 7?

   c What is the largest multiple of 5 that can be made from these digits?

10 Which of these is the same as 40^2?

   A 20   **B** 80   **C** 1600   **D** 16   **E** 160

11 Find two prime numbers which add to give a square number.
    Find as many answers as possible.

12 Test whether these are prime numbers. Write Yes or No.

   a 37   **b** 49   **c** 91   **d** 79

13 a If $11^2 = 121$, what is $1 \cdot 1^2$?   **b** If $\sqrt{289} = 17$, what is $\sqrt{2 \cdot 89}$?

Mental calculation

1 Do these multiplications mentally.

  a 4×9 **b** 8×7 **c** 5×11 **d** 9×6 **e** 20×8

  f 6×40 **g** 800×8 **h** 500×9 **i** 900×4 **j** 7×600

2 Find the answers to these mentally.

  a $54 \div 9$ **b** $32 \div 8$ **c** $36 \div 4$ **d** $81 \div 9$ **e** $720 \div 8$

  f $210 \div 7$ **g** $2400 \div 6$ **h** $4200 \div 7$ **i** $1200 \div 4$ **j** $1800 \div 6$

3 Find the missing numbers mentally.

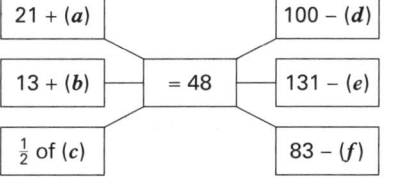

$21 + (a)$		$100 - (d)$
$13 + (b)$	$= 48$	$131 - (e)$
$\frac{1}{2}$ of (c)		$83 - (f)$

4 What numbers go in the boxes?

  a $87 + \boxed{} = 100$ **b** $100 - \boxed{} = 34$ **c** $2 \cdot 7 + \boxed{} = 10$ **d** $0 \cdot 3 + \boxed{} = 1$ **e** $1 - \boxed{} = 0 \cdot 63$

5 $1 \cdot 47 + 3 \cdot 69 = 5 \cdot 16$.   What is the answer to $5 \cdot 16 - 3 \cdot 69$?

6 Five friends buy tickets to a concert at £2·40 each.
   How much do the tickets cost altogether?

7 Calculate these mentally.

  a $3 + 6 \times 2$ **b** $8 \times 4 - 2$ **c** $\dfrac{24}{4 \times 3}$ **d** 2×3^2

  e $3(6 + 2)$ **f** $8^2 - 6^2$ **g** $\dfrac{1^2 + 2^2}{5}$ **h** $5(7^2 - 5^2)$

8 Find the answers to these mentally.

  a $2 + 7 + 19 + 26$ **b** $70 - 13 - 14$ **c** $86 - 71$ **d** $124 - 77$ **e** $84 + 29 - 73$

9 a June had two scones and a cup of tea.

   i How much did this cost?

   ii How much change did she get from £5?

  b Ben bought two teacakes and two cups of coffee. He was charged £3·10.
     How can you tell this is wrong without doing the calculation?

Teacake	£0·45
Gateau	£1·20
Coffee	£0·90
Tea	£0·70
Scone	£0·80

10 Find the answers to these mentally.

  a $0 \cdot 7 + 0 \cdot 5$ **b** $0 \cdot 8 + 0 \cdot 6$ **c** $1 \cdot 9 - 0 \cdot 8$ **d** $2 \cdot 4 + 0 \cdot 7$ **e** $8 \cdot 3 - 1 \cdot 2$

11 Calculate these mentally.

  a $2 \times 3 \times 8$ **b** 130×4 **c** 20×70 **d** 63×11

  e $1 \cdot 3 \times 4$ **f** $8 \times 0 \cdot 6$ **g** $1 \cdot 8 \times 50$

12 Find the answers to these mentally.

  a $270 \div 3$ **b** $846 \div 2$ **c** $8 \cdot 1 \div 9$ **d** $3 \cdot 2 \div 4$

13 Calculate these mentally.

  a $\sqrt{68 - 43}$ **b** $\sqrt{24 + 25}$ **c** $\sqrt{8^2 - 15}$ **d** $\sqrt{6^2 + 8^2}$

14 Find the answers to these mentally.

  a $3 \times 7 \div 3$ **b** 34×3 **c** 80×20 **d** 18×21 **e** 14×12

NUMBER

NUMBER

Estimating

1 Which is the best approximation for these calculations?

a 207 + 398   **A** 200 + 300   **B** 300 + 400   **C** 200 + 400   **D** 300 + 300

b 29 × 13   **A** 30 × 20   **B** 30 × 10   **C** 20 × 10   **D** 20 × 20

c 31·7 – 12·86   **A** 30 – 10   **B** 30 – 20   **C** 31 – 12   **D** 317 – 128·6

d 2·89 × 41·6   **A** 2 × 41   **B** 2 × 40   **C** 289 × 416   **D** 3 × 40

2 Estimate the answers to these.

a 487 + 618   b 294 – 89   c 18 × 43   d 478 ÷ 5·21

e 7·92 + 1·07   f 11·21 – 4·88   g 3·147 × 82

3 Lisa bought 18 roses at £6·99 each. She calculated the cost to be £141·82. Explain how you can tell that Lisa cannot be right.

4 Write down two ways to find an estimate for these.

a 25·2 – 14·6   b 8·47 × 1·98   c 207 + 349

5 Estimate the answers to these. Then find the accurate answer.

a 2·3 × 7   b 8·9 × 3   c 12·14 × 6   d 20·8 × 4

e 147·8 × 5   f 203·8 × 8

Order of operations   except question 4.

1 Write the answers to each of the following.

a 2 + 5 × 3   b (2 + 5) × 3   c 2 × 3 + 5

d 2 + 8 ÷ 4   e (2 + 3) × 5   f 5 + 2 × 3

2 Work out the answers to these.

a (6 + 4) × (7 – 3)   b 4 × (11 – 5)   c 2 + 6 ÷ 3 – 1   d 12 ÷ (21 ÷ 7)

3 Put brackets in these calculations to make the answers correct.

a 2 + 7 × 4 = 36   b 4 + 3 + 2 × 10 = 90   c 8 – 3 × 2 + 7 = 45   d 4 + 5 + 2 × 3 = 25

 4 Use your calculator to find the answers to these to 1 d.p.

a 19·7 – 6·2 × 1·8   b (3·6 + 7·8) × 1·4   c $\dfrac{4·9}{8·43 × 0·71}$ d $\dfrac{8·47 + 3·95}{13·72 – 7·18}$

Written calculation

1 Work these out.

| **a** 27 +58 | **b** 168 +144 | **c** 76 −25 | **d** 142 − 28 | **e** 208 − 49 | **f** 300 −176 |

2 Calculate these.

a 219 + 736   **b** 421 − 185   **c** 27 + 125 + 218   **d** 202 − 84   **e** 706 − 299

3 Copy and complete these multiplications.

| **a** 26 × 3 | **b** 472 × 4 | **c** 196 × 7 | **d** 274 × 9 |

4 Find the answers to these.

a 179×6 **b** $18{\cdot}4 \times 8$ **c** $29{\cdot}8 \times 3$ **d** $11{\cdot}29 \times 5$

5 Copy these and find the answer. Write any remainders as R ___ .

a $3\overline{)86}$ **b** $6\overline{)143}$ **c** $4\overline{)762}$ **d** $8\overline{)651}$

6 Find the answers to these. If there is a remainder write it as R ___ or as a fraction.

a $294 \div 7$ **b** $896 \div 5$ **c** $188 \div 9$ **d** $377 \div 4$

7 Find the answers to these.

| **a** 27 ×13 | **b** 243 × 24 | **c** 176 × 17 | **d** 529 × 36 |

8 **a** Find the sum of 76 and 138.

b Find the difference between 279 and 423.

c Find the product of 42 and 27.

9 Find the answers to these. Estimate first.

a $4{\cdot}6 \times 3$ **b** $12{\cdot}7 \times 9$ **c** $1{\cdot}42 \times 8$ **d** $15{\cdot}9 \times 7$

Fractions

1 What fraction of each of these is shaded?

a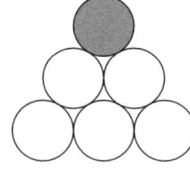
b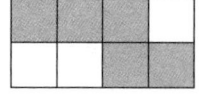
c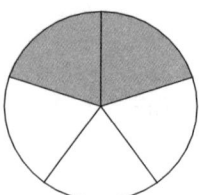

2 Find

a $\frac{1}{4}$ of £24   **b** $\frac{1}{2}$ of 80 m   **c** $\frac{1}{5}$ of 35 cm   **d** $\frac{1}{3}$ of 60p   **e** $\frac{1}{10}$ of £700.

3 Which fractions from the box are equivalent to $\frac{2}{5}$?

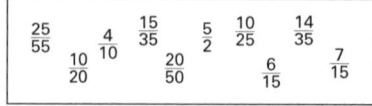

$\frac{25}{55}$ $\frac{4}{10}$ $\frac{15}{35}$ $\frac{5}{2}$ $\frac{10}{25}$ $\frac{14}{35}$ $\frac{10}{20}$ $\frac{20}{50}$ $\frac{6}{15}$ $\frac{7}{15}$

4 Shahid had 24 golf balls. He gave $\frac{1}{4}$ of them to his friend Jay.
He gave $\frac{1}{3}$ of the rest to his sister. How many golf balls did Shahid have left?

5 Find the missing numbers.

a $\frac{2}{3} = \frac{8}{\square}$ **b** $\frac{3}{4} = \frac{\square}{28}$ **c** $\frac{3}{7} = \frac{15}{\square}$ **d** $\frac{5}{12} = \frac{15}{\square}$

e $\frac{8}{9} = \frac{48}{\square}$ **f** $\frac{36}{48} = \frac{3}{\square}$ **g** $\frac{49}{63} = \frac{\square}{9}$ **h** $\frac{18}{81} = \frac{2}{\square}$

6 What fraction of

a 1 metre is 23 centimetres   **b** 1 minute is 24 seconds?

7 Use cancelling to write these fractions in their lowest terms.

a $\frac{9}{12}$ **b** $\frac{15}{35}$ **c** $\frac{21}{27}$ **d** $\frac{64}{72}$ **e** $\frac{44}{64}$ **f** $\frac{25}{40}$ **g** $\frac{24}{90}$

8 a I am equivalent to $\frac{3}{4}$. My denominator is 6 more than my numerator. What fraction am I?

b I am equivalent to $\frac{28}{63}$. My numerator and denominator are square numbers. What fraction am I?

9 Find these mentally.

a $\frac{1}{5}$ of £55   **b** $\frac{1}{4}$ of 120 cm   **c** $\frac{3}{4}$ of 20 m

d $\frac{3}{5}$ of 35 ℓ **e** $\frac{5}{8}$ of 64 kg   **f** $\frac{2}{7}$ of 42p

10 On Joel's farm, 85 of the 125 piglets had straight tails.

a What fraction had straight tails?

b What percentage had straight tails?

11 Write the equivalent fraction that has a denominator of 60.

a $\frac{1}{2}$ **b** $\frac{1}{4}$ **c** $\frac{3}{4}$ **d** $\frac{7}{10}$ **e** $\frac{5}{6}$ **f** $\frac{7}{12}$ **g** $\frac{2}{15}$

12 Calculate these.

a $\frac{1}{3} \times 27$ **b** $40 \times \frac{5}{8}$ **c** $\frac{3}{5} \times 45$ **d** $1\frac{1}{3} \times 33$ **e** $7 \times \frac{2}{3}$ **f** $2\frac{1}{2} \times 7$

Decimals, fractions and percentages

 Except for questions 7 and 8.

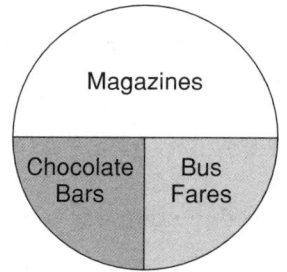

1 The pie chart shows how Hannah spends her pocket money one week.

   a What fraction of her pocket money does Hannah spend on magazines?

   b What percentage does she spend on chocolate bars?

2 Write these as decimals.

   a $\frac{1}{2}$ **b** $\frac{3}{4}$ **c** $\frac{3}{10}$ **d** $\frac{37}{100}$ **e** $2\frac{1}{4}$ **f** $\frac{3}{100}$

   g 20%   **h** 43%   **i** 65%   **j** 7%   **k** 150%   **l** 109%

3 Write these as fractions in their simplest form.

   a 0·7   **b** 0·25   **c** 0·4   **d** 0·36   **e** 0·85   **f** 0·08

   g 0·72   **h** 0·94   **i** 0·22

4 In a survey of 240 pupils, only 15% were left-handed.
How many left-handed pupils were questioned in the survey?

5 Find the answers to these mentally.

   a 30% of 150   **b** 75% of 80   **c** 50% of 48

   d 10% of 40   **e** 15% of 40   **f** 21% of 600

6 Write these as percentages.

   a $\frac{2}{5}$ **b** $\frac{9}{10}$ **c** $\frac{1}{4}$ **d** $\frac{18}{25}$ **e** $\frac{7}{20}$

   f $2\frac{1}{2}$ **g** 0·85   **h** 0·1   **i** 0·99   **j** 1·26

 7 Use a calculator to find these.

   a 65% of £258   **b** 18% of 144 kg   **c** 28% of 65 m

 8 Cheeky Charlie's and Sam's Stores both sell dog food.

   Which store sells the dog food with the highest percentage of protein?

Ratio and proportion

1   When Daniel tiles his bathroom wall, he uses 3 blue tiles for every 5 white tiles.
    If he uses 15 blue tiles in one area, how many white tiles does he use?

2   If 2 bus fares cost £1·24, how much will 8 bus fares for the same journey cost?

3   A potting compost is made up of 6 parts peat to 4 parts sharp sand.

   a What is the ratio of peat to sharp sand in the compost?

   b What proportion of the compost is sharp sand? Give your answer as a percentage.

   c What proportion of the compost is peat? Give your answer as a fraction.

4   Dillon's aviary has 24 canaries and 16 lovebirds.
    What is the ratio in its simplest form of canaries to lovebirds?

5   Grant and Carla share 72 hologram cards between them in the ratio 5 : 4.
    How many cards does Carla get?

6   The ratio of black beads to purple beads in a necklace is 4 : 3

    How many black beads are in a necklace with this number of purple beads?

   a 18   b 27   c 33   d 60

Ex 3   Functions

Equation	Expression	Formula	Function
a $6m + 3 = 18$	$5q + 3$	$L = 4M + 50$	$y = 5x - 4$
b $4(x + 2) = 24$	$2(3m - p)$	$p = 4 - 3q$	$y = 4 - \frac{1}{2}x$
c $3 - 2m = 5$	$\dfrac{7x - 4}{5}$	$v^2 - u^2 = 2as$	$y = \frac{x}{3} + 2$
d $\dfrac{2a + 4}{3} = 5$	$\dfrac{p^2 - 4}{3p}$	$p = 4s$	$y = \dfrac{4 - 3x}{7}$

Ex 4   Rules for writing expressions

1 a $2m$ **b** $8q$ **c** $\frac{x}{8}$ **d** $4(x + 7)$ **e** $\dfrac{4 + p}{5}$

   f $2ab$ **g** $7pq$ **h** $\dfrac{6m}{5}$ **i** $2a^3$ **j** $4x^2$

2 a mn **b** $\frac{y}{2}$ **c** $4m + 3$ **d** $\dfrac{a + b}{3}$ **e** $\dfrac{cd}{4}$ **f** $5(s + t)$

Homework Sheet 8.2

Ex 1   More understanding algebra — algebraic operations

1 a $2x$ **b** m^2 **c** $x + 7$
   d $7 - t$ **e** y^2 **f** $a + 4$
2 a True   **b** False   **c** True
3 $26 + 29$ is not equal to $26 + 20$. It should say
$a + b = 26 + 29$
$ = 26 + 20 + 9$
$ = 46 + 9$
$ = 55$

Ex 2   Inverse operations

1 a $p = q + 3, \quad q = 3 - p$

   b $2b = a, \quad \frac{a}{b} = 2$

   c $x = \frac{y}{4}, \quad 4 = \frac{y}{x}$

   d $n = 6 \times m, \quad 6 = \dfrac{n}{m}$

   e $5 - r = t, \quad 5 - t = r$

2 a Adding 9
   b Dividing by 4
   c Multiplying by 2, then subtracting 3
   d Adding 10, then multiplying by 6
   e Dividing by 5, then adding 4

3 a C $\dfrac{y - 3}{2} = x$

   b B $m = \frac{n}{4} - 3$

Homework Sheet 8.3

Simplifying expressions — multiplying and dividing

1 a $18m$ **b** $20a$ **c** ^-12q **d** ^-18t
2 a c^5 **b** x^3 **c** e^8
3 a x^3 **b** p^4 **c** $6y^2$ **d** $14m^2$
4 a p **b** y^2 **c** r^2 **d** 1
   e $4q^2$ **f** $3b^3$ **g** $2m$
5 a $4x$ **b** $3y$ **c** $6n$
6 A SNEEZE IS AS FAST AS AN EXPRESS TRAIN.

Homework Sheet 8.4

Brackets

1 a $2a + 14$ **b** $11x - 22$ **c** $12 - 4q$
   d $5y - 30$ **e** $^-3m - 24$ **f** $^-4j + 12$
   g $6p - 24$ **h** $14m + 35$ **i** $24 - 16m$
   j $6a - 12b$ **k** $^-35t + 21r$ **l** $336a - 294b$

Homework Sheet 8.5

Ex 1   Collecting like terms

1 THE YOUNGEST POPE WAS ELEVEN
2 $x + y + 2x - y + 3x + 4 + 3y - 2 + x + 2y$
   $= 7x + 5y + 2$
3 a $x + y$ **b** $3x - 2y$ **c** $7x - y$

Ex 2   Multiplying out brackets, then simplifying

1 a $2x + 6 + 5x + 5 = 7x + 11$
   b $6a + 24 + 3a - 6 = 9a + 18$
   c $4m - 28 + 5m + 20 = 9m - 8$
   d $18y + 36 + 2y = 20y + 36$
   e $6 + 4k + 28 = 4k + 34$
   f $5 - x - 2 = 3 - x$
   g $11 - 4m - 6 = 5 - 4m$
   h $24p - 6q - 3q + 2p = 26p - 9q$

Homework Sheet 8.6

Substituting

1 COCA-COLA WAS ORIGINALLY GREEN
2 a 7   **b** 9   **c** 18   **d** 39
   e 6   **f** 4   **g** 0   **h** 5
3 a 10   **b** 55

Homework Sheet 8.7

Ex 1   Substituting into formulae

1 a $T = 210$ minutes   **b** $T = 270$ minutes
   c $T = 350$ minutes
2 a $c = £360$ **b** $c = £600$
3 a $A = 72 \text{ m}^2$ **b** $A = 106 \cdot 4 \text{ m}^2$
4 a $A = 300 \text{ cm}^2$ **b** $A = 481 \text{ cm}^2$ **c** $A = 320 \text{ cm}^2$

Ex 2   Substituting into a formula

1 a $A = 28 \text{ cm}^2$ **b** $h = 10$ cm
2 a $s = 180$ m   **b** $t = 9$ seconds

Homework Sheet 8.8

Writing expressions, formulae and equations

1   **a** $12m$ pounds   **b** $(2m - 1)$ pounds   **c** $\dfrac{2m - 1}{2}$ pounds

2   $6x + 4y$

3   **a** $t - 3$ **b** $2(t - 3)$

Homework Sheet 8.9

More writing and simplifying algebraic expressions

1   **a, b**   Two possible expressions are
$x + x + x + x + y + y + y$ and $4x + 3y$
  c $x + x + x + x + y + y + y = 4x + 3y$

2   $12m - 28$, $5m + 7(m - 4)$, $12(m - 4) + 20$
$5m + 7(m - 4) = 5m + 7m - 28$
$= 12m - 28$
$12(m - 4) + 20 = 12m - 48 + 20$
$= 12m - 28$

3   Let the first number be n. The other numbers are $n + 1$,
$n + 2, n + 3, n + 4, n + 5, n + 6$.
The sum $= n + n + 1 + n + 2 + n + 3 + n + 4 + n + 5 + n + 6$
$= 7n + 21$
$7n$ is divisible by 7 and 21 is divisible by 7, so the sum
of the consecutive numbers is divisible by 7.

Homework Sheet 8.10

Writing equations

1   **a** $k + t = 840$ **b** $t = 3k$ **c** $k + 160 = \dfrac{t}{2}$

2   **a** $n - 19 = h$ **b** $n + h = 32$ **c** $n - 3 = 4h$
  d $3(n - h) = 52$

Homework Sheet 8.11

Writing and finding formulae

1   **a** $y = \dfrac{f}{3}$ **b** $c = 4d + 7$ **c** $s = 6p + 16$
  d $b = \dfrac{LW}{5}$

Homework Sheet 8.12

Ex 1   Writing and solving equations

1   **a** $x = 3\frac{1}{2}$ or $3\cdot5$ **b** $x = 6$ **c** $y = 5$ **d** $m = 5$
  e $a = 8$ **f** $m = 4$

2   **a** $3(2p + 7) = 57$ **b** $9y - 17 = 163$
       $p = 6$ $y = 20$
  c $7c + 16 = 61\cdot5$
       $c = £6\cdot50$

3   **a** $x = 3$ **b** $m = 11$ **c** $b = 6$
  d $m = 4$ **e** $x = 1\cdot8$ **f** $y = 1\cdot2$

Ex 2   More solving equations

1   **a** $x = 3$ **b** $n = 8$ **c** $x = 2$ **d** $k = 4$
  e $p = 5$ **f** $c = 2$

2   $2(6h - 3) = 162$
      $h = 14$

3   **a** $8m + 340 = 2500$
        $m = 270$ g
  b $n + n - 6 = 50$
        $n = 28$ chocolates
  c $\dfrac{4k + 5}{3} = 11$
        $k = 7$

Homework Sheet 8.13

Ex 1   Solving equations by transforming both sides

1   **a** $x = 7$ **b** $a = 3$ **c** $k = 32$ **d** $y = 4$
  e $m = 8$ **f** $t = 4$

2 

Ex 2   Solving equations

1   **a** $x = 4$ **b** $p = 4$ **c** $y = 3$ **d** $m = 7$

2   $6y + 4 = 9y - 14$
      $y = 6$

3   $5n + 6 = 4n + 18$
      $n = 12$

Homework Sheet 8.14

Solving equations using a graph

1   **a**

Centimetres	5	10	15	20	25
Inches	2	4	6	8	10

  b   It is always 5 : 2

  c   The number of inches is directly proportional to the
number of centimetres, because they are always in
the same ratio.

  d   Inches versus centimetres

Inches versus centimetres

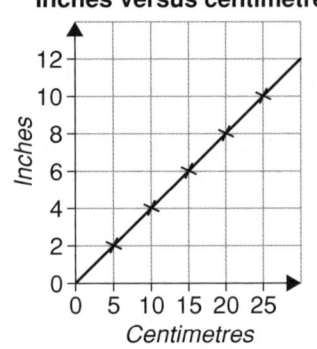

  e $y = \frac{2}{5}x$
  f   **i**   26 inches
       ii   32 inches

2 a

Number of laps	3	6	9	12	15
Sponsor money in £	20	40	60	80	100

b

Number of laps versus sponsor money

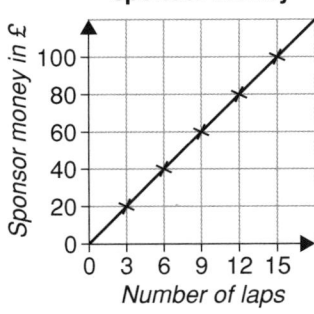

The points lie in a straight line.

c $y = \frac{20}{3}x$

d i £140 ii £200

Homework Sheet 9.1

Writing sequences from flow charts
1 a 99, 92, 85, 78, 71, 64, 57, 50, 43
  b 3, 6, 12, 24, 48, 96, 192, 384, 768

Homework Sheet 9.2

Generating sequences by multiplying and dividing
1 a 5, 10, 20, 40, 80, 160   b 1, ⁻4, 16, ⁻64, 256, ⁻1024
  c 120, 60, 30, 15, $7\frac{1}{2}$, $3\frac{3}{4}$ d ⁻32, 16, ⁻8, 4, ⁻2, 1
2 a 250, 1250   b $\frac{1}{3}$, $\frac{1}{9}$ c 0·8, 0·16

Homework Sheet 9.3

Counting forwards or backwards in increasing or decreasing steps
1 a 3, 5, 8, 12, 17, 23
  b 0, 2, 6, 12, 20, 30
  c 70, 69, 66, 61, 54, 45
2 a 20, 21, 23, 26, 30, 35
  b 65 plates

Homework Sheet 9.4

Arithmetic sequences
1 a 4, 10, 16, 22, 28, 34   b 5, 14, 23, 32, 41, 50
  c 14, 9, 4, ⁻1, ⁻6, ⁻11
2 a $a = 9, d = 4$ b $a = 16, d = ⁻7$
  c $a = ⁻8, d = 3$ d $a = 0·8, d = ⁻0·5$

Homework Sheet 9.5

Predicting the next few terms of a sequence
1 a 20, 8, ⁻4, [C]   b 39, 46, 53 [A]
  c 21, 28, 36 [B]   d 45, 33, 19 [D]
2 3, 6, 12, 21, 33, 48, ... **Rule**: add on 3, 6, 9, 12, 15, ...
  3, 6, 12, 24, 48, 96, ... **Rule**: multiply by 2

Homework Sheet 9.6

Writing sequences from term-to-term rules and rules for the *n*th term
1 a 10 000, 2000, 400, 80, 16   b 2, ⁻6, 18, ⁻54, 162
  c 7, 9, 13, 19, 27   d 1, 4, 5, 9, 14
  e 3, 4, 6, 10, 18
2 a 9, 11, 13, 15, 17   b 101, 96, 91, 86, 81
  c 3·5, 6·5, 9·5, 12·5, 15·5   d 0·3, 0·6, 0·9, 1·2, 1·5
3 a 37 b 31 c 45·5 d 4·5
4 a *1st term* 9, *rule* add 2
  b *1st term* 101, *rule* subtract 5
  c *1st term* 3·5, *rule* add 3
  d *1st term* 0·3, *rule* add 0·3
5 Possible answers are
  a *First term* 8 or 16 or any multiple of 8 *rule* add 8
  b *First term* 11 or 19 or 3 more than any multiple of 8 and *rule* add 8
  c *First term* any even number and *rule* add any even number
  d *First term* any number and the *rule* multiply by any negative number

Homework Sheet 9.7

Describing linear sequences
1 a Any value of *b* that is a multiple of 8 or 0
  b The sequence starts at 6 and ascends in steps of 8. Each term is two less than a multiple of 8.
2 a The ascending multiples of 5 starting at 5
  b The ascending multiples of 5 starting at 10
  c The ascending multiples of 5 starting at 25
  d The sequence which starts at 6 and ascends in steps of 5. Each term is one more than a multiple of 5
  e The sequence which starts at 4 and ascends in steps of 5. Each term is one less than a multiple of 5.
3 a The descending multiples of 9 (9 times table backwards) starting at 90.
  b Possible answers are
    i $T(n) = 4n + b$ where *b* is any number
    ii $T(n) = a - 3n$ where *a* is any number

Homework Sheet 9.8

Sequences in practical situations

1 a 1, 4, 7, 10, 13, ...
The next few terms are 16, 19, 22, 25, ...
The next pattern has three more crosses than the pattern before.
The first pattern has just 1 cross and then three crosses are added each time.

b $n + 2$ $(n - 1) = 3n - 2$ or
  ↑   ↑
crosses crosses in two
in middle sets of
row   verticals

There are three crosses times the (pattern number minus 1) plus the cross that is in the first pattern
$3(n - 1) + 1 = 3n - 3 + 1$
$= 3n - 2$

2 a 8, 12, 16, 20, ...
The next few terms are 24, 28, 32, 36, ...
The sequence is the result of adding the 4 corner tiles to four times the length of one side of the square.
$4 + 4(1) = 8,$ $4 + 4(2) = 12,$ $4 + 4(3) = 16,$
$4 + 4(4) = 20$

b $4 + 4(n) = 4 + 4n$
  ↑   ↖
corner tiles round 4 sides
tiles   of square

3 a 1, 3, 6, 10, ... These are triangular numbers.
b 1, 4, 9, 16, ... These are square numbers.

Homework Sheet 9.9

Finding the rule for the *n*th term
1 a $3n - 1$ **b** $4n + 3$ **c** $69 - 6n$ **d** $5 - 3n$
2 a 53, 62, 71
  b Add on 9
  c 134
  d $9n - 1$

Homework Sheet 9.10

Functions
1 a

Input	Output
7	32
20	97
1·5	4·5
$4\frac{1}{2}$	$19\frac{1}{2}$

Input	Output
8	4
29	11
5·6	3·2
$2\frac{3}{8}$	$2\frac{1}{8}$

2 a

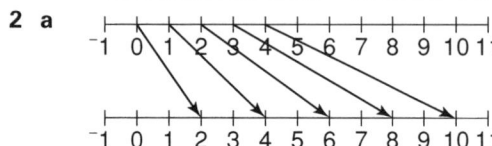

2 b

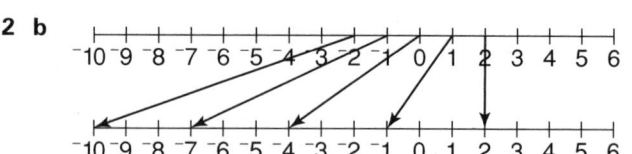

2 b

Input	Output
$^-2$	$\frac{1}{2}$
$^-1$	1
0	$1\frac{1}{2}$
1	2
2	$2\frac{1}{2}$

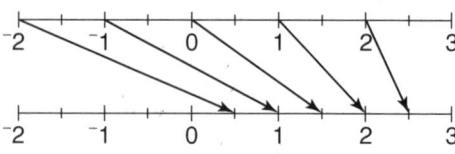

4 a True **b** True

Homework Sheet 9.11

Finding the function given the input and output
1 a $x \rightarrow 3x + 2$ **b** $x \rightarrow 5x - 6$
  c $x \rightarrow 2x - 4$ or $x \rightarrow 2(x - 2)$ **d** $x \rightarrow 4x + 1$

Homework Sheet 9.12

Properties of functions when combining number operations
1 a [add 11] **b** [multiply by 12] **c** [subtract 3] **d** [multiply by 4]
2 $x \rightarrow 3(x - 2)$ or $x \rightarrow 3x - 6$
3 a Subtract 5, multiply by 2 **b** Add 8, divide by 6
  c Multiply by 10, subtract 4
4 a 11 **b** 6 **c** 14, 8, 28 **d** 9, 5, 2
5 a $x \rightarrow \frac{x + 4}{6}$
  b $x \rightarrow 3(x + 2)$
  c $x \rightarrow \frac{x}{3} - 4$
  d $x \rightarrow 2x - 7$

Homework Sheet 10.1

Graphing functions
1 c, d
2 a, c, e

3 a

x	⁻1	0	3
y	⁻7	⁻4	5

b (⁻1, ⁻7), (0, ⁻4), (3, 5)

c

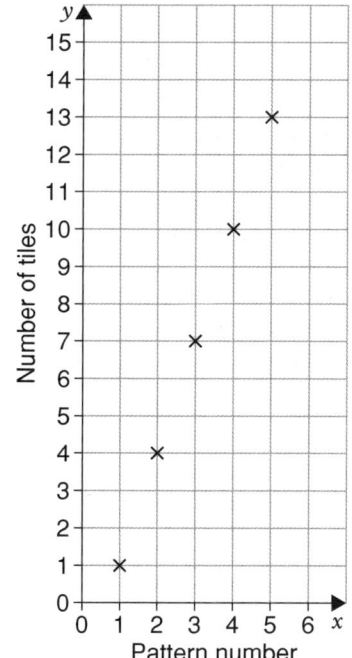

d Some possible answers are
(⁻2, ⁻10), (1, ⁻1), (2, 2), (4, 8)

e Yes, because $x = 8$ and $y = 20$ satisfy the equation
$y = 3x - 4$
$20 = (3 \times 8) - 4$

f No, because $x = ⁻11$ and $y = ⁻29$ do not satisfy the
equation $y = 3x - 4$
$⁻29 \neq (3 \times ⁻11) - 4$

4 a The points do lie on a straight line.

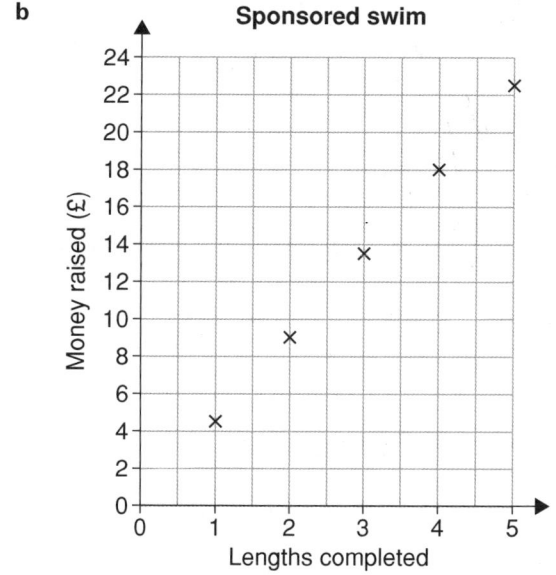

b Yes, because 25 would lie on the imagined straight
line drawn through the points.

c No, because there are no values between these
points. Each pattern has three tiles more than the
last and patterns cannot have fractional numbers.

Homework Sheet 10.2

Equations of straight line graphs

1 a i $y = 4x + 1$ **ii** $y = 4$ **iii** $y = ⁻2x$ **iv** $x = ⁻2$
  b $y = x + 4$, $y = x - 3$
2 a E   **b** D   **c** C   **d** F   **e** A   **f** B
3 a $y = ⁻5x$ **b** $y = x + 4$
4 a 2   **b** 3   **c** 2

Homework Sheet 10.3

Reading and plotting real-life graphs

1 a 220 g   **b** 90 days old   **c** 440 g
  d 360 g   **e** 86 days old
  f The puppy increased in weight far faster than the
  kitten.

2 a

Number of laps completed	1	2	3	4	5
Money raised in £	4·50	9	13·50	18	22·50

b

Sponsored swim

Money raised (£) vs *Lengths completed*

c Yes, the relationship is linear.
d No, because Luc will not raise any money for
incomplete lengths, so the gaps between the points
have no meaning.

3 a (10, 21), (20, 26), (30, 31), (40, 36), (100, 66)

b

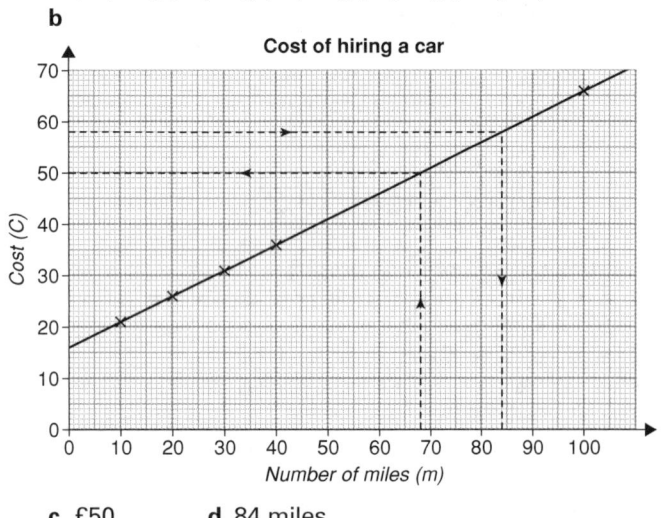

Cost of hiring a car

c £50   **d** 84 miles

b The number of cars stayed the same for the first day or so, then decreased gradually to reach a low by late on Wednesday. New stocks were brought in on Thursday increasing the numbers rapidly. Then sales saw the number of cars gradually decrease again.

3

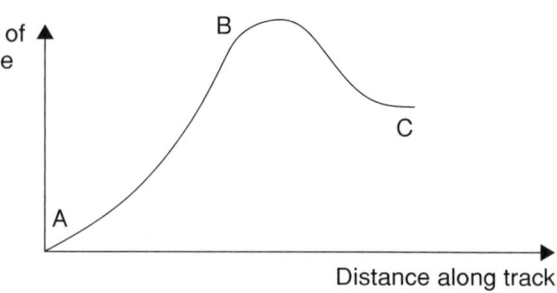

Homework Sheet 10.4

Distance/time graphs

1

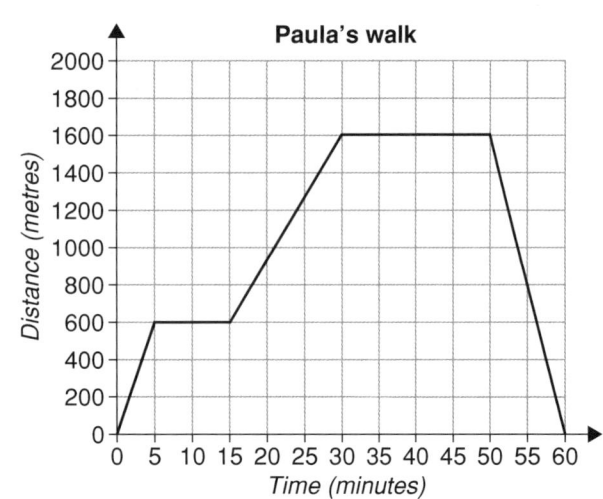

Paula's walk

2 a 20 km   **b** 45 minutes or $\frac{3}{4}$ hour   **c** $2\frac{1}{4}$ hours
   d 15 km   **e** $1\frac{1}{4}$ hours

Homework Sheet 10.5

Interpreting and Sketching real-life graphs

1 a 0830 and 1400   **b** 1100   **c** 1300
   d Because no drinks are bought from the machine before 0830 or after 1800 because the factory is closed then.

2 a The temperature of the water in the kettle increased as the kettle was put on to boil. It stayed at boiling point for a little while, then the kettle was switched off and the water temperature decreased. The kettle was switched back on and the temperature increased to boiling point again, then decreased as it cooled.

Homework Sheet 11.1

Angles made with intersecting and parallel lines

1 a Any pair from a, c, e, g or any pair from b, d, f, h
   b Any pair containing one of a, c, e, g and one of b, d, f, h
   c Any from a, e b, f c, g d, h
   d Either of: d, f or c, e

2 a 104° — corresponding angles are equal
   b 53° — alternate angles are equal
   c 73° — alternate angles are equal

3 a b, d, h **b** e, f, g **c** d, h, i

4 A and C

5

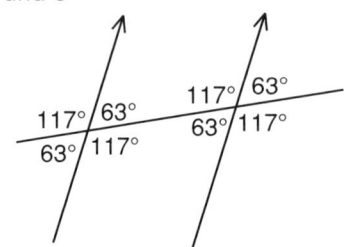

Homework Sheet 11.2

Using geometrical reasoning to find angles

1 m = 68° corresponding angles
   p = 103° vertically opposite angles are equal
   q = 180° − 103° angles on a straight line add to 180°
     = **77°**
   n = 77° alternate angles

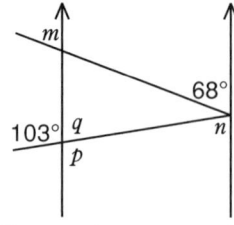

2 a $5x + 8° = 7x − 32°$ corresponding angles are equal
   $8 = 2x − 32°$ subtracting $5x$ from both sides
   $40° = 2x$ adding 32 to both sides
   $x = 20°$ dividing both sides by 2

b $4x - 17° + 6x + 13° + 124° = 360°$ angles at a point add to 360°

$10x + 120° = 360°$ collecting like terms

$10x = 240°$ subtracting 120 from both sides

$x = 24°$ dividing both sides by 10

c $x + 19° + 4x + 26° = 180°$ angles on a straight line add to 180°

$5x + 45° = 180°$ collecting like terms

$5x = 135°$ subtracting 45 from both sides

$x = 27°$ dividing both sides by 5

3 AB and CD are not parallel, because, if they were 65° and 125° would add up to 180° and they do not.

4

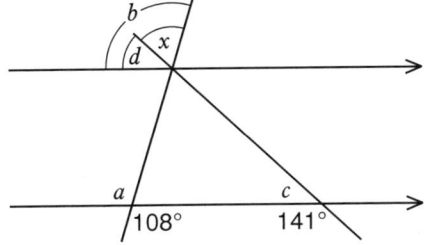

One method is:

$a = 108°$ vertically opposite angles are equal

$b = 108°$ corresponding angles are equal

$c = 180° - 141°$

$= 39°$ angles on a straight line add to 180

$d = 39°$ corresponding angles are equal

So $x = b - d$

$= 108° - 39°$

$= 69°$

Homework Sheet 11.3

Angles in triangles

1

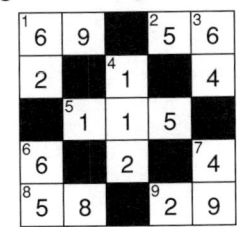

2 a $a = 57°, b = 57°, c = 123°, d = 61°, e = 61°, f = 119°,$
$g = 62°,$

b $h = 114°, i = 66°, j = 66°, k = 47°, m = 67°, n = 47°$

3 One method is:

$a = 180° - 81° = 99°$ angles on a straight line add to 180°

$x = 180° - (47° + 99°)$ angle sum of a $\triangle$ = 180°

$= 34°$

$b = 81°$ (base angles of an isos. $\triangle$ are equal)

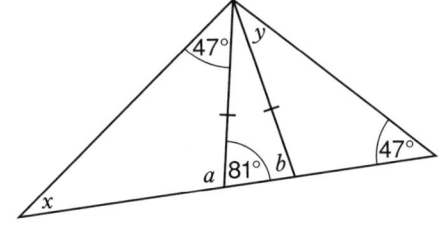

$81° = 47° + y$ so
$y = 81° - 47°$ (ext. angle of a $\triangle$ is equal to two opp. int. angles)
$= 34°$
So angle x = angle y

4 a $4x - 12° + 3x + 1° = 6x + 11°$ ext. angle of a triangle

$7x - 11° = 6x + 11°$ collecting like terms

$x - 11° = 11°$ subtracting $6x$ from both sides

$x = 22°$ adding 11 to both sides.

b $7x + 6° + 2x - 1° + 5x - 7° = 180°$ angle sum of a triangle

$14x - 2° = 180°$ collecting like terms

$14x = 182°$ adding 2 to both sides

$x = 13°$ dividing both sides by 14

Homework Sheet 11.4

Angles in quadrilaterals
1 a $x = 76°$ **b** $x = 98°$ **c** $x = 58°$

Homework Sheet 12.1

Visualising and sketching 2-D shapes
1 Hexagon

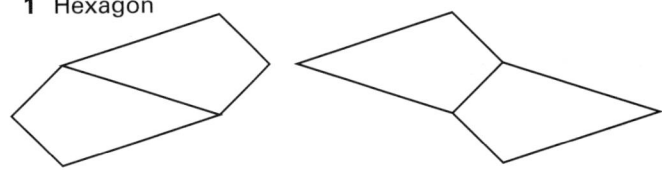

or pentagon, if the kite contains a right angle.

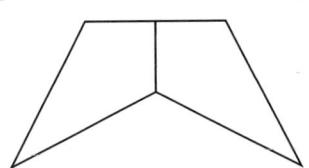

Homework Sheet 12.2

Properties of triangles, quadrilaterals and polygons
1 a Rectangle, rhombus
  b Parallelogram, rectangle, rhombus
  c Square, rhombus, kite, arrowhead
2 a Parallelogram **b** Kite
3 a True, a square is a special sort of rectangle with all 4 sides equal.
  b False, some rhombuses are squares because a square is a special sort of rhombus with 4 right angles, but some rhombuses have no right angles.
  c True, a rhombus is a special sort of parallelogram, with 4 equal sides.

4

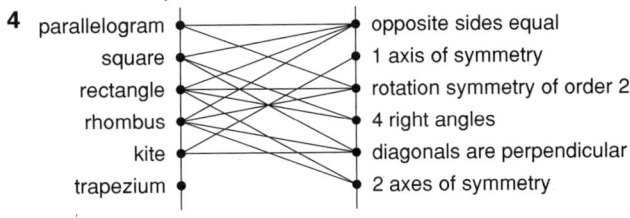

5 a Isosceles **b** Trapezium

Homework Sheet 12.3

Congruence
1 **a** C, G　　**b** E, N, B　　**c** L　　**d** D　　**e** T
2 F, L, P
3 **a** FE, HI, JK, OM
　b D, G, L, N
　c BC, DE, GI, JL, MN

Homework Sheet 12.4

Describing and sketching 3-D shapes. Constructing nets
1 **a i** 8　　**ii** 4　　**iii** 4
　b i 5　　**ii** 8

2 **a** 　**b**　**c**

3 **a** 　**b**　**c**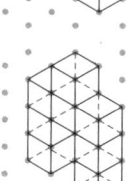

Homework Sheet 12.5

Plans and elevations
1 **a**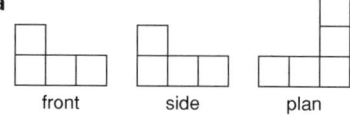
front　side　plan

b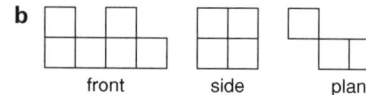
front　side　plan

c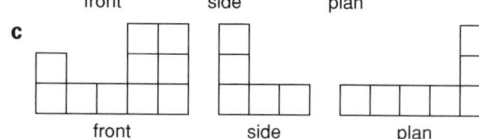
front　side　plan

2 **a** 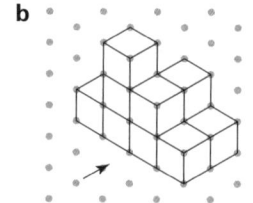　**b**

Homework 12.6

Construction
1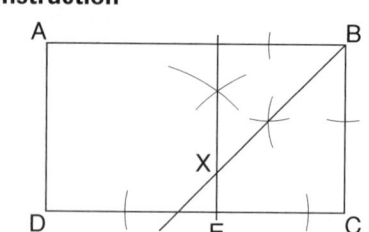

AX = 5·1 cm

2 **a and b**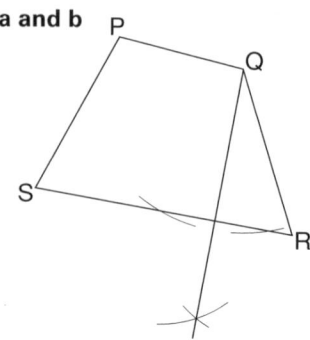

3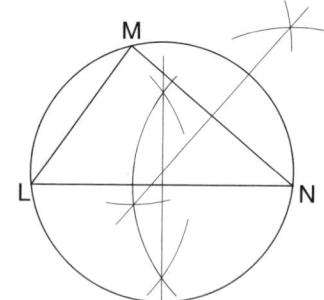

The circle should pass through the three vertices of the triangle, L, M and N.

4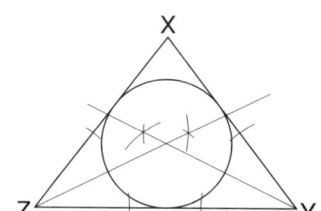

Homework Sheet 12.7

Constructing triangles and quadrilaterals
1 Dotted line is 9·6 cm
2 Longest diagonal is 9·1 cm

4　a Possible answers are given. The nets are not full size.

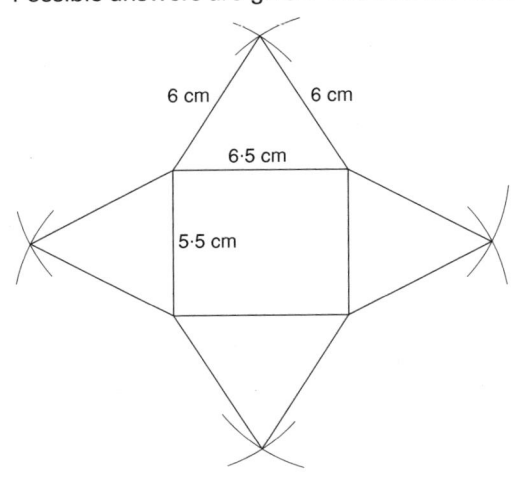

6 cm　6 cm
6·5 cm
5·5 cm

b

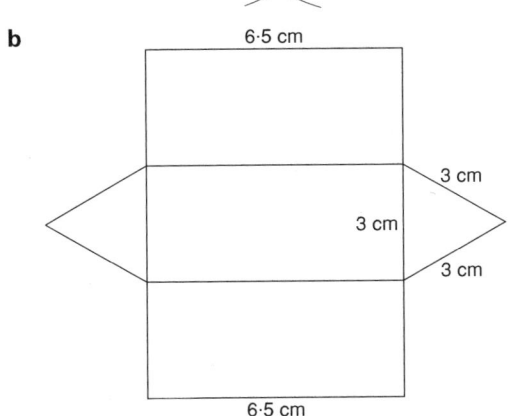

6·5 cm
3 cm
3 cm
3 cm
6·5 cm

c

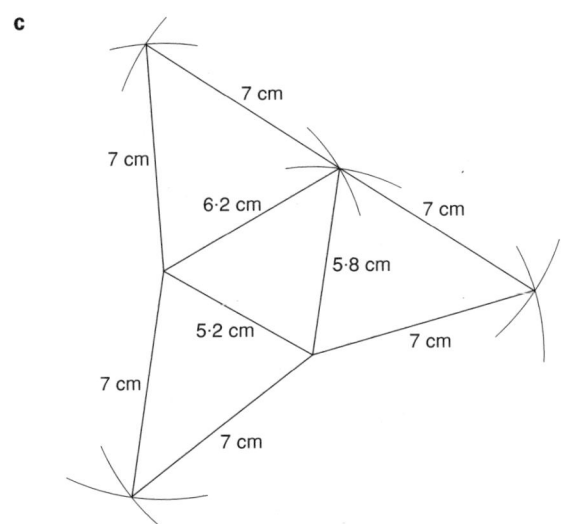

7 cm
7 cm
6·2 cm　7 cm
5·8 cm
5·2 cm　7 cm
7 cm
7 cm

Homework Sheet 12.8

Locus

1　a Flash's head will travel along lines parallel to the ground, then the sides and top of the ramp.

b The locus will be a circle, centre the stake, radius the length of the rope.

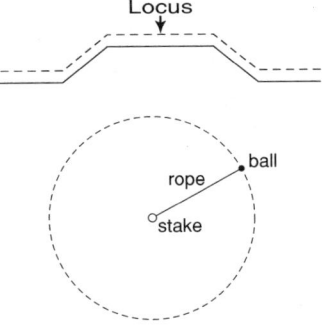

Locus
rope　ball
stake

c The locus will be a line, parallel to the rails and halfway between them.

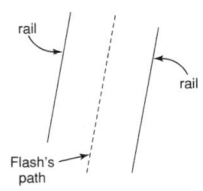

rail
rail
Flash's path

d The locus will be the diagonal of the square arena. This bisects both angles at the corners of the arena where the buckets are.

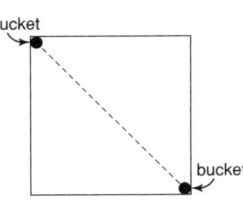

bucket
bucket

Homework Sheet 13.1

Coordinates and transformations

1　a (2, 1), (5, 1), (5, 2), (4, 2), (4, 3), (3, 3), (3, 4), (2, 4)

b i (⁻2, 1), (⁻5, 1), (⁻5, 2), (⁻4, 2), (⁻4, 3), (⁻3, 3), (⁻3, 4), (⁻2, 4)

ii (1, ⁻2), (1, ⁻5), (2, ⁻5), (2, ⁻4), (3, ⁻4), (3, ⁻3), (4, ⁻3), (4, ⁻2)

iii (⁻4, ⁻4), (⁻1, ⁻4), (⁻1, ⁻3), (⁻2, ⁻3), (⁻2, ⁻2), (⁻3, ⁻2), (⁻3, ⁻1), (⁻4, ⁻1)

iv (2, ⁻3), (5, ⁻3), (5, ⁻4), (4, ⁻4), (4, ⁻5), (3, ⁻5), (3, ⁻6), (2, ⁻6)

Homework Sheet 13.2

Combinations of transformations

1

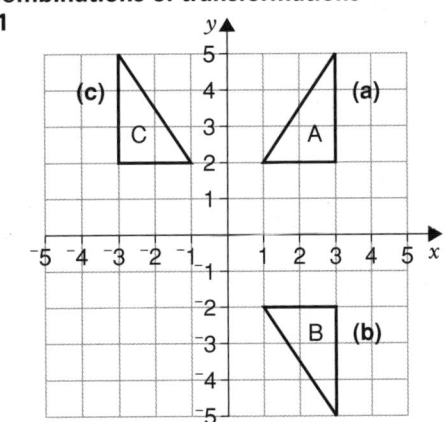

(c)　C　A　(a)
B　(b)

d Reflection in the y-axis

2

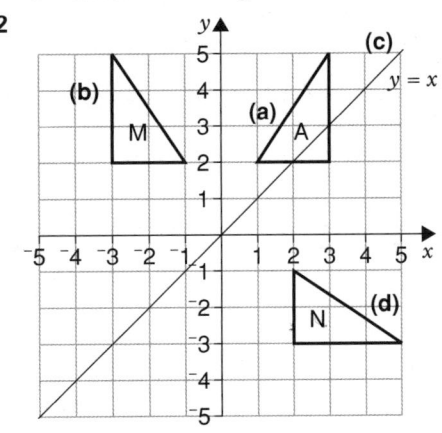

(b)　M　A　(c)　$y = x$　(a)
N　(d)

e Rotation of 90° about the origin

3　a C　**b** D　**c** A　**d** B

4　a C　**b** B　**c** A

Homework Sheet 13.3

Symmetry
1 a 1 line of symmetry; no rotation symmetry
  b No lines of symmetry; rotation symmetry of order 3
  c 5 lines of symmetry; rotation symmetry of order 5
  d 4 lines of symmetry; rotation symmetry of order 4
2 a A possible answer is
  b A possible answer is
  c A possible answer is

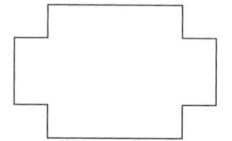  

There are many answers to this question.
3 Possible answers are

a 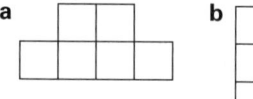 **b** 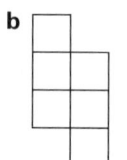 **c**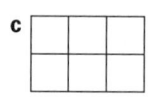

Homework Sheet 13.4

Ex 1 Enlargement
1 a 2 **b** 4 **c** 5
2 a 4 **b** 2 **c** 3

Ex 2 More enlargement
1 a

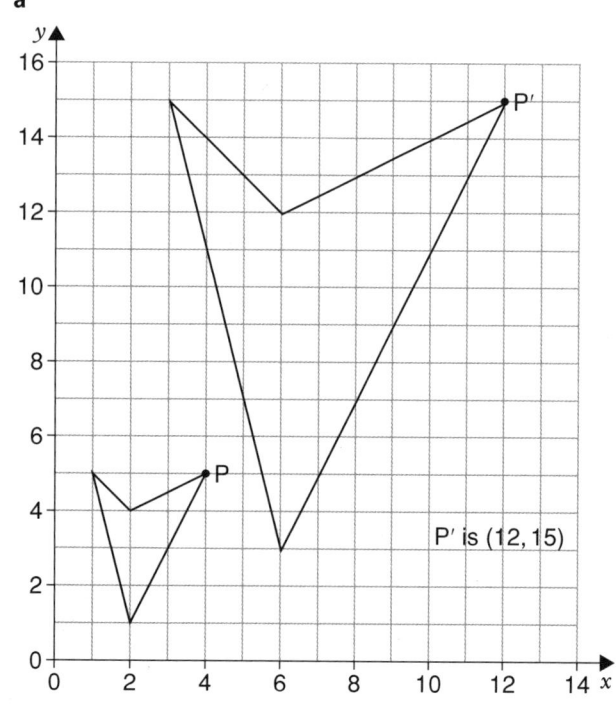

P' is (12, 15)

b

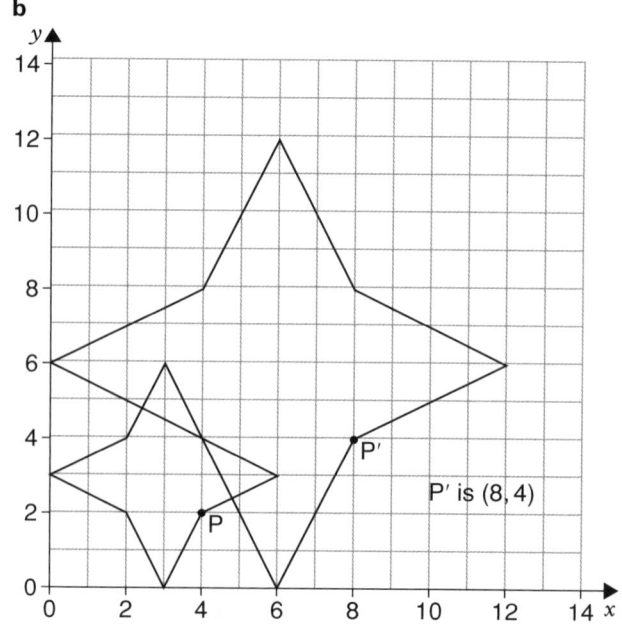

P' is (8, 4)

c

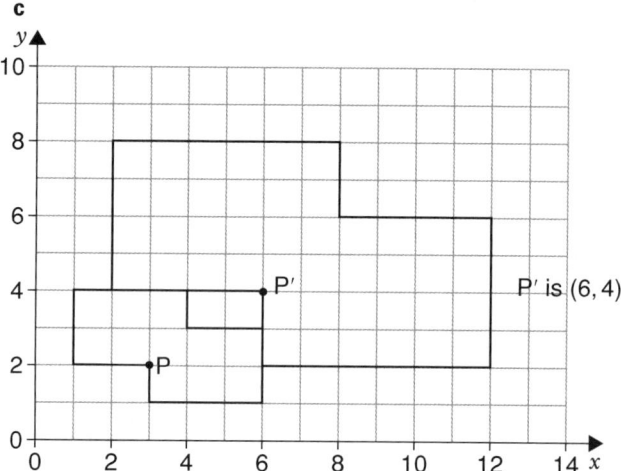

P' is (6, 4)

2

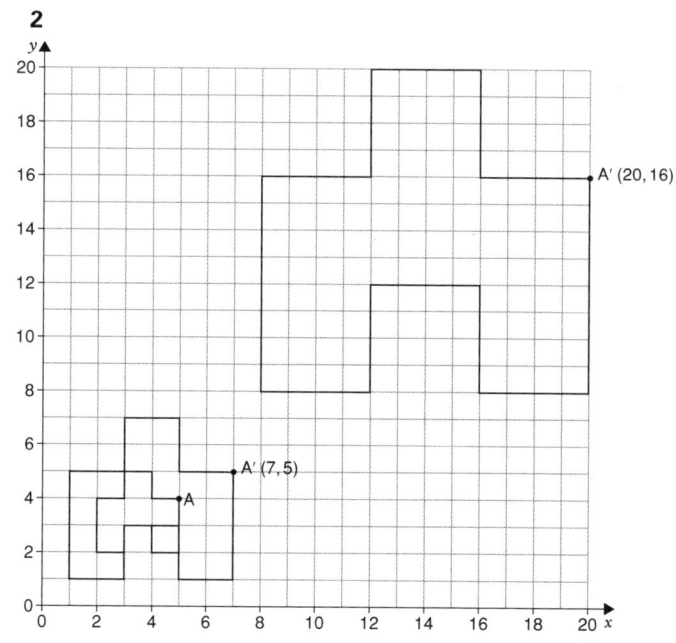

A' (20, 16)
A' (7, 5)

3

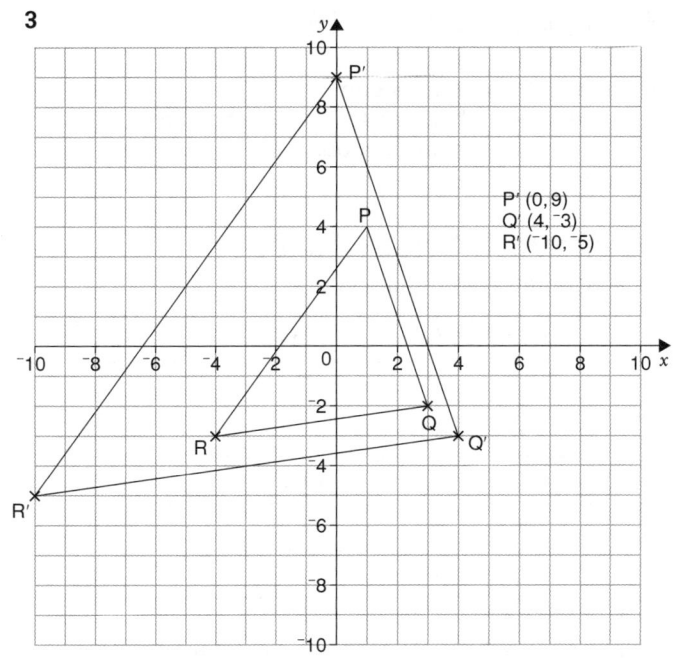

P' (0, 9)
Q' (4, ⁻3)
R' (⁻10, ⁻5)

2 B is the point (⁻1, ⁻8)

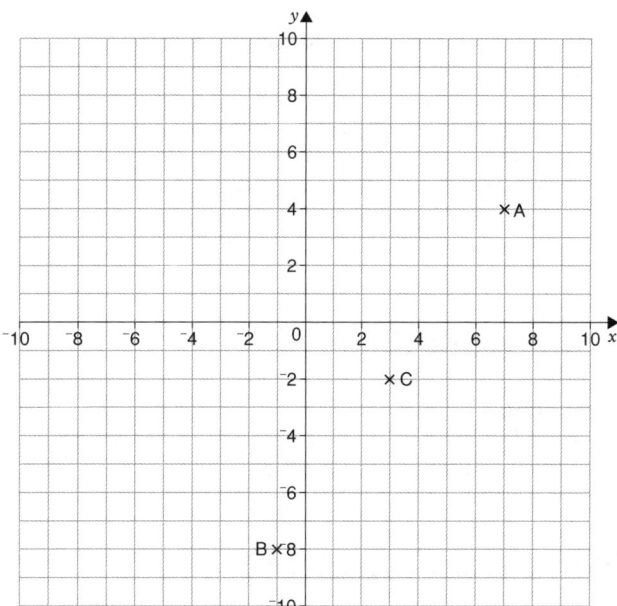

Homework Sheet 13.5

Ex 1   Scale drawing
1 6–7 m
2 **a** 12 m   **b** 22·5 m   **c** 7·5 m
3 1 cm represents 4 m
4 1 cm represents 2 km
5 1 cm represents 5 m
6 **a** 150 m   **b** 425 m   **c** 60 m

Ex 2   Making a scale drawing
1

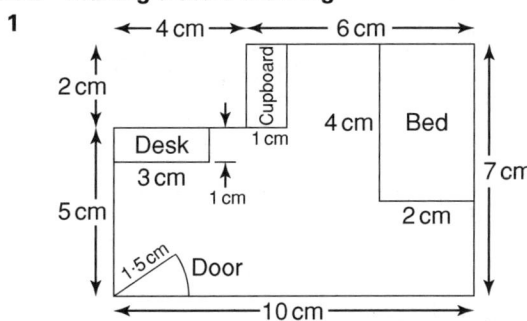

Homework Sheet 13.6

Ex 1   Finding the mid-point of a vertical or horizontal line
1 **a** (2, 8)   **b** (3, 4)   **c** (1, 3)   **d** (⁻2, ⁻5$\frac{1}{2}$)   **e** (1, 2$\frac{1}{2}$)

Ex 2   Finding the mid-point of a line segment
1 **a** (3, 3$\frac{1}{2}$)   **b** (1, 1)   **c** (⁻5$\frac{1}{2}$, 4$\frac{1}{2}$)   **d** (⁻1$\frac{1}{2}$, 3)

Homework Sheet 14.1

Ex 1   Metric measurements
1 4 hrs 36 mins
2 7 weeks 6 days
3 4 years 8 months
4 **a** 640 000 m^2 **b** 4·6 hectares
5 2930 cm
6 6·185 tonnes
7 6·72 litres
8 Thursday 16th May at 8 minutes past 7 in the evening (1908)

Ex 2   Capacity
1 CATERPILLARS HAVE TWELVE EYES
2 **a** 40 cℓ **b** 25 cm^3

Homework Sheet 14.2

Metric and imperial equivalents
1 96 km
2 5·375 m
3 63 pints
4 210 g
5 33 pounds
6 31·5 litres

Homework Sheet 14.3

Units, measuring instruments and accuracy
1 a Nearest km   b Nearest mm
   c Nearest tonne   d Nearest minute
   e Nearest mℓ or 100 mℓ   f Nearest g
   g Nearest cm
2 C nearest 10 m or D nearest 100 m
   b A nearest centimetre
   c D nearest 100 m

Homework Sheet 14.4

Estimating
1 a 190–220 cm   b 140–180 g   c 40–60 cm
   d 160–250 mℓ   e 12–22 cm   f 8–12 ℓ

Homework Sheet 14.5

Bearings
1 a C 180°   b D SW
2 a i 115°   ii 208°   iii 303°
   b i 304°   ii 099°   iii 234°
3 a Loksi   b 135° (±3°)   c 286° (±3°)
   d 142° (±3°)   e 322° (±3°)

Homework Sheet 14.6

Ex 1   Perimeter and area
1 a 27 cm^2 b 44 cm^2 c 560 mm^2
2 a 13·68 m^2 b 43·52 cm^2 c 88 m^2
3 a 15·1 m   b 27·8 cm   c 39·5 m
4 1·4 m
5 675 cm^2

Ex 2   More area
1 a 10 units^2 b 13·5 units^2 c 14 units^2

Homework Sheet 14.7

Surface area and volume
1 Volume   a 5376 cm^3 b 1176 cm^3 c 283·5 cm^3
   Surface area   2192 cm^2 700 cm^2 292·5 cm^2
2 1·404 m^3
3 36 dominoes
4 a 360 000 cm^3 or 0·36 m^3 b 38 400 cm^2 or 3·84 m^2
5 a 1, 1, 10 cm;   1, 2, 9 cm;   1, 3, 8 cm;   1, 4, 7 cm;
     1, 5, 6 cm;   2, 2, 8 cm;   2, 3, 7 cm;   2, 4, 6 cm;
     2, 5, 5 cm;   3, 3, 6 cm;   3, 4, 5 cm;   4, 4, 4 cm
   b 10 cm^3, 18 cm^3, 24 cm^3, 28 cm^3, 30 cm^3, 32 cm^3,
     42 cm^3, 48 cm^3, 50 cm^3, 54 cm^3, 60 cm^3, 64 cm^3
     The 4 × 4 × 4 cuboid has the greatest volume. It is a
     cube.

Homework Sheet 15.1

Discrete and continuous data
1 a Discrete   b Discrete   c Continuous
   d Discrete   e Continuous

Homework Sheet 15.2

1 a   **Graphing continuous data**

Time taken (secs)	Tally	Frequency
$40 < t \leqslant 50$	ЖЖ	5
$50 < t \leqslant 60$	III	3
$60 < t \leqslant 70$	ЖЖ II	7
$70 < t \leqslant 80$	III	3
$80 < t \leqslant 90$	II	2

  b 7   c 5   d 8
  e Because they show the spread of times better. If
    you used the intervals $25 < t \leqslant 50$, there would only
    be a frequency of 5 in the first class interval, all the
    remaining 13 in the second and two in the third.

Homework Sheet 15.3

Two-way tables
1 a 24   b 58   c 82
   d The most popular language chosen by the boys was
     German, whereas for the girls, it was French. About
     the same proportion of girls and boys chose to
     study Spanish.
2   A possible answer is:

	Television	Video player	DVD player
Cash			
Cheque			
Credit card			

Homework Sheet 15.4

Surveys
1 Possible answers are:
   a i   Do you intend to study in Year 12?
     ii   Do you intend to go to university?
     iii   What kind of career do you intend to follow?
     iv   Do you think liking a job, or earning a lot of
       money is more important?
   b i   70% may want to study in year 12.
     ii   60% may want to go to university.
     iii   55% may want to follow a professional career.
     iv   65% may think that liking a job is more
       important than earning a lot of money.

c Gender, Year 12 intentions, university intentions, career intentions, opinions on job satisfaction, marriage intentions, travel intentions, home ownership intentions

d Tick the relevant boxes in this survey about your intentions after Year 12 .

Are you male or female?

    male □   female □

Do you intend to study in Year 12?

    yes □   no □   not sure □

Do you intend to go to university?

    yes □   no □   not sure □

Which kind of career do you see for yourself in the future?

    Professional [doctors, lawyers, bankers, ...] □

    Professional [teachers, police, fire service, ...] □

    Caring profession [nurses, childminders, ...] □

    White collar [retail, secretarial, IT specialist, ...] □

    Manual

        [bricklayers, electricians, car mechanics, ...] □

    Leisure-based

        [sports professional, hotel management, ...] □

    Other □

Do you intend to get married

    before the age of 20 □   between 20 and 30 □

    over 30 □   never? □

Which do you think is more important?

    job satisfaction □   high wages □

Do you intend to travel abroad in your career?

    Yes, definitely □   Maybe □

    Yes, in a gap year □   No □

Can you see yourself by the age of 25

    living with parents □

    renting a flat/house □

    owning your own home □

e A suitable sample size would be 20% of your year group.

You could choose this by putting names in alphabetical order and choosing every fifth person.

f The data is a primary source.

Homework Sheet 16.1

Mode and range

1 **a** $10 < t \leqslant 12$ **b** 8·6 seconds.

Homework Sheet 16.2

Ex 1 **Mean**

1 $\frac{738}{153} = 4\cdot8$ cars

Ex 2 **Assumed means**

1 11·85 cm

2 63·3 drawing pins

Homework Sheet 16.3

Median, mean, mode, range

1 **a** £24·50 **b** £24 **c** £31·58 to the nearest penny

    d £81

2 **a** £30·06 to the nearest penny **b** £42

    c The pupils in Kathleen's class raised more, on average, than the pupils in Jed's class.

    d The pupils in Kathleen's class had the greatest range of amounts raised.

3 She could have scored 4 and 9 on the other throws.

4 **a** 2 **b** 5 **c** 3 **d** 5 or 0

Homework Sheet 16.4

Finding the median, range and mode from a stem-and-leaf diagram

1 **a**

Number of phonecalls per hour at Brentons											
1	7	9	9								
2	0	1	5	5	6	8	8				
3	0	1	2	3	4	5	6	7	8	9 9 9 9
4	0	2	4	6	7	8					

  b 6 times

  c Median 34   mode 39   range 31

  d Median 30   mode 23   range 35

  e Fashion Extra received fewer phonecalls per hour than Brentons, shown by the lower median and mode. Fashion Extra also had a greater variance in the number of phonecalls per hour, shown by the higher range.

Homework Sheet 16.5

Comparing data

1 **a** Mean 247·8 g   median 236 g   range 350 g

  b Mean 235·4 g   median 238 g   range 215 g

  c On average, the organic fertiliser produced the heavier crop of raspberries, shown by the higher mean, although it is not that much higher. The median mass for each type of fertiliser was about the same. The chemical fertiliser, however, produced a much more consistent crop mass, shown by the much lower range.

  d A possible answer is:

    Hester will probably recommend the chemical fertiliser, because its crop yield was not much less than for the organic fertiliser, but the yield from cane to cane did not vary as much.

Homework Sheet 16.6

Compound bar charts and line graphs

1 a

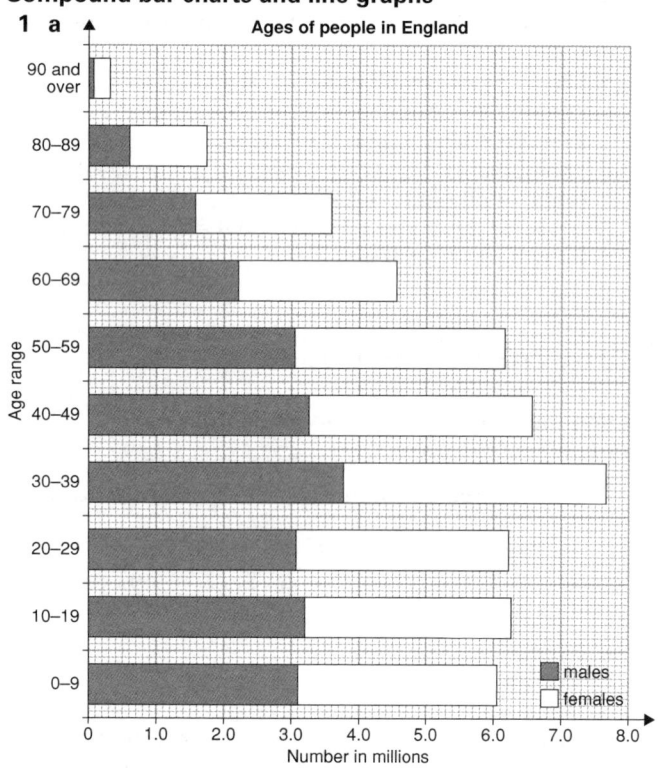

Ages of people in England

b There are more males and females in the ranges 30–39 and 40–49 than any other age range. The gender distribution is about equal up to the age of 70, but there are more females than males older than 70. This implies that men die, in general, at a younger age than women.

2 a

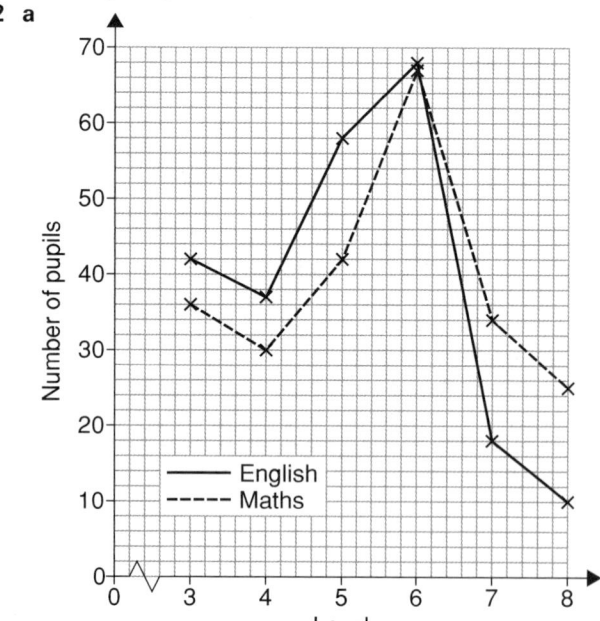

b More pupils gained levels 3, 4 or 5 in English than in maths, but more pupils gained levels 7 and 8 in maths than in English. Generally the maths results were better than the English results. The modal level for both maths and English was 6.

Homework Sheet 16.7

Frequency diagrams

1 a

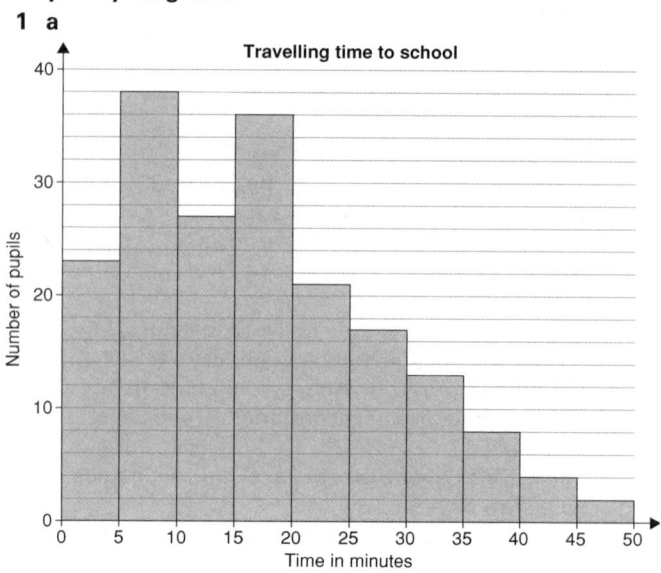

Travelling time to school

b 88   **c** 27

d No, because the 36 pupils who took 15 minutes but less than 20 minutes could all have taken exactly 15 minutes, or none might have taken exactly 15 minutes. We don't know how many of the 36 pupils took exactly 15 minutes.

Homework Sheet 16.8

Drawing pie charts

1 Gym $\frac{8}{40} \times 360° = 72°$

Swimming $\frac{5}{40} \times 360° = 45°$

Badminton $\frac{12}{40} \times 360° = 108°$

Aerobics $\frac{15}{40} \times 360° = 135°$

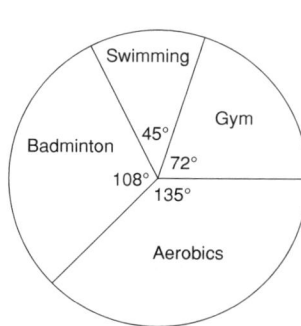

2 Dog $\frac{17}{60} \times 360° = 102°$

Cat $\frac{14}{60} \times 360° = 84°$

Hamster $\frac{2}{60} \times 360° = 12°$

Sheep $\frac{22}{60} \times 360° = 132°$

Cow $\frac{5}{60} \times 360° = 30°$

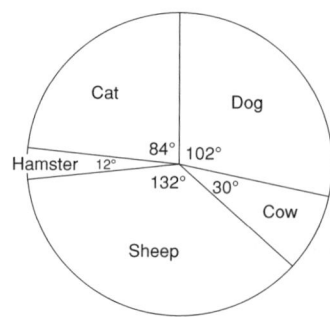

Homework Sheet 16.9

Scatter graphs

1 a

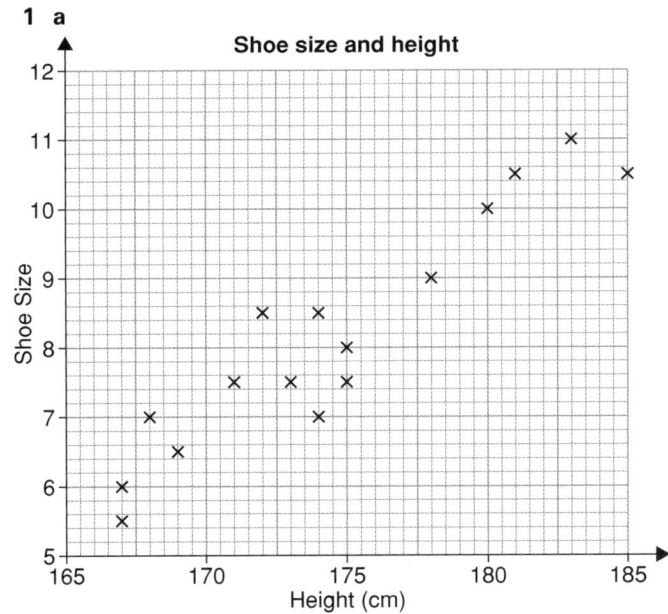

Shoe size and height

b Yes, someone who is taller generally has larger feet.

2 a

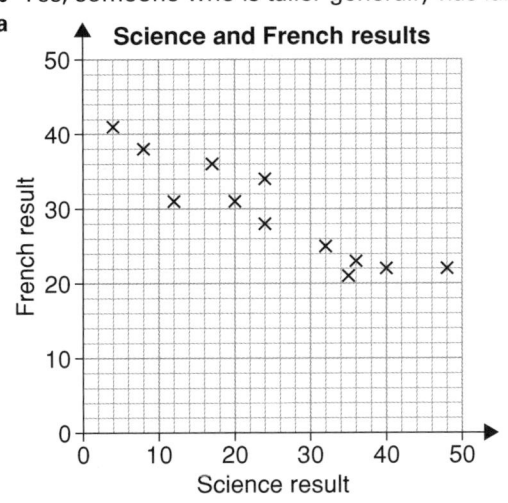

Science and French results

b The graph shows that someone who is good at science tends to be not so good at French.

Homework Sheet 16.10

Interpreting graphs and tables

1 a It has gradually decreased overall, although there was a slight rise again in 2001.
b It has gradually increased after a fall in 1996. There was a slight fall in 1999.
c Possible answers are:
Medical research showing the dangers of smoking.
Higher price of tobacco.
Smoking becoming antisocial or unfashionable.

2 a Generally the number of overweight children is increasing.
b The age groups 6–11 and 12–19 vary in being more prone to being overweight.

3 a Food and clothes
b Household services
c Possible answers are:
Higher cost of cars/petrol.
More than one car per household.
Higher number of miles travelled.
Higher cost of maintenance/tax/insurance.

Homework Sheet 17.1

Language of probability

1 2, because there are more 2s than any other number.
2 Box A, because with Box A, she has an $\frac{8}{21}$ (0·38) chance of picking a dark chocolate; with Box B, she only has a $\frac{13}{27}$ (0·35) chance of picking a dark chocolate.
3 The circle round the bullseye, because it has the greatest area.

Homework Sheet 17.2

Calculating probability

1 $\frac{5}{9}$
2 a $\frac{4}{12}$ or $\frac{1}{3}$ **b** $\frac{8}{12}$ or $\frac{2}{3}$ **c** $\frac{7}{12}$ **d** $\frac{5}{12}$ **e** $\frac{11}{12}$
3 a $\frac{1}{12}$ **b** $\frac{11}{12}$ **c** $\frac{6}{12}$ or $\frac{1}{2}$ **d** $\frac{5}{12}$
   e $\frac{4}{12}$ or $\frac{1}{3}$ **f** $\frac{3}{12}$ or $\frac{1}{4}$ **g** $\frac{9}{12}$ or $\frac{3}{4}$ **h** $\frac{7}{12}$
4 a 3 red, 12 yellow and 15 green
   b i 1
   c ii 0

Homework Sheet 17.3

Calculating probabilities by listing outcomes

1 a

Aidan	Aidan	Aidan	Brian	Brian	Brian	Colin	Colin	Colin
Daphne	Ellen	Fiona	Daphne	Ellen	Fiona	Daphne	Ellen	Fiona

2 a

Dice

	X	1	2	3	4	5	6
Spinner	1	1	2	3	4	5	6
	2	2	4	6	8	10	12
	3	3	6	9	12	15	18
	4	4	8	12	16	20	24

Homework Sheet 17.4

Estimating probability from experiments

1 $\frac{37}{100}$
2 a i $\frac{31}{120}$ **ii** $\frac{13}{120}$
   b No, because when an experiment is repeated the results are very likely to be different.
   c Yes, because the larger the amount of data collected, the more accurate the result.

3

Starter Resource Sheets

Photocopy onto card, cut out.

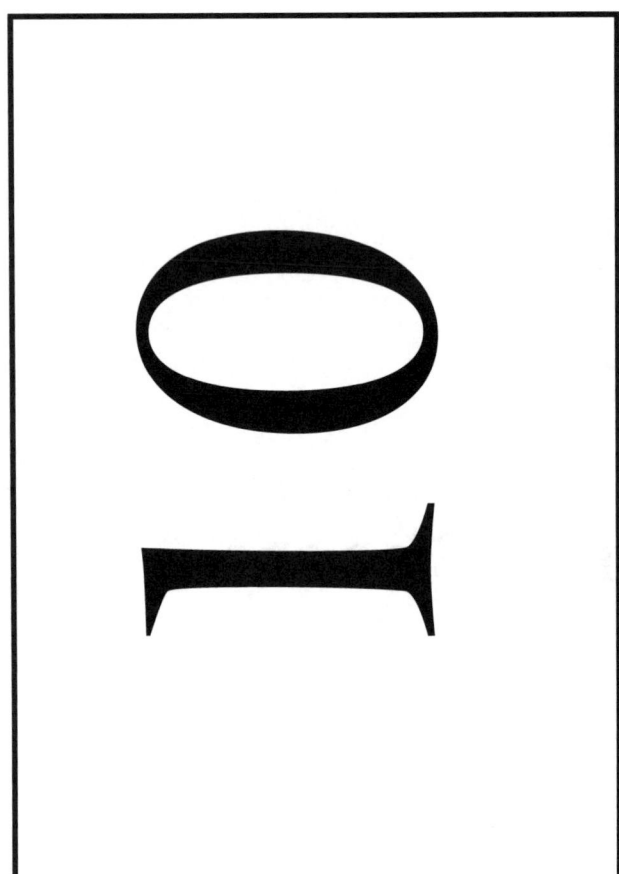

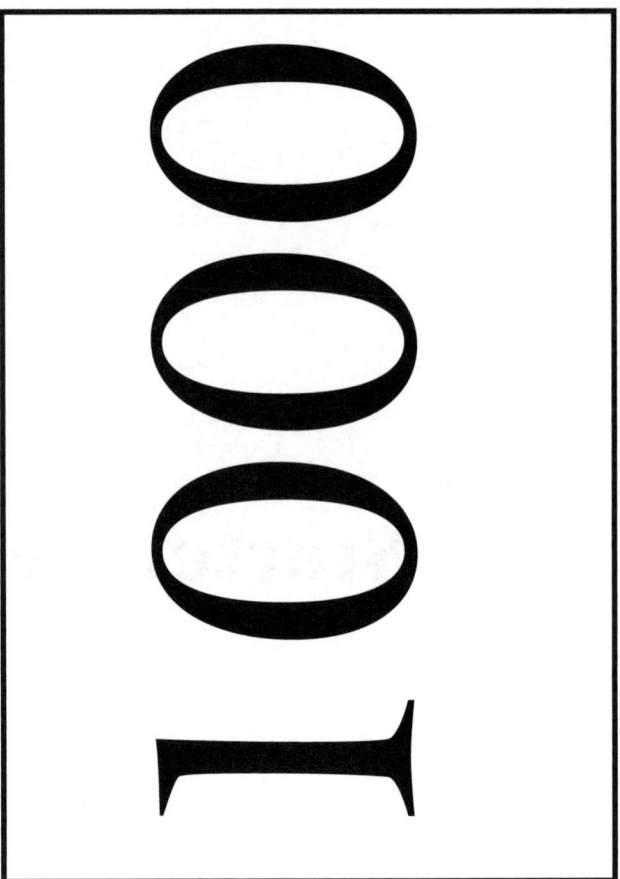

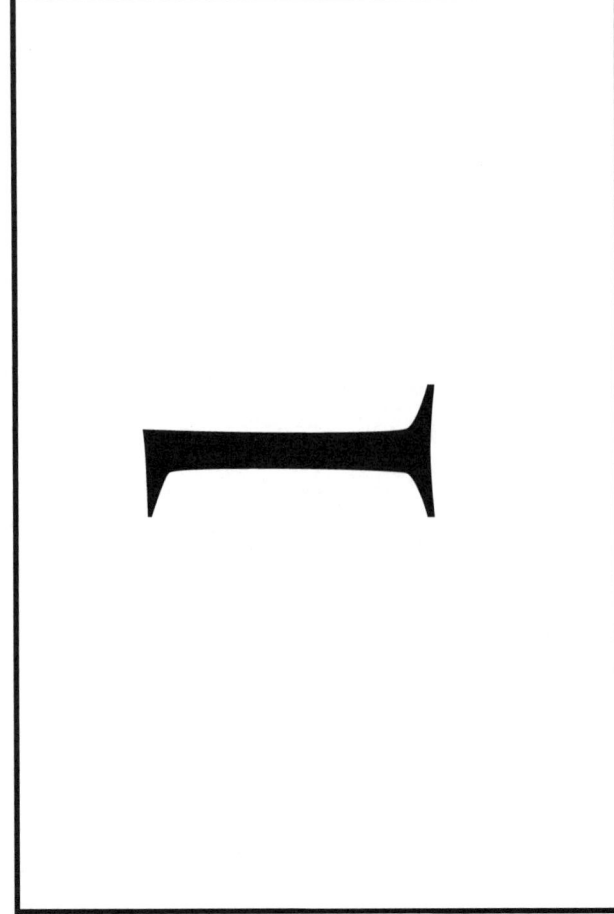

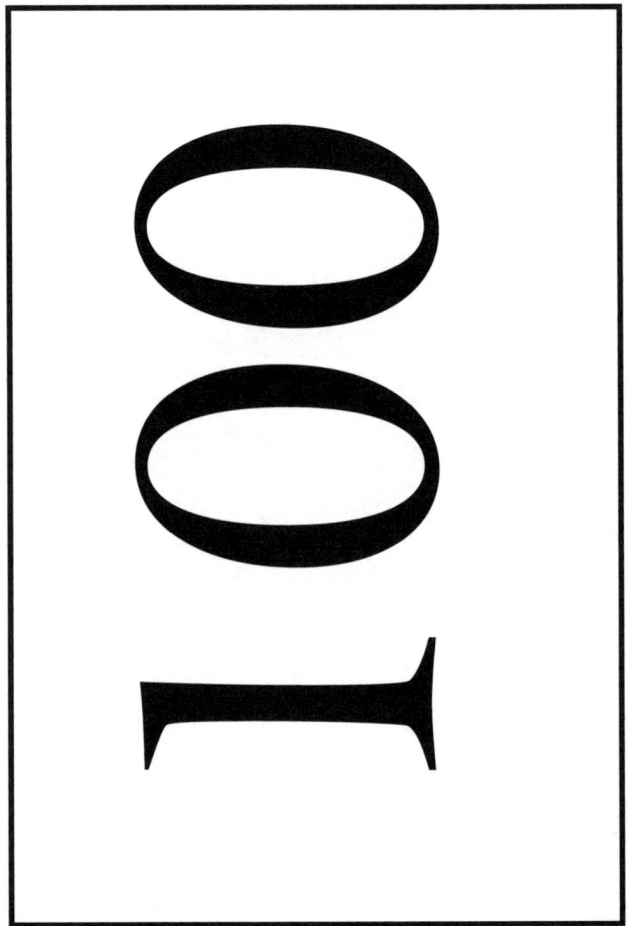

Photocopy onto card, cut out.

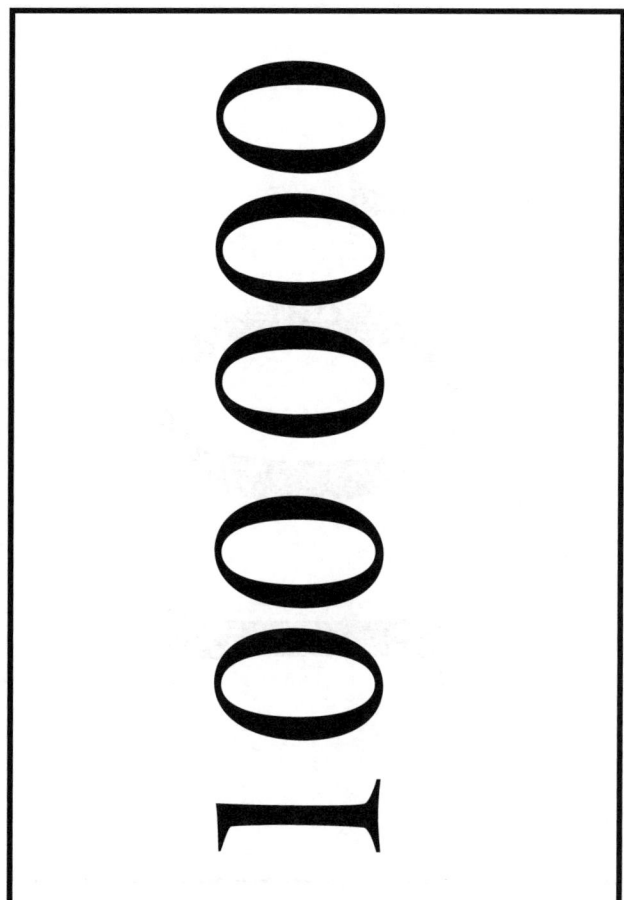

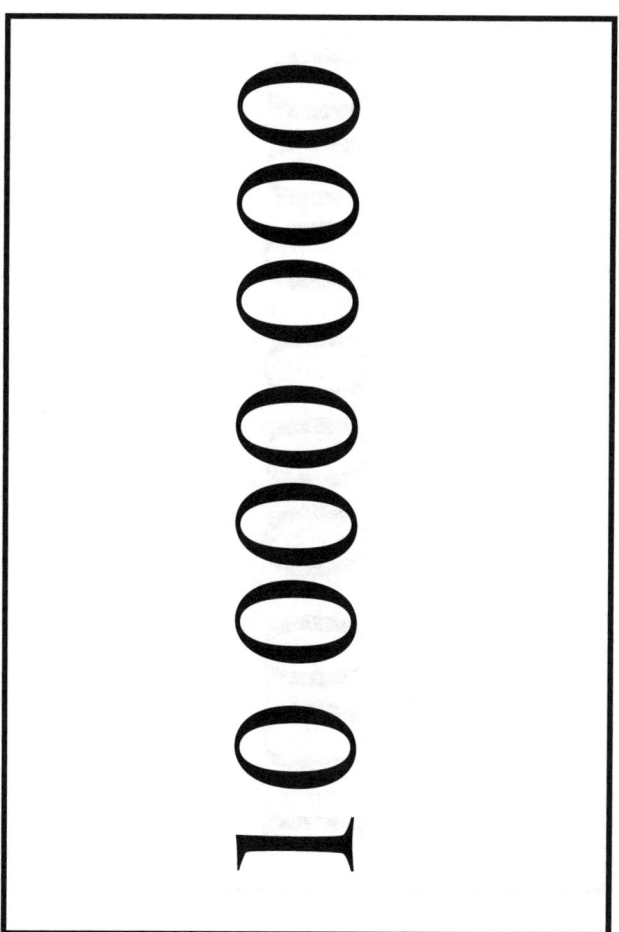

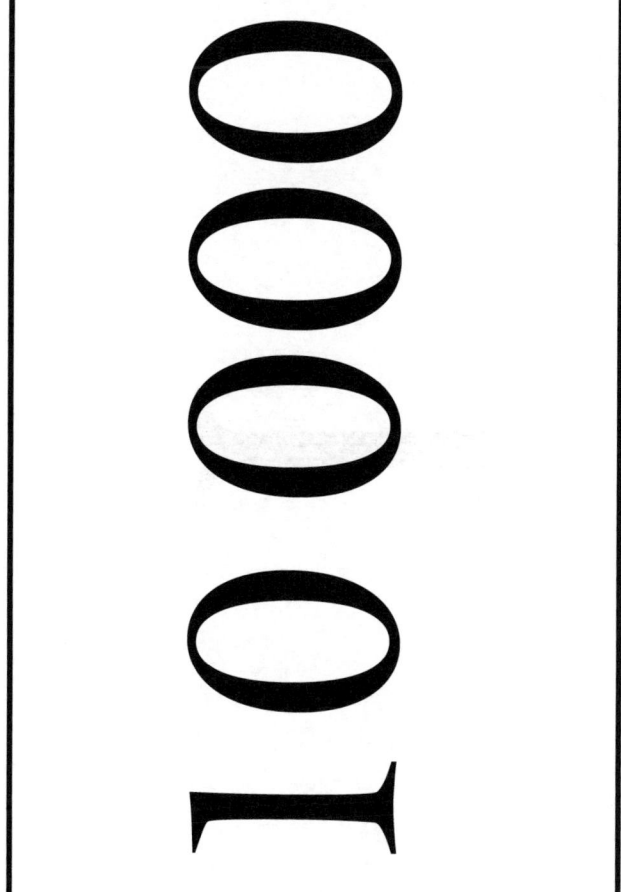

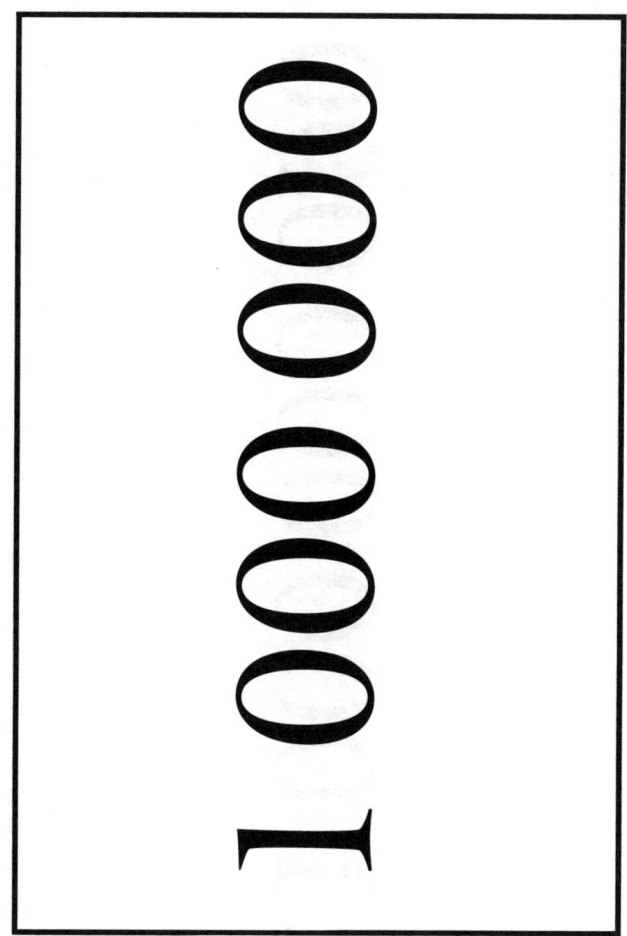

Photocopy onto card, cut out.

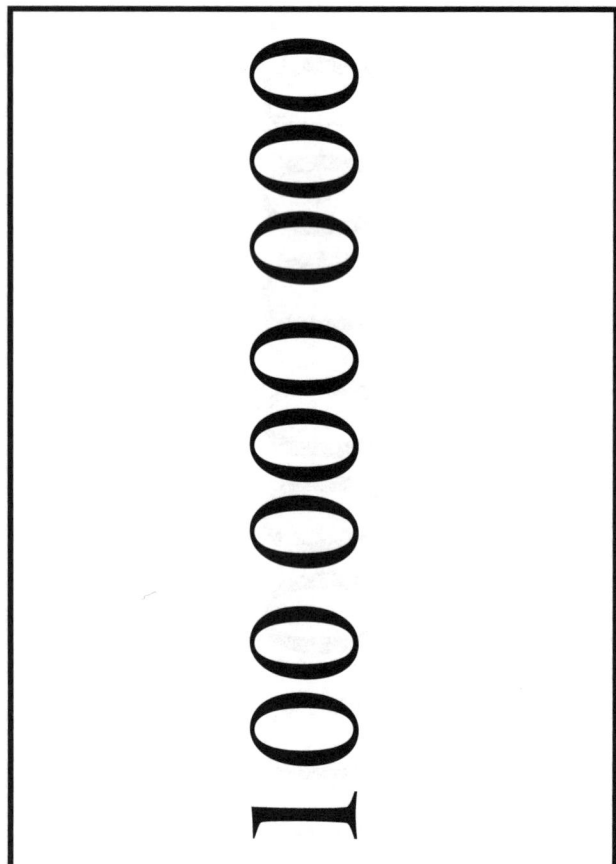

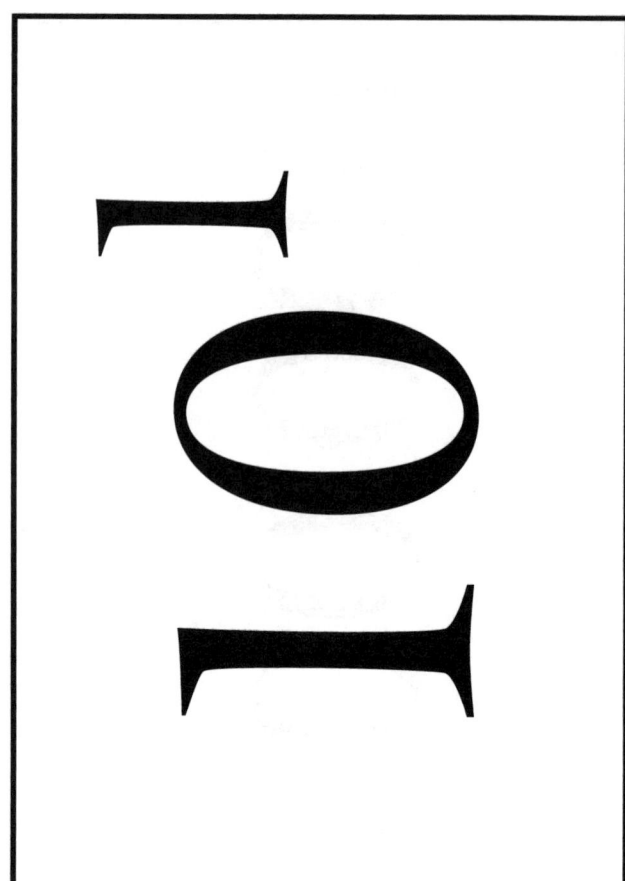

1000000000

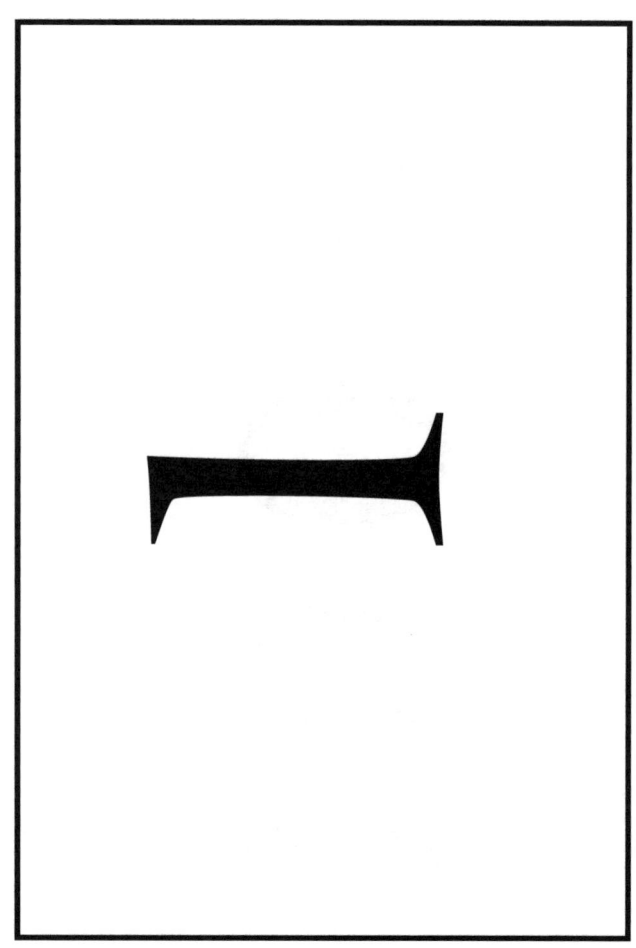

Photocopy onto card, cut out.

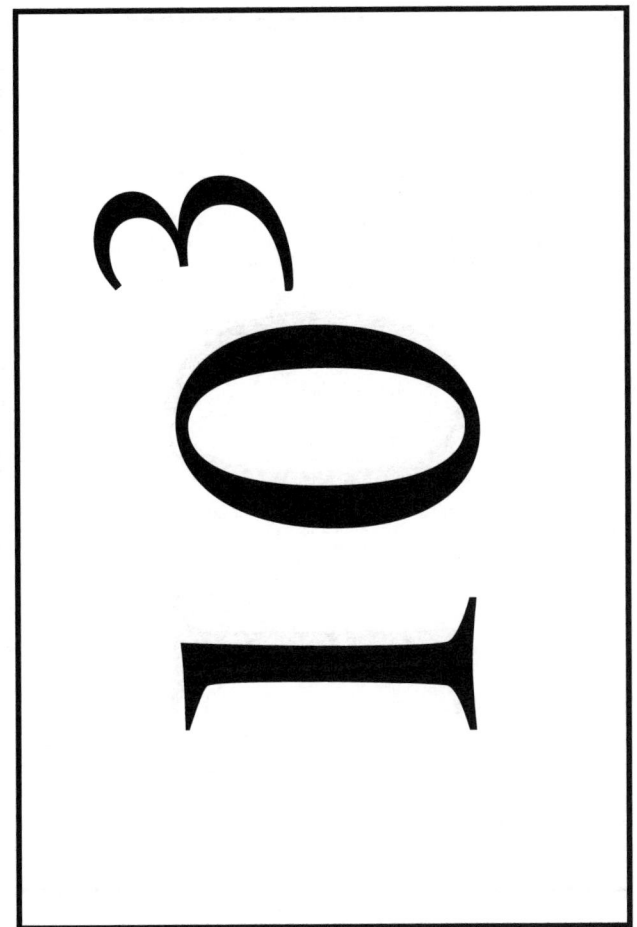

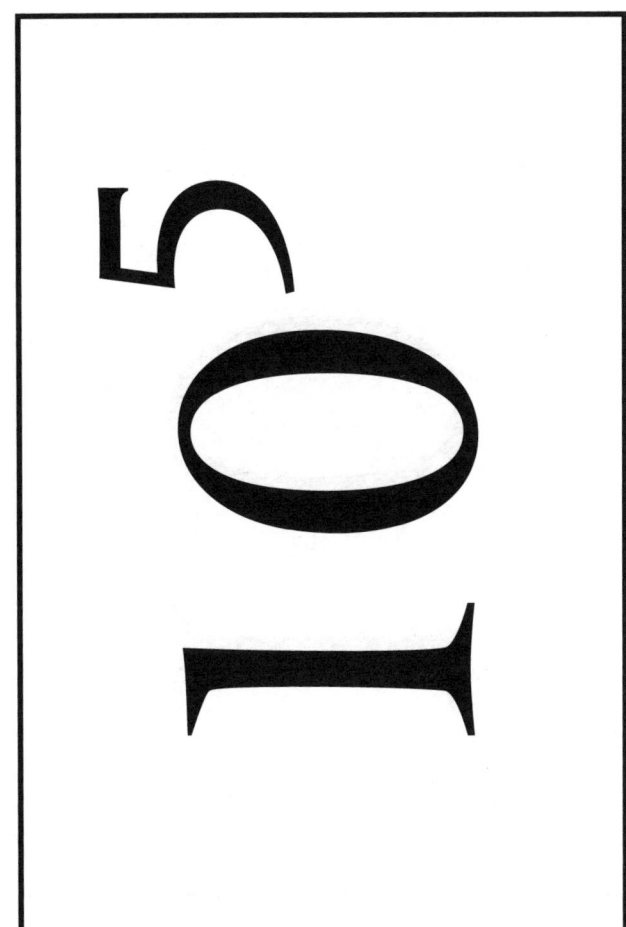

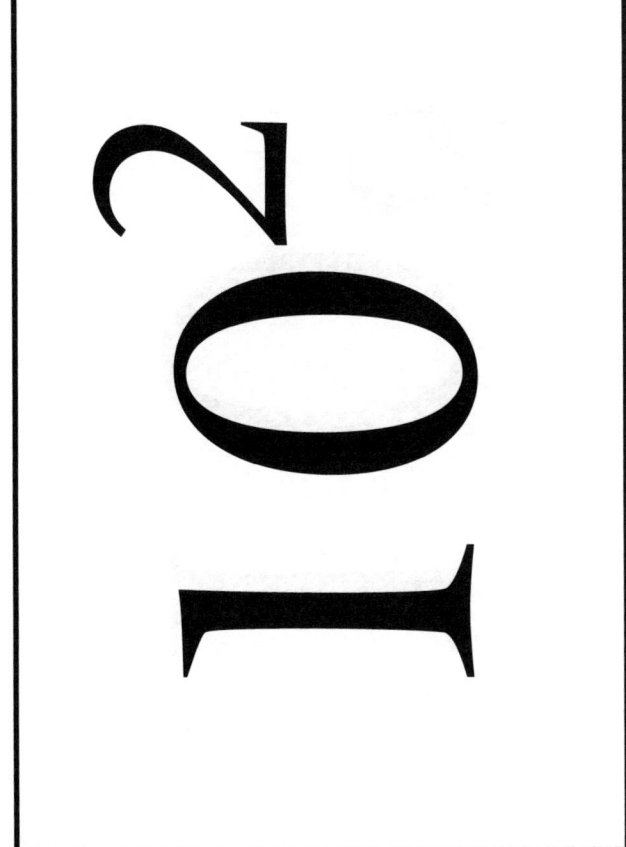

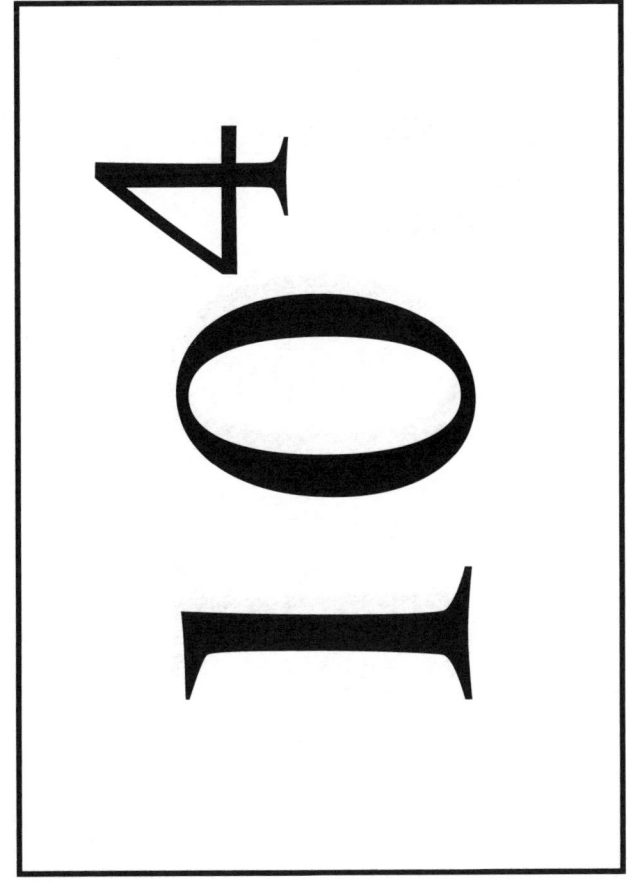

Photocopy onto card, cut out.

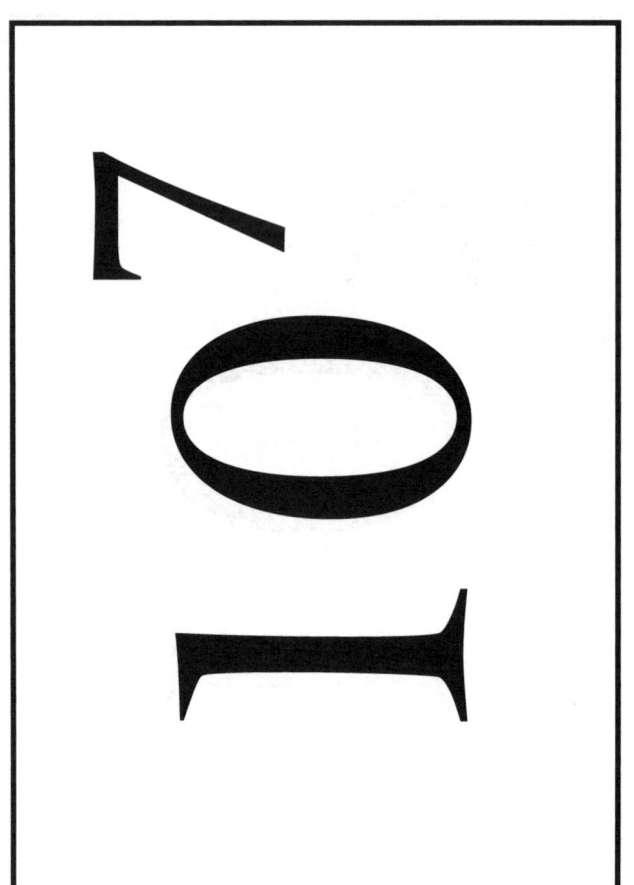

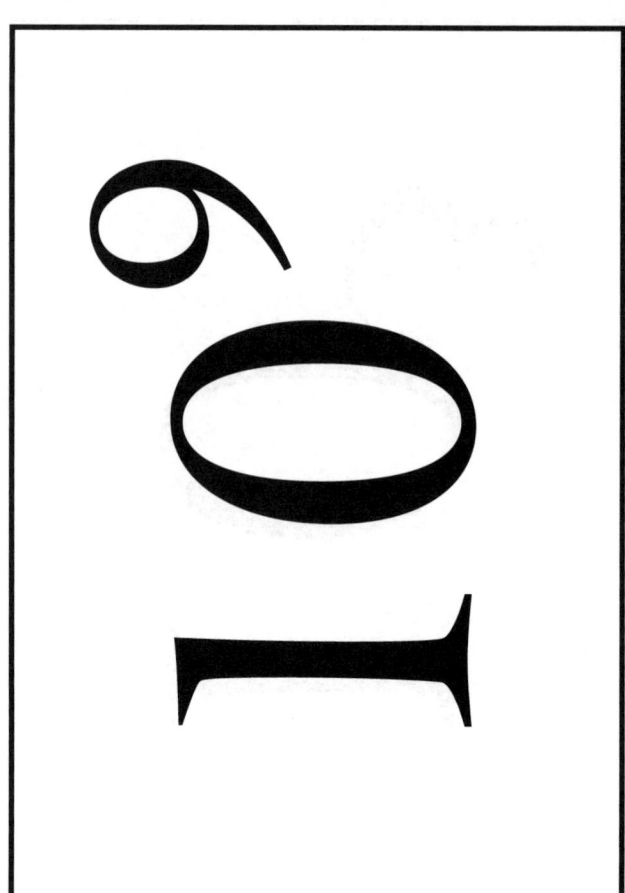

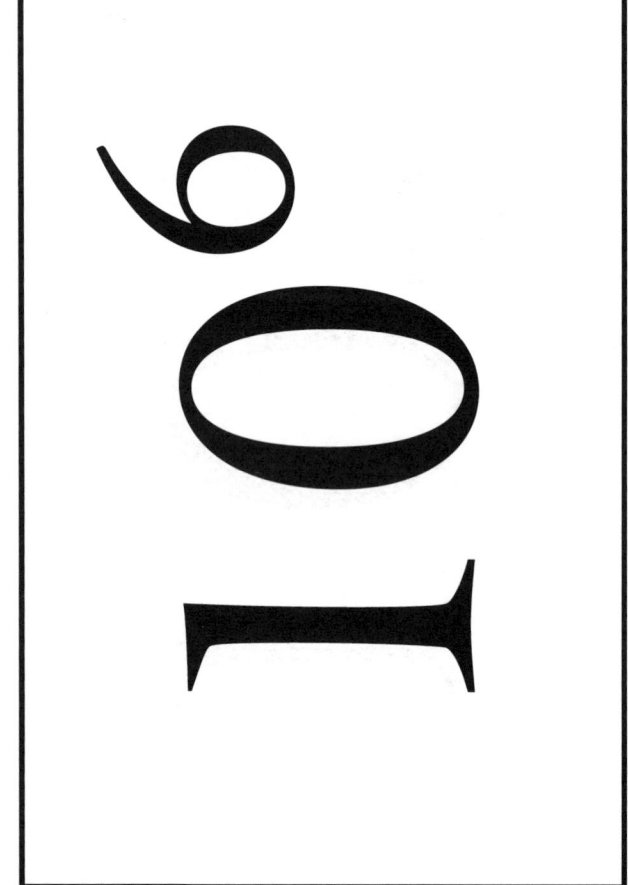

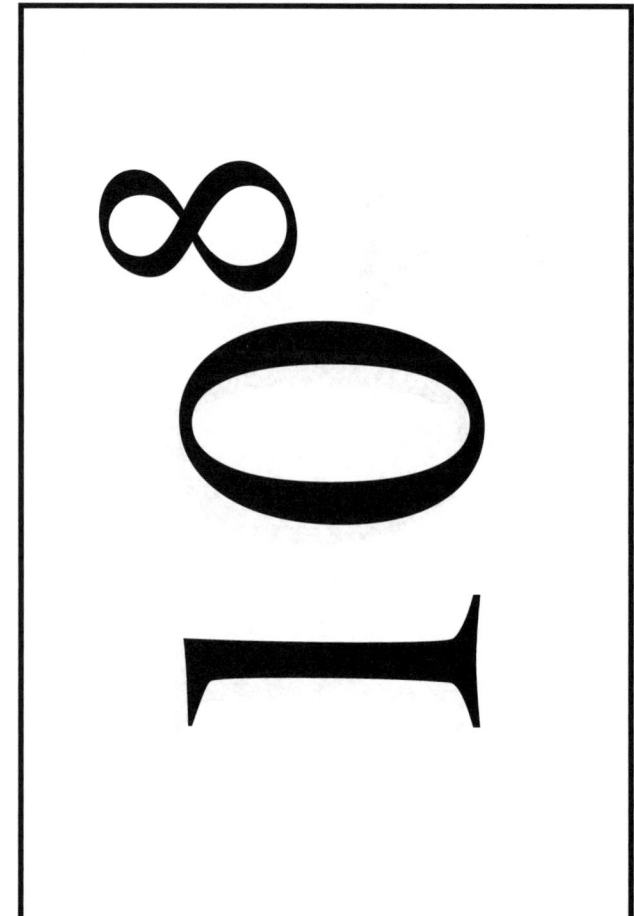

Photocopy onto card, cut out.

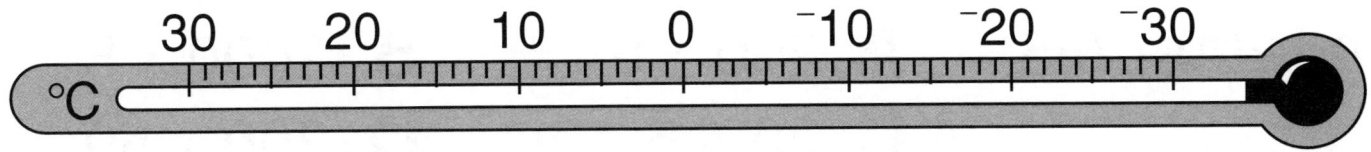

Which of A, B, C or D is equivalent to the calculation given?

a $32 \div 0 \cdot 4$ **A** $32 \div 0 \cdot 04$ **B** $3 \cdot 2 \div 4$
C $32 \div 4$ **D** $320 \div 4$

b $48 \div 0 \cdot 6$ **A** $48 \div 6$ **B** $480 \div 0 \cdot 6$
C $480 \div 60$ **D** $480 \div 6$

c $200 \div 0 \cdot 5$ **A** $200 \div 5$ **B** $2000 \div 5$
C $20 \div 5$ **D** $200 \div 50$

d $936 \div 0 \cdot 03$ **A** $936 \div 3$ **B** $93\,600 \div 3$
C $93 \cdot 6 \div 3$ **D** $9360 \div 30$

e $96 \div 2 \cdot 3$ **A** $96 \div 23$ **B** $960 \div 2 \cdot 3$
C $960 \div 23$ **D** $9600 \div 23$

f $420 \div 2 \cdot 6$ **A** $4200 \div 26$ **B** $420 \div 26$
C $42\,000 \div 26$ **D** $4200 \div 26$

Estimate what fraction of these is shaded.

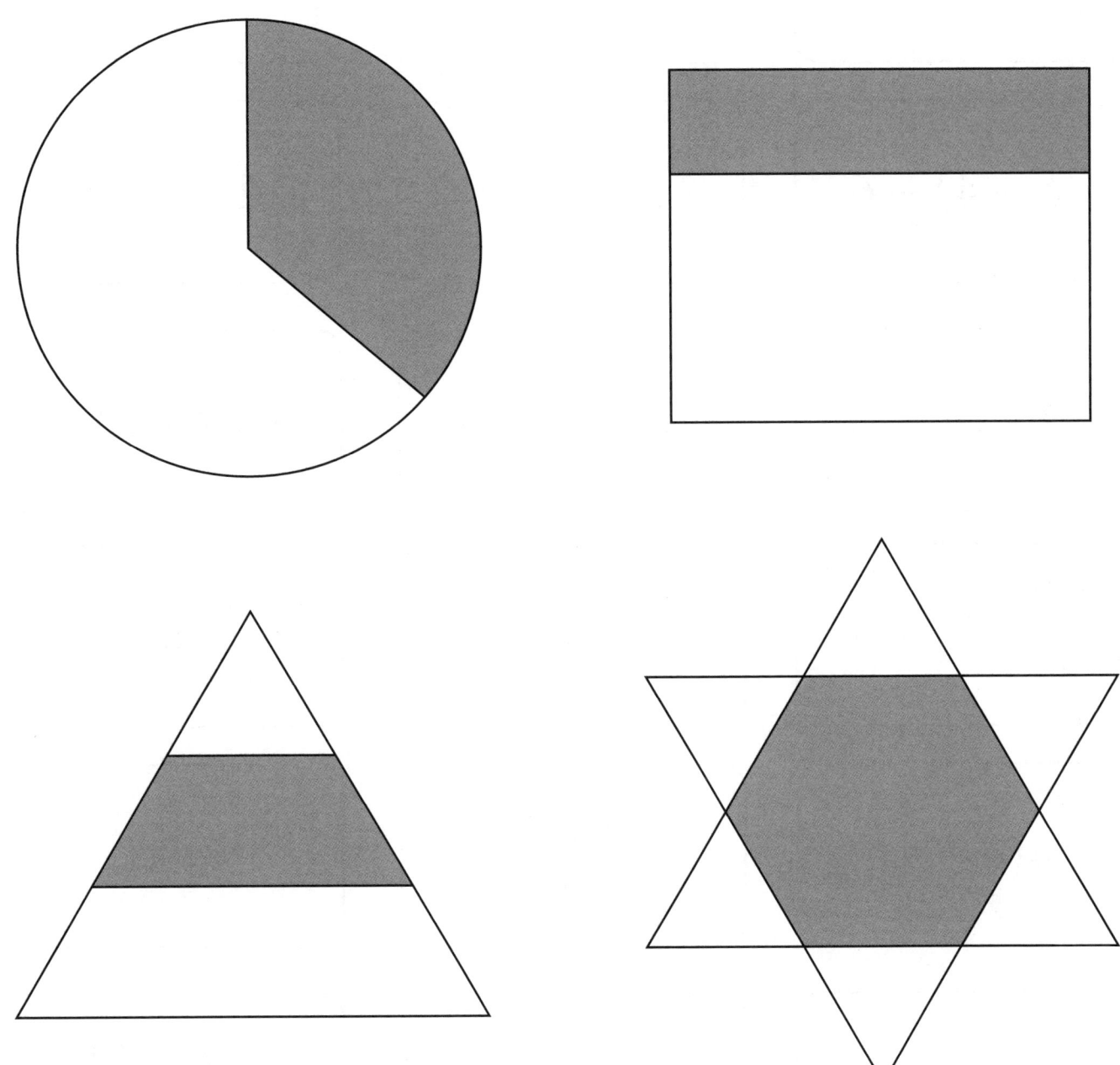

$3 \times a$

$a \times b$

$4 \times (a + b)$

$(b + 4) \div 6$

$a \times b \times c$

$4 \times a \times b$

$^-2 \times (a + b)$

$(a - 2) \div 4$

$3 \times (a + 2) \div 7$

$(a - b) \times 5$

$3a$

ab

$4(a + b)$

$\dfrac{b + 4}{6}$

abc

$4ab$

$^-2(a + b)$

$\dfrac{3(a + b)}{7}$

$\dfrac{a - 2}{4}$

$5(a - b)$

$$x + y + z = xyz$$

$$x + y + z = x + yz$$

$$a + b = b + a$$

$$abc = acb$$

$$y \times y = 2y$$

$$x + y + x = 2x + y$$

$$x + x = x^2$$

$$\frac{ab}{c} = \frac{bc}{a}$$

$$m \times m \times 3 = 3m^2$$

$$m + m = 2m$$

$$x + y - z = x - z + y$$

$$2b^2 = (2b)^2$$

$2(x + 4)$

$3x + 9$

$3(x - 3)$

$2x - 4$

$2(x - 4)$

$2x + 4$

$3(x + 3)$

$2x + 6$

$2(x + 2)$

$3x - 3$

$2(x - 2)$

$2x - 8$

$3(x + 1)$

$3x + 3$

$3(x - 1)$

$3x - 9$

Photocopy onto card, cut out.

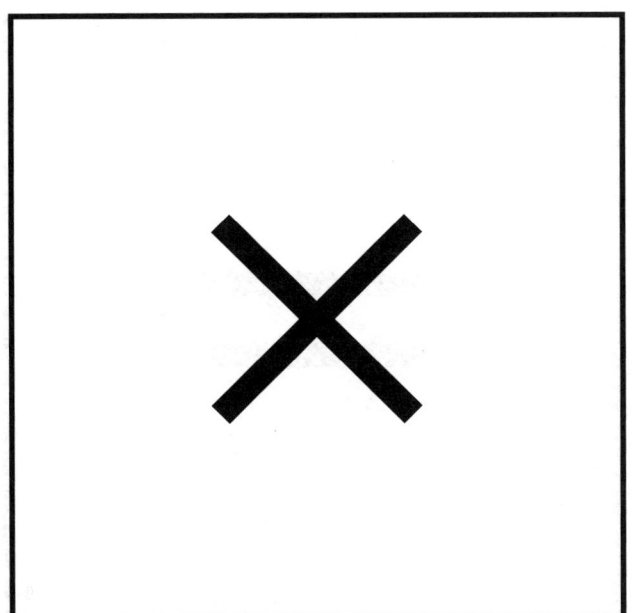

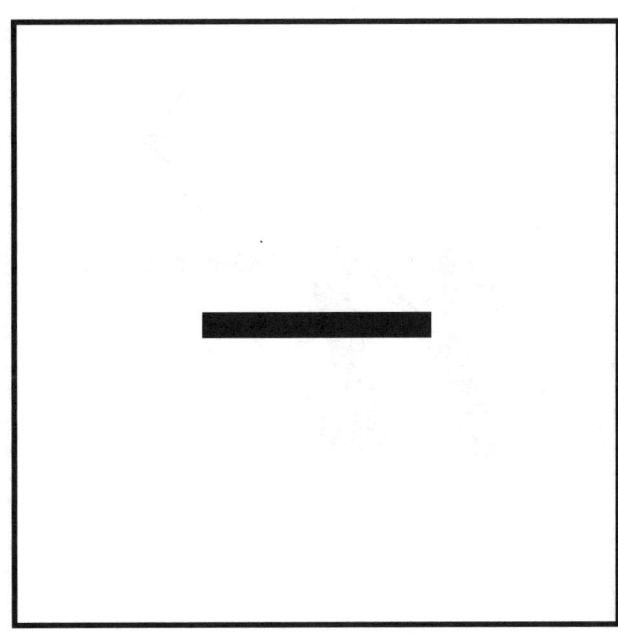

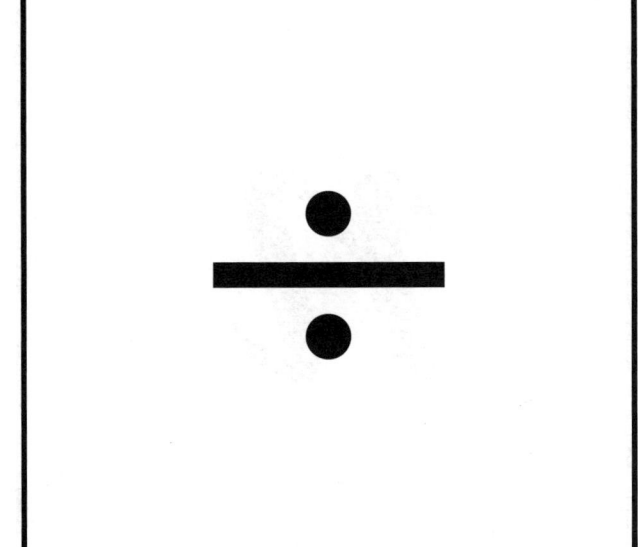

Photocopy onto card, cut out.

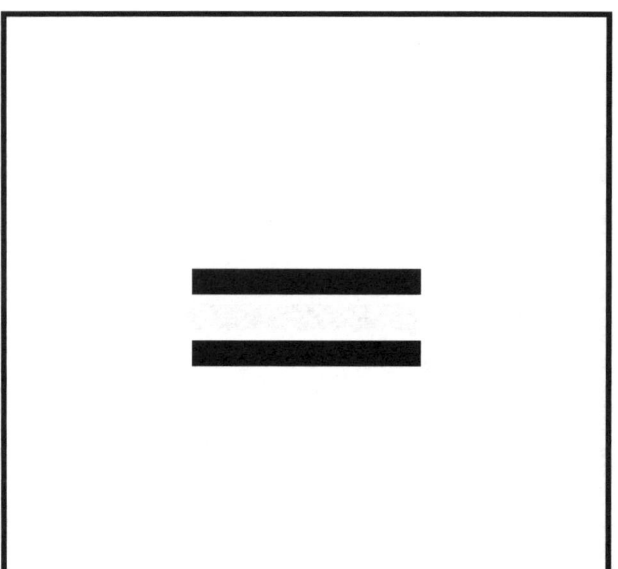

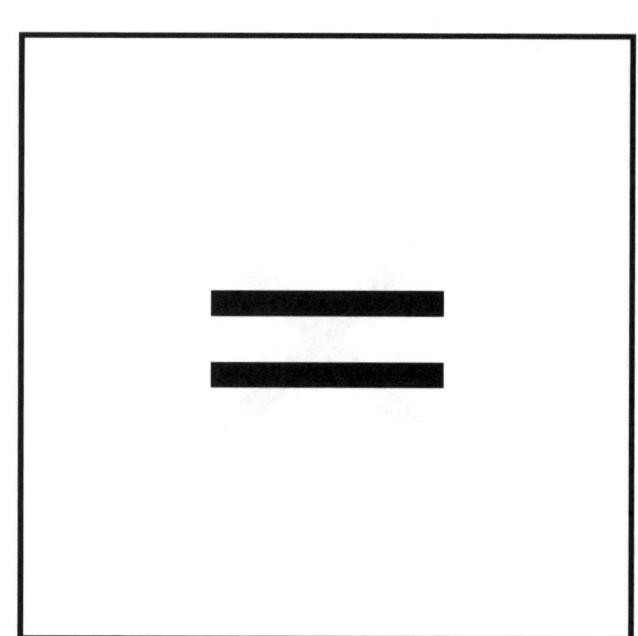

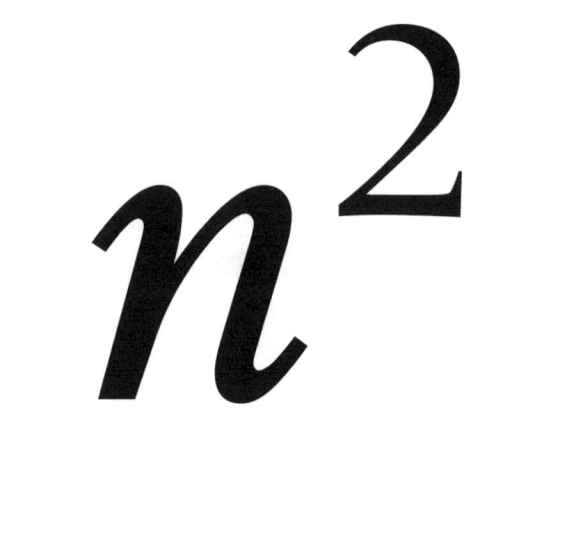

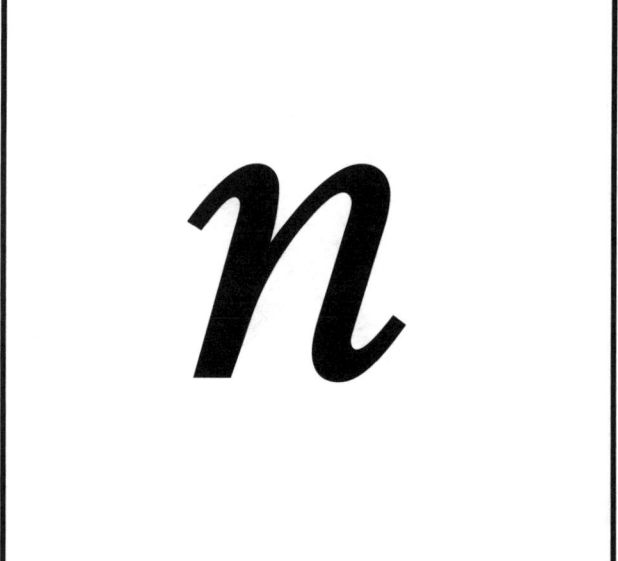

Photocopy onto card, cut out.

n	3
12	5
3	4

Photocopy onto card, cut out.

2

0

1

2

3

4

Photocopy onto card, cut out.

5

6

7

8

9

⁻1

Photocopy onto card, cut out.

$$^-2$$

$$^-3$$

$$^-4$$

$$^-5$$

$$^-6$$

$$^-7$$

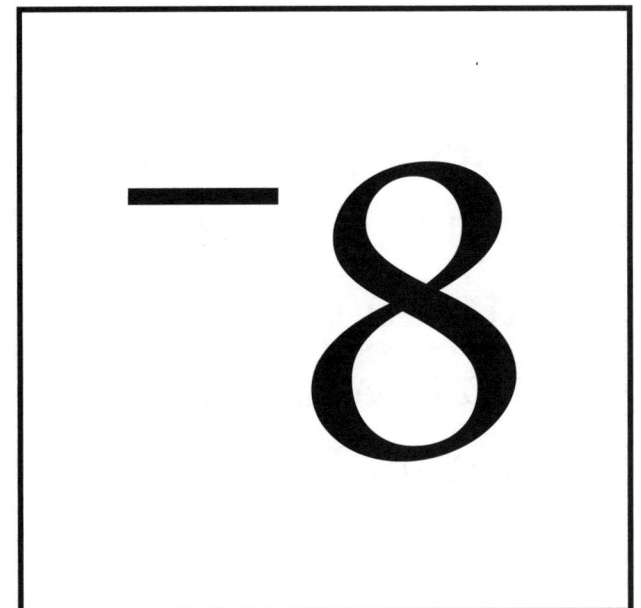

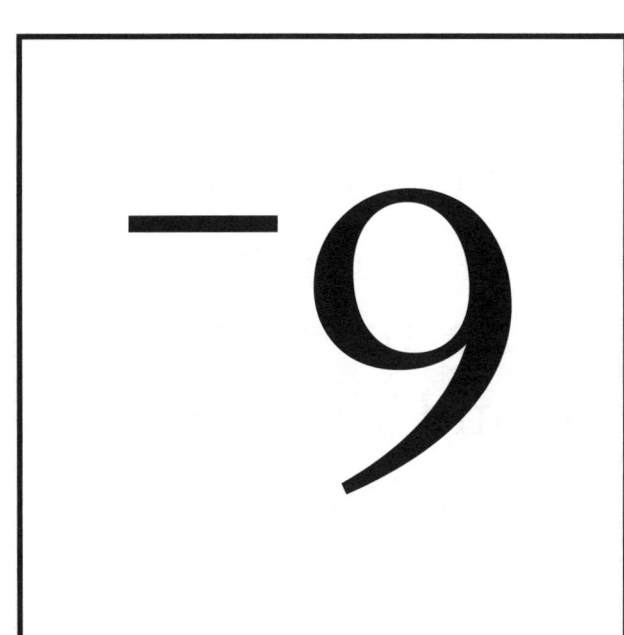

Photocopy onto card, cut out.

Jo has *a* pounds.
Tim has *b* pounds.
Write an equation if
Jo and Tim together
have £25.

Jo has *a* pounds.
Tim has *b* pounds.
Write an equation if
Jo has twice as
much as Tim.

Jo has *a* pounds.
Tim has *b* pounds.
Write an equation if
Tim gives Jo £4 so
they then both have
the same amount.

Jo has *a* pounds.
Tim has *b* pounds.
Write an equation if
half of Tim's
amount is equal to
three times Jo's
amount.

Write an equation
for this.
Multiply a number
by 2.
Subtract 3.
The answer is 7.

Write an equation
for this.
Add 5 to a number.
Divide the answer
by 2.
The answer is 15

The cost, C, of n muffins at £2 each is .... $C = 2n + 3$

The cost, C, of n cakes at £3 each is .... $C = 3n + 2$

The cost, C, of a CD for £n and a CD case for £2 is .... $C = 3n$

The cost, C, of a pie for £n and a cake for £3 is .... $C = n + 3$

The total cost, C, of a hiring a bike for n hours at £3 per hour plus a £2 lock charge is .... $C = 2n$

The total cost, C, of a hiring a ladder for n days at £2 per day plus a £3 cleaning charge is .... $C = n + 2$

Photocopy onto card, cut out.

n

$+$

$n + 16$

$+$

$2(n + 16)$

$=$

144

© **New National Framework Mathematics 8** Nelson Thornes Ltd

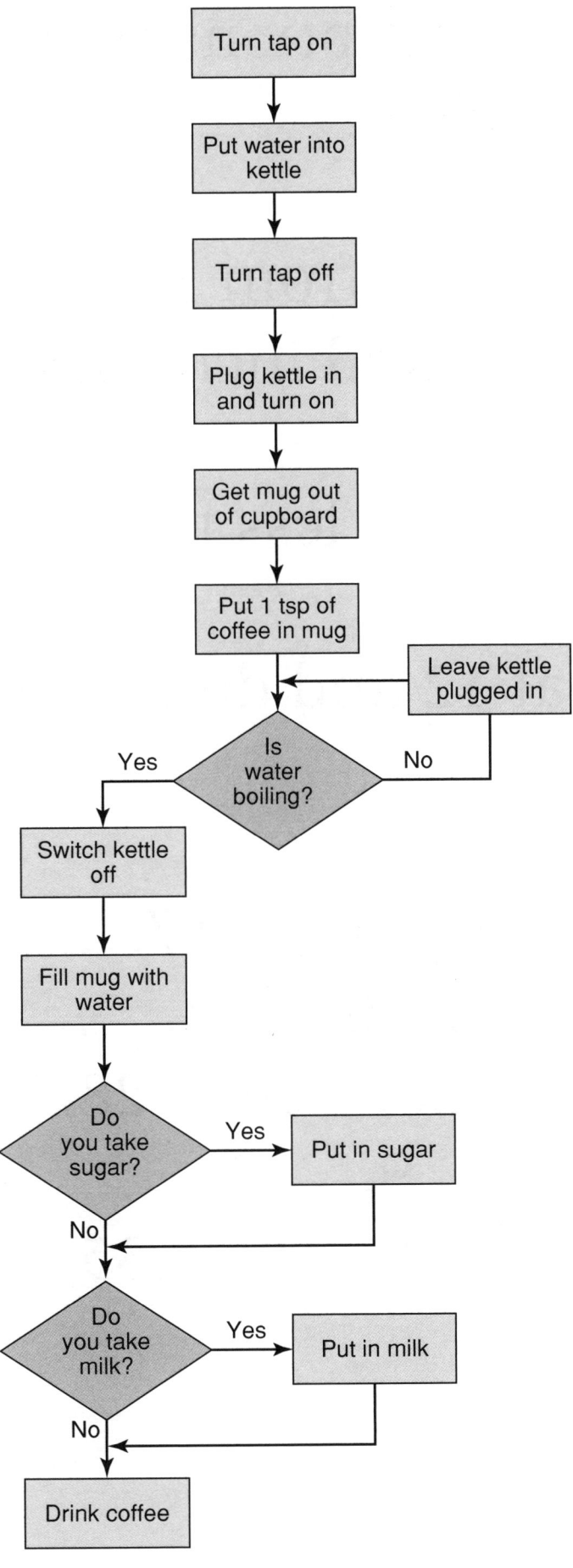

Match the descriptions with the sequences below.

a odd numbers from 3 to 11

b begins at 5 and increases in steps of 3

c each term is 1 less than a multiple of 6

d begins at 12 and decreases in steps of 4

A 5, 8, 11, 14, 17, ...

B 12, 8, 4, 0, $^-$4, ...

C 12, 16, 20, 24, 28, ...

D 3, 5, 7, 9, 11, ...

E 11, 17, 23, 29, 35, ...

Part 1

→ +4 → +3 →

→ +5 → −2 →

→ −4 → −6 →

→ ×4 → ×2 →

→ ×6 → ÷2 →

Part 2

x → ×3 → +6 →

x → +2 → ×3 →

Part 3

→ ×4 → −3 →

→ −3 → ×4 →

Part 4

1, 2, 3 → +5 → [ ] → 1, 2, 3

10, 3, 7 → ×2 → +1 → ×2 → +1 → [ ] → [ ] → 10, 3, 7

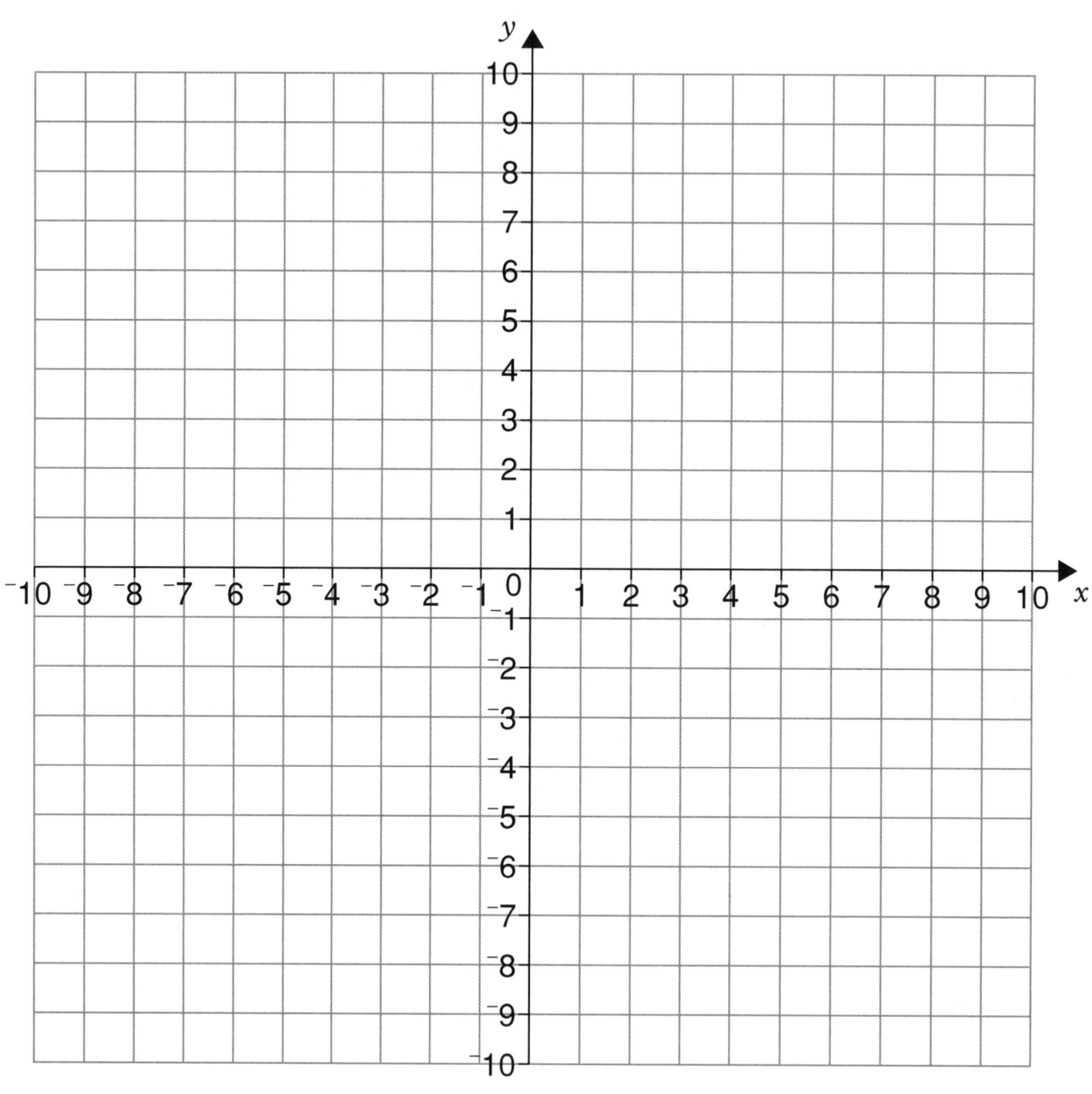

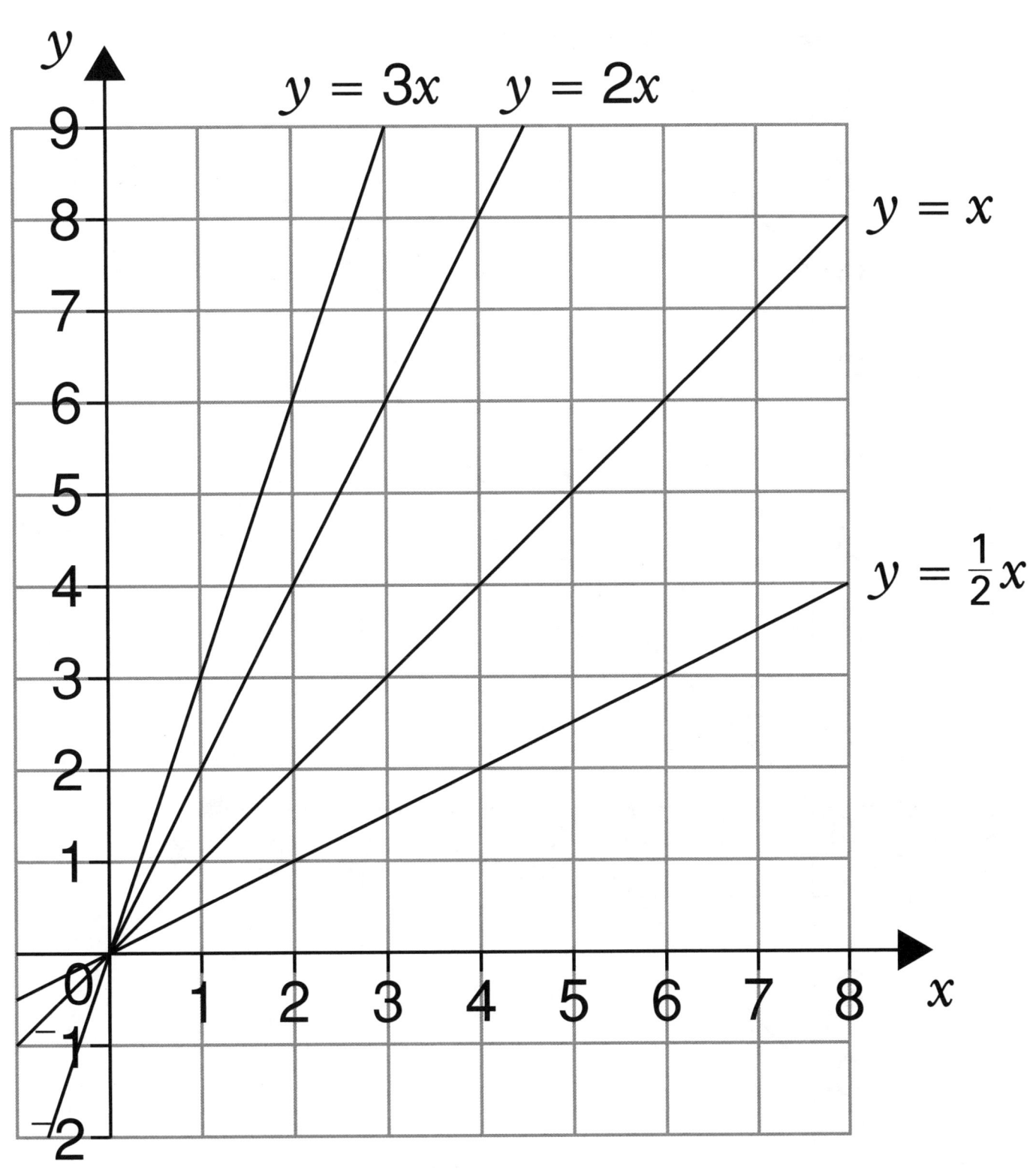

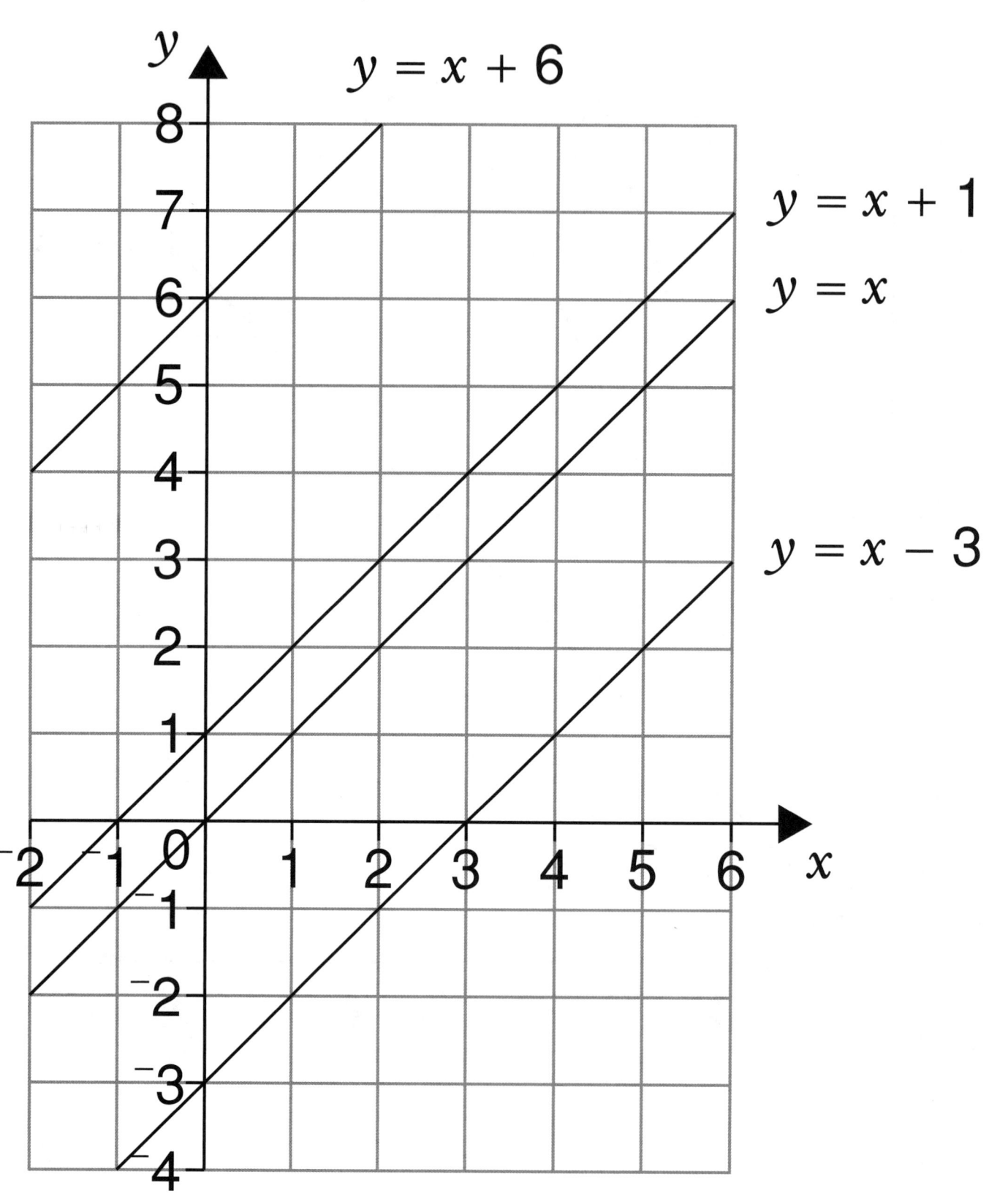

$s = 20 + 5t$ s is the distance in metres
t is the time in seconds

t (seconds)				
s (metres)				

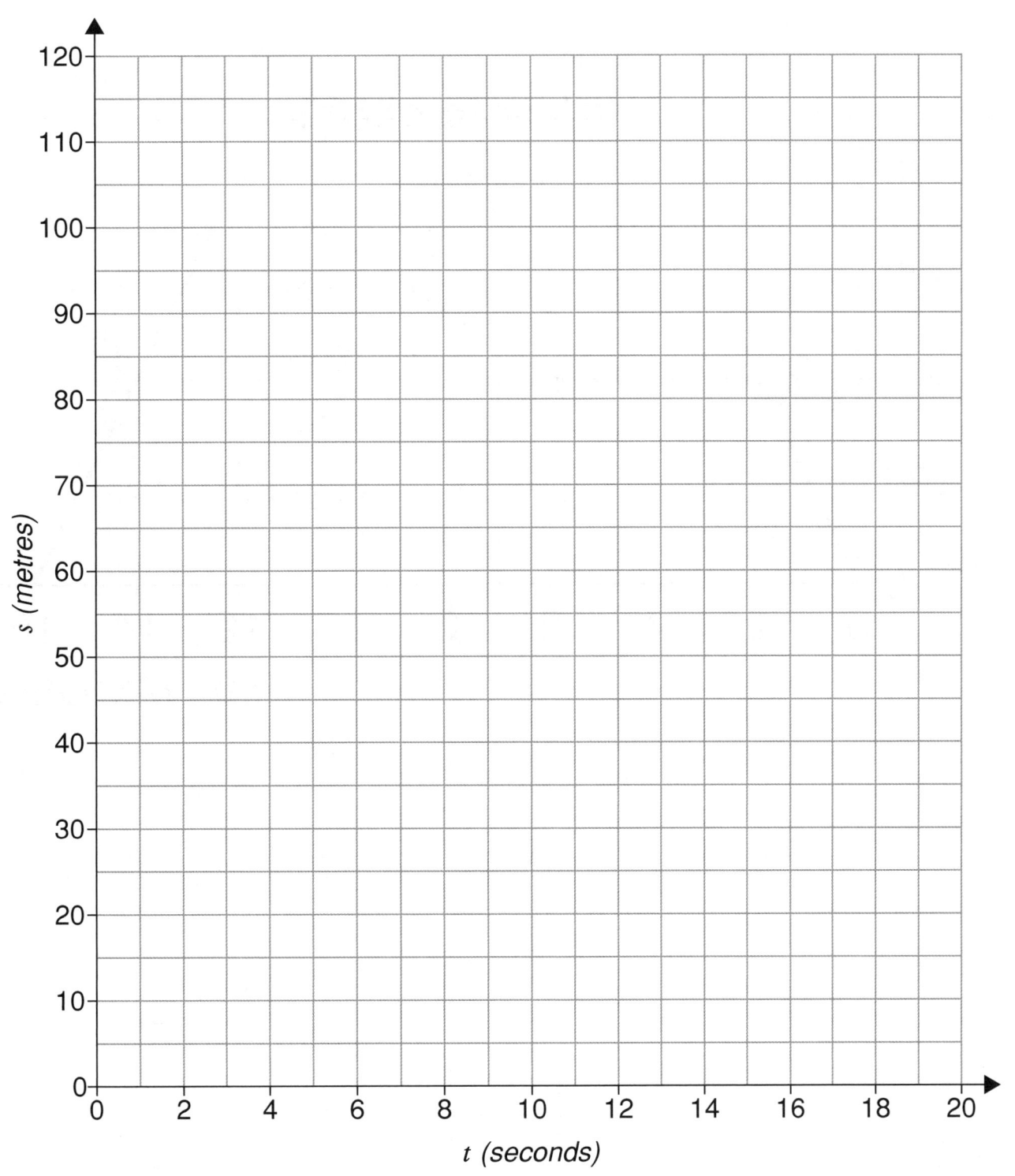

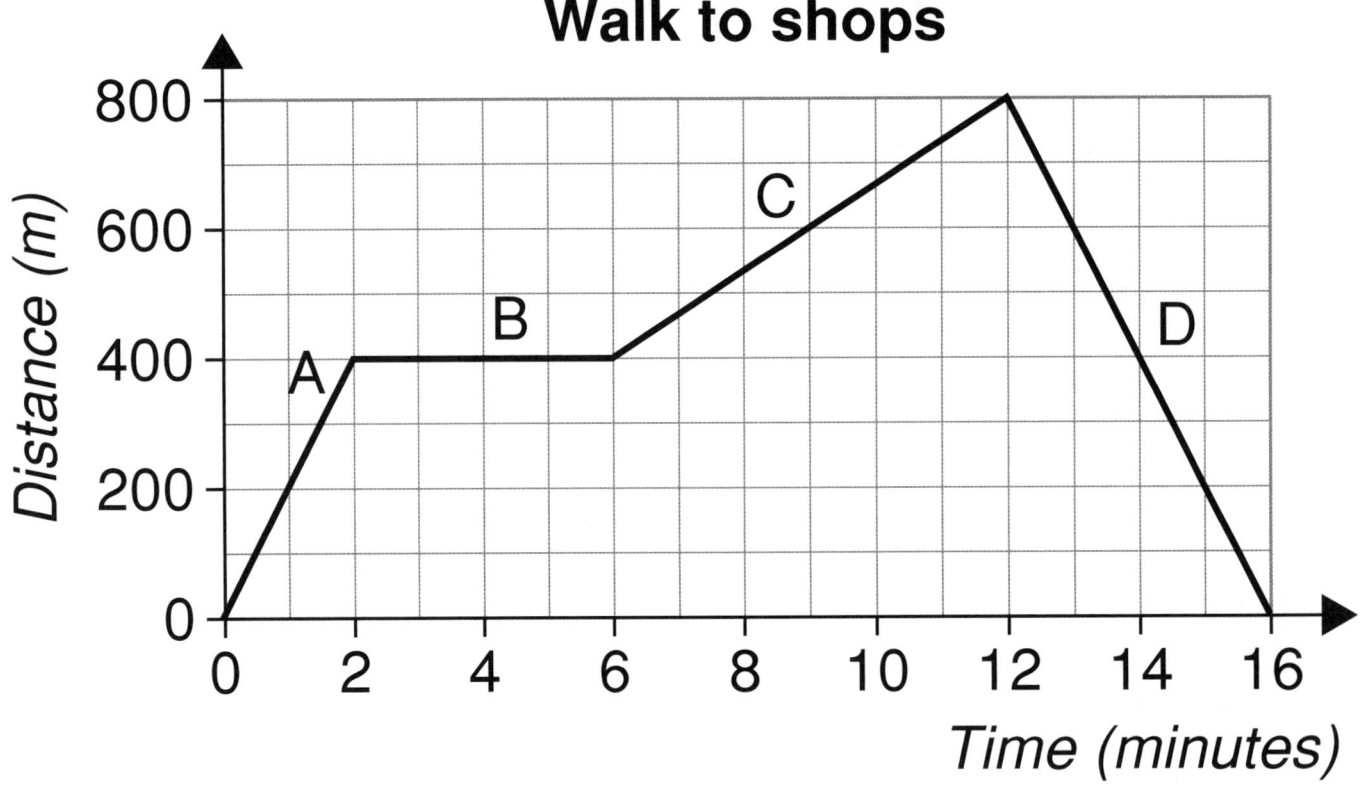

Walk to shops

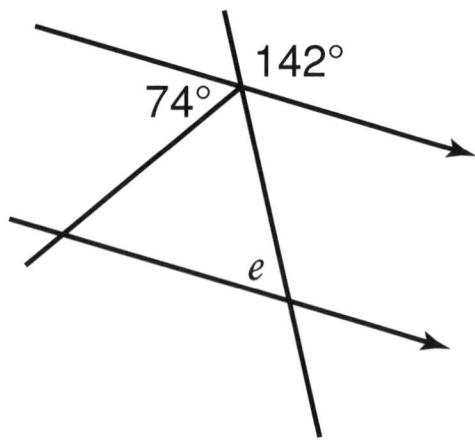

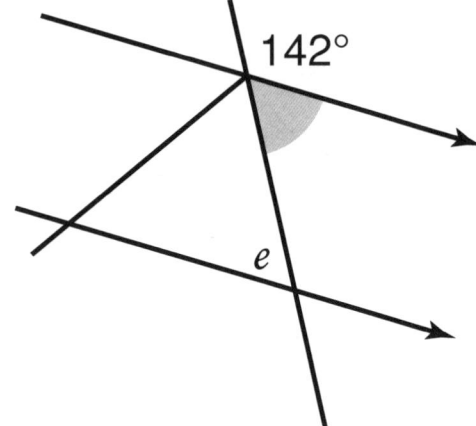

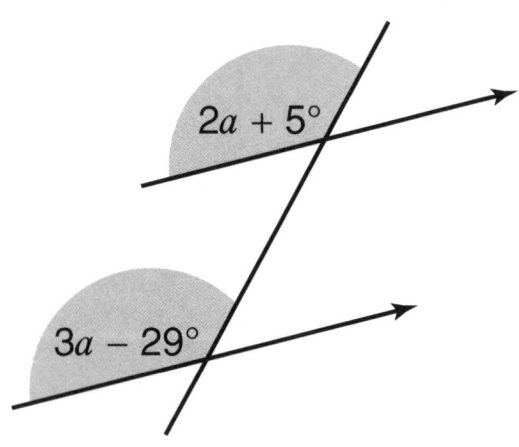

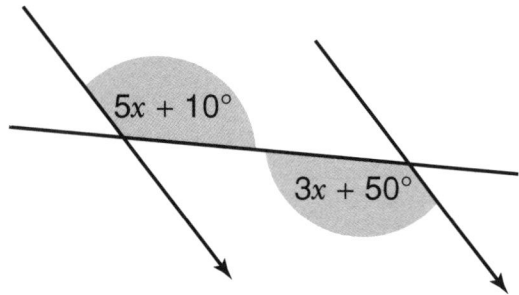

Cut out the shapes

Cut out the shapes

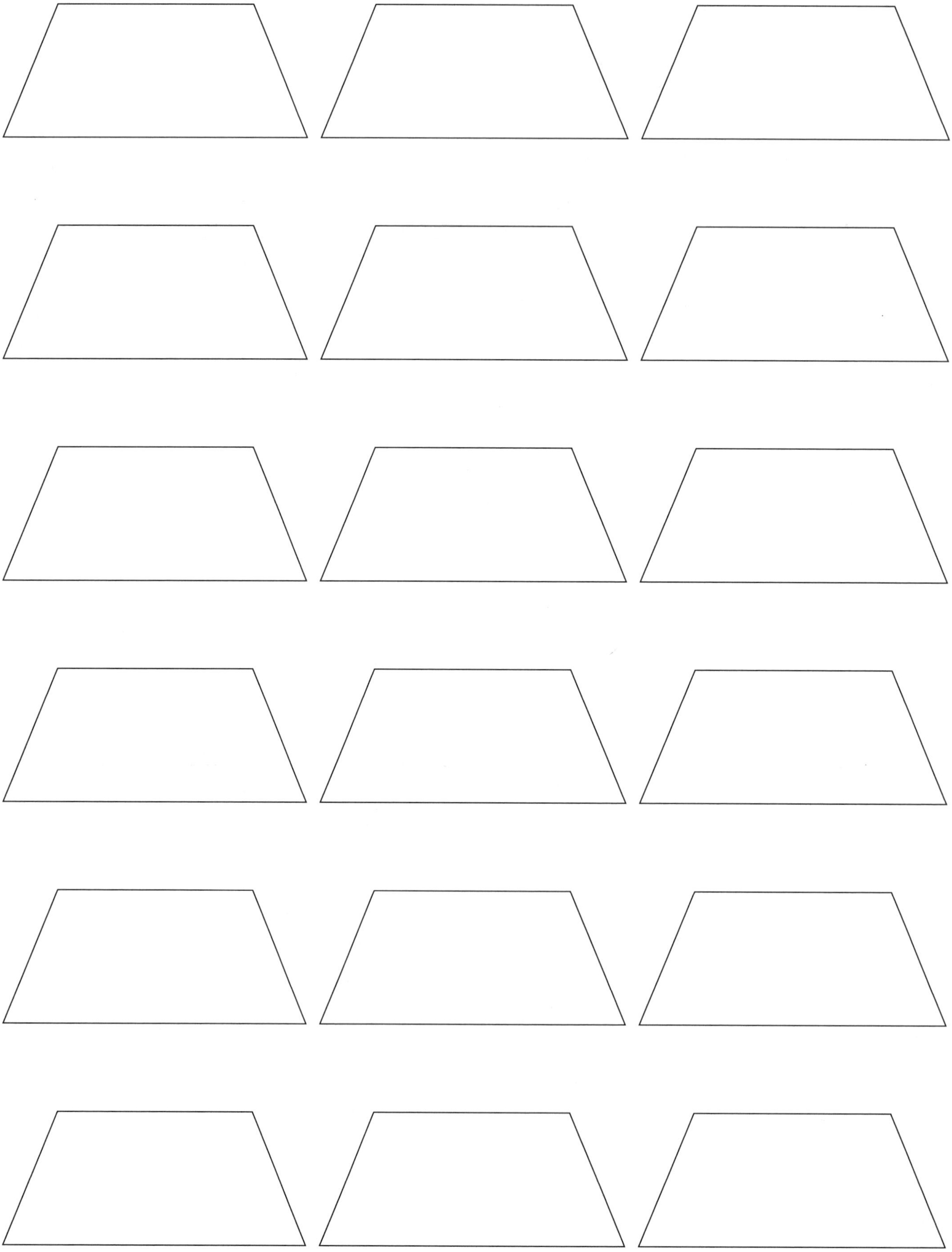

Cut out the shapes

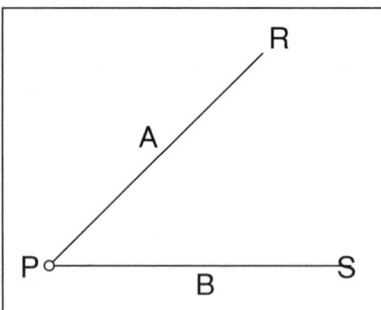

Open out the compasses to a length less than PR or PS. With the point on P, draw an arc as shown. Label the points A and B.

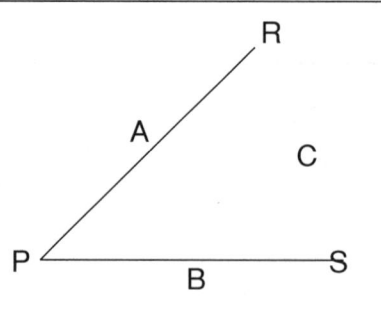

With the point first on A and then on B, draw two arcs to meet at C.

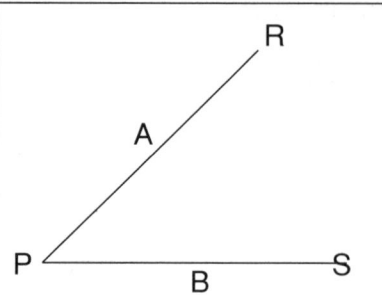

Draw the line from P through C. This line, PC, is the bisector of the angle P.

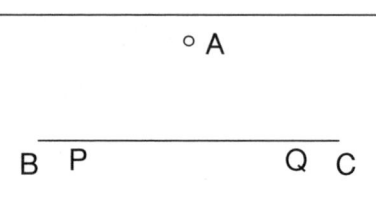

Open out the compasses.
With the point on A, draw an arc to cross BC at P and Q.

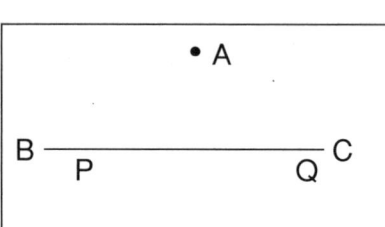

With the point first on P, then on Q, draw two arcs to meet at R.

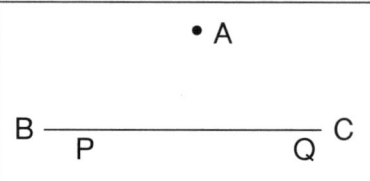

Join A and R. AR is the perpendicular from A to the line segment BC.

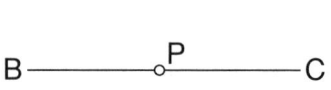

Open out the compasses to less than half the length of BC.
With the point on P, draw arcs, one on each side of P. Label where they cross BC as S and T.

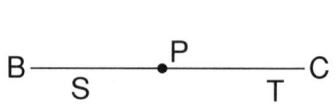

Open out the compasses a little more.
With the point first on S and then on T, draw arcs so they cut at Q and R.

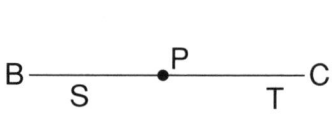

Draw the line through Q and R. QR is the perpendicular from P on the line segment BC.

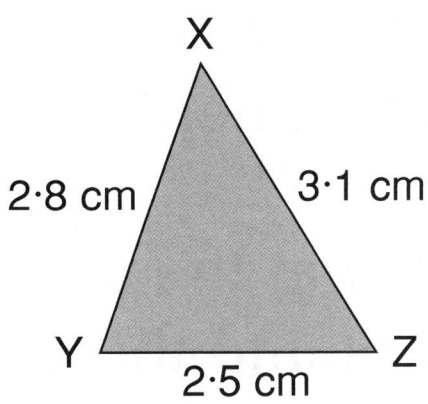

Given
Three sides (called SSS)

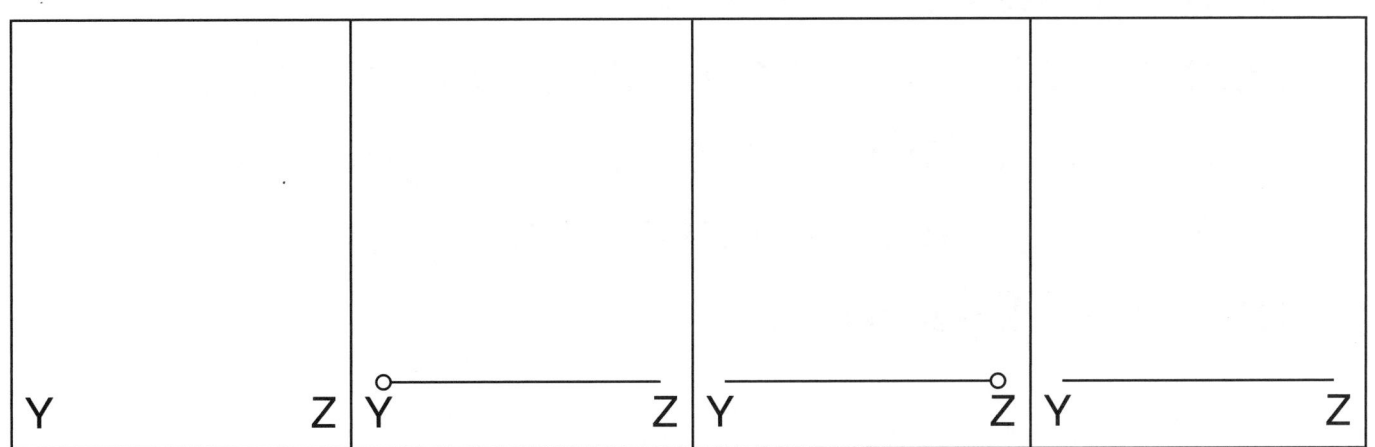

Draw a line 2.5 cm long.

Open the compasses out to 2.8 cm. With the point of the compasses on Y, draw an arc.

Open the compasses out to 3.1 cm. With the point of the compasses on Z, draw an arc to cross the first arc.

Complete the triangle.

Discussion

● Mary walked from her home to a local fair. She walked out of her house and turned left.

She turned right at the church.

She turned right into the park and walked straight to the other side.

She turned left at the park gate.

She turned left at the book shop.

The local fair is at the square on the right at the end of the road.

Mary met her brother at the fair. He asked her to give him directions to get home. What directions might Mary give her brother? **Discuss.**

9

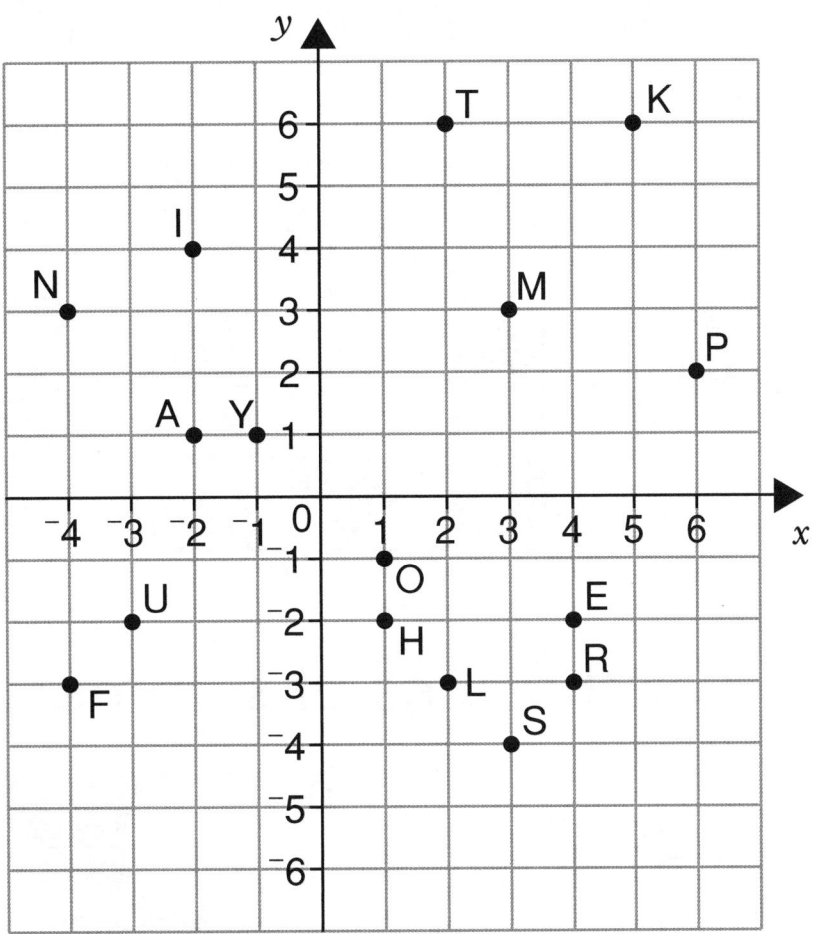

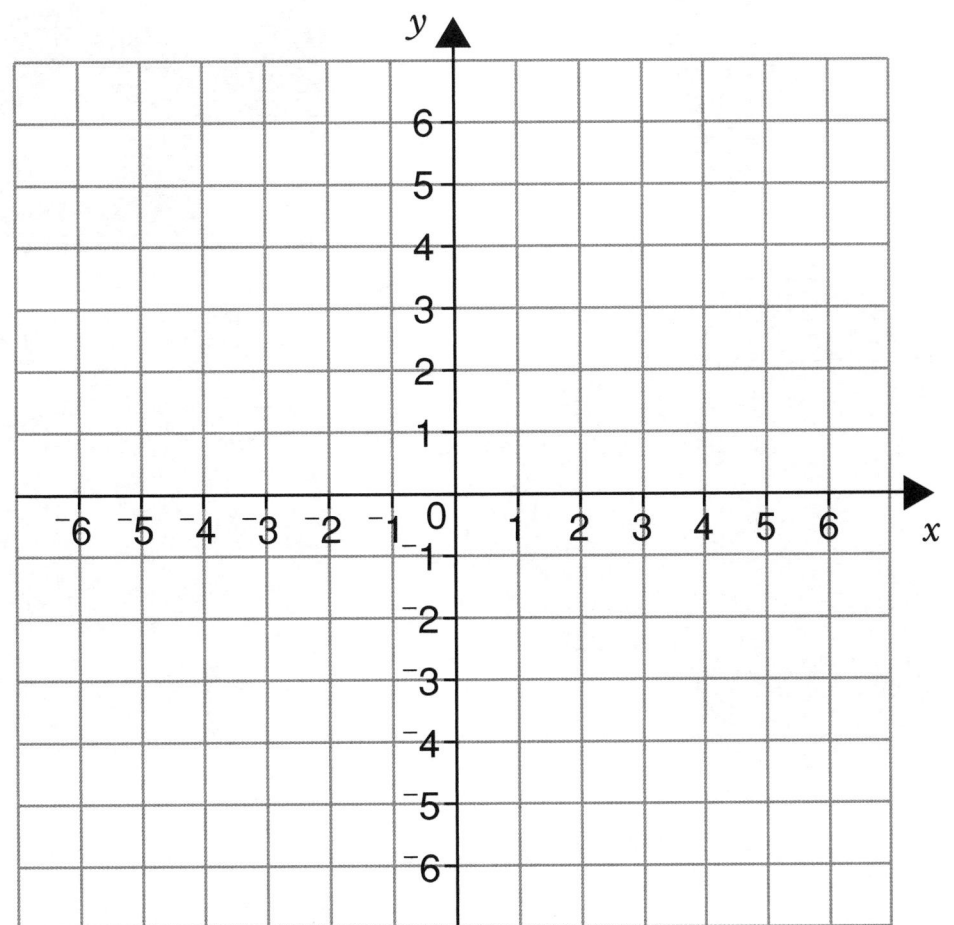

5　Do these shapes have rotational symmetry?

a　　　**b**　　　**c**

6　Draw on all the lines of symmetry.

a

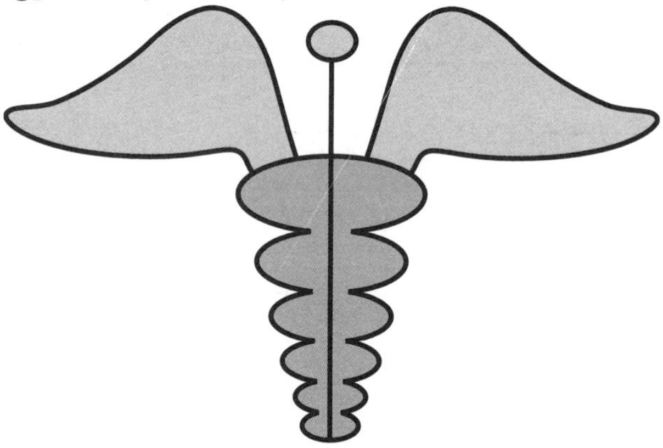

b

c

d

13 What is the order of rotation symmetry of these?

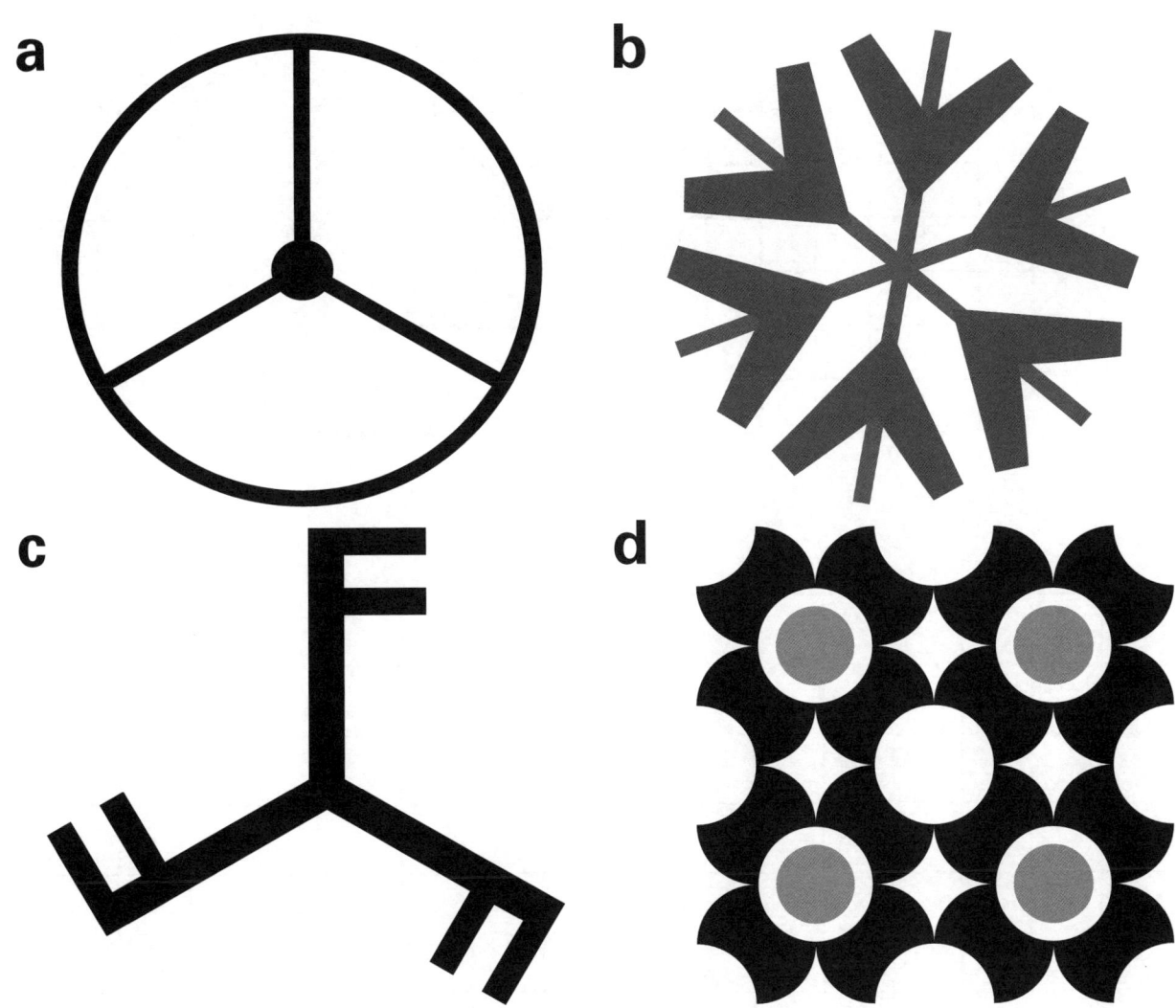

a

b

c

d

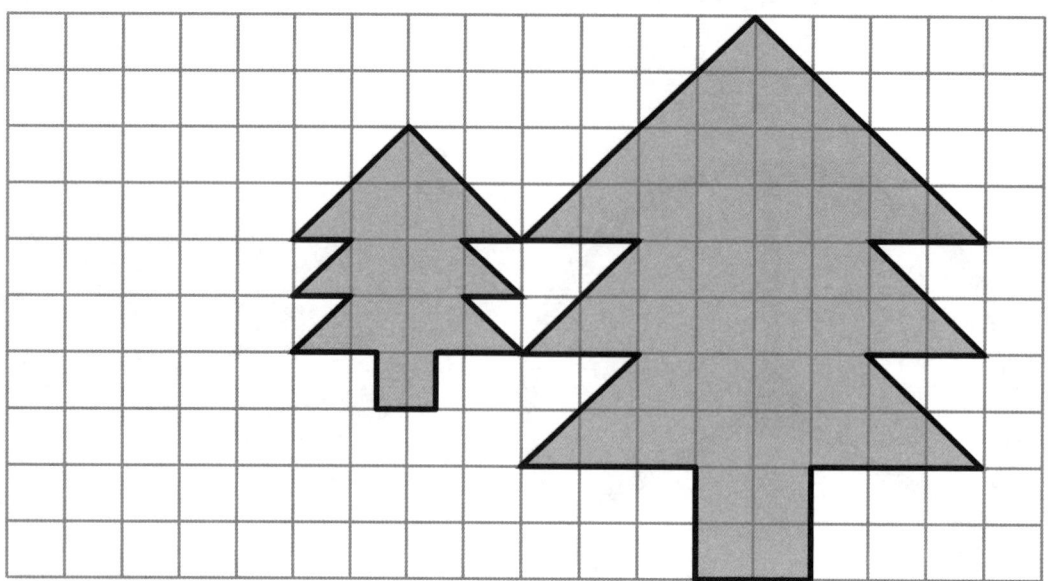

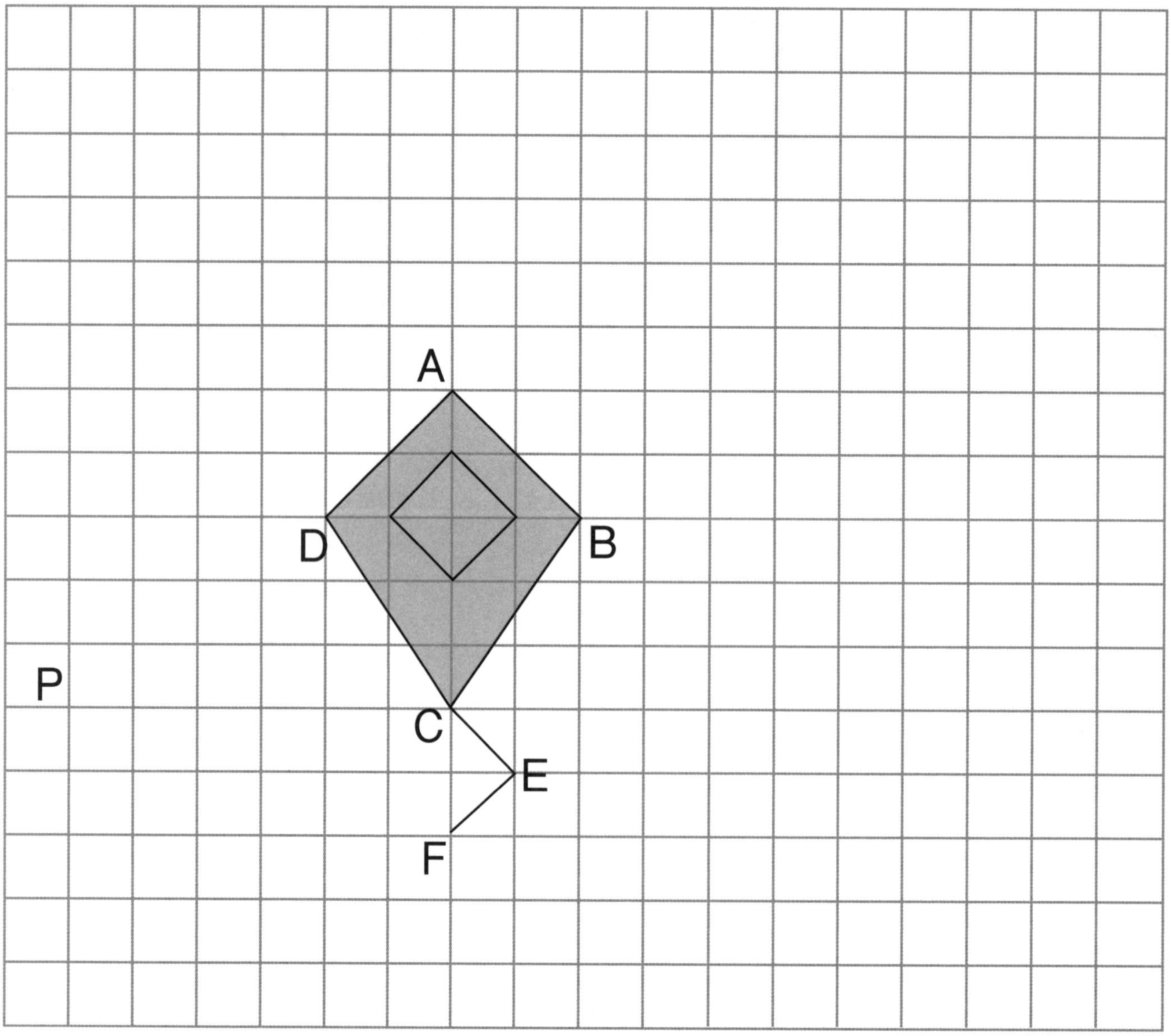

A

D · B

C

P

E

F

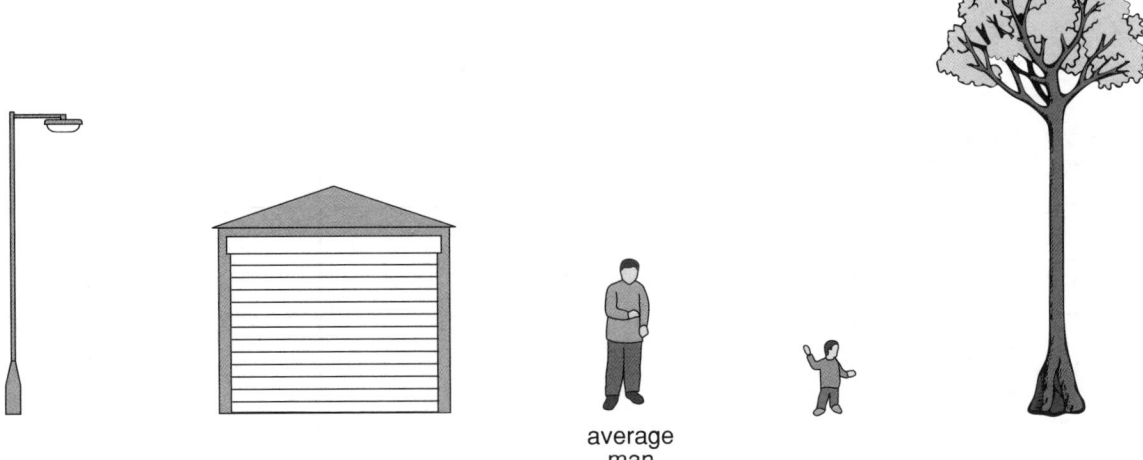

average
man

Amount spent	Tally	Frequency
£0·01–£2	IIII	5
£2·01–£4	IIII II	7
£4·01–£6	IIII I	6
£6·01–£8	III	3
£8·01–£10	IIII	4
over £10	I	1

Time (t min)	Tally	Frequency
$10 < t \leqslant 12$	I	1
$12 < t \leqslant 14$	II	2
$14 < t \leqslant 16$	IIII IIII II	12
$16 < t \leqslant 18$	IIII IIII IIII I	16
$18 < t \leqslant 20$	IIII	4

Comparing data

Remember

To **compare data** we sometimes use the range and one or more of the mode, median or mean.

Example This data gives the number of hours two different brands of electric jug lasted.

Brand A	608	635	612	585	683	697	641	610	604	664
	607	638	601	636	615	642	701	618	625	620
Brand B	437	861	735	824	632	321	532	682	614	732
	486	324	507	461	913	912	586	671	469	893

The mean of brand A $= \dfrac{\text{sum of data values}}{\text{number of values}} = 632 \cdot 1$

The median of brand A $= 622 \cdot 5$

The range of brand A $= 701 - 585 = 116$

The mean of brand B $= 629 \cdot 6$

The median of brand B $= 623$

The range of brand B $= 913 - 321 = 592$.

Tennis club membership

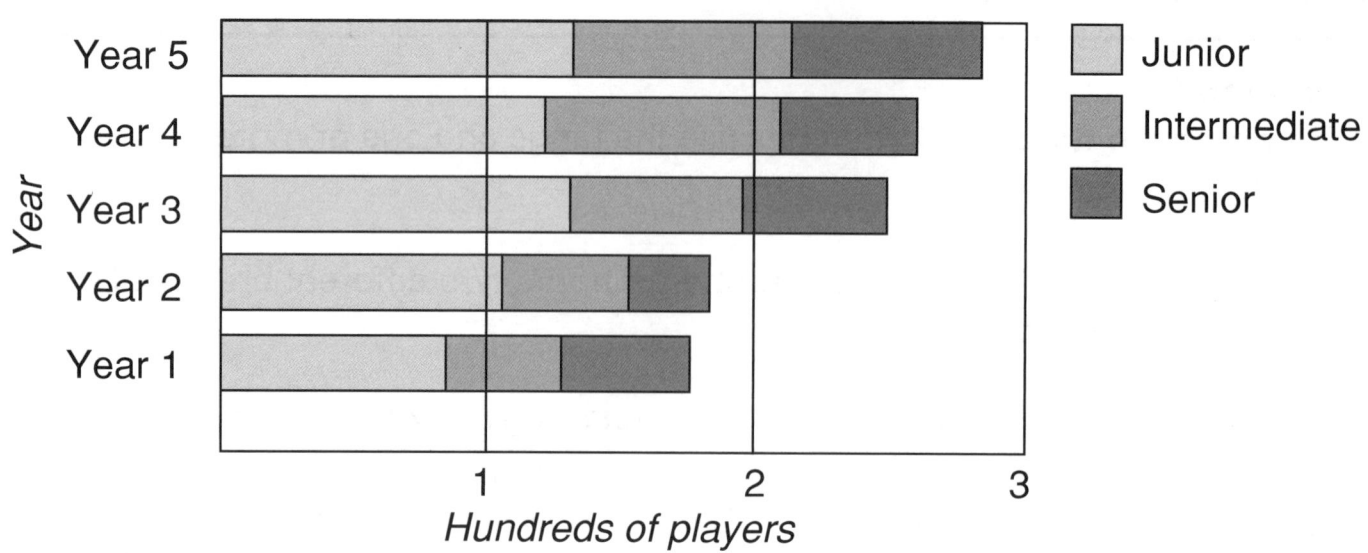

	Meat	Dairy products, eggs and fats and spreads	Vegetables	Cereals	Other

Year of 2nd birthday	'89	'90	'91	'92	'93	'94	'95	'96	'97	'98	'99
Percentage MMR	80	84	87	90	92	91	91	92	92	91	88
Percentage whooping cough	75	79	84	88	92	93	93	94	94	94	94

Percentage immunised by 2nd birthday

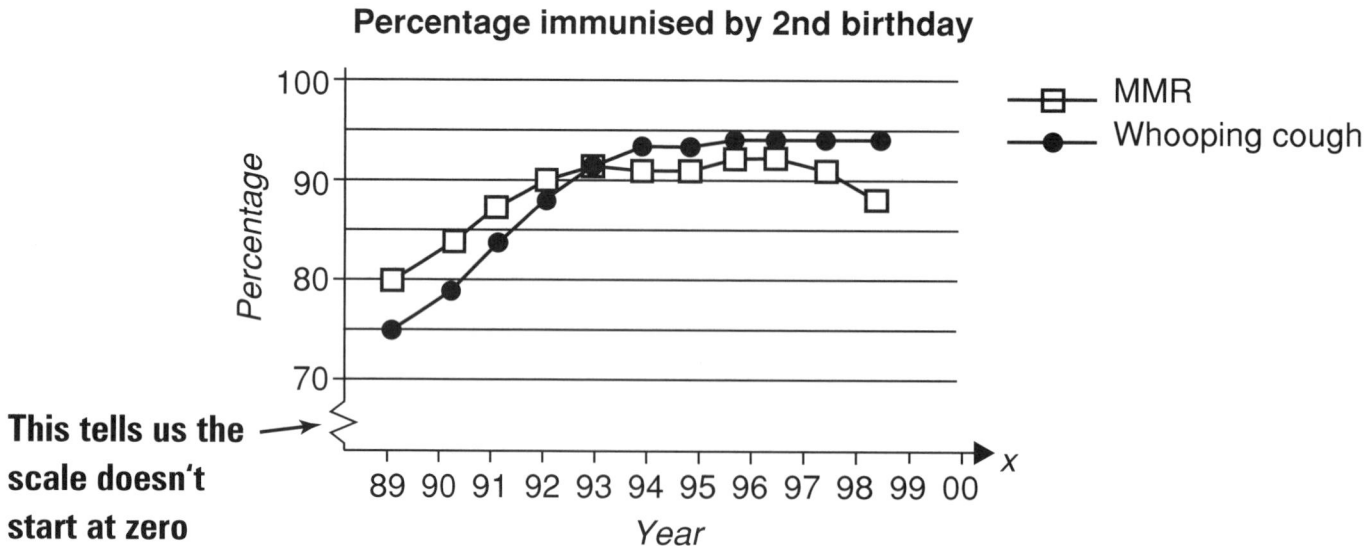

This tells us the scale doesn't start at zero

Class interval (l in mm)	Frequency
$30 \leqslant l < 35$	2
$35 \leqslant l < 40$	2
$40 \leqslant l < 45$	4
$45 \leqslant l < 50$	9
$50 \leqslant l < 55$	10
$55 \leqslant l < 60$	2

Thumb lengths

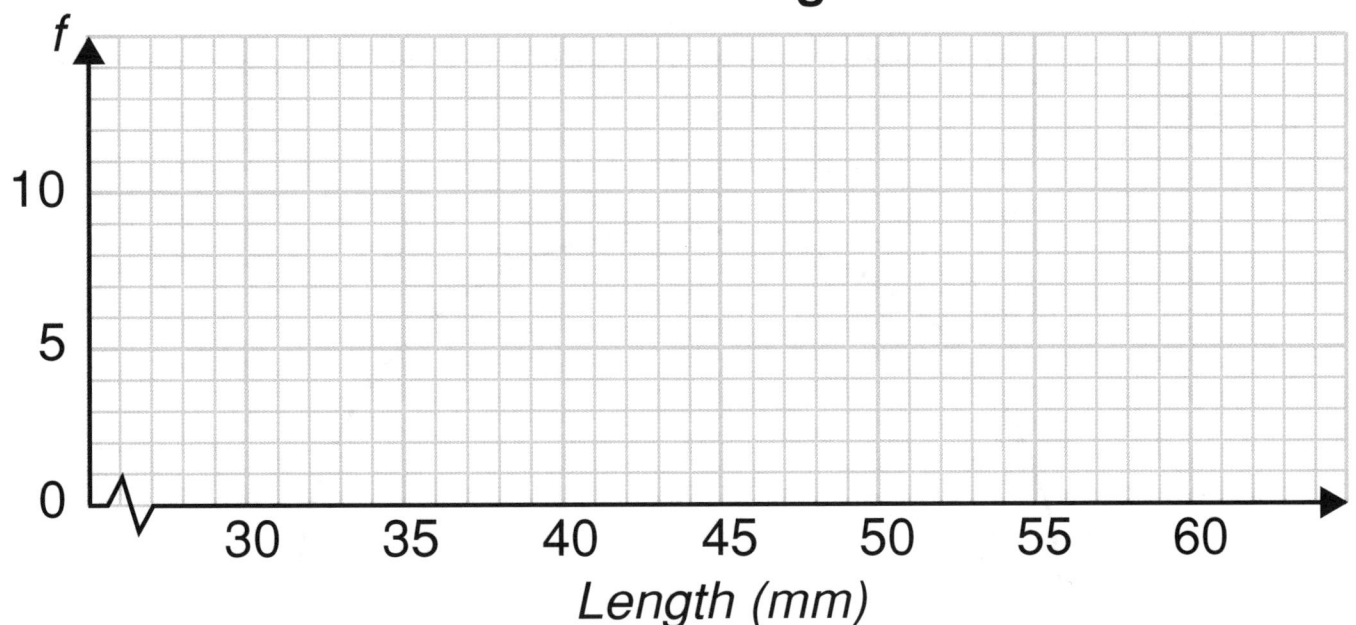

Number of videos watched	Frequency	Fraction of pie chart	Angle
0			
1			
2			
3			
4			
5			
>5			

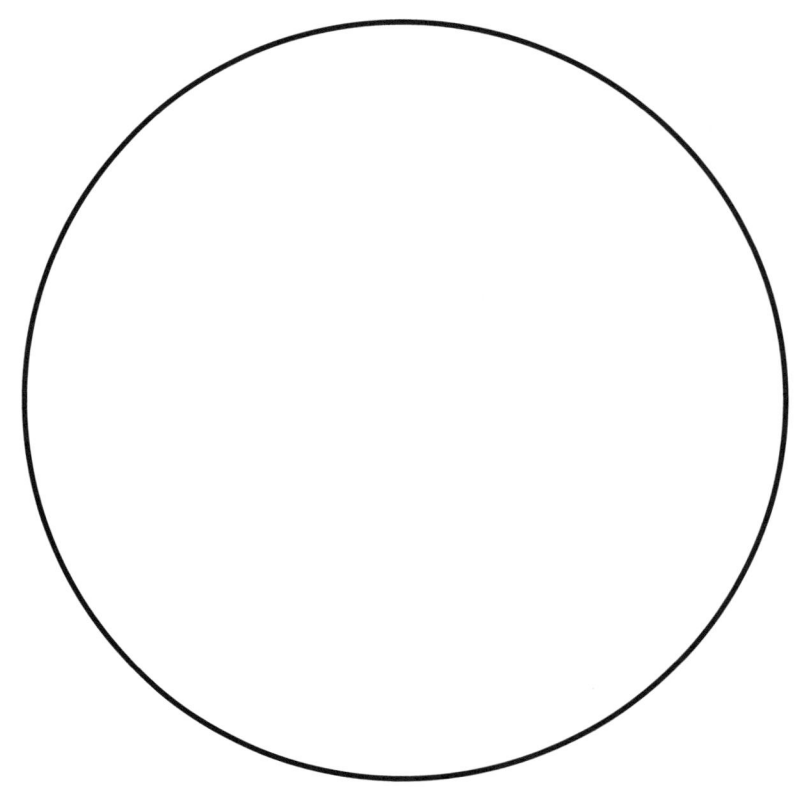

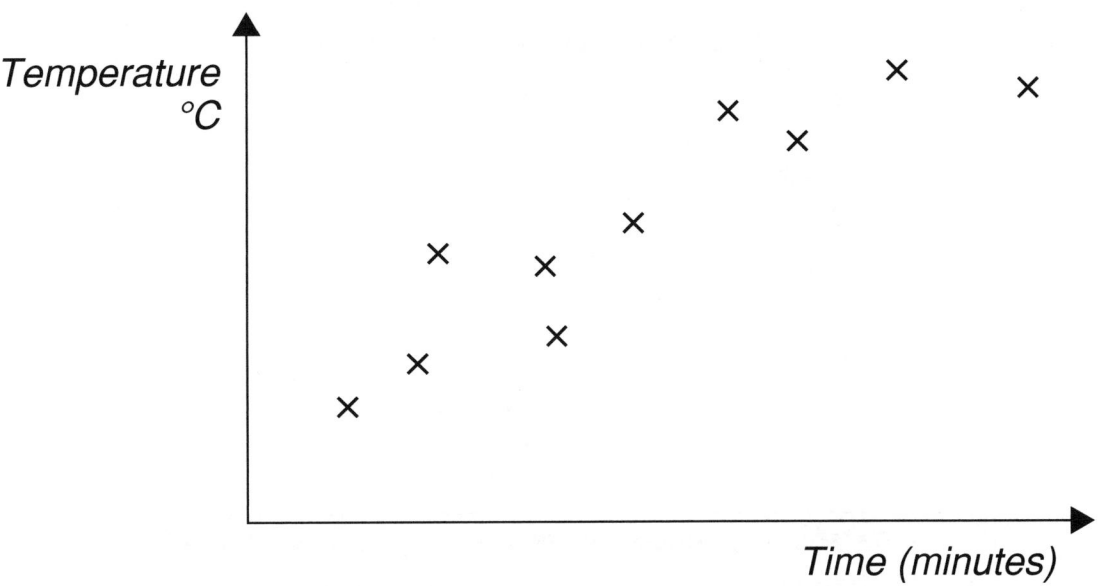

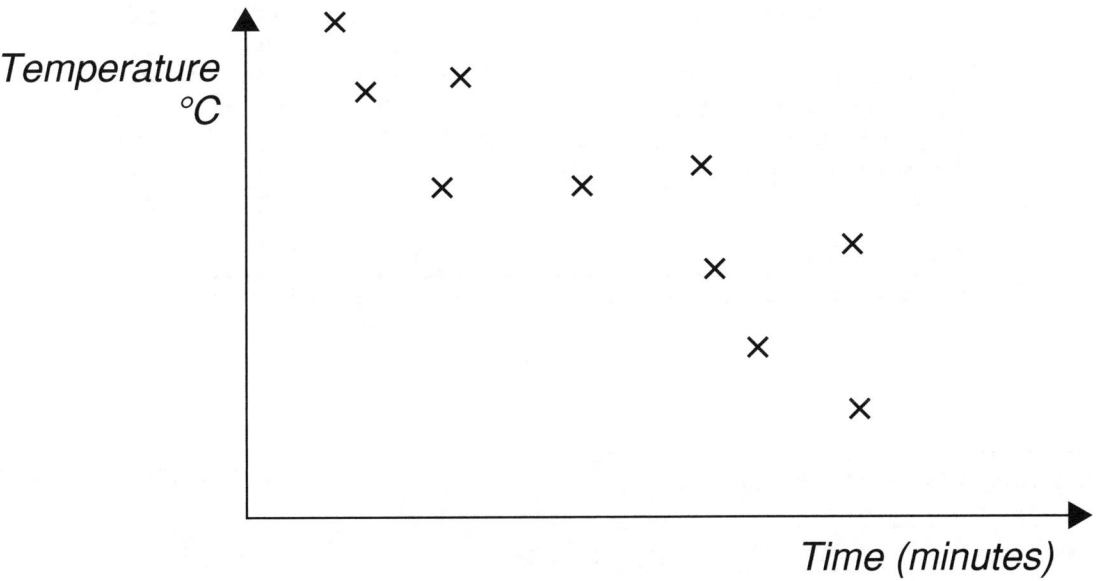

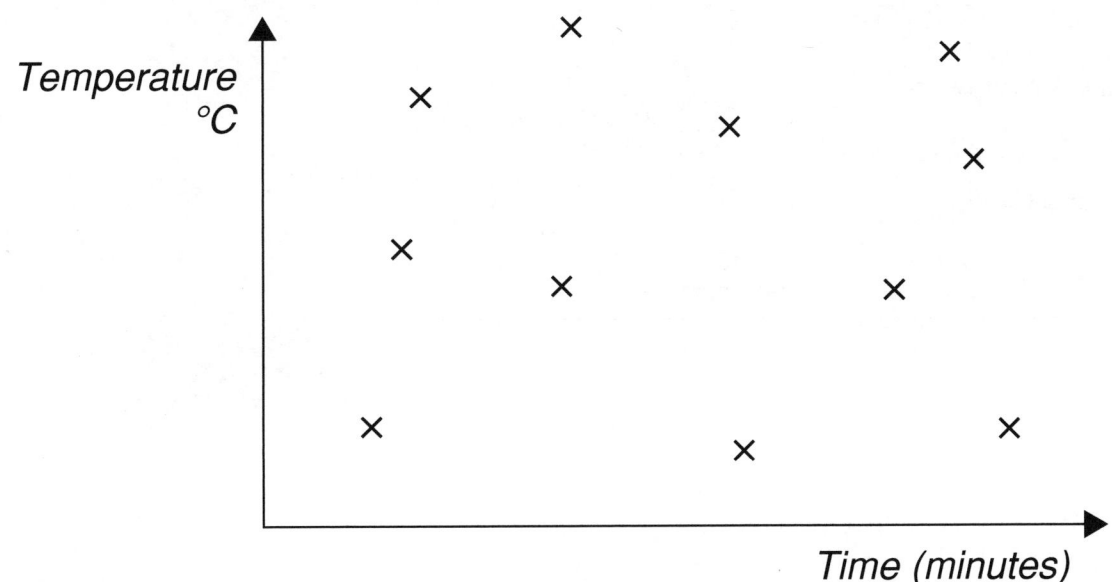

Fat consumption in Great Britain

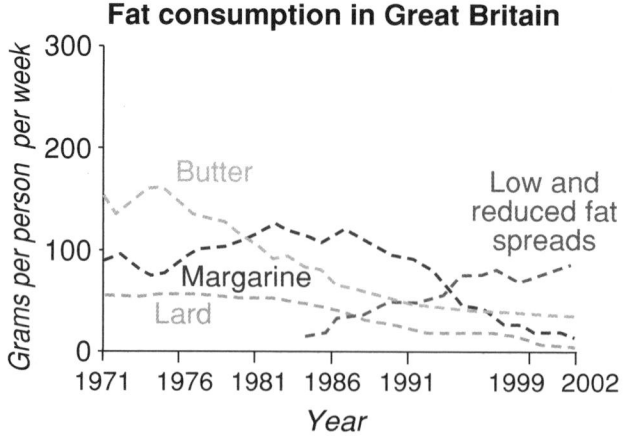

Reason why school pupils tried to give up smoking			
Current smokers			*England*
Why tried to give up	Boys	percentage Girls	Total
Worried about my health	47	55	52
Cost	28	35	32
To make me feel fitter	32	19	24
My family/friends persuaded me	12	10	11
Smoking made me smell or look nasty	7	10	9
Did not like/enjoy it	4	9	7
Other	12	13	13
Bases (=100%)	137	210	347

Percentages total more than 100 because some pupils gave more than one answer

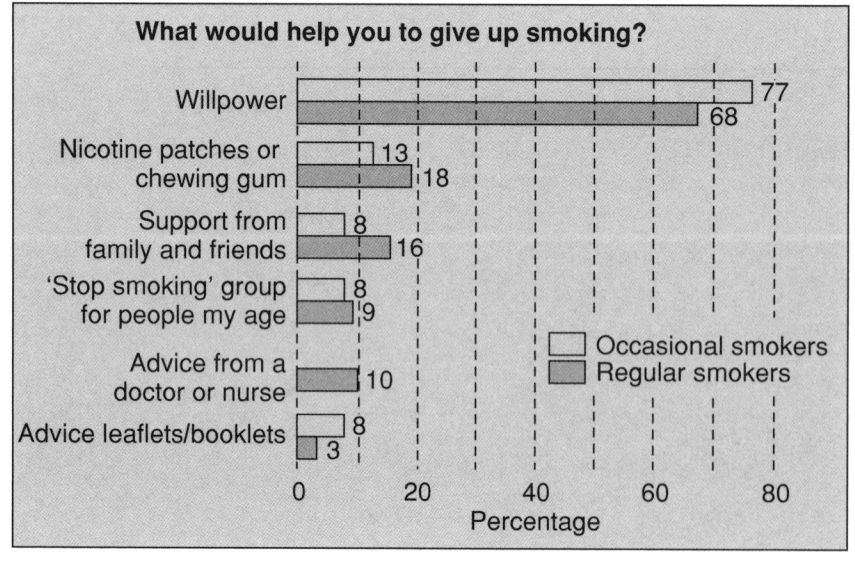

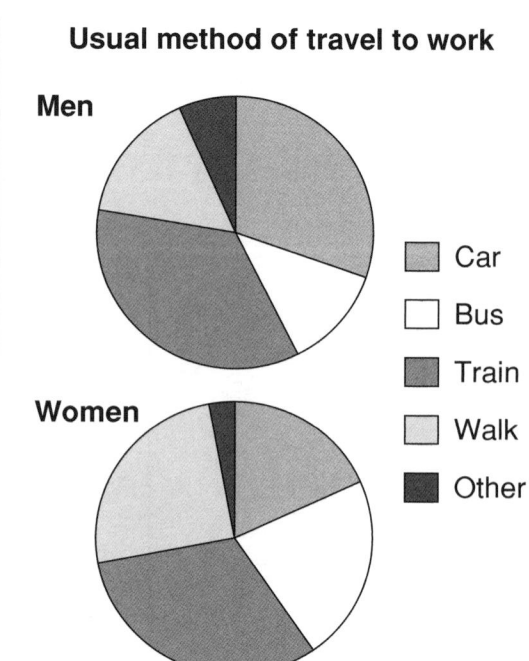

Usual method of travel to work

14

Key
G Green
R Red
B Blue

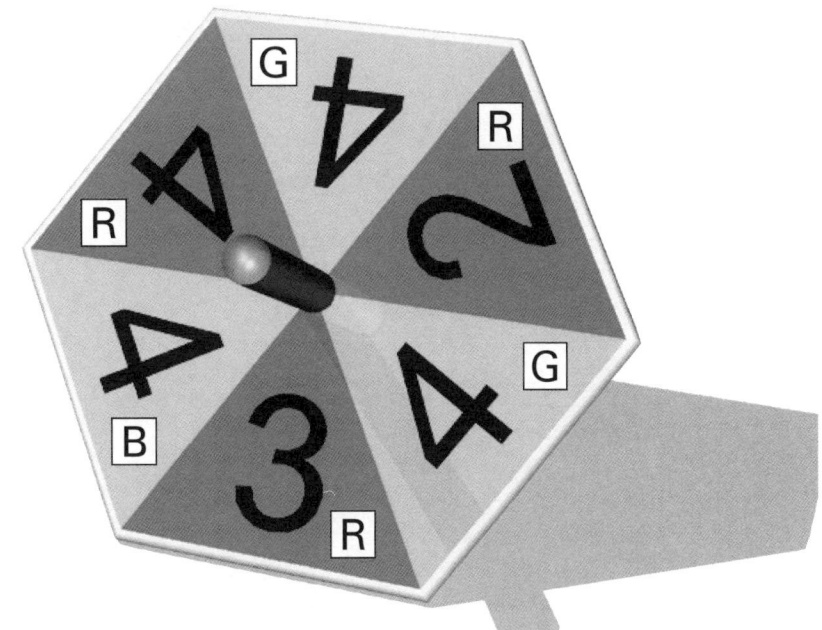

This spinner is spun.
Use one of 'certain', 'better than even chance', 'even chance', 'less than even chance', 'impossible' to describe the likelihood of the spinner stopping on
a Red **b** Green **c** Yellow **d** 4 **e** Red 4 **f** Green 2
g a number less than 5.

18

Key
G Green
B Blue
P Purple
R Red

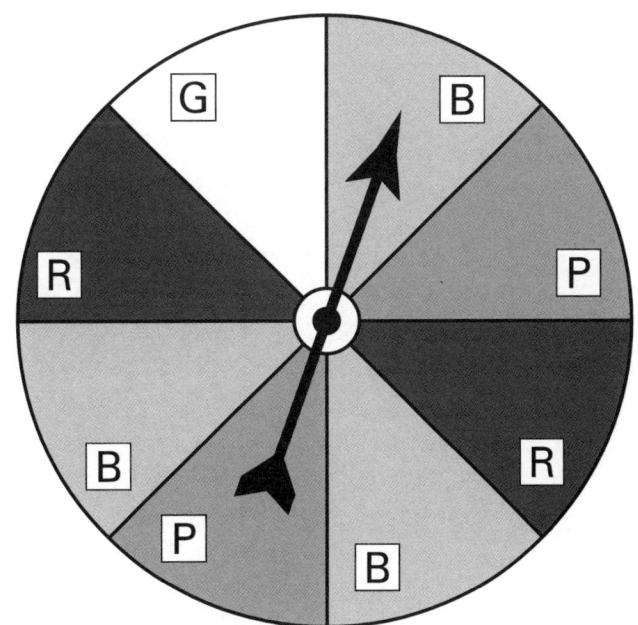

This spinner is spun.
a Which colour is it least likely to stop on?
Give a reason for your answer.
b Copy this scale.

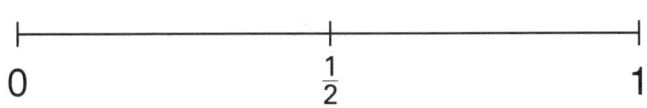

0          $\frac{1}{2}$ 1

Mark with an X the probability that the colour will be blue.
c Write down the probability that the colour will be red.

New National **Framework**

MATHEMATICS **8**

Teacher Resource Pack

M. J. Tipler   K. M. Vickers

nelson thornes

New National Framework

MATHEMATICS 8

Teacher Resource
Pack

Jocelyn Douglas

Maureen Hayes

David Miller

Peter Sherran

Maryanne Tipler

Kathy Vickers

Published in 2003 by:
Nelson Thornes Ltd
Delta Place
27 Bath Road
CHELTENHAM
GL53 7TH
United Kingdom

03 04 05 06 07 / 10 9 8 7 6 5 4 3 2 1

A catalogue record for this book is available from the British Library

ISBN 0 7487 7886 1

Page make-up by Mathematical Composition Setters Ltd.

Printed and bound in Great Britain by Antony Rowe Ltd.

Acknowledgements
The author and publisher are grateful for the following contributions to this Teacher Resource Pack.
Worksheets and Chapter Reviews, Maureen Hayes
ICT Support, David Miller
Graphical Calculator Support, Peter Sherran

Contents

Introduction

This *Teacher Resource Pack* has been developed to support and extend material covered in the *New National Framework Mathematics 8 Core* book. It provides you with everything you need to resource your work to implement the requirements of the *National Curriculum, Framework* and *Medium-Term Plans*.

Worksheets

These can be used to provide additional consolidation and practice to support work in the four Support sections of the book. If pupils are having difficulty with the practice questions in these chapters, further practice material is provided in these worksheets.

Chapter Reviews

This section provides a complete set of photocopiable write-on question sheets that can be used for additional practice, extension and consolidation work from the chapters. They could also be set for homework or used as assessment material.

ICT Support

The ICT section of this file gives full and comprehensive teachers' notes on the use of spreadsheets, dynamic geometry packages (Geometer's SketchPad) Logo (MswLogo) and a graphing package (Omnigraph). Worksheets are also provided.

Graphical Calculator Support

The Graphical Calculator section of this file gives notes on the use of the three most popular graphic calculators from Casio, Sharp and Texas. In the pupil books, some of the practical exercises require the use of a graphical calculator.

Pupil Book Answers

Full answers for *New National Framework Mathematics 8 Core* are provided.

3 $y = \boxed{}\, x$. What number could go in the box to give the graph of a line which is steeper than $y = 3x$?

4 Match these equations with the lines drawn on the grid.

  a $x = {}^-2$

  b $y = 3$

  c $y = {}^-2$

  d $y = x$

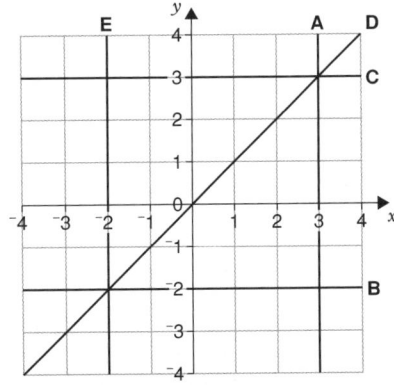

Graphs of real-life situations

1 A car-hire firm charges £25, plus £8 for every 100 miles driven.

  a Copy and complete this table.

Miles driven	0	100	400
Charge (£)			

  b On a copy of this grid, plot the points from the table.
     Draw a straight line through the points.

  c Show how to use your graph to find the charge if the hire
     car was driven 250 miles.

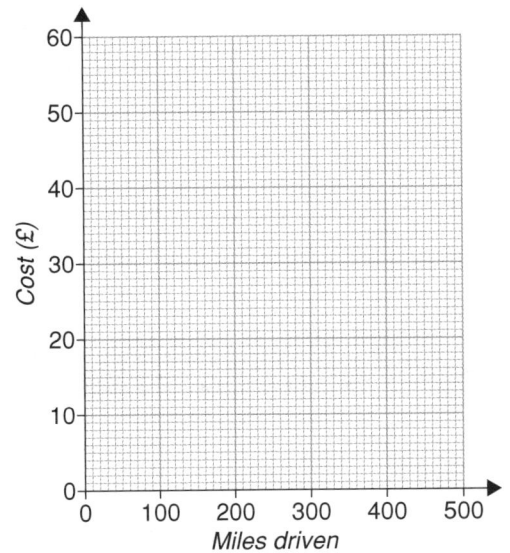

2 Bernie and Max are painting fence panels. Bernie paints 2 panels in 18 minutes.
   Max paints 3 panels every 40 minutes.

  a Copy and complete these tables.

Number of panels painted by Bernie	2	10	20
Time in minutes			

Number of panels painted by Max	3	15	30
Time in minutes			

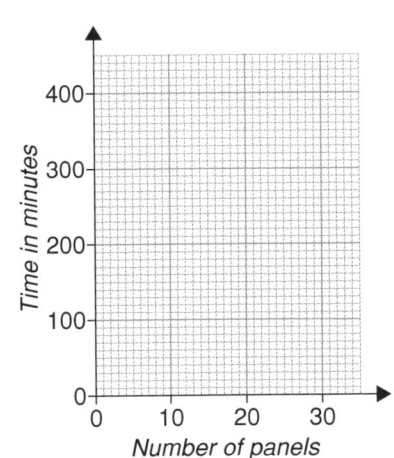

  b Draw and label the graphs showing how long it takes to
     paint the panels for both Bernie and Max.

  c Use the graphs to find how long it takes each of them to
     paint 25 panels.

  d Find how many panels they each paint in 100 minutes.

Lines

1 **a** Estimate the lengths of these lines to the nearest centimetre. Then measure them accurately.

b Name two pairs of lines that are perpendicular.

c Name a pair of lines that are parallel.

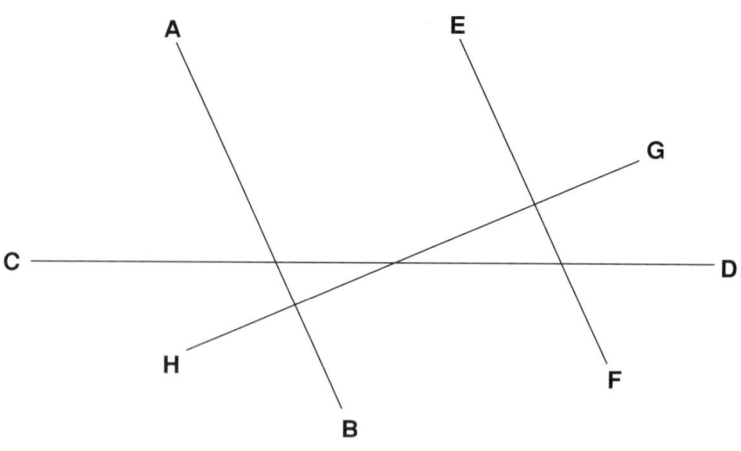

2 **a** Draw a line, AB, 7 cm long.

b Use your ruler and set square to draw a line through B, 4 cm long and perpendicular to BA. Label the line BC.

c Draw a line through C, parallel to AB, 5 cm long. Label it CD.

d Join AD. What shape have you drawn?

3 Look at the diagram and write down all the lines that are

a horizontal

b vertical

c parallel but not horizontal or vertical

d perpendicular but not horizontal or vertical.

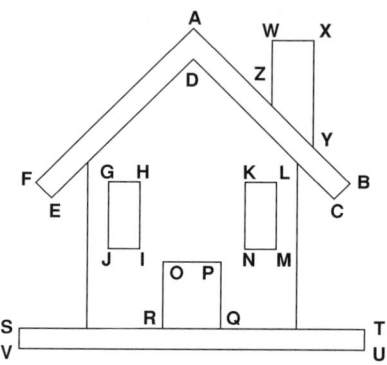

Angles

1 Name the angles shown, using one letter for each.

a 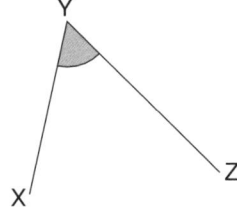   **b** 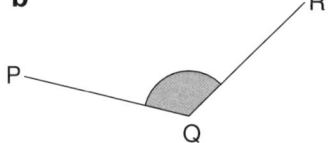   **c**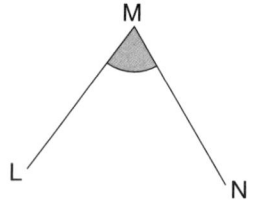

2 Name the angles shown in **question 1** using three letters for each.

3 From the diagram, write down all the named angles that are

a obtuse   **b** acute

c reflex   **d** right angles.

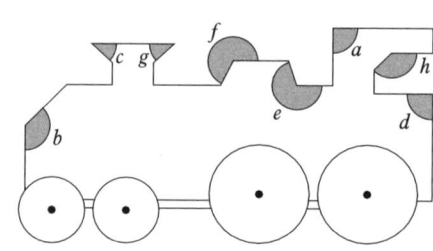

4 In the diagram in **question 3**, measure the size of angles a, b, c and e.

5 Name the shaded angles.

a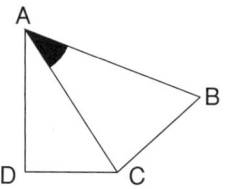

b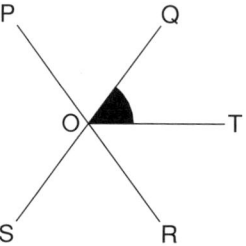

6 Measure these angles using a protractor.

a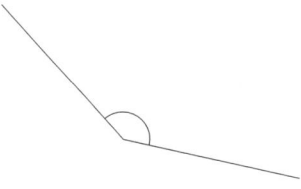

b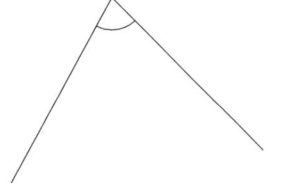

c

7 Draw these angles.

a 47°   **b** 134°   **c** 222°

8 **a** Using a ruler, draw a line LM, 5·5 cm long.

b Using a protractor, draw a line LN, such that angle NLM is 36° and LN is 4·5 cm long.

c Join NM. Measure and write down the size of angle NML.

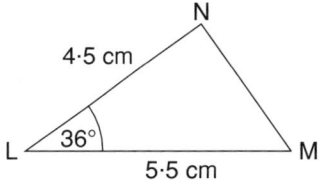

Calculating angles

1 Find the angles marked with letters.

a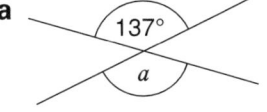

b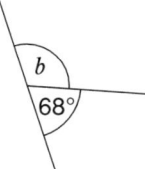

c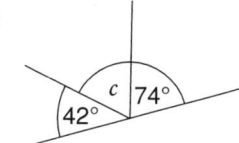

d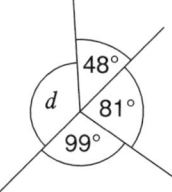

e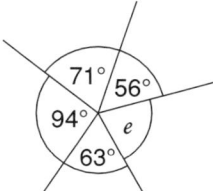

f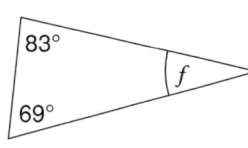

g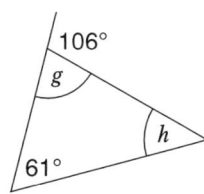

h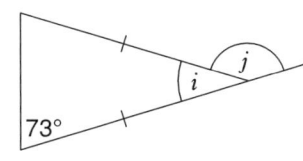

i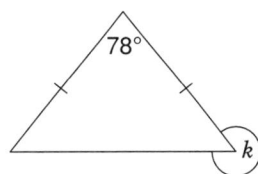

2 Find the value of the angles marked with letters.

a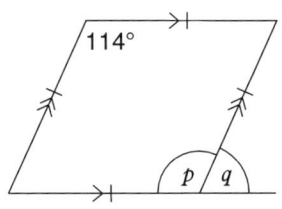

b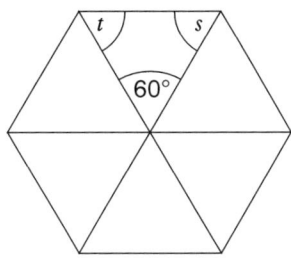

Regular hexagon

SHAPE, SPACE AND MEASURES

2-D shapes

The box contains:

A Square
B Isosceles triangle
C Parallelogram
D Equilateral triangle
E Rectangle
F Scalene triangle
G Rhombus
H Kite
I Right-angled triangle
J Trapezium

1 Choose the best name from the box for each of these.

a 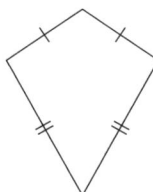   **b** 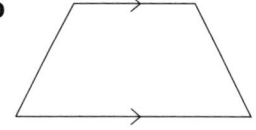   **c**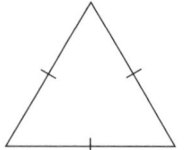

d 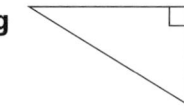   **e** 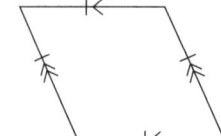   **f**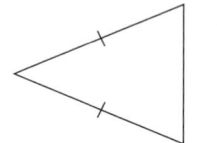

g    **h** 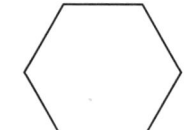   **i** 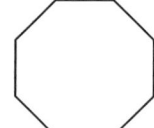   **j**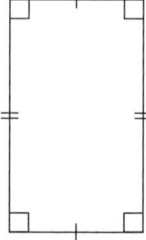

2 Name these polygons.

a 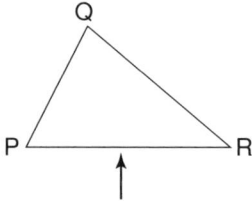   **b** 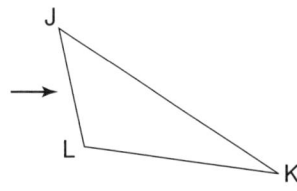   **c**

d Which of **a**, **b** and **c** are regular polygons?

3 Name these triangles, using capital letters.

a   **b**

4 Using a lower-case letter, name the side indicated by the arrow in each triangle in **question 3**.

SHAPE, SPACE AND MEASURES

Constructing triangles

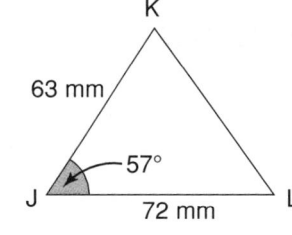

1 Use compasses, ruler and protractor to accurately construct this triangle.
On your drawing measure the size of ∠JLK. Give your answer to the
nearest degree.

2 Use compasses, ruler and protractor to accurately construct this
triangle.
On your drawing measure the length of AC. Give your answer to the
nearest millimetre.

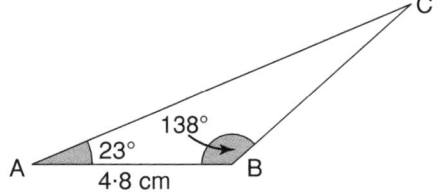

3 The diagram shows the shape of a company logo.

Use a ruler, set square and protractor to draw the logo accurately.

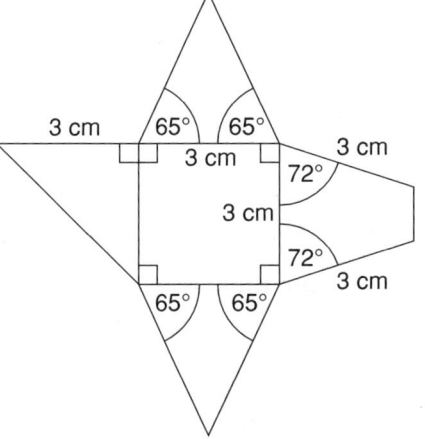

Coordinates

1 Draw a grid, with each axis labelled from ⁻6 to +6.

  a Plot these points and join them, in order, to make a shape.

    (3, 1) (1, ⁻2) (⁻1, 1) (1, 4)

  b Name the shape.

  c How many axes of symmetry does it have?

  d (⁻4, 6) (⁻5, ⁻2) and (3, ⁻3) are three vertices of a square.
Plot them and write down the coordinates of the fourth vertex.

SHAPE, SPACE AND MEASURES

3-D shapes

1 Name these shapes.

a 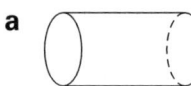 **b**  **c** 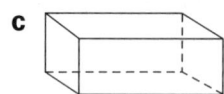 **d**

e 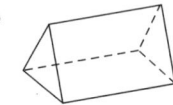 **f**

2 **a** How many faces does this shape have? **b** How many edges does it have?

c How many vertices does it have? **d** Name the shape.

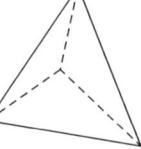

3 Imagine four cubes. Two cubes are green and two cubes are yellow.
Put them together in a line, with the two green cubes in the middle.

a How many yellow faces are showing?

b How many green faces are showing?

Now put the cubes together in a square, with the two green cubes at opposite corners.

c How many yellow faces are showing now?

4 **A** 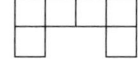 **B** 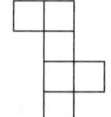 **C**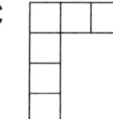

a Which of these nets will fold to make a cube?

b Draw another net that will fold to make a cube.

5 Imagine you are flying above a building shaped like a triangular prism. What is the maximum number of faces you can see?

6 Use a ruler, set-square and protractor to draw accurately a net for each of these 3-dimensional shapes.

a 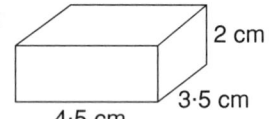 **b** 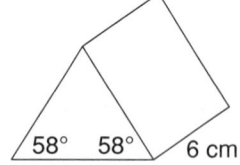 **c**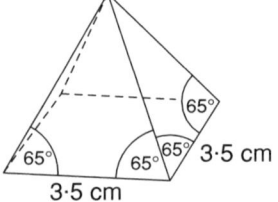

Transformations

1 Draw the reflections of these shapes in the dotted mirror lines.

a

b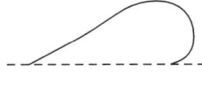

c

2 Which of the diagrams below show this shape 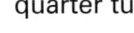 rotated through a

a half turn

b full turn

c quarter turn?

A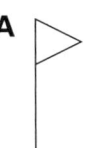

B

C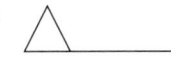

3 Draw the image after the translation 7 units left and 6 units up.

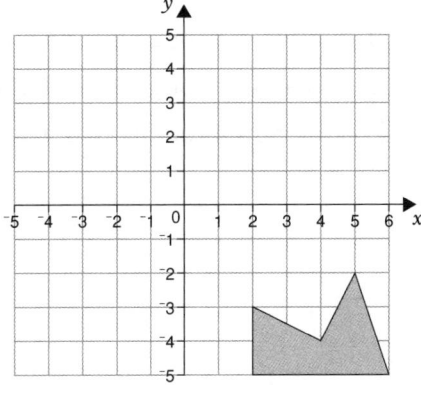

4 What is the inverse of the translation given in **question 3**?

5 Which of the diagrams below shows this shape after

a a rotation of 90° clockwise about (0, 0)

b a rotation of 180° about (0, 0)

c a translation 4 units left and 3 units down?

A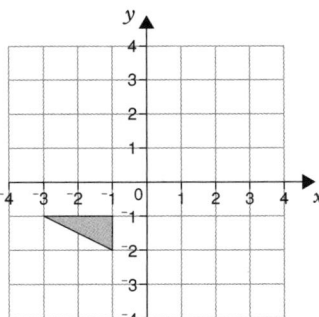

B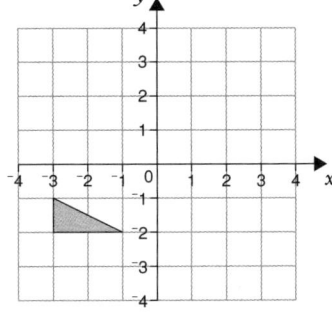

C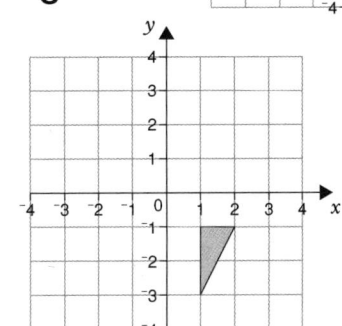

SHAPE, SPACE AND MEASURES

6 The centre of rotation and angle of rotation are given.
Draw the image shapes.

a

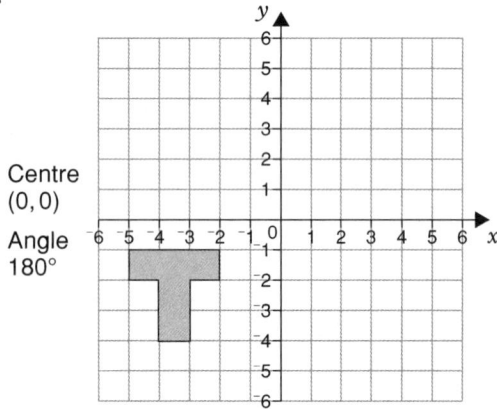

Centre
(0, 0)

Angle
180°

b

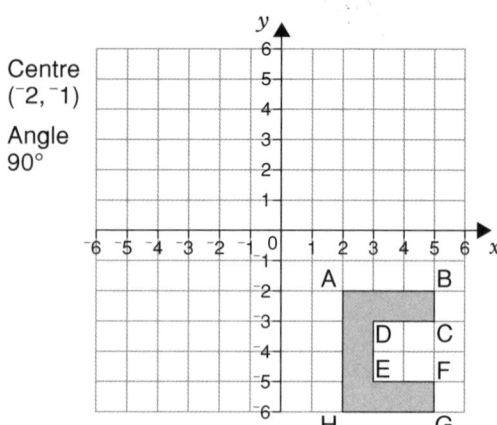

Centre
(⁻2, ⁻1)

Angle
90°

7 A rotation through 90° about the origin maps P onto P'.

Which of these maps P' onto P?

A a rotation through 270° clockwise about (0, 0)

B a rotation through 90° clockwise about (0, 0)

C a rotation through 270° about (0, 0)

D a rotation through 90° anticlockwise about (0, 0)

Symmetry

1 Do these shapes have rotation symmetry?

a

b

c

2 On each of them draw all the lines of symmetry.

a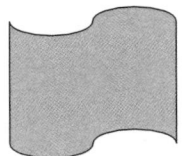

b

c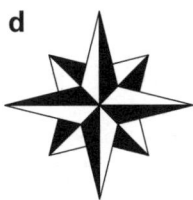

3 What is the order of rotation symmetry of these?

a

b

c

d

4 **a** I have a rectangle made out of paper. The rectangle measures 16 cm by 24 cm.
I fold the rectangle in half to make a smaller rectangle. What size could the smaller
rectangle be? Is this the only answer?

b I have a piece of paper in the shape of an equilateral triangle of side 12 cm.
If I fold it once along a line of symmetry, describe the shape I then have, giving any
lengths and angles you know.

5 Reece folded a square piece of paper in half, then in half again.
He made two straight cuts.
He opened the sheet of paper and it looked like this.
Use a copy of this folded sheet of paper.
Draw on it to show where Reece made his cuts.

SHAPE, SPACE AND MEASURES

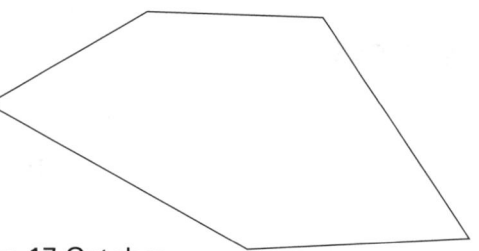

Measures

1 Use a ruler to measure the sides of this shape and add them to find its perimeter.

2 **a** John and Isobel went on holiday on 5 September and returned on 17 October. How many days long was their holiday, including these two dates?

 b Mohammed's favourite television programme begins at quarter to 7 in the evening and ends at 8:25 p.m. How long is the programme?

3 Find the missing numbers.

 a 2·78 ℓ = ___ mℓ **b** 429 g = ___ kg **c** 2·83 km = ___ m **d** 18·3 cm = ___ mm

 e 1·7 kg = ___ g **f** 2·9 ℓ = ___ cℓ **g** 176 cm = ___ m **h** 846 m = ___ km

 i 3284 mℓ = ___ ℓ **j** 84 cℓ = ___ ℓ **k** 42 mm = ___ cm **l** 8·3 m = ___ mm

4

___	___	___	___	___	___	___	___	___
1·3 cm	5 ℓ	40 mℓ	1 kg	48 km	8·5 m	8·5 m	5 ℓ	50 kg

L	___	___	___	___	___	___	___
1 mm	43 sec	1 kg	1·5 ℓ	1·3 cm	5 ℓ	40 mℓ	3·2 kg

___	**L**	___	___	___
75 cm	1 mm	11·5 cm	5 ℓ	160 g

Put the letter beside each measurement above its matching estimate in the box.

L Thickness of a protractor ≈ 1 mm   **A** Distance from Chester to Manchester

K Capacity of a lemonade bottle   **Y** Width of your thumbnail

W Mass of an apple   **I** Time to swim the length of a pool

R The mass of a new born baby   **E** The distance round a waist line

U The capacity of an egg cup   **N** The length of a garage

B The height of a cola can   **T** The mass of a sack of potatoes

O The capacity of a watering can   **C** The mass of a bag of sugar

5 Find the missing numbers.

 a 238 mm + 49 cm = ___ cm **b** 2·4 kg − 1862 g = ___ kg

 c 472 cℓ − 3·28 ℓ = ___ ℓ **d** 84·2 cm + 394 mm = ___ m

6 Find the measurements given by pointers **A**, **B** and **C**. In **d** you will need to estimate.

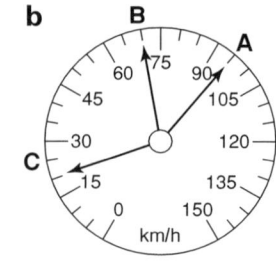

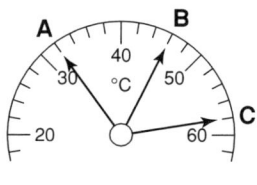

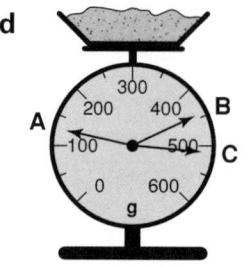

SHAPE, SPACE AND MEASURES

7   Choose the best range.

   a **A**   0 cm ⩽ Height of a dining table ⩽ 50 cm

   B   50 cm ⩽ Height of a dining table ⩽ 100 cm

   C   100 cm ⩽ Height of a dining table ⩽ 150 cm

   b **A**   0 kg ⩽ Mass of the Yellow Pages phonebook ⩽ 1·5 kg

   B   1·5 kg ⩽ Mass of the Yellow Pages phonebook ⩽ 3 kg

   C   3 kg ⩽ Mass of the Yellow Pages phonebook ⩽ 4·5 kg

8   **a** Give the rough metric equivalent for these.

   i 2 pints   **ii** 25 miles   **iii** 11 lb   **iv** 5 gallons

   b Give the rough imperial equivalent for these.

   i 4 ℓ   **ii** 36 ℓ   **iii** 64 km   **iv** 3 kg

9   Jane is organising a barbecue.

   a She needs 15 kg of charcoal. The charcoal is packed in 750 g bags.
      How many bags does she need?

   b She buys 20 litres of cola. Each can contains 800 mℓ. How many cans does she buy?

   c She needs 36 m of paper cloth to cover the tables. Each roll has 450 cm.
      How many rolls does she need?

   d She buys 12 kg of spare ribs. About how many pounds is this?

Compass directions

1   The diagram shows a network of paths in a local park.
   There is a fountain in the centre at I. Each junction of paths is labelled
   with a letter.

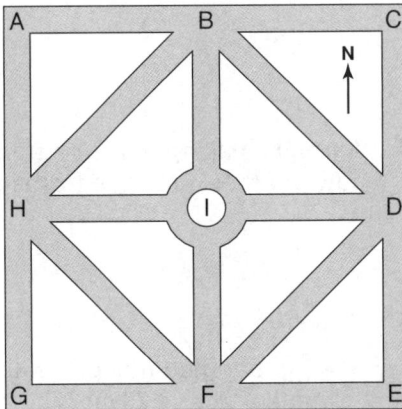

   a Luke walks from the fountain to D. In what direction does he walk?

   b When he arrives at D,

   i   what junction is north of him?

   ii what junction is south of him?

   c Lanni walks from F to H. In what direction is she walking?

   d Theresa is standing at junction B.

   i   If she looks SW, what junction does she see?

   ii If she looks SE, what junction does she see?

   e Damon starts at junction A and walks south to the next junction.
      Then he walks NE to the next junction. Finally he walks SE to the next junction.
      Where is he now?

2   **a** Beth was facing south. She turned a $\frac{1}{2}$-turn clockwise, then a $\frac{3}{4}$-turn anticlockwise,
      then a $\frac{1}{4}$-turn clockwise. What direction is she facing now?

   b Joseph was facing west. He turned a $\frac{1}{4}$-turn clockwise, then a $\frac{1}{2}$-turn anticlockwise
      then a $\frac{3}{4}$-turn anticlockwise and finally a $\frac{1}{2}$-turn clockwise. What direction is he facing now?

HANDLING DATA

Worzheet 34 8 Core

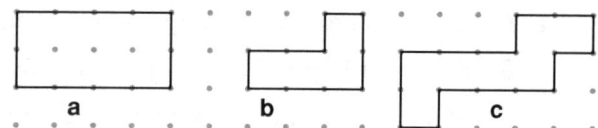

Perimeter, area and surface area

1 The distance between dots is 1 cm.
Without measuring, find the perimeters of the shapes.

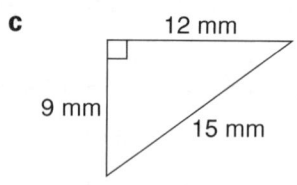

2 What is the area of each of the shapes in **question 1**?

3 Find the perimeters of these.

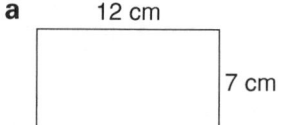   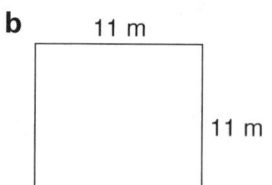

a 12 cm, 7 cm

b 11 m, 11 m

c 12 mm, 9 mm, 15 mm

4 Find the area of each of the shapes in **question 3**.

5 Choose a suitable unit from the box to estimate the area of these.

mm²   cm²   m²

  a The area of a page from a book

  b The area of a shirt button

  c The area of a car park

6 Find the surface area of each of these shapes.

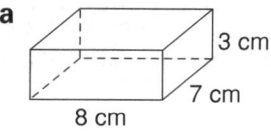

a 3 cm, 7 cm, 8 cm

b 13 cm, 12 cm, 8 cm, 5 cm

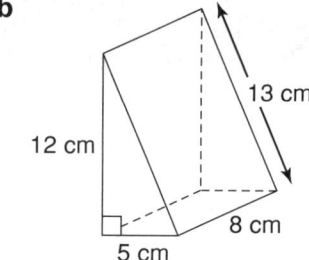

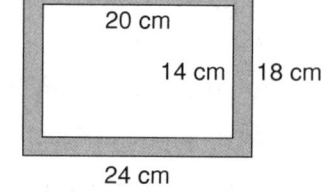

7 Aaron is making a picture frame. He begins with a rectangular piece of card and
cuts another rectangle from it. What is the area of the picture frame?

20 cm, 14 cm, 18 cm, 24 cm

8 Draw

  a a right-angled triangle, with one side AB
    and an area of 20 cm²

  b Make a copy of the grid and draw an
    isosceles triangle, with one side AB and
    an area of 12 cm².

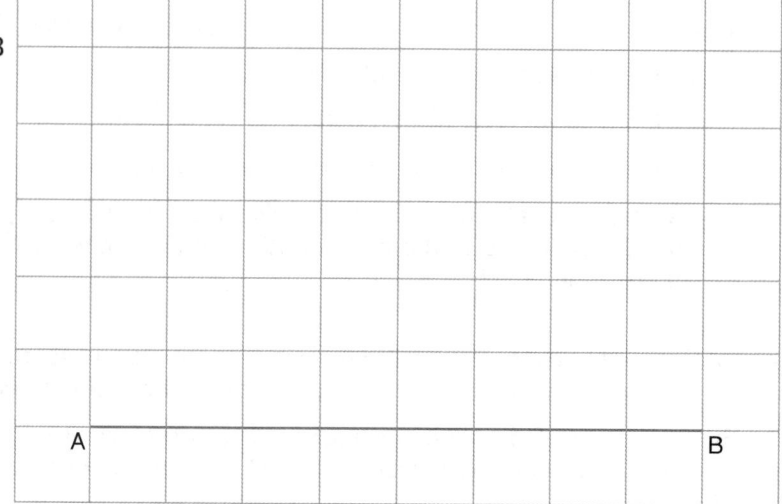

Planning and collecting data

1 Bindya collected this data to find out which would be the most popular place for a Year 8 day out.

   a Fill in the frequency column.

   b What was the most popular place?

Place	Tally	Frequency
Alton Towers	₩₩ ₩₩ ₩₩ I	
Blackpool	₩₩ ₩₩ III	
Castleton Caves	₩₩ III	8
Jorvik Viking Museum	₩₩ ₩₩ I	
Llandudno	IIII	

2 Ross recorded the number of letters in each word in a paragraph of his book.

```
3  5  2  1  3  4  7  2  1  5

8  3  1  4  3  3  2  6  5  4

1  3  4  3  5  2  7  8  3  4
```

Ross drew a tally chart to show the number of letters in each word. Finish Ross's chart.

Number of letters	Tally	Frequency
1		
2		
3		
4		
5		
6		
7		
8		

3 For each of these questions, choose one of these three methods of collecting the data.

   A Questionaire or data collection sheet

   B Experiment

   C Secondary source, such as website, book, newspaper, CD-ROM ...

   a Do more male pupils like PE lessons than female pupils?

   b Do more people die of heart attacks in city areas than in the country?

   c Do pupils have a shorter memory in the morning than in the afternoon?

4 Aidan wanted to know if Year 8 boys can walk backwards faster than Year 8 girls.

   a Write down two things Aidan might find out from his survey.

   b What data does Aidan need to collect?

   c Design a collection sheet for this data.

   d How could Aidan collect the data?

Displaying and interpreting data

1 This bar-line graph shows sales of shoes from Thompson's shoe shop one day.

   20 of size 7 were sold.

   26 of size 8 were sold.

   21 of size 9 were sold.

   a Finish this bar-line graph.

   b How many size 5 were sold?

   c How many size 4 were sold?

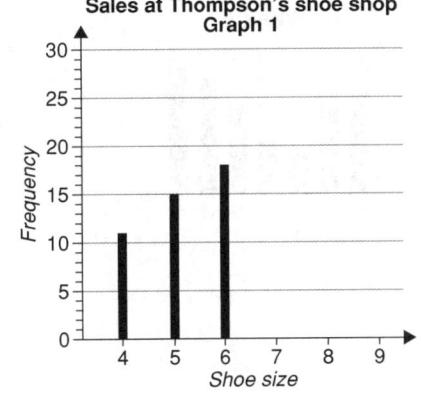

Sales at Thompson's shoe shop
Graph 1

HANDLING DATA

This bar-line graph shows sales of shoes from the same shoe shop another day.

The data was collected on two days, one at the beginning of June and another at the end of August.

d Look at the two graphs.

Which do you think represents sales on a day at the end of August?

How can you tell this from the graphs?

Sales at Thompson's shoe shop
Graph 2

2 This diagram shows how many pupils in Daryl's class like tea and coffee.

a How many like coffee, but not tea?

b How many like coffee and tea?

c How many do not like either coffee or tea?

d How many pupils are there in Daryl's class?

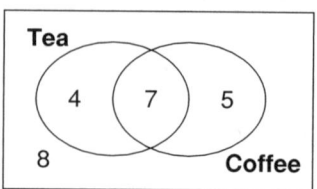

3 These pictograms show the numbers of different types of animals treated on one day at two different veterinary practices.

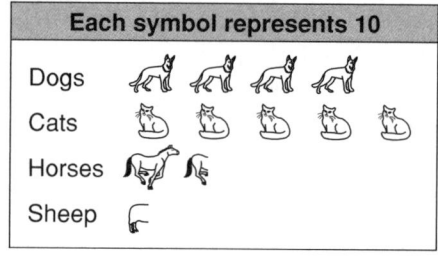

Owley Grange

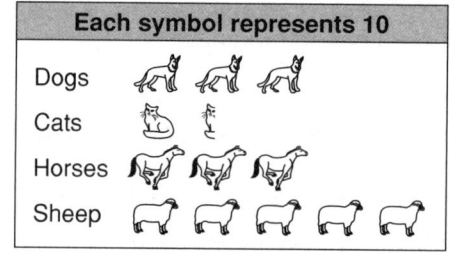

Manor House

a How many dogs were treated at Owley Grange?

b How many more cats were treated at Owley Grange than at Manor House?

c One of the veterinary practices was in a country village and the other was in a large town. Which do you think was which? Explain.

4 These graphs shows sales from a bookshop on three different weeks of one year.

Match these statements to the bar charts.

a Sales were good early in the week, then fell off so much that Mr Jasper did not open the shop at all on Sunday.

b Sales dipped in the middle of the week, but picked up considerably by the weekend.

c Sales were fairly consistent throughout the week.

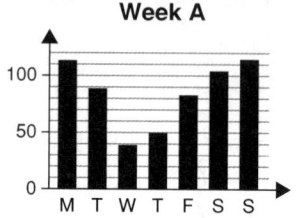

Week A

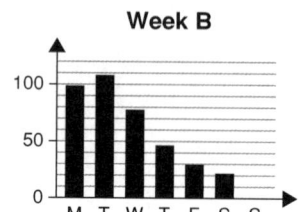

Week B

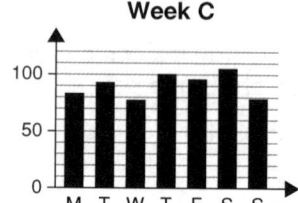

Week C

HANDLING DATA

5 Bryony did a survey to find out what pupils in her school did for lunch one day.

Her results are shown in the pie chart.

a What did most pupils do for lunch?

b What did fewest pupils do?

c Are these sentences true? Write Yes or No.

   A About 25% had a school dinner.

   B More pupils had a packed lunch or went to the fish and chip shop than had lunch at home.

   C About twice as many pupils had lunch at home as those who went to the fish and chip shop.

   D More than half the pupils had a school dinner or a packed lunch.

6 Kirsty did a survey to investigate whether more men than women wear glasses.

She summarised the data in this Carroll diagram.

	Men	Women	
	73	49	**Wearing glasses**
	86	64	**Not wearing glasses**

a How many people were surveyed altogether?

b How many of those surveyed were wearing glasses?

c How many men were in the survey?

d How many women who were not wearing glasses did Kirsty record?

7 This table shows part of a database about drinks.

Name	Volume	Can/bottle	Fizzy/still	Cost	Value for money rating (1–10)
Zipfoam	200 mℓ	Bottle	Fizzy	75p	4
Chocmilk	180 mℓ	Bottle	Still	59p	7
Dr Pippas	230 mℓ	Can	Fizzy	45p	5
Orangerine	210 mℓ	Can	Fizzy	70p	9
Currantina	200 mℓ	Bottle	Still	90p	6
Fizzpop	160 mℓ	Bottle	Fizzy	36p	8

a List the drinks that cost more than 60p.

b List the drinks that contain less than 200 mℓ.

c List the bottled drinks with a value for money rating of more than 5.

8 This graph shows the air temperature taken every 4 hours one day.

a What time did the temperature first register as 6 °C?

b What was the temperature at midday?

c Estimate the temperature at 10 p.m.

   Explain why this may not be accurate.

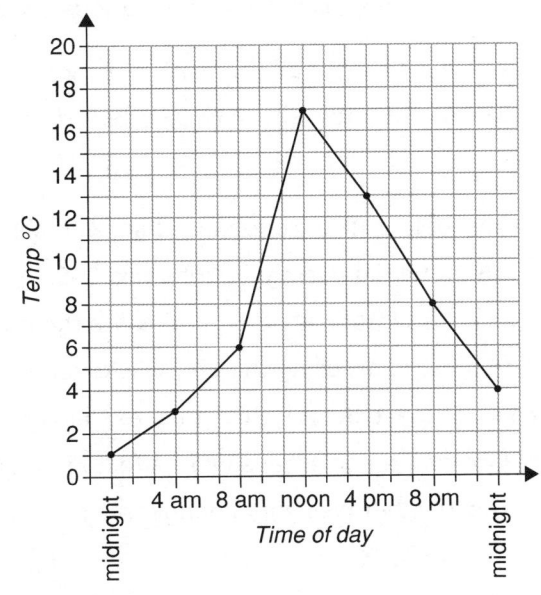

HANDLING DATA

9 The bar chart shows the amount spent one month by an English and a French family on household expenses.

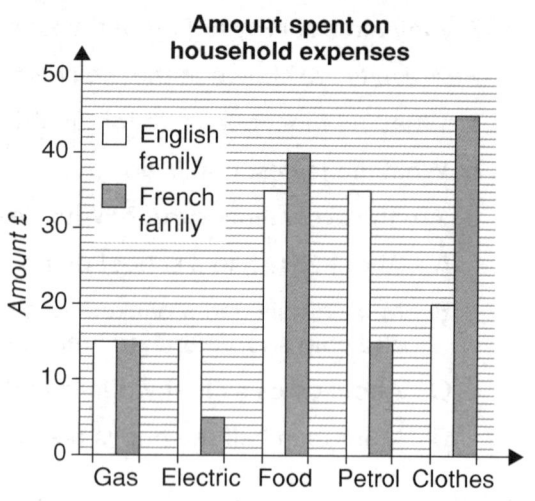

Amount spent on household expenses

☐ English family
▨ French family

Gas Electric Food Petrol Clothes

Amount £

a On which household expenses did the French family spend more than the English family?

b The English family spent less on food than the French family.

Which of these is the most likely reason?

A The English family don't eat very much.

B Food is generally cheaper in England.

C There are fewer children in the English family than the French family.

c Use the graph to write two sentences which compare the amounts spent by the two families.

10 Some boys and girls were asked what kind of takeaway food they preferred. These pie charts show the results. Morwenna looked at these charts and decided that more girls preferred Chinese takeway compared to the number of boys. Explain why she may not be right.

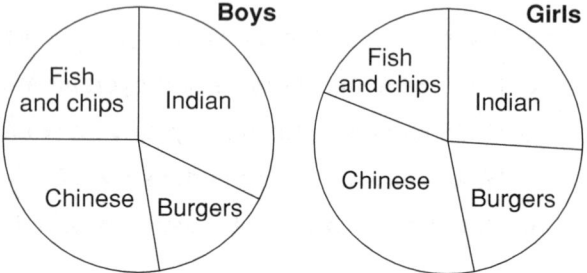

Boys

Fish and chips | Indian | Chinese | Burgers

Girls

Fish and chips | Indian | Chinese | Burgers

11 a This frequency chart shows the number of sunny days per month for six months in 1974 and 1998. Draw a bar chart to show the data.

Month	May	June	July	August	September	October
1974	7	15	14	12	8	9
1998	5	11	9	16	10	11

b Write a few sentences comparing 1974 and 1998.

Grouped data

1 The number of apples picked from various trees in a fruit farm were:

27 32 14 19 28 37 16 24 17 20
34 18   6 12 18 22 26 33 19 16
14 38 29 24 23 13 17 21   3 18

a Put this data onto a frequency chart with class intervals 1–5,   6–10,   11–15,   16–20, ...

b Regroup the data using the intervals 0–9,   10–19,   20–29,   30–39.

c Which class intervals do you think are the most useful? Explain why.

d Draw a bar chart using the intervals given in **b**.

e Which of these do you think best explains the shape of the graph?

A Two of the apple trees were affected by blight.

B The weather was very changeable during the growing season.

HANDLING DATA

Mode, median, mean, range

1 Mr Pritchard drew this table to show the number of cucumbers produced by his cucumber plants one summer.

a What is the modal number of cucumbers per plant?

b What is the range of the number of cucumbers?

c Why might Mr Pritchard want to know the modal number of cucumbers per plant?

Number of cucumbers	Tally	Frequency
0	II	2
1	III	3
2	︳IIII IIII	10
3	IIII IIII III	13
4	IIII IIII	9
5	IIII I	6

2 Mr Shah recorded the number of caravans staying overnight at his site during the summer months.

a What is the modal class?

b Draw a bar chart to show the data.

	Tally	Frequency
0–9	IIII II	7
10–19	IIII IIII III	13
20–29	IIII IIII IIII IIII IIII II	27
30–39	IIII IIII IIII IIII II	22
40–49	IIII IIII IIII II	17

3 Find the mean, median and range of each of these sets of data.

a 2, 13, 6, 4, 9, 8, 17, 7, 11, 14

b 511 kg, 862 kg, 713 kg, 476 kg, 622 kg, 413 kg, 825 kg, 570 kg

c £2·30, £1·96, £8·42, £4·17, £1·90

4 These are the masses of 7 oranges.

162 g, 147 g, 153 g, 164 g, 168 g, 142 g, 149 g

a What is the mean mass of an orange?

b What is the range of the masses?

c What is the median mass?

5 Gus counted the number of seconds six of his friends could stand on one leg without losing their balance. These are his results.

| 8 | 10 | 6 | 8 | ? | ? |

If Gus got a range of 7 and a mean of 9 for his data, what are the two missing numbers?

6 This table shows the number of children in the families of the pupils in class 8PF.

Find the mean number of children per family. Give your answer to 1 d.p.

Number of children per family	Number of families	Total
1	IIII	
2	IIII IIII I	
3	IIII II	
4	IIII	
5	II	
6	I	

7 This table gives the points scored by two golfers, Monica and Betty, over eight matches.

	Match 1	Match 2	Match 3	Match 4	Match 5	Match 6	Match 7	Match 8
Monica	34	37	35	36	38	35	36	39
Betty	36	38	32	39	42	37	34	40

a Find the range of scores for each golfer.

b Calculate the mean for each player. Give your answer to 1 d.p.

c Which golfer, Monica or Betty, would you choose for the team? Explain why.

HANDLING DATA

Probability

1 Put these events in order of likelihood. Put the most likely first.

A You will win an olympic medal one day.

B You will eat something today.

C You will get 100% in every test this year.

D A baby will be born somewhere in the world tomorrow.

E You will see a squirrel today.

2 Decide if each of these is certain, very likely, likely, unlikely, very unlikely or impossible.

a You will watch television tonight.

b Boiling water will reach a temperature of 200 °C.

c Someone in Britain will eat a boiled egg today.

d You will grow to a height of 200 cm in the future.

e Next week Tuesday will follow Monday.

f You will see a red car on your way home from school today.

3 Write no chance, poor chance, even chance, good chance or certain for each of these events.

a Someone in your road will win the lottery next Saturday.

b The next time you toss a coin, it will land on heads.

c You pick a red disc out of a bag containing 10 red discs and 1 blue disc.

d The next person you meet will have purple hair.

e You will get homework sometime this year.

4 The numbers from 1 to 10 are written on cards and put in a box.

One card is chosen without looking.

a Copy and finish this list of possible outcomes.   1, 2, 3, ...

b Does each number have an equal chance of being chosen?

c Is it equally likely an odd number or an even number will be chosen? Explain.

5 Mr Foster threw an unbiased coin three times and it landed on heads each time. If he throws the coin again, is it more likely to land on heads, tails or is there an even chance of either? Explain your answer.

6 A box contains these cards that are used in a game.

a If a card is chosen at random, which shape (circle, triangle or diamond) is most likely to be picked? Give a reason for your answer.

b Copy this scale.

```
|---------------------|---------------------|
0                     1                     1
                      2
```

Mark with an X the probability that the card will be a triangle.

c What is the probability that a diamond will be picked?

HANDLING DATA

7 This spinner is spun. What is the probability that it lands on

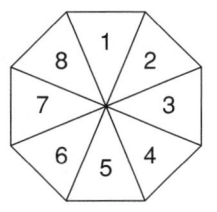

a an even number

b a number bigger than 3

c a number less than 9

d a 4 or an 8

e a number less than 1?

8 There are 20 cards in a box with letters of the alphabet on them. Cerys takes one out
without looking. She writes down the letter, then puts the card back in the box.
She does this 20 times. Cerys records her results in a chart.

letter	frequency
A	4
C	6
F	1
H	3
T	4
X	2

a Cerys thinks there must be equal numbers of cards with A and T on, because there
are equal numbers in the chart. Explain why she could be wrong.

b Cerys thinks that there are more cards with C on than with F on. Is she right? Explain.

c Cerys also decides that there cannot be a card in the bag with the letter M on because
there are none on the chart.

Explain why she could be wrong.

HANDLING DATA

Worksheet 1

1 **a** 4 hundreds or 400   **b** 4 ten thousands or 40 000
   c 4 tenths, $\frac{4}{10}$ or 0·4   **d** 4 thousandths, $\frac{4}{1000}$ or 0·004
   e 4 hundredths, $\frac{4}{100}$ or 0·04
2 **a** 2508·36   **b** 814 011·615   **c** 0·13   **d** 83·021
3 **a** 9778   **b** 25 785   **c** 897
4 **a** Add 1 g   **b** Subtract 0·1 g   **c** Add 0·03 g
   d Subtract 0·01 g   **e** Subtract 0·5 g
5 **a** Thirty-six thousand, eight hundred and two
   b Six hundred and five thousand and four point three seven
   c Two million, forty-one thousand, six hundred and eighty-nine point zero four eight.
6 **a** 2945   **b** 0·45   **c** 2·5   **d** 4·95
7 **a** 3·28   **b** 18·46   **c** 7·05   **d** 3·195
8 **a** Largest 8421, smallest 1248
   b Largest 975·3, smallest 3·579

Worksheet 2

1 **a** 430   **b** 1600   **c** 12 000   **d** 67
   e 830   **f** 14   **g** 1·289   **h** 0·02
   i 0·3074   **j** 860
2 **a** 24 cm   **b** 240 cm   **c** £100
3 **a** 100 tins   **b** £3700
4 **a** 100   **b** 1000   **c** 10   **d** 10   **e** 10

Worksheet 3

1 **a** False   **b** True   **c** True
2 **a** 2687, 2867, 6278, 6782, 7286, 8276
   b 0·46, 4·06, 4·6, 6·004, 6·04, 40·6, 46
3 284 762 miles
4 **a** =   **b** <   **c** >   **d** <   **e** >
5 **a** 2457
   b i 52   **ii** 27   **iii** 524   **iv** 2754
6 John 37·09 sec, Darren 37·4 sec, Philip 37·43 sec, Paul 38·04 sec, Daniel 38·2 sec, Hassan 39·55 sec, Malcolm 39·6 sec, Silvio 40·68 sec, Kirk 40·8 sec, Nigel 41·62 sec
7 0·407 kg, 0·418 kg, 0·43 kg, 0·48 kg, 0·5 kg, 0·509 kg, 0·527 kg, 0·56 kg, 0·602 kg
8 **a** <   **b** >   **c** >
9 **a** 643·1   **b** 641·3   **c** 1·364

Worksheet 4

1 **a** 780, 1250, 1000, 350   **b** 800, 1200, 1000, 300
2 285
3 **a** 96 500   **b** 97 000
4 **a** £7·74 to the nearest penny
   b 169 g to the nearest gram
5 **a** 7 615 000 km^2 **b** 7 625 000 km^2

Worksheet 5

1 A Trapezium
2 **a** 645, 1335
   b 1044, 612   **c** 216, 656

Worksheet 6

1

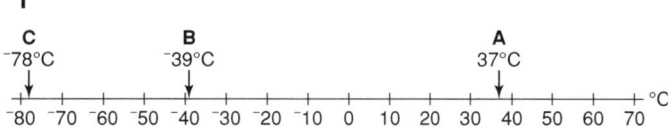

2 **a** $^-8\,°C$, $^-5\,°C$, $^-2\,°C$, $0\,°C$, $3\,°C$, $4\,°C$
   b 2·7, 1·9, 0·75, $^-$0·4, $^-$3·6, $^-$4·1
3 **a** £15   **b** $^-$£15
4 **a** 14 °C   **b** 4 °C   **c** 11 °C   **d** 25 °C
5 19 °C
6 $^-$75 m
7 $^-$34 m
8 **a** $\boxed{^-3}$ **b** $\boxed{^-3}$ + $\boxed{2}$

9

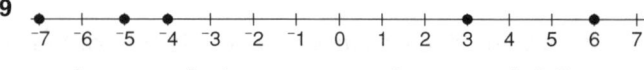

   a $^-6$ **b** 1   **c** $^-2$ **d** 0·5
10 **a** $2 + {}^-5 = {}^-3$ **b** $2 + {}^-9 = {}^-7$
       $2 + {}^-6 = {}^-4$
       $2 + {}^-7 = {}^-5$

11

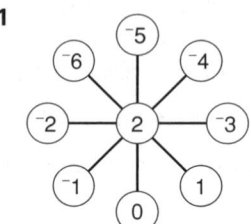

Worksheet 7

1 **a** 4, 8, 12, 16, 20   **b** 7, 14, 21, 28, 35
2 **a** 1, 12   2, 6   3, 4   **b** 1, 36   2, 18   3, 12   4, 9   6, 6
   c 1, 84   2, 42   3, 28   4, 21   6, 14   7, 12
3 2, 3, 5, 7, 11, 13
4 **a** 4, 16   **b** 2, 7, 11, 19   **c** 1, 2, 15, 30
   d 1, 4, 9, 16
5 **a** 1, 2, 3, 4, 6, 9, 12, 18, 36
   b 1, 2, 3, 4, 5, 6, 10, 12, 15, 20, 30, 60
   c 1, 2, 3, 4, 6, 12
   d 12
6 **a** 1, 3, 6, 10, 15, 21, 28, 36, 45, 55   **b** 36, 49
7 X
8 **a** 8   **b** 5   **c** 9   **d** 1   **e** 11
9 **a** 375, 573, 537, 735   **b** 753   **c** 735
10 C 1600
11 Possible answers are: 2 + 7 = 9, 5 + 11 = 16, 2 + 23 = 25, 3 + 13 = 16, 5 + 11 = 16,   5 + 31 = 36, 7 + 29 = 36 13 + 23 = 36, 17 + 19 = 36,   2 + 47 = 49
12 **a** Yes   **b** No   **c** No   **d** Yes
13 **a** 1·21   **b** 1·7

Worksheet 8

1 **a** 36   **b** 56   **c** 55   **d** 54   **e** 160
   f 240   **g** 6400   **h** 4500   **i** 3600   **j** 4200
2 **a** 6   **b** 4   **c** 9   **d** 9   **e** 90
   f 30   **g** 400   **h** 600   **i** 300   **j** 300
3 **a** 27   **b** 35   **c** 96   **d** 52   **e** 83
   f 35
4 **a** 13   **b** 66   **c** 7·3   **d** 0·7   **e** 0·37
5 1·47
6 £12·00
7 **a** 15   **b** 30   **c** 2   **d** 18   **e** 24
   f 28   **g** 1   **h** 120
8 **a** 54   **b** 43   **c** 15   **d** 47   **e** 40
9 **a i** £2·30   **ii** £2·70
   b Two teacakes cost less than £1 and each coffee costs less than £1, so the total should be less than £3.
10 **a** 1·2   **b** 1·4   **c** 1·1   **d** 3·1   **e** 7·1
11 **a** 48   **b** 520   **c** 1400   **d** 693   **e** 5·2
    f 4·8   **g** 90

12 a 90 **b** 423 **c** 0·9 **d** 0·8
13 a 5 **b** 7 **c** 7 **d** 10
14 a 7 **b** 102 **c** 1600 **d** 378 **e** 168

Worksheet 9

1 a C 200 + 400 **b** B 30 × 10
c A 30 − 10 **d** D 3 × 40
2 a 500 + 600 = 1100 **b** 300 − 90 = 210
c 20 × 40 = 800 **d** 500 ÷ 5 = 100
e 8 + 1 = 9 **f** 11 − 5 = 6
g 3 × 80 = 240
3 20 × £7 = £140, but both 18 and £6·99 have been rounded up for this estimation, so the real cost will be less than £140.
4 a 30 − 10 = 20 or 25 − 15 = 10
b 8 × 2 = 16 or 8·5 × 2 = 17
c 200 + 300 = 500 or 200 + 350 = 550
5 a 2 × 7 = 14, 16·1 **b** 9 × 3 = 27, 26·7
c 10 × 6 = 60, 72·84 **d** 20 × 4 = 80, 83·2
e 150 × 5 = 750, 739 **f** 200 × 8 = 1600, 1630·4

Worksheet 10

1 a 17 **b** 21 **c** 11 **d** 4
e 25 **f** 11
2 a 40 **b** 24 **c** 3 **d** 4
3 a (2 + 7) × 4 = 36 **b** (4 + 3 + 2) × 10 = 90
c (8 − 3) × (2 + 7) = 45 **d** 4 + (5 + 2) × 3 = 25
4 a 8·5 **b** 16·0 **c** 0·8 **d** 1·9

Worksheet 11

1 a 85 **b** 312 **c** 51
d 114 **e** 159 **f** 124
2 a 955 **b** 236 **c** 370 **d** 118
e 407
3 a 78 **b** 1888 **c** 1372 **d** 2466
4 a 1074 **b** 147·2 **c** 89·4 **d** 56·45
5 a 28 R 2 **b** 23 R 5 **c** 190 R 2 **d** 81 R 3
6 a 42 **b** 179 R 1 or $179\frac{1}{5}$
c 20 R 8 or $20\frac{8}{9}$ **d** 94 R 1 or $94\frac{1}{4}$
7 a 351 **b** 5832 **c** 2992 **d** 19 044
8 a 214 **b** 144 **c** 1134
9 a 5 × 3 = 15, 13·8 **b** 13 × 10 = 130, 114·3
c 1 × 8 = 8, 11·36 **d** 16 × 10 = 160, 111·3

Worksheet 12

1 a $\frac{1}{6}$ **b** $\frac{5}{8}$ **c** $\frac{2}{5}$
2 a £6 **b** 40 m **c** 7 cm **d** 20p
e £70
3 $\frac{4}{10}, \frac{20}{50}, \frac{10}{25}, \frac{6}{15}, \frac{14}{35}$
4 12
5 a 12 **b** 21 **c** 35 **d** 36
e 54 **f** 4 **g** 7 **h** 9
6 a $\frac{23}{100}$ **b** $\frac{2}{5}$
7 a $\frac{3}{4}$ **b** $\frac{3}{7}$ **c** $\frac{7}{9}$ **d** $\frac{8}{9}$
e $\frac{11}{16}$ **f** $\frac{5}{8}$ **g** $\frac{4}{15}$
8 a $\frac{18}{24}$ **b** $\frac{4}{9}$
9 a £11 **b** 30 cm **c** 15 m **d** 21 ℓ
e 40 kg **f** 12p
10 a $\frac{17}{25}$ **b** 68%
11 a $\frac{30}{60}$ **b** $\frac{15}{60}$ **c** $\frac{45}{60}$ **d** $\frac{42}{60}$
e $\frac{50}{60}$ **f** $\frac{35}{60}$ **g** $\frac{8}{60}$
12 a 9 **b** 25 **c** 27 **d** 44
e $4\frac{2}{3}$ **f** $17\frac{1}{2}$

Worksheet 13

1 a $\frac{1}{2}$ **b** 25%
2 a 0·5 **b** 0·75 **c** 0·3 **d** 0·37
e 2·25 **f** 0·03 **g** 0·2 **h** 0·43
i 0·65 **j** 0·07 **k** 1·5 **l** 1·09
3 a $\frac{7}{10}$ **b** $\frac{1}{4}$ **c** $\frac{2}{5}$ **d** $\frac{9}{25}$
e $\frac{17}{20}$ **f** $\frac{2}{25}$ **g** $\frac{18}{25}$ **h** $\frac{47}{50}$
i $\frac{11}{50}$
4 36
5 a 45 **b** 60 **c** 24 **d** 4
e 6 **f** 126
6 a 40% **b** 90% **c** 25% **d** 72%
e 35% **f** 250% **g** 85% **h** 10%
i 99% **j** 126%
7 a £167·70 **b** 25·92 kg **c** 18·2 m
8 Sam's Stores

Worksheet 14

1 25 white tiles
2 £4·96
3 a 3 : 2 **b** 40% **c** $\frac{3}{5}$
4 3 : 2
5 32
6 a 24 **b** 36 **c** 44 **d** 80

Worksheet 15

1

2 a 8n **b** 6x **c** 4(y + 3) **d** 2(10 − m)
e p^2 **f** $4t + q^2$ **g** $7(r^2 + 6)$
3 a n + 8 **b** n − 7 **c** 4 − n **d** 10n
e $\frac{n}{3}$ **f** 2n + 5 **g** 5(n − 4) **h** 3(n + 9)
i $n^2 − 1$
4 a m − 6 **b** y + 8 **c** p − 2 **d** 2x

Worksheet 16

1 a x + y = y + x **b** mn = nm **c** abc = cab
d t(u + v) = t(v + u) **e** a + (b + c) = (a + b) + c
2 a 3b **b** a **c** 9x **d** 10y
e q **f** 4m **g** 10t
3 a 1 **b** 1 **c** 3 **d** 4
e 11 **f** 2a **g** 4r **h** 3b
i 3f **j** 4
4 a 6 **b** 6 **c** 1 **d** 7
e 8 **f** 15 **g** 5
5 a 8 **b** 2 **c** 9 **d** 1
e 28 **f** 15 **g** 34 **h** 10

Worksheet 17

1 a £42 **b** £62
2 a 2 hours **b** 6 hours
3 a £4·60 **b** £10·50
4 a 28 cm **b** 100 cm
5 a £24·60 **b** £28

Worksheet 18

1 a 6 **b** 11 **c** 15 **d** 5
e 3 **f** 6
2 a C 2t + 6 = 30 **b** £12

3 a $x + 5 = 7$　**b** $x - 6 = 8$　**c** $4x = 28$
　d $4x = 260$　**e** $2x + 3 = 45$
4 a $n = 5$　**b** $x = 10$　**c** $y = 7$
　d $a = 18$　**e** $m = 8$　**f** $t = 10$
　g $n = 7$　**h** $q = 40$　**i** $x = 0$
5 a $n + 16 = 28$　　　**b** $12n + 40 = 760$
　　　$n = 12$　　　　　　　$n = 60$
　　　　　The monthly payments were £60.

　c $2x + 8 = 30$
　　　$x = 11$ cm

Worksheet 19

1 a 3, 7, 11, 15, 19, 23　　**b** 40, 37, 34, 31, 28, 25
2 a 37　　**b** 76　　**c** 37　　**d** 37
3 a 4, 7·5, 11, 14·5, 18, 21·5
　b 10, 4, ⁻2, ⁻8, ⁻14, ⁻20
4 a 4, 11, 18, 25, 32　　**b** 6, 3·5, 1, ⁻1·5, ⁻4
　c 1, 1·4, 1·8, 2·2, 2·6　　**d** 2, 1·7, 1·4, 1·1, 0·8
5 a D　　**b** A　　**c** B　　**d** C
6 a

Pattern number	1	2	3	4	5	6
Number of counters	5	8	11	14	17	20

　b Starts at 5 and increases in steps of 3.
　　For each new pattern, three counters are added,
　　one to each arm. The first pattern has two extra
　　counters.
　c The number of counters is 3 × pattern number + 2.
　　So the nth pattern has $3n + 2$ counters.
7 a 6, 7, 8, 9, 10　　**b** ⁻1, 0, 1, 2, 3
　c 4, 8, 12, 16, 20　　**d** 7, 10, 13, 16, 19
　e 6, 5, 4, 3, 2

Worksheet 20

1 a ⁻4, 2, ⁻2, ⁻1　　　　**b** 20, 32, 12, 44
2 a

x	2	6	9	10
y	4	16	25	28

　b i $y = 3x - 2$
　ii $x \rightarrow 3x - 2$

3 a

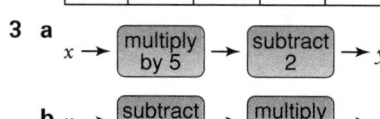

$x \rightarrow$ [multiply by 5] $\rightarrow$ [subtract 2] $\rightarrow y$

　b $x \rightarrow$ [subtract 2] $\rightarrow$ [multiply by 5] $\rightarrow y$
　　$x \rightarrow 5(x - 2)$
　c i 33　**ii** 25

Worksheet 21

1 a

x	⁻2	0	3
y	⁻8	0	12

b

x	⁻4	0	1
y	2	6	7

c

x	0	2	5
y	4	2	⁻1

2 a

x	⁻3	0	4
y	⁻4	⁻1	3

　b (⁻3, ⁻4), (0, ⁻1), (4, 3)

c

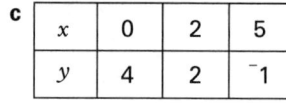

　d Yes
　e Some possible answers are:
　　(⁻2, ⁻3), (⁻1, ⁻2), (1, 0), (3, 2).

3 $y = 4x$ or $y = 5x$ or $y = 6x$ ...
4 a E　　**b** C　　**c** B　　**d** D

Worksheet 22

1 a

Miles driven	0	100	400
Charge (£)	25	33	57

b

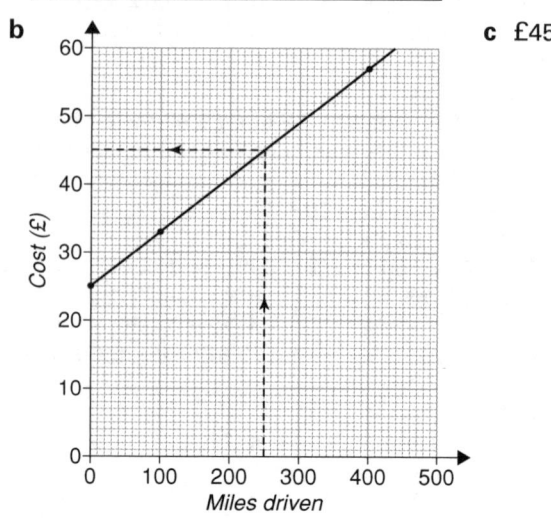

c £45

2 a

Number of panels painted by Bernie	2	10	20
Time in minutes	18	90	180

Number of panels painted by Max	3	15	30
Time in minutes	40	200	400

b

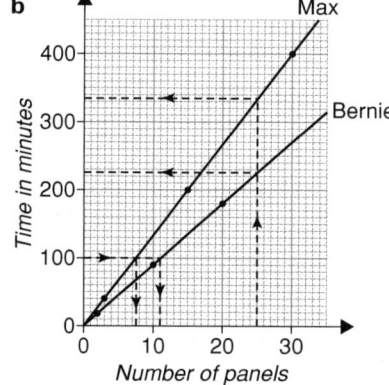

c It takes Bernie about 225 minutes and Max about
　335 minutes to paint 25 panels.
d Max paints $7\frac{1}{2}$ panels and Bernie paints about
　11 panels in 100 minutes.

Worksheet 23

1 a AB = 5·3 cm, CD = 8·9 cm, EF = 4·7 cm, GH = 6·4 cm
　b AB and GH, EF and GH
　c AB and EF
2

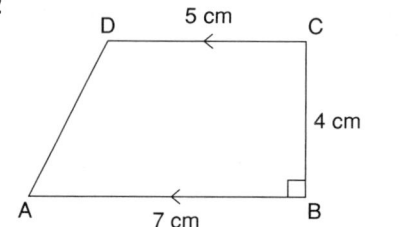

A trapezium

3 a WX, GH, KL, JI, NM, OP, RQ, ST, VU
  b WZ, XY, GJ, HI, KN, LM, OR, PQ, SV, TU
  c AF and DE and BC, AB and DC and FE
  d AF and AB, AF and FE, FE and ED, ED and DC,
     DC and CB, CB and BA

Worksheet 24

1 a Angle Y **b** Angle Q **c** Angle M
2 a Angle XYZ or ZYX **b** Angle PQR or RQP
  c Angle LMN or NML
3 a b, h **b** c, g **c** e, f **d** a, d
4 $a = 90°$, $b = 135°$, $c = 43°$, $e = 250°$ [allow ± 3°]
5 a Angle CAB or BAC **b** Angle QOT or TOQ
6 a 143° **b** 72° **c** 322° [allow ± 3°]
7 a   **b**   **c**

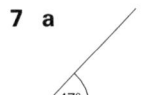   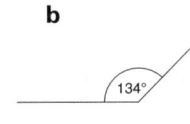   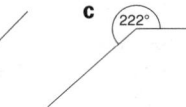

8

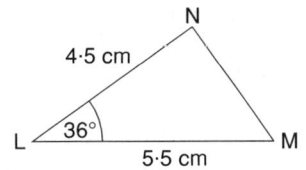

Angle NML = 55°

Worksheet 25

1 a $a = 137°$ **b** $b = 112°$ **c** $c = 64°$ **d** $d = 132°$
  e $e = 76°$ **f** $f = 28°$ **g** $g = 74°$, $h = 45°$
  h $i = 34°$, $j = 146°$ **i** $k = 309°$
2 a $p = 114°$, $q = 66°$ **b** $s = 60°$, $t = 60°$

Worksheet 26

1 a H Kite **b J** Trapezium
  c D Equilateral triangle **d A** Square
  e F Scalene triangle **f C** Parallelogram
  g I Right-angled triangle **h G** Rhombus
  i B Isosceles triangle **j E** Rectangle
2 a Pentagon **b** Hexagon **c** Octagon
  d b and **c** are regular polygons.
3 a Triangle PQR **b** Triangle JKL
4 a q **b** k

Worksheet 27

1 ∠JLK = 54°
2 AC = 9.9 cm

Worksheet 28

1 a

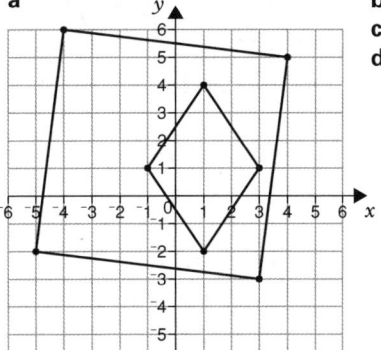

  b Rhombus
  c 2 axes of symmetry
  d (4, 5)

Worksheet 29

1 a Cylinder **b** Square-based pyramid **c** Cuboid
  d Cone **e** Triangular prism **f** Sphere
2 a 4 **b** 6 **c** 4 **d** Tetrahedron
3 a 10 **b** 8 **c** 8
4 a Only B.
  b One possible answer:

  There are other possible answers.

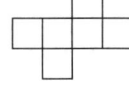

5 3
6 a Possible answers, these nets are not drawn full size.

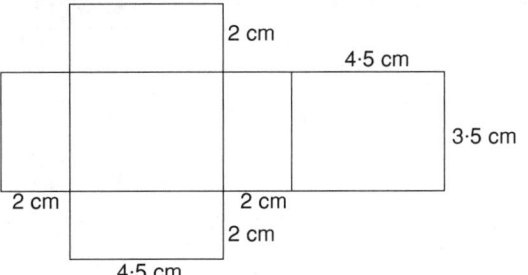

  b

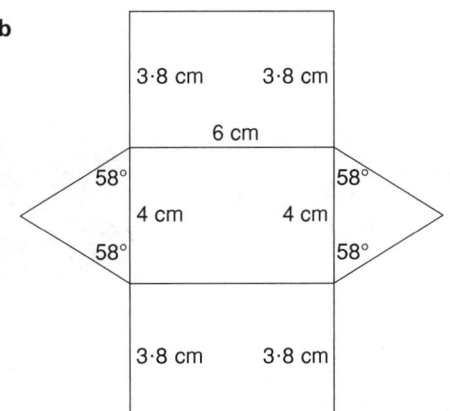

  c

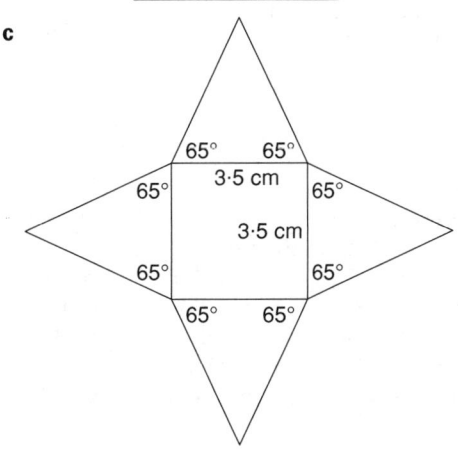

Worksheet 30

1 a 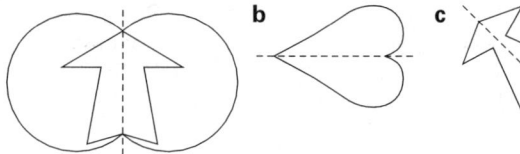 **b** **c**

2 a B **b** A **c** C

3
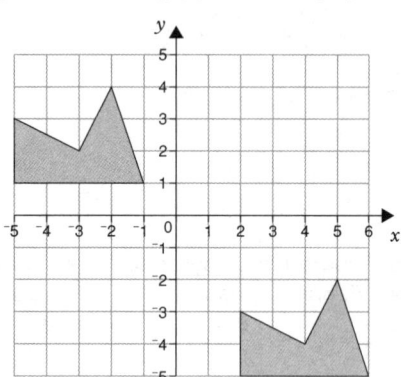

4 7 units right and 6 units down

5 a C **b** A **c** B

6 a
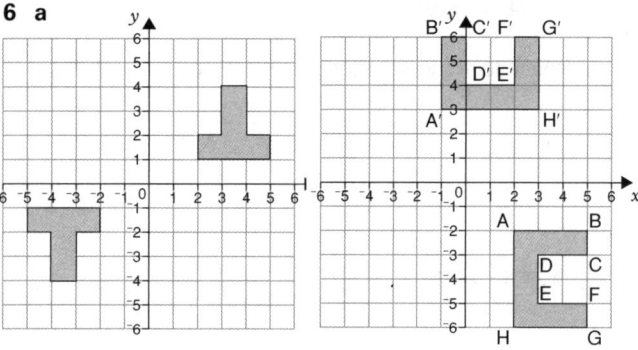

7 B and C

Worksheet 31

1 a Yes **b** No **c** Yes

2 a **b** **c**

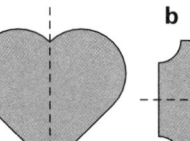 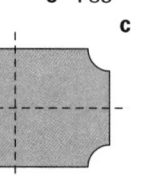

3 a 2 **b** 8 **c** 3 **d** 4

4 a 16 cm by 12 cm or 8 cm by 24 cm

   b A right-angled triangle with angles of 30°, 60° and 90°, the shortest side 6 cm and the longest side 12 cm.

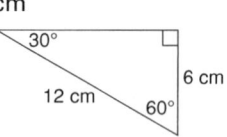

5
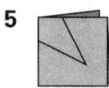

Worksheet 32

1 15 cm (± 3 mm)

2 a 43 days **b** 1 hour 40 minutes

3 a 2780 **b** 0·429 **c** 2830 **d** 183

   e 1700 **f** 290 **g** 1·76 **h** 0·846

   i 3·284 **j** 0·84 **k** 4·2 **l** 8300

4 YOU CANNOT LICK YOUR ELBOW

5 a 72·8 cm **b** 0·538 kg **c** 1·44 ℓ **d** 1·236 m

6 a A 8·5 cm **b** A 95 km/h **c** A 32 °C **d** A 120–140 g

   B 9·25 cm B 70 km/h B 46 °C B 430–450 g

   C 6·75 cm C 20 km/h C 58 °C C 510–520 g

7 a B **b** B

8 a i 1200 mℓ **ii** 40 km **iii** 5 kg **iv** 22·5 ℓ

   b i 7 pints **ii** 8 gallons **iii** 40 miles **iv** 6·6 lb

9 a 20 bags **b** 25 cans **c** 8 rolls

   d 26·4 or $26\frac{1}{2}$ lb.

Worksheet 33

1 a East **b i** C **ii** E **c** NW

   d i H **ii** D **e** D

2 a South **b** East

Worksheet 34

1 a 12 cm **b** 10 cm **c** 16 cm

2 a 8 cm^2 **b** 4 cm^2 **c** 7 cm^2

3 a 38 cm **b** 44 m **c** 36 mm

4 a 84 cm^2 **b** 121 m^2 **c** 54 mm^2

5 a cm^2 **b** mm^2 **c** m^2

6 a 202 cm^2 **b** 300 cm^2

7 152 cm^2

8 a
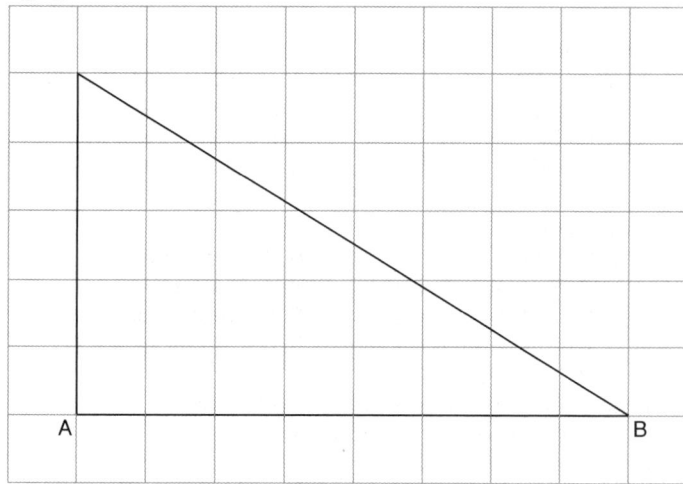

b
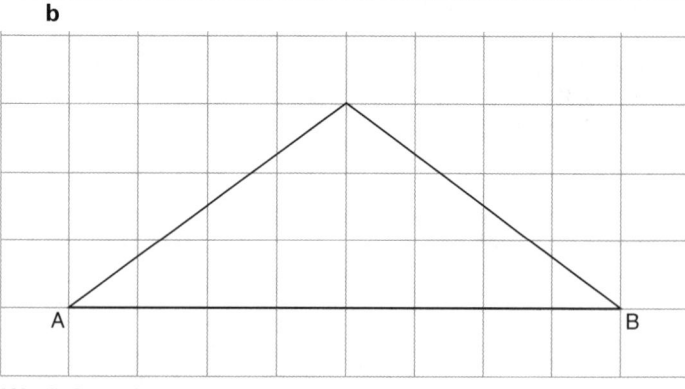

Worksheet 35

1 a

Place	Tally	Frequency
Alton Towers	⦀⦀ ⦀⦀ ⦀⦀ I	16
Blackpool	⦀⦀ ⦀⦀ III	13
Castleton Caves	⦀⦀ III	8
Jorvik Viking Museum	⦀⦀ ⦀⦀ I	11
Llandudno	IIII	4

b Alton Towers

2

Number of letters	Tally	Frequency
1	IIII	4
2	IIII	4
3	ΤΗΗ III	8
4	ΤΗΗ	5
5	IIII	4
6	I	1
7	II	2
8	II	2

3 a A   **b** C   **c** B

4 a Possible answers are
  i Aidan might find out that Year 8 boys can walk backwards faster than Year 8 girls.
  ii Aidan might find a greater spread of times to walk backwards for the Year 8 girls than for the Year 8 boys.
  b Aidan needs to record whether they are a boy or a girl and the time to walk a certain distance backwards.
  c A possible answer is

	Boys			Girls		
	Time (sec)	Tally	Frequency	Time (sec)	Tally	Frequency
	$10 \leqslant t < 12$			$10 \leqslant t < 12$		
	$12 \leqslant t < 14$			$12 \leqslant t < 14$		
	$14 \leqslant t < 16$			$14 \leqslant t < 16$		
	$16 \leqslant t < 18$			$16 \leqslant t < 18$		

  d Aidan would need to time about 30 to 40 pupils to walk backwards over a set distance. He would need to have equal numbers of boys and girls.

Worksheet 36

1 a

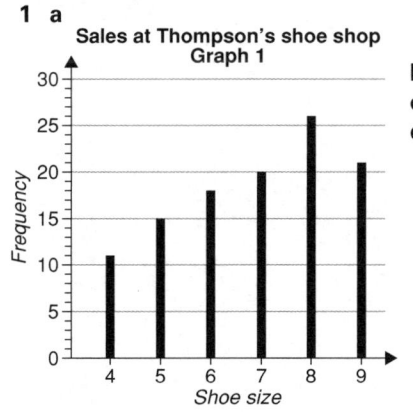

Sales at Thompson's shoe shop
Graph 1

  b 15
  c 11
  d Graph 2 represents sales at the end of August because more smaller sizes are sold. This is the time when new shoes are bought for school children ready to return to school.

2 a 5   **b** 7   **c** 8   **d** 24
3 a 40   **b** 35
  c Manor House was in a country village because more horses and sheep were treated than at Owley Grange.
4 a Week B   **b** Week A   **c** Week C
5 a Lunch at home   **b** Packed lunch
  c A Yes B No C Yes D No
6 a 272   **b** 122   **c** 159   **d** 64
7 a Zipfoam, Orangerine, Currantina
  b Chocmilk, Fizzpop
  c Chocmilk, Currantina, Fizzpop
8 a 8 a.m.   **b** 17 °C
  c 6 °C. Because the line joining the points only gives an estimate — the temperature was not measured then.

9 a Food and clothes
  b C There are fewer children in the English family than the French family.
  c Possible answers are:
    They spent the same amount on gas.
    The French family spent more on clothes.
    The English family spent more on electricity.
    Petrol is cheaper in France/the French family do not use their car as much.
10 The pie charts only show the proportion of the boys and the proportion of the girls who preferred the different takeaway foods. A higher proportion of the girls preferred Chinese, but there may have been many more boys in the survey, so more boys in total may have chosen Chinese.

11 a

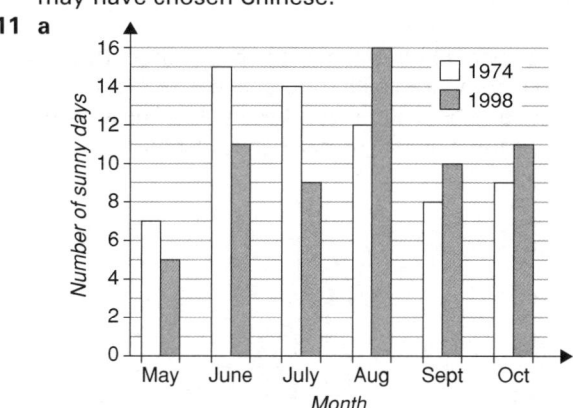

  b In May, June and July, it was sunnier in 1974 than in 1998.
    1998 had a sunnier August, September and October than 1974.
    Overall there was only a small difference in the number of sunny days between May and October in both years.

Worksheet 37

1 a

Number of apples	Tally	Frequency
1–5	I	1
6–10	I	1
11–15	IIII	4
16–20	ΤΗΗ ΤΗΗ	10
21–25	ΤΗΗ	5
26–30	IIII	4
31–35	III	3
36–40	II	2

b

Number of apples	Tally	Frequency
0–9	II	2
10–19	ΤΗΗ ΤΗΗ III	13
20–29	ΤΗΗ ΤΗΗ	10
30–39	ΤΗΗ	5

  c The class intervals in **b** because they show the spread of the data better.

d

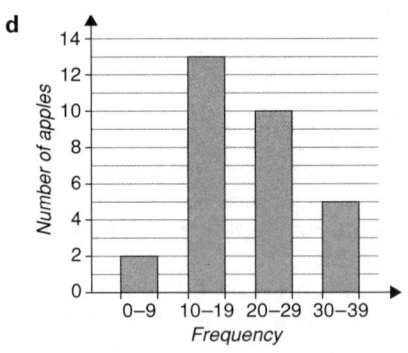

e A Two of the apple trees were affected by blight.

Worksheet 38
1 a 3   **b** 5

   c To decide whether that is the best variety to plant again compared with others he has grown and maybe to give him a good estimate of how many cucumbers he will produce in a season.

2 a 20–29

   b

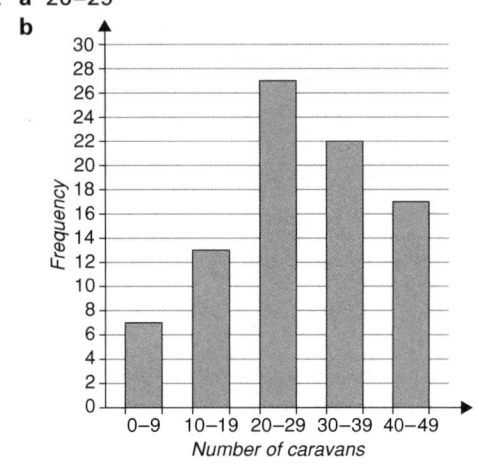

3 a Mean 9·1, Median 8·5, Range 15

   b Mean 624 kg, Median 596 kg, Range 449 kg

   c Mean £3·75, Median £2·30, Range £6·52

4 a 155 g   **b** 26 g   **c** 153 g

5 13 and 9

6 2·7 (1 d.p.) children per family

7 a Range for Monica 5,   Range for Betty 10

   b Mean for Monica 36·3,   Mean for Betty 37·3

   c *Either*   Betty because she has the highest mean score although she is less consistent, shown by her higher range of scores.

   or   Monica because she is more consistent, shown by her lower range of scores, although her mean score was slightly lower.

Worksheet 39
1 Possible answer is: **D, B, E, A, C**

2 Possible answers are:

   a Very likely   **b** Impossible   **c** Very likely

   d Very unlikely   **e** Certain   **f** Likely

3 Possible answers are:

   a Poor chance   **b** Even chance   **c** Good chance

   d Poor chance   **e** Certain

4 a

| 1 | 2 | 3 | 4 | 5 | 6 | 7 | 8 | 9 | 10 |

   b Yes

   c Yes, it is equally likely, because there are the same number of odd numbers as even numbers.

5 There is an even chance of either because each time he throws the coin there is an equal chance of it landing on heads or tails.

6 a Circle, because there are more circles than the other shapes.

   b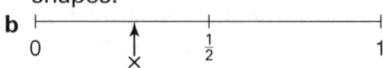

   c $\frac{2}{10}$ or $\frac{1}{5}$

7 a $\frac{4}{8}$ or $\frac{1}{2}$ **b** $\frac{5}{8}$ **c** 1   **d** $\frac{2}{8}$ or $\frac{1}{4}$ **e** 0

8 a Cerys may have taken out a card with A or T on more than once, so she cannot be certain that there are the same number of each.

   b It is likely that there are more Cs than Fs, but Cerys could have picked the same C card out more than once, so she cannot be certain.

   c Cerys could be wrong, because there could be a card with the letter M that she missed everytime she picked a card.

2

Chapter Reviews

CHAPTER 1

Name ....

1 Write these as a power of ten.

 a a thousand   **b** a hundred thousand   **c** a million

 ....

2 Your heart beats 3×10^9 times in a lifetime. Write 3×10^9 in words.

 ....

3 Write, in figures, the number that is one less than two and a half million.

 ....

4 a $2 \cdot 934 + 0 \cdot 001$ ....   **b** $11 \cdot 258 - 0 \cdot 001$ ....

 c $4 \cdot 689 + 0 \cdot 001$ ....   **d** $8 \cdot 470 - 0 \cdot 001$ ....

 e $6 \cdot 357 + 0 \cdot 005$ ....   **f** $12 \cdot 423 - 0 \cdot 006$ ....

5 What must be added to $6 \cdot 2843$ to get $6 \cdot 2849$? ....

6 Which of these is the same as $2 \cdot 15 \times 0 \cdot 01$?

 A $2 \cdot 15 \div 10$ **B** $2 \cdot 15 \times 10$ **C** $2 \cdot 15 \div 100$ **D** $2 \cdot 15 \times 100$

 ....

7 Find the answers to these.

 a $4 \cdot 73 \times 0 \cdot 01$ ....   **b** $2 \cdot 8 \div 0 \cdot 1$ ....

 c $3 \cdot 94 \times 0 \cdot 1$ ....   **d** $16 \cdot 25 \div 0 \cdot 01$ ....

8 Write < or > in the space for each of these.

 a $2 \cdot 985$ .... $2 \cdot 979$ **b** $6 \cdot 309$ .... $6 \cdot 313$ **c** $^-2 \cdot 48$ .... $^-2 \cdot 51$

9 Rewrite these in descending order.

 a $3 \cdot 642$, $3 \cdot 246$, $3 \cdot 624$, $3 \cdot 4$, $3 \cdot 64$ ....

 b $0 \cdot 285$, $0 \cdot 52$, $0 \cdot 08$, $0 \cdot 582$, $0 \cdot 208$ ....

10 Which number is halfway between these?

 a $3 \cdot 2$ and $3 \cdot 3$ ....   **b** $1 \cdot 06$ and $1 \cdot 07$ ....   **c** $^-0 \cdot 4$ and $^-0 \cdot 5$ ....

11 Ben did a survey on the heights of the pupils in his class. This shows part of his frequency table.

Height (cm)	Tally	Frequency
$150 < h \leqslant 155$	**[A]**	
$155 < h \leqslant 160$	**[B]**	
$160 < h \leqslant 165$	**[C]**	

 a Explain what $155 < h \leqslant 160$ means.

 ....

 ....

 b Ben's friend, Paula, is 155 cm tall. In which box, **A**, **B** or **C**, should Ben put the tally mark?

 ....

12 Round these to:

	the nearest thousand	the nearest ten thousand	the nearest million
a 6 478 936			
b 1 846 523			
c 4 079 046			

13 A newspaper reported the number of people attending Notting Hill Carnival one year as 950 000, rounded to the nearest thousand.

    a What is the smallest number of people that could have attended?

    ....

    b What is the largest number of people that could have attended?

    ....

14 Approximate each of these to the given number of decimal places.

    a 16·437 (2 d.p.) ....   **b** 1·8426 (1 d.p.) ....

    c 12·408 (1 d.p.) ....   **d** 12·408 (2 d.p.) ....

    e 2·035 (1 d.p.) ....   **f** 4·895 (2 d.p.) ....

15 Write the answers to these as a recurring decimal or round to 2 d.p.

    a John weighed 9 test tubes in science. Their total weight was 560 g.

    About how much did each test tube weigh?   ....

    b Veronica cut a 6 metre length of ribbon into 7 equal pieces.

    About how long was each piece?   ....

    c Glen ran 50 metres in 6 seconds.

    About how far did he run each second?   ....

    d Liz made 11 rock cakes weighing, in total, 1500 g.

    About how heavy was each rock cake?   ....

Name ....

except for question **9**.

1 Fill in the numbers to find the number in the top circle.

a Two numbers are added to get the number above.

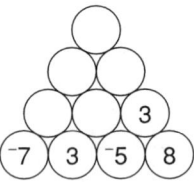

b The number on the right is subtracted from the number on the left to get the number above.

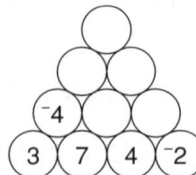

c Two numbers are multiplied to get the number above.

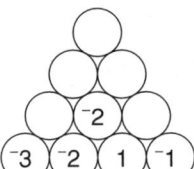

d The number on the left is divided by the number on the right to get the number above.

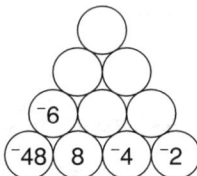

2 a $^-4(^-6 + ^-2) =$ ....

b $^-7 + (3 - ^-5) =$ ....

c $\dfrac{^-4 \times ^-7}{2 \times ^-2} =$ ....

3 If $4 \times ^-3 = ^-12$, write down one more multiplication fact and two division facts using the numbers 4, $^-3$ and $^-12$.

....

4 Fill in these addition and multiplication squares.

a

+	$^-15$	8	$^-4$
$^-11$			
$^-9$			
13			

b

×	6	$^-5$	$^-2$
$^-4$			
8			
$^-3$			

5 Fill in the missing number in each box.

a $^-7 + 19 =$ ☐

b $^-4 - 16 =$ ☐

c $8 + ^-10 =$ ☐

d $^-17 - ^-11 =$ ☐

e $^-6 -$ ☐ $= 1$

f $^-1 +$ ☐ $= ^-5$

6 Which of 558, 1035, 1260, 2376 and 6048 are divisible by

a 12

....

b 15

....

c 18?

....

7 a Use a table or factor tree to write 378 as a product of prime factors.

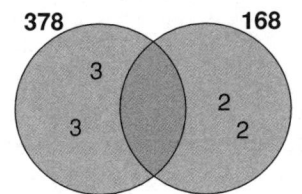

b Complete this diagram for the prime factors of 378 and 168.

c Find the HCF of 378 and 168.   ....

d Find the LCM of 378 and 168.   ....

8 Write down the answers to each of these.

a $(^-3 + ^-4)^2$ **b** $\dfrac{(4 - ^-2)^2}{^-2 \times 3}$ **c** 5^3 **d** $^3\sqrt{64}$

....

9 Give the answer to each of these, rounded to 2 d.p. where necessary.

a $4\cdot3^2$ ....   **b** $\sqrt{11\cdot2}$ ....   **c** $2\cdot1^3$ ....

d $(^-8)^3$ ....   **e** $^3\sqrt{2744}$ ....   **f** $\sqrt{15^2 - 12^2}$ ....

10 Which is bigger, the square of 20 or the cube of 8?

....

11 29 and 61 are 2-digit prime numbers that can be written as the sum of three different square numbers.

$29 = 4 + 9 + 16 = 2^2 + 3^2 + 4^2$

$61 = 9 + 16 + 36 = 3^2 + 4^2 + 6^2$

There are more 2-digit primes that can be written like this.

Find three of them.

....

Name ....

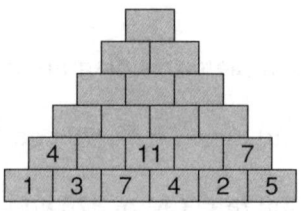

1 a $2 + {}^-5 - 4 - {}^-3 + 6 = $ ....

  b $11 + {}^-8 - 5 + 2 - {}^-4 - 7 = $ ....

  c $0{\cdot}2 - 0{\cdot}5 + {}^-0{\cdot}8 - {}^-0{\cdot}3 + 0{\cdot}9 = $ ....

2 In this pyramid, each number is the sum of the two numbers below it. What number goes in the top square?

....

3 Find x in each of these.

  a $200 - x = 163$

    $x = $ ....

  b $x + 3{\cdot}74 = 10$

    $x = $ ....

  c $1000 = x + 348$

    $x = $ ....

  d $100 - x = 27$

    $x = $ ....

4 Fill in this magic square.

1·0		1·4
		0·7
		1·2

5 a $620 + 190 = $ ....

  b $430 - 170 = $ ....

  c $224 + 416 = $ ....

  d $738 - 219 = $ ....

6 a $20 \times 60 = $ ....

  b $1200 \div 40 = $ ....

  c $240 \times 3 = $ ....

  d $18 \times {}^-7 = $ ....

7 Fill in the missing numbers in this number chain.

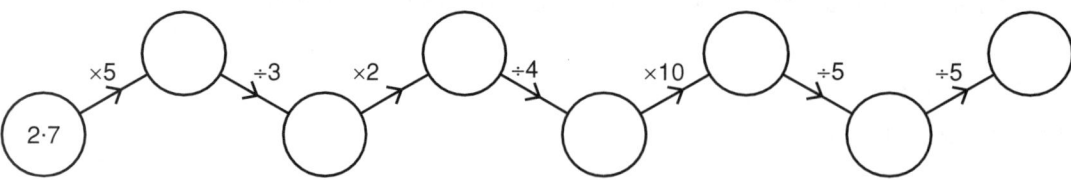

8 a $1{\cdot}8 \times 2{\cdot}5$

  ....

  b ${}^-3{\cdot}6 \times 6{\cdot}5$

  ....

9 a $4 \times 0{\cdot}9 = $ ....

  b $0{\cdot}2 \times 0{\cdot}8 = $ ....

  c $0{\cdot}3 \div 6 = $ ....

  d $0{\cdot}5 \div 4 = $ ....

10 a $260 \div 5 = $ ....

  b $156 \div 4 = $ ....

  c ${}^-18{\cdot}9 \div 3 = $ ....

  d $372 \div 12 = $ ....

CHAPTER 3

11 a How many metres in 3·4 km? ....

    b Two angles of a triangle are 36° and 67°. What is the size of the third angle? ....

    c What is the total cost of 7 apples at 13p each and 5 oranges at 15p each? ....

    d A bag contains red, yellow and green discs. A disc is drawn out at random. The probability of getting a red disc is 0·25. The probability of getting a green disc is 0·32. What is the probability of getting a yellow disc? ....

    e The square of a number is 81. What is the number? ....

    f Find the mean of 19, 34 and 43. ....

12 Jack was given £50 by his uncle. He put £15 into savings, gave £7 to his brother and divided the remainder into five equal amounts to spend over the next five weeks. How much did he have to spend each week? ....

13 The organisers of a pop concert need to know how many people will fit into the field where the concert will take place. Do they need to know an exact number or an estimate? Explain.

....

....

14 Estimate the answers to these. Show how you found your estimate.

    a $218 \div 47$ **b** $42·6 \times 3·85$ **c** $\dfrac{8·9 \times 4·18}{5·76}$ **d** $\dfrac{14·7 - 2·78}{1·92 + 2·03}$

....

15 a Paul estimates the cost of 58 cream teas at £4·85 each as £300. Is he correct? Show how you found your answer.

....

....

    b Is the estimated cost of the cream teas higher or lower than the actual cost? Explain why, without calculating the actual cost.

....

....

CHAPTER 3

Name ....

 Except for questions **16**, **17** and **18**.

1 a $4{\cdot}68 + 0{\cdot}72 + 3 + 13{\cdot}89$ **b** $4{\cdot}6 - 0{\cdot}28 - 1{\cdot}4 + 11{\cdot}7$

....

2 Beppe bought a book for £6·95, a CD for £8·49 and a magazine for £1·36. How much change would he get from £20?

....

3 Petrol cost 79p a litre. Sian puts 26·4 litres of petrol into her car.

    a How much will it cost, to the nearest penny?

....

    b How much change will she get from £30?

....

4 a $16{\cdot}7 \times 0{\cdot}6$ **b** $24 \times 4{\cdot}8$

....

5 654 cartons of yoghurt are packed in boxes of 24.

    a How many boxes can be filled?

....

    b How many cartons are left?

....

6 Work out the answers to these, as decimals.

a 296 ÷ 25

b 596 ÷ 16

....

....

7 Give the answers to these, rounded to 1 d.p.

a 427 ÷ 13

b 1236 ÷ 22

....

....

8 334·4 cm of string can be wound round a circular bobbin exactly 19 times. How far is it once round the bobbin?

....

9 Which is cheapest, a doughnut or an iced bun?

....

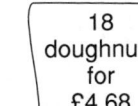
18 doughnuts for £4.68

23 iced buns for £6.05

10 a 162 ÷ 0·4

b 372 ÷ 0·06

c 432 ÷ 1·2

....

....

....

11 Give the answers to these to 1 d.p.

a 67 ÷ 0·7

b 85 ÷ 0·13

....

....

12 Raoul worked out the total cost of 18 boxes of frozen chips at £4·96 a box as £93·42. Is this answer sensible? Explain why.

....

....

13 Tiffany wrote down 2·52 as the answer to 7·56 ÷ 0·3. Explain how you know her answer cannot be correct.

....

....

CHAPTER 4

14 a Write down an equivalent calculation you
could do to check the answer to $3 \cdot 6 \times 1 \cdot 8$. ....

b Find the answer to $3 \cdot 6 \times 1 \cdot 8$. ....

15 a What calculation, using inverse operations,
could you do to check the answer to $634 \div 8 = 78$? ....

b Do the calculation to check if the answer was correct.
Write Yes or No in the answer space

....

16 a $(16 - 3 \cdot 8) \times 1 \cdot 5 + 4 \cdot 9 =$

b $8 \cdot 6 + 1 \cdot 2 \div 0 \cdot 4 - 8 \cdot 3 =$

....

....

17 Give the answers to these to 1 d.p.

a $\dfrac{7 \cdot 2 + 1 \cdot 4}{8 \cdot 3 - 3 \cdot 6}$

b $\dfrac{2 \cdot 9}{1 \cdot 8 \times 2 \cdot 4}$

c $\sqrt{18 - 3 \cdot 1^2}$

d $\dfrac{9 \cdot 6 \div 0 \cdot 28}{1 \cdot 4 + 26 \cdot 9}$

....

18 Use the calculator memory to find the answer.

At the garden centre, Charlie buys

7 heathers at £1·95 each

4 shrubs at £6·48 each

2 roses at £4·76 each

3 bags fertiliser at £2·49 each.

What is the total cost?

....

Name ....

 Except for question **13f**.

1 In the flag shown here, estimate the fraction of the flag that is

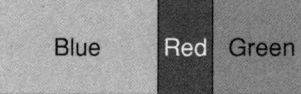

a blue   b green.

....

2 a What fraction of 36 is 9?   b What fraction of 120 is 60?

....

3 Alex counted the animals in a boarding kennels one day.
There were 18 dogs and 12 cats. What fraction of the animals were cats?   ....

4 Write these as fractions, in their lowest terms.

a 0·075   b 0·288

....

5 Write these as decimals, rounding to 2 d.p. where necessary.

a $\frac{19}{20}$ b $\frac{7}{25}$ c $\frac{5}{6}$

....

d $\frac{33}{6}$ e $5\frac{3}{4}$

....

6 What fraction of this shape is shaded? Give your answer as a decimal.

....

7 Write these as recurring decimals.

a $\frac{2}{3}$ b $\frac{9}{11}$

....

8 Write < or > in the box.

a $\frac{3}{7} \square \frac{19}{42}$ b $\frac{13}{40} \square \frac{3}{8}$ c $\frac{7}{12} \square \frac{4}{7}$

9 Stacey scored 14 out of 20 in the first round of the school quiz and 22 out of 30 in
the second round. In which round did she do better?

....

CHAPTER 5

10 Put these in order from smallest to largest.

$\frac{1}{4}$, $\frac{5}{8}$, $\frac{7}{24}$, $\frac{2}{3}$, $\frac{5}{6}$, $\frac{1}{2}$

....

11 On the number line, show each of these fractions with an arrow.

$\frac{12}{30}$, $\frac{28}{20}$, $\frac{36}{40}$, $\frac{18}{15}$

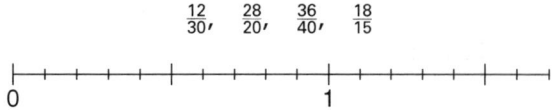

12 Find the fraction halfway between $\frac{1}{4}$ and $\frac{11}{36}$.

....

13 a $\frac{2}{7} + \frac{3}{7} =$ **b** $\frac{7}{8} - \frac{1}{4} =$ **c** $\frac{1}{2} + \frac{2}{5} =$

....

d $\frac{11}{12} - \frac{5}{9} =$ **e** $\frac{2}{3} + \frac{1}{5} - \frac{3}{10} =$ **f** $\frac{4}{5} - \frac{3}{7} + \frac{1}{2} =$

....

14 Give the answers to these as mixed numbers. Cancel, if possible.

a $\frac{2}{3}$ of 11 litres   **b** $\frac{5}{8}$ of 18 km   **c** $\frac{3}{11}$ of 16 m   **d** $\frac{2}{9}$ of 21 kg

....

15 Jarred has 72 videos in his collection. Kenny has $\frac{5}{8}$ as many videos as Jarred. How many videos does Kenny have?

....

16 a Write a division for 'How many sixths in 8?'   ....

     b Find the answer.   ....

17 Use this diagram to find the answer to $6 \div \frac{3}{5}$.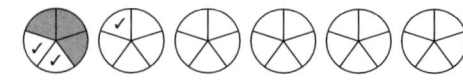

....

© New National Framework Mathematics 8 Nelson Thornes Ltd

Name ....

 Except for questions **1, 6** and **8**.

1 Complete this table.

Decimal	Percentage	Fraction
0.45		
	60%	
		$\frac{4}{25}$
1.2		
	175%	

2 Write these as percentages to 2 d.p.

  a $\frac{2}{3}$ = ...
  b $\frac{4}{11}$ = ...
  c $\frac{7}{19}$ = ...

3 You have 650 muscles in your body and you use 200 of them when you walk. What percentage is this? Give your answer to 1 d.p.

....

4 Put these in order, from smallest to largest.

0·5,   47%, $\frac{13}{30}$,   0·42, $\frac{4}{9}$

....

5 Jade did a survey about healthy eating. 48 of the 230 Junior School children said they avoided fatty foods, 0·23 of the Secondary School children said the same and 97 of the 512 adults also agreed. Which age group had the highest percentage that said they avoided fatty foods?

....

6 Find the answers to these without using a calculator. Show your method.

  a 15% of 360
       b 21% of 700

  ....
      ....

7 The tallest person who ever lived, Robert Wadlow, was 2·72 m tall. The smallest person who ever lived, Gul Mohammed, was about 21% of Robert Wadlow's height. How tall was Gul Mohammed? Give your answer in centimetres.

....

8 Use a written method to find these. Do not use a calculator.

  a 13% of 60 m
       b 42% of 80 g

  ....
      ....

  c 39% of £150
       d 161% of 85 litres

  ....
      ....

CHAPTER 6

9 Nathan bought a bike for £185 and sold it three years later at a loss of 15%.
How much did he sell it for?

....

10 The retail cost of an SR30 camera was £162. Happy Snaps were offering 12% off in
their sale and Top Prints were offering the camera in their sale at £20 off.
In which shop was the sale price lowest?

....

11 A salesperson is paid £360 each month plus 8% commission on the value of the goods he
sells that month. How much will his total wage be in a month in which he sells goods
with a total value of £9840?

....

12 This pie chart shows the proportions of the different types of
monkeys in the ape house at the zoo.

There are 18 chimpanzees in the ape house.

a Estimate how many marmosets are there?

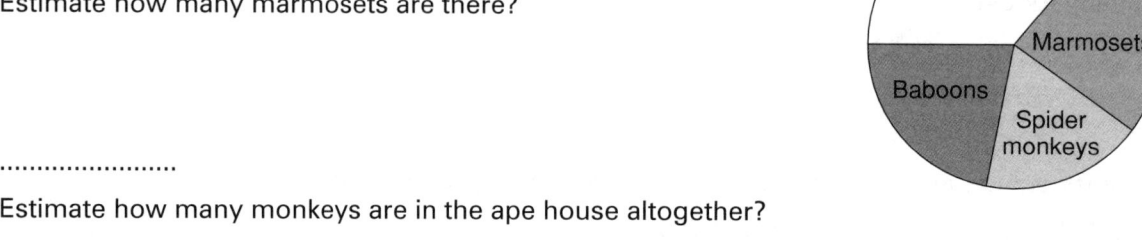

....

b Estimate how many monkeys are in the ape house altogether?

....

13 Fingernails grow about 3 mm a month. The hairs on your head grow about 433% faster
than fingernails. How much do the hairs on your head grow each month?

....

14 A day on the planet Venus lasts for 243 Earth days. A day on Mercury lasts for 24% of a
day on Venus. How many Earth days, to the nearest day, does a day on Mercury last?

....

Name ....

🖩 Except for questions **3** and **6a–d**.

1 a 4 gerbils cost £11·00.

What will 9 gerbils cost?

....

b 7 budgies cost £27·65.

What would 3 budgies cost?

....

2 5 trains every hour pass through Newgill Station and 7 trains every hour pass through Sandlock Station. If, in a set period of time, 450 trains pass through Newgill Station, how many trains will have passed through Sandlock Station?

....

3 Write these ratios in their simplest form.

a 24 : 18 : 42   **b** 20 : 55 : 75   **c** $5\frac{1}{4}$: 7   **d** 4·8 : 2

....

4 Julie gets £3·60 pocket money and Paula gets 90p. Write the ratio of Julie's pocket money to Paula's as a ratio in its simplest form.

....

5 Write 350 g : $2\frac{1}{2}$ kg in its simplest form.

....

6 A game uses discs with circles, squares, triangles or crosses on them. These are the discs used in the game.

a Write the ratio of circles to crosses.

....

b Write the ratio of triangles to crosses to squares.

....

c What fraction of the discs are circles?

....

d What percentage of the discs are squares?

....

e Give the proportion of triangles as a decimal to 1 d.p.

....

CHAPTER 7

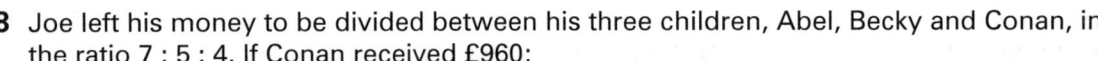
7 The ratio of sheep to cows to hens on a farm is 3 : 2 : 5.
If the farm has 40 hens, how many sheep does it have?

....

8 Joe left his money to be divided between his three children, Abel, Becky and Conan, in the ratio 7 : 5 : 4. If Conan received £960:

a How much did Becky receive?

....

b How much did Joe leave altogether?

....

9 This pie-chart shows the proportion of 900 people in a survey who wear glasses. How many wear glasses?

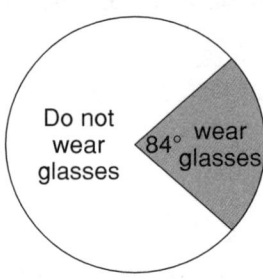

....

10 A recipe for 4 people used 280 g of flour. What mass of flour is needed for 9 people?

....

11 Divide 117 in the ratio 2 : 5 : 6.

....

12 The ratio of boys to girls to teachers in a school is 13 : 15 : 1.
There are 812 people altogether in the school.

a How many teachers are there?

....

b How many girls are there?

....

13 A metal alloy is made up of iron, zinc and copper in the ratio 7 : 3 : 2.
How much of each is in 1128 g of the alloy?

....

© New National Framework Mathematics 8 Nelson Thornes Ltd

CHAPTER 7

14 Write a formula for this:

The number of litres, L, of yellow paint needed to paint the lines on the road is found by multiplying the number of metres of lines, M, by 2 and adding 4.

....

15 Write and solve an equation for each of these.

a Sheba, the sheepdog, is twice as heavy as her puppy, Flash. If Sheba and Flash together weigh 27 kg, how heavy is Flash?

....

b I think of a number, multiply it by 3 and add 6. I get the same answer as I do if I multiply the number by 5 and subtract 4. What is the number?

....

c Adding 4 to a number and multiplying the result by 3 gives the same answer as multiplying the number by 5. What is the number?

....

d Jane and Simon try to win cash prizes on a fairground game. Jane wins 3 prizes and then spends 30p. Simon wins 6 prizes and then spends 50p. They end up with £1·90 between them. How much was each cash prize?

....

16 Solve these equations.

a $6m = 5$ **b** $4(y + 3) = 40$ **c** $\dfrac{14}{p} = 2$

d $3(a + 2) + 2(a - 4) = 33$ **e** $11x - 6 = 7x + 10$

CHAPTER 8

17 Calculate the size of x.

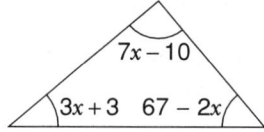

$x =$ ....

18 One inch is about 2·5 centimetres.

a Fill in this table.

Number of inches (i)	10	20	30	40	50
Number of centimetres (c)	25				

b Is the number of centimetres directly proportional to the number of inches?

....

c What is the ratio of centimetres : inches?   ....

d On the grid plot the graph of centimetres against inches.

Is it possible to draw a
straight line through the points?   ....

e Write a formula for the relationship between centimetres
(c) and inches (i).

....

f Find the length in centimetres that is equivalent to
36 inches.

....

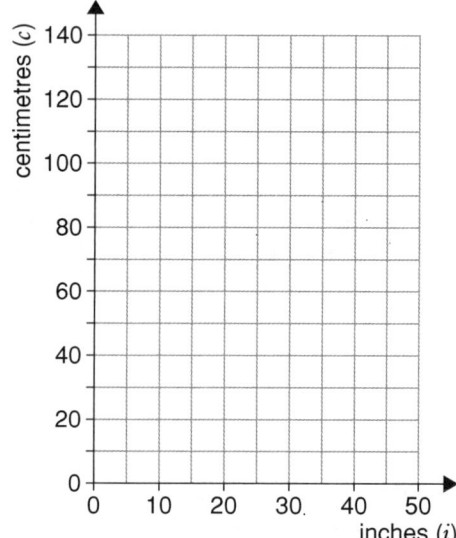

Name ....

Flow chart:

Start → Write down 0·4 → Add 0·2 → Write down the answer → Is the answer more than 1·2? — No (loops back to Add 0·2) / Yes → Stop

1 Write down the sequence given by this flow chart.

....

2 Write down the first five terms of these sequences.

   a **1st term** 243, **rule** divide by 3

   ...., ..., ..., ..., ...

   b **1st term** 2, **rule** multiply by ⁻2

   ...., ..., ..., ..., ...

3 Write down the first six terms of these sequences

   a Start at 3 and count forwards by 1, 3, 5, 7, ...

   ...., ..., ..., ..., ..., ...

   b Start at 16 and count backwards by 1, 2, 3, 4, ...

   ...., ..., ..., ..., ..., ...

4 Write down the first six terms of these arithmetic sequences.

   a $a = 5, d = 3$ ...., ..., ..., ..., ..., ...

   b $a = 3, d = ⁻2$ ...., ..., ..., ..., ..., ...

   c $a = ⁻9, d = 4$ ...., ..., ..., ..., ..., ...

5 Predict the next three terms of these sequences.

   a 4, 5, 7, 10, ..., ..., ...

   b 84, 77, 70, 63, ..., ..., ...

   c 4, 9, 16, 25, ..., ..., ...

   d 1458, 486, 162, 54, ..., ..., ...

6 How might the sequence 3, 4, 7, ... continue?

   Give two possible answers

   ...., ..., ...   and   ...., ..., ...

7 Fill in the missing terms

   a 7, ☐ , ☐ , 19, ☐   **term-to-term rule**: add consecutive even numbers starting with 2.

   b 1, 5, ☐ , ☐ , 17, ☐   **term-to-term rule**: add the two previous terms.

   c 1, 2, ☐ , ☐ , 120, ☐   **term-to-term rule**: multiply by consecutive numbers starting with 2.

8 The *n*th term of a sequence is given. Write the first five terms.

a $T(n) = 2n + 7$ ...., ..., ..., ..., ...

b $T(n) = 48 - 3n$ ...., ..., ..., ..., ...

c $T(n) = 1 \cdot 5n$ ...., ..., ..., ..., ...

9 Describe the sequences in **question 8** using a term-to-term rule.

a ....

b ....

c ....

10 $T(n) = 4n + 1$ generates a sequence of ascending numbers with a difference of 4 which are all one more than a multiple of 4. It starts at 5. Describe these sequences in a similar way.

a $T(n) = 5n - 2.$

b $T(n) = 77 - 7n.$

11 Stacey made these patterns with square tiles.

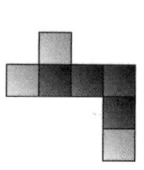

Pattern 1   Pattern 2   Pattern 3

a Draw pattern 4.

How many square tiles are in this pattern?   ....

b Complete this table

Pattern number	1	2	3	4	5
Number of tiles					

c Describe the sequence made by the number of tiles.

....

....

d Explain how you would find the number of tiles in the *n*th pattern. Refer to the diagram.

....

....

....

....

e Write an expression for the *n*th term. ....

f Will one of the patterns have 30 tiles? ....

Give a reason for your answer.

....

....

....

....

12 Find an expression for the *n*th term of these.

a 3, 9, 15, 21, 27, ...

....

b 8, 15, 22, 29, 36, ...

....

c 66, 62, 58, 54, 50, ...

....

13 Find the output for these.

a 6, 2, ⁻5, 1·6 → | multiply by 5 | → | add 7 | → —, —, —, —

b 3, ⁻8, 1, 4·5 → | subtract 3 | → | multiply by 4 | → —, —, —, —

14 Fill in the mapping diagram for this function machine.

⁻3, ⁻2, ⁻1, 0, 1, 2 → | multiply by 2 | → | add 4 | →

15 a If you drew a mapping diagram for $x \rightarrow x - 2$, what would you notice about the mapping arrows?

....

b If you drew a mapping diagram for $y = 3x$ and extended the mapping arrows back to the zero line, what would you notice?

....

CHAPTER 9

16 Write down the functions for each of these.

a 1, 3, 2, 5, 4 → [?] → [?] → 5, 9, 7, 13, 11 $y =$ ....

b 2, 5, 1, 3, 4 → [?] → [?] → 0, 9, ⁻3, 3, 6 $y =$ ....

17 What single operation could replace the two given?

a $x →$ [multiply by 8] → [divide by 4] → y

   $x →$ [   ] → y

b $x →$ [subtract 10] → [add 3] → y

   $x →$ [   ] → y

18 Dale drew this function machine.

$x →$ [multiply by 2] → [subtract 5] → y

He then drew this function machine.

$x →$ [subtract 5] → [multiply by 2] → y

The function machines contain the same operations in different orders.
Will Dale get the same answer if he puts 7 into both machines?

....

Explain your answer.

19 Find the inverse function of these.

a $x → 5x + 2$

....

b $x → 2(x - 3)$

....

c $x → \dfrac{x}{3} + 4$

....

d $x → \dfrac{x + 6}{3}$

....

e $x → 4(x + 8)$

....

f $x → x^2 - 3$

....

20 Write down the missing inputs.

a —, —, — → [add 5] → [multiply by 3] → 24, 3, 39

b —, —, — → [subtract 1] → [divide by 2] → 3, ⁻3, 1.5

Name ....

1 Complete these tables of coordinate pairs for the given rules.

a $y = 2x + 4$

x	$^-3$	0	2
y			

b $y = 3x - 1$

x	$^-1$	1	4
y			

c $y = 4 - 3x$

x	$^-2$	0	4
y			

2 a Write down the coordinate pairs from the table in **question 1c** for $y = 4 - 3x$.

....

b On the grid, plot the three points. Draw and label the line with equation $y = 4 - 3x$.

c Does the point $(2, ^-2)$ lie on the line?

....

d Write down the coordinate pair of a different point which lies on the line.

....

Does the coordinate pair satisfy the equation $y = 4 - 3x$?

....

e Do these points lie on the line?

i $(1\frac{1}{2}, ^-\frac{1}{2})$ .... **ii** $(2·5, ^-1·5)$ ....

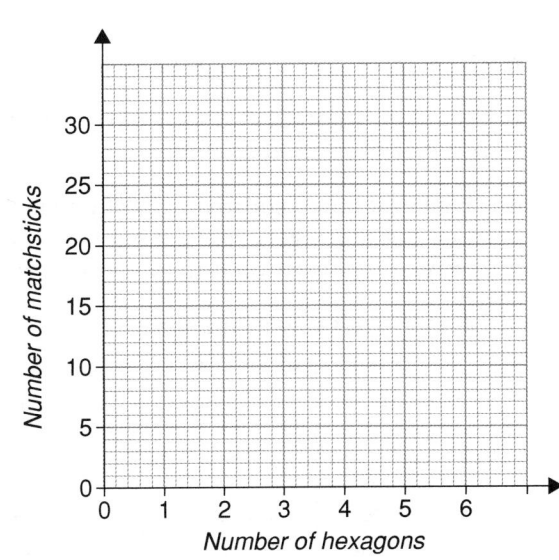

3 Ella was using matchsticks to make hexagon patterns.

1 hexagon **2 hexagons** **3 hexagons** **4 hexagons**

a Complete this table

Number of hexagons	1	2	3	4	5	6
Number of matchsticks	6	11				

b Plot the sequence on this grid. Do the points lie in a straight line? ....

c Is 24 a term of the sequence? ....

Explain.

d Would it be sensible to draw a straight line through the points? ....

Explain.

4 Choose an equation from the box to match each of these. You may use some equations more than once.

A $y = 5x - 3$	
B $y = {}^-2x + 1$	
C $y = x + 2$	

  a It has a negative gradient.   ....

  b It crosses the y-axis at $^-3$.   ....

  c Its gradient is steeper than $y = 3x + 1$.   ....

  d It is parallel to $y = 5x + 2$.   ....

  e It cuts the y-axis at (0, 2).   ....

5 This graph gives the masses of two boys, Martin and Lee.

  a How heavy was Martin when he was $2\frac{1}{2}$ years old?

    ....

  b How heavy was Lee when he was 5 years old?

    ....

  c How heavy was Martin when he was born?

    ....

  d Between which ages did Martin overtake Lee in mass?

    ....

  e About how much heavier was Martin than Lee when they were 8 years old?

    ....

Masses of Martin and Lee

Mass in kg

Age in years

6 One pint is about 0·55 litres.

  a Write a formula to show the relationship between litres (ℓ) and pints (p).   ....

  b Complete this table.

p	0	10	20
ℓ			

  c On the grid, draw the graph of litres versus pints.

  d Use your graph to find the approximate number of litres in 12 pints.   ....

  e This chart gives approximate equivalents between pints and litres, for a brewery making beer. The amounts are given to the nearest 5 pints and the nearest 5 litres.

    Fill in the missing amounts.

70 pints	.... pints	.... pints
=	=	=
.... litres	120 litres	100 litres

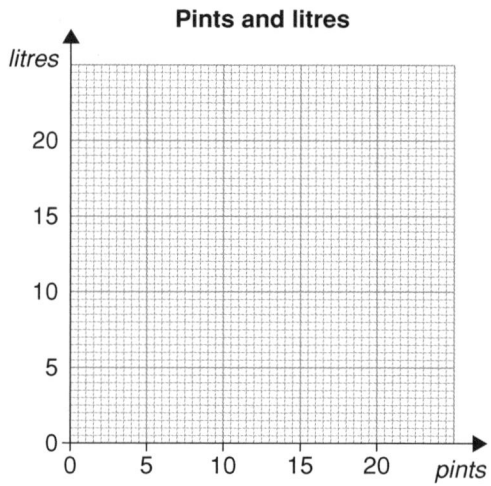

Pints and litres

litres

pints

7 Peter and Oliver go for a cycle ride from Peter's house one afternoon. They left Peter's house at 1400 hours and cycled to a wood 2 kilometres away. This took 15 minutes.

They climbed trees in the wood for 30 minutes, then they cycled another 3 kilometres away from Peter's house, to a stream. This took 30 minutes.

They stayed at the stream for 15 minutes, then cycled straight back to Peter's house, arriving there at 1600 hours.

On the grid, draw the distance/time graph for Peter and Oliver's cycle ride.

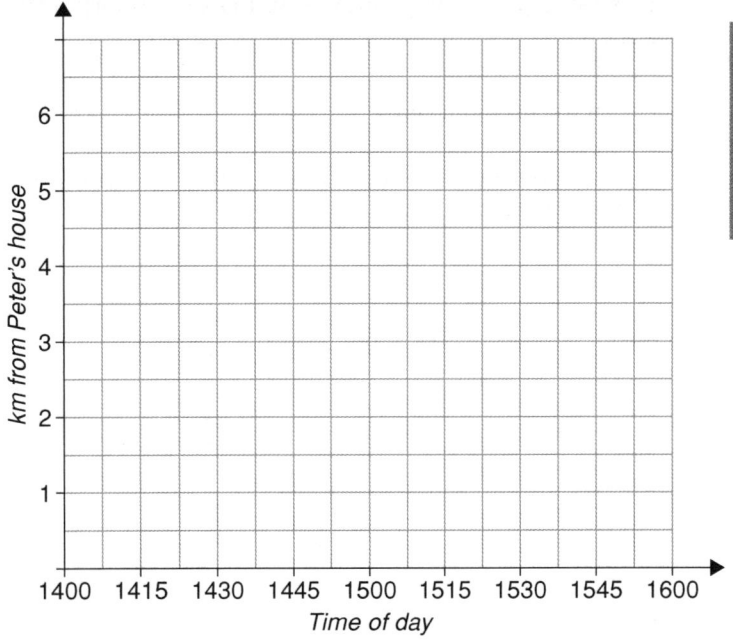

8 Explain these graphs.

a Watering plants

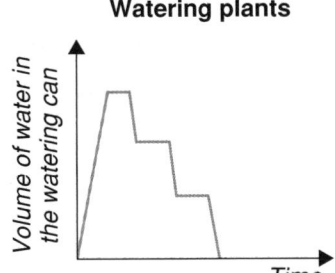

b Car journey

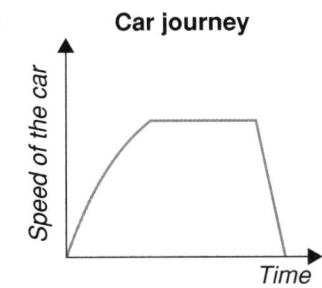

9 The graph shows the ice cream sales from Mr Luigi's ice cream van during the four weeks of August.

One week was much hotter than the other weeks. Which week do you think this was?

....

Ice cream sales

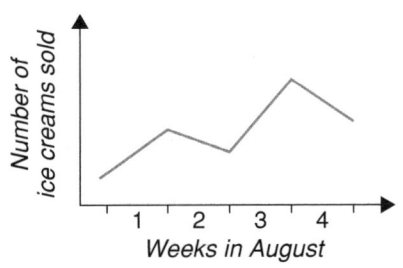

10 Water is poured steadily into two containers. The graphs show the depth of water against time for each container.

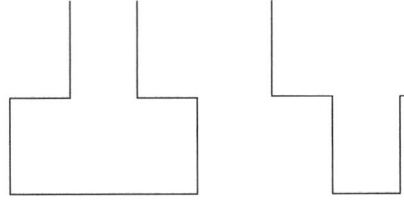

Container A   **Container B**

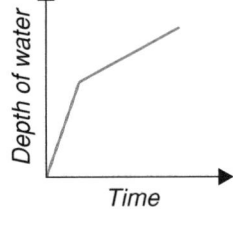

Graph 1

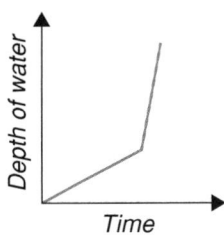

Graph 2

Which graph is for which container?

Graph 1 is for container ....

Graph 2 is for container ....

Explain.

11 This table gives the profits of Benny Boy Incorporated from January to June.

Month	Jan	Feb	Mar	Apr	May	Jun
Profits (in thousands)	7	6	7.5	8	9	10

a Plot the points on each of the grids shown.

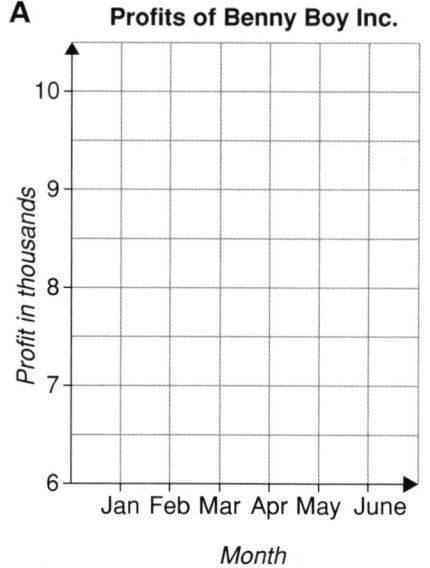

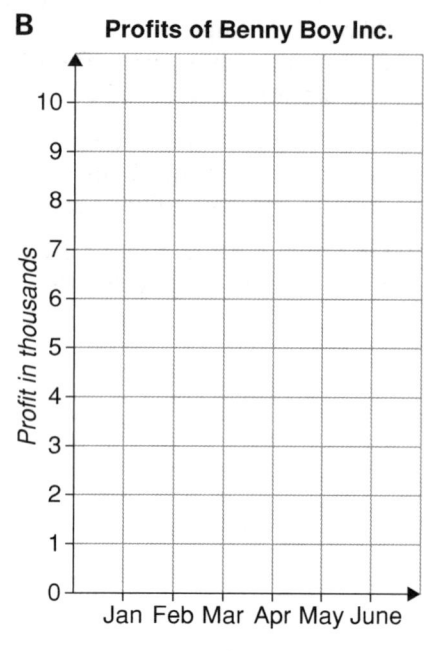

b Which graph do you think the managing director would use to show how well his company is doing?

....

Explain.

Name ....

1 a Which of the following pairs of angles are complementary angles?

    A 79° and 101°   **B** 37° and 53°   **C** 68° and 32°

    ....

  b Which of the following pairs of angles are supplementary angles?

    A 23° and 147°   **B** 47° and 43°   **C** 116° and 64°

    ....

2 Find the angles marked with letters. Explain your reasoning.

a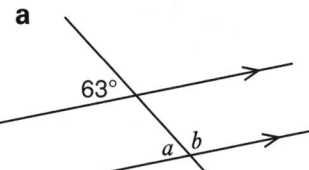

b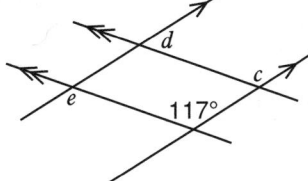

c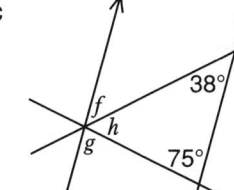

a = ...

Reason:

c = ...

Reason:

f = ...

Reason:

b = ...

Reason:

d = ...

Reason:

g = ...

Reason:

e = ...

Reason:

h = ...

Reason:

3 Find the angles marked with letters.

a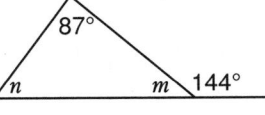

b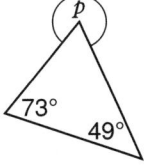

c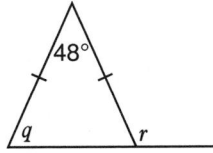

m = ...

n = ...

p = ...

q = ...

r = ...

d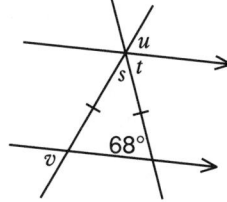

e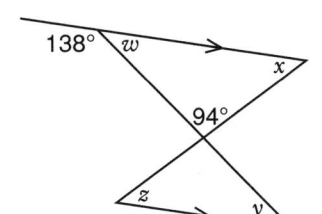

s = ... t = ...

u = ... v = ...

w = ...

x = ...

y = ...

z = ...

CHAPTER 11

4 Calculate the value of k.

a

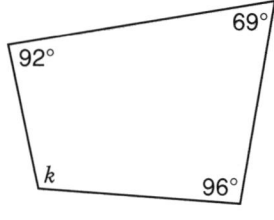

92° 69° k 96°

$k = \ldots\ldots\ldots\ldots\ldots\ldots$

b

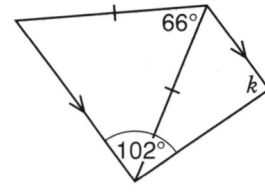

66° k 102°

$k = \ldots\ldots\ldots\ldots\ldots\ldots$

5 Calculate the value of x. Show your working.

a

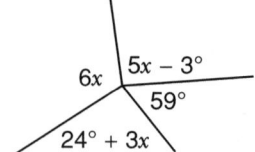

$6x$ $5x - 3°$ $59°$ $24° + 3x$

$x = \ldots\ldots\ldots\ldots\ldots\ldots$

b

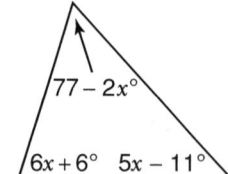

$77 - 2x°$ $6x + 6°$ $5x - 11°$

$x = \ldots\ldots\ldots\ldots\ldots\ldots$

c

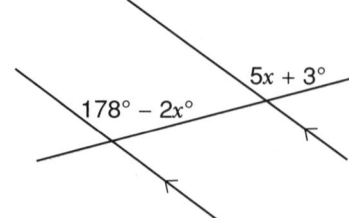

$5x + 3°$ $178° - 2x°$

$x = \ldots\ldots\ldots\ldots\ldots\ldots$

Name ....

1 a Imagine a special quadrilateral with opposite sides equal.

What could it be? ....

b Suppose it has no lines of symmetry. What could it be now?

....

2 Write down all the special quadrilaterals

  a with exactly two lines of symmetry ....

  b with diagonals that are perpendicular to each other

  ....

  c with no rotational symmetry. ....

3 Two identical triangular prisms have ends that are right-angled triangles.

Imagine painting the triangular ends on both prisms blue and all the rectangular faces red.

Glue the two prisms with their largest rectangular faces together.

  a What shape is the end of the new prism?

  ....

  b How many red faces does the new prism have?

  ....

  c At how many edges do a blue and a red face meet?

  ....

4 Lesley is tiling a floor with tiles like the shape shown in the diagram.

Draw seven more tiles to show how they tessellate.

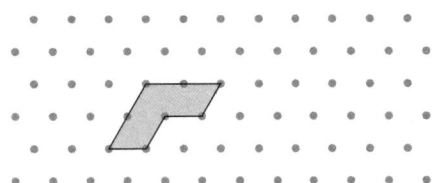

5 Triangles ABC and DEF are congruent.

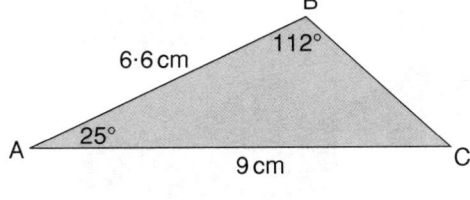

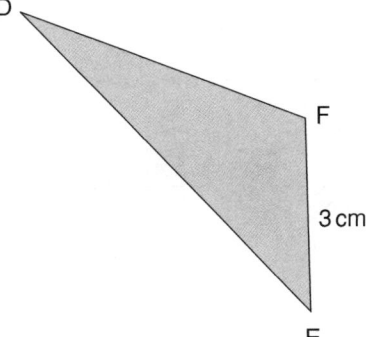

  a What is the size of angle F? ....

  b What is the length of DE? ....

CHAPTER 12

6 a Draw this shape on the isometric paper.
Using dashed lines, show the position
of the hidden edges.

plan view

front view   side view

b Draw the plan view, front elevation and side elevation of the shape.

Plan view   **Front elevation**

Side elevation

7 This shows a model made with 9 cubes, 5 black and 4 white.

These drawings show the four side views.

Write the letter of each side view under the correct drawing.

a   **b**   **c**   **d**

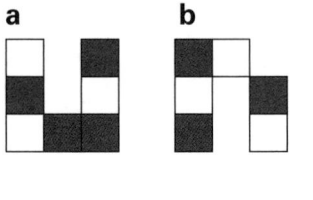 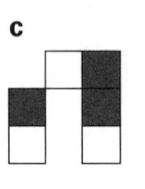 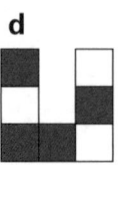

....

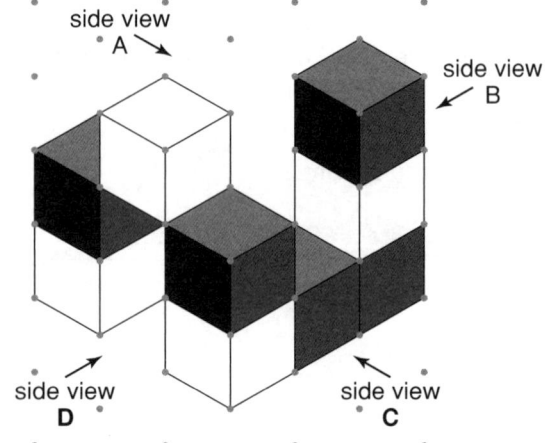

side view
A

side view
B

side view
D

side view
C

CHAPTER 12

8 On the isometric paper, draw the shape that this represents.
Draw it from the view given by the arrow.

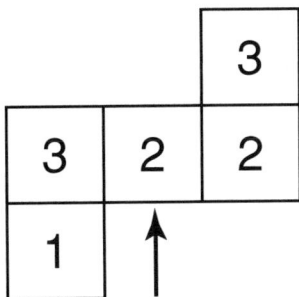

9 a Using compasses, construct a line through X that is perpendicular to AB.

  b Name the point where this line meets BC as M.

  c Bisect angle ABC.

  d Name the point where this bisector meets XM as N.

  e Measure and write down the length of MN to the nearest millimetre.

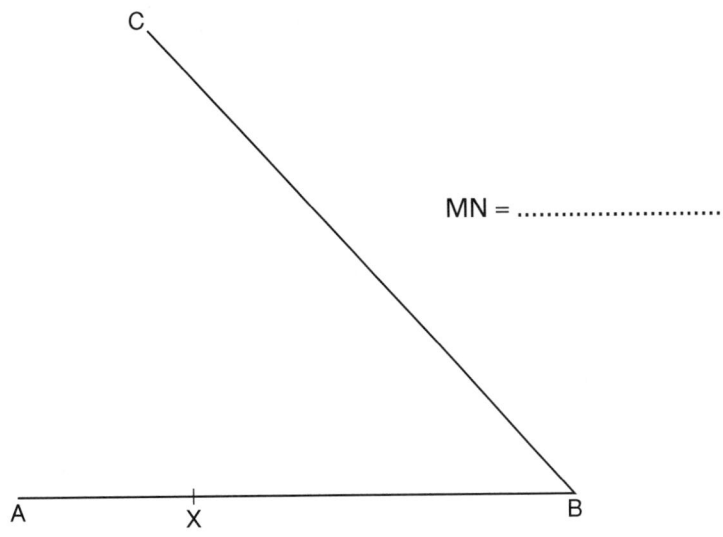

MN = ...

CHAPTER 12

10 Use a ruler and compasses to draw the net for this pyramid.

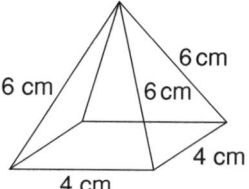

6 cm

6 cm

6 cm

6 cm

4 cm

4 cm

b Which of the combinations of transformations will map

i Q to R   ....

ii R to P   ....

iii S to Q?   ....

A Rotation 90° about the origin followed by translation 3 units left and 8 units down.

B Reflection in *y*-axis followed by rotation 180° about (2, 1).

C Translation 2 units left and 1 unit down, followed by a reflection in $y = {}^-2$.

D Reflection in the *y*-axis followed by a translation 2 units left.

E Translation 1 unit left and 1 unit up, followed by a reflection in the line $y = x$.

4 Describe the reflection and rotation symmetry of each of these.

a    ....
     ....
     ....
     ....

b 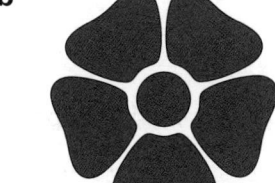   ....
     ....
     ....
     ....

c    ....
     ....
     ....
     ....

d    ....
     ....
     ....
     ....

5 Join these shapes together to make a
single shape with both
reflection and rotation symmetry.

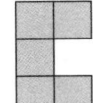  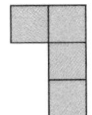

6 **a** On the grid below, enlarge the boat by
a scale factor of 2, centre of enlargement (0, 0).
Write down the coordinates of P′.

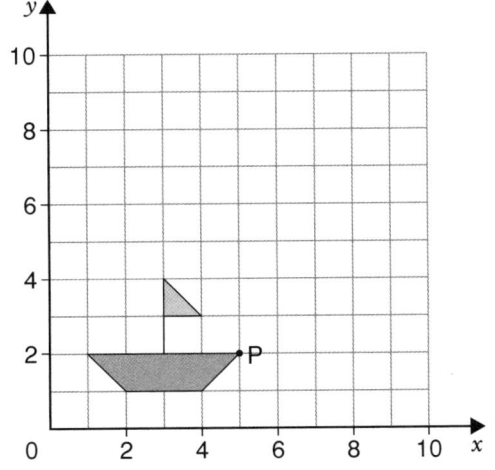

P′ is ....

b On the grid below, enlarge the boat by
a scale factor of 2, centre of enlargement (2,1).
Write down the coordinates of P′.

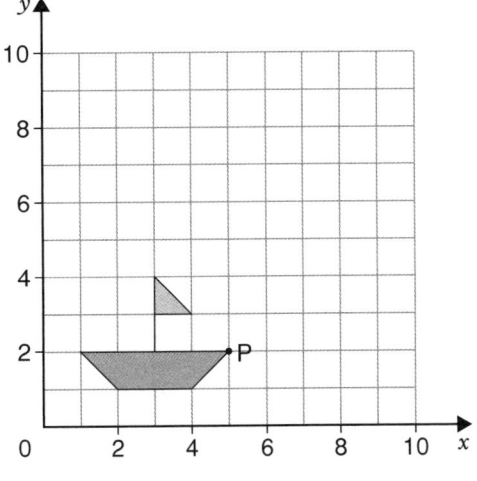

P′ is ....

CHAPTER 13

7 Estimate the length of the train in this picture.

The man is about 1·8 m tall.

.... m

8 Claudia is making a scale drawing of her village.

She is using the scale 1 cm represents 50 m.

How long in real life is a length that, on the scale drawing, is

a 3 cm   **b** 11 cm   **c** 8·5 cm?

....

9 Write down the coordinates of the midpoint of PQ.

a P (2, 4)   Q (8, 12)   **b** P (3, ⁻5)   Q (⁻6, 9)

....

Name ....

1 Fill in the missing numbers.

   a 14 cm = ... mm   **b** 2500 mℓ = ... ℓ **c** 2·9 tonnes = ... kg

   d 586 m = ... km   **e** 1·2 ℓ = ... cℓ **f** 47 cm^3 = ... mℓ

   g 1·6 ℓ = ... cm^3 **h** 3460 ℓ = ... m^3 **i** 27000 m^2 = ... ha

2 Convert these.

   a 470 seconds to minutes and seconds   **b** 6·4 hours to hours and minutes

      ....

   c 375 weeks to years and weeks   **d** 2580 minutes to days and hours

      ....

3 A packing box weighs 1·36 kg. When it is filled with 16 cans of paint it weighs 28·88 kg. How much does each can of paint weigh?

                                                    ....

4 A rectangular field has an area of 36 hectares. One side is 900 m. How long is the other side?

                                                    ....

5 The longest any game has been played non-stop was a game of Monopoly that lasted for 11 days, 2 hours and 35 minutes. If the players started playing at 3:47 p.m. on 24 December 1974, when did they finish?

                                                    ....

CHAPTER 14

6 Nathan used a 65 c*l* jug to fill his fish tank with water. He needed 17 jugfuls of water to fill the tank. How many litres does the fish tank hold?

....

7 a Insert an equivalent quantity in metric units in the brackets after each imperial unit quantity:

Hetty drove about 12 miles (..........................) to the supermarket to buy 8 ounces

(..........................) of ground almonds, 18 inches (..........................) of lace trim and

1 pint (..........................) of milk. She then drove 100 yards (..........................) to the

garage to buy 5 gallons (..........................) of petrol.

b Insert an equivalent imperial quantity in the brackets after each quantity given in metric units:

Rupert weighed out 150 g (..........................) of bird seed and put it on the bird table,

which was 50 cm (..........................) long and 35 cm (..........................) wide. He then

poured 450 m*l* (..........................) of water into the water bowl.

8 To what degree of accuracy do you think these are given? Choose from the box.

a Jarred completes the cross-country run in 48 minutes.

....

b Laila says she has 360 cm of string.

....

c Chrissie says her school bag weighs 1400 g.

....

d Kirk says he can drink 600 m*l* in one go.

....

to the nearest		
mm	cm	10 cm
m	10 m	km
sec	min	30 min
m*l*	c*l*	*l*
g	100 g	kg

9 Estimate these, using metric units. Give a range for each.

a The length of a ballpoint pen

....

b The capacity of a teacup

....

c The mass of a can of fizzy drink

....

CHAPTER 14

10 The diagram shows the map of an island, with several features marked on it.

Measure and write down the bearing of these.

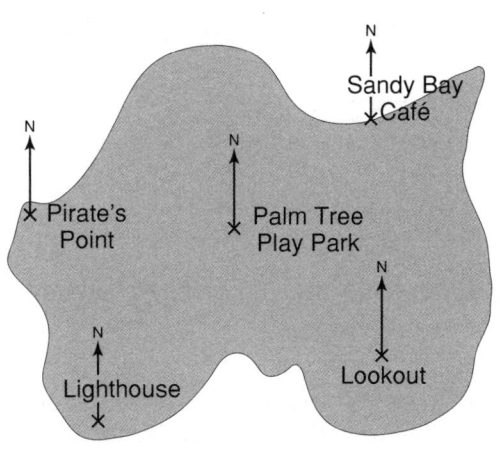

a Palm Tree Play Park from Lighthouse

....

b Lookout from Pirate's Point

....

c Pirate's Point from Sandy Bay Café

....

d Palm Tree Play Park from Lookout

....

11 Draw lines to link the compass bearings with their three-figure bearings. The first one has been done for you.

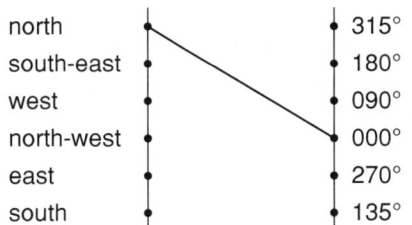

12 Find the areas of these triangles.

a

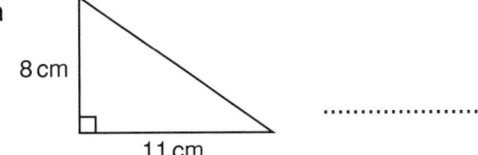

....

b

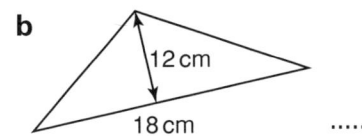

....

13 Find the areas of these tiles.

a

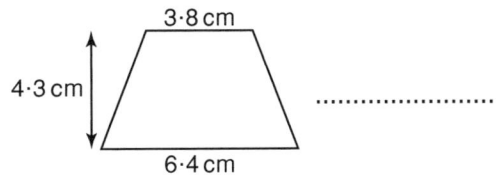

....

b

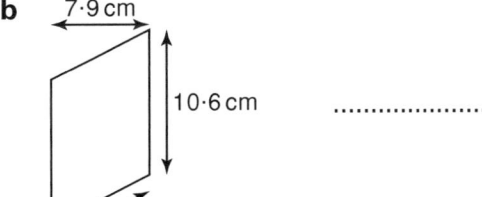

....

14 Find the area of each of these shapes. The grid represents cm².

a

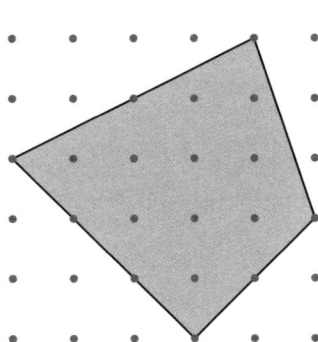

b

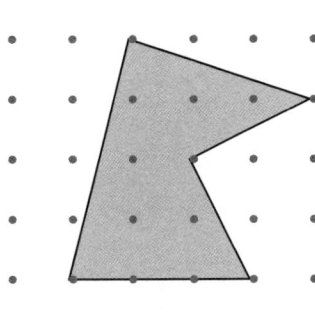

....

....

15 This diagram shows two cuboids that have the same volume.

 a What is the volume of Cuboid A?

 ....

 b What is the length marked x?

 ....

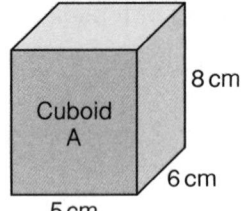 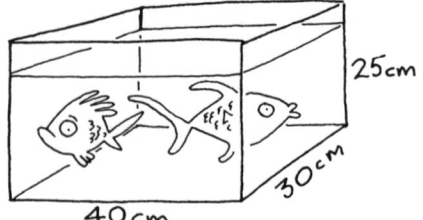

16 a What is the volume of this fish tank?

 ....

 b What volume of water is in the tank if it is
 filled to within 5 cm of the top?

 ....

17 A pencil sharpener measures 2·4 cm by 1·2 cm by 0·9 cm. What is the largest number
 of pencil sharpeners that can be packed into a box 16 cm by 12 cm by 3 cm?

 ....

CHAPTER 14

Name ....

1   Write in the box a D if you think the data is discrete or a C if you think it is continuous data.

   a The time taken to swim the Channel ☐

   b The number of fish in a tropical tank ☐

   c Ladies' dress sizes ☐

   d The heights of the pupils in a class ☐

2   Laila measured the distances the pupils in her class could throw a tennis ball.
These are the results, in metres.

78	43	81	56	44	72	50	39	48	57
36	75	49	60	38	66	82	74	69	55
70	80	46	56	52	64	49	53	58	62

   a Fill in this frequency table.

   b How many pupils threw the ball 40 metres or more but less than 50 metres?

   ....

   c How many pupils threw it 60 metres or more ?

   ....

Distance (m)	Tally	Frequency
$30 \leqslant d < 40$		
$40 \leqslant d < 50$		
$50 \leqslant d < 60$		
$60 \leqslant d < 70$		
$70 \leqslant d < 80$		
$80 \leqslant d < 90$		

   d If Laila had chosen to use class intervals of $30 \leqslant d < 50$, $50 \leqslant d < 70$, $70 \leqslant d < 90$, would they have been more useful? Explain.

   ....

   ....

   ....

3   This table shows what treatment customers to a hair salon had in one day.

   a How many females had colour?

   ....

	Dry trim	Trim/blow wave	Colour
Female	9	15	7
Male	13	6	1

   b How many males went to the hair salon that day? ....

   c How many customers had a dry trim? ....

   d Comment on the difference between treatments required by females and males.

   ....

   ....

CHAPTER 15

4   You are doing a survey to find out how much time teenagers spend connected to the Internet.

   a Write down three questions related to this that you could ask.

     i ....

     ii ....

     iii ....

   b Write down some possible results for each of these questions.

     i ....

     ii ....

     iii ....

   c Write down what data you would need to collect.

   d Suggest a suitable sample size.

   e Explain how you would ensure an unbiased sample.

   f Would the data collected be from a primary or secondary source?

5   You are doing a survey to find out how much the pupils in your year spend on clothes.

Design a questionnaire to collect the data. Include at least 6 questions in your questionnaire.

Name ....

1 The following table gives the length in metres of the best throws for the Men's Javelin in the 2000 Sydney Olympics.

Length of throw (m)	$65 < L \leqslant 70$	$70 < L \leqslant 75$	$75 < L \leqslant 80$	$80 < L \leqslant 85$	$85 < L \leqslant 90$	$90 < L \leqslant 95$
Frequency	2	5	11	10	8	1

a What is the modal class?

....

b Why might it be useful to know this modal class?

2 Grant wrote down the times it took him to get by taxi from the station to the airport on a Friday and on a Sunday.

These are his results.

Time in minutes – Friday	73	47	68	77	52
Time in minutes – Sunday	49	36	44	41	38

a Compare the range of these times for a Friday and a Sunday.

b Why might it be useful for Grant to know these ranges for his future travel?

3 The table shows the number of goals scored by football teams in a county league through one season.

Number of goals scored	0	1	2	3	4	5	6	7
Frequency	17	38	26	19	13	9	6	2

Use a calculator to find the mean number of goals scored. Give your answer to 2 d.p.

....

4 The heart rate in beats per minute was measured for sixteen students.
These are the results.

66   71   78   65   77   62   59   66
81   76   72   66   70   63   68   72

Use an assumed mean to find the mean heart rate. Give your answer to 1 d.p.

....

5 Kelly and Dean were trying to see how many steps they could take, walking on their hands. They decided to find the mean of three tries.

Kelly's numbers of steps	2	11	14
Dean's numbers of steps	□	9	□

They both had the same mean. Dean's range was half of Kelly's. What are Dean's two missing numbers of steps?

....

6 The table gives the expenses of a sales representatitve for 30 days. Find the mean expense per day.

Petrol	£170
Lunches	£148
Telephone	£46
Internet	£221

....

7 Find the median, mode and range from this stem-and-leaf diagram.

Number of cars per 15 minutes passing over the Menai Bridge

Median = ...

Mode = ...

Range = ...

```
0 | 0 1 5 7 8 8 9
1 | 0 0 1 1 2 4 4 4 5 5 6 8 8 8 9 9 9 9
2 | 0 2 2 2 3 5 5 6 6 7 7 8 9 9 9
3 | 1 3 3 3 5 6 6 7 7 7 8 9 9
4 | 0 2 2 4 5 5 6 7 7 8
5 | 1 3 3 5 7 8 8
```

8 The table shows Megan's test marks through the year for French and German.

French marks	7	4	6	8	9	2	9	8	7	8	6	5
German marks	7	8	6	7	5	9	6	5	7	6	7	5

Find the mean, median, mode and range for each set of marks. If Megan has to choose which language to study next year, based on these results, which do you think she should choose? Explain your reasoning.

French   Mean = ...   Median = ...

           Mode = ...   Range = ...

German   Mean = ...   Median = ...

           Mode = ...   Range = ...

Megan should choose .... because

9 This table gives the average monthly rainfall in The Gambia, on the west coast of Africa, and Kenya, on the east coast of Africa

	J	F	M	A	M	J	J	A	S	O	N	D
Rainfall in The Gambia (cm)	1	1	0	0	1	6	28	50	31	11	2	1
Rainfall in Kenya (cm)	3	3	6	19	32	11	9	6	6	9	10	6

a On the axes, draw a line graph with both sets of data.

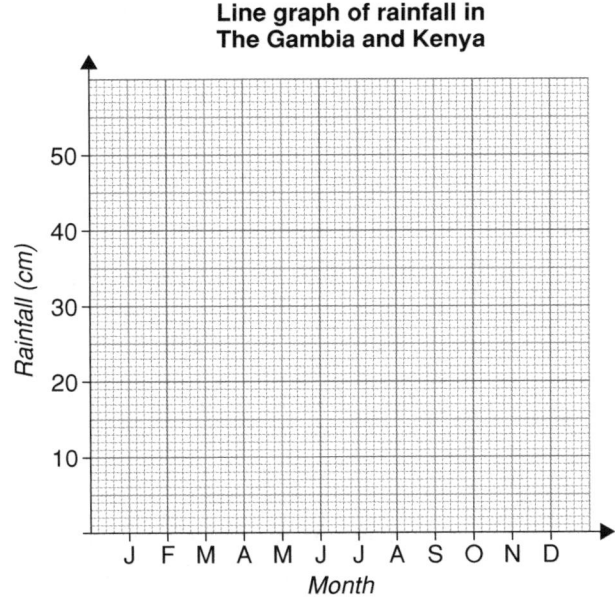

b Do the points in between the ones plotted have meaning? ....

Explain:

c Use your graph to compare the rain fall in The Gambia and Kenya.

10 This table gives the number of men, women and children attending a concert.

	Monday	Tuesday	Wednesday	Thursday	Friday	Saturday
Men	38	25	31	42	45	43
Women	47	36	58	26	67	61
Children	19	28	17	4	32	36

Draw a compound bar graph on the grid below to show this data.

11 Alisha timed how many seconds the pupils in her class could stand on one leg without losing balance. The table shows the grouped frequency distribution of the results.

Time in seconds	0–	5–	10–	15–	20–
Frequency	5	13	11	4	1

a Draw a frequency diagram.

b How many pupils could stand on one leg for 15 seconds or more?

....

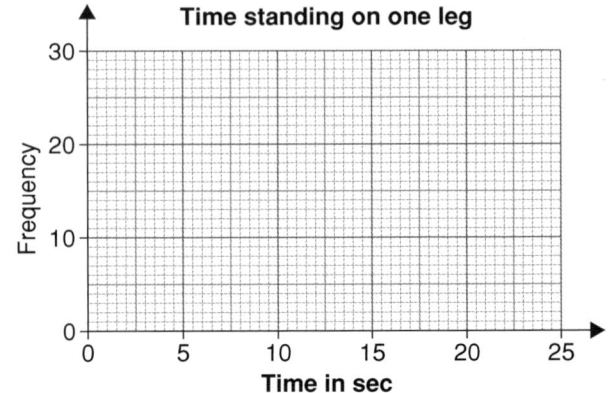

12 The total price of £1200 for a package holiday is made up of these costs:

Category	Flight	Hotel accommodation	Meals	Insurance	Excursions
Cost	£540	£420	£140	£40	£60

a Work out the pie chart angle for each sector.

b Draw the pie chart to show the data.

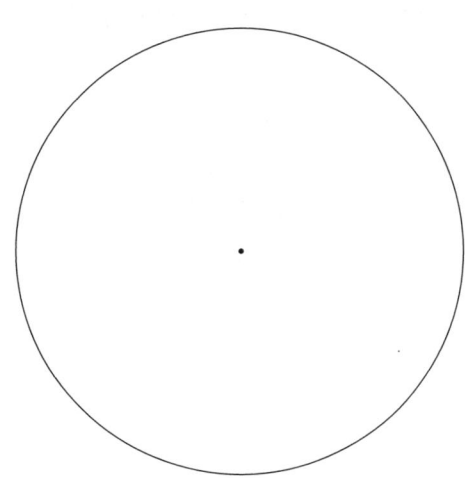

13 Rachid collected data about the heights and shoe sizes of 12 people.

Height in cm	183	167	169	172	175	174	180	178	181	173	168	175
Shoe size	11	6	$6\frac{1}{2}$	$8\frac{1}{2}$	8	7	10	9	$10\frac{1}{2}$	$7\frac{1}{2}$	7	$7\frac{1}{2}$

a Draw a scatter graph for this data.

b Does the scatter graph show that taller people tend to have larger feet?

....

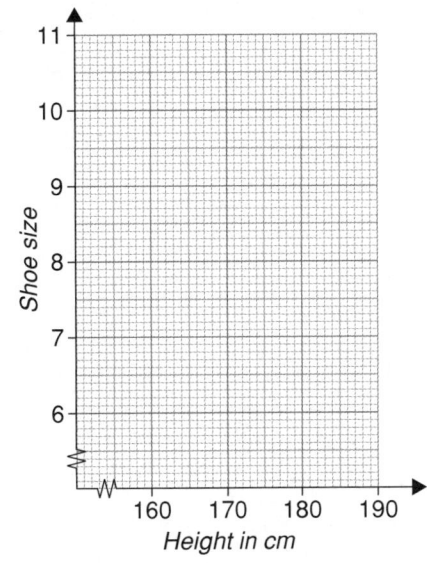

CHAPTER 16

14 This graph shows the percentage of children aged 3 to 17 who had access to a home computer and who used the Internet at home from 1984 to 2001.

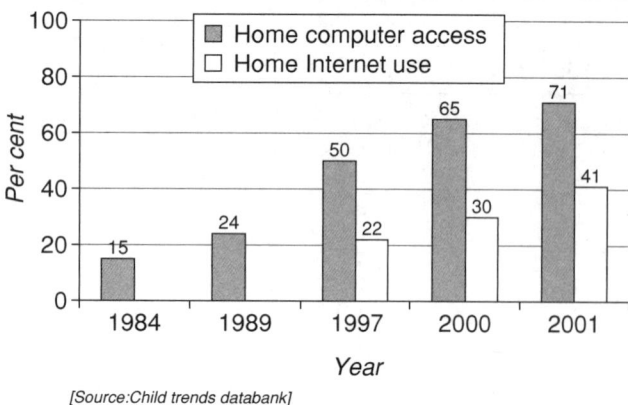

[Source:Child trends databank]

a How has home computer and Internet use changed?

b Using the data in the graph, predict what you think might happen by 2010.

Name ....

1 Mandy has three boxes of mints. Some mints are white imperial mints and some are striped everton mints.

A    **B**    **C**

Which box, A, B or C is she most likely to pick an everton mint from, if she takes a mint without looking?

....

2 Edwin is designing a game where you win by landing on a ✳ on a spinner. He has made three different spinners.

A 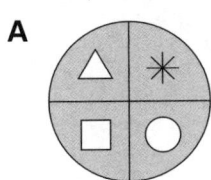   **B** 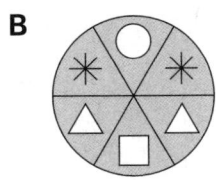   **C**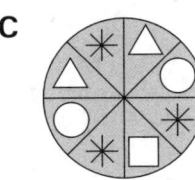

Which spinner would give you the greatest chance of a win?

....

Explain why:

3 The probability of a car engine failing in the first six months is 0·17. What is the probability of the car engine not failing?

....

4 Jessica draws a card from an ordinary pack of 52 playing cards. What is the probability that it is

a a 4

....

b a number from 6 to 9 inclusive

....

c a picture card (Jack, Queen or King)

....

d not an Ace?

....

5 Jack tossed a six-sided and a twelve-sided dice and added the scores together.

a Complete the table below to show all the possible answers.

12-sided dice

+	1	2	3	4	5	6	7	8	9	10	11	12
1												
2												
3												
4												
5												
6												

6-sided dice

b Find the probability of getting a total of:

i 12

....

ii less than 6

....

iii 16 or more

....

6 Liam has 4 discs numbered 1, 2, 3 and 4. He puts them in a bag, draws one out at random and notes its number. He then puts it back in the bag and draws a second disc out at random, noting its number.

a Draw a sample space to show the numbers on the two discs.

b Find the probability that:

i both discs were number 3

....

ii both discs were the same number

....

iii the discs had different numbers on

....

iv the numbers on the discs added up to 3

....

v only one disc had the number 4 on

....

vi at least one disc had the number 2 on

....

vii no discs had the number 3 on.

....

7 Spiros spun a spinner 60 times.

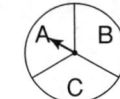

It landed on A 17 times.

a Estimate the probability that the next time it is spun, it will land on A again.

....

b He spun it another 60 times. It landed on A 19 times. Use all 120 results to estimate the probability of it landing on A.

....

c Which probability is more accurate? ....

Explain why.

8 Geena has a bag containing one red, one blue, one green, one yellow and one black counter. She takes one out at random.

a What is the theoretical probability that it is black? ....

b Describe an experiment you could do to estimate the experimental probability that the counter is black.

c Would the experimental probability be more likely to be closer to the theoretical probability if you did 50 trials or 500 trials?

....

Chapter 1 Review

1 a 10^3 **b** 10^5 **c** 10^6

2 Three times ten to the power nine

3 2 499 999

4 a 2·935   **b** 11·257   **c** 4·69 or 4·690
    d 8·469   **e** 6·362   **f** 12·417

5 0·0006

6 C $2·15 \div 100$

7 a 0·0473   **b** 28   **c** 0·394   **d** 1625

8 a 2·985 > 2·979   **b** 6·309 < 6·313   **c** ⁻2·48 > ⁻2·51

9 a 3·642, 3·64, 3·624, 3·4, 3·246
    b 0·582, 0·52, 0·285, 0·208, 0·08

10 a 3·25   **b** 1·065   **c** ⁻0·45

11 a h is greater than 155 cm and less than or equal to 160 cm
    b [A]

12

a	6 479 000	6 480 000	6 000 000
b	1 847 000	1 850 000	2 000 000
c	4 079 000	4 080 000	4 000 000

13 a 949 500   **b** 950 499

14 a 16·44   **b** 1·8   **c** 12·4
    d 12·41   **e** 2·0   **f** 4·90

15 a 62·29 or 62·22 g   **b** 0·857142 or 0·86 m
    c 8·3 or 8·33 m   **d** 136·36 or 136·36 g

Chapter 2 Review

1 a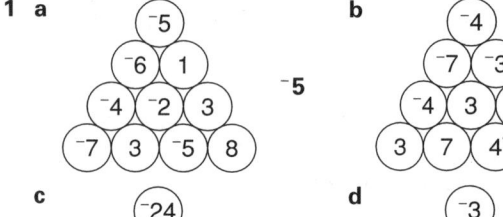

    c 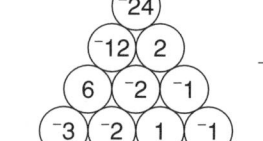   **d**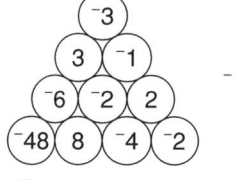

2 a 32   **b** 1   **c** ⁻7

3 ⁻3 × 4 = ⁻12, ⁻12 ÷ 4 = ⁻3, ⁻12 ÷ ⁻3 = 4

4 a

+	⁻15	8	⁻4
⁻11	⁻26	⁻3	⁻15
⁻9	⁻24	⁻1	⁻13
13	⁻2	21	9

    b

×	6	⁻5	⁻2
⁻4	⁻24	20	8
8	48	⁻40	⁻16
⁻3	⁻18	15	6

5 a 12   **b** ⁻20   **c** ⁻2   **d** ⁻6
    e ⁻7   **f** ⁻4

6 a 1260, 2376, 6048   **b** 1035, 1260
    c 558, 1260, 2376, 6048

7 a $2 \times 3 \times 3 \times 3 \times 7$ **b** 378   168

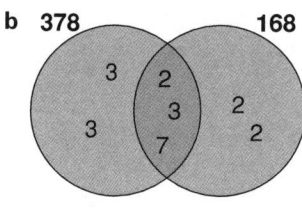

    c 42   **d** 1512

8 a 49   **b** ⁻6   **c** 125   **d** 4

9 a 18·49   **b** 3·35   **c** 9·26   **d** ⁻512
    e 14   **f** 9

10 The cube of 8

11 Any three from 41, 53, 59, 83 and 89.
    Possible answers are:
    $41 = 1^2 + 2^2 + 6^2$ $53 = 1^2 + 4^2 + 6^2$ $59 = 1^2 + 3^2 + 7^2$
    $83 = 3^2 + 5^2 + 7^2$ $89 = 2^2 + 6^2 + 7^2$ or $3^2 + 4^2 + 8^2$

Chapter 3 Review

1 a 2   **b** ⁻3   **c** 0·1

2 141

3 a $x = 37$ **b** $x = 6·26$ **c** $x = 652$ **d** $x = 73$

4

1·0	0·9	1·4
1·5	1·1	0·7
0·8	1·3	1·2

5 a 810   **b** 260   **c** 640   **d** 519

6 a 1200   **b** 30   **c** 720   **d** ⁻126

7

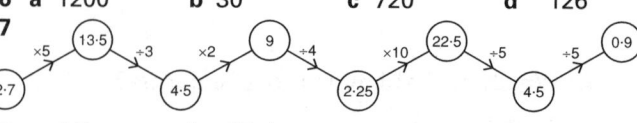

8 a 4·5   **b** ⁻23·4

9 a 3·6   **b** 0·16   **c** 0·05   **d** 0·125

10 a 52   **b** 39   **c** ⁻6·3   **d** 31

11 a 3400   **b** 77°   **c** £1·66   **d** 0·43
    e 9   **f** 32

12 £5·60

13 An estimate, because the number that will fit into the field will vary, depending on how close together they sit.

14 a $200 \div 50 = 4$ **b** $40 \times 4 = 160$
    c $\dfrac{9 \times 4}{6} = 6$ **d** $\dfrac{15 - 3}{2 + 2} = \dfrac{12}{4} = 3$

15 a Yes, he is correct. $58 \times £4·85 \approx 60 \times £5 = £300$
    b Higher, because both 58 and £4·85 were rounded up for the estimated cost.

Chapter 4 Review

1 a 22·29   **b** 14·62

2 £3·20

3 a £20·86   **b** £9·14

4 a 10·02   **b** 115·2

5 a 27   **b** 6

6 a 11·84   **b** 37·25

7 a 32·8   **b** 56·2

8 17·6 cm

9 A doughnut is cheapest.

10 a 405   **b** 6200   **c** 360

11 a 95·7   **b** 653·8

12 No – because $8(\underline{18}) \times 6(£4·9\underline{6}) = 48$, so the answer should end in 8.
    (Accept other reasonable explanations.)

13 Because when you divide by a decimal between 0 and 1, your answer should be bigger than 7·56.

14 a $3·6 \times 3 \times 0·6$ or (other equivalent)
    b $= 10·8 \times 0·6 = 6·48$

15 a 78×8 **b** 624   **No**

16 a 23·2   **b** 3·3

17 a 1·8   **b** 0·7   **c** 2·9   **d** 1·2

18 £56·56

Chapter 5 Review

1 a $\frac{1}{2}$ Blue   **b** $\frac{1}{3}$ Green

2 a $\frac{1}{4}$ **b** $\frac{1}{2}$

3 $\frac{2}{5}$ cats

4 a $\frac{3}{40}$ **b** $\frac{36}{125}$

5 a 0·95   **b** 0·28   **c** 0·83   **d** 5·5
    e 5·75

6 $\frac{3}{8} = 0·375$

7 a 0·6̇   **b** 0·8̇1̇

8 a <   **b** <   **c** >

9 The second round.

10 $\frac{1}{4}, \frac{7}{24}, \frac{1}{2}, \frac{5}{8}, \frac{2}{3}, \frac{5}{6}$

e ....

Except for questions **16**, **17** and **18**.

a 4·68 + 0·72 + 3 + 13·89   **b** 4·6 − 0·28 − 1·4 + 11·7

....

Beppe bought a book for £6·95, a CD for £8·49 and a magazine for £1·36.
How much change would he get from £20?

....

Petrol cost 79p a litre. Sian puts 26·4 litres of petrol into her car.
a How much will it cost, to the nearest penny?

....

b How much change will she get from £30?

....

a 16·7 × 0·6   **b** 24 × 4·8

....

554 cartons of yoghurt are packed in boxes of 24.
a How many boxes can be filled?

....

b How many cartons are left?

....

6 Work out the answers to these, as decimals.

   a 296 ÷ 25   **b** 596 ÷ 16

   ....

7 Give the answers to these, rounded to 1 d.p.

   a 427 ÷ 13   **b** 1236 ÷ 22

   ....

8 334·4 cm of string can be wound round a circular bobbin exactly 19 times. How far is it once round the bobbin?

                                                                                                       ....

9 Which is cheapest, a doughnut or an iced bun?

18 doughnuts for £4.68

23 iced buns for £6.05

   ....

10 a 162 ÷ 0·4   **b** 372 ÷ 0·06   **c** 432 ÷ 1·2

   ....

11 Give the answers to these to 1 d.p.

   a 67 ÷ 0·7   **b** 85 ÷ 0·13

   ....

12 Raoul worked out the total cost of 18 boxes of frozen chips at £4·96 a box as £93·42. Is this answer sensible? Explain why.

   ....

   ....

13 Tiffany wrote down 2·52 as the answer to 7·56 ÷ 0·3. Explain how you know her answer cannot be correct.

   ....

   ....

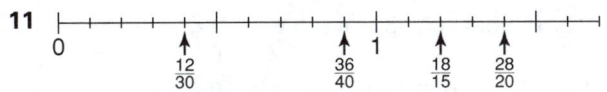

11
```
├──┼──┼──┼──┼──┼──┼──┼──┼──┼──┼──┤
0        ↑        ↑   1    ↑    ↑
        12/30     36/40   18/15 28/20
```

12 $\frac{5}{18}$

13 a $\frac{5}{7}$ b $\frac{5}{8}$ c $\frac{9}{10}$ d $\frac{13}{36}$ e $\frac{17}{30}$ f $\frac{61}{70}$

14 a $7\frac{1}{3}$ litres b $11\frac{1}{4}$ km c $4\frac{4}{11}$ m d $4\frac{2}{3}$ kg

15 45 videos

16 a $8 \div \frac{1}{6}$ b 48

17 10

Chapter 6 Review

1

Decimal	Percentage	Fraction
0·45	45%	$\frac{9}{20}$
0·6	60%	$\frac{3}{5}$
0·16	16%	$\frac{4}{25}$
1·2	120%	$1\frac{1}{5}$
1·75	175%	$1\frac{3}{4}$

2 a 66% b 36% c 36·84%

3 30·8%

4 0·42, $\frac{13}{30}$, $\frac{4}{9}$, 47%, 0·5

5 Secondary School children

6 a 10% = 36 b 10% = 70
     5% = 18   20% = 140
     15% = 54   1% = 7
                   21% = 147

7 57·12 cm

8 a 10% = 6 m b 10% = 8 g
     1% = 0·6 m   40% = 32 g
     3% = 1·8 m   1% = 0·8 g
     13% = 7·8 m   2% = 1·6 g
                    42% = 33·6 g

   c 10% = £15 d 10% = 8·5 litres
     40% = £60   60% = 51 litres
     1% = £1·50   1% = 0·85 litre
     39% = £58·50 161% = 85 + 51 + 0·85 litres
                        = 136·85 litres

9 £157·25

10 Top Prints

11 £360 + £787·20 = £1147·20

12 a about 12 b 48–54

13 12·99 mm or 13 mm

14 58 Earth days

Chapter 7 Review

1 a £24·75 b £11·85

2 630 trains

3 a 4 : 3 : 7 b 4 : 11 : 15 c 3 : 4 d 12 : 5

4 4 : 1

5 7 : 50

6 a 3 : 5 b 5 : 10 : 7 c $\frac{3}{14}$ d 25% e 0·2

7 24

8 a £1200 b £3840

9 210 wear glasses

10 630 g

11 18 : 45 : 54

12 a 28 teachers b 420 girls

13 iron 658 g, zinc 282 g, copper 188 g

Chapter 8 Review

1 A Expression B Function C Equation D Formula

2 a $3xy$ b $5m^2$ c $2c^2$ d $16k^2$
   e $\frac{a+b}{c}$ f $20n$

3 True, True, False, True, True

4 a B b C

5 a $35m$ b ^-6x c ^-24y d $12k$
   e p^5 f $3n$ g b^3 h $2t^2$
   i r^3

6 a $4h + 32$ b $2m - 12$ c $^-5c - 15$ d $^-3x + 12$
   e $3p + 6q$

7 a $6x + 6y$ b $9p + 10$ c $5a^2 - 3a$ d $7x + 5$
   e $11 - m$

8 a 1 b 5·8 c 2 d $^-1·4$

9 a $^-7$ b 10 c 2 d 13

10 a $v = 28$ m/s b $v = 72$ m/s

11 $A = 9·3$ cm^2

12 a $x - 5$ b $2x + 2$ c $\frac{2x + 2}{3}$ d $x + 7$

13 $77 - 3m$, $7(11 - m) + 4m$, $3(11 - m) + 44$
   $7(11 - m) + 4m$ $3(11 - m) + 44$
   $= 77 - 7m + 4m$ $= 33 - 3m + 44$
   $= 77 - 3m$ $= 77 - 3m$
   They all equal $77 - 3m$.

14 $L = 2M + 4$

15 a $2h + h = 27$, Flash weighs 9 kg
   b $3n + 6 = 5n - 4$, the number is 5
   c $3(x + 4) = 5x$, the number is 6
   d $3p - 30 + 6p - 50 = 190$, each cash prize was 30p

16 a $m = \frac{5}{6}$ b $y = 7$ c $p = 7$ d $a = 7$
   e $x = 4$

17 $x = 15°$

18 a

Number of inches (i)	10	20	30	40	50
Number of centimetres (c)	25	50	75	100	125

   b Yes
   c 5 : 2
   d

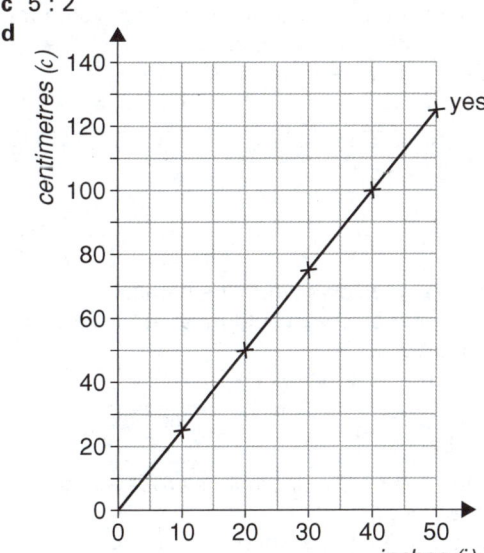

   e $c = 2·5i$
   f 90 centimetres

Chapter 9 Review

1 0·4, 0·6, 0·8, 1·0, 1·2

2 a 243, 81, 27, 9, 3 b 2, $^-4$, 8, $^-16$, 32

3 a 3, 4, 7, 12, 19, 28 b 16, 15, 13, 10, 6, 1

4 a 5, 8, 11, 14, 17, 20 b 3, 1, $^-1$, $^-3$, $^-5$, $^-7$
   c $^-9$, $^-5$, $^-1$, 3, 7, 11

5 a 14, 19, 25 b 56, 49, 42
   c 36, 49, 64 d 18, 6, 2

6 11, 18, 29 and 12, 19, 28
   (add two previous terms) (add, 1, 3, 5, 7, ...)

7 a 9, 13, 27 b 6, 11, 28 c 6, 24, 720

8 a 9, 11, 13, 15, 17   **b** 45, 42, 39, 36, 33
   c 1·5, 3, 4·5, 6, 7·5

9 a 1st term 9, **rule** add 2
  b 1st term 45, **rule** subtract 3
  c 1st term 1·5, **rule** add 1·5

10 a A sequence of ascending numbers with a difference of 5 which are all 2 less than a multiple of 5. It starts at 3.
  b A sequence of descending numbers with a difference of 7 which are all multiples of 7. It starts at 70.

11 a

Pattern 4
16 tiles

b

Pattern number	1	2	3	4	5
Number of tiles	7	10	13	16	19

c The sequence starts at 7 and increases by 3 each time.
d Each arm has n, the shape number, squares. There are always 4 dark grey squares in each shape. So in the nth shape there are $3 \times n$ squares in the arms plus 4 dark grey squares.
e $3n + 4$
f No, because there is no whole number value for n that will make a $3n + 4$ equal to 30.

12 a $6n - 3$ **b** $7n + 1$ **c** $70 - 4n$

13 a 37, 17, ⁻18, 15   **b** 0, ⁻44, ⁻8, 6

14

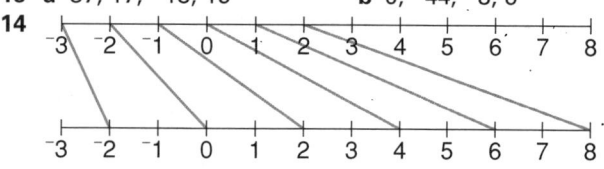

15 a The mapping arrows would be parallel.
  b The mapping arrows would meet at the zero line if extended back.

16 a $y = 2x + 3$ **b** $y = 3(x - 2)$

17 a $x \to \boxed{\text{multiply by 2}} \to y$ **b** $x \to \boxed{\text{subtract 7}} \to y$

18 No, because when the operations are reversed you get a different function.
The first is $y = 2x - 5$, so when $x = 7$, $y = 9$.
The second is $y = 2(x - 5)$, so when $x = 7$, $y = 4$.

19 a $x \to \dfrac{x - 2}{5}$ **b** $x \to \dfrac{x}{2} + 3$ **c** $x \to 3(x - 4)$
   d $x \to 3x - 6$ **e** $x \to \dfrac{x}{4} - 8$ **f** $x \to \sqrt{x + 3}$

20 a 3, ⁻4, 8   **b** 7, ⁻5, 4

Chapter 10 Review

1 a

x	⁻3	0	2
y	⁻2	4	8

b

x	⁻1	1	4
y	⁻4	2	11

c

x	⁻2	0	4
y	10	4	⁻8

2 a (⁻2, 10) (0, 4) (4, ⁻8)   **b**

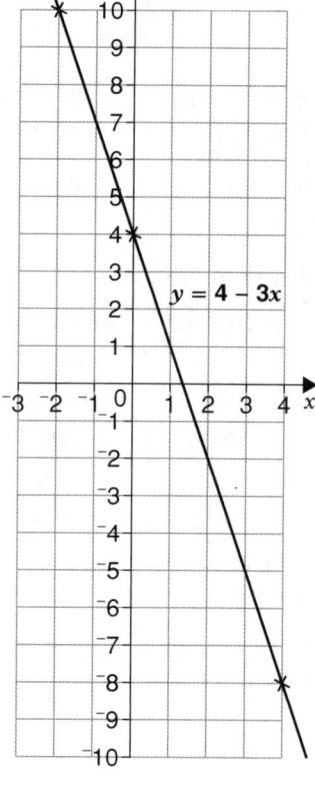

c Yes, (2, ⁻2) does lie on the line.
d Possible answers are (⁻1, 7) (1, 1) (3, ⁻5) Yes the coordinate pair satisfies the equation $y = 4 - 3x$.
e i Yes   **ii** No

3 a

Number of hexagons	1	2	3	4	5	6
Number of matchsticks	6	11	16	21	26	31

b The points do all line in a straight line.

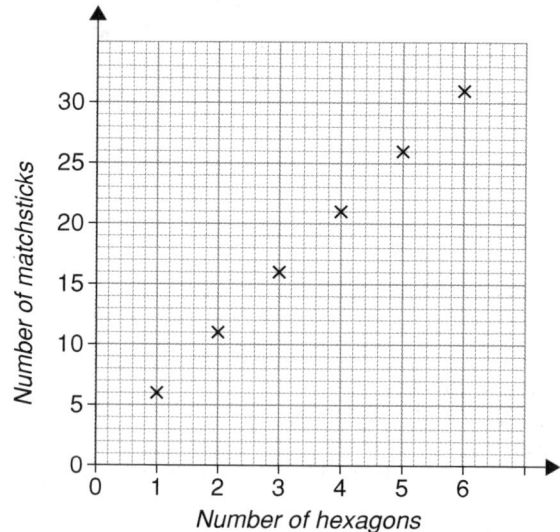

c 24 is not a term of the sequence because all the terms of the sequence are 1 more than multiples of 5 and 24 is not.
d It would not be sensible to draw a straight line because the points between the plotted ones are meaningless. You can't have non whole numbers of hexagons.

4 a B   **b** A   **c** A   **d** A   **e** C

5 a 14 kg   **b** 20 kg   **c** 3 kg   **d** 6 to 8 years
  e about 2 kg

6 a $l = 0.55p$

b

p	0	10	20
l	0	5·5	11

c

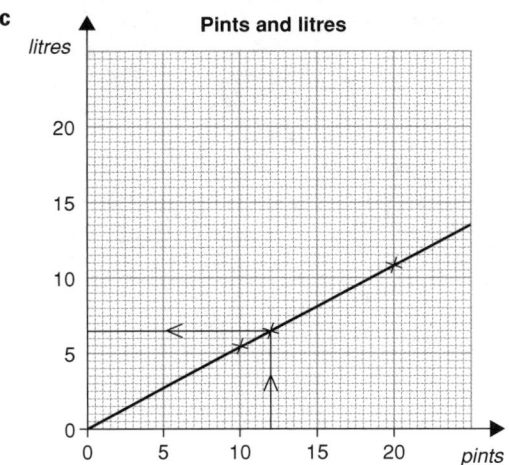

Pints and litres

d about 6·6 litres

e 70 pints   220 pints   180 pints
  =   =   =
  40 litres   120 litres   100 litres

7

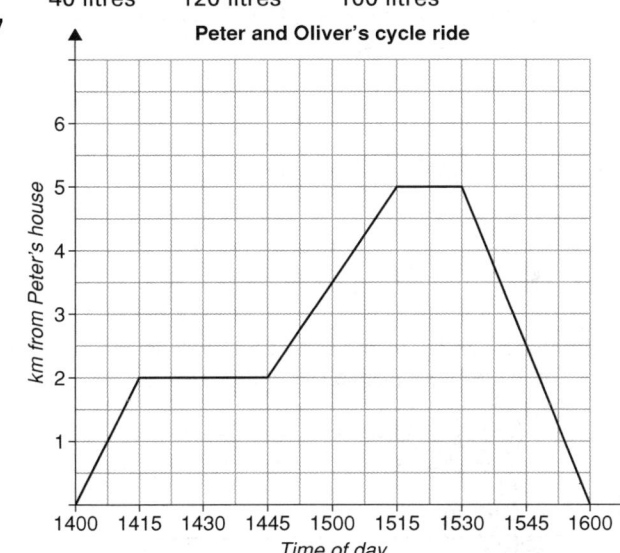

Peter and Oliver's cycle ride

8 a The watering can starts empty but is filled up. The volume of water remains constant until the first plant is watered, when the volume decreases. It then stays constant again until the second plant is watered, when it decreases again. This then happens again for a third plant by which time the watering can is empty again.

b The car starts from rest and its speed increases rapidly at first, then more gradually until it reaches a certain speed. It then continues at a constant speed for a while, before braking suddenly to be brought to a rapid halt.

9 Week 3

10 Graph 1 is for container B because the narrow lower part of the container means it fills more rapidly than for the wider upper part. So the graph has a steep section, then a less steep section.
Graph 2 is for container A, because it is the other way round from container B: wide first, then narrow, so the graph is less steep first, then steeper in the second section.

11 a A

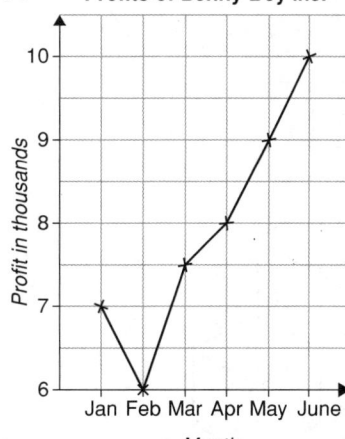

Profits of Benny Boy Inc.

B

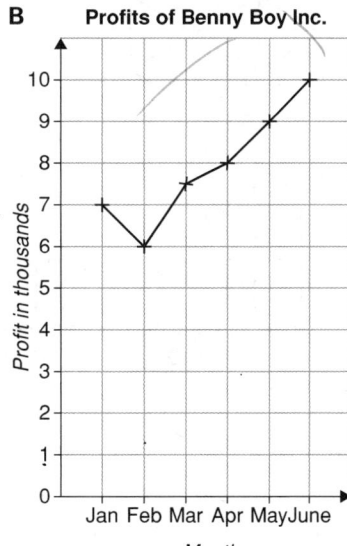

Profits of Benny Boy Inc.

b Graph A, because the scale used makes the graph appear to show a much steeper rise in profits — it is a misleading graph.

Chapter 11 Review

1 a B   **b** C

2 a Possible reasons are given.
  $a = 63°$ (corresponding angles)
  $b = 180° - 63°$ (angles on a straight line)
   $= 117°$

b $c = 117°$ (corresponding angles)
  $d = 63°$ (alternate angles)
  $e = 117°$ (alternate angles)

c $f = 38°$ (alternate angles)
  $g = 75°$ (alternate angles)
  $h = 180° - (38° + 75°)$ (angles on a straight line)
   $= 67°$

3 a $m = 36°$, $n = 57°$ **b** $p = 302°$ **c** $q = 66°$, $r = 114°$
  d $s = 44°$, $t = 68°$, $u = 68°$, $v = 68°$
  e $w = 42°$, $x = 44°$, $y = 42°$, $z = 44°$

4 a $103°$ **b** $78°$

5 a $x = 20°$ **b** $x = 12°$ **c** $x = 25°$

Chapter 12 Review

1 a Square, rectangle, parallelogram, rhombus
  b Parallelogram

2 a Rectangle, rhombus
  b Square, rhombus, kite, arrowhead
  c Kite, trapezium, arrowhead

3 a A square   **b** 4 red faces   **c** 8 edges

4 Any correct tessellation.

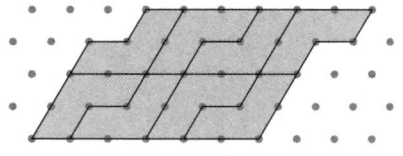

5 a 43°   **b** 6·6 cm

6 a

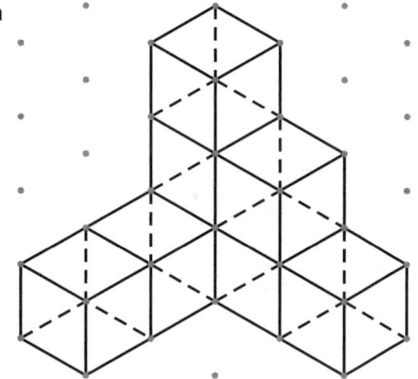

  b   **c**   **d**

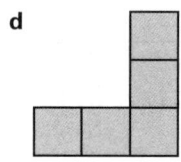

Plan view   Front elevation   Side elevation

7 a C   **b B**   **c D**   **d A**

8

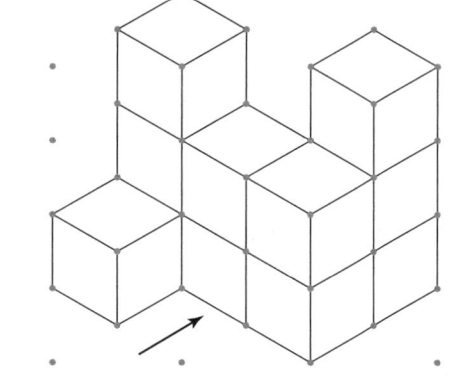

9

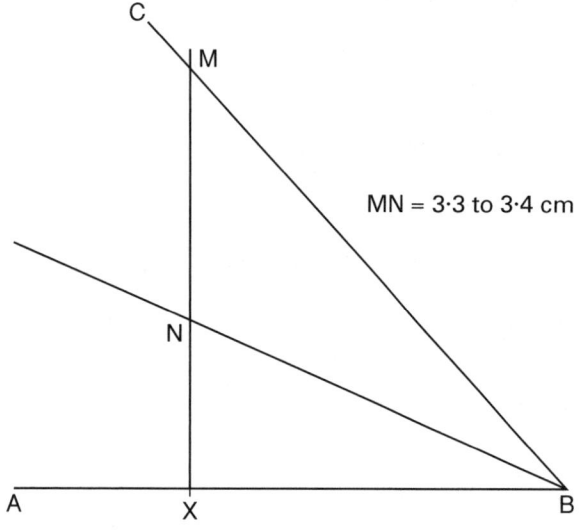

MN = 3·3 to 3·4 cm

10

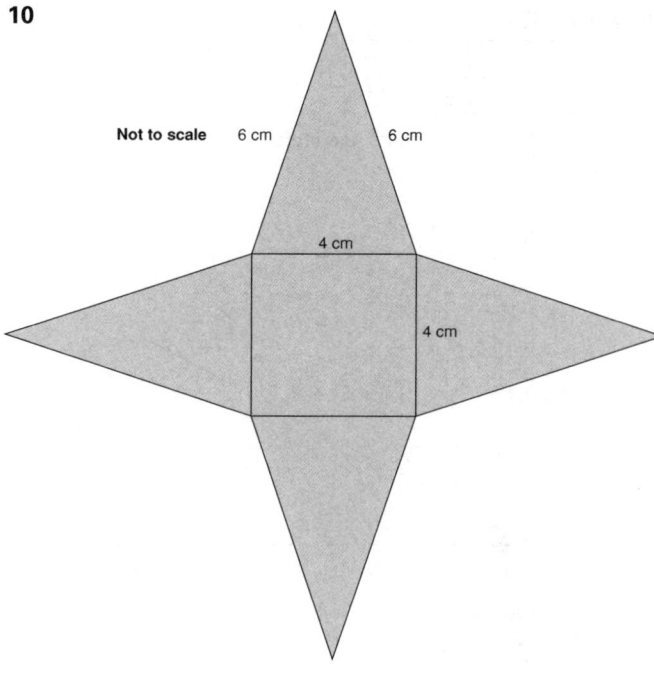

Not to scale   6 cm   6 cm

4 cm

4 cm

11 a A circle with the hand as its centre and the radius of the circle will be the length of the string

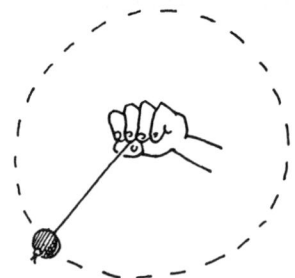

  b A line parallel to the top of the wall and the height of the cat's head above the wall.

  c The perpendicular bisector of the line joining the two telegraph poles.

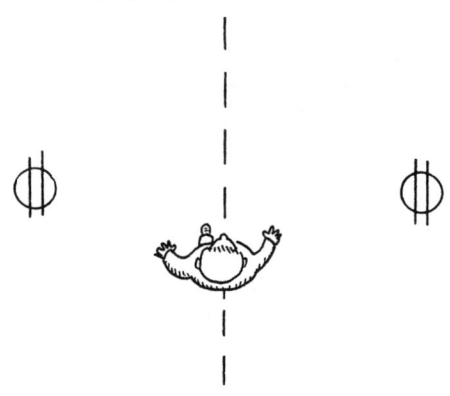

Chapter 13 Review

1 a A′($^-$2, 1)   B′($^-$5, 2)   **b** A′($^-$2, $^-$1)   B′($^-$5, $^-$2)
      C′($^-$3, 5)   D′($^-$1, 4)   C′($^-$3, $^-$5)   D′($^-$1, $^-$4)
 c A′(0, $^-$6)   B′(3, $^-$5)
      C′(1, $^-$2)   D′($^-$1, $^-$3)

2 a

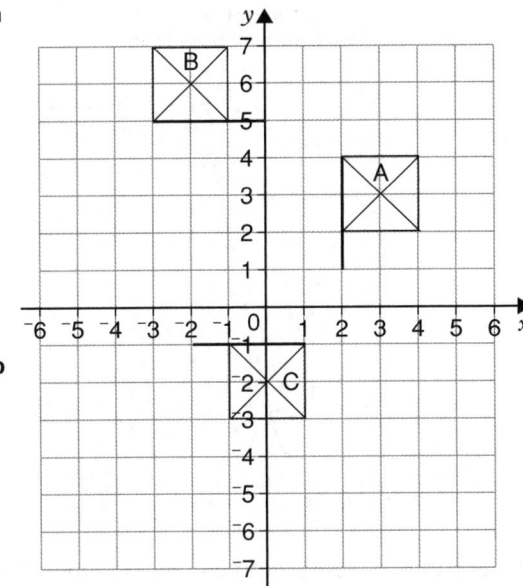

b

c Rotation 90° clockwise about (⁻1, 2) or Rotation 270° about (⁻1, 2)

3 a i B   **ii** E   **iii** A
   b i D   **ii** E   **iii** A

4 a 1 line of symmetry; no rotational symmetry
   b 5 lines of symmetry; rotational symmetry of order 5
   c No lines of symmetry; rotational symmetry of order 4
   d 4 lines of symmetry; rotational symmetry of order 4

5

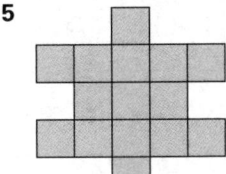

6

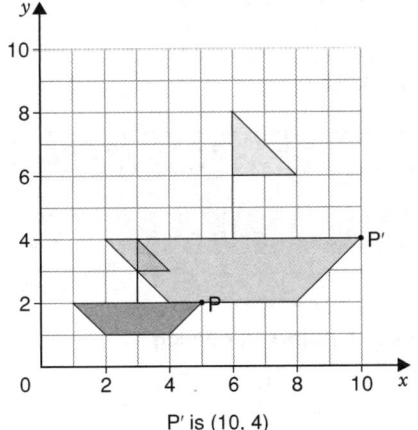

P′ is (10, 4)

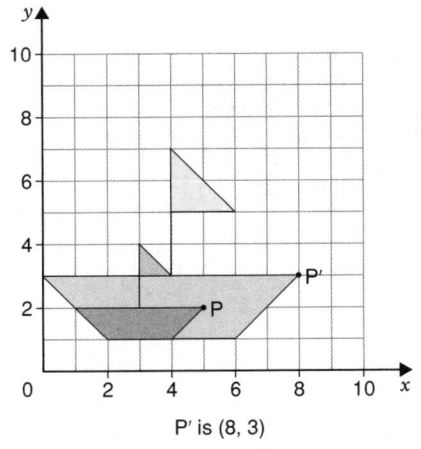

P′ is (8, 3)

7 34–40 m
8 a 150 m   **b** 550 m   **c** 425 m
9 a (5, 8)   **b** (⁻1½, 2)

Chapter 14 Review

1 a 140   **b** 2·5   **c** 2900
   d 0·586   **e** 120   **f** 47
   g 1600   **h** 3·46   **i** 2·7
2 a 7 minutes 50 seconds   **b** 6 hours 24 minutes
   c 7 years 11 weeks   **d** 1 day 19 hours
3 1·72 kg
4 400 m
5 6:22 p.m. on 4 January 1975
6 11·05 litres
7 a 19·2 km, 240 g, 45 cm, 600 mℓ, 100 m, 22·5 ℓ
   b 5 oz, 20 inches, 14 inches, 0·75 pints
8 a to the nearest minute
   b to the nearest 10 cm
   c to the nearest 100 g
   d to the nearest cℓ
9 a 13–17 cm   **b** 200–250 mℓ   **c** 300–400 g
10 a 034°   **b** 113°   **c** 254°   **d** 312°
   (Allow ±3°)

11

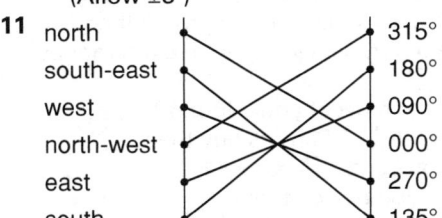

north	315°
south-east	180°
west	090°
north-west	000°
east	270°
south	135°

12 a 44 cm²   **b** 108 cm²
13 a 21·93 cm²   **b** 83·74 cm²
14 a 13 cm²   **b** 8·5 cm²
15 a 240 cm³   **b** 6 cm
16 a 30 000 cm³ **b** 24 000 cm³
17 195 pencil sharpeners

Chapter 15 Review

1 a C   **b** D   **c** D   **d** C
2 a

Distance (in m)	Tally	Frequency
30 ⩽ d < 40	III	3
40 ⩽ d < 50	ЖЖ I	6
50 ⩽ d < 60	ЖЖ III	8
60 ⩽ d < 70	ЖЖ	5
70 ⩽ d < 80	ЖЖ	5
80 ⩽ d < 90	III	3

   b 6   **c** 13
   d No, because the data would not be distributed widely enough.
3 a 7   **b** 20   **c** 22
   d More females than males visited altogether. The majority of the females wanted a trim and blow wave rather than just a dry trim or colour. Most of the men wanted just a dry trim.

4 Possible answers are:
  a i Do you have access to the Internet at home?
    ii Do you only have access to the Internet for a restricted time?
    iii Estimate how many times you log on to the Internet each week.
  b i 65% have access to the Internet at home.
    ii 83% can only use the Internet after 6 p.m.
    iii Most pupils used the Internet service at least three times a week.
  c For each person the data needed is, whether they use the Internet at home or in school, how often, at what times of the day and for how long.
  d A good sample size would be 30–50 people.
  e Choose equal numbers of boys and girls, a range of ages and choose using a random method.
  f primary
5 Possible answers are:
  i Are you male ☐ female ☐?
  ii How much pocket money do you get each week?
    Less than £1 ☐ At least £1 but less than £4 ☐
    At least £4 but less than £8 ☐
    At least £8 but less than £12 ☐
    At least £12 but less than £16 ☐ More than £16 ☐
  iii Do you have a part-time job to earn extra money?
    Yes ☐ No ☐
  iv How do you get most of your clothes?
    Buy them myself ☐ Parents buy them ☐
    Get them as presents ☐ Other ☐
  v How often do you buy clothes?
    Every week ☐ Every month ☐
    Less often than that ☐
  vi How much do you usually spend on clothes on average each month?
    Less than £10 ☐ At least £10 but less than £20 ☐
    At least £20 but less than £30 ☐
    At least £30 but less than £50 ☐ More than £50 ☐

Chapter 16 Review
1 **a** $75 < L \leqslant 80$
  b A possible answer is:
    More officials could stand near that area to measure the distances of the throws.
2 **a** Friday: 30 minutes   Sunday: 13 minutes
  b A possible answer is:
    To plan what time he should leave home to be certain of getting to the airport on time, even when the traffic is really heavy.
3 2·26 goals (2 d.p.)
4 69·5 beats per minute
5 6 and 12
6 £19·50 each day
7 Median = 27, Mode = 19, Range = 58
8 French: Mean = 6·58, Median = 7, Mode = 8, Range = 7
  German: Mean = 6·5, Median = 6·5, Mode = 7, Range = 4
  A possible answer is:
  French, because the mean and median marks are higher than German, although the range is greater, meaning the marks vary more.
  She could choose German because her marks have a smaller range meaning they are more consistent and the mean and median are only a little less than for French.

9 **a**

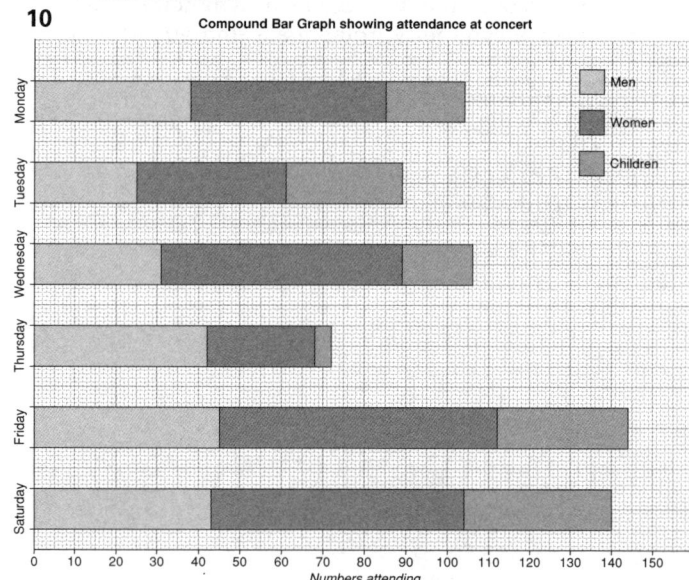

b No, because the data gives the average monthly rainfall and we do not know when in the month the rain fell.
c The rainfall varies more in The Gambia, with very low rainfall for most of the year, but high from July to September. In Kenya the highest rainfall is in April and May, but not as high as the highest in The Gambia.

10

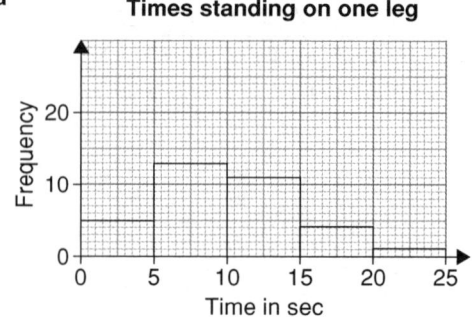

11 **a**

Times standing on one leg   **b** 5

12 a Flight   162°
    Hotel accommodation   126°
    Meals   42°
    Insurance   12°
    Excursions   18°

b

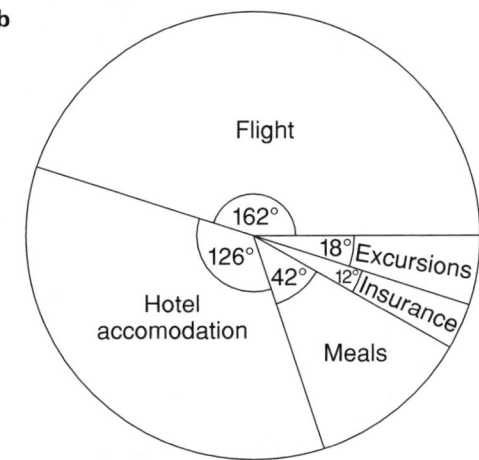

13 a

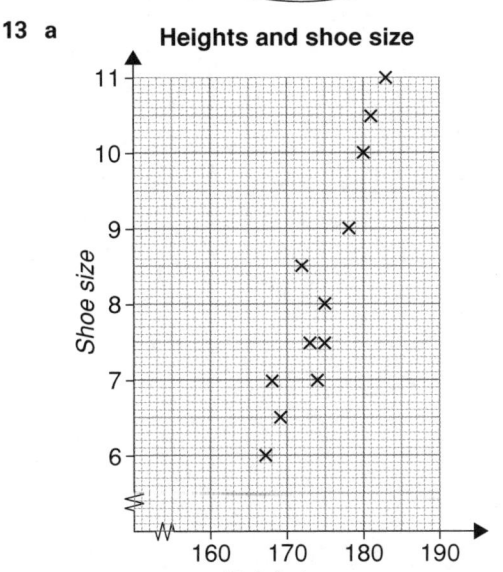

b Yes

14 a There has been a marked increase in the number of children who have access to a home computer since 1984, from less than a quarter of homes to almost three-quarters. No children had Internet access at home in 1989 because it wasn't available, but in the nineties the number of children with Internet access at home has steadily risen, with about $\frac{2}{5}$ of them accessing it in 2001.

  b If the trend continues, nearly all children will have access to a home computer by 2010 and most will be able to access the Internet.

Chapter 17 Review

1 Box B

2 Spinner C because it has the highest probability.
Spinner A: $\frac{1}{4}$ or 0·25   Spinner B: $\frac{2}{6}$ or 0·3$\dot{3}$
Spinner C: $\frac{3}{8}$ or 0·375

3 0·83

4 a $\frac{4}{52}$ or $\frac{1}{13}$ **b** $\frac{16}{52}$ or $\frac{4}{13}$ **c** $\frac{12}{52}$ or $\frac{3}{13}$ **d** $\frac{48}{52}$ or $\frac{12}{13}$

5 a

+	1	2	3	4	5	6	7	8	9	10	11	12
1	2	3	4	5	6	7	8	9	10	11	12	13
2	3	4	5	6	7	8	9	10	11	12	13	14
3	4	5	6	7	8	9	10	11	12	13	14	15
4	5	6	7	8	9	10	11	12	13	14	15	16
5	6	7	8	9	10	11	12	13	14	15	16	17
6	7	8	9	10	11	12	13	14	15	16	17	18

  b i $\frac{6}{72}$ or $\frac{1}{12}$ **ii** $\frac{10}{72}$ or $\frac{5}{36}$ **iii** $\frac{6}{72}$ or $\frac{1}{12}$

6 a

	1	2	3	4
1	1, 1	1, 2	1, 3	1, 4
2	2, 1	2, 2	2, 3	2, 4
3	3, 1	3, 2	3, 3	3, 4
4	4, 1	4, 2	4, 3	4, 4

  b i $\frac{1}{16}$ **ii** $\frac{4}{16}$ or $\frac{1}{4}$ **iii** $\frac{12}{16}$ or $\frac{3}{4}$ **iv** $\frac{2}{16}$ or $\frac{1}{8}$
     v $\frac{6}{16}$ or $\frac{3}{8}$ **vi** $\frac{7}{16}$ **vii** $\frac{9}{16}$

7 a $\frac{17}{60}$ **b** $\frac{36}{120}$ or $\frac{3}{10}$
  c The second estimate is more accurate because the more trials you do, the more accurate the result.

8 a $\frac{1}{5}$
  b Draw the counters out, one at a time, replacing when the colour has been noted. Repeat this for 200 times altogether.
  c More likely to be closer for 500 trials.

3

ICT Support

The use of ICT with New National Framework 8 Core

The aim of this resource is to help you and your pupils use ICT in meaningful mathematical activities. As a natural consequence of this, you and your pupils should become proficient users of the software.

Software used is as follows:

Excel 97, 2000 or later (a spreadsheet)
Geometer's SketchPad (a dynamic geometry package, Cabri-géomètre is an alternative)
MswLogo (there are other alternatives)
Omnigraph for Windows (a graph plotter: Autograph is an alternative)
Smile for Windows (short programs)

Other resources that users may find useful are:

Eagle, Ruth, *Dynamic Geometry with SketchPad* (Keele Mathematical Education Publications, 2000)
Key Maths ICT CD-ROM Year 8 (Nelson Thornes, 2002)
Miller, Dave, *Logo for Windows* (Keele Mathematical Education Publications, 1997)
Internet

Activities are provided for specific chapters where the use of ICT is considered appropriate. However it is likely that with experience you will find more activities yourself where the use of ICT is appropriate.

The layout of the resource

Information is included about spreadsheet, Geometer's SketchPad and MswLogo skills and techniques found in 7 Core, 7 Plus and 8 Core. It is assumed that pupils have followed through the work in Year 7, though some reminders are given.

For each chapter the layout is consistent. The initial content is described briefly in terms of software required — this is deliberate since it is assumed that you will have a broad view of the content of the chapter and therefore be able to identify where to place the activities in the sequence of the chapter.

First page of the chapter

ICT-based activities	short programs (including Smile for Windows)
Excel spreadsheet	spreadsheet
Geometer's SketchPad	dynamic geometry package
Internet	links to sites
MswLogo	programming language (turtle graphics)
Omnigraph for Windows	graph plotter

Following pages of the chapter

More detail follows under these headings, in the order listed on the first page of the chapter.

Worksheets

These follow at the end of the appropriate chapter.

Use of the materials

Particular emphasis is attached to a number of ideas:

- you and your pupils' ICT skills will improve with use of the materials
- the uses here will help pupils' understanding of mathematics
- you will try out the materials first before using them with pupils
- pupils need time with the software and the activity if it is to help them better understand the underlying mathematics
- liaison with colleagues in the ICT department is important, particularly if they are able to help improve pupils' basic skills with the software — you should know what pupils are learning in any generic ICT lessons

- good teaching is still required when ICT is used
- sometimes the focus of the lesson may appear to involve learning 'software' techniques, but this will help with mathematical learning at a later point
- non-classroom use of ICT, at home or in a school homework/mathematics ICT club should be encouraged — though attention should be given to an entitlement of all pupils to access ICT equipment (mathematics departments and schools need to address this issue)
- interactive whiteboards or computers with data projectors are becoming increasingly available and should be used wherever relevant and possible.

Basic assumptions in the terminology

It is assumed that your pupils know how to use a mouse. The following terminology is used:

Click to click with left-hand button on mouse, occasionally left-click is used for emphasis.

Right-click to click with right-hand button on mouse — in Excel and most Microsoft software the right-click brings up a special menu that changes according to where the mouse is positioned (a very powerful and timesaving feature).

Click and drag to click with left-hand mouse button **AND** keep the button depressed until some desired event has happened (when you let go of the mouse button).

Drag see Click and drag.

Highlight for a spreadsheet this will mean to click with left-hand mouse button **AND** keep the button depressed until all of the named cells are 'selected' or 'highlighted'. For other software this will mean to do the same thing but the process may be slightly different.

Cell one of the rectangular grid shapes given in the body of a spreadsheet.

Array a rectangle of cells, such as from A1 to B4 often referred to as A1:B4 in spreadsheet terms.

Format menu this refers to one of the menus usually found at the very top of the program (usually under the blue coloured top line). '*Format menu, Cells, Font tab*' means click on the **Format menu** then click on the word **Cells** that appears on this drop-down menu, then click on the **Font** 'tab' that you should now see.

Resources/suppliers

Geometer's SketchPad
QED Books (previously in York), Room 1, Stonehills House, Stonehills, Welwyn Garden City, AL8 6NH
Tel: 0345 402275, Fax: 01707 334233

MswLogo, Dynamic Geometry with SketchPad, Logo for Windows
Keele Mathematical Education Publications (KMEP), Department of Education, Keele University, Keele, Staffordshire, ST5 5BG
Tel: 01782 583124, Fax: 01782 583555

Omnigraph for Windows
SPA, PO Box 59, Tewkesbury, GL20 6AB
Tel: 01684 833700, Fax: 01684 833718

Key Maths ICT CD-ROM Year 8
Nelson Thornes, Delta Place, 27 Bath Road, Cheltenham GL53 7TH
Tel: 01242 267100, Fax: 01242 221914

Smile for Windows
SMILE Mathematics, Isaac Newton Centre, 108A Lancaster Road, London, W11 1QS
Tel: 020 7221 8966, Fax. 020 7243 1570:

There are also other places where you can find most of these resources.

Spreadsheet use in the ICT Support section of the Year 7 Teacher Resource Packs

The following skills and techniques are covered in the ICT Support section of the Teacher Resource Packs for 7 Core and 7 Plus. A few of these are mentioned only in the teachers' notes and are not explained on the pupil worksheets.

Number	Spreadsheet skills
Core Chapter 1	
Core Chapter 1 Teacher notes	**ROUND** function: **ROUND**(A2, −2). **IF** function: **IF**(A2 = A1, "Good", "Bad"). Making a simple interactive worksheet.
1	Undo entry or mistake. Embolden text. Centre text. Widen columns.
2	Enter a formula into a cell. Copy a formula down a column and what to do if this does not happen. Delete things from an array of cells (a rectangular block of cells). Know that * is used for multiply. Know that / is used for divide.
3	Centre text vertically. Change font size. Colour connected or unconnected cells. Colour fonts. Put borders around cells.
Plus Chapter 1	Using a spreadsheet for multiplying and dividing by powers of 10. The spreadsheet shows pupils how to enter a formula into cells and how to copy a formula down a column. Also shows how to solve problem of 'cell drag and drop' not being available.
Plus Chapter 1 Teachers' notes	How to find a ÷ sign.
Core Chapter 2	
Core Chapter 2 Teachers' notes	Create a hundred square. The **OR** function reports True or False depending on whether one or more condition(s) is (are) true (gives True) or all are false (gives False): **OR**(A1 = "", B1 = 7, C5 = "No"). Use of the Sigma icon: click to find a sum. Use of the **INT** function: **INT**(A1/B1) takes integer part of (A1/B1). Use of the **SUM** function: **SUM**(A1, B1:B4, C7) adds values in cells A1, B1, B2, B3, B4 and C7.
4	Create a spiral grid of numbers using *Copy* and *Paste* (by right-clicking). Drag formulas to left, right, up and down.
Core Chapter 8	
Core Chapter 8 Teachers' notes	Use a spreadsheet to evaluate an expression such as $3x + y$ so that it equals, for example, 48. Use a spreadsheet to generate a conversion between, for example, gallons and pints.
Core Chapter 9	
Core Chapter 9 Teachers' notes	Create grids and then sum cells based on the grid, leading to number sequences 'displayed' in another grid. Extending the idea to different patterns and more grids.
5	Creating number sequences using position-to-term and term-to-term formulae in the spreadsheet cells. Consider advantage of each.
Core Chapter 10	
Core Chapter 10 Teachers' notes	The worksheets in this chapter look at drawing a chart and using right-click on these charts to bring up a relevant menu. Importance of using *XY(Scatter)* and not *Line* for suitable charts and of 'de-selecting' the *Autoscale* feature on *Format menu, Chart area, Font*.
6	Drawing a chart for a graph using the scatter graph option (*all mathematical graph work should use this not the line graph option*). Using the chart wizard and selecting the appropriate chart.

Number	Spreadsheet skills
7	Drawing a chart for a graph using the scatter graph option, the chart title, labelling the axes and the legend (key).
8	Drawing a chart for a graph using the scatter graph option, placing gridlines.
9	Right-click on the chart to find all the menus available: format axis, format chart title, format axis title, format chart area, format gridlines, format data series and format plot area are explained.
10	Format chart options explained.
11	Format *y*-axis explained (format *x*-axis is the same).
Core Chapter 16	
Core Chapter 16 Teachers' notes	What to do for a bar chart when the *x*-axis is numerical data.
12	Making a bar chart for two sets of data (straightforward case).
13	Making a pie chart for one set of data (straightforward case).
Plus Chapter 16 Teachers' notes	Use of **AVERAGE** function to find a mean.
Core Chapter 18	
Core Chapter 18 Teachers' notes	The difference between automatic and manual calculation. ***Tools menu, Options, Calculation***, select ***Manual***. Nothing will change on the spreadsheet until you press the **F9** key. Use of ***Copy***, followed by ***Paste special*** choosing the ***values*** option, so that formulas are replaced by values.
14	Use of **RAND( )** and = **INT**(**RAND( )** $* 6 + 1$) to generate 'computer random' numbers between 1 and 6. Use of the **FREQUENCY** array function to count frequency distribution in a given part of the spreadsheet.

Spreadsheet use in the ICT Support section of this Pack

The following skills and techniques are covered in the ICT Support section of this Pack. A few of these are mentioned only in the teachers' notes and are not explained on the pupil worksheets.

	Spreadsheet skills
Chapter 1	
Teachers' notes	How to find a ÷ sign.
Chapter 6	
Teachers' notes	The **ROUND** function is used again. **LOG10** is used to find an answer. **Absolute referencing** is explained, as is use of the **F4** key.
Pupil worksheets	Use of ***Format menu, Cells, Number tab, Number*** then ***Currency*** for display of numbers in pounds and pence. Use of $ symbol for cell referencing in formulas.
Chapter 7	
Pupil worksheets	***Insert menu, Name, Define*** a cell, which defines a name for a cell. This is another way to use **absolute referencing**. ***Format menu, Cells, Number tab, Currency*** for dollars.
Chapter 9	
Pupil worksheets	***Format menu, Cells, Alignment*** to centre text and make the text so that it will fit in the cells (***Shrink to fit***). Use of the ***Ctrl*** key to highlight unconnected cells. ***View menu, Toolbars, Drawing*** to find the ***Drawing menu*** which is usually put at the bottom of the screen. Used to draw the fraction line and the arrow. Using the ***Drawing menu***. ***Format Autoshape*** to change an item from the ***Drawing menu***.
Chapter 10	
Pupil worksheets	Adding a second graph to a chart by using the ***Source data*** option on the chart.

Number	Spreadsheet skills
Chapter 16	
Pupil worksheets	Find the mean on a spreadsheet from a frequency table. Use of the **AVERAGE** function. Use a spreadsheet as a database, and how to extract information from the database. *Sort menu, Autofilter* feature of excel to search the database and **SUBTOTAL** to find the mean of 'filtered' data. *Custom* feature of *Autofilter* to search the database and **MEDIAN** function. How to draw a bar chart. How to draw a line graph. *Format axis, Scale tab* to replace the 'default' scale with one of your choice. How to draw a pie chart. How to draw a scatter graph.
Chapter 18	
Teachers' notes	The difference between automatic and manual calculation. *Tools menu, Options, Calculation*, select *Manual*. Nothing will change on the spreadsheet until you press the **F9** key. Use of *Copy*, followed by *Paste special* choosing the *values* option, so that formulas are replaced by values.
Pupil worksheets	Use of **RAND()** and = **INT(RAND()** * 6 + 1) to generate 'computer random' numbers between 1 and 6. Use of the **FREQUENCY** array function to count frequency distribution in a given part of the spreadsheet.

Geometer's SketchPad use in the ICT Support section of the Year 7 Teacher Resource Packs

The following skills and techniques are covered in the ICT Support section of the Teacher Resource Packs for 7 Core and 7 Plus. A few of these are mentioned only in the teacher notes and are not explained on the pupil worksheets.

Number	Geometer's SketchPad skills
Chapter 11	
1	Use of all tools except the circle drawing tool. Changing to the different versions of the *Line drawing tool*. Use of *Construct menu*, for both *Parallel line* and *Point on object*. Use of *Measure menu* for both *Length* and *Angle*. Drawing parallel lines and finding the minimum distance between them.
2	Use of *Construct menu*, for both *Perpendicular line* and *Point at intersection*. Drawing perpendicular lines and considering the relationship to a parallel line.
3	Use of features from 1 and 2. Drawing intersecting lines and a line parallel to one of the lines and considering angle properties.
Plus 1	Use of *Measure menu, Calculate*.
Chapter 13	
4	Use of *Transform menu, Mark vector A → G*, followed by *Transform menu, Translate* (by a marked vector should be 'shown' in the window) to move AG to GG' where AGG' is a straight line. Use of right-click, *Display, Hide Line* on a highlighted line to hide it. Use of *Transform menu, Mark mirror*, followed by surrounding a shape then *Transform menu, Reflect* to reflect the shape in the given mirror line.
5	Use of *Line style, Thick* to make one line stand out from another.
6	Use of *Transform menu, Mark centre*, followed by surrounding a shape then *Transform menu, Rotate* to rotate a shape by an angle that you have to name — it rotates the angle anticlockwise.
Plus 1	Use of *colour*, by right-click on object.
Plus 3	Use of *Construct menu, Point at midpoint*.
Plus 5	Use of *Graph menu* with *Show grid, Hide axes* and *Snap to grid*. Use of *Transform menu, Dilate*, followed by **Scale factor** (*Dilate* = US for *Enlargement*).

Geometer's SketchPad use in the ICT Support section of this Pack

The following skills and techniques are covered in the ICT Support section of this Pack. A few of these are mentioned only in the teachers' notes and are not explained on the pupil worksheets.

	Geometer's SketchPad skills
Chapter 11	
Teachers' notes	Differences between versions 3 and 4 in highlighting objects.
Chapter 12	
Pupil worksheet	Shows how to construct a line of fixed length (by relating it to another line).

MswLogo use in the ICT Support section of the Year 7 Teacher Resource Packs

The following skills and techniques are covered in the ICT Support section of the Teacher Resource Packs for 7 Core and 7 Plus.

	MswLogo skills
Core Chapter 12	
1	Starting MswLogo and the different parts of the screen. **FORWARD**, **RIGHT**, **LEFT** and **BACK** instructions. Drawing rectangles.
2	Abbreviations for the instructions **FORWARD**, **RIGHT**, **LEFT** and **BACK**. The importance of a space after each command. Use of **REPEAT**. Use of *Setmenu, Pencolor*. Use of **Penup (PU)**, **Pendown (PD)**, **Penerase (PE)** and **Penpaint (PPT)**. Drawing squares and other regular polygons.
Core Chapter 13	
3	Use of a nested **REPEAT** instruction (one **REPEAT** inside another) to create a shape with rotational symmetry.

Omnigraph use in the ICT Support section of the Year 7 Teacher Resource Packs

The following skills and techniques are covered in the ICT Support section of the Teacher Resource Packs for 7 Core and 7 Plus.

Number	Omnigraph skills and activities
Chapter 10	
1	Entering equations (different formats allowed), powers, pi and the square root symbol, transformations, a new shape, zoom menu and re-scale and finding the mouse position.
2	Lines of the form $y = mx$ where $m > 0$.
3	Lines of the form $y = x + c$ where $^-4 < c < 4$.
4	Lines of the form $x = a$ and $y = b$.
5	Asks pupils to find out for themselves about equations of the form $y = mx + c$ (can be used in place of 2–4 above).
6	Sheet where pupils can write equation and complete a table of values for each equation.

ICT-based activities

There are three Smile for Windows programs that can be used with this chapter. The first is suggested for use in Year 7 but could be used again.

Decimals

Multiply and divide decimals by powers of 10
Smile for Windows: Numeracy: Tenners
For use with '× and ÷ by multiples of 10, 100 and 1000' (Pupil Book page 19).

Order numbers with up to three places of decimals
Smile for Windows: Sense of Number: BoxD
For use with *'Putting decimals in order'* (Pupil Book page 24).

Estimate number on a number line
Smile for Windows: Numeracy: NumberLines
For use with *'Putting decimals in order'* (Pupil Book page 24).

Excel spreadsheet or equivalent activities

Bigger or smaller

For use with the Investigation '× and ÷ by 0·1 and 0·01' (Pupil Book page 20).

In this worksheet pupils have to consider when multiplying and dividing by two different numbers can have the same effect.

The spreadsheet reminds pupils how to enter a formula into cells and how to copy a formula down a column. The teachers' notes also tell you how to find a ÷ sign.

Activities

Smile for Windows: Numeracy: Tenners 'provides practice at multiplying and dividing by multiples of 10'.

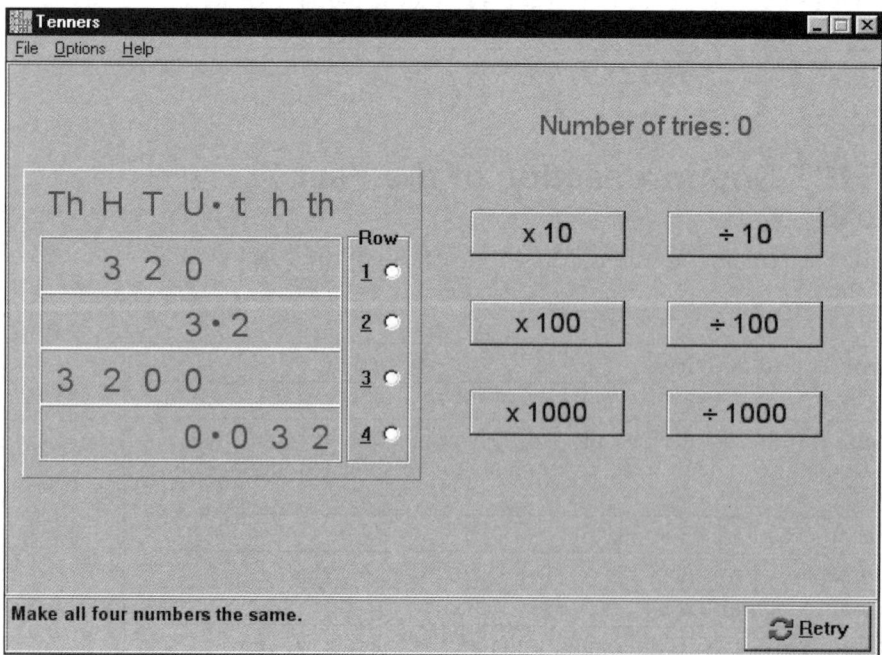

Pupils have to make all the numbers the same by changing (ideally) three of the numbers given 'into the fourth'. In the example above all numbers can be made into 3·2: for row 1 click on the dot next to the 1 and then click on ÷ 100, the rest are left to you.

Smile for Windows: Sense of Number: BoxD 'helps develop an understanding and confidence with place value' with numbers with up to three decimal places.

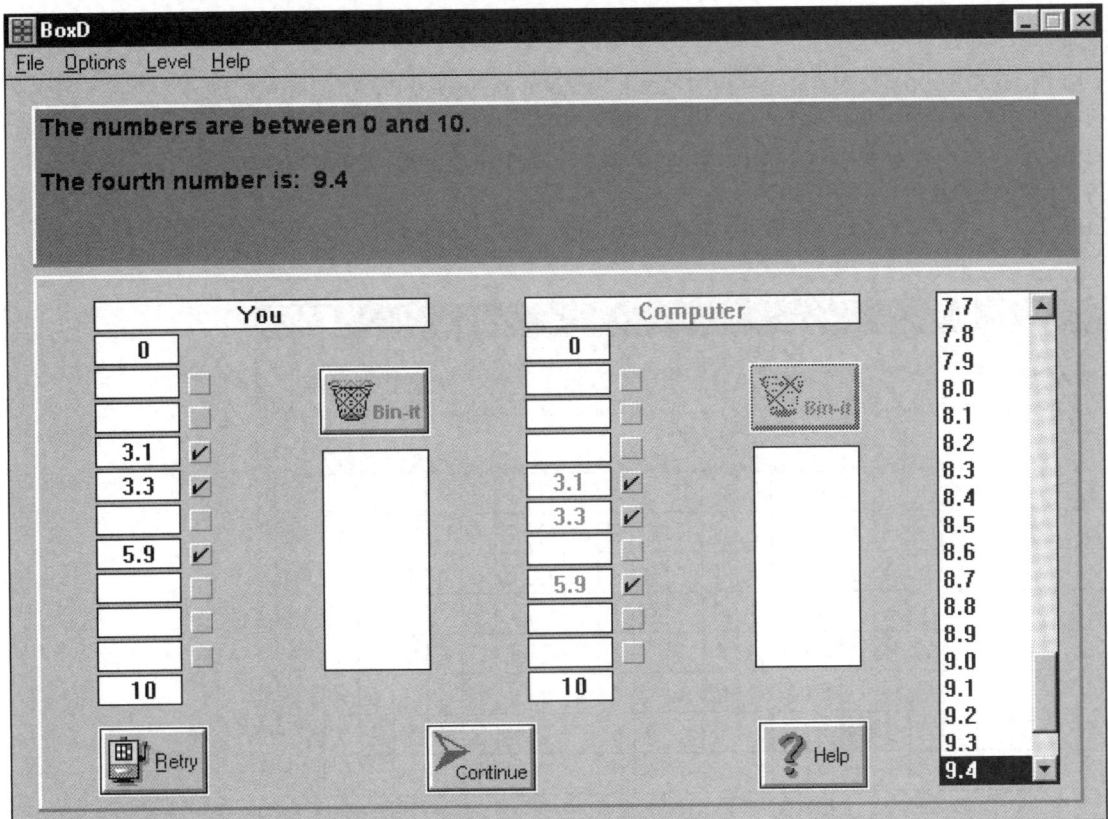

Smile for Windows: Numeracy: NumberLines

NumberLines provides students with opportunities, which involve decimals, to develop their estimation skills and strategies. There are four levels of difficulty.

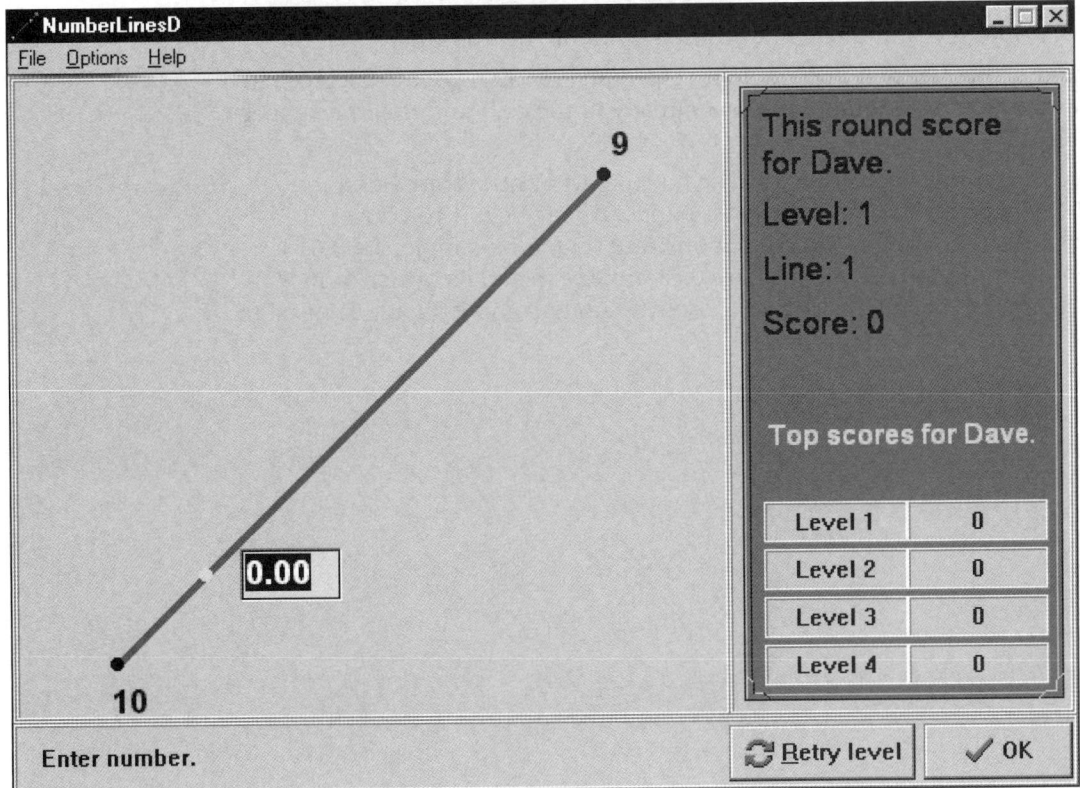

Spreadsheet ideas

Bigger or smaller

Comments and advice

You may need to refer to the ICT Support section of the 7 Core Teacher Resource Pack, Chapter 1 if pupils have forgotten the spreadsheet basics used in this worksheet.

The drag and drop feature may not be available. To restore this you need to go into *Tools menu, Options, Edit* and click on *Allow cell drag and drop*.

The quickest way to find the ÷ symbol is to use Word, and then click on *Insert menu, Symbol* and you will see the window below.

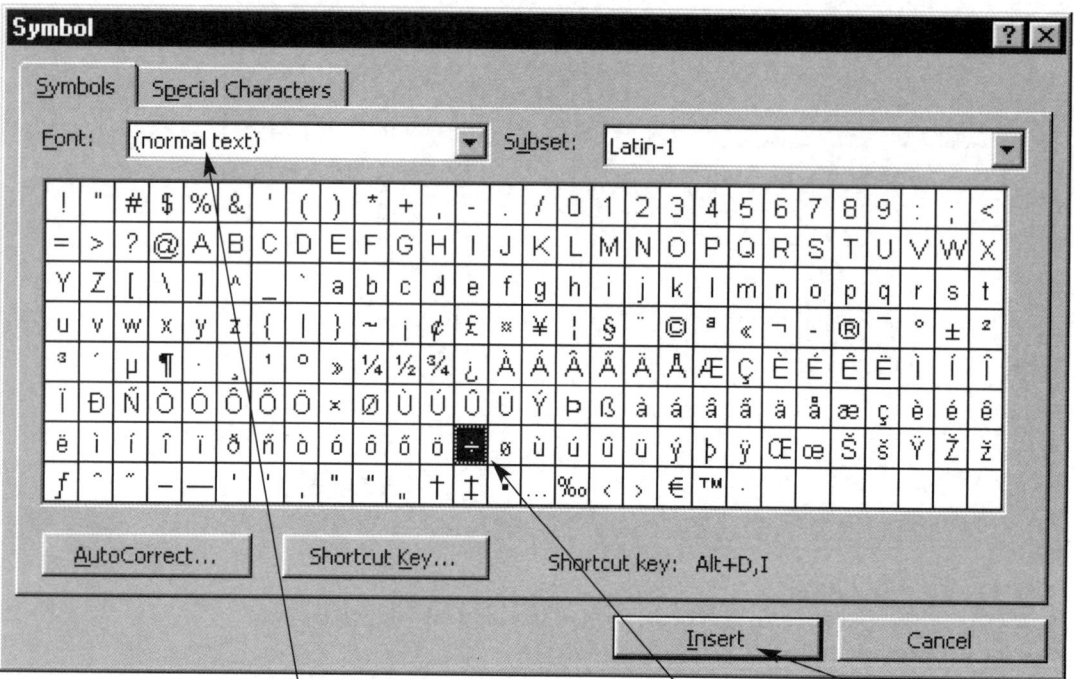

Make sure that you see *normal text* and then click on the ÷ symbol, then on *Insert*.

Highlight the ÷ sign, *Copy* it then *Paste* it into the four cells in Excel. Then go into each cell in turn, press F2 (which moves you to the end of the text already in the cell) and then press space and the relevant number.

Pupils may have trouble with **questions 9** and **10**. The fourth part of **question 9** and its equivalent for 10 and 0·1 will be much harder for most pupils. It may help if you have millimetre graph paper so that you can physically demonstrate that, for example, 4 ÷ 0·01 is 400 since it is equivalent to 'how many squares of size 0·01 are in a rectangle of size 4 cm²?'. Many pupils need this degree of concrete work and those who understand it should be used to help those who don't.

In this worksheet you are going to find out when multiplying and dividing by two different numbers can have the same effect.

Your spreadsheet

You are going to make the spreadsheet below.

	A	B	C	D	E	F	G	H	I
1	**Number**	× 10	÷ 0.1	× 100	÷ 0.01	÷ 10	× 0.1	÷ 100	× 0.01
2	4.7								
3	0.8								
4	0.92								
5	16								
6	12.3								
7	572								
8	46.23								
9	5271								

1 Type the text into cells A1 to I1. If you do not know how to find the ÷ symbol use / instead.

2 Type the numbers into cells A2 to A9.

3 Into cell B2 type the formula: **= A2 * 10**

4 Press the Enter key after typing the formula in cell B2 and you should see 47 in cell B2.

5 Do the same thing for cells C2 to I2, but use the correct formula based on the rule given in row 1.

6 Using the cell drag and drop features of Excel, highlight cells B2 to I2 and drag them all down to row 9. You should now see all the answers in your spreadsheet.

7 Which columns are the same and which columns are different? What does this mean?

8 Change some of the numbers in column A. What do you notice?

9 Here are a number of statements. Use your spreadsheet results to help you choose the correct response.

Statement	True	False
When you multiply a number, such as 4·7, by any other positive number it always gets bigger.		
When you divide a number, such as 0·8, by any other positive number it always gets smaller.		
Multiplying by 0·1 is the same as dividing by 10.		
Multiplying by 100 is the same as dividing by 0·01.		

10 Explain your **True** answers and give a counter-example for your **False** answers.

ICT-based activities

For use with '*Fractions and decimals*' (Pupil Book page 102).

You will need a computer-based calculator for this work. On many computers the program is called calc.exe and is found in the windows directory. It is usually found by clicking on **Start Menu, Programs, Accessories, Calculator**. The advantage of using this calculator, rather than a normal or scientific calculator or even a spreadsheets, is that it has 'extended precision'. This is explained in the 'Me' version of windows by:

'Understanding Extended Precision

Extended Precision, a feature of Calculator, implies that all operations are accurate to at least 32 digits. Calculator also stores rational numbers as fractions to retain accuracy. For example, 1/3 is stored as 1/3, rather than 0·333. However, errors accumulate during repeated operations on irrational numbers. For example, Calculator will truncate pi to 32 digits, so repeated operations on pi will lose accuracy as the number of operations increases.'

The 32-digit accuracy is good enough for the work here, but normal calculator accuracy is not.

Spreadsheet ideas

Investigating fractions

For use with '*Fractions and decimals*' (Pupil Book page 102).
If you cannot find the computer calculator do a search to find it — ask your ICT coordinator for help.

You could do this with a normal calculator, but it is rarely obvious when a fraction recurs.

Terminating proper fractions will be those of the form $\dfrac{a}{2^m 5^n}$ where a is any positive number less than the denominator and m and n are integers $\geqslant 0$, i.e. the denominator has factors of 2 or 5 or both.

Note that a fraction with denominator n recurs with at most $n - 1$ digits and in many cases many fewer — this is to do with the different remainders when the numerator is divided by n (there can only be $n - 1$ of these). Many fractions with denominator n recur in fewer than $n - 1$ digits: the number of digits is always a factor of $n - 1$.

In this activity you are going to use a computer calculator to work out fractions as decimals. The advantage of the computer calculator is that it works with more digits than a normal calculator or spreadsheet.

1 Start the calculator on the computer (usually ***Start Menu, Programs, Accessories, Calculator***). You should see the calculator below.

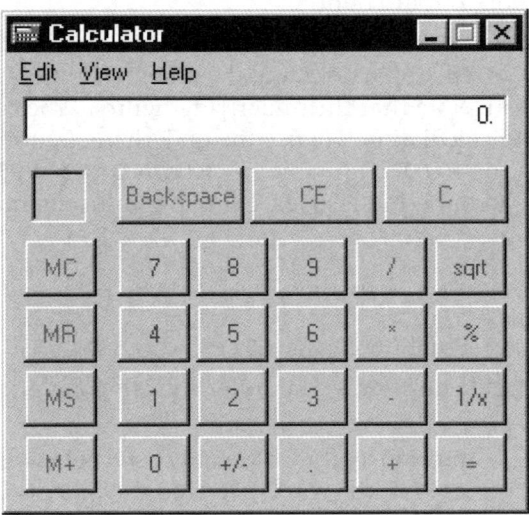

2 Here are five fractions

$$\frac{3}{32}, \frac{3}{26}, \frac{1}{15}, \frac{3}{11}, \frac{3}{14}$$

and five calculator displays with their corresponding recurring decimals (in four cases).

0·272727272727272727272727272727	0·2̇7̇
0·214285714285714285714285714571	0·2̇14285̇7̇
0·066666666666666666666666666667	0·06̇
0·115384615384615384615384615462	0·1̇15384̇6̇
0·09375	

Match the fractions with the correct display.

3 Explain how the recurring decimal for each of the fractions is worked out.

4 Work out some more fractions for yourself (keep to denominators less than 100) and write down the fraction as a terminating decimal or as a recurring decimal.

5 Which fractions terminate and which give a recurring decimal?

6 Explain.

Excel spreadsheet or equivalent activities

Discount deals

For use with the Practical (Pupil Book page 131).
In this worksheet pupils use a spreadsheet to work out discounts.

The spreadsheet does not tell pupils what formula to enter; it is assumed that you will do this. Pupils are shown how to format cells to show Currency (£) correcting the amount in the display to the nearest penny (two decimal places). Note that the underlying figure in the cell is not changed, so any use of the figure in a later calculation will use the underlying figure not the rounded figure. You can solve this by using the **ROUND** function: = **ROUND(A2*0·95,2)**. In cell B2 this would not only show the correct amount, but also leave it there for use in future calculations.

Population increase

For use with the Investigation *'Population Increase'* (Pupil Book page 131).

In this worksheet pupils use a spreadsheet to decide how long it takes for a given population to double in size.

Pupils are introduced to **absolute referencing** — previously they have only used **relative referencing** in a spreadsheet. These terms are not used in the worksheet and the process is not fully explained in the worksheet.

Activities

Discount deals

The new price based on 5% of £9·95 is £9·4525, which rounds to £9·45 in a mathematics lesson and would be the same in a shop. However, if the new price was £9·4575 in a mathematics lesson it would be £9·46, but in a shop would have to be £9·45 (otherwise the discount is less than 5% — a trade descriptions matter). Although we are encouraged to make mathematics more like real life, this is probably something that you will wish to ignore.

Population increase

Pupils are often surprised with how quickly things can double (at 4% pupils might think 25 years is a sensible answer, since 4% of 1000 is 40 and $40 \times 25 = 1000$). However, the actual answer is 18.

Working out the answers quickly
If you want the spreadsheet to work out the answers quickly, put the following formula into a cell (such as cell F1):

$$= (LOG10(2))/(LOG10(1+B1/100))$$

This uses A-level work and the spreadsheet function **LOG10** that gives the value of the logarithm to base 10 of the number in the brackets. The cell gives the solution to the equation of the problem, which is:

$$2000 = 1000 \times (1·04)^x$$

where 2000 is the final population, 1000 the starting population, 4% is the rate of increase and x is the number of years for the population to rise from 1000 to 2000 at this rate. The solution is 17·67 years (2 decimal places).

Absolute and relative referencing

The spreadsheet assumes that when you use a formula in a cell, you usually will want the reference to move as the cell moves. For example, look at the way the formulas change in the cells that follow:

	A	B	C	D
1	**Rate of population increase**	**4**	%	
2				
3	**Population at start**			1000
4	Population after	1	year	=D3*(100+$B$1)/100
5	Population after	=B4+1	years	=D4*(100+$B$1)/100
6	Population after	=B5+1	years	=D5*(100+$B$1)/100
7	Population after	=B6+1	years	=D6*(100+$B$1)/100

The formula in cell B5 is interpreted by the spreadsheet to mean 'take the value in the cell one above this (i.e. cell B4) and add 1 to it'. When copied '**relative referencing**' is maintained.

The formula in cell D4 also refers to the cell above (the D3 part) but the dollar signs tell the spreadsheet to always use the value in B1 ('**absolute referencing**') as the cell is copied.

When entering the formula in cell D4 just type:

$$=\textbf{D3}*(\textbf{100}+\textbf{B1}$$

If you then press the *F4 key* before entering the close of bracket and making sure that you don't type any dollar signs, then the dollar signs will be added. If you press the **F4 key** again it changes it again and cycles around these four options:

B1 $B$1   B$1 $B1

There will be times when you may want to use the other options as well. The B$1 option allows the column (the B part) to change as the formula is copied, but will always refer to row 1. It is similar for $B1, which fixes the column but allows the row to move.

In this activity you are going to use a spreadsheet to help you work out discounts on prices.

1 Set up a spreadsheet as below.

	A	B	C
1	**Original price**	**Price less 5%**	**Price less 10%**
2	9.95		
3	17.95		
4	14.95		

2 Enter a formula into cell B2 that will work out the new price based on a 5% discount. If you see 9·4525 you have probably entered the correct formula. What price will the shop charge the customer?

3 Copy the formula in cell B2 down into cells B3 and B4, where you will see prices that have to be rounded to pounds and pence.

4 Do the same to produce the new prices in column C for a discount of 10%.

5 You could make it so that the spreadsheet shows the rounded price straight away. You do this by formatting columns A, B and C so that the spreadsheet knows that all three columns are in pounds and pence. To do this highlight the grey letters A, B and C above the columns and then click on *Format menu, Cells, Number tab, Currency* and then select the following:

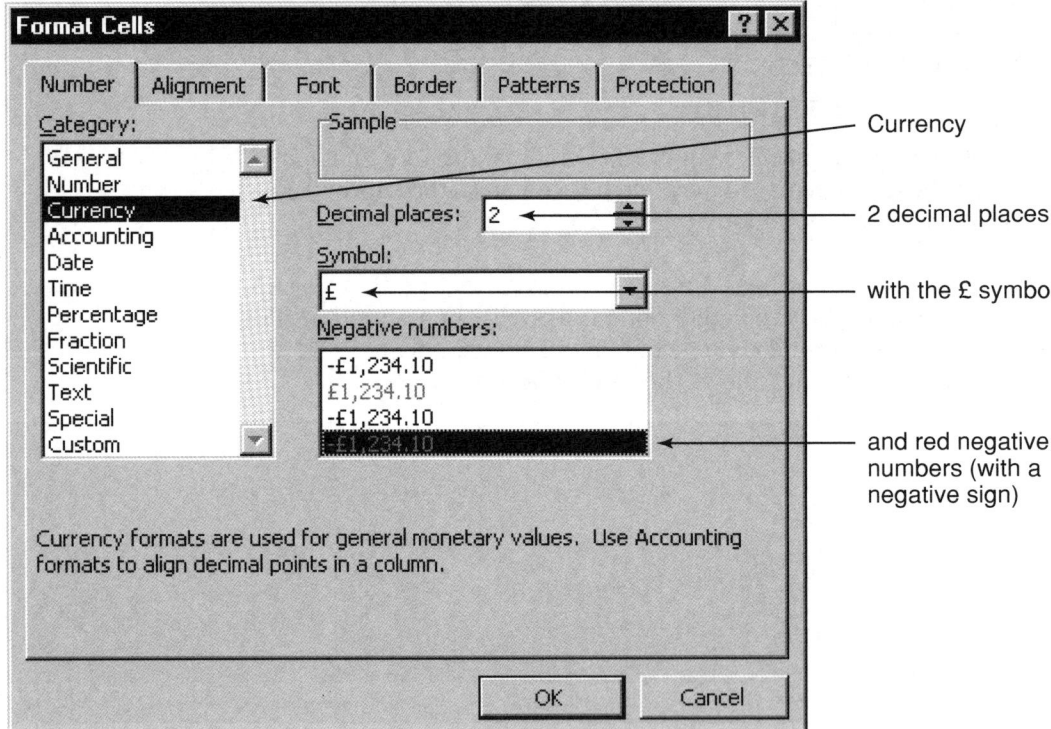

6 Extend your spreadsheet to include more discounts and add more rows.

In this activity you are going to use a spreadsheet to help you work out how long it takes different populations to increase by different amounts.

1 You are told that the annual rate of increase of a population is 4%. Guess how many years it will take at this rate for a population of 1000 to double. Why did you guess this number?

2 You are going to see how long it takes a population of 1000 to double in size when the annual rate of increase is 4%. Set up a spreadsheet as below, making sure that you use a formula in cell B5. Copy it down to row 30.

	A	B	C	D
1	**Rate of population increase**	4	%	
2				
3	**Population at start**			1000
4	Population after	1	year	
5	Population after	2	years	
6	Population after	3	years	
7	Population after	4	years	

3 You want to put a formula into cell D4 so that it uses the population in the year before (here cell D3) and the population rate in cell B1. Type the following formula into cell D4:

$$= D3 * (100 + \$B\$1)/100$$

4 Copy this formula down to row 30. Check the formula in row 30. It should be:

$$= D29 * (100 + \$B\$1)/100$$

The dollar signs in front of the B and the 1 mean that the B1 part of the formula does not change.

5 How many years does it take before the population of 1000 is doubled? What is the population at this time (to the nearest whole number)?

6 Repeat this for different starting populations. What do you notice?

7 Put 1000 back as your starting population. Change the rate to 5%. How many years does it take before the population of 1000 is doubled? What is the population at this time (to the nearest whole number)?

8 Repeat this for different starting populations. What do you notice?

9 Repeat this process again and write up what you have found out. Use examples or printouts to help explain what you have written.

10 You could also look at population decrease (use ⁻4 instead of 4). At this rate how long does it take for the population to decrease to half its size?

Excel spreadsheet or equivalent activities

Constant multiplier

For use with Practical (Pupil Book page 139) question 1.

In this worksheet pupils use a spreadsheet to make a constant multiplier table of values.

Pupils are shown how to *Name* a cell, which is another way to use *absolute referencing*.

Currency converter

For use with Practical (Pupil Book page 139) question 2.

In this worksheet pupils use a spreadsheet to convert pounds to dollars and dollars to pounds. They need to use *Format menu, Cells, Number tab, Currency* for dollars.

Activities

Constant multiplier

To·use the name in other cells pupils either have to type the name into a formula or click in the cell to put it into the formula. If they type in at the keyboard the cell reference, here B2, the formula will go wrong when copied down since relative referencing will be used and the formula in subsequent rows will not refer to cell B2, but to the cell above. An alternative method is to use $B$2.

Currency converter

The formula used in cell H2 is:

$$= 1/D2$$

Possible answers to questions 3 and 4 are given in the appropriate cells below:

	C	D	E	F	G	H
5		=D2*B5				=H2*D5

There are many currency conversion websites that give daily rates. They can be found by searching the Internet. One is:

http://travel.guardian.co.uk/currencyconverter/

In this activity you are going to use a spreadsheet to make a 'constant multiplier' — a table that shows multiples of a single priced item.

1 Set up a spreadsheet as below so that column B is in currency format.

	A	B
1	**Constant multiplier**	
2	Cost of one item	£1.75
3	2	
4	3	
5	4	
6	5	
7	6	
8	7	
9	8	
10	9	
11	10	

2 Put the cursor in cell B2 and click on *Insert menu, Name, Define*. You should see the window below:

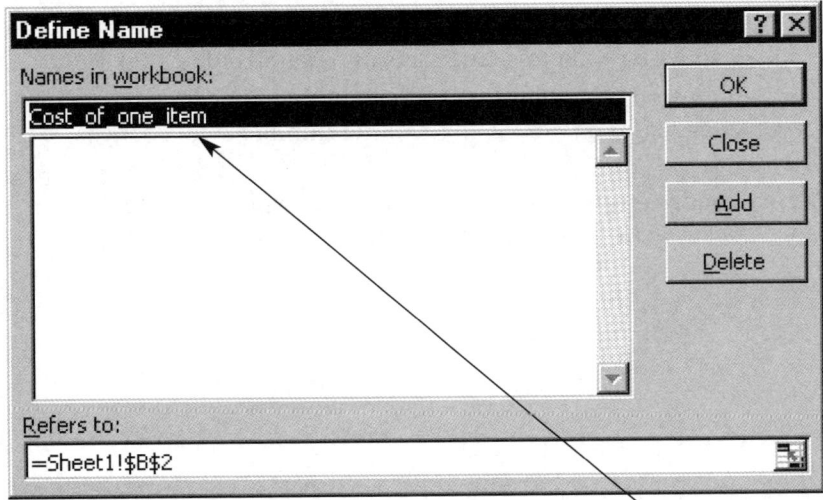

This lets you name cell B2 and suggests that you call it the name shown (which is taken from cell A2, the one next to it, but it replaces spaces with _). You can decide to keep this name or change it to something easier, such as 'Cost'.

3 Into cell B3 you are going to put a formula linked to the cell B2, the named cell, but you must not type in B2. Instead when you see B2 in the formula below you must click into cell B2 and you will see the name of the cell in the formula. Do this and then in cell C2 you should see:

= **A3∗Cost_of_one_item**

OR   the name of your cell B2 instead of **Cost_of_one_item**.

4 Copy this formula down to cell B11.

5 Make another constant multiplier to show prices for up to 20 items costing £2·40 each.

In this activity you are going to use a spreadsheet to make a currency converter so that you can change pounds into a currency of your choice and also change the currency back into pounds.

1 At the start of 2003 the British pound would buy $1·61 (US dollars). Make the spreadsheet shown below. You need to make cells B2, H2, B5 and H5 so they are in currency format for the £ and cells D2, F2, D5 and F5 so they are in currency format for the $ (English) United States.

	A	B	C	D	E	F	G	H
1	**Currency converter**							
2	Exchange rate	£1.00	equals	$1.61	so	$1.00	equals	
3								
4			Pounds to dollars conversion table			Dollars to pounds conversion table		
5			equals				equals	
6								

2 You know that £1 is worth $1·61, so you now need to work out the rate in reverse, i.e. $1 is worth how many British pounds? This figure should be found by placing a formula in cell H2. Discuss this with your friends. If you use the correct formula you should see $1 equals £0·62.

3 To be able to use the converter you have to enter an amount in pounds into cell B5 and a formula in cell D5. Do this now.

4 Enter an amount in dollars into cell F5 and a formula to change this into pounds in cell H5.

5 Change the figures so that you have an up-to-date rate for the US dollar. If you have made your spreadsheet correctly, when you change the rate in B2 all the other numbers will change (except B5 and F5).

6 Make another version of the currency converter for a different country. You will need to find the currency rate from a newspaper, travel agent or the Internet.

Excel spreadsheet or equivalent activities

Inverses

For use with Practical (Pupil Book page 163).

In this worksheet pupils use a spreadsheet to substitute values of x into an equation to find a corresponding value of y and then substitute the resulting y into the inverse equation.

True or false

For use with Chapter 8 Exercise 5 **question 3** (Pupil Book page 161).

In this worksheet pupils set up two sides of an equation to test whether the left-hand side equals the right-hand side.

Brackets

For use with Practical (Pupil Book page 168).

In this worksheet pupils compare expressions to see which are the same. Examples taken are those involving brackets.

A formula game

For use with **question 1** of the Investigation 'Substituting' (Pupil Book page 174).

In this worksheet pupils use a spreadsheet to make up a formula for another pupil to find.

Formulae

For use with Practical (Pupil Book page 184).

In this worksheet pupils use a spreadsheet to make up a table of values and then 'draw' a graph for the table of values. They use the graph to read off values.

Activities

Inverses

You may decide to give your pupils the formulae for cells B6 and C6 at the start, but some pupils should be able to do this for themselves.

The formula needed for cell B6 is: $= A6/2+3$

The formula needed for cell C6 is: $= 2*(B6-3)$

You substitute the value of B6 into cell C6, not the value of A6. Some pupils may get this wrong.

Note that this work is intended to show that when you follow this process through, you get back to the original value of x. Pupils are likely to say that you always get back to the same value, but clearly this will only be based on a finite number of cases, and so is not a proof, but an indication that it is highly likely to always be true.

True or false

The first two (including the example used) are always true, the second two are generally false, but occasionally there are numbers that make the equation true (and these are given as the suggested start values). Some pupils may believe that since you can find a value when the equation is true, then the equation is generally true.

You may wish to ask pupils if there are many other start values which will make the equations true.

Brackets

This worksheet deliberately uses expressions that show common misconceptions.

Pupils need to have completed the Chapter 7 worksheet 'Constant multiplier' to know how to name a cell.

A formula game

The best tactic is to make values of *a* and *b* zero in turn to work out the rule.

Formulae

In this worksheet pupils use a spreadsheet to make up a table of values and a graph for the table of values. This is very similar to the work found in the ICT section of Chapter 10 for Year 7, where pupils are asked to make a conversion graph. Once they have completed this task they should be given similar and related tasks which should also involve solving problems based on the information.

In this activity you are going to use a spreadsheet to substitute values into an equation and then put that answer into the inverse equation to see what happens.

1 Set up a spreadsheet as below ready for the equation $y = \frac{x}{2} + 3$.

	A	B	C
1	Equation	y = x/2 + 3	
2			
3	Inverse	x = 2(y – 3)	
4			
5	Value of x	y value from Equation	x value from inverse
6	1		
7	2		
8	3		
9	25		

2 Explain why $x = 2(y - 3)$ is the inverse equation of the equation $y = \frac{x}{2} + 3$.

3 Put a formula into cell B6 so that it substitutes the value in cell A6 into the equation $y = \frac{x}{2} + 3$.

4 The answer in cell B6 should be 3·5. Why? If you are correct, copy the formula into cells B7:B9, if not try again and discuss your answer with a friend. Check that these answers are what you expect. If they are not, try again and discuss it with your teacher.

5 Put a formula into cell C6 so that it substitutes the value in cell B6 into the inverse equation $x = 2(y - 3)$.

6 The answer in cell C6 should be 1. Why? If you are correct, copy the formula into cells C7:C9, if not try again and discuss your answer with a friend. Check that these answers are what you expect. If they are not, try again and discuss it with your teacher.

7 What do you notice?

8 Put new starting numbers into column A. Make sure that you use decimals and negative numbers. What do you notice?

9 Delete the values in cells B1, B3 and B6:C9. Place a new equation into cell B1 and its inverse into cell B3. Complete **questions 3** to **8** for your new equation.

In this activity you are going to use a spreadsheet to check your answers to Chapter 8 Exercise 5 **question 3** on page 161.

1 Set up a spreadsheet as below. Place a formula in each of cells G5, H5 and I5 so that each one is linked to the correct value in one of C5, D5 or E5.

	A	B	C	D	E	F	G	H	I
1	**Left-hand side**	**x + y + z**							
2	**Right-hand side**	**y + z + x**							
3									
4	**LHS value**	**RHS value**	x	y	z		y	z	x
5			1	2	3				

2 Place a formula in cell A5, based on cells C5, D5 and E5, so that it gives you the left-hand side value (make sure that you add x, y and z in the correct order).

3 Do the same thing in cell B5 to give you the right-hand side value.

4 What do you notice?

5 Change the numbers in cells C5:E5. Does it make a difference? How?

6 You are now going to do the same sort of thing to test which equations always give the same left-hand side and right-hand side values. Make a new spreadsheet to test each of these: True of False?

$pqr = qpr$ (start with $p = 1$, $q = 2$, $r = 3$)

$x + y + z = xyz$ (start with $x = 1$, $y = 3$, $z = 2$)

$n + m + p = nm + p$ (start with $n = 2$, $m = 2$, $p = 50$)

In this activity you are going to use a spreadsheet to compare different expressions to see which ones are the same.

1 Set up a spreadsheet so that you can compare expressions 2, 3 and 4 to see which one is the same as the expression given in cell A1 (expression 1).

	A	B	C	D
1	3(a + b)	a value	1	
2		b value	0	
3				
4	Expression 1	3(a + b)		
5	Expression 2	3a + b		
6	Expression 3	3a + b x 3		
7	Expression 4	3a + 3b		
8				

2 Click into cell C1 and name the cell (using *Insert menu, Name, Define*). Do the same for cell C2.

3 Use these names to put a formula that follows the rule in cell B4, $3(a + b)$, into cell C4.

4 Do the same sort of thing for each of the cells C5:C7.

5 Put your own values into cells C1 and C2. Which of the expressions 2, 3 and 4 is always the same as expression 1?

6 Are the others ever the same as expression 1?

7 Do the same sort of thing for expressions like:

$4(a + b)$	$3(a + 5)$	$5(a - b)$	$6(a - 2b)$
$4a + b$	$3a + 5$	$5a - b$	$6a - 2b$
$4a + b \times 4$	$3a + 15$	$5a - b \times 5$	$6a - b \times 6$
$4a + 4b$	$3a + 8$	$5a - 5b$	$6a - 12b$

8 Choose some expressions of your own to test.

In this activity you are going to use a spreadsheet to make up a formula which your friends have to guess.

1 Set up a spreadsheet so that it looks like the spreadsheet below, making sure that all three cells B4:B6 contain a formula so that when you change the values in cells B1 and B2 all the other cells in column B change.

	A	B
1	a value	7
2	b value	5
3		
4	4a	28
5	3b	15
6	Sum	43
7		

2 You should see from this spreadsheet that the answer in cell B6 is always 4a + 3b. However, you are now going to make up your own version and see if your friends can guess the rule. Make sure that your friends can't see your screen and write your own values into cells A4 and A5. Make sure that the formula in cells B4 and B5 then matches what you have put into cells A4 and A5.

3 Highlight all four cells A4:B5. Then go to the Font Colour icon (or *Format menu, Cells, Font tab, Colour*) and make the colour of the text white (so it won't be seen). The value in B6 should still be seen.

4 Now ask your friends to change the values in cells B1 and B2 to work out your rule. When they have guessed it correctly make the colour of the text in cells A4:B5 a different colour, so they can see the rule.

In this activity you are going to use a spreadsheet to help solve a problem. The problem is solved by making a table of values and then 'drawing' a spreadsheet graph from the table of values.

Miles and kilometres

1 You may already know that 5 miles is approximately 8 kilometres. Use this information to complete the table below.

Number of miles	0	5	10	15	20	25	30	35	40	45
Number of kilometres	0	8								

2 Put this table into a spreadsheet, making sure that you use two formulae so you don't have to type all the numbers into the spreadsheet.

3 Highlight the whole table and use the chart wizard to make the chart shown below.

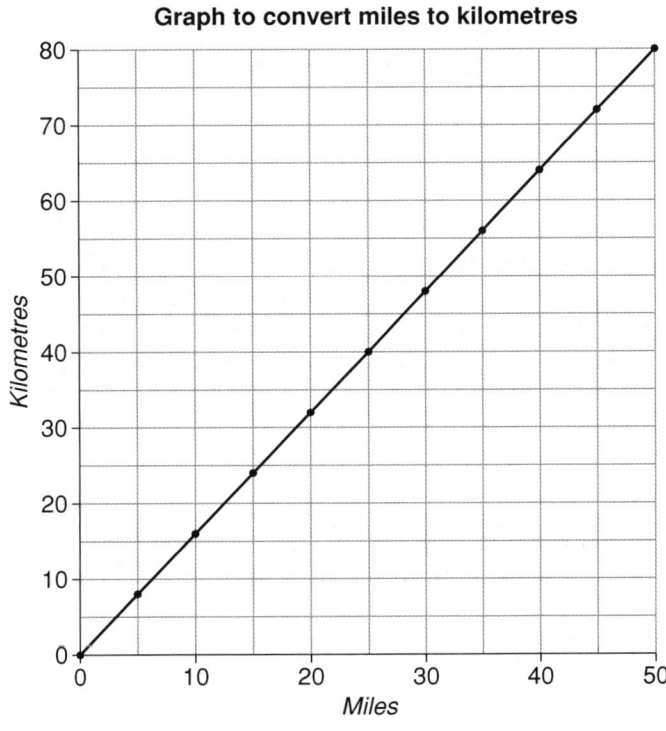

Graph to convert miles to kilometres

Your chart will probably look different to the one shown here. It may have the gridlines in different places and it may have different shading. It is important that you know how to change things on the chart. It must be easy to read information from the chart and it should look interesting. Make your chart look as similar as possible to the one above and make it look colourful.

4 Use your graph to read off how far 50 kilometres is in miles. How could you check this?

5 Make up more questions like **question 4** for a friend to answer.

6 Find other problems that you can use a spreadsheet to help solve in the same sort of way.

Excel spreadsheet or equivalent activities

Pascal's or the Chinese triangle

For use with the Investigation 'Pascal's Triangle' (Pupil Book page 205).
In this worksheet pupils use a spreadsheet to work out Pascal's or the Chinese triangle.

Pupils use **Format menu, Cells, Alignment tab** to centre text and make the text so that it will fit in the cells (**Shrink to fit**) and use the **Ctrl** key to highlight unconnected cells.

Generating sequences

For use with **question 2** of the Practical (Pupil Book page 211).
In this worksheet pupils learn how to use a spreadsheet to generate sequences using a position-to-term formula.

Sequence differences

For use with **question 2** of the Practical (Pupil Book page 211).
In this worksheet pupils use a spreadsheet to investigate the difference between terms in a sequence.

Functions

For use with Practical (Pupil Book page 221).
In this worksheet pupils use a spreadsheet to produce input and output values for a mapping.

Pupils use **View menu, Toolbars, Drawing** to find the **Draw menu** which is usually put at the bottom of the screen. To draw an arrow click on the arrow on the **Draw menu** at the bottom of the screen and then draw it in place (by clicking with the mouse and dragging the other end to where you want it). To change it right-click on it and then on **Format AutoShape**.

Activities

Pascal's or the Chinese triangle

Pupils may need to be reminded how to make columns to a certain width (see Chapter 1, in 7 Core Teacher Resource Pack, Spreadsheet basics 3). Highlighting unconnected cells is also dealt with on the same worksheet.

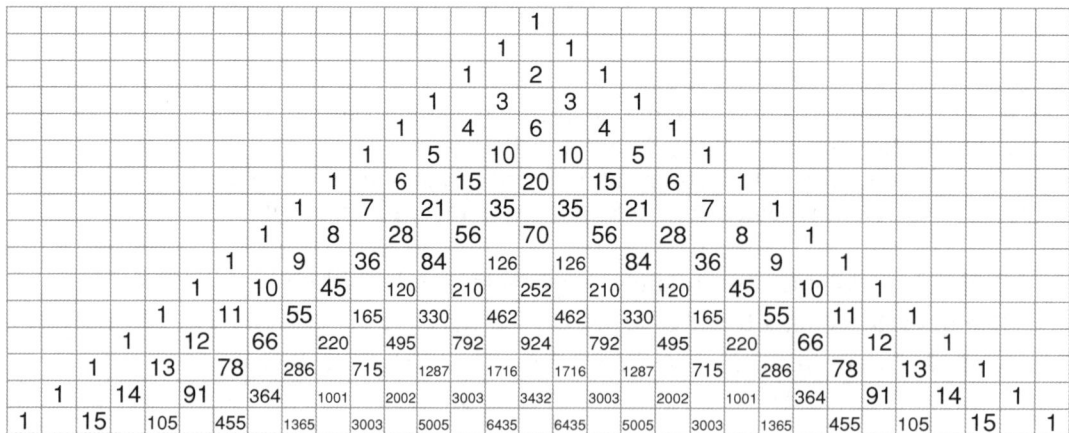

A copy of the triangle down to row 16 is shown above.

Generating sequences

You will need to check pupils' spreadsheet formulae to make sure that they are not using a term-to-term formula (in column B) or just typing in numbers (in column A).

This and the 'Functions' worksheet involve similar types of spreadsheet skills and activities.

Sequence differences

This introduces pupils to first row differences for linear sequences. The idea can be adapted for any sequence by having more terms to start with and more difference rows, though not all will give a final common difference.

Functions

In this worksheet pupils use a spreadsheet to produce input and output values for a mapping.

In this activity you are going to use a spreadsheet to make a copy of Pascal's triangle and use it to help investigate number patterns. Blaise Pascal was a French mathematician and physicist who lived in the 17th century. He is well known for his work on probability and for the 'triangle' named after him. This triangle of figures was also known to Chinese mathematicians before the time of Pascal, so many people prefer to call it the Chinese triangle.

1 Set up a spreadsheet like this, starting with the number 1 in cell P1 and continue placing 1s down the two diagonals as shown to row 4.

	K	L	M	N	O	P	Q	R	S	T	U
1						1					
2					1		1				
3				1				1			
4			1						1		

2 Highlight all of the spreadsheet by clicking in the top left hand grey corner. —————

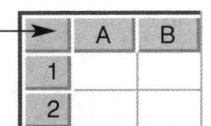

3 Make the font size of all text 10 and make the column width 2·57.

4 With the highlight on all the spreadsheet click on *Format menu, Cells, Alignment tab* and you will see the window here.

Make these both *Center*.

Also make sure that *Shrink to fit* is ticked. Click on OK.

5 Continue placing the number 1 in the spreadsheet down the diagonals until you get to cells A16 and AE16.

6 In cell P3 put this formula: = O2 + Q2

7 Copy cell P3 (right-click on the cell and then click *Copy*) and paste the formula into cells O4 and Q4.

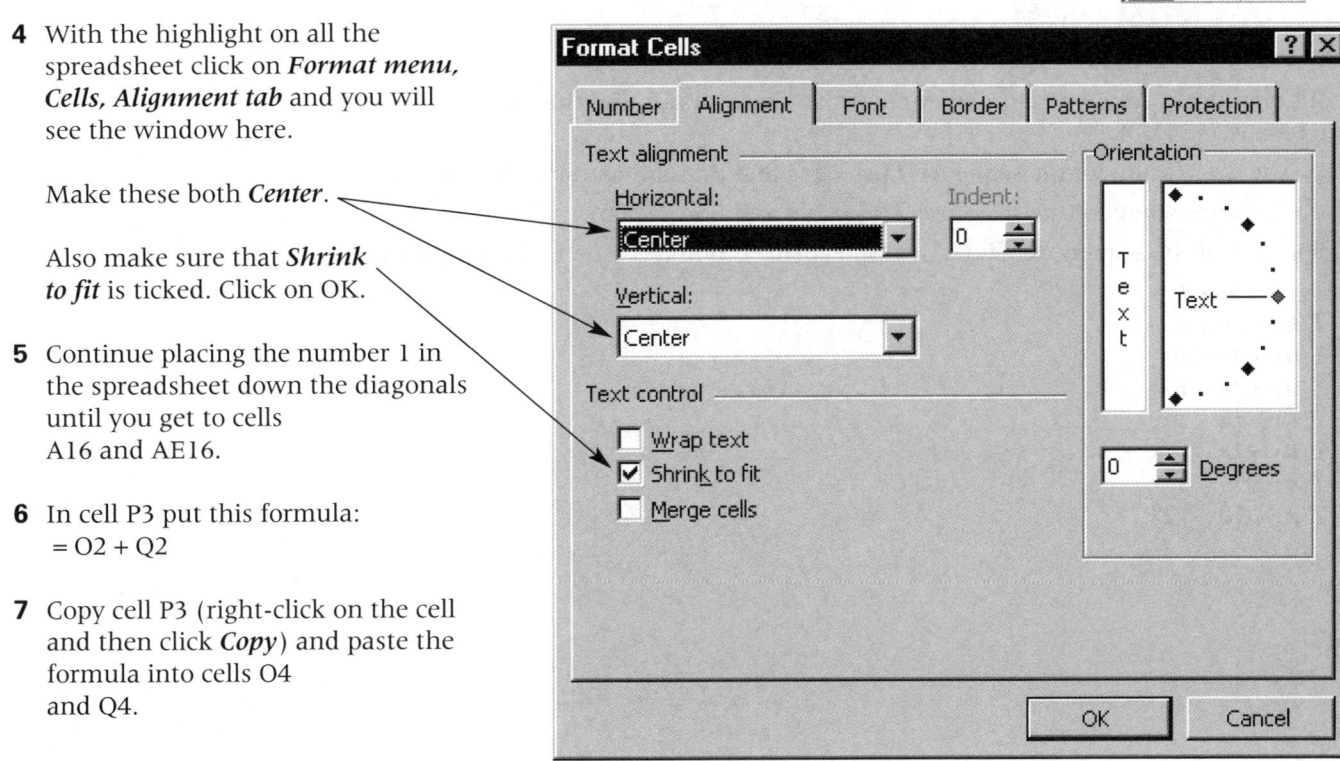

8 You are now going to fill in a lot of cells quickly in the same way to make a 'brick wall type' pattern. Go to cell P3 and copy it. Click into cell N5, then hold down the **Control key** (marked **Ctrl**) and keep it held down while you click into all these cells one at a time: N5, P5, R5, M6, O6, Q6 and S6. Still with the Ctrl key held down right-click over any one of these marked cells and click on *Paste*. The formula is now in all the cells, which now contain numbers.

9 Do this two rows at a time until you get down to row 16. You should notice that as the numbers get higher they shrink so they can still be seen.

10 How is each number, except those on the outside, worked out?

11 To look at patterns you can colour cells by clicking in them. To colour a lot of cells in one go, highlight them by using the **Ctrl** key as you did to copy the formula. Your book makes suggestions about what you should look at.

In this activity you are going to use a spreadsheet to help you with work on sequences.

1 The first spreadsheet you are going to make is one based on the sequence generated by the rule $T(n) = 15n$. You should make the sheet below by using a formula in cell A6 and copying it down to A9. You should also use a formula in cell B5 that is linked to cell A5.

	A	B	C
1	**Spreadsheet sequences**		
2	**Rule of the sequence T(n) = 15n**		
3			
4	**Term number**	**Term**	
5	1	15	
6	2	30	
7	3	45	
8	4	60	
9	5	75	

2 How else could you have generated the terms in column B?

3 Highlight cells A9:B9 and drag them down so that you can find the 24th multiple of 15, which is $T(24)$.

4 Do the same thing for the sequence $T(n) = 24n$ and find the 18th multiple of 24.

5 Do the same thing for the sequence $T(n) = 29n$ and find $T(100)$.

6 For each of the sequences below use a spreadsheet to find the first 20 terms of the sequence:

$T(n) = 3n + 1$
$T(n) = 3n + 4$
$T(n) = 2n - 1$
$T(n) = 2n + 1$
$T(n) = 4n + 2$
$T(n) = 4n - 6$
$T(n) = 7n - 3$
$T(n) = 10n + 2$

In this activity you are going to use a spreadsheet to help look at the differences between consecutive terms in a sequence.

1 The first spreadsheet you are going to look at is one based on the sequence generated by the rule $T(n) = 3n + 1$. Make the sheet below by using a formula in cell D4 and copying it across into cells F4, H4, J4 and so on until cell T4. You should also use a formula in cell B5 that is linked to cell B4 and copy this into the cells shown and beyond to cell T5.

	A	B	C	D	E	F	G	H	I	J	K	L	M	N
1	**Sequence rule**													
2	**T(n) = 3n + 1**													
3														
4	Term number	1		2		3		4		5		6		7
5	Term	4		7		10		13		16		19		22
6	Difference in terms													

2 Type this formula into cell C6:

$$= D5 - B5$$

3 Copy this formula into cells E6, G6, I6, etc. up to cell S6.

4 What do you notice about all the differences?

5 For each of the sequences below, generate the sequence in a similar way and look at the differences.

$T(n) = 3n + 4$
$T(n) = 2n - 1$
$T(n) = 2n + 1$
$T(n) = 4n + 2$
$T(n) = 4n - 6$
$T(n) = 7n - 3$
$T(n) = 10n + 2$

6 What have you found out? Can you explain why it works?

7 You have found that the differences between consecutive terms of a sequence is always 5. Give three different possible sequences in the form $T(n) = ...$ that could generate this set of differences.

In this activity you are going to use a spreadsheet to help show mapping diagrams and work out an input–output table. Later on you could use the same process to produce a graph of the mapping diagram.

1 Make the spreadsheet as shown below. To make the area A1:C11 white (or another colour), highlight it all and colour it white (or any other colour) — this makes it easier to see the arrow and the fraction line.

	A	B	C
1	**Mapping diagram**		
2			
3	x ⟶		x + 3
4			4
5			
6	Input	Output	
7			
8			
9			
10			
11			

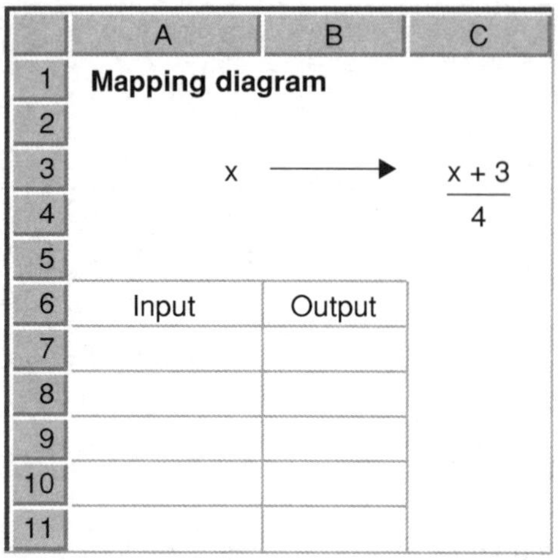

2 To add the arrow and the fraction line you need to have the **Drawing menu** available. To do this go to **View menu, Toolbars, Drawing** (if **Drawing** is already ticked it means it is already there). The **Drawing menu** is usually put at the bottom of the screen. To draw an arrow click on the arrow on the **Drawing menu** at the bottom of the screen and then draw it in place (by clicking with the mouse and dragging the other end to where you want it). To change it right-click on it and then on **Format AutoShape**.

3 Do the same sort of thing to make the fraction line.

4 Choose inputs for the mapping and use a formula based on the mapping to generate the outputs.

5 Adapt the spreadsheet for the following mappings:

$x \longrightarrow \frac{x}{10} + 3$

$x \longrightarrow 3(x - 4)$

$x \longrightarrow 6x - 4$

$x \longrightarrow 0{\cdot}6x - 4$

6 Adapt the spreadsheet to show the following:

$y = 5(x - 1)$

$y = 10x - 16$

$y = 4 - 3x$

Excel spreadsheet or equivalent activities

Drawing a graph on a spreadsheet

For use with the Investigation *'Features of Graphs'* (Pupil Book page 236).
In this worksheet pupils have to plot a table of values on a spreadsheet and then plot the corresponding graph. They then add four more graphs and compare the graphs.

The worksheet shows pupils how to add a second graph to a chart by using the *Source data* option on the chart.

Omnigraph or equivalent

Drawing graphs with a graph plotter

For use with the Investigation *'Features of Graphs'* (Pupil Book page 236).
In this worksheet pupils plot graphs using a graph plotter such as Omnigraph or Autograph.

Activities with spreadsheets

Drawing a graph on a spreadsheet

The worksheets 'Spreadsheet basics 6–10' in the ICT section of the 7 Core Teacher Resource Pack, Chapter 10 Teachers' Notes explain how to use the chart wizard and format a chart so that it looks like a graph suitable for a mathematical purpose.

The lines drawn will all be parallel. Pupils may not mention the intercept property, but should be encouraged to look for more than 'the lines are all parallel'.

Activities with Omnigraph

Drawing graphs with a graph plotter

In this worksheet pupils plot graphs using a graph plotter such as Omnigraph or Autograph.

Instructions on how to use Omnigraph are found in the ICT section of the 7 Core Teacher Resource Pack, Chapter 10 Teachers' Notes, where the emphasis is on using the graph plotter to make a table of values.

In this worksheet you are going to use a spreadsheet to complete a table of values for an equation and then draw the graph of this equation.

1 Make the spreadsheet below, using a formula for the numbers in row 3 from C3 onwards.

	A	B	C	D	E	F	G	H	I	J
1	**The graph of y = 2x**									
2										
3	x		−4	−3	−2	−1	0	1	2	3	4
4	y = 2x									
5										

2 Type the numbers into cells B4 to J4 using a formula based on the numbers in row 3.

3 Highlight cells A3:J4 and use the chart wizard to draw a graph of $y = 2x$. Make sure that you format the chart so that it is properly labelled and displays the information clearly. It helps if the axes are approximately the same scale.

4 Adapt the spreadsheet so that row 5 has the y-coordinates for the graph $y = 2x + 1$ (use a formula based on the x-values in row 3).

5	**y = 2x + 1**									

5 To add the graph to the chart already drawn, right click on the chart and then click on *Source data*. Click on *Series* and you will see the window shown here.

6 Click on *Add*.

7 Make the section showing **Name**, **X Values** and **Y Values** so that it looks like that shown below.

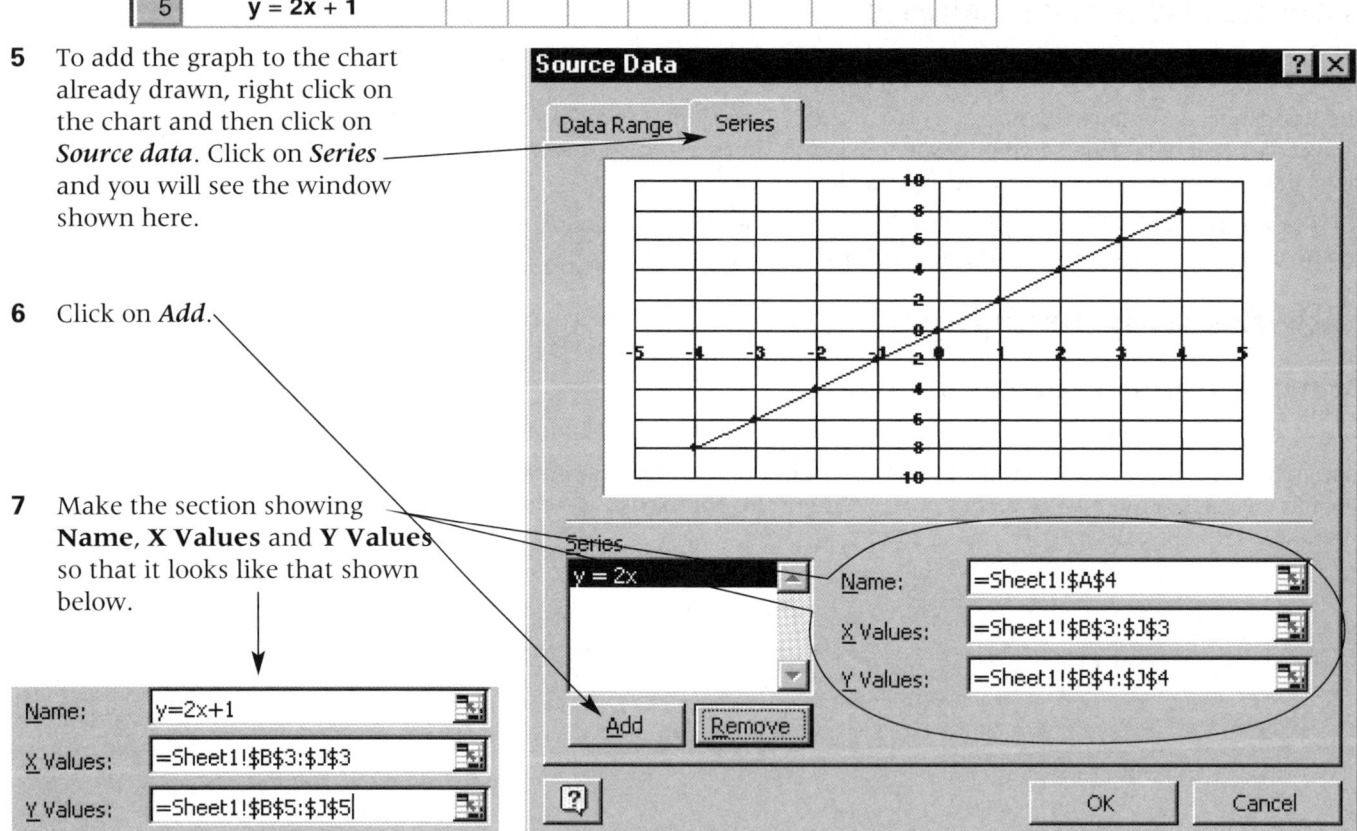

8 You should then see the two graphs on the same grid. What do you notice?

9 Add three more lines in the same way: $y = 2x + 5$, $y = 2x − 3$, $y = 2x − 1$. What, if anything, is the same about all the graphs and what is different about them? How can you tell which graph belongs to which equation?

10 You could use the skills you have learnt to complete the investigation 'Features of graphs' on page 236 of your book.

In this worksheet you are going to use a graph plotter to investigate graphs of the form $y = mx + c$ where m and c are numbers.

1 Put the equation $y = x$ into a graph plotter. Your screen should then look something like the screen below.

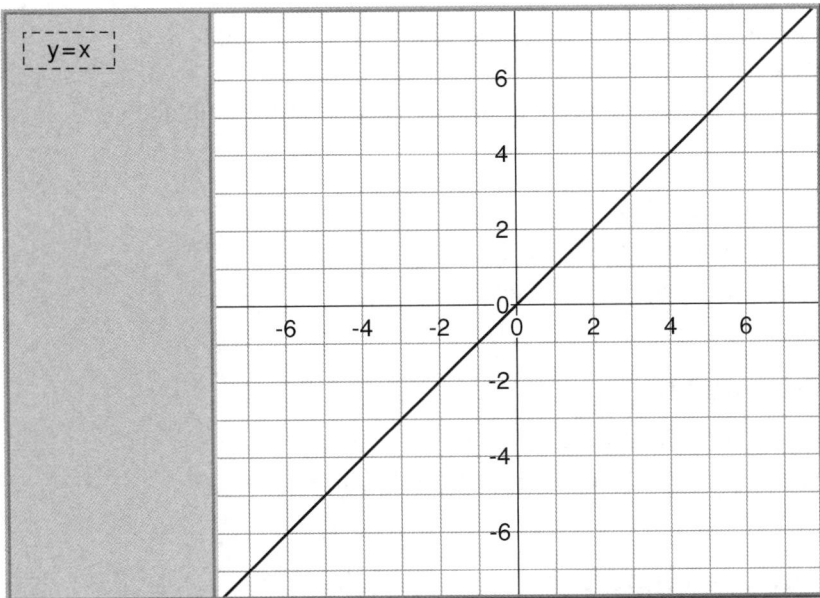

2 Enter all these equations into the graph plotter and describe what you notice.

$y = x + 1, \ y = x - 2, \ y = x + 3, \ y = x - 4$

3 Clear the screen and put $y = x$ into the graph plotter. Enter all these equations into the graph plotter and describe what you notice.

$y = 2x, \ y = 3x, \ y = 4x, \ y = 0 \cdot 5x$

4 You could use the skills you have learnt to complete the investigation 'Features of graphs' on page 236 of your book.

Geometer's SketchPad or equivalent activities

Three intersecting lines

For use with **question 1** of the Practical (Pupil Book page 272).

In this worksheet pupils investigate angles at a point — in particular, angles on a line and vertically opposite angles (terminology is not used).

Shows how to draw lines, construct a point of intersection, measure a given angle and add angles.

Parallel lines

For use with **question 2** of the Practical (Pupil Book page 272).

In this worksheet pupils investigate angles between parallel lines.

Shows how to construct a parallel line.

A quadrilateral between parallel lines

For use with **question 3** of the Practical (Pupil Book page 272).

In this worksheet pupils investigate angles 'inside' two pairs of parallel lines.

Shows how to construct a polygon interior.

Angles in a triangle and on a straight line

For use with the Practical (Pupil Book page 277).

In this worksheet pupils investigate and compare the sum of the angles in a triangle and the sum of the angles on a straight line.

Dynamic Geometry with SketchPad

Eagle, Ruth, *Dynamic Geometry with SketchPad* (Keele Mathematical Education Publications, 2000)

This book provides a good introduction to the use of Geometer's SketchPad. It includes ideas for classroom use, worksheets and 32 ready-prepared files that can be used to help pupils learn mathematics.

Geometer's Sketchpad ideas

See also Chapter 11 of the ICT Support section in the 7 Core Teacher Resource Pack.

If you have version 3 of Geometer's SketchPad, using the shift key to highlight more than one point sometimes creates problems. The other way to do this is to use the pointer tool to 'part surround' objects. If it includes part of an object then the whole object is selected. Note that if a line is selected it does not necessarily mean that the end points of the line are selected, since the points at the end are different entities, though if the points have defined the line they will move with the line (see what happens when you select just the line, copy it and then paste it).

If you have version 4 of Geometer's SketchPad, (late 2002 onwards) selecting objects is easier: you click on the white screen to clear all, then select those in the order that you want them (they all stay selected until you click on the white screen), though you can turn each one 'off' by clicking on it again.

The most common mistake is to use the wrong tool and add extra points (for example). There is an *Undo* feature on the *Edit menu*. The other common mistake is not to construct objects correctly (so that expected constraints are lost when one of the parts of the diagram is moved).

The **Line drawing tool** allows you to draw three different sorts of lines. Click on the tool itself and hold the mouse button down. You should see the new 'sub-options' shown here. The one with an arrow at both ends is a line without end and the one with dots at each end is a line segment. The first one is the one currently in use.

Activities

Three intersecting lines

Pupils find vertically opposite angles a difficult concept.

Parallel lines

This is also dealt with in the Year 7 Teacher Resource Pack.

A quadrilateral between parallel lines

Shading the area helps pupils 'see' the parallelogram.

Angles in a triangle and on a straight line

It helps if pupils see that these two sums are the same. It is important that pupils see how the same totals arrive (i.e. a proof that one implies the other, and that this could be either way round).

Question 8 is concerned with the external angle. Pupils find it difficult to prove why the angle equals the sum of the other two.

In this worksheet you are going to use Geometer's SketchPad to investigate the angles made by three lines intersecting at a single point.

1 Make the diagram below on Geometer's SketchPad where all three lines pass through the point A. The instructions for doing this are given below (2–3). If you know how to do this, do it, then go to **5** to measure the angles about point A.

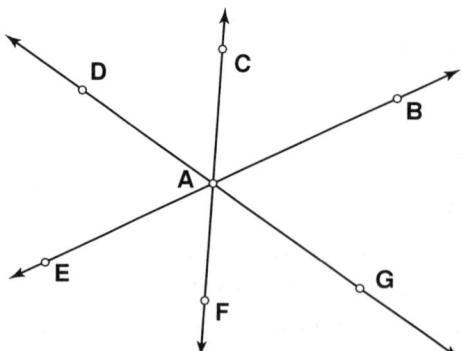

2 Make sure that you are in the **Line drawing (Straightedge) tool** and that it has an arrow at both ends. Click on the drawing screen and then drag the line out to a point. Go to the **Text tool** and click on the two marked points, which now should be labelled A and B.

3 Change back to the **Line drawing tool**, click again on A and then drag the line out to another point, which you should label C. Repeat for another line starting at A and a point D.

4 Use the **Point tool** to make points E, F and G as shown on the diagram above. Label them.

5 You now need to measure the angles. To do this change to the **Pointer (Selection arrow) tool** and click anywhere on the screen, not on a point or a line. Click in order on points B, A and C. [If they are not all surrounded with pink or black you need to click anywhere on the screen not on a point or a line, hold down the Shift key, then click in order on B, A and C – keeping the Shift key held down. Release the Shift key.]

6 Go to *Measure menu, Angle* and you should see the size of angle BAC given on the screen.

7 Do the same thing for each angle: CAD, DAE, EAF, FAG and GAB.

8 Which angles are the same and which are different?

9 What happens to the angles as you move one of B, C or D about point A, but make sure that the points B, C, D, E, F and G stay in the same order about A?

10 Do the same pairs of angles always stay the same as each other? Explain.

11 Click on the white part of the screen so nothing is highlighted. Click on *Transform menu, Calculate*. You should now see a calculator on the screen. Click on **<mBAC** on the screen, then on the **+** on the calculator, then on **<mCAD**, then on the + on the calculator, then on **<mDAE** then on **OK** on the calculator. What happens and what do you notice?

12 Repeat and add up the other three angles. What do you notice? Explain.

In this activity you will use Geometer's SketchPad to find out about angles between parallel lines.

1 Make the screen below where lines AB and CD intersect at E.

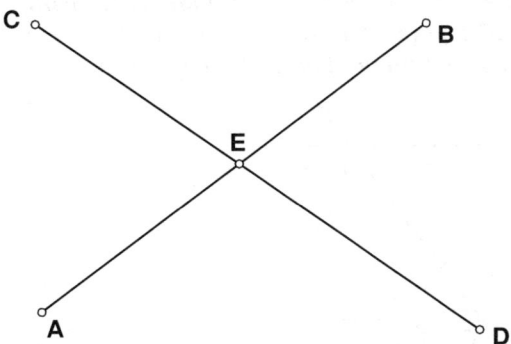

2 To make a line parallel to AB, you need to place a point on line CD. To do this click on the *Point making tool* then move it to the line CD and then click to make the point F (so that it is on the line CD).

3 Highlight F and then line AB at the same time (using the Shift key) and then click on *Construct menu, Parallel line*.

4 Place two points G and H so they are on this line.

5 The diagram on the right shows part of the screen above, but with shading around the angles at point E. Which of the angles around the point F are equal to the shaded angles around point E? Give reasons.

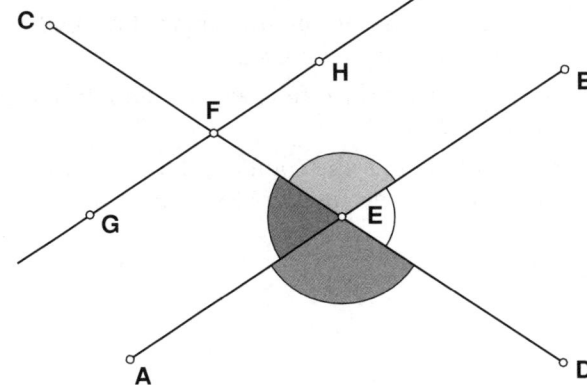

6 Use the *Measure menu, Angle* to find the size of the angles around point E.

7 Do the same thing to find the size of the angles around point F. What do you notice?

In this activity you will use Geometer's SketchPad to find out about angles between pairs of parallel lines.

1 Make line AB using the **Line drawing tool** with arrows at both ends. Then make a line AC starting from A. Then highlight point C and the line AB, then click on **Construct menu, Parallel line**. In the same way highlight line AC and point B to make the line CD. Finally highlight these two new lines and click on **Construct menu, Point at intersection**.

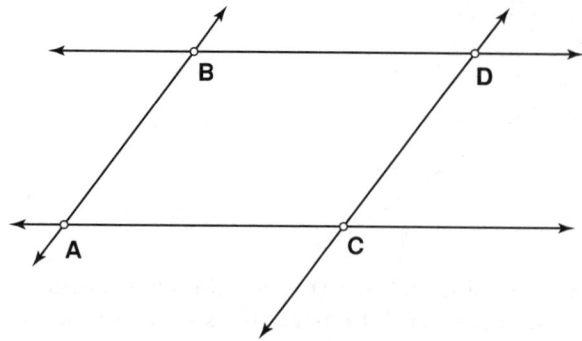

2 You now need to make it so that the four angles of quadrilateral ABDC are measured by Geometer's SketchPad.

3 What do you notice about the angles?

4 Move the points of the quadrilateral around. What happens to the angles?

5 Use the calculator to sum the angles BAC and ACD. What do you notice? Move the points around. What do you notice?

6 Do the same for the other pair of angles. What do you notice? Move the points around. What do you notice?

7 Highlight in order the points A, B, D, C. Click on **Construct menu, Polygon interior**. What shape is this interior? Explain.

In this activity you will use Geometer's SketchPad to find out about angles between pairs of parallel lines.

1 Make the diagram below.

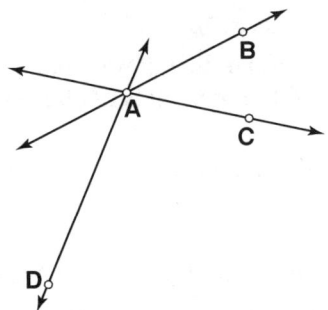

2 Highlight line AB and point C to construct a parallel line through C and parallel to AB, to give the figure below.

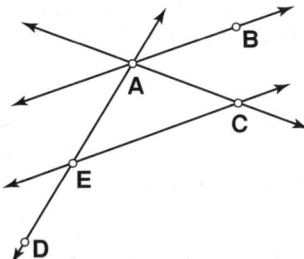

3 Construct point E at the intersection of lines AD and the new parallel line through C.

4 Put an extra point F on line BA so that it is the other side of A from B. Do the same so that G is beyond E on CE and so that H is beyond C on CE.

5 Measure now and compare all the angles BAC, CAE, EAF, AEC, ACE, GEA and HCA. Which are the same and which are different? What changes as you move the lines around?

6 Use the calculator to add up the three angles inside the triangle. Do the same for the three angles at point A. What do you notice? What changes as you move the lines around?

7 Explain what this all means.

8 You should also have found one angle that equals the sum of two others. Where are these angles? Explain what this means.

Geometer's SketchPad or equivalent activities

Different parallelograms

For use with the Investigation *'Properties'* (Pupil Book page 287).

In this worksheet pupils investigate the different types of parallelograms that can be made from two 'base' lines. They also look at the properties of the diagonals of all these types of parallelograms.

Shows how to construct parallel lines and hide parts of a diagram.

Tessellations

For use with **question 3** of the Investigation *'Tessellations'* (Pupil Book page 292).

Rotations are used to make tessellations of quadrilaterals. Mid-points of lines are found.

Constructions on a line

For use with the Practical (Pupil Book page 306).

In this worksheet pupils investigate the perpendicular bisector of a line by constructing two circles, each has one end of the line as a centre with the radius equal to the length of the line.

Constructing a triangle with three known lengths

For use with the Practical (Pupil Book page 306).

In this worksheet pupils learn how to use Geometer's SketchPad to construct a triangle with three known lengths.

Shows how to construct a line of fixed length (by relating it to another line). Translation of a point by a vector.

MswLogo or equivalent

Making shapes with Logo

For use with the Practical (Pupil Book page 291).

In this worksheet pupils investigate the different types of parallelograms (squares, rectangles and rhombuses) that can be made by using a simple Logo procedure.

Investigating loci

For use with the Practical (Pupil Book page 309).

In this worksheet pupils work out instructions to produce example shapes. They then make more of their own.

MswLogo for Windows

Miller, Dave, *Logo for Windows* (Keele Mathematical Education Publications, 1997)

This book provides a good introduction to the use of MswLogo and contains many worksheets and ideas. It includes a copy of the MswLogo program that can be used in all educational establishments.

Internet

Tessellations (SketchPad)

http://mathforum.com/sum95/suzanne/tess.gsp.tutorial.html

Tessellations (Paint or Paintbrush)

http://www.wsd1.org/bitsbytes/9798/bboct97/default.htm#STORY4

Activities with Geometer's SketchPad

Different parallelograms

One emphasis should be that all the shapes below are types of parallelogram, i.e. they are all included in the set of parallelograms. In the same way, all squares are rectangles.

Type of parallelogram	Properties of sides	Properties of angles
Parallelogram	Opposite pairs equal	Opposite pairs equal
Square	All four equal	All four equal (90°)
Rectangle	Opposite pairs equal	All four equal (90°)
Rhombus	All four equal	Opposite pairs equal

In the table below, note that technically 'Never' is not correct, rather that the response should be:

Type of parallelogram	Diagonals equal	Diagonals at right angles	Diagonals bisected
Parallelogram	Sometimes	Sometimes	Always
Square	Always	Always	Always
Rectangle	Always	Sometimes	Always
Rhombus	Sometimes	Always	Always

Tessellations

Note that this worksheet is found in Chapter 13 of 7 Plus Teacher Resource Pack, ICT Support section.

Two websites are also suggested for work related to tessellations. One uses Geometer's SketchPad, the other makes use of a painting program. Both are fun and produce Escher-type tessellations.

1 Tessellations (SketchPad)

http://mathforum.com/sum95/suzanne/tess.gsp.tutorial.html
This site tells you how to construct a tessellation based on a parallelogram. You can also download a SketchPad file that produces a tessellation based on a parallelogram. Click to follow the Java applet – it's fun.

2 Tessellations (Paint or Paintbrush)

http://www.wsd1.org/bitsbytes/9798/bboct97/default.htm#STORY4
Describes how to make an Escher-type tessellation with a painting program.

Constructions on a line

The special case is used to find the perpendicular bisector of the line AB. The generic case, where two identical circles with radius r such that $\frac{r}{2} < AB < r$, is not used here, but could be completed after the worksheet 'Constructing a triangle with three known lengths'. You could fix a line first and then make two circles that have their radii equal to this fixed length.

Constructing a triangle with three known lengths

A standard method to fix a line to a given length. Note that this can be considered as an equivalent to a straight edge and compasses construction. In this example the two lines GJ and HL are in place so that the activity is much more concrete and this may help many pupils understand the underlying construction. You are really only looking for the point where they meet, but this makes it harder for some pupils to understand. (Technically a circle should be drawn to make the first line GH, in order that GH can also be in any direction.)

Activities with MswLogo

Making shapes with Logo

The first part looks at making parallelograms, and pupils are asked to note that the procedure (here referred to as 'instructions') requires two angles that add up to 180 to make a parallelogram. The special cases are the rectangle when both angles are 90 and the rhombus when both the lengths are the same.

A rectangle procedure is, for example, REPEAT 2 [FD 30 RT 90 FD 40 RT 90] and a rhombus procedure is, for example, REPEAT 2 [FD 30 RT 50 FD 30 RT 130]. A procedure for the square is, for example, REPEAT 2 [FD 30 RT 90 FD 30 RT 90], which can be written more efficiently as REPEAT 4 [FD 30 RT 90].

The Framework (page 15) suggests that a kite can be made by adapting this procedure. This is not possible at this level since pupils need a knowledge of trigonometry to do this with Logo.

The second part lets pupils explore more complex and interesting patterns. Note that to make a closed shape such as the hexagon the product of the number of repeats and the angle should make 360 (or a multiple of 360 if the shape is allowed to be traced out more than once). In a similar way the repeat number and the angle, shown in bold below, have to do the same for a closed 'design'.

REPEAT **5** [REPEAT 8 [FD 50 RT 45] RT **72**]

Investigating loci

The shapes shown are made by the following procedures (reading across and then down).

REPEAT 5 [FD 60 RT 144]
REPEAT 8 [FD 60 RT 135]

REPEAT 3 [FD 20 LT 90 FD 20 RT 90 FD 20 RT 90 FD 20 LT 90 FD 20 RT 120]
REPEAT 4 [FD 20 RT 60 FD 20 LT 120 FD 20 RT 60 FD 20 LT 90]
REPEAT 3 [FD 20 RT 60 FD 20 LT 120 FD 20 RT 60 FD 20 LT 120]

In this worksheet you are going to use Geometer's SketchPad to investigate the different parallelograms that can be made from two lines. You will also look at the properties of the diagonals of these parallelograms.

1 You are going to make parallelogram ABCD. Start by drawing two lines, AB and BC, and constructing parallel lines based on these two lines to make the diagram shown below. The instructions are given after the diagram.

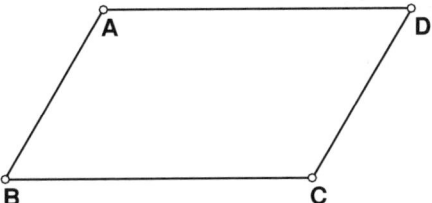

2 Highlight line BC and point A and click on *Construct menu, Parallel line*.

3 Do the same for line AB and point C.

4 Construct D, the point of intersection of the two parallel lines (highlight both, *Construct menu, Point of intersection*).

5 Highlight the two lines you have constructed and hide them (*Display menu, Hide lines*). Now join CD and DA using the **Line drawing tool**.

6 Highlight AB and measure its length. Do the same for BC and for angle ABC. What do you know about the other two lengths and the other three angles? Explain.

7 Move the points A, B and C around. How many different types of parallelograms can you make? What are they?

8 Copy and complete the table below, filling in all the gaps:

Type of parallelogram	Properties of sides	Properties of angles
Parallelogram		
Square		

9 Now join the diagonals of the parallelogram and measure them. You will also need to construct the point of intersection so that you can find the angle between the diagonals. Finally, move the points A, B and C around so that you can complete the table below using **Always**, **Sometimes** or **Never** in the 3 'diagonal columns'.

Type of parallelogram	Diagonals equal	Diagonals at right angles	Diagonals bisected
Parallelogram			
Square			

In this worksheet you have to make a tessellation of a trapezium shape and then tessellate shapes of your own choice.

1 Make trapezium ABCD shown here where AB and DC are parallel lines. To do this, first make AB and BC. Then highlight AB and the point C, click on *Construct menu, Parallel line*. Construct a point on this parallel line, then hide the line. Join CD and DA.

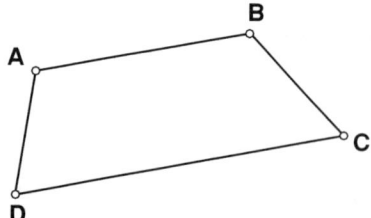

2 Highlight CD then *Construct menu, Point at midpoint*. Highlight this midpoint and click on *Transform menu, Mark centre E*. Highlight all the shape, then click on *Transform menu, Rotate* and make the angle 180°.

3 The diagram below shows ABCD rotated by 180° about E, then all of this shape is rotated 180° about F and then all of this shape is rotated 180° about G. E, F and G are all midpoints of lines. Do this.

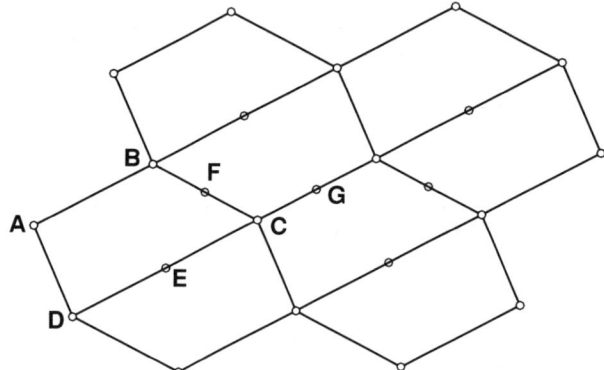

4 Continue in this way to make a larger tessellation.

5 What do you notice about the lines and angles?

6 Move points A, B and C. What do you notice?

7 Choose a quadrilateral of your own and follow the same sort of process. You should always be able to make a tessellation using this method.

In this activity you are going to use Geometer's SketchPad to investigate constructions made on a single line.

1 Construct a line AB and colour this line red (highlight just the line, **Display menu, Colour**). Label all points as you go through these instructions.

2 Change to the **Circle drawing tool** and draw a circle with centre A and dragging the mouse to point B, so AB is the radius of the circle. Move point B. If the circle moves so that it is still through B you have done this correctly. If not, delete the circle and try again.

3 Make the circle blue.

4 Make another blue circle in the same way so that it has centre B and radius BA.

5 Highlight both circles to find the two points of intersection (**Construct menu, Point of intersection**). Colour the two points of intersection black. Join these two points with a line, CD. Make this line green.

6 Highlight the lines AB and CD and find the point of intersection, point E.

7 Highlight both circles and make them disappear (**Display menu, Hide circles**). You should now see the diagram below.

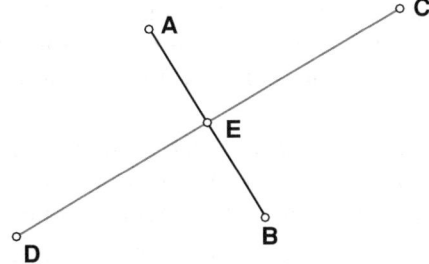

8 Move points A and B around to see what happens.

9 How is point E linked to AB? Explain.

10 How is line CD linked to AB? Explain.

11 How is triangle ABC linked to AB? Explain.

12 Draw the construction using a straight edge and a pair of compasses.

In this activity you are going to use Geometer's SketchPad to draw a triangle with sides of fixed lengths.

1 Make three lines, AB, CD and EF, in the top left-hand corner of the screen and measure their lengths. Make them the same lengths as those in the diagram.

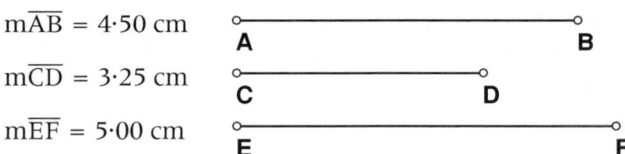

$m\overline{AB} = 4\cdot50$ cm

$m\overline{CD} = 3\cdot25$ cm

$m\overline{EF} = 5\cdot00$ cm

2 Highlight the point A and then the point B in that order. Click on **Construct menu, Mark vector A → B**. Make a new point, G, in the middle of the screen. Highlight just G then click on **Transform menu, Translate** and then make sure that the new window says '**Vector from A to B**' and that '**By marked vector**' is selected, then click on **OK**. You will see a new point, label it H.

3 Join line GH and this is one side of your triangle. It will always be the same length as AB and parallel to it.

4 You are now going to do a similar thing at each end of the line GH, to make one line equal in length to CD and the other equal in length to EF. However, you will have to follow one step more since these two lines have to be made so that they can take any direction. The instruction is given in **5**.

5 Highlight the point C and then D in that order. Click on **Construct menu, Mark vector C → D**. Highlight just G then click on **Transform menu, Translate** with '**Vector from C to D**' and '**By marked vector**' selected, then click on **OK**. You will see a new point, label it I.

6 Now make a circle centre G with radius GI. Put another point J anywhere on this circle. Join GJ to make a line and then hide point I and the circle.

7 Highlight the point E and then F in that order. Click on **Construct menu, Mark vector E → F**. Highlight just H then click on **Transform menu, Translate** with '**Vector from E to F**' and '**By marked vector**' selected, then click on **OK**. You will see a new point, label it K.

8 Now make a circle centre H with radius HK. Put another point L anywhere on this circle. Join HL to make a line and then hide point K and the circle.

9 You should now see a diagram like the one on the right. Move points J and L until you make a triangle. What are the lengths of the triangle? Explain.

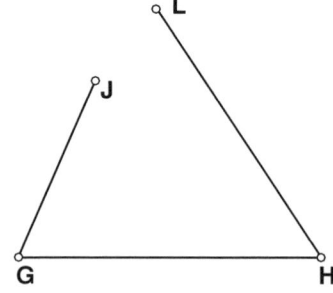

10 Change the lengths of AB, CD and EF to see what happens. Can you always make a triangle? If not why not? Explain.

11 Make sure that you have a triangle on your screen. Click on **Display menu, Show all hidden objects**. What do you notice? Explain.

In this activity you are going to learn how to use MswLogo to draw parallelograms and then adapt this to make other quadrilaterals based on parallelograms. You will then look at patterns that can be drawn in Logo.

1 Type this into the instruction window:

REPEAT 2 [FD 30 RT 50 FD 40 RT 130]

2 What shape do you get?

3 What are the internal angles of the shape? Explain, and use a diagram to show the shape and its angles.

4 Make another of the same type of shape with different angles.

5 If you want to keep making the same shape, what is important about the two angles?

6 At what angles (> 0) could the shape have a different name?

7 How does changing the lengths make a difference? Explain.

8 What if the lengths are the same? What new shapes can now be made?

9 Adapt the procedure above to make a rectangle and the other possible shapes that could be made by the instruction above. Test your instructions and draw some examples using them.

Logo experiments

1 Copy this instruction into Logo.

REPEAT 8 [FD 50 RT 45]

2 What shape does it make?

3 What is important about the number of repeats and the angle?

4 Copy this instruction into Logo.

REPEAT 5 [REPEAT 8 [FD 50 RT 45] RT 72]

5 What shape does it make?

6 What is important about the number of repeats and the angle?

7 Make more shapes of your own like this.

In this activity you are going to use Logo to make some shapes and then adapt your work to make shapes of your own. In each case you can see the shape made by a point following the instructions you are typing into the computer.

1 Here are two shapes with rotational and line symmetry of the same order (but different to each other). Make these shapes and then make more of your own based on the same idea.

 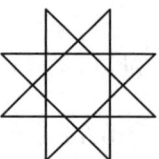

2 Here are three more shapes also with rotational and line symmetry of the same order (but different to each other). Make these shapes and then make more of your own based on the same idea.

3 Make more shapes of your own design using ideas inspired by the work in this activity.

Geometer's SketchPad or equivalent activities

Enlargement

For use with **part C** of the Practical (Pupil Book page page 323).

In this worksheet pupils investigate what happens when a quadrilateral AEFG is made from quadrilateral ABCD with E, F and G as the mid-points of AB, AC and AD respectively.

Activities

Enlargement

This is the general case. For some pupils you may want to start with either a square (easiest) or a parallelogram. In both these cases when you make the three translations of quadrilateral AEFG you completely match and fill quadrilateral ABCD (with no overlapping).

When you move the four vertices A, B, C and D to make it so that the overlap of the translations disappears, the shape is a parallelogram (and therefore could be a rectangle, rhombus or square).

In all cases AEFG is similar to ABCD with lengths in the ratio 1 : 2.

An enlargement scale factor 2 centre A will take AEFG onto ABCD and an enlargement scale factor 0·5 centre A will take ABCD onto AEFG.

In this worksheet you are going to make two linked quadrilaterals and look at the connections between them.

1 Draw a quadrilateral ABCD and join one diagonal AC. Construct E to be the mid-point of AB. Do the same thing for F, the mid-point of AC and G, the mid-point of AD.

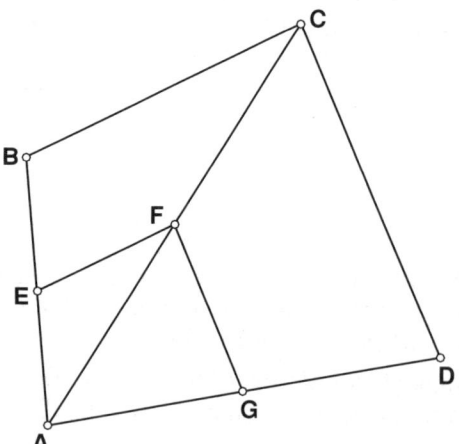

2 You now have the two quadrilaterals ABCD and AEFG. What do you notice? Explain.

3 What link is there between quadrilateral ABCD and quadrilateral AEFG? Explain.

4 Highlight in order all of points A, E, F and G. Click on *Construct menu, Polygon interior*.

5 Highlight points A and E in this order and click on *Transform menu, Mark as vector*. Translate all of quadrilateral AEFG and its interior by this vector. Repeat for points A and F and translate quadrilateral AEFG and its interior. Repeat again for points A and G and translate quadrilateral AEFG and its interior.

6 Colour two opposite interiors light yellow and the other two light red. You should now have the diagram shown here.

7 What can you say about the lengths and areas of the two quadrilaterals ABCD and AEFG? Explain.

8 Move A, B, C and D until there is no overlap of yellow and red areas. What shape is the quadrilateral? Explain.

9 Move A, B, C and D so you have a general quadrilateral. Make an enlargement (called a dilation in Geometer's SketchPad) of AEFG so that it is exactly the same size as ABCD. Explain what this means.

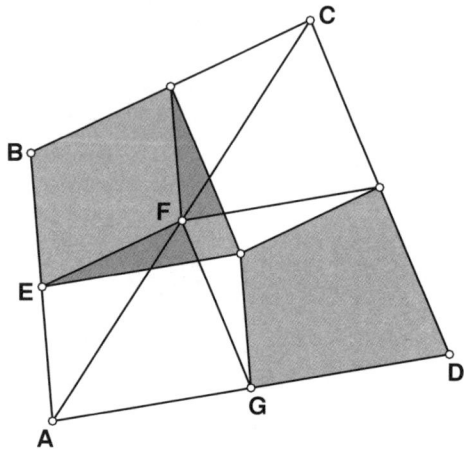

Excel spreadsheet or equivalent activities

Finding means with a spreadsheet

For use with '**Mean**' and Exercise 2 (Pupil Book page 383).

Find the mean on a spreadsheet from a frequency table.

Use of the **AVERAGE** function.

Interrogating a database 1

For use with the Practical (Pupil Book page 391).

In this worksheet pupils use a spreadsheet table as a database and they learn how to extract information from the database.

They use the *Data menu, Filter, Autofilter* feature of Excel to search the database and use the **SUBTOTAL** function to find the mean of 'filtered' data.

Interrogating a database 2

For use with the Practical (Pupil Book page 391).

In this worksheet pupils use a spreadsheet table as a database and they learn more about how to extract more information from the database.

They use the *Custom* feature of *Autofilter* to search the database and use the **MEDIAN** function to find the median of data taken from the spreadsheet (**MEDIAN** cannot be found easily on filtered data).

Bar charts

For use with '**Compound bar charts and line graphs**' (Pupil Book page 391).

In this worksheet pupils are shown how to draw a bar chart.

Drawing a line graph

For use with '**Compound bar charts and line graphs**' (Pupil Book page 391).

In this worksheet pupils learn how to draw a line graph.

Use *Format axis, Scale tab* to replace the 'default' scale with one of your choice.

Pie charts

For use with the Practical (Pupil Book page 399).

In this worksheet pupils are shown how to draw a pie chart.

Scatter graphs

For use with the Practical (Pupil Book page 402).

In this worksheet pupils are shown how to draw a scatter graph.

Spreadsheet ideas

We would suggest that Excel, or an equivalent, should always be used to produce a report on a statistical survey. It should be used to make the tables and any charts that arise from the tables. These should be copied and pasted into a word processor such as Word. As with handwritten work, the charts and tables should be properly titled and labelled and should present the data in an interesting and informative manner. Pupils should use appropriate charts and not 'unusual' charts. Blocked and 3-D charts are usually not appropriate since they distort the data.

One common error is to produce a chart where the text is large and the chart is small — this means that Autoscale on the Font menu has not been turned off. Charts like this would not be acceptable if hand-drawn and should therefore not be accepted from the spreadsheet. You will need to remind pupils of this often.

The more pupils use spreadsheets and word processors for writing reports the better they become at it.

Note that it is not possible to use Excel easily to draw a frequency diagram for a table in this form:

Class interval (Length in mm)	Frequency
$30 \leqslant L < 35$	2
$35 \leqslant L < 40$	2
$40 \leqslant L < 45$	4
$45 \leqslant L < 50$	9
$50 \leqslant L < 55$	10
$55 \leqslant L < 60$	2

This is because Excel will place the class interval as shown on the chart below:

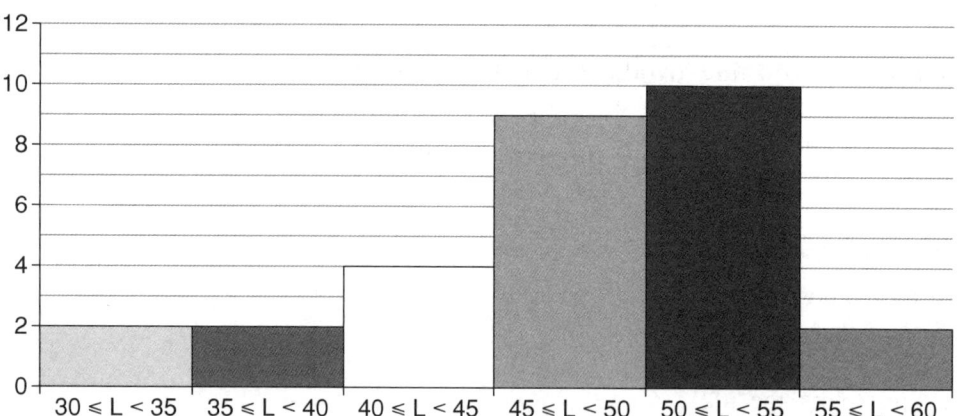

To place the figures at the end of the bars would involve creating a special column of information in addition to that above and then finding a way to make the label move to one end of the bar.

Activities

Finding means with a spreadsheet

Note that it is possible to find a mean directly from a spreadsheet using the **AVERAGE** function. However, this is only appropriate if you have the raw data rather than data in a tabular form.

AVERAGE(A2:A10) would find the mean of the values given in cells A2:A10.

Interrogating a database 1

You will probably need to collect these data in advance of the lesson and could have pupils input their own. As you do this you need to ensure that they all use the same way of describing gender ('F' and 'M' are used here for female and male) and that you agree on allowed eye colours (are 'brown' and 'hazel' the same, should 'blue-green' and 'blue green' both be allowed?). Too many colours on a small database is not a good idea, but pupils do like this sort of information to be accurate.

You may extend this database to include more information, but some personal data will clearly not be appropriate.

You will then need to check the data to make sure that it is 'reasonably correct'. Typical mistakes are that pupils enter a measurement in the wrong units and that they add spaces at the end of words. Spaces become evident when you use the **Autofilter** feature of Excel and you see two or more entries of, for example, female gender where one is F and the other is F followed by a space. Where this happens the easiest way to solve it is to use a find and replace on the relevant column: find F followed by a space and replace it with F. Errors like this could affect the mode.

Pupils need to take care when looking for means. Typically, they do things in a wrong order and so may find the mean of an incorrect subset.

The help menu tells you what **SUBTOTAL** can do in addition to finding the mean — most uses are not relevant to the 11–16 curriculum.

Interrogating a database 2

This continues the work on the same database, making use of the **Custom** feature of **Autofilter**.

The median does not work on 'filtered' data.

Bar charts

The bar chart that uses days of the week for the *x*-axis is straightforward. However, if there were numbers across the *x*-axis things would be more problematic. Here you will need to go to Step 2 of 4 of the chart wizard (so you need to press **Next** and not **Finish** for part 3 of the worksheet). You will then see this window.

Notice how it has three sets of bars (it has read the first set of figures as representing a bar). To change this you need to click on the **Series tab** near the top of the window.

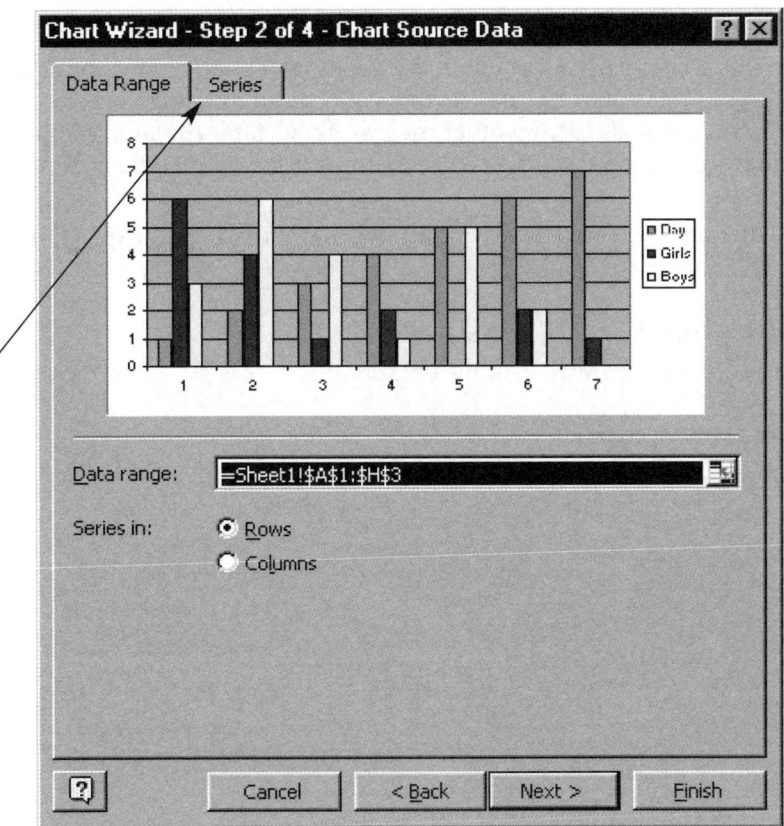

You will then see this window.

You need to make sure that the highlight is on the Day set of data (shown in the Series box).

Then you have to click on *Remove*.

Then click into the *Category (X) axis labels* box.

Highlight cells B1 to H1 and you should see

= Sheet1!$B$1:$H$1

placed in the box.

Click on Finish and then improve the chart.

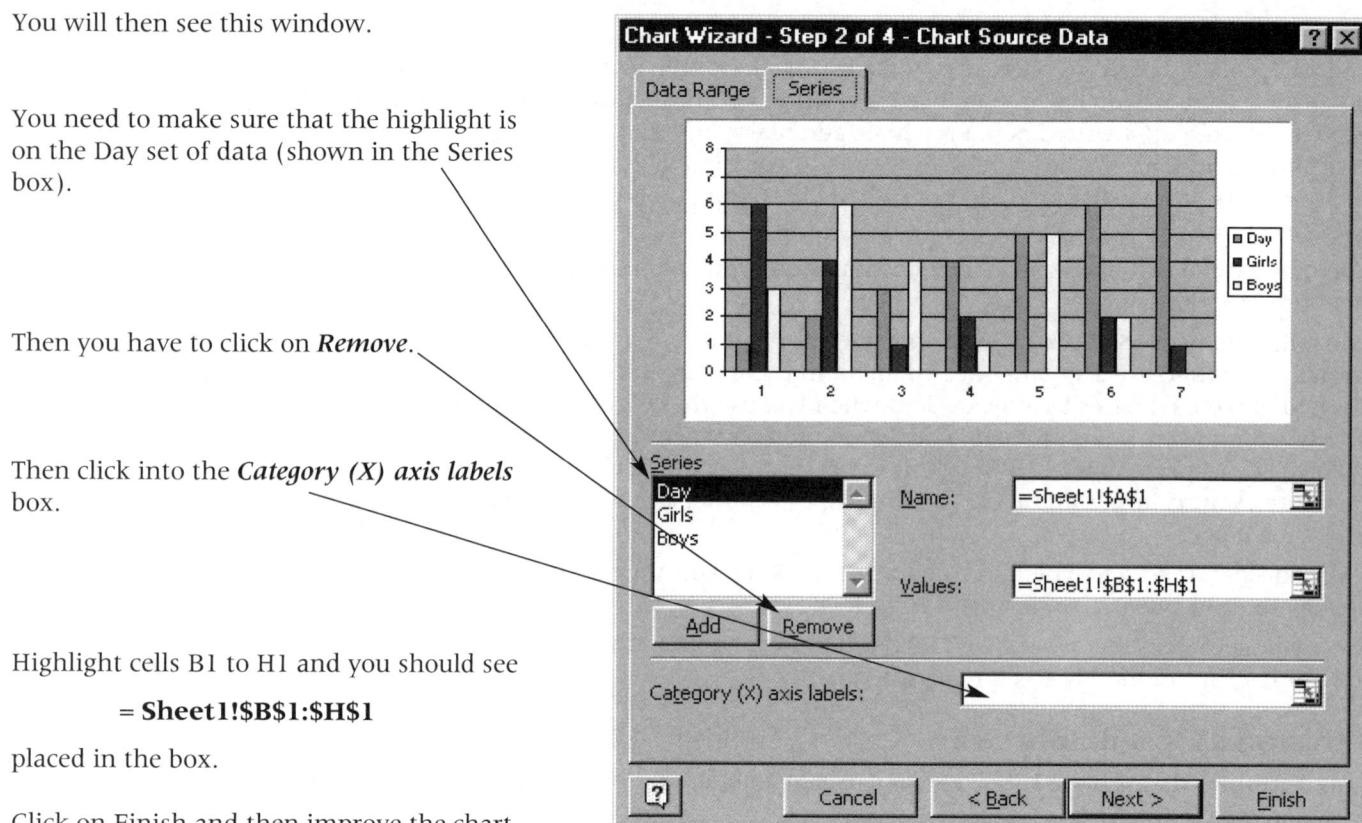

Drawing a line graph

The problem with a line graph is evident when data is not correctly ordered and when the *x*-axis data is not regularly spaced. A line graph in Excel takes numerical data in the top row as its *x*-axis and data from other rows as its '*y*' data. (It could do this for columns as well.) It will not interpret any values on the *x* data — it just places the *x* data against consecutive 'marks' on the *x*-axis.

Generally we would advise pupils not to use the **Line graph** option, but to use the **XY(Scatter)** instead.

Pie charts

Straightforward way to draw a pie chart.

Scatter graphs

Straightforward way to draw a scatter graph.

In this activity you are going to make a table of frequencies on a spreadsheet then use the spreadsheet to calculate the mean.

1 The table below shows the scores of 285 people in a quiz. Make the table in a spreadsheet.

	A	B	C	D	E	F	G	H	I	J	K	L
1	Score	1	2	3	4	5	6	7	8	9	10	Totals
2	Number of people	8	16	23	26	31	87	41	32	17	4	
3	Total score											
4												
5	Mean											

2 You want to count the total number of people and the total score. To count the total number of people you need to add up all the figures in cells B2 to K2. The quickest way to do this is to use the **SUM** function. Into cell L2 type the formula:

$$= \textbf{SUM}(B2{:}K2)$$

3 How many people were there altogether?

4 The total score from the people who scored 1 is 8. The total score from the people who scored 2 is 32 (2×16) and so on. The spreadsheet can do this for you. All you have to do is put the correct formula into cell B3 and then copy it across to cell K3.

5 Place your formula into cell B3 and copy this formula across to cell K3 (if you get 40 in K3 you probably have the correct formula).

6 You already have a formula in cell L2 that adds the values in B2:K2, so you can copy this down into cell L3.

7 What is the total score for all the 285 people?

8 To find the mean you must place the correct formula in cell B5. This formula is:

$$= L3/L2$$

9 What is the mean score?

10 Do some more examples like this for yourself.

In this worksheet you will use a spreadsheet as a database, having collected the following pieces of information about a group of people.

1 Make or load a database that contains the information given below for at least 20 people, making sure that you leave rows 1 and 2 empty. It is assumed that your database is in cells A3:H22, so you will need to change the H22 below if your database is larger.

	A	B	C	D	E	F	G	H
1								
2								
3	Name	Gender	Age (months)	Height (cm)	Foot length (cm)	Arm span (cm)	Wrist circumference (cm)	Eye colour
4								

2 Highlight the complete database, cells A3:H22. Click on **Data menu, Filter, Autofilter**. You will see drop-down arrows added to the cells A3:H3 (see right for an example).

Name ▼	Genc ▼	Age (montl ▼
Sayyad Iqbal	M	150
Mary O'Neill	F	148
Sue Wood	F	150

3 Click on the drop-down arrow by gender and you will see a menu like the one shown here.

Name ▼	Genc ▼	Age (montl ▼
Sayyad Iqb	(All)	150
Mary O'Ne	(Top 10...) (Custom...)	148
Sue Wood	F	150
Everton Gr	M	147
Anita Chop	(Blanks) (NonBlanks)	155

4 Click on the M. What happens?

5 Click on the age drop-down arrow (in cell C3) and click on one of the figures. How many males in your group are that age in months?

6 Click on the two blue drop-down arrows and for each click on **All**. You should now be back to the start and have all the data showing for all the group.

7 Into cell A1 type **MEAN**. You are now going to find out how to find the mean of a filtered list, so that as the list changes the mean changes as well. To begin with you want to find the mean age of all the people. You could use the **AVERAGE** function, but this will not work when you only want to look at some of the group and not all the group, so instead you use the **SUBTOTAL** function.

8 Type = **SUBTOTAL**(1, C4:C22) into cell C1. (If you have more/less data replace 22 with the bottom row of your data.) The 1 at the start is the code for the mean. You should see a figure that will be around 150 (if this is completed for a Year 8 group). Now click so only the females are in the list. What is the mean age of the females?

9 What is the mean age of the males? Which group has a larger mean? Is this what you would expect?

10 Do the same thing to find the mean height of all the people, then of the females and then of the males. Which group has a larger mean? Is this what you would expect?

11 Repeat for foot length, arm span and wrist circumference.

12 Why can't you do the same for eye colour?

13 Save your database. You will need it for Worksheet: Interrogating a database 2.

In this worksheet you will continue to use a spreadsheet as a database, having collected information about a group of people. This follows the worksheet 'Interrogating a database 1'.

1 Load the same database and make sure that *Autofilter* is selected.

2 Click on the arrow by **Age** and then on *Custom* on the drop down menu and you will see a window like the one shown below:

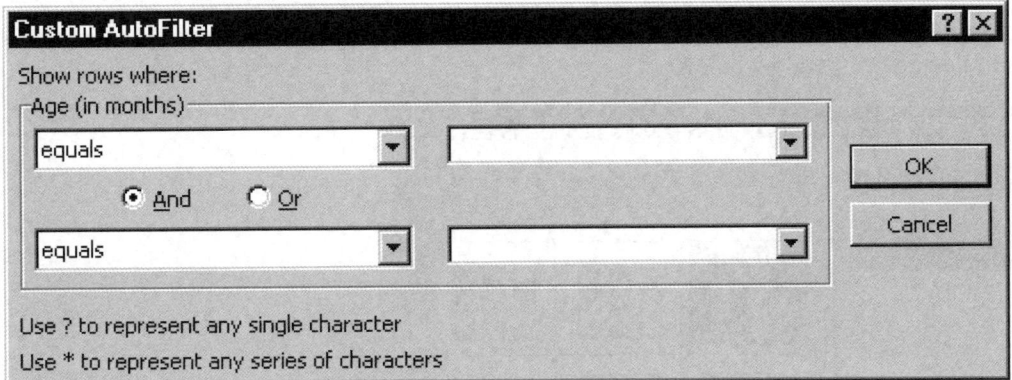

3 This lets you put two conditions onto the column, connecting them by **And** or by **Or**. In this case you want to look at just the people who are aged 148 months or more *and* less than 151 months. To do this make the window below:

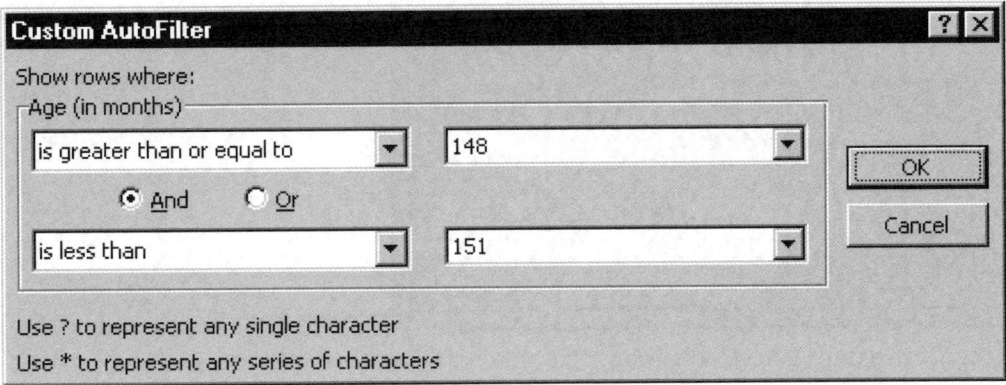

4 To find the median of any of the data in the list is not quite as easy as finding the mean. To find the median of any column or part column you have to first filter the list so it just contains the people that you want (for example Females). You then need to copy that list of people and paste it to another worksheet. Then you click on the cell beneath the pasted data and type into the cell:

= **MEDIAN**(cell list)

where you type the cell list inside the brackets — this would be something like A4:A15.

5 The mode function in Excel only works on numerical data and only finds one mode (even if there are several), so it is best not to use this spreadsheet function.

In this worksheet you are reminded how to create a spreadsheet bar chart from a table.

	A	B	C	D	E	F	G	H
1	**Day**	**Monday**	**Tuesday**	**Wednesday**	**Thursday**	**Friday**	**Saturday**	**Sunday**
2	**Girls**	6	4	1	2	0	2	1
3	**Boys**	3	6	4	1	5	2	0

1 Make the table above in your spreadsheet. Highlight the table from A1 to H3. Click on the Chart wizard and you should see the window below.

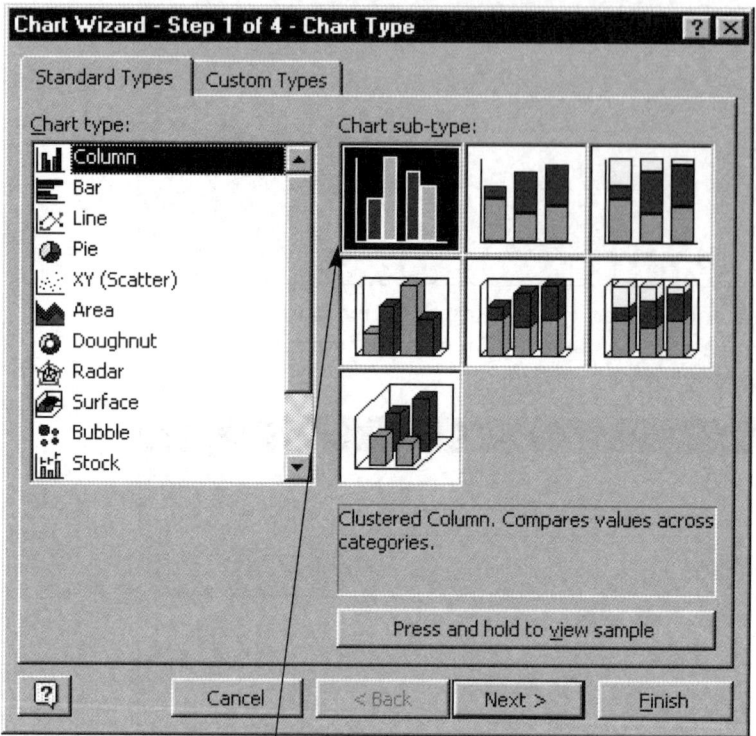

2 Although you want a bar chart the spreadsheet calls this a column chart (use the pictures and not the words to guide you).

3 You want the marked chart sub-type, so once you have this click on *Finish* and you should see the chart given below.

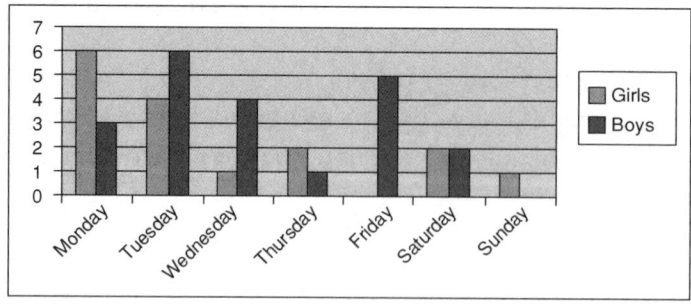

4 Format this chart so that it has a title, is properly labelled and is easy to read with sensible sized text.

In this worksheet you learn how to draw a line graph. For this it is best to follow the *XY(Scatter)* option and not the *Line graph* option, since it is quicker and easier.

	A	B	C	D	E	F	G	H	I	J	K	L
1	Year of second birthday	1989	1990	1991	1992	1993	1994	1995	1996	1997	1998	1999
2	Percentage MMR	80	84	87	90	92	91	91	92	92	91	88
3	Percentage whooping cough	75	79	84	88	92	93	93	94	94	94	94

1 Make the table above in your spreadsheet. Highlight the table from A1 to L3. Click on the Chart wizard and follow the option for *XY(Scatter)* and the picture for joined with a curved line and click on *Next*.

2 You should then see the diagram below.

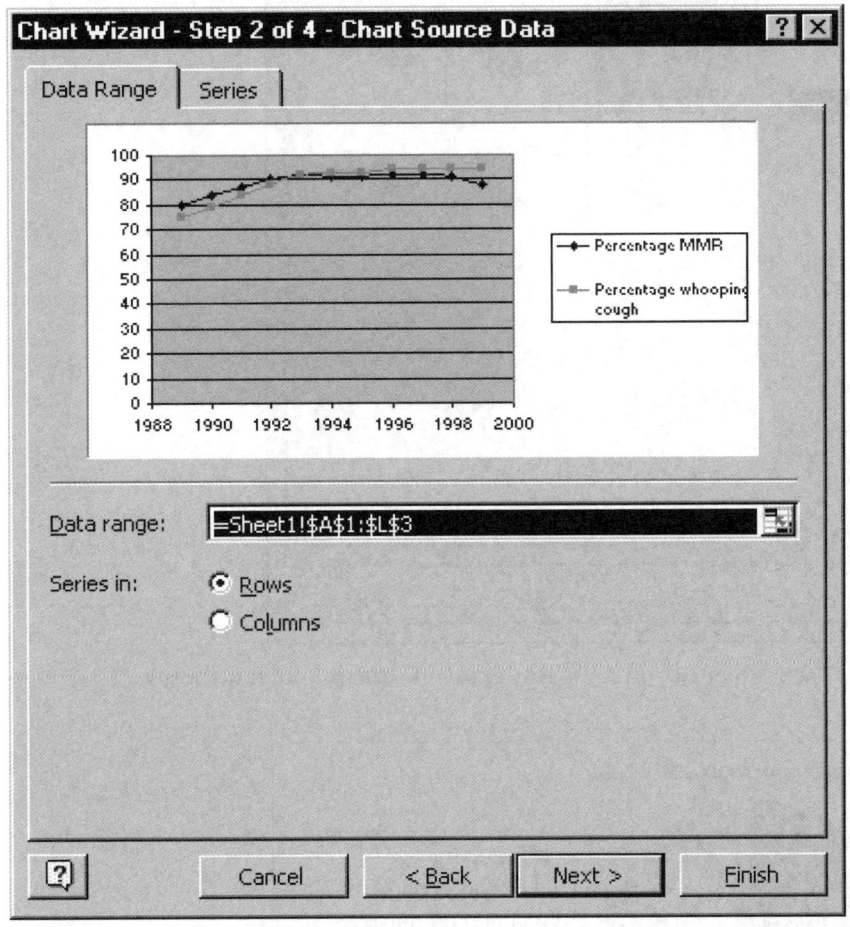

3 Give your line graph a title, label the axes and delete the Legend.

4 You will probably see that your graph is all grouped at the top (as in the diagram above). This is not helpful so you need to change it. To do this right-click on the *y*-axis and then click on *Format axis*, *Scale tab* and type 70 in the **Minimum** box, then click on *OK*.

5 Remember that the text of the chart should not get too big. If it does, right-click in the top-right of the chart, click on *Format chart area, Font* and 'turn off' the **Autoscale** and make the font size 12 or less.

In this worksheet you will learn how to create a spreadsheet pie chart from a table.

	A	B	C	D	E	F
1	Activity	Sleeping	Playing sport	Doing homework	Eating	With friends
2	Hours spent on activities	9	5	2	1	7

1 Make the table above in your spreadsheet. Highlight the table from A1 to F2. Click on the Chart wizard and you should see the window below. Highlight the *Pie* option as shown.

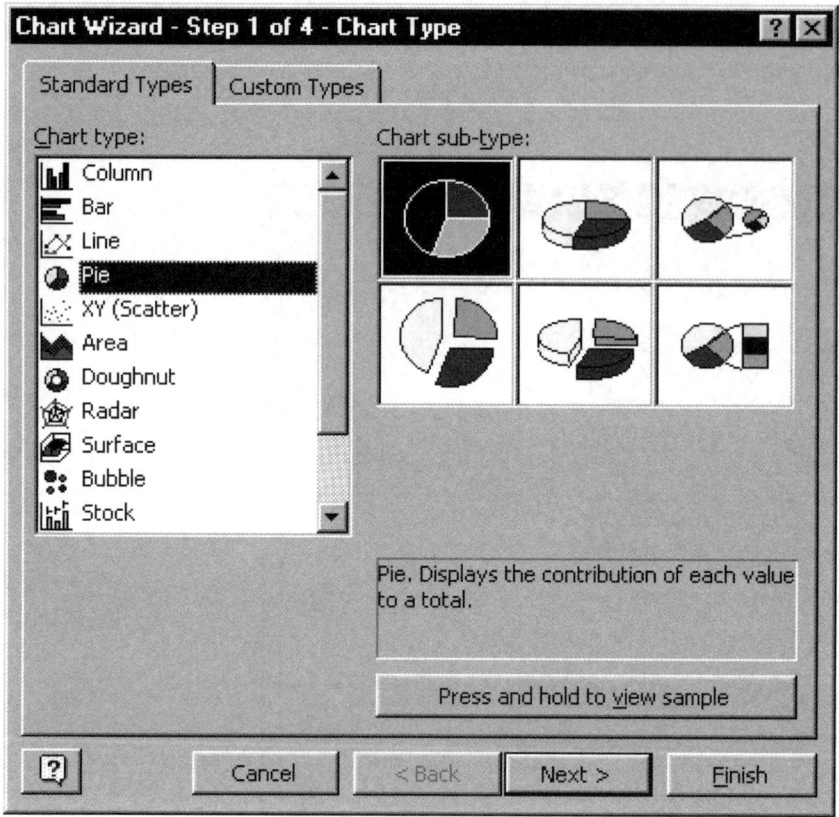

2 You want the marked chart sub-type, so once you have this click on *Finish* and you should see the chart given below.

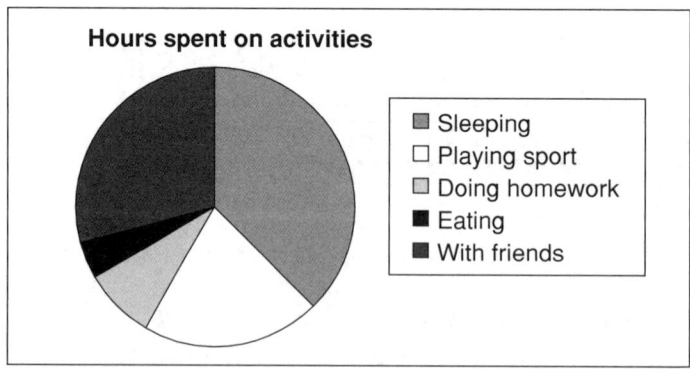

3 Format this chart so that it is easy to read.

4 Why is this a good example for a pie chart? Make another pie chart for an appropriate set of data.

In this worksheet you will learn how to create a spreadsheet scatter graph from a table.

	A	B	C	D	E	F	G	H	I	J	K	L	M
1	Age (years)	2	6	8	10	7	3	5	4	8	7	9	6
2	Height (cm)	80	110	122	136	114	95	104	100	126	120	130	112

1 Make the table above in your spreadsheet. Highlight the table from A1:M2. Click on the *Chart wizard* and follow the option for *XY(Scatter)* and the picture where the dots are not joined. Click on *Next*.

2 Make your scatter graph so it is easy to read the information it contains. It should look something like the one shown below.

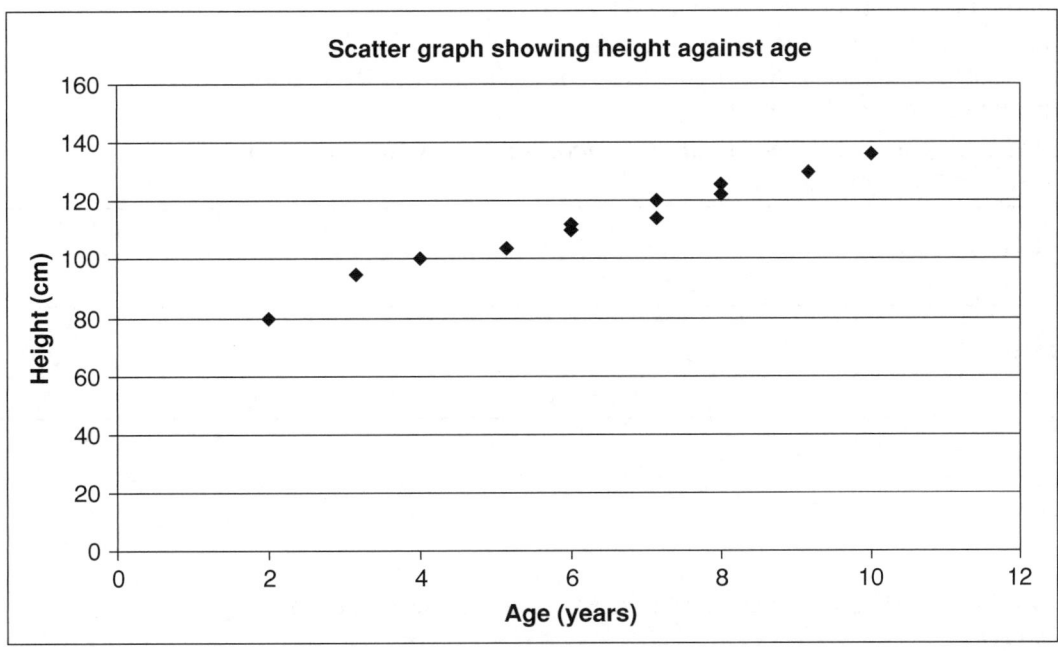

3 What does the scatter graph tell you about any link between age and height?

4 If the chart contained data about more people in this age range what do you think it would show?

5 What if it contained information about people of all ages?

6 Draw some more scatter graphs of your own.

Excel spreadsheet or equivalent activities

Throwing dice 1

For use with **part B** of the Practical (Pupil Book page 423).

In this worksheet pupils are asked to construct a spreadsheet that generates 60 throws of a dice. From this pupils construct an automatic frequency count.

Use of **FREQUENCY** function.

Throwing dice 2

For use with **part B** of the Practical (Pupil Book page 423).

In this worksheet pupils add a bar chart to the spreadsheet in 'Throwing dice 1', then increase the number of throws to 600. A bar chart is linked to a frequency table for this data so that pupils can compare the data in the two bar charts.

Right-click on one of the chart bars to use *Format data, Options tab* to vary the colours of the bars and change the distance between the bars.

Internet

The lottery website

www.lottery.co.uk

Activities

Throwing dice 1

This is almost the same as the worksheet for Chapter 18 in the ICT Support section of the 7 Core Teacher Resource Pack.

The first part shows how to use the **INT** and **RAND( )** functions to generate random numbers for dice and the second part shows how to count the frequency distribution using the **FREQUENCY** function (which is not easy).

The dice random throw generator is:

$$= \textbf{INT}(\textbf{RAND}(\,) * 6 + 1)$$

Note that **RAND( )** generates a random number between 0 and 1, including 0 but **NOT** including 1. Hence **RAND( )** * 6 generates a random number between 0 and 6, including 0 but **NOT** including 6. Taking the integer part would therefore give you integers 0 to 5.

This can be adapted for a dice with any number of sides: change the 6 to 8 for an octahedral dice.

Throwing dice 2

The first part adds a bar chart to the spreadsheet above. The second part asks pupils to extend the data to 600 throws, make a new table and new chart by copying the old ones — you might wish to have pupils construct these from scratch.

The *y*-axis on the chart for 600 values has to be changed so that the minimum is 0, otherwise Excel assumes that you only want to see that part of the chart where the data can be found. This gives a misleading picture and should be discussed with pupils.

You may wish to make the minimum on both charts 0 and the maximum equal to 20 and 200 respectively (one-third of the total number of throws). This may make the difference between the two charts more evident.

The chart for 600 throws is likely to be 'flatter' than that for 60 throws. In percentage terms the difference of each frequency from the 'theoretical average' (of 100 and 10) will be less for the 600 throws — this could also be examined and discussed.

Spreadsheet ideas

Throwing dice 2 (*continued*)

Keeping results

The random number generator changes every time the spreadsheet changes. This means that it is difficult to keep a set of results. There are two ways round this. The first involves setting the spreadsheet so that values are only calculated when you want. To do this use **Tools menu, Options, Calculation tab**, select **Manual**. This means that nothing will change on the spreadsheet until you press the **F9** key.

The other method is to highlight all of the spreadsheet from A1:J12 and then use **Copy**, move to, for example, cell A14 followed by **Paste special** choosing the **values** option, so the formulas are all replaced by values.

In this activity you use a spreadsheet to simulate throwing a dice 60 times.

1 The array A3:F12 below has been generated by the computer. You will now make your own version of this spreadsheet. Type in all parts of the table below **except** for cells A3:F12.

	A	B	C	D	E	F	G	H	I	J
1	Throwing a dice									
2										
3	5	4	5	6	6	2				
4	6	4	5	6	3	4				
5	4	1	2	5	3	1			Dice score	Frequency
6	6	2	3	2	1	3			1	
7	2	2	3	1	5	4			2	
8	1	2	5	4	2	2			3	
9	3	3	1	2	4	1			4	
10	2	3	1	1	5	1			5	
11	6	3	4	3	1	3			6	
12	3	6	3	4	1	2				

2 Into cell A3 type the formula:

$$= \mathbf{INT}(\mathbf{RAND}()*6+1)$$

3 What happens? Copy it into cells B3 to F3. What happens? Copy A3:F3 down to F12.

The automatic frequency count

1 Highlight cells J6 to J11. Click on *Insert menu, function* and you will see the window here. Click on *Statistical*.

2 Then click here until you see the word **FREQUENCY** in the right-hand column.

3 Click on *FREQUENCY* and then click on *OK*.

4 The window here will now open.

5 In here type: **A3:F12**.

6 In here type: **I6:I11**.

7 **Don't click** on OK.

8 **Instead hold down the Shift and Ctrl keys and press Enter at the same time**.

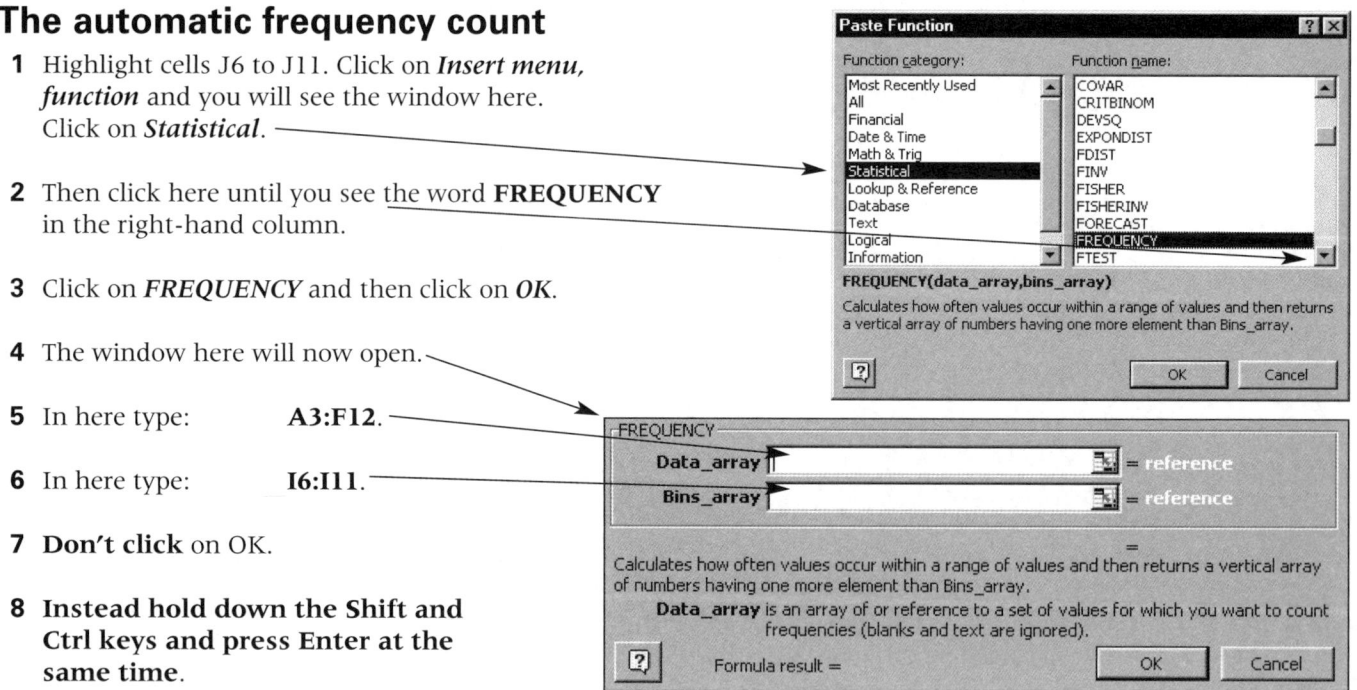

9 You should now see how many times each number from 1 to 6 appears in the cells A3:F12.

10 If you press the **F9** key, you should get a new set of values. Do this. What do you notice?

© New National Framework Mathematics 8 Nelson Thornes Ltd

In this activity you will use the spreadsheet from 'Throwing Dice 1' and add a bar chart linked to the table. You will also increase the number of dice throws to see what happens when you throw the dice 600 times.

1   You need to open the spreadsheet for the worksheet 'Throwing Dice 1' and find the table in cells I5:J11.

2   Highlight the table, including the words, and click on the **Chart wizard** and then on the **Column chart** option as though you were going to make a normal bar chart. Click on **Next** to move to **Step 2 of 4** (which should be on the top blue line). Click on the **Series tab** near the top, making sure that you are still at **Step 2 of 4** shown on the window below.

3   At the moment there are two sets of bars here, which is wrong.

4   Click on **Remove** (so that **Dice scores** is taken off the chart and so you have just one set of bars showing).

5   Click into the **Category (X) axis labels** box and then highlight on the sheet cells I6:I11. Click on **Finish**.

6   Right-click on the chart, and go to **Chart options**. Label your chart properly and remove the Legend.

7   Right-click on one of the chart bars and then **Format data series, Options tab**. Change the gap to zero and tick **Vary colours by point**.

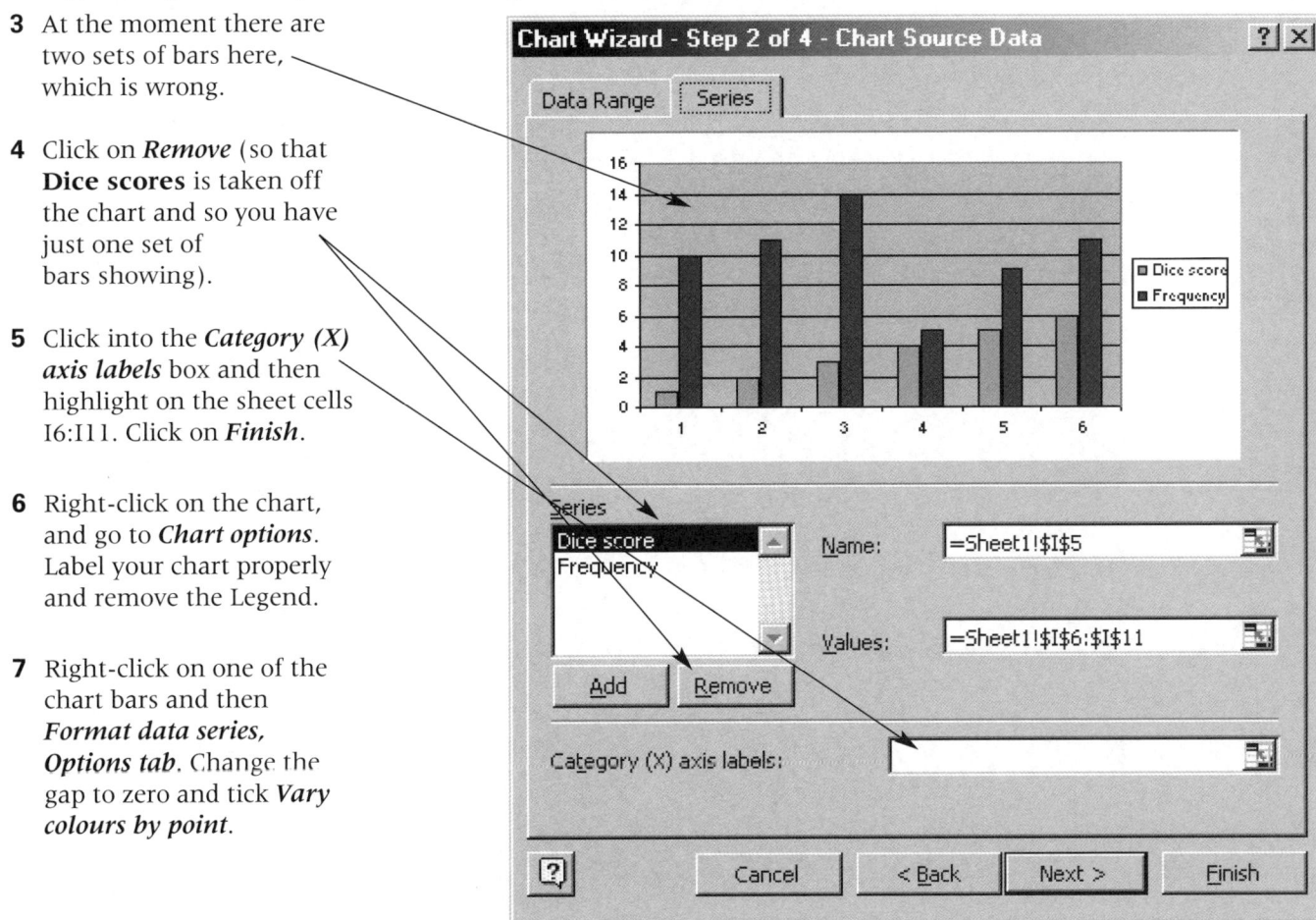

8   You will now add more throws and make another table and bar chart so that you can compare 60 throws with 600 throws. Highlight the data in cells A12:F12 and copy it down to A102:F102.

9   Copy cells I5:J11 and paste into cells I15:J21. Highlight cells J16:J21 and delete the **Frequency count**. Make a new **Frequency count** so that the **Data array** reads **A3:F102** and the **Bins array** is **I16:I20**. Press the **Shift and Ctrl keys together and press Enter at the same time**.

10   Copy and paste the graph, so that you have two copies. Right-click on the new one and change the source data so that the new chart takes its data from the table at I15:J21. Click on the y-axis and on **Format axis, Scale tab** and change the **Minimum** to 0.

11   Press the F9 key to change the values and note how your two bar charts change. How are they the same? How are they different? Explain.

4

Graphical Calculator Support

Introduction

The graphical calculator support materials provide calculator command sequences for Casio, Sharp and Texas machines.

These sequences include such things as Int, Random, Lbl and Goto that are accessed from the various menus of calculator commands. It is important to note that simply spelling out these commands using the keypad *will not work*. Each command is automatically written onto the calculator display once selected from the appropriate menu.

Advice on navigating through the menus to find the commands that you need is provided for the Casio *fx-7400G PLUS*, the Casio *fx-9750G PLUS* the Sharp EL9400 and the Texas TI-83. The **command sequences** for each of these makes will apply to all models, though there will be some variation in the **key sequences** required to produce the commands on screen.

Casio *fx-7400G PLUS*

The various command menus may be accessed by selecting either RUN or PRGM from the main menu screen. This screen is shown whenever the machine is switched on and can also be displayed at any other time by pressing the MENU button.

The difference between selecting RUN and PRGM is that commands written in RUN mode cannot be saved for future use, whereas commands written in PRGM mode are saved automatically as a program, once a program name is entered.

This diagram shows where the commands are located. You may find it helpful to refer to it when writing commands into the machine. In the diagram, the use of block capitals indicates that there is a corresponding sub-menu.
The F keys are located directly under the screen and are aligned with menu commands.

The ▶ key is located to the right of the F keys.

PRGM (key)

COM	CTL	JUMP	▶	?	◢	CLR	DISP	▶	REL	I/O	:
F1	F2	F3		F1	F2	F3	F4		F1	F2	F3

COM								
If	Then	Else	IfEnd	▶	For	To	Step	Next
F1	F2	F3	F4		F1	F2	F3	F4

CTL			
Prog	Return	Break	Stop
F1	F2	F3	F4

JUMP			
Lbl	Goto	Isz	Dsz
F1	F2	F3	F4

CLR		
ClrText	ClrGraph	ClrList
F1	F2	F3

REL						
=	≠	>	<	▶	≥	≤
F1	F2	F3	F4		F1	F2

OPTN (key)

LIST	CALC	STAT	PROB	▶	NUM	ANGLE
F1	F2	F3	F4		F1	F2

PROB

x!	nPr	nCr	Ran#
F1	F2	F3	F4

NUM

Abs	Int	Frac	Rnd	▶	Intg
F1	F2	F3	F4		F1

Examples

To display	Press				
?	(Shift) PRGM	▶	F1		
Goto	(Shift) PRGM	F3	F2		
Ran#	OPTN	PROB	F4		
Int	OPTN	▶	F1	F2	

Casio *fx-9750G PLUS*

As for the Casio *fx-7400G PLUS*, the various command menus may be accessed by selecting either RUN or PRGM from the main menu screen. This screen is shown whenever the machine is switched on and can also be displayed at any other time by pressing the MENU button. The main difference is that the Casio *fx-9750G PLUS*, has six F keys at the top of the keyboard instead of four and so more commands are available from each screen.

The difference between selecting RUN and PRGM is that commands written in RUN mode cannot be saved for future use, whereas commands written in PRGM mode are saved automatically as a program, once a program name is entered.

This diagram shows where the commands are located. You may find it helpful to refer to it when writing commands into the machine. In the diagram, the use of block capitals indicates that there is a corresponding sub-menu. Use the EXIT key to leave a main or sub-menu. The F keys are located directly under the screen and are aligned with menu commands.

PRGM (key)

COM	CTL	JUMP	?	◪	▷
F1	F2	F3	F4	F5	F6

CLR	DISP	REL	I/O	:	▷
F1	F2	F3	F4	F5	F6

COM

If	Then	Else	IfEnd		▷
F1	F2	F3	F4	F5	F6

For	To	Step	Next
F1	F2	F3	F4

CTL

Prog	Return	Break	Stop
F1	F2	F3	F4

JUMP

Lbl	Goto	⇒	Isz	Dsz
F1	F2	F3	F4	F5

CLR

Text	Grph	List
F1	F2	F3

DISP

Stat	Grph	Dyna	F-Tbl	R-Tbl
F1	F2	F3	F4	F5

REL

=	≠	>	<	⩾	⩽
F1	F2	F3	F4	F5	F6

OPTN (key)

LIST	MAT	CPLX	CALC	STAT	▷		HYP	PROB	NUM	ANGLE
F1	F2	F3	F4	F5	F6		F2	F3	F4	F5

PROB

x!	nPr	nCr	Ran#
F1	F2	F3	F4

NUM

Abs	Int	Frac	Rnd	Intg
F1	F2	F3	F4	F5

Examples

To display	Press			
?	(Shift) PRGM	F4		
Goto	(Shift) PRGM	F3	F2	
Ran#	OPTN	F6	F3	F4
Int	OPTN	F6	F4	F2

Sharp EL9400

Commands that are mathematically based are accessed by pressing the MATH key. You can use the vertical scroll keys to move the highlighting up or down through the main menu list, on the left, to display the corresponding sub-menu items on the right.

The diagram below shows some examples of how this works.

A CALC
B NUM
C PROB
D CONV
E ANGLE
F INEQ
G LOGIC

1 abs (
2 round (
3 ipart
4 fpart
5 int
6 min (

A CALC
B NUM
C PROB
D CONV
E ANGLE
F INEQ
G LOGIC

1 random
2 nPr
3 nCr
4 !

A CALC
B NUM
C PROB
D CONV
E ANGLE
F INEQ
G LOGIC

| 1 = |
| 2 ≠ |
| 3 > |
| 4 ⩾ |
| 5 < |
| 6 ⩽ |

For example, to select **random**, highlight PROB and then press 1.

The PRGM key does two things:

- When first pressed, it offers the options to Execute an existing program, Edit an existing program or create a New program.

- When editing, or writing a new program, it provides access to the commands that you may need.

The menu system accessed after pressing the PRGM key operates in the same way as the one accessed by pressing the MATH key, but with a different set of options. Note that the MATH key may still be used to access the maths commands when writing or editing a program.

A sample screen accessed by pressing the PRGM key is shown below.

A PRGM
B BRNCH
C SCRN
D I/O
E COORD
F FORM
G S_PLOT
H COPY

| 1 Label |
| 2 Goto |
| 3 If |
| 4 Gosub |
| 5 Return |

This shows, for example, that highlighting BRNCH on the left and then pressing 4 will write Gosub onto the screen.

Texas TI-83

Commands that are mathematically based are accessed by pressing the MATH key. You can use the horizontal scroll keys to move the highlighting through the main menu list across the top of the screen, to display the corresponding sub-menu items underneath.

The diagram below shows some examples of how this works.

MATH	NUM	CPX	PRB
1 : .Frac			
2 : .Dec			
3 : ^{3}			
4 : $^3\sqrt{(}$			
5 : $^x\sqrt{(}$			
6 : fMin (			
7 : ↓fMax (			

MATH	NUM	CPX	PRB
1 : abs (			
2 : round (			
3 : iPart			
4 : fPart			
5 : int (			
6 : min (			
7 : ↓max			

For example, highlighting NUM and pressing 5 writes **int (** onto the display.

The ↓ symbol shown in the 7th item indicates that the list extends downwards. The additional items are accessed by pressing the downward scroll key.

The PRGM key does two things:

- When first pressed, it offers the options to Execute an existing program, Edit an existing program or create a New program.
- When editing, or writing a new program, it provides access to the commands that you may need.

The menu system accessed from the PRGM key operates in the same way as the one accessed by pressing the MATH key, but with a different set of options. Note that the MATH key may still be used to access the maths commands when writing or editing a program.

A sample screen accessed by pressing the PRGM key is shown below.

CTL	I/O	EXEC

1 : If
2 : Then
3 : Else
4 : For (
5 : While
6 : Repeat
7 : ↓ End

This shows, for example, that highlighting CTL and then pressing 4 will write **For (** onto the display.

Core ref	Description	Detail
Chapter 1 Practical: 'Zooming in on a line' (Pupil Book page 28)	Command sequences for generating random numbers between 0 and 1 displayed to 3 d.p.	The commands that you need to enter into your machine are given below. See the introduction for guidance about where the commands are found in the menu system for each make of calculator. Once you have entered the commands, press ENTER or EXE repeatedly to produce the numbers. **Casio** Int(999Ran#+1)÷1000 **Sharp** Int(999random+1)/1000 **Texas** Int(999rand+1)/1000
Chapter 2 'Adding and subtracting integers' (Pupil Book page 36)	A program to generate two columns of 20 random integers and a third, initially hidden, column containing the sum of corresponding values.	**Casio** Select PRGM from the main menu, give the program a suitable name and enter the following commands. 20 → Dim List 1 20 → Dim List 2 20 → Dim List 3 For 1 → J To 20 Int(21Ran#−10) → List1[J] Int(21Ran#−10) → List2[J] List 1 + List 2 → List 5 Next **Using the program** Press EXIT then highlight your program name in the program list and press F1. Press the MENU key and select LIST to see the columns of numbers. Pupils can enter their answers into the third column and then check by scrolling across to column 5. **Sharp** 20 ⟹ dim(L1) 20 ⟹ dim(L2) 1 → J Label 1 int(21random − 10) ⟹ L1(J) int(21random − 10) ⟹ L2(J) J+1 ⟹ J If J<20Goto1 L1+L2 ⟹ L4 **Using the program** Press 2ndF QUIT then press PRGM and select EXEC. Highlight your program name in the program list and press ENTER. This will execute the program, sending the information to the columns. Press the STAT key then select EDIT and press ENTER to see the columns of numbers. Pupils can enter their answers into the third column and then check by scrolling across to column 4.

(continued)

Core ref	Description	Detail
Continued		**Texas** $20 \rightarrow \dim(L_1)$ $20 \rightarrow \dim(L_2)$ $1 \rightarrow J$ Lbl 1 RandInt$(-10,10) \rightarrow L_1(J)$ RandInt$(-10,10) \rightarrow L_2(J)$ IS $>$ (J,20) Goto 1 $L_1 + L_2 \rightarrow L_4$ **Using the program** Press 2nd QUIT then press PRGM and select EXEC. Highlight your program name in the program list and press ENTER twice. This will execute the program, sending the information to the columns. Press the STAT key then select EDIT and press ENTER to see the columns of numbers. Pupils can enter their answers into the third column and then check by scrolling across to column 4.
Ex5, Q6 (Pupil Book page 45)	A program to check whether a given whole number is prime or not.	Up to a point there is value in doing this without the aid of a graphical calculator. However, in order to extend the question, the following programs may be helpful. **Casio** Select PRGM mode from the main menu and give the program a name such as 'PRIME'. Please note that words such as "VALUE", that are written inside quotation marks, are keyed in letter by letter. The quotation marks must be included and these are entered by pressing F6 and F2. "VALUE"? $\rightarrow$ N $2 \rightarrow$ C Lbl 1 C $>$ √N $\Rightarrow$ Goto 3 Frac (N ÷ C) = 0 $\Rightarrow$ Goto 2 C + 1 $\rightarrow$ C Goto 1 Lbl 2 "NOT PRIME" "A FACTOR IS" C Goto 4 Lbl 3 "PRIME" Lbl 4 **Using the program** Key in a number and press EXE when the calculator displays VALUE? The calculator will then respond with PRIME or NOT PRIME depending on the value entered. In the case where the number is not prime, the calculator will also display its smallest prime factor. Press EXE again to test another value.

(continued)

Core ref	Description	Detail
Continued		**Sharp** Press the PRGM key and select NEW then press ENTER. Give the program a name such as 'PRIME'. Please note that words such as "NOT PRIME", that are written inside quotation marks, are keyed in letter by letter. The quotation marks must be included and these are entered by pressing the PRGM key, highlighting PRGM and pressing 2. Input N $2 \Rightarrow C$ Label 1 If $C > \sqrt{N}$ Goto 3 If fpart $(N/C) = 0$ Goto 2 $C + 1 \Rightarrow C$ Goto 1 Label 2 Print "NOT PRIME Print "A FACTOR IS Print C Goto 4 Label 3 Print "PRIME" Label 4 **Using the program** Key in a number and press ENTER when the calculator displays N = ? The calculator will then respond with PRIME or NOT PRIME depending on the value entered. In the case where the number is not prime, the calculator will also display its smallest prime factor. Press ENTER again to test another value. **Texas** Press the PRGM key and select NEW then press ENTER. Give the program a name such as 'PRIME'. Please note that words such as "NOT PRIME", that are written inside quotation marks, are keyed in letter by letter. The quotation marks must be included and these are entered by pressing ALPHA then +. Prompt N $2 \rightarrow C$ Lbl 1 If $C > \sqrt{(N)}$ Goto 3 $N/C \rightarrow Y$ If fPart $(Y) = 0$ Goto 2 $C + 1 \rightarrow C$ Goto 1 LBl 2 Disp "NOT PRIME" Disp "A FACTOR IS", C Goto 4 LBl 3 Disp "PRIME" Lbl 4 *(continued)*

Core ref	Description	Detail
Continued		**Using the program** Key in a number and press ENTER when the calculator displays N = ? The calculator will then respond with PRIME or NOT PRIME depending on the value entered. In the case where the number is not prime, the calculator will also display its smallest prime factor. Press ENTER again to test another value.

Ex 8, Q11 (Pupil Book page 53)	Finding square roots without using the square root key.	The process of squaring lots of values in order to find the square root of a number can be made more efficient by using the table facility of a graphical calculator.
		
		 Casio
		 Select TABLE from the main menu.
		 Set $Y1 = X^2$.
		
		 Press F5 and enter values for Start, End and pitch as below.
		
		 Start : 3
		 End : 4
		 pitch : 0·01
		
		 Press EXE to return to the previous screen and then press F6 to see the table. You can now scroll through the table to find values close to $\sqrt{11}$.
		 Greater accuracy may be achieved by reducing the size of the pitch, but this may use too much memory and so you should correspondingly reduce the difference between the Start and End values.
		
		 Sharp
		 Press the Y = key and set $Y1 = X^2$.
		
		 Press 2ndF TBLSET. Leave Input set to Auto and enter values for TBLStrt and TBLStep as below.
		
		 TBLStart =
		             3
		 TBLStep =
		             0·01
		
		 Press TABLE to see the table of values. You can now scroll through the table to find values close to $\sqrt{11}$. Greater accuracy may be achieved by reducing the size of the TBL step, but you need to make the Start value closer to the target figure or it can take a long time to scroll through the results.
		
		 Texas
		 Press the Y = key and set $Y1 = X^2$.
		
		 Press 2nd TBLSET. Leave Indpnt set and Depend set to Auto and enter values for TblStart and ΔTBL as below.
		
		 TblStart = 3
		 ΔTBL = 0·01
		
		 Press 2nd TABLE to see the table of values. You can now scroll through the table to find values close to $\sqrt{11}$. Greater accuracy may be achieved by reducing the size of ΔTBL, but you need to make the Start value closer to the target figure or it can take a long time to scroll through the results.

Core ref	Description	Detail
Chapter 8 Practical (Pupil Book page 171)	Understanding an algebraic identity.	Pupils may be convinced that $(x + 6) - (x - 2) = 8$ for any given value of x by evaluating the left-hand side for many different values. A good way to do this, once again, is to use a table. As a follow-up it may be useful for pupils to devise and check some algebraic identities of their own. **Casio** Select TABLE from the main menu. Set Y1 = X + 6, Y2 = X − 2 and Y3 = (X + 6) − (X − 2) Press F5 and choose values for Start, End and pitch. It's probably best to choose positive whole numbers throughout initially. Press EXE to return to the previous screen then press F6 to see the table. You can then scroll through the values to show that the value in the Y3 column is always 8. **Sharp** Press the Y = key. Set Y1 = X + 6, Y2 = X − 2 and Y3 = (X + 6) − (X − 2) Press 2ndF TBLSET and choose values for TBLStrt and TBLStep. It's probably best to choose positive whole numbers throughout initially. Press TABLE to see the results. You can then scroll through the values to show that the value in the Y3 column is always 8. **Texas** Press the Y = key. Set Y1 = X + 6, Y2 = X − 2 and Y3 = (X + 6) − (X − 2) Press 2nd TBLSET and choose values for TblStart and ΔTBL. It's probably best to choose positive whole numbers throughout initially. Press 2nd TABLE to see the results. You can then scroll through the values to show that the value in the Y3 column is always 8.
Chapter 9 Practical (Pupil Book page 202).		The key sequence given in the Pupil Book makes use of the Ans key and may easily be adapted to answer questions 2 and 3. It may be worth answering the first two parts of question 4 without a calculator, initially, in order to examine the process. <table><tr><td>Day</td><td>Logs to sell</td></tr><tr><td>1</td><td>10</td></tr><tr><td>2</td><td></td></tr><tr><td>3</td><td></td></tr><tr><td>4</td><td></td></tr><tr><td>…</td><td>…</td></tr></table>

(continued)

Core ref	Description	Detail
Continued		To find the missing value in the second row involves adding 15 and then dividing by 2. The problem is that there are three ways to proceed: $10 + 15 = 25$ $25 \div 2 = 12{\cdot}5$ Do we assume that the woodcutter sells 12 logs, 13 logs or 12·5 logs? Does it make any difference to the *long-term* behaviour of the values in the second column? Does it make any difference to the woodcutter? This could be a good discussion point. It turns out that the long-term behaviour is the same, regardless of which approach is used. The values in the second column converge to 15 and so the woodcutter eventually makes $15 \times 67p = £10{\cdot}05$ per day. You can use the Ans key to produce the figures in the second column, assuming that the logs don't have to be sold in whole numbers. **Casio** Select RUN from the main menu. 10 EXE (Ans + 15) ÷ 2 EXE EXE EXE ... **Sharp** 10 ENTER (Ans + 15)/2 ENTER ENTER ENTER ... **Texas** 10 ENTER (Ans + 15)/2 ENTER ENTER ENTER ... The advantage of using the calculator is that it is easy to see what happens if the start number is changed. Pupils will be surprised to find that, regardless of the starting value, the sequence will always converge to 15.
Chapter 9 Ex6, Q6 (Pupil Book page 209)	Generating a sequence using a **term-to-term rule**.	Sequences of the form described in **question 6** may be generated using the Ans key. For example, to generate the sequence with first term 5 and rule Add 2 you can key in: **Casio** 5 EXE + 2 EXE EXE EXE ... The sequences for Sharp and Texas are the same, but replace EXE with ENTER.
Chapter 9 Practical (Pupil Book page 211)	Generating a sequence using a **term-to-term rule**.	The Ans key can also be used to generate sequences when the term-to-term rule involves subtraction, multiplication or division. For example, to generate the sequence with first term 8 and rule Divide by 2 you can key in: **Casio** 8 EXE ÷ 2 EXE EXE EXE ... The sequences for Sharp and Texas are the same, but replace EXE with ENTER.

Core ref	Description	Detail
Chapter 9 Ex9, Q4 (Pupil Book page 218)	Generating a sequence using a **formula**.	You can use a table to generate a sequence using a formula for the nth term. For example, the nth term of the sequence 3, 5, 7, 9 ... is given by $2n + 1$. The following instructions may be adapted for different sequences by changing the formula. **Casio** Select TABLE from the main menu. Set $Y = 2X + 1$ Press F5 and choose a suitable Start and End number. Set the pitch value to 1. Note that the end number does not have to belong to the sequence. It just sets an upper limit for the values calculated. Press EXE then F6 to see the table. The first column of the table gives the position of each term and this makes it easy to locate the term in a given position. **Sharp** Press the Y = key and set $Y1 = 2X + 1$. Press 2ndF TBLSET and choose a suitable start number. Set the TBLStep value to 1. Press TABLE to see the table of values. The first column of the table gives the position of each term and this makes it easy to locate the term in a given position. **Texas** Press the Y = key and set $Y1 = 2X + 1$. Press 2nd TBLSET and choose a suitable start number. Set the ΔTBL value to 1. Press 2nd TABLE to see the table of values. The first column of the table gives the position of each term and this makes it easy to locate the term in a given position.
Chapter 10 Investigation (Pupil Book page 236) Practical (Pupil Book page 239).	Features of graphs	A graphical calculator is an ideal tool for exploring the properties of graphs of simple equations. Pupils will get most out of the investigation by being systematic in the way that they change the values of m and c. **Casio** Select GRAPH from the main menu. Enter an equation such as $Y1 = X$ and press EXE. You can enter up to 20 equations at one time. Press SHIFT F3 to select the view window for the graphs and then press F1 to use the initial settings. These settings use the same scale on each axis which is useful at this stage. Also, if you use the TRACE function with these settings, the cursor will move in steps of 0·1 You can use the arrow keys to scroll up, down, left and right. Press EXE to return to the previous screen and then F6 to draw the graph. Press EXIT to return to the screen where you can enter more equations or edit ones already there.

(continued)

Core ref	Description	Detail
Continued		**Sharp** Press the Y = key and enter an equation such as Y1 = X and press ENTER. You can enter up to 10 equations at one time. Press ZOOM and then 5 to select the default settings for the axes. The graph will be drawn automatically. You can adjust each of the axis settings individually by pressing WINDOW. Press Y = again to make any changes to the equations. Now press GRAPH to return to the graph screen. **Texas** Press the Y = key and enter an equation such as Y1 = X and press ENTER. You can enter up to 10 equations at one time. Press ZOOM and then 5 to select the Zsquare settings for the axes. The graph will be drawn automatically. You can adjust each of the axis settings individually by pressing WINDOW. Press Y = again to make any changes to the equations. Now press GRAPH to return to the graph screen.

5

Resource Sheets

Number Support

Practice Questions (pages 7, 10 and 11)

3

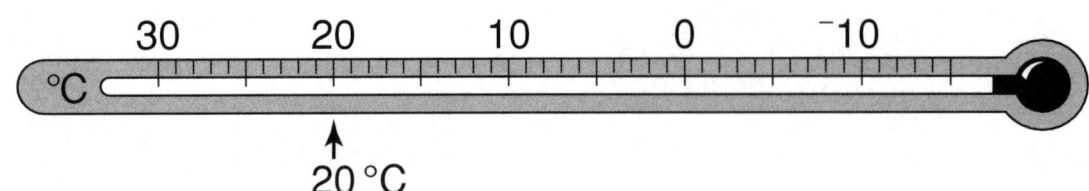

20 °C

24

65	100	76	82	92	44
87	29	15	8	42	99
35	6	28	60	11	56
300	21	17	7	30	33
77	4	1	63	72	14
110	55	74	40	200	76

33

2373	6840
7970	2470
7947	3856
1265	5830
7295	8312
4864	4836
1031	5177
3587	2869

1 Place Value, Ordering and Rounding

Exercise 2 (pages 18 and 19)

4

12·655	4·03	4·69
0·012	5·39	5·399
3·04	6·832	4·71
6·742	4·719	4·680
4·05	0·048	12·682

Exercise 2 (pages 18 and 19) continued

Review 1

<table>
<tr><td>2·404</td><td>4·996</td><td></td><td>2·404</td><td>4·996</td><td>6·991</td><td></td><td>C
5·84</td><td>2·404</td><td>4·996</td></tr>
<tr><td>4·69</td><td>4·698</td><td>8·316</td><td>6·991</td><td></td><td>8·316</td><td>4·698</td><td>8·316</td><td>6·991</td><td>0·042</td></tr>
<tr><td>6·991</td><td>4·698</td><td>4·36</td><td>13·681</td><td>5·841</td><td></td><td>4·698</td><td>6·991</td><td>5·841</td><td></td></tr>
<tr><td>4·684</td><td>7·378</td><td>4·996</td><td></td><td>7·378</td><td>13·681</td><td>4·698</td><td>4·923</td><td>7·134</td><td>6·991</td></tr>
</table>

Write the letter beside each calculation above its answer in the box.

C 5·83 + 0·01 = **5·84**   **H** 7·234 − 0·1   **O** 4·683 + 0·001   **G** 4·924 − 0·001
I 4·697 + 0·001   **Y** 0·043 − 0·001   **E** 13·683 − 0·002   **S** 5·834 + 0·007
M 4·361 − 0·001   **W** 7·382 − 0·004   **F** 8·311 + 0·005   **L** 4·689 + 0·001
A 2·399 + 0·005   **N** 5·000 − 0·004   **T** 7 − 0·009

Investigation — × and ÷ by 0·1 and 0·01 (page 20)

Number	× 10	÷ 0·1	× 100	÷ 0·01	÷ 10	× 0·1	÷ 100	× 0·01
4·7								
0·8								
0·92								
16								
12·3								
572								
46·23								
5271								

Exercise 4 (page 24)

Review 2

56·3	0·42	87		56·3	0·56	4·8	87	0·563	0·56	13 700

420	0·56	0·0563	4·2	0·56	0·0563	4·8	13·7	870	0·56	0·0563	0·137	87

U									
0·08	870	87	4·2		13·7	870	13·7	5·6	0·56	0·56

Put the letter beside each question above its answer in the box.

U	$0.8 \times 0.1 = 0.08$	C	13.7×0.01	E	$8.7 \div 0.1$	F	$137 \div 0.01$	H	42×0.01
A	$0.137 \div 0.01$	S	$8.7 \div 0.01$	D	$0.42 \div 0.1$	N	5.63×0.01	L	$42 \div 0.1$
R	5.63×0.1	T	$5.63 \div 0.1$	W	8×0.6	O	0.8×0.7	Z	8×0.7

Puzzle (page 27)

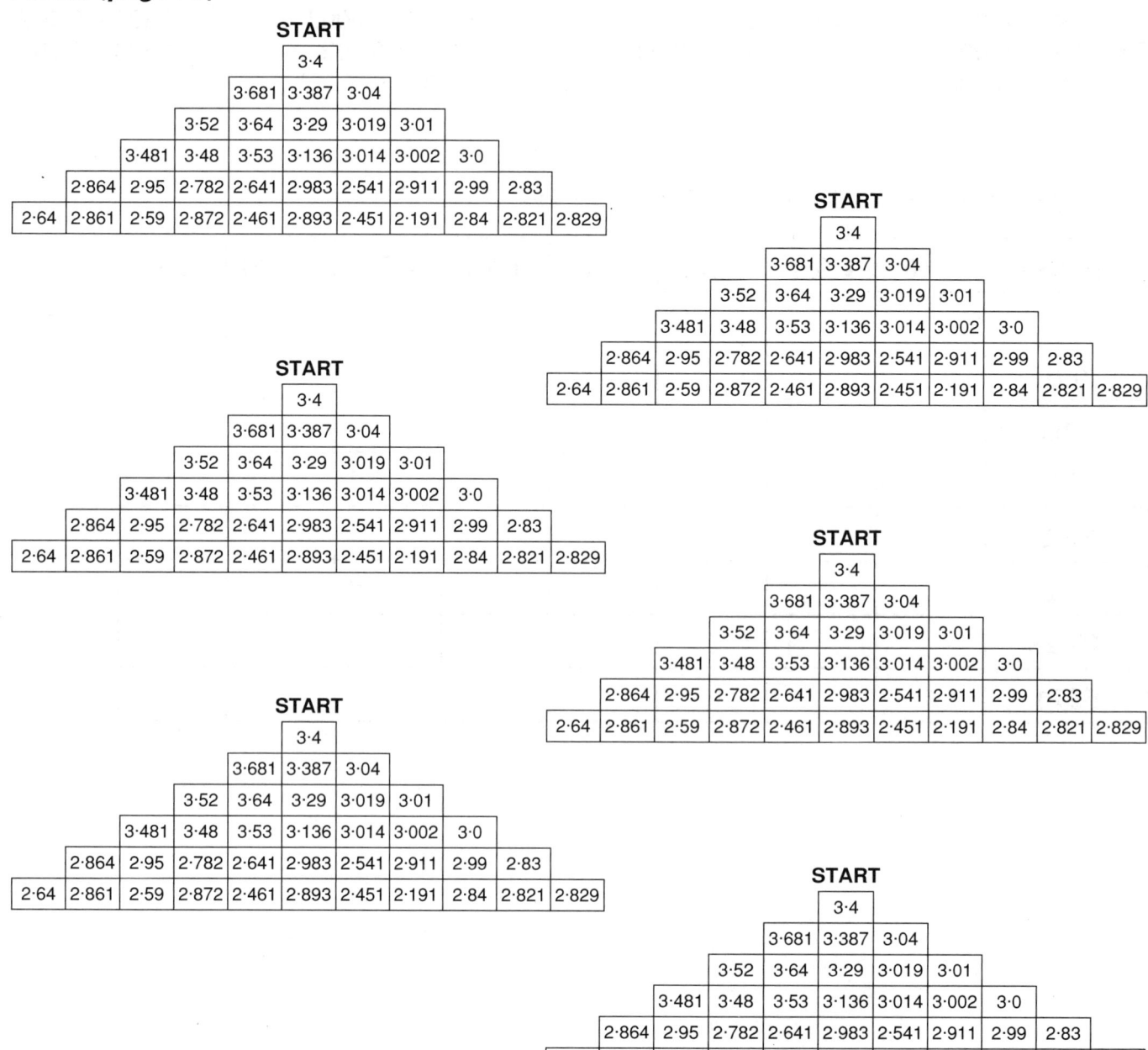

Ladders — a game for a group (page 27)

  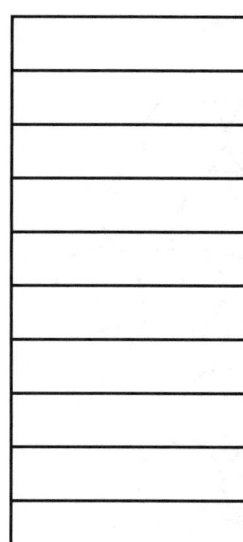

2 Integers, Powers and Roots

Exercise 1 (pages 37 and 38)

1

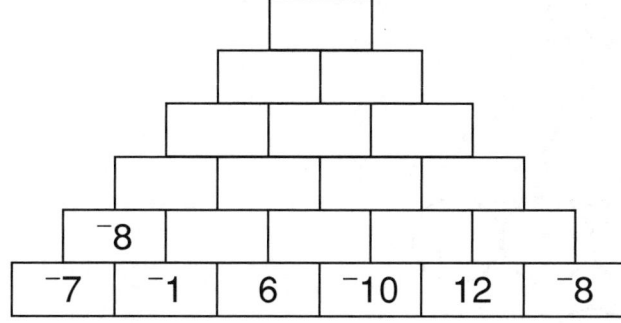

2

3

a	⁻2	3	⁻1	4	⁻3	5	⁻6	1	⁻7	⁻4
b	3	⁻1	4	⁻2	5	⁻3	2	⁻4	⁻2	⁻5
a−b	⁻5	4								

Exercise 1 (pages 37 and 38) continued

Review 1

a
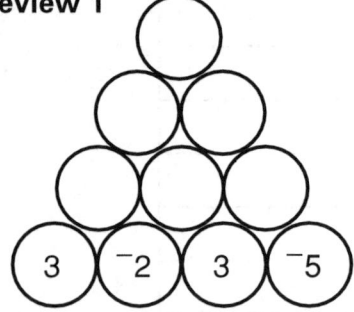

b
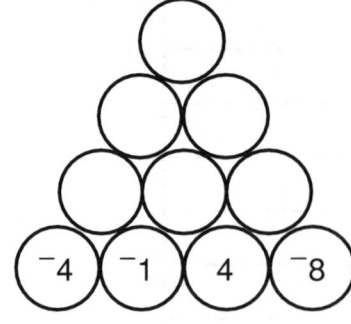

c
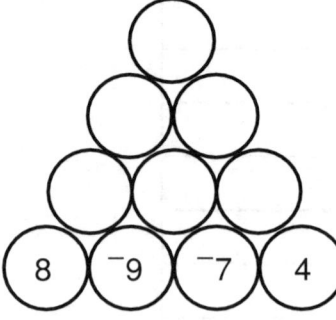

Review 2

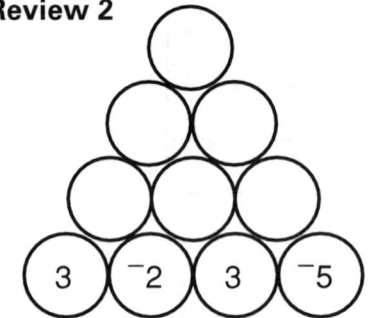

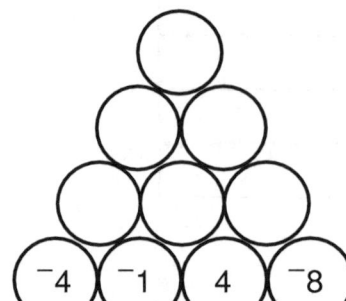

Discussion (page 39)

×	⁻3	⁻2	⁻1	0	1	2	3
3	⁻9	⁻6	⁻3	0	3	6	9
2				0	2	4	6
1				0	1	2	3
0				0	0	0	0
⁻1							⁻3
⁻2							⁻6
⁻3							⁻9

Exercise 2 (pages 40, 41 and 42)

3 a
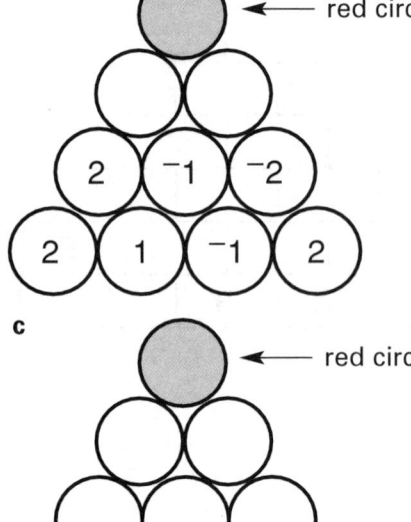
← red circle

b
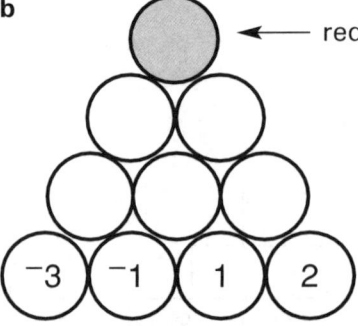
← red circle

c
← red circle

d
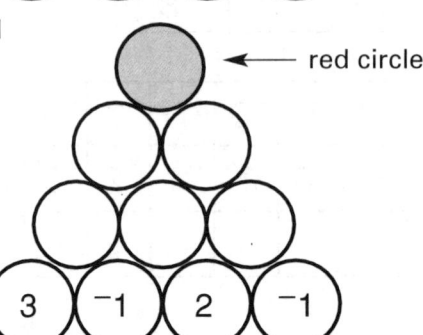
← red circle

Exercise 2 (pages 40, 41 and 42) continued

4 a

×	2	⁻5	⁻3	4
⁻1	⁻2	5		
3	6			
⁻2				
5				

b

×	⁻1	2	3	⁻4
3				
⁻1				
⁻4				
⁻6				

c

×	⁻6	2	⁻1	8
⁻4				
7				
⁻6				
⁻1				

d

×	⁻3	7	⁻2	⁻8
6				
⁻1				
⁻7				
2				

8 a

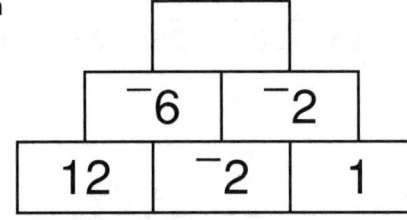

b

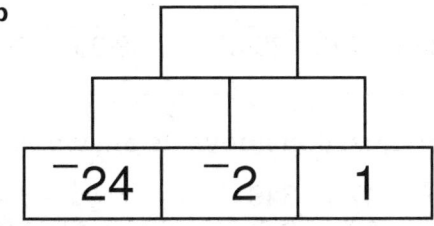

c

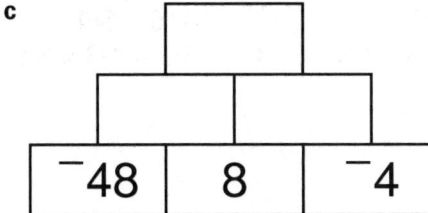

d

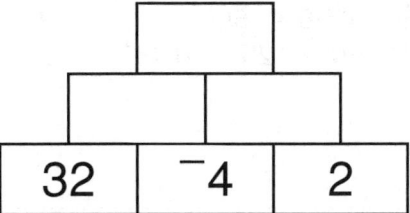

10

×				
	⁻10			
		2		
			⁻24	
				30

Review 3

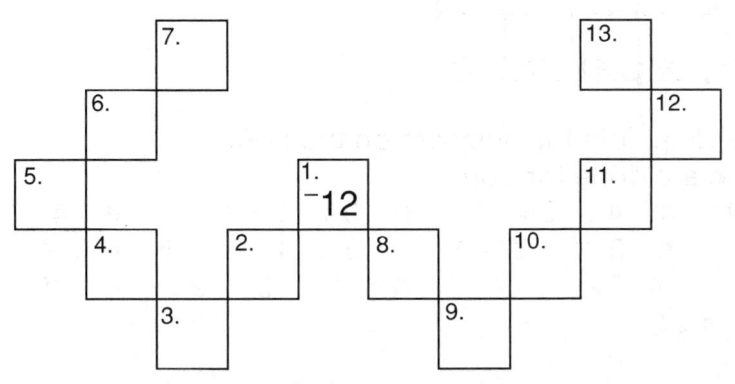

Exercise 3 (page 43)

3 a

+	⁻84	253	⁻461	514
⁻198				
⁻257				
346				
⁻98				

b

×	47	⁻14	⁻28	8·6
⁻6·4				
⁻72				
95				
⁻69				

Review

									F
⁻140	⁻102	475	24·9	⁻140	8·25	25·62	⁻65	⁻26	326

24·9	⁻140	8·25	1620	⁻1827	25·62	⁻1827	⁻140	⁻65	⁻1827	24·9

3844	⁻24·9	338·4	338·4	475	25·62	⁻8·25	⁻26	182·7	⁻26	338·4	⁻26	⁻2·6	⁻65

Write the letter beside each question above its answer in the box.

F ⁻84 + ⁻242 = ⁻326   **H** 247 + ⁻349   **E** 348 − ⁻127   **I** ⁻29 × 63
C ⁻31 × ⁻124   **L** ⁻94 × ⁻3·6   **O** 1456 ÷ ⁻56   **M** ⁻792 ÷ 96
N ⁻29 × ⁻6·3   **G** 145·6 ÷ ⁻56   **U** ⁻792 ÷ ⁻96   **Y** ⁻416 ÷ 6·4
T ⁻84 + ⁻98 − ⁻42   **P** 144 × ⁻25 × ⁻0·45   **S** ⁻8·3 × 3·9 ÷ ⁻1·3   **A** ⁻8·3 ÷ 1·3 × 3·9
D ⁻4·2 × ⁻6·1

Exercise 4 (page 44)

Review

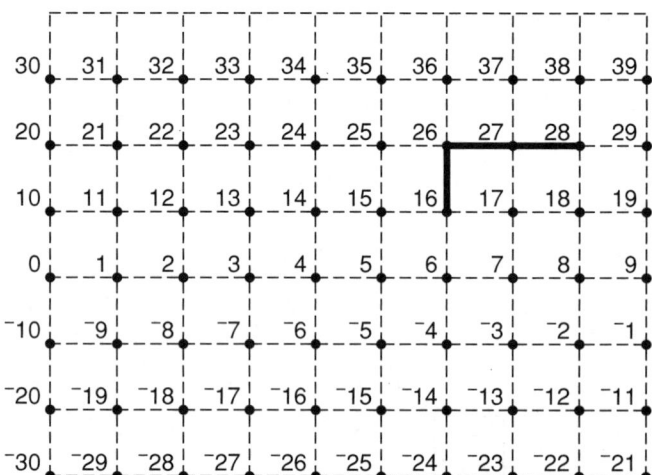

Do the calculations in order from **a** to **q**. Join the answers on the grid.
What picture do you get? **a**, **b** and **c** are done for you.

a ⁻6 × ⁻5 + ⁻2   **b** ⁻2(⁻14 + 1)   **c** 4 + ⁻3 × ⁻4   **d** 9 + ⁻1 × 2   **e** 4 × ⁻5 + 7
f ⁻2(4 − ⁻8)   **g** ⁻4 × 8 + 5   **h** 3(⁻5 + 2) + 1   **i** 2 × ⁻1 − 3 × ⁻5   **j** ⁻1 × 3 − 4 × ⁻2
k ⁻6 × ⁻2 + 4 × ⁻4   **l** $\frac{3(6 - ⁻2)}{4}$ **m** $\frac{6 × ⁻5}{⁻2}$ **n** ⁻8 × 2 − ⁻9   **o** 40 − ⁻1 × ⁻4
p ⁻4 × ⁻9 − ⁻1   **q** ⁻7 × ⁻5 + ⁻4 × 2

Exercise 7 (page 51)

Review 3

Across

1. LCM of 21 and 28
2. LCM of 84 and 126
4. HCF of 720 and 648
6. HCF of 300 and 990
7. HCF of 168 and 196
10. HCF of 132 and 330
11. LCM of 264 and 72

Down

1. LCM of 120 and 168
2. HCF of 110 and 132
3. HCF of 144 and 120
4. LCM of 150 and 125
5. HCF of 216 and 420
8. HCF of 900 and 792
9. HCF of 756 and 1485

Test Yourself (pages 59 and 60)

1 a
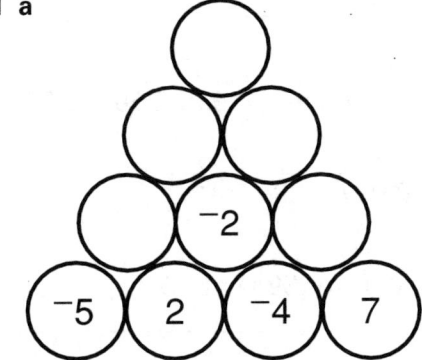

b
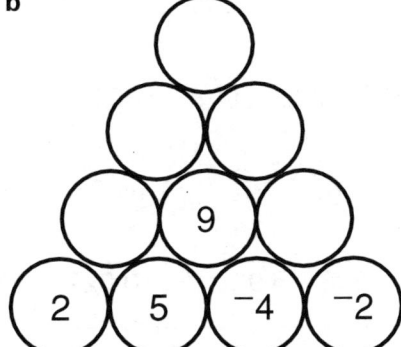

c
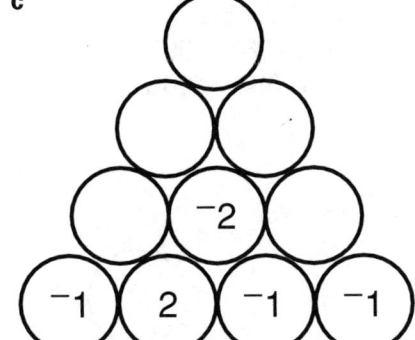

d
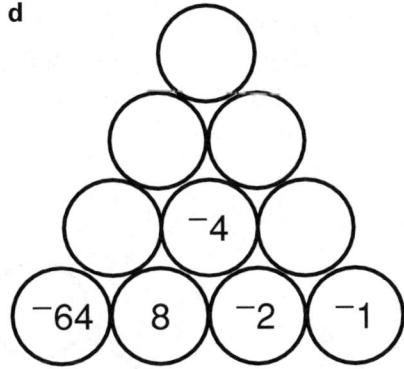

4 a

+	$^-$109	79	$^-$212
346			
$^-$127			
$^-$449			

b

×	$^-$7·8	42	$^-$4·5
$^-$34			
8·6			
$^-$3·8			

3 Mental Calculation

Exercise 1 (page 65)

Review 3

3	8		18
		7	4
			5
6	9		15

13		24	10
		21	
19	17	14	20
			25

Review 4

	H											
$\overline{}$ 2400	$\overline{}$ 1140	$\overline{}$ 394	$\overline{}$ 4199	$\overline{}$ 766	$\overline{}$ 11·8	$\overline{}$ 2400	$\overline{}$ 0·027	$\overline{}$ 766	$\overline{}$ 4199	$\overline{}$ 4199	$\overline{}$ 766	$\overline{}$ 5·92

$\overline{}$ 3·7 $\overline{}$ 766 $\overline{}$ 11·3 $\overline{}$ 4199 $\overline{}$ 766 $\overline{}$ 3·7 $\overline{}$ 0·027 $\overline{}$ 3·8 $\overline{}$ 5·92 $\overline{}$ 0·027 $\overline{}$ 394 $\overline{}$ 11·3

$\overline{}$ 906 $\overline{}$ 11·8 $\overline{}$ 11·8 $\overline{}$ 0·9 $\overline{}$ 906 $\overline{}$ 5·92 $\overline{}$ 0·027 $\overline{}$ 3·8 $\overline{}$ 5·92 $\overline{}$ 0·027 $\overline{}$ 394 $\overline{}$ 11·3

Write the letter beside each calculation above its answer in the box.

H 770 + 370 = **1140**   **T** 7300 – 4900   **S** 8·2 + 3·6   **A** 9·7 – 5·9
I 526 + 380   **F** 8·3 – 4·6   **O** 468 + 298   **M** 5206 – 1007
E 690 – 239 – 57   **N** 8·42 – 2·5   **K** 4·6 – 3·7   **C** 0·084 – 0·057
R 8·6 + 3·2 – 4·1 – ⁻3·6

Exercise 2 (pages 68 and 69)

7 a

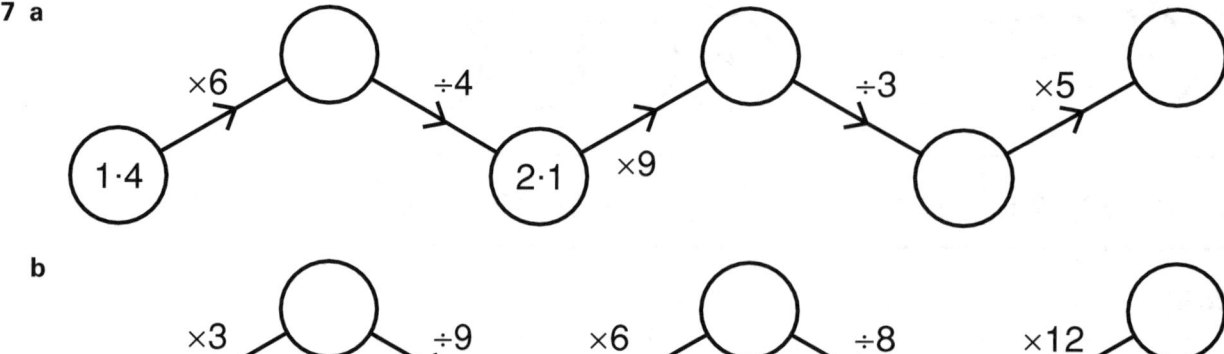

b

Exercise 2 (pages 68 and 69) continued

Review 1

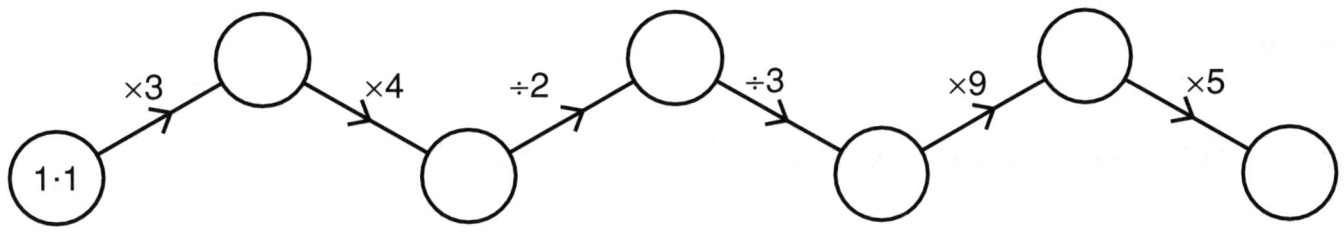

Review 4

2·8	45		2	1560	45		91	45	15·6	2·8	15	0·02	103	20·3	‾400	520

					H						
0·02	45	45	20·3		2	1600	‾400		520	103	1200	‾400

Write the letter beside each calculation above its answer in the box.

H $80 \times 20 = 1600$ **M** 30×40 **T** $600 \div 300$ **S** 130×4 **W** 312×5

E $^-16 \times 25$ **R** $5·2 \times 3$ **O** $^-2·5 \times ^-18$ **N** $4 \times 0·7$ **L** $0·08 \div 4$

C $455 \div 5$ **A** $618 \div 6$ **F** $270 \div 18$ **K** $8·12 \times 2·5$

Test Yourself (page 79)

4

1·4		1·6
	1·7	
		2·0

7

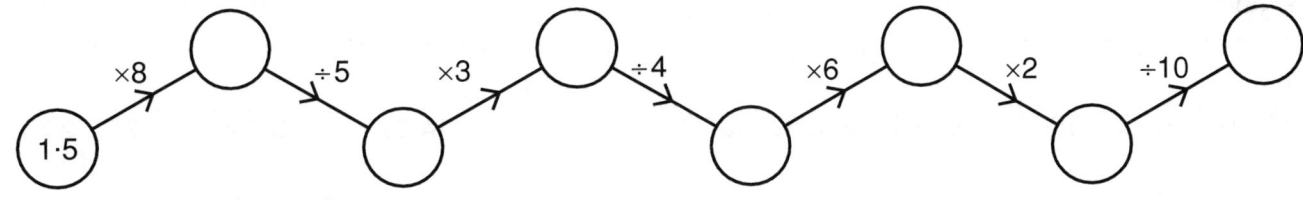

4 Written and Calculator Calculation

Exercise 1 (page 82)

Review 1

$\overline{35\cdot25}$	$\overline{18\cdot21}$	$\overline{23\cdot85}$	$\overline{157\cdot11}$	$\overline{14\cdot09}$	$\overline{3\cdot54}$	$\overline{10\cdot65}$	$\overline{4\cdot86}$

E

$\overline{14\cdot09}$ $\overline{7\cdot04}$ $\overline{10\cdot65}$ $\overline{18\cdot67}$ $\overline{105\cdot74}$ $\overline{35\cdot25}$ $\overline{10\cdot65}$ $\overline{7\cdot04}$ $\overline{223\cdot43}$ $\overline{6\cdot6}$ $\overline{4\cdot86}$ $\overline{29\cdot42}$

   E   E

$\overline{14\cdot09}$ $\overline{157\cdot11}$ $\overline{105\cdot74}$ $\overline{105\cdot74}$ $\overline{10\cdot65}$ $\overline{7\cdot04}$ $\overline{14\cdot09}$ $\overline{157\cdot11}$ $\overline{10\cdot65}$ $\overline{14\cdot09}$

   E   E

$\overline{14\cdot09}$ $\overline{157\cdot11}$ $\overline{105\cdot74}$ $\overline{105\cdot74}$ $\overline{34\cdot35}$ $\overline{6\cdot6}$ $\overline{10\cdot65}$ $\overline{14\cdot09}$ $\overline{223\cdot43}$ $\overline{7\cdot04}$

     E   E

$\overline{14\cdot67}$ $\overline{105\cdot74}$ $\overline{18\cdot67}$ $\overline{105\cdot74}$ $\overline{4\cdot86}$ $\overline{10\cdot65}$ $\overline{4\cdot86}$ $\overline{29\cdot42}$ $\overline{10\cdot65}$

           E

$\overline{157\cdot11}$ $\overline{10\cdot65}$ $\overline{35\cdot25}$ $\overline{76\cdot38}$ $\overline{14\cdot09}$ $\overline{18\cdot21}$ $\overline{8\cdot71}$ $\overline{105\cdot74}$ $\overline{14\cdot67}$

     E   E

$\overline{11\cdot09}$ $\overline{105\cdot74}$ $\overline{7\cdot04}$ $\overline{14\cdot67}$ $\overline{105\cdot74}$ $\overline{3\cdot54}$ $\overline{223\cdot43}$ $\overline{4\cdot86}$ $\overline{29\cdot42}$

Put the letter beside each question above its answer in the box.

E $98\cdot71 + 2\cdot8 + 4\cdot23 = $ **105·74**   **O** $52\cdot6 + 147\cdot2 + 23\cdot63$ **T** $5\cdot07 + 3 + 0\cdot82 + 5\cdot2$
F $64\cdot7 + 33\cdot4 - 21\cdot72$ **C** $5\cdot36 + 2\cdot9 - 4\cdot72$ **P** $16 + 0\cdot82 - 5\cdot73$
U $16\cdot48 - 5\cdot2 - 4\cdot68$ **N** $15 - 3\cdot8 - 6\cdot34$ **D** $57 + 3\cdot84 - 31\cdot42$
I $53\cdot76 - 40\cdot25 + 4\cdot7$ **L** $41\cdot36 + 2\cdot08 - 3\cdot96 - 4\cdot23$
Q $18 - 3\cdot57 + 16\cdot72 - 4\cdot3 + 7\cdot5$ **A** $18\cdot63 + 0\cdot42 - 12 + 3\cdot6$
R $24 - 18\cdot62 - 3\cdot4 + 5\cdot7 - 0\cdot64$ **H** $186\cdot4 - 13\cdot89 + 2\cdot5 - 17\cdot9$
S $52\cdot7 + 3\cdot6 - 15\cdot82 - 32\cdot61 + 6\cdot8$ **G** $7\cdot94 + 13\cdot82 - 4\cdot09 - 0\cdot52 + 6\cdot7$
M $83\cdot72 - 68\cdot4 - 0\cdot89 - 5\cdot72$ **V** $4\cdot3 + 18\cdot04 - 6\cdot58 - 0\cdot09 + 3$

Exercise 2 (page 84)

Review 1

                          I

$\overline{62\cdot56}$ $\overline{967\cdot2}$ $\overline{20\cdot13}$ $\overline{76\cdot941}$ $\overline{476\cdot52}$ $\overline{8\cdot736}$ $\overline{130\cdot2}$ $\overline{14\cdot82}$ $\overline{8\cdot736}$ $\overline{51\cdot72}$ $\overline{20\cdot13}$

                             I

$\overline{33\cdot18}$ $\overline{52\cdot08}$ $\overline{29\cdot12}$ $\overline{3\cdot249}$ $\overline{20\cdot13}$ $\overline{52\cdot08}$ $\overline{22\cdot568}$ $\overline{29\cdot12}$ $\overline{130\cdot2}$ $\overline{14\cdot82}$ $\overline{51\cdot72}$

$\overline{103\cdot25}$ $\overline{33\cdot18}$ $\overline{967\cdot2}$ $\overline{62\cdot56}$ $\overline{8\cdot736}$ $\overline{51\cdot72}$ $\overline{76\cdot941}$ $\overline{8\cdot736}$

                          I

$\overline{20\cdot13}$ $\overline{52\cdot08}$ $\overline{62\cdot56}$ $\overline{14\cdot82}$ $\overline{130\cdot2}$ $\overline{52\cdot08}$ $\overline{8\cdot736}$ $\overline{51\cdot72}$ $\overline{20\cdot13}$

$\overline{29\cdot12}$ $\overline{20\cdot13}$ $\overline{8\cdot736}$ $\overline{8\cdot736}$ $\overline{20\cdot13}$ $\overline{967\cdot2}$ $\overline{14\cdot82}$ $\overline{476\cdot52}$ $\overline{8\cdot736}$

Write the letter beside each calculation above its answer in the box.

I $186 \times 0\cdot7 = $ **130·2**   **Y** $3\cdot61 \times 0\cdot9$ **N** $14 \times 3\cdot72$ **H** $4\cdot31 \times 12$ **S** $2\cdot6 \times 5\cdot7$
E $18\cdot3 \times 1\cdot1$ **O** $1\cdot58 \times 21$ **D** $2\cdot72 \times 23$ **W** $41\cdot3 \times 2\cdot5$ **L** $16 \times 1\cdot82$
G $4\cdot03 \times 5\cdot6$ **A** $9\cdot27 \times 8\cdot3$ **R** $186 \times 5\cdot2$ **T** $3\cdot64 \times 2\cdot4$ **M** $83\cdot6 \times 5\cdot7$

Exercise 6 (page 93)

Review 1

| $\overline{37{\cdot}68}$ | $\overline{7}$ | $\overline{7}$ | $\overline{16}$ | | $\overline{16{\cdot}74}$ | **A** $\overline{1{\cdot}5}$ | $\overline{15}$ | $\overline{7}$ | | $\overline{484{\cdot}6}$ | $\overline{3{\cdot}5}$ | $\overline{15}$ | $\overline{7}$ | | $\overline{7}$ | $\overline{1}$ | $\overline{7}$ | $\overline{16}$ |

Put the letter that is beside each calculation above its answer in the box.

A $4{\cdot}7 - (8{\cdot}3 + 3{\cdot}7) + (5{\cdot}8 + 3) = 1{\cdot}5$
V $\sqrt{(17^2 - 8^2)}$
S $160 \div \{18 - (12 - 4)\}$

H $(18{\cdot}7 - 9{\cdot}4) \times (8{\cdot}7 - 6{\cdot}9)$
F $7 \times 8{\cdot}32^2$ to 1 d.p.
E $140 \div \{4 \times (3 + 2)\}$

I $\frac{15 + 24}{17 - 6}$ to 1 d.p.
Y $\frac{(5+7)^2}{(21 - 9)^2}$
B $(14{\cdot}6 - 8{\cdot}32) \times \{18 \div (9 - 6)\}$

5 Fractions

Exercise 3 (page 102)

Review

| $2\frac{19}{500}$ | $\frac{6}{125}$ | $\frac{19}{20}$ | $\frac{3}{500}$ | $\frac{181}{200}$ | $\frac{19}{50}$ | $2\frac{3}{10}$ | $2\frac{19}{50}$ | | $\frac{81}{200}$ | $\frac{19}{50}$ | $2\frac{48}{125}$ | $\frac{2}{25}$ | | $\frac{17}{40}$ | $\frac{7}{8}$ | **S** $\frac{1}{20}$ | $2\frac{3}{100}$ | $2\frac{3}{10}$ | $\frac{19}{50}$ |

| $\frac{19}{20}$ | $2\frac{19}{50}$ | | $2\frac{3}{100}$ | $\frac{6}{125}$ | $2\frac{3}{10}$ | | **S** $\frac{1}{20}$ | $\frac{12}{125}$ | $\frac{19}{50}$ | $\frac{19}{20}$ | $2\frac{19}{50}$ | $\frac{81}{200}$ |

Convert these to fractions in their simplest form.
Write the letter besides each question above its answer in the box.

S $0{\cdot}05 = \frac{1}{20}$ **I** $0{\cdot}95$ **W** $0{\cdot}08$ **R** $0{\cdot}380$
G $0{\cdot}405$ **A** $0{\cdot}875$ **F** $0{\cdot}425$ **D** $0{\cdot}905$
L $0{\cdot}006$ **H** $0{\cdot}048$ **P** $0{\cdot}096$ **E** $2{\cdot}3$
N $2{\cdot}38$ **T** $2{\cdot}03$ **C** $2{\cdot}038$ **O** $2{\cdot}384$

Exercise 5 (page 107)

10 a

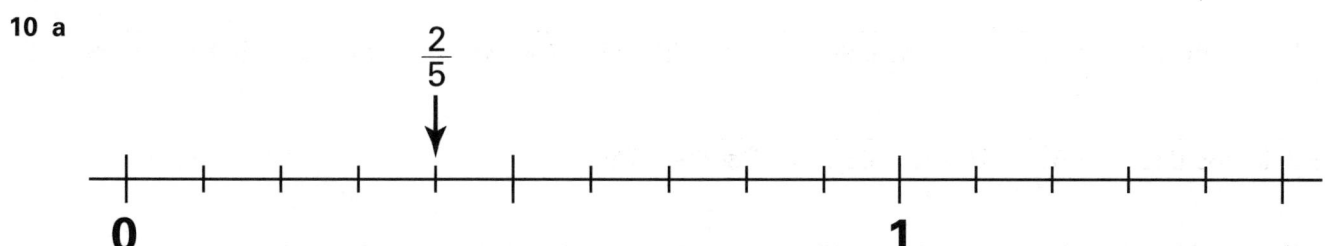

b

Exercise 5 (page 107) continued

Review 4

a

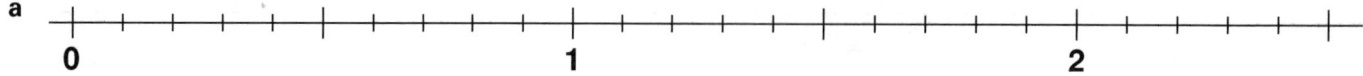

b

6 Percentages, Fractions, Decimals

Exercise 1 (pages 120 and 122)

1

Fraction	Decimal	Percentage	Fraction	Decimal	Percentage
$\frac{2}{5}$			$\frac{4}{25}$		
	0·8				3%
		60%		0·06	
	0·65			0·125	
		55%	$1\frac{3}{4}$		
$\frac{9}{20}$					136%
	0·35				$33\frac{1}{3}$%

Review 4

$\overline{66·7\%}$　　$\overline{8·7\%}$　$\overline{65·0\%}$　$\overline{66·7\%}$　$\overline{21·1\%}$　$\overline{76·6\%}$　$\overline{39·0\%}$　$\overline{8·7\%}$　$\overline{71·9\%}$

$\overline{73·2\%}$　$\overline{66·7\%}$　$\overline{70·6\%}$　　$\overline{65·0\%}$　$\overline{16·7\%}$　$\overline{21·1\%}$　$\overline{70·6\%}$　　$\overline{39·0\%}$　$\overline{65·0\%}$　$\overline{8·7\%}$

$\overline{8·7\%}$　$\overline{65·0\%}$　$\overline{41·5\%}$　$\overline{70·3\%}$　$\overline{66·7\%}$　$\overline{73·2\%}$　$\overline{71·9\%}$

**E**

$\overline{39·0\%}$　$\overline{70·6\%}$　$\overline{8·7\%}$　$\overline{39·0\%}$　$\overline{30·4\%}$　$\overline{71·4\%}$　　$\overline{41·5\%}$　$\overline{16·7\%}$　$\overline{65·0\%}$

Write the letter beside each fraction above its answer in the box.
Write these as percentages to 1 d.p.

E $\frac{5}{7} = 71·4\%$　　**U** $\frac{1}{6}$　　　**N** $\frac{12}{17}$　　　**A** $\frac{2}{3}$　　　**O** $\frac{34}{82}$

R $\frac{4}{19}$　　　　**D** $\frac{7}{23}$　　　**I** $\frac{16}{41}$　　　**M** $\frac{26}{37}$　　　**S** $\frac{19}{218}$

C $\frac{71}{97}$　　　　**H** $\frac{64}{89}$　　　**F** $\frac{164}{214}$　　　**T** $\frac{152}{234}$

Test Yourself (page 134)

1

Decimal	Percentage	Fraction
0·85		
	35%	
		$\frac{13}{20}$
0·08		

Decimal	Percentage	Fraction
		$1\frac{2}{5}$
		$\frac{1}{3}$
	148%	
2·48		

7 Ratio and Proportion

Investigation — Triangle Paths (page 147)

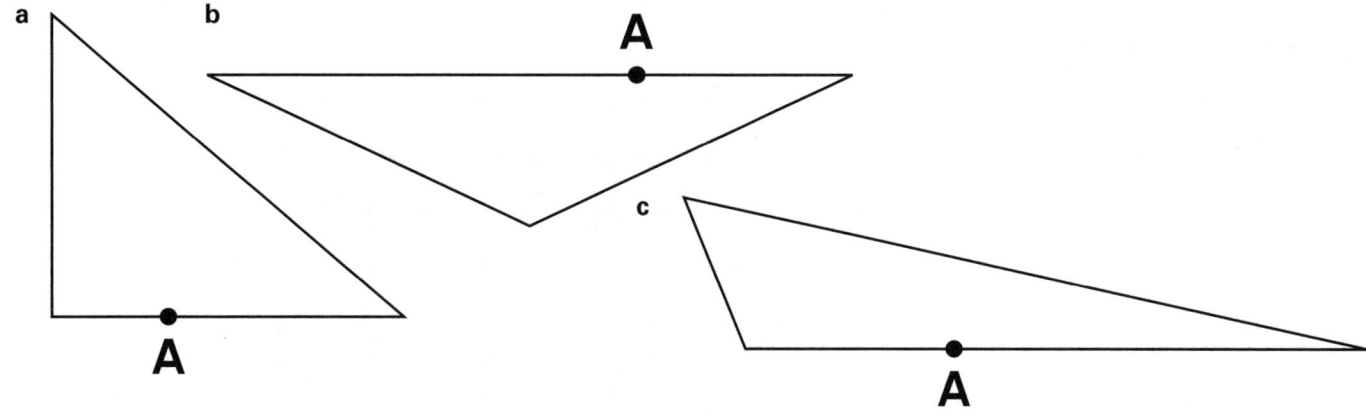

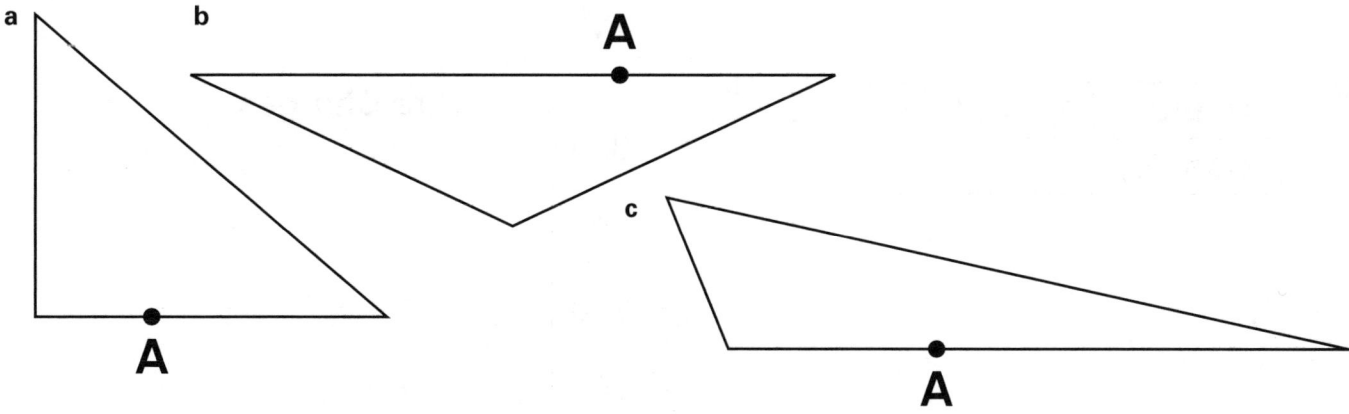

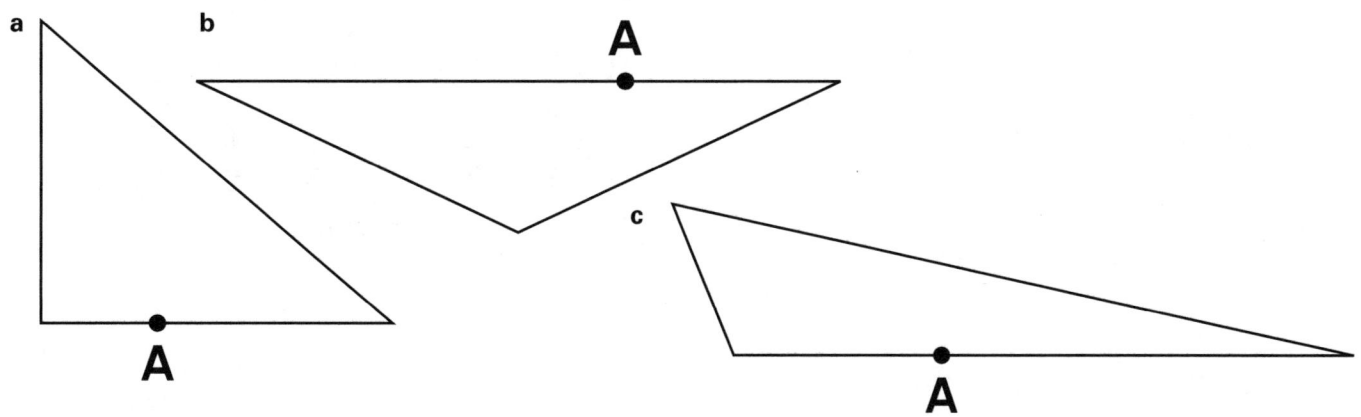

Algebra Support

Practice Questions (pages 154, 155 and 156)

3

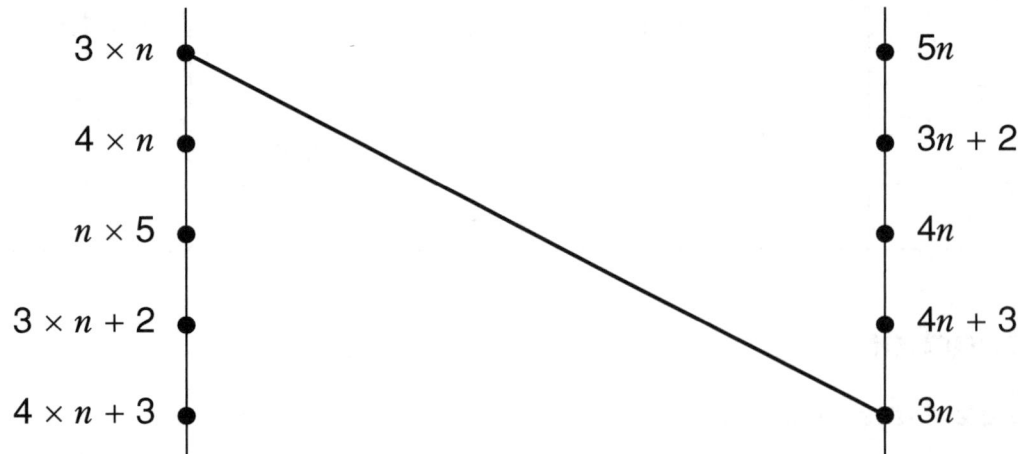

| $3 \times n$ |
| $4 \times n$ |
| $n \times 5$ |
| $3 \times n + 2$ |
| $4 \times n + 3$ |

| $5n$ |
| $3n + 2$ |
| $4n$ |
| $4n + 3$ |
| $3n$ |

13 a

x	−3	0	1
y			

c

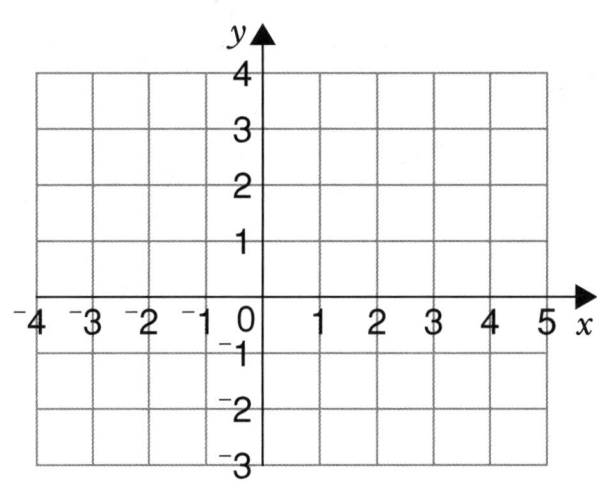

22 a

Hours	1	4	8	10
Cost (£)				

b

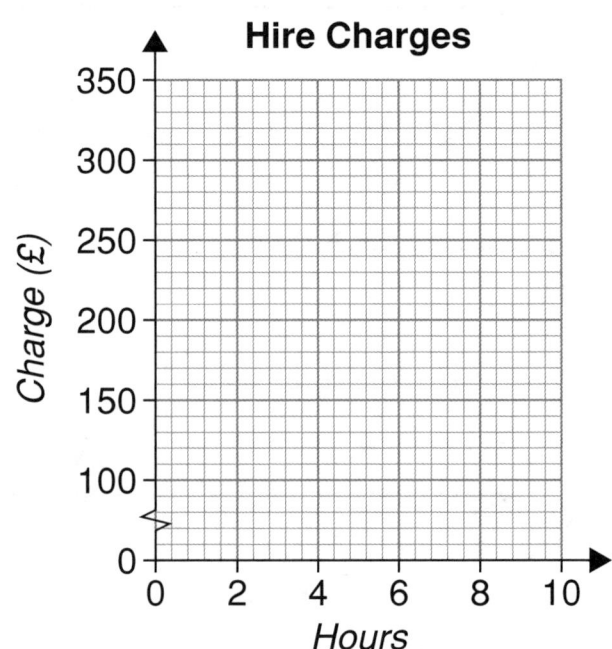

Hire Charges

8 Expressions, Formulae and Equations

Exercise 7 (page 166)

Review 3

I		I				I							
$\overline{a^3}$	$\overline{a^4}$	$\overline{a^3}$	$\overline{\frac{5a}{3}}$	$\overline{\frac{4b}{3}}$	$\overline{2a}$	$\overline{a^3}$	$\overline{4a}$	$\overline{\frac{4b}{3}}$	$\overline{\frac{5a}{3}}$	$\overline{30}$	$\overline{5b^2}$	$\overline{12n^2}$	$\overline{4a}$

| $\overline{1}$ | $\overline{a^4}$ | $\overline{b^2}$ | $\overline{30}$ | $\overline{\frac{5a}{3}}$ | $\overline{\frac{3b^2}{4}}$ | $\overline{5b^2}$ | $\overline{a^4}$ | $\overline{\frac{5a}{3}}$ | $\overline{30}$ | $\overline{2a}$ | $\overline{a^3}$ | $\overline{a}$ | $\overline{5b^2}$ |

| $\overline{a^2}$ | $\overline{1}$ | $\overline{\frac{5a}{3}}$ | $\overline{\frac{5a}{3}}$ | $\overline{b^2}$ | $\overline{5b^2}$ | $\overline{\frac{4b}{3}}$ | $\overline{1}$ | $\overline{2a}$ | $\overline{\frac{3b^2}{2}}$ | $\overline{1}$ | $\overline{\frac{3b^2}{2}}$ | $\overline{12n^2}$ | $\overline{\frac{3b^2}{2}}$ | $\overline{1}$ | $\overline{b^2}$ | $\overline{12n^2}$ |

| $\overline{\frac{4b}{3}}$ | $\overline{1}$ | $\overline{b^2}$ | $\overline{\frac{8a}{7}}$ |

Simplify these expressions.
Write the letter beside each expression above its answer in the box.

I $a \times a^2 = a^3$ **N** $a^2 \times a^2$ **F** $\frac{8a}{4}$ **R** $\frac{20a}{5}$ **Y** $\frac{30b}{b}$

D $\frac{64a}{56}$ **B** $\frac{a^5}{a^3}$ **V** $a^6 \div a^5$ **E** $\frac{10b^2}{2}$ **W** $\frac{9b^2}{12}$ **T** $\frac{45a}{27}$

S $\frac{72b}{54}$ **L** $b^4 \div b^2$ **O** $a^3 \div a^3$ **C** $\frac{21b^2}{14}$ **A** $3n \times 4n$

Brackets — a game for a group (page 167)

$3n$

$2n$

$3x$

$2x$

Brackets — a game for a group (page 167) continued

^-2n	^-2x
$^-5$	2
6	$^-6$
$5n$	$5x$
5	$2(1-n)$

Brackets — a game for a group (page 167) continued

$$2(x+3)$$

$$2(3-x)$$

$$3(n+2)$$

$$2(n-3)$$

$$2(n+3)$$

$$3(n-2)$$

$$3(x+2)$$

$$3(x-2)$$

$$5(n+1)$$

$$5(x-1)$$

Exercise 9 (pages 169 and 170)

7 a

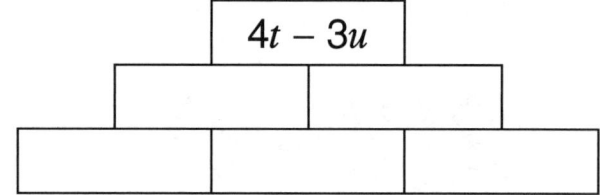

$$4t - 3u$$

b

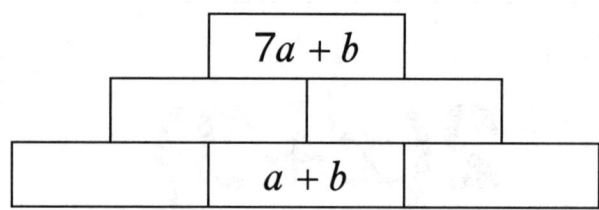

$$7a + b$$

$$a + b$$

Review 1

| $\overline{11a - 6}$ | $\overline{3a + b}$ | $\overline{11a + b}$ | | $\overline{4a^2}$ | $\overline{9b + 8}$ | $\overline{11a + b}$ | $\overline{7a + 1}$ | $\overline{4a^2}$ | $\overline{6a + 2}$ | $\overline{11a + b}$ |

| $\overline{3a + b}$ | $\overline{11a + b}$ | $\overline{8a^2}$ | $\overline{6a + 2}$ | $\overline{3a + b}$ | $\overline{11a - 6}$ | | $\overline{6a + 4b}$ | $\overline{10a + b}$ | | $\overline{4a^2}$ |

N   **N**

$\overline{6a^2}$ $\overline{4a^2}$ $\overline{6a + 9b}$ $\overline{8a^2}$ $\overline{6a + 9b}$ $\overline{11a - 6}$ $\overline{3a + b}$ $\overline{11a + b}$

$\overline{6a^2}$ $\overline{8a^2}$ $\overline{a - 2}$ $\overline{a - 2}$ $\overline{15b + 1}$ $\overline{11a + b}$ $\overline{4a^2}$ $\overline{6a + 2}$ $\overline{11a + b}$ $\overline{5a + 5b}$

$\overline{5a + 3b}$ $\overline{4a^2}$ $\overline{5a + 5b}$ $\overline{10a + b}$ $\overline{8a^2}$ $\overline{9b + 8}$ $\overline{11a + b}$

$\overline{10a + b}$ $\overline{11a + b}$ $\overline{11a + b}$ $\overline{11a - b}$ $\overline{5a + 5b}$ $\overline{8a^2}$ $\overline{3b - 1}$

N

$\overline{8a^2}$ $\overline{6a + 9b}$ $\overline{a + 11}$ $\overline{3a + b}$ $\overline{11a + b}$ $\overline{5a + 5b}$

Simplify these. Write the letter beside each, above its answer in the box.

N $4a + 3b + 2a + 6b = 6a + 9b$ **S** $9a + 3b - 4a + 2b$ **W** $3a + 4b + 2a - b$ **F** $8a + 4b + 2a - 3b$
O $5a + 6b + a - 2b$ **H** $7a + 2b - 4a - b$ **R** $5a + 3 + 2a - 2$ **V** $3b + 11 + 5b - 3 + b$
C $9a + 7 - 8a + 4$ **L** $12b - 7 + 3b + 8$ **G** $4a + 6 + 3a - 4 - a$ **E** $5a + 3b + 6a - 2b$
T $6a - 2 + 5a - 4$ **X** $4b + 9 - b - 10$ **D** $4a + 5 - 3a - 7$ **A** $9a^2 - 5a^2$
I $6a^2 + 2a^2$ **M** $5a^2 + 3a^2 - 2a^2$

Exercise 11 (page 174)

Review 1

$\overline{2}$	$\overline{1}$	$\overline{^-1\cdot5}$	$\overline{16}$	$\overline{^-1}$	$\overline{^-1}$	$\overline{^-5\cdot5}$	$\overline{50\cdot5}$		$\overline{0}$	$\overline{18}$	$\overline{6}$ $\overline{10\cdot5}$

$\overline{50\cdot5}$ $\overline{6}$ $\overline{^-5\cdot5}$ $\overline{^-5\cdot5}$ $\overline{13}$ $\overline{^-1}$ $\overline{0}$ $\overline{^-1\cdot5}$ $\overline{^-1}$ $\overline{1}$ $\overline{7\cdot5}$ $\overline{^-5\cdot5}$

U
$\overline{52}$ $\overline{1}$ $\overline{18}$ $\overline{32}$ $\overline{128}$ $\overline{^-5\cdot5}$ $\overline{50\cdot5}$ $\overline{16}$ $\overline{61}$ $\overline{16}$ $\overline{10\cdot5}$

If $n = 4$, $m = 2\cdot5$ and $p = ^-3$, evaluate these. Write the letter that is beside each above its answer in the box.

U $2n^2 = 32$ **M** $3n^2 + 4$ **T** $2n^3$ **D** $4n^2 - 3$ **L** $\frac{n^3 - 4}{10}$

V $3m$ **P** $4m + 3$ **E** $2 - 3m$ **Y** $7(m - 1)$ **N** $4(m + 2)$

O $3p + 9$ **S** $3n^2 + m$ **F** $2m + 2p$ **A** $4m - 2p$ **R** $\frac{3p + 6}{2}$

I $\frac{2p + 3}{p}$ **G** $\frac{p - 1}{p + 1}$

Exercise 20 (page 190)

Review

Use a copy of the cross number. Fill it in by solving the equations.

Across
1. $n - 7 = 4$
2. $\frac{n}{7} = 3$
5. $n - 20 = 12$
9. $n - 10 = 3$
10. $\frac{n}{3} = 14$
13. $2n - 1 = 41$
15. $\frac{n}{3} + 4 = 5$

Down
1. $\frac{3n}{4} + 1 = 10$
3. $n + 3 = 16$
4. $3n = 12$
6. $\frac{2n}{3} = 16$
7. $2 + n = 15$
8. $n - 1 = 13$
11. $2n = 44$
12. $1 + 2n = 21$
14. $2n - 3 = 19$

Discussion (page 192)

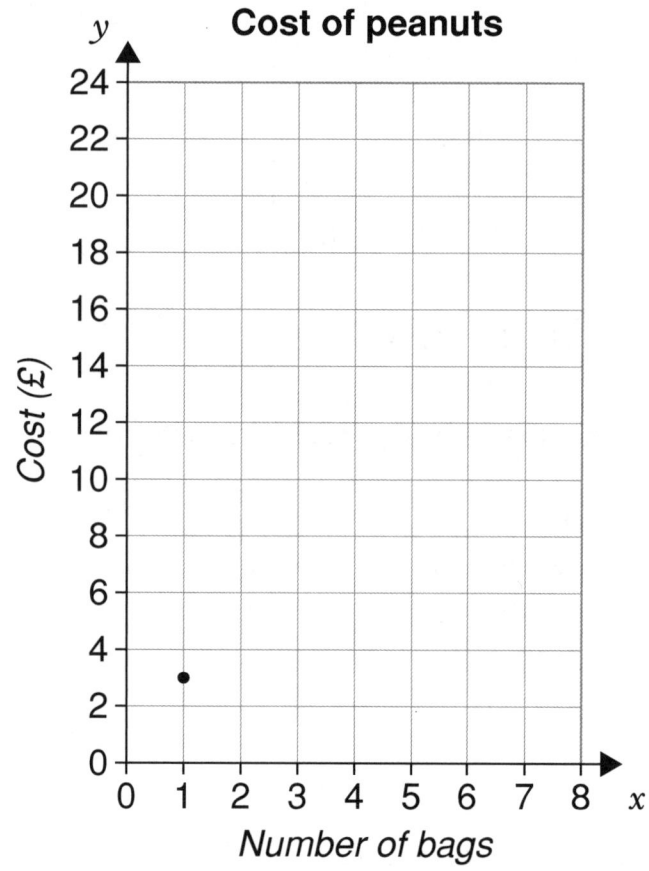

Cost of peanuts

Number of bags

Exercise 22 (pages 192 and 193)

1 a

Number of pizzas	1	2	3	4
Cost (£)				

d

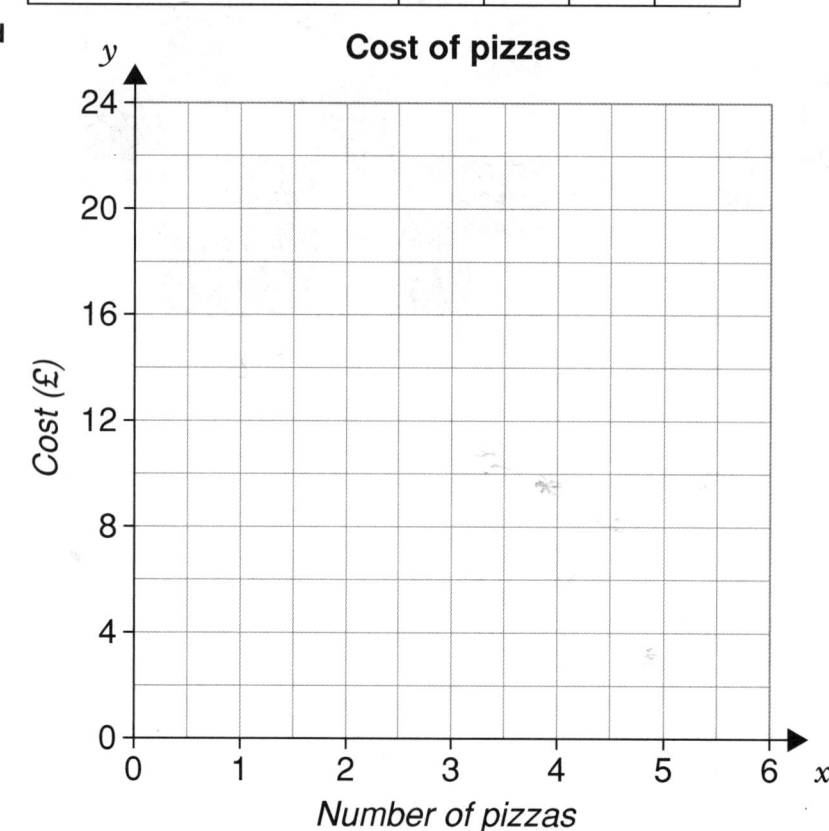

Cost of pizzas

Number of pizzas

Exercise 22 (pages 192 and 193) continued

Review

a

Miles	5	10	15	20	25
Kilometres	8				

d

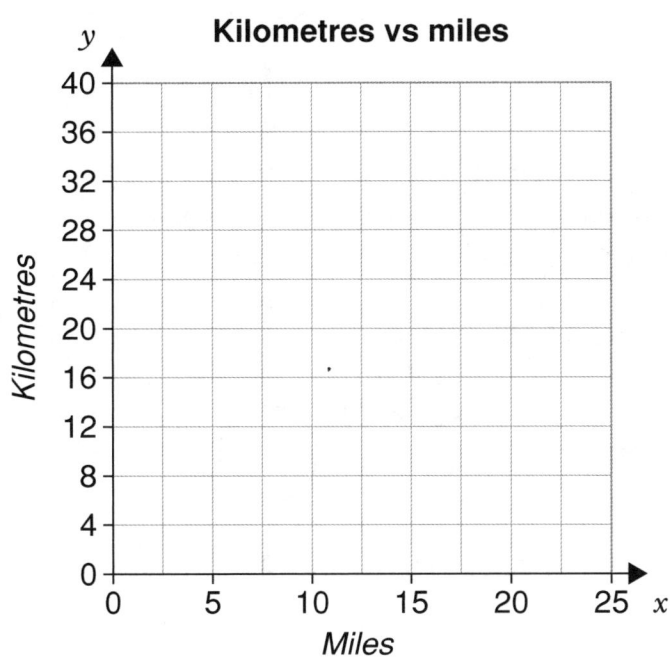

Kilometres vs miles

Test Yourself (page 198)

21 a

British pounds (*p*)	10	20	30	40	50
Euro (*e*)	15				

d

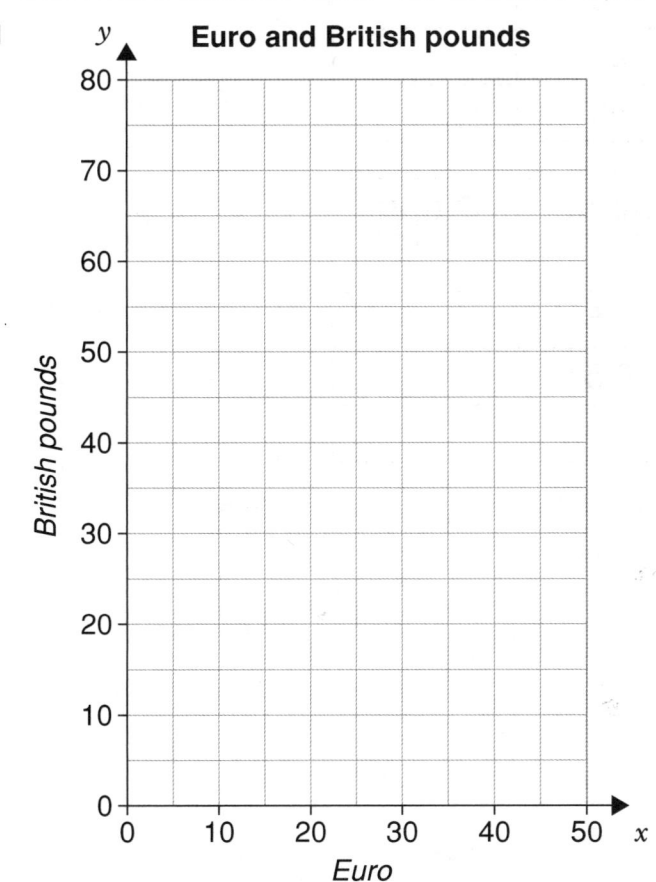

Euro and British pounds

9 Sequences and Functions

Exercise 4 (page 206)

7

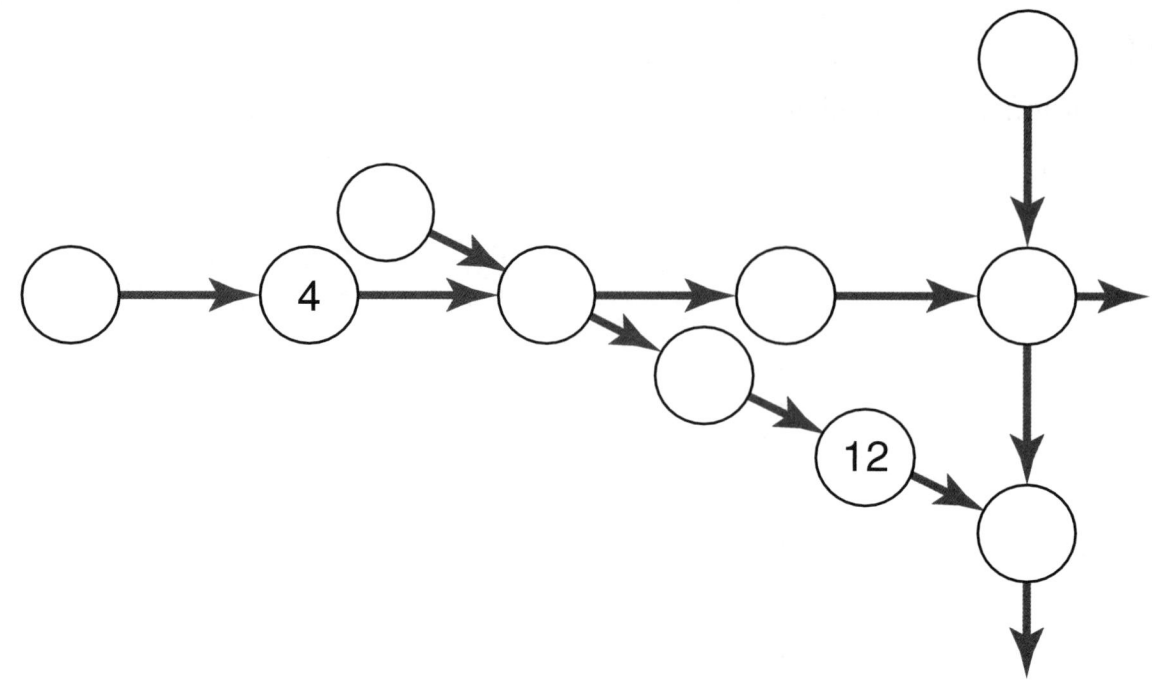

Exercise 10 (pages 219, 220 and 221)

1 a
Input	Output
5	
2	
0	
12	

b
Input	Output
4	
10	
24	
68	

c
Input	Output
11	
21	
⁻2	
⁻5	

d
Input	Output
7	
1·5	
0·5	
⁻1	

Exercise 10 (pages 219, 220 and 221) continued

3 a

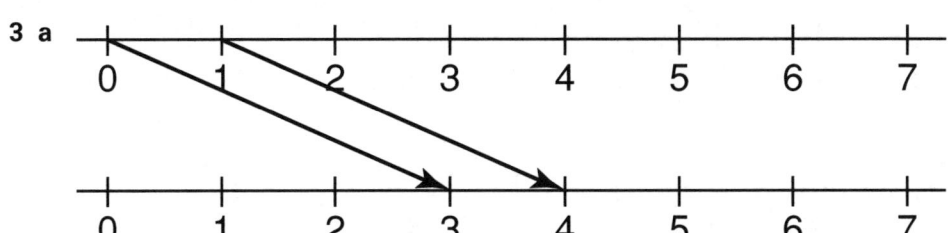

b

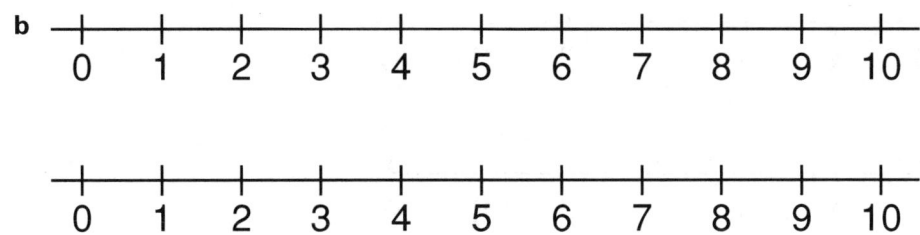

4 a

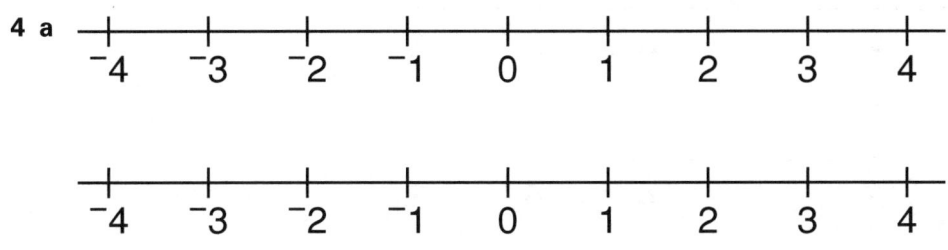

b

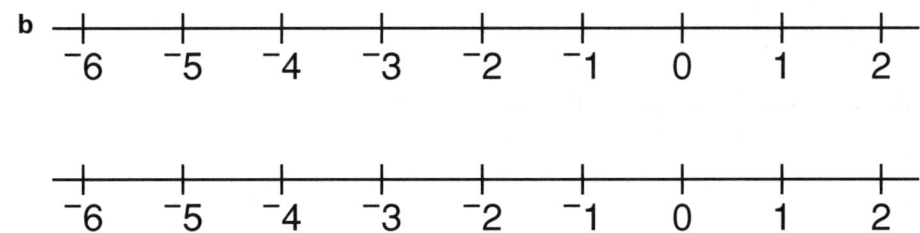

c

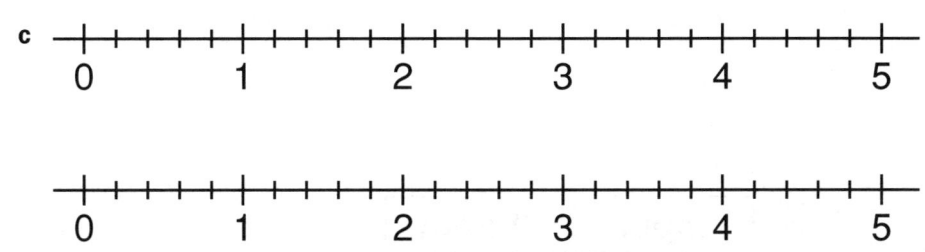

Exercise 10 (pages 219, 220 and 221) continued

6 a i
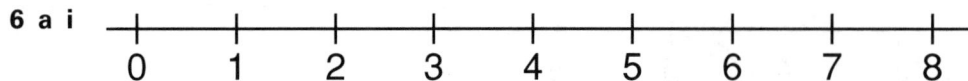

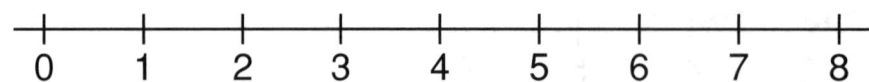

b ii
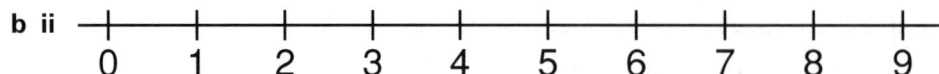

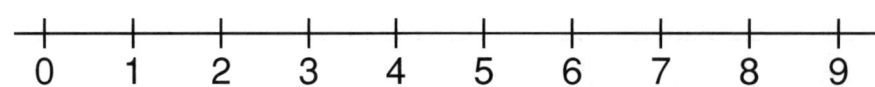

c iii
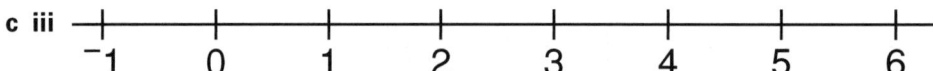

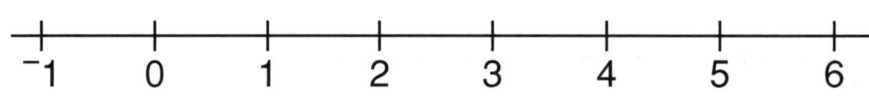

7 a
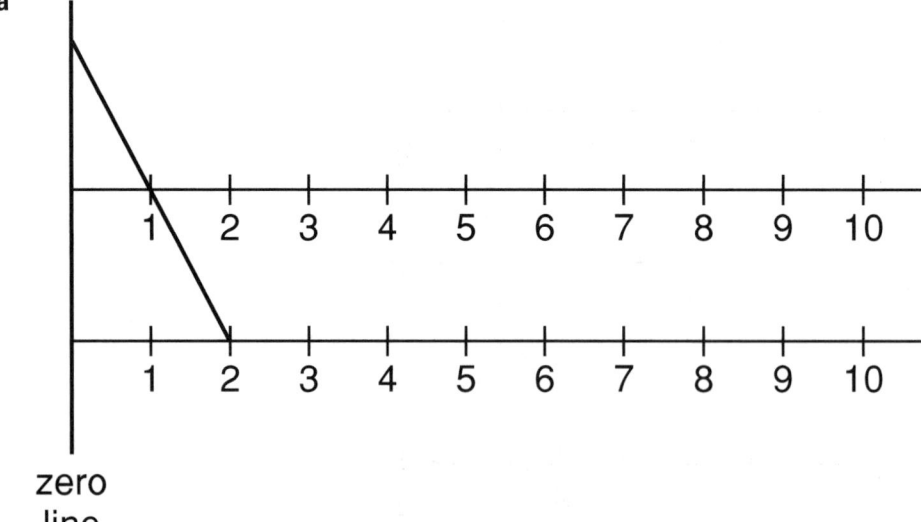

zero
line

Review 1

a

Input	Output
2	
15	
2·5	
$\frac{1}{2}$	

b

Input	Output
3	
8	
2·4	
$2\frac{1}{2}$	

Exercise 10 (pages 219, 220 and 221) continued

Review 3

a
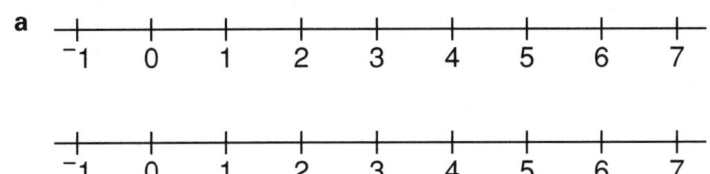

b
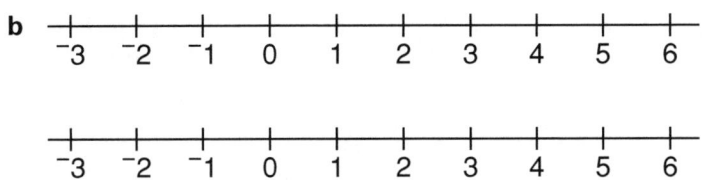

c
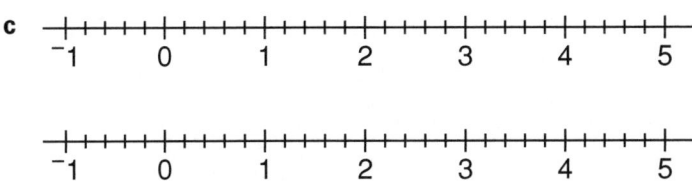

Exercise 12 (page 226)

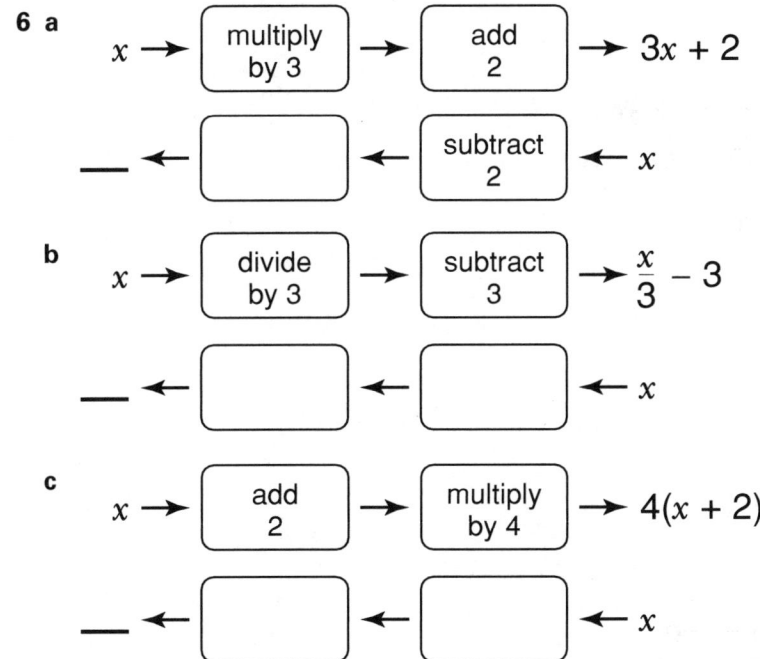

6 a

$x \rightarrow$ multiply by 3 $\rightarrow$ add 2 $\rightarrow 3x + 2$

$\underline{} \leftarrow \leftarrow$ subtract 2 $\leftarrow x$

b

$x \rightarrow$ divide by 3 $\rightarrow$ subtract 3 $\rightarrow \dfrac{x}{3} - 3$

$\underline{} \leftarrow \leftarrow \leftarrow x$

c

$x \rightarrow$ add 2 $\rightarrow$ multiply by 4 $\rightarrow 4(x + 2)$

$\underline{} \leftarrow \leftarrow \leftarrow x$

Test Yourself (page 231)

14
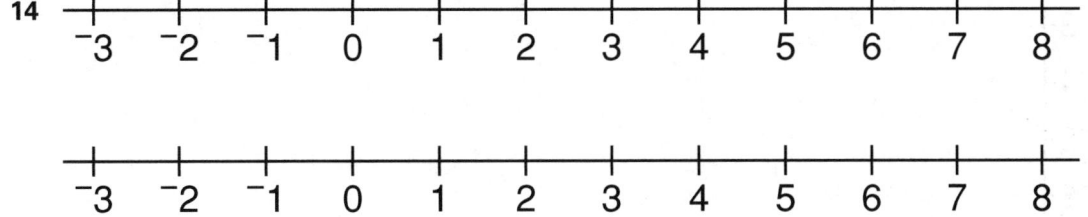

10 Graphs

Exercise 1 (pages 235 and 236)

7 b

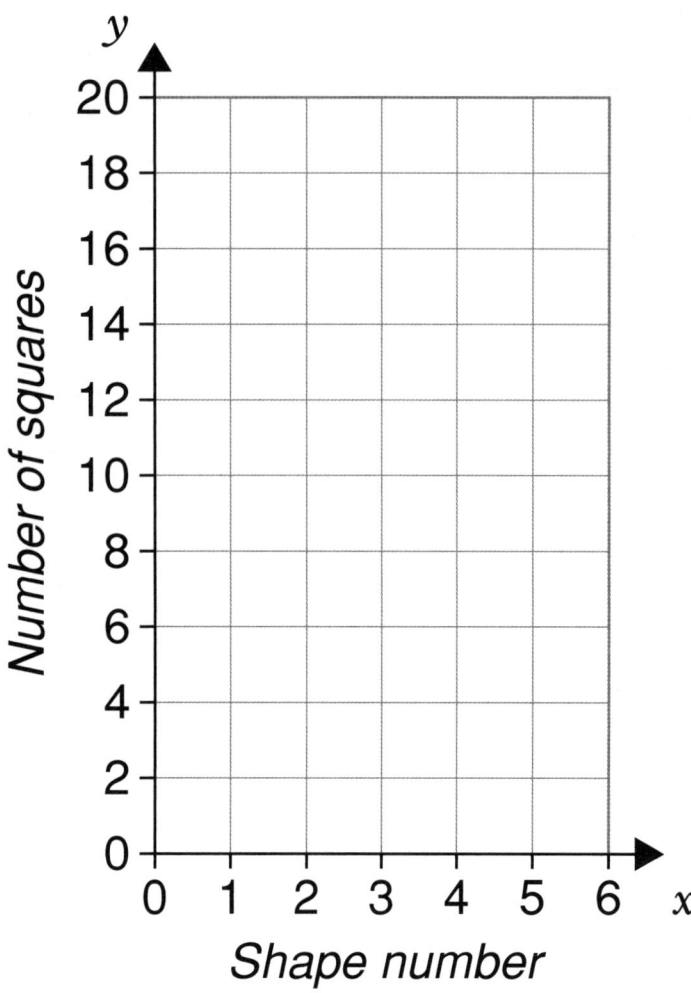

Review 2

c

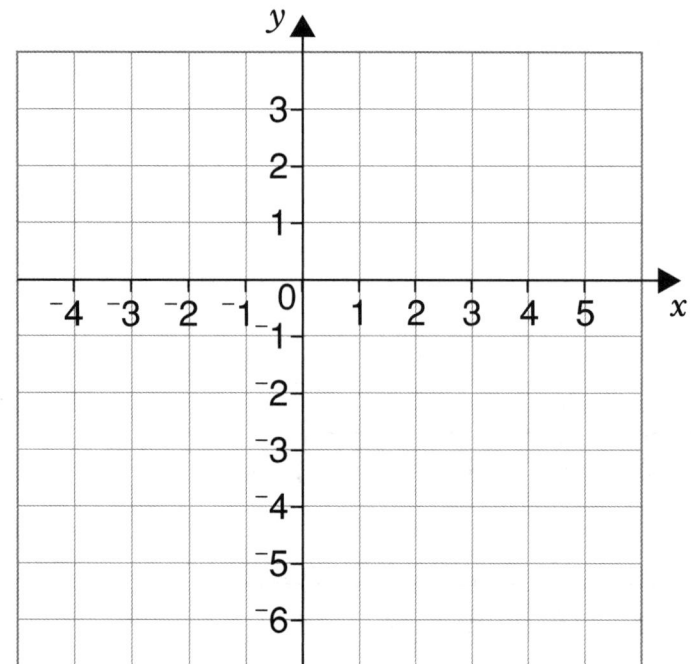

Exercise 3 (pages 242, 243, 244 and 245)

2 c

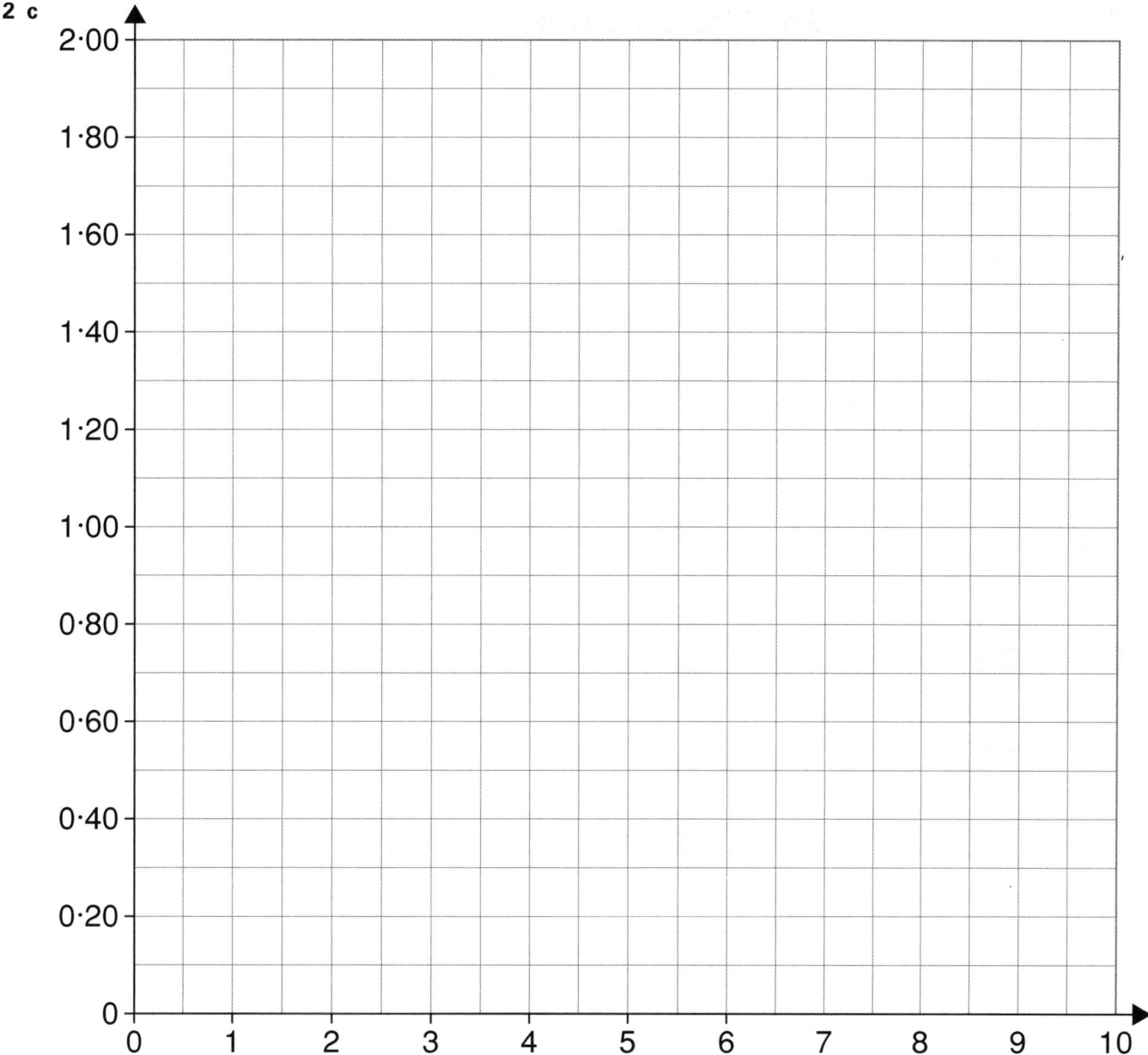

3 d

Age of girl (in years)	Height in cm at start of year (approximate)	Height in cm at end of year (approximate)	Approximate growth in cm
1 to 2	74	86	12
2 to 3	86		
3 to 4			

Exercise 3 (pages 242, 243, 244 and 245) continued

4 b

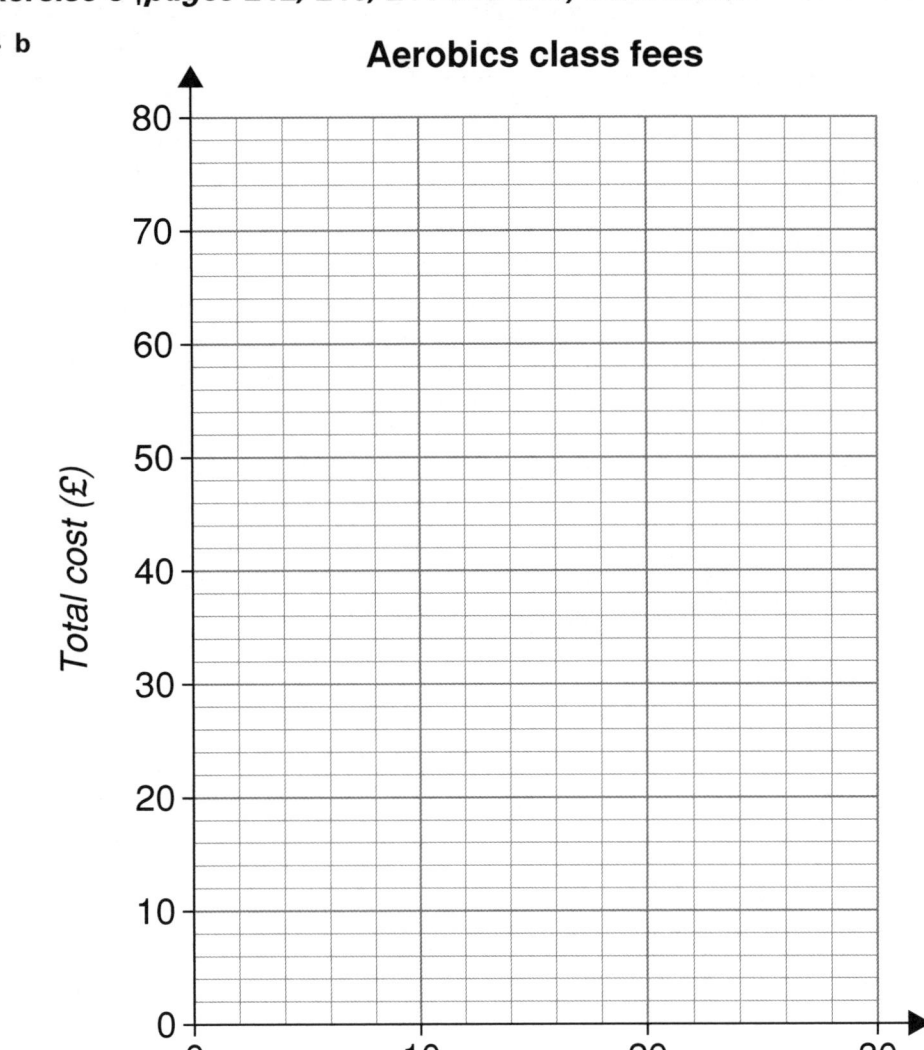

Aerobics class fees

Total cost (£) (y-axis, 0 to 80)

Number of classes (x-axis, 0 to 30)

Exercise 3 (pages 244 and 245) continued

7 a

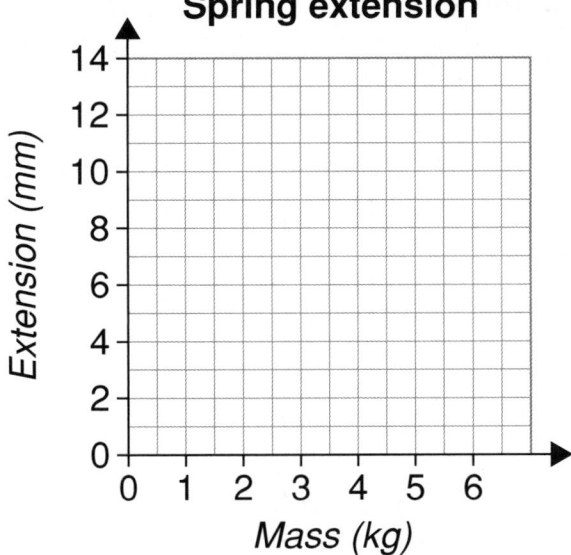

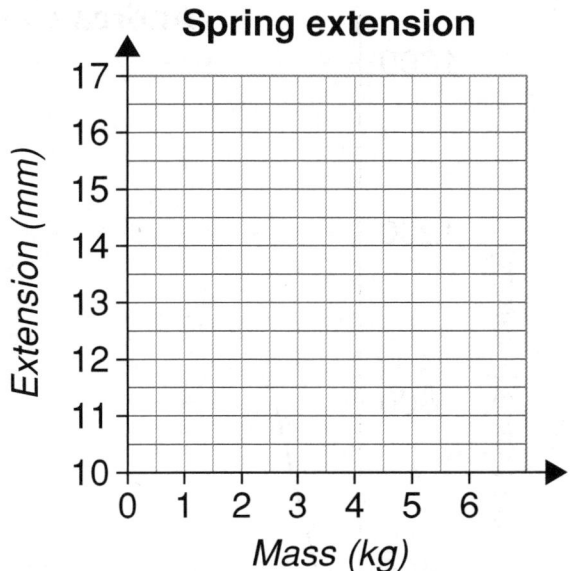

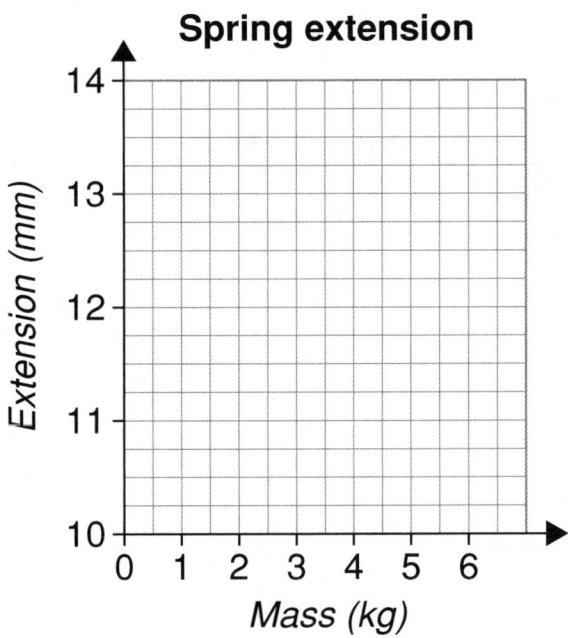

Review 2

b

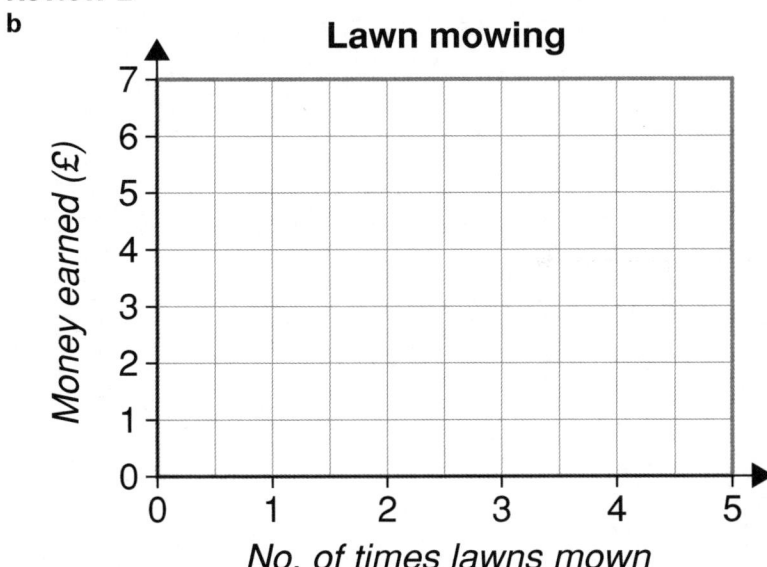

Exercise 4 (pages 246 and 247)

1

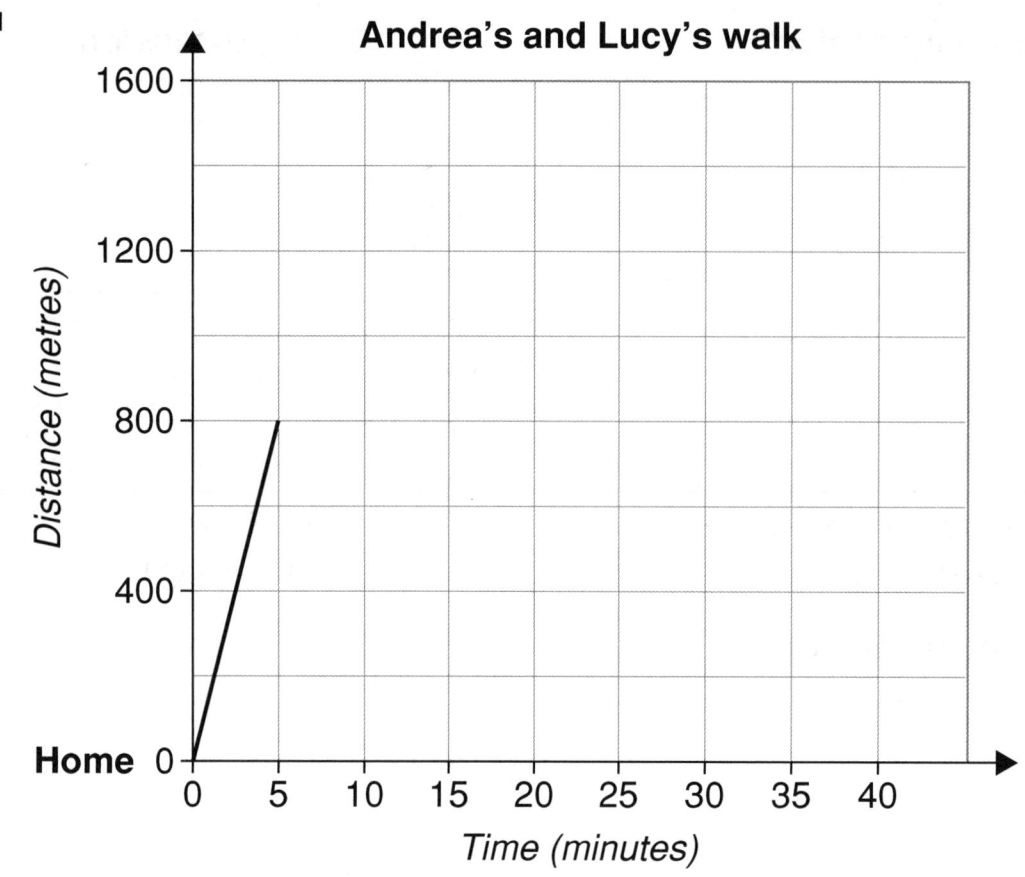

Andrea's and Lucy's walk

2

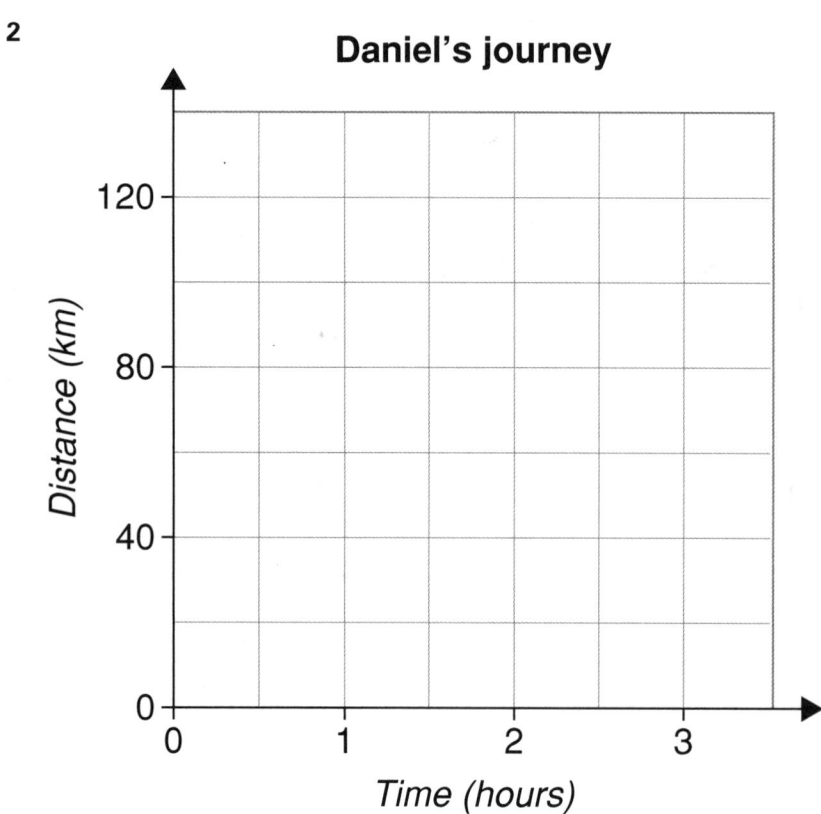

Daniel's journey

Exercise 4 (pages 246 and 247) continued

Review

Training run

Distance (km)

Time (hours)

Exercise 5 (pages 251 and 252)

9

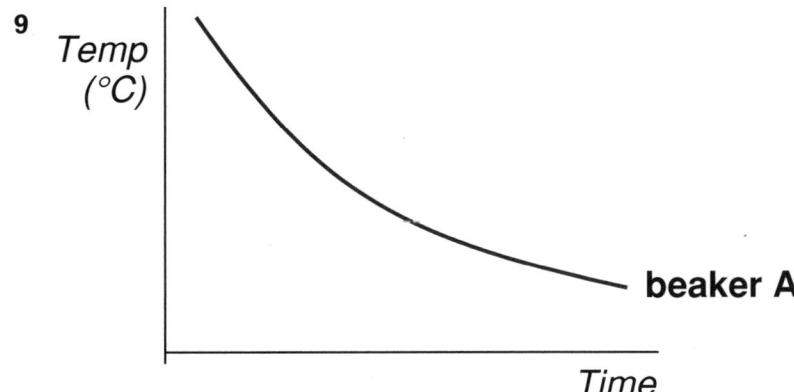

Temp (°C)

beaker A

Time

Review 4

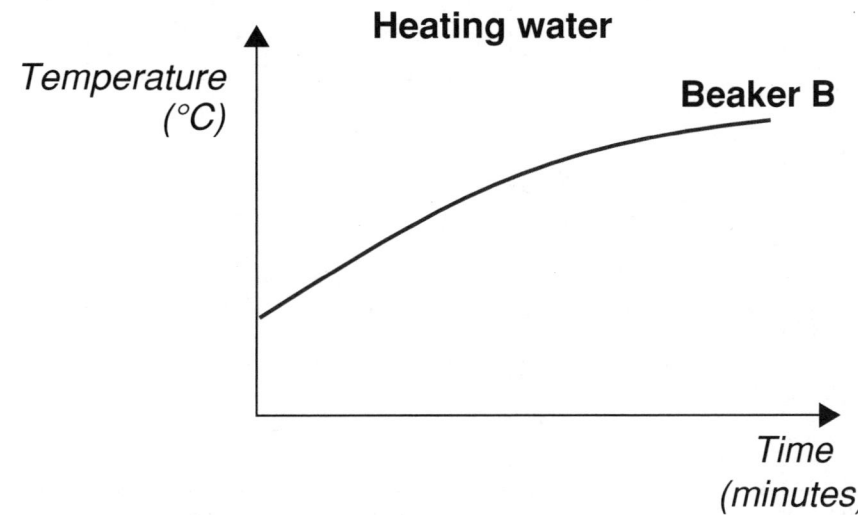

Heating water

Temperature (°C)

Beaker B

Time (minutes)

Test Yourself (pages 255, 256 and 257)

1 a $y = 3x + 3$

x	-2	0	2
y			

b $y = 2x - 4$

x	-1	2	4
y			

c $y = 3 - 2x$

x	-1	1	3
y			

2 b

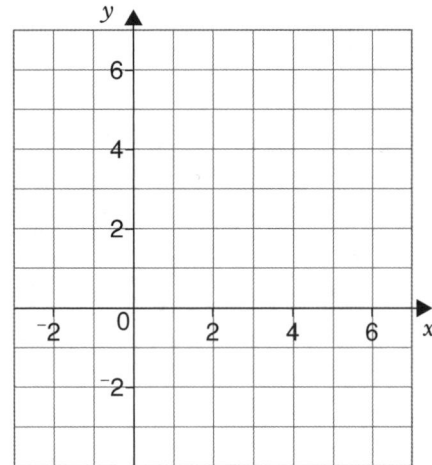

7

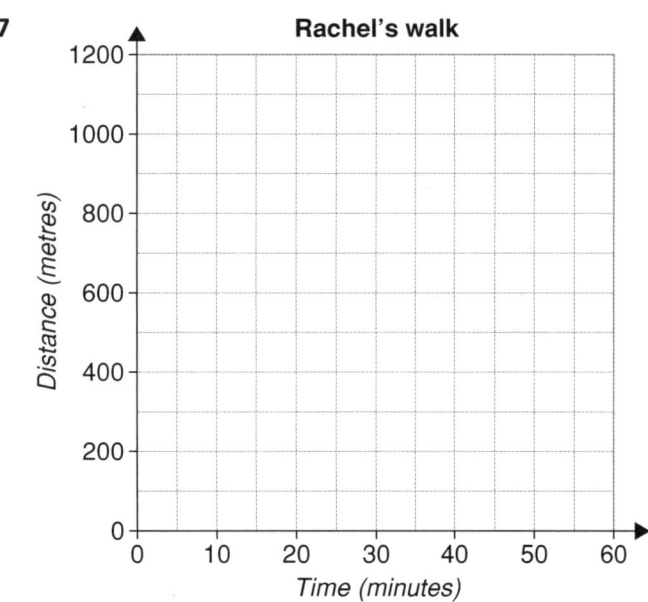

Rachel's walk

11 a i

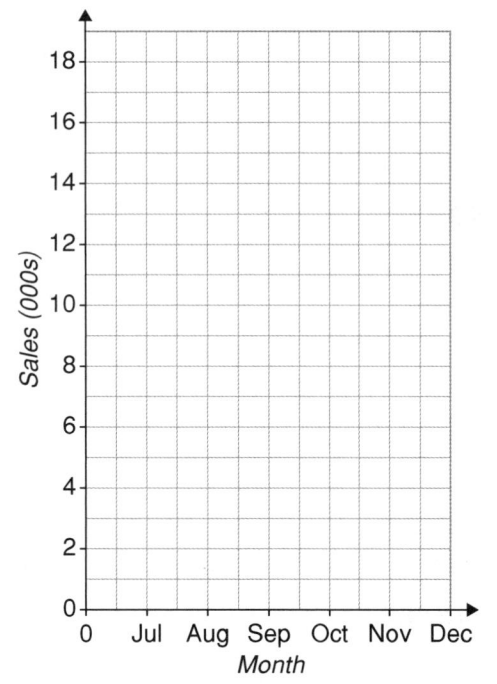

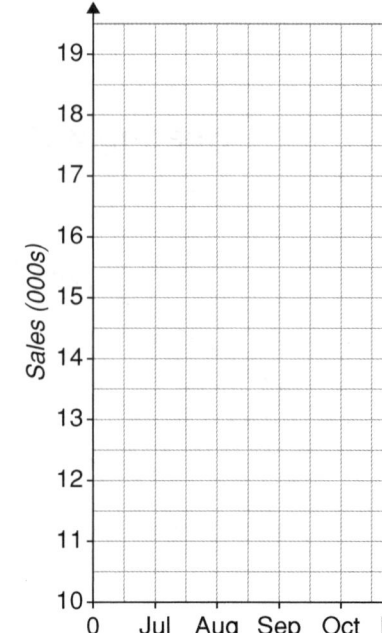

Shape, Space and Measures Support

Practice Questions (pages 264, 266, 267, 269 and 270)

6 a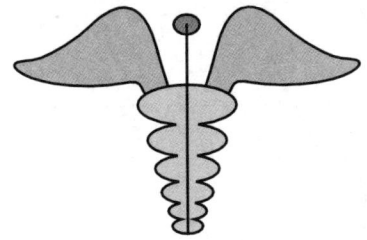

b

c

d

16

							V	
2 mℓ	200 mℓ	5 mℓ	2 kg		5 g	20 kg	1·5 m	2 cm

75 cm	200 mℓ	200 ℓ	6 mm

2 kg	10 ℓ	200 mℓ	25 m	20 kg	2 mℓ	5 g	2 kg

Put the letter beside each measurement above its matching estimate in the box.

V length of a table ≈**1·5 m**   **H** mass of a pencil
F width of a television   **O** capacity of a glass
S mass of a small puppy   **U** capacity of a paddling pool
A mass of a full suitcase   **T** capacity of a bucket
M length of a driveway   **W** capacity of a teaspoon
E length of a pencil sharpener   **C** volume of an eye dropper
R width of a paper clip

Practice Questions (pages 264, 266, 267, 269 and 270) continued

19 a

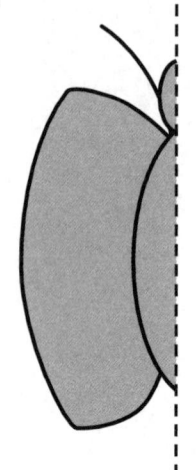

b

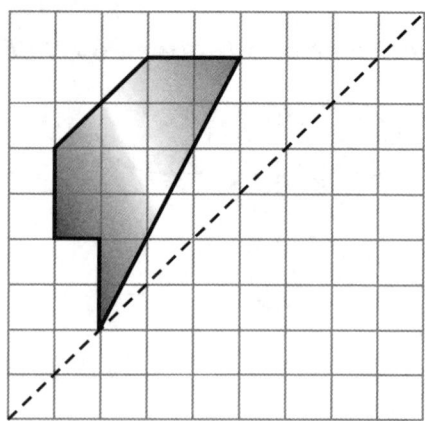

c

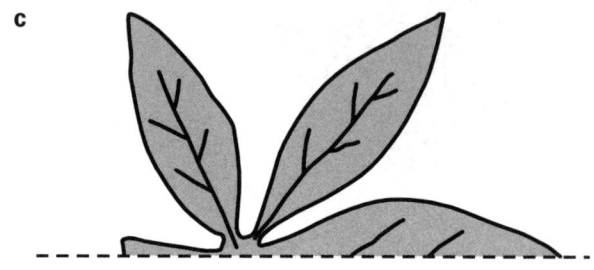

22

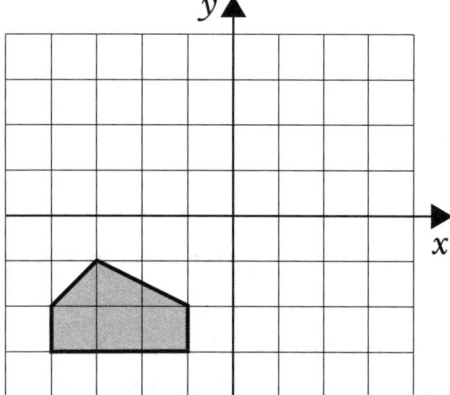

33 a

(0, 0)
90°

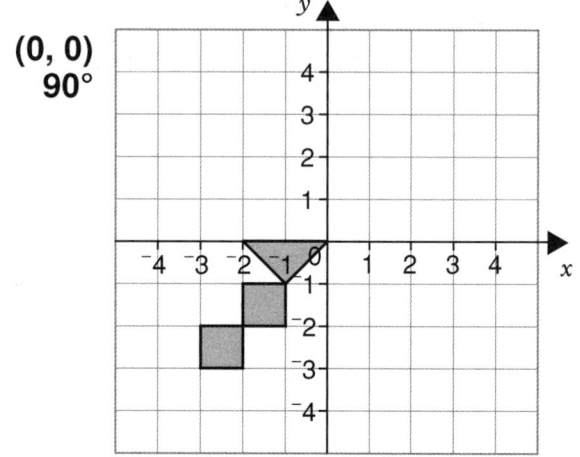

b

(⁻1, 1)
180°

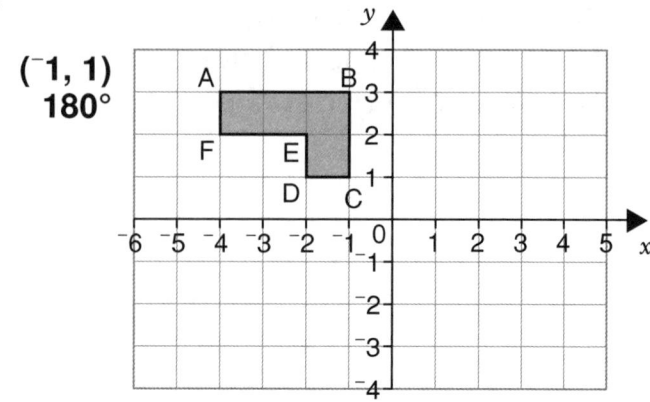

Practice Questions (pages 264, 266, 267, 269 and 270) continued

40

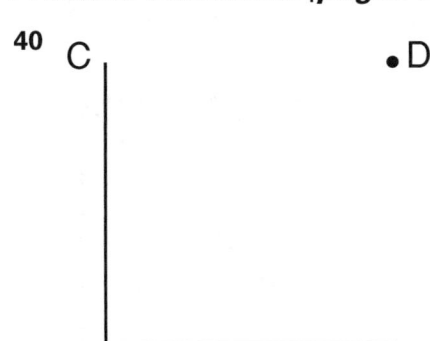

11 Lines and Angles

Exercise 3 (pages 279, 280 and 281)

3

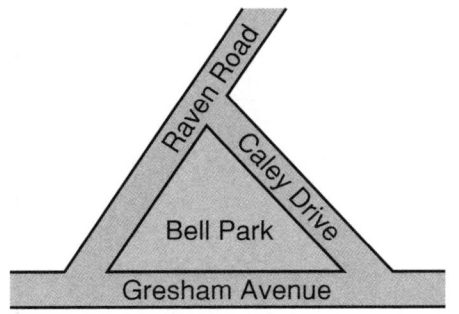

9

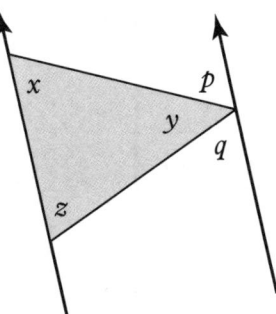

Review 1

Across	Down
1 c	**1** h
2 d	**3** e
5 g	**4** f
8 b	**6** a
9 i	**7** j

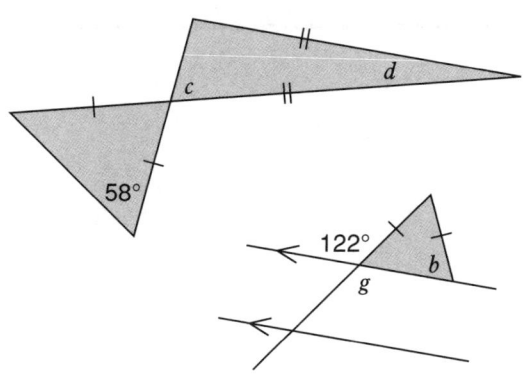

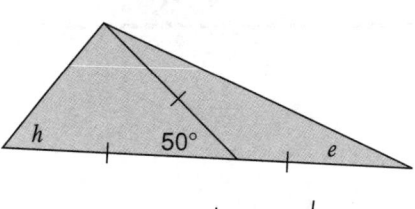

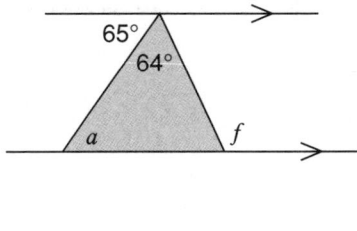

Review 3

Fill in the gaps.   **reason**

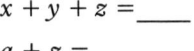 _____   _____

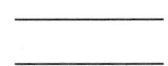 _____   _____

How can you finish this to prove $a = x + y$?

12 Shape, Construction and Loci

Piecing it Together (page 285)

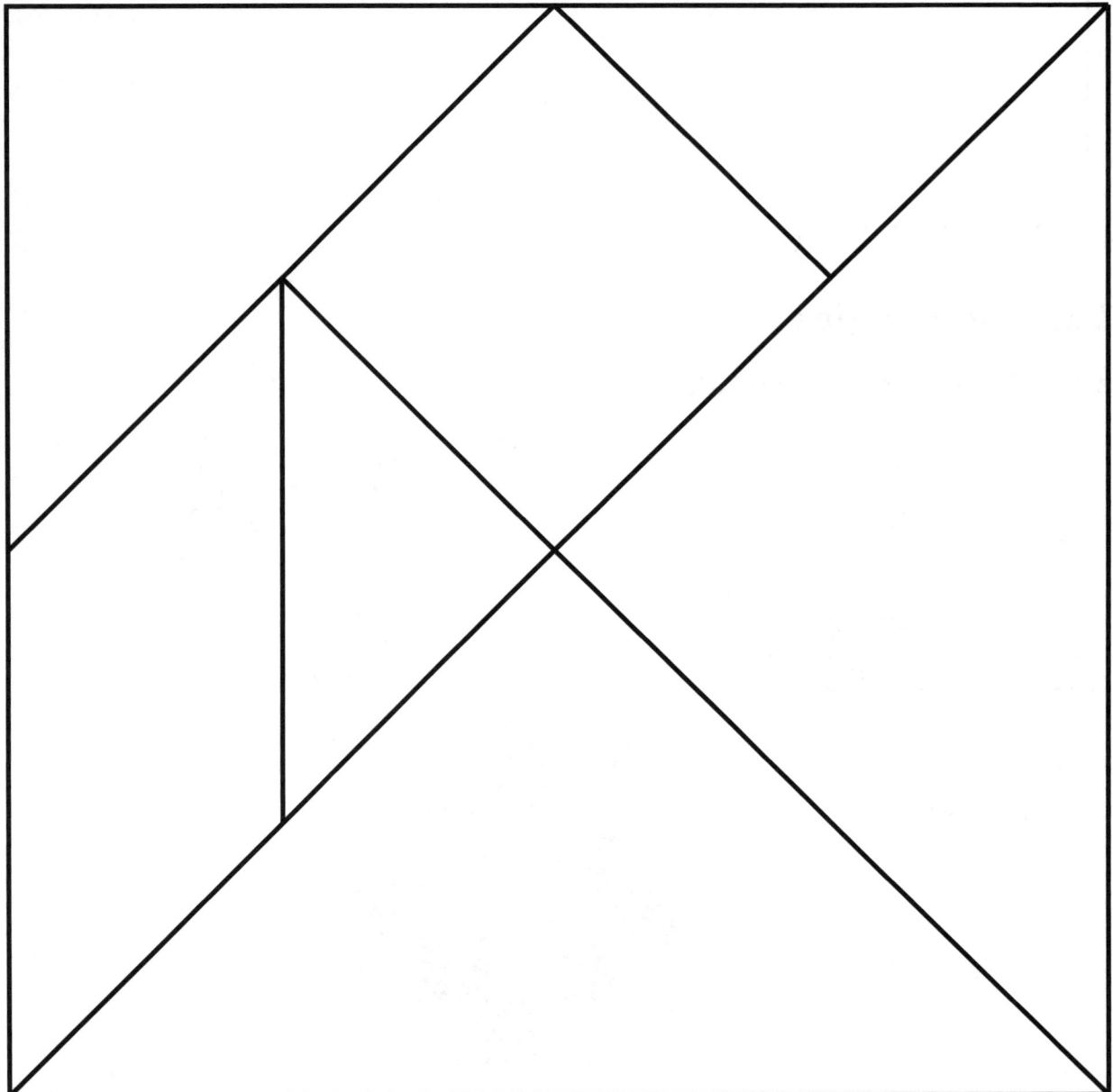

Investigation — Properties of Shapes (pages 287 and 288)

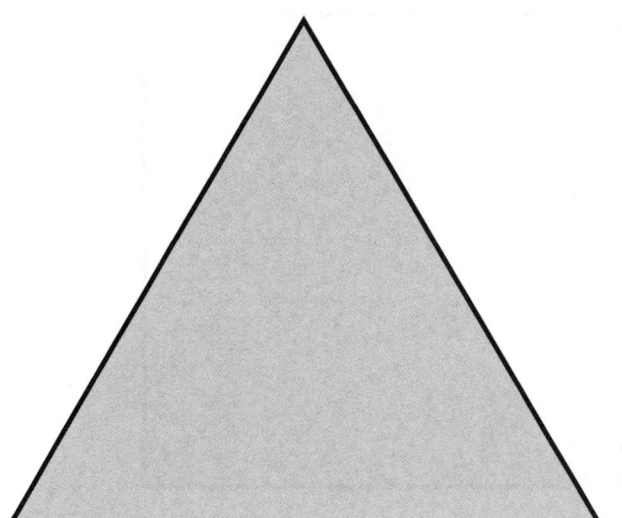
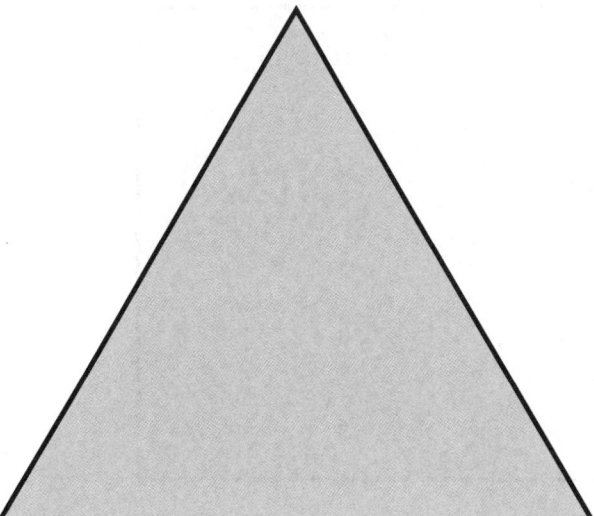
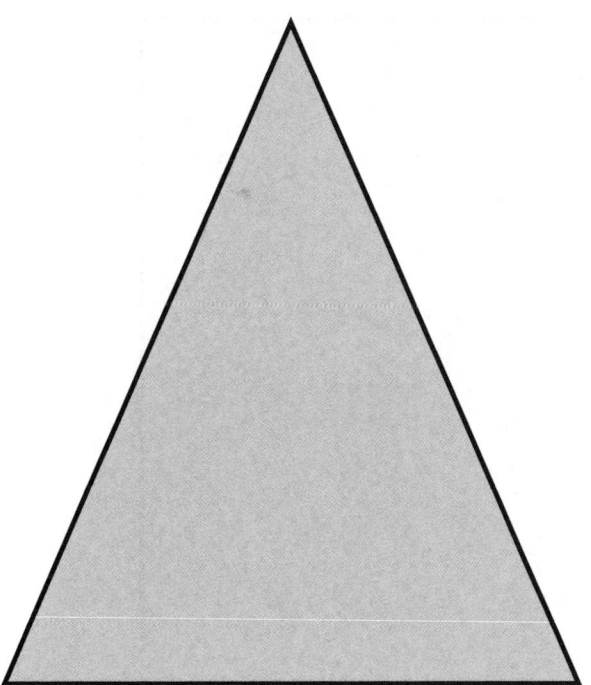
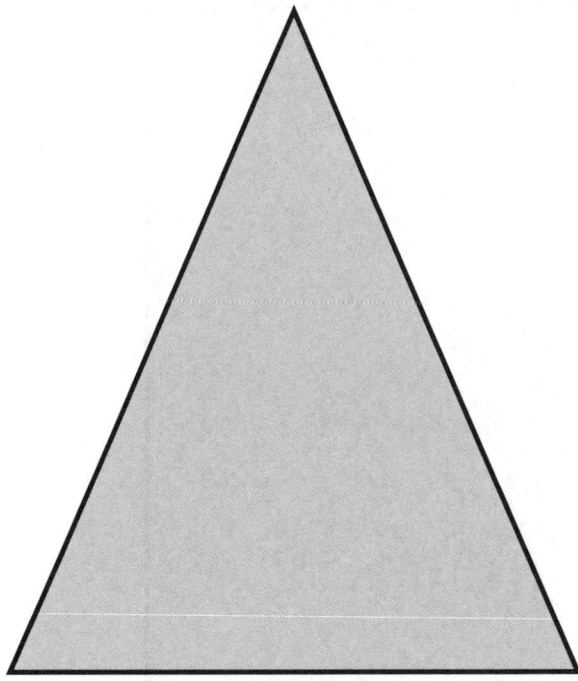

Investigation — Properties of Shapes (pages 287 and 288) continued

Investigation — Properties of Shapes (pages 287 and 288) continued

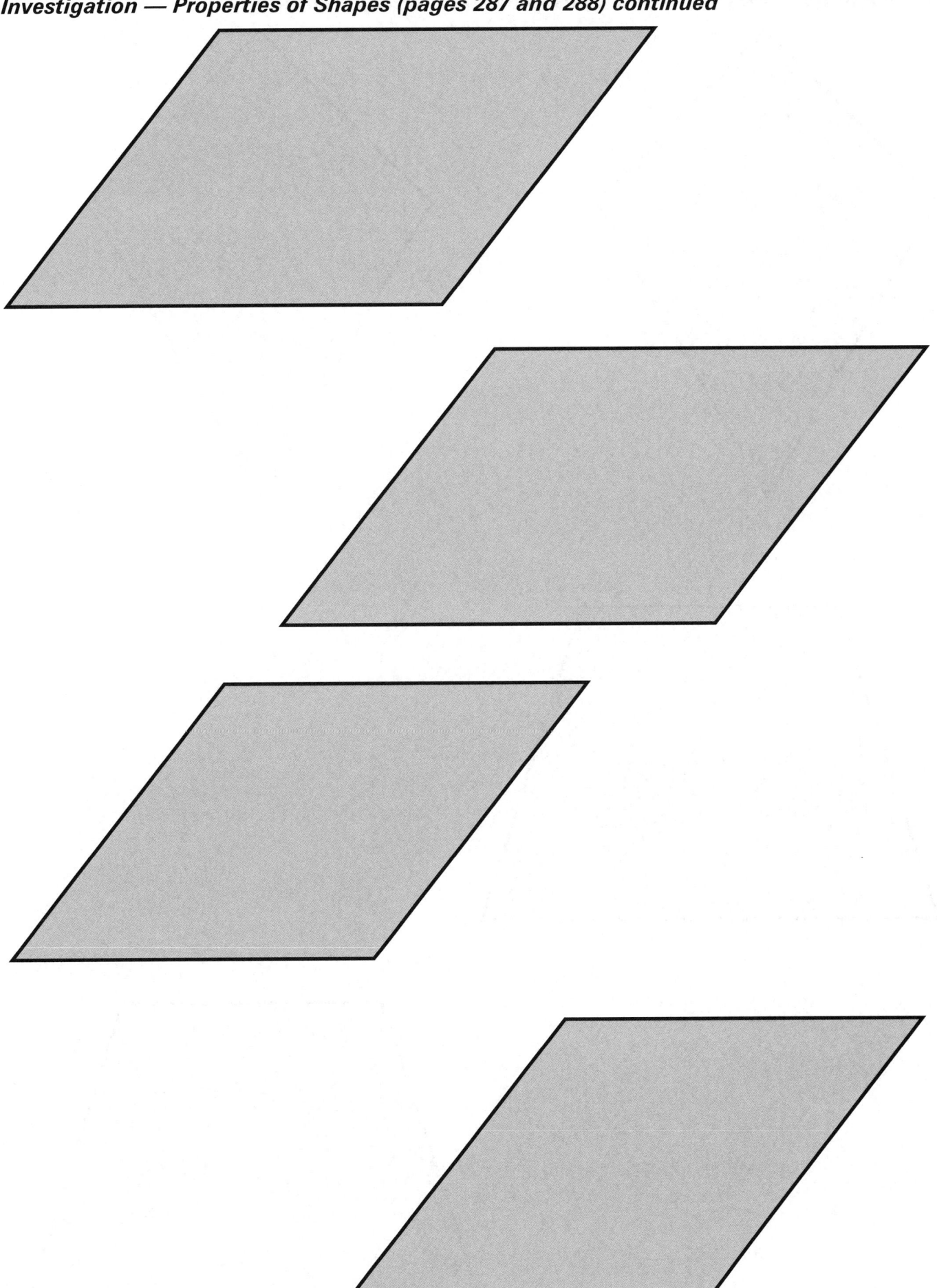

Investigation — Properties of Shapes (pages 287 and 288) continued

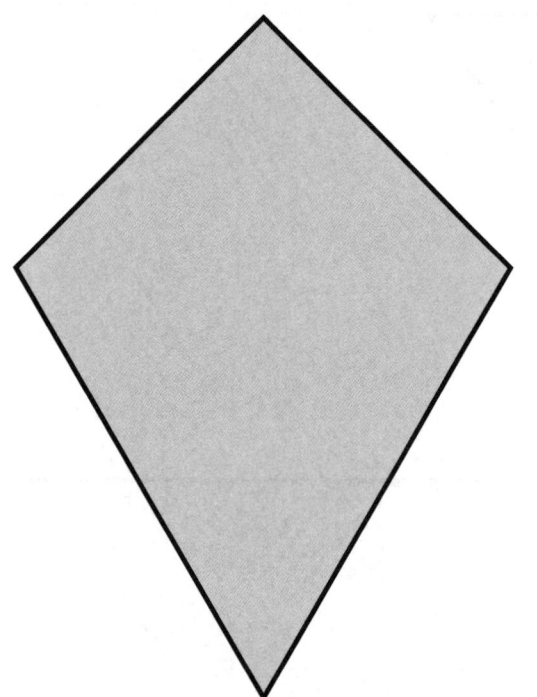

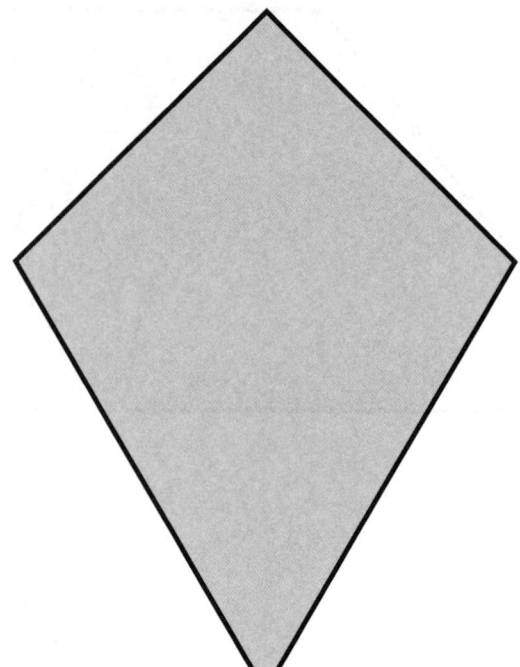

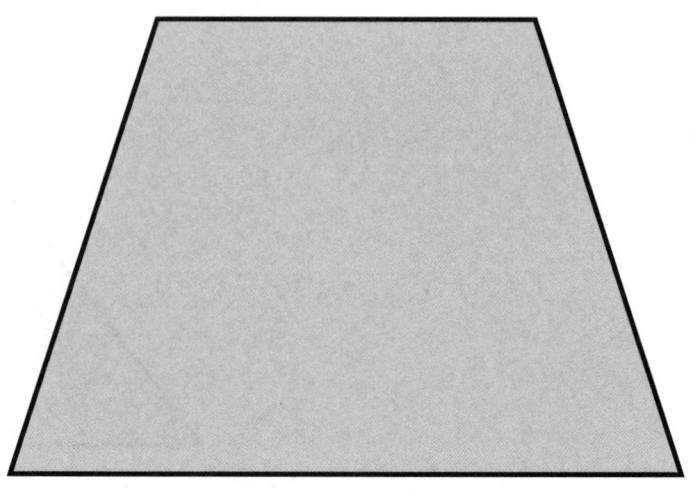

Investigation — Properties of Shapes (pages 287 and 288) continued

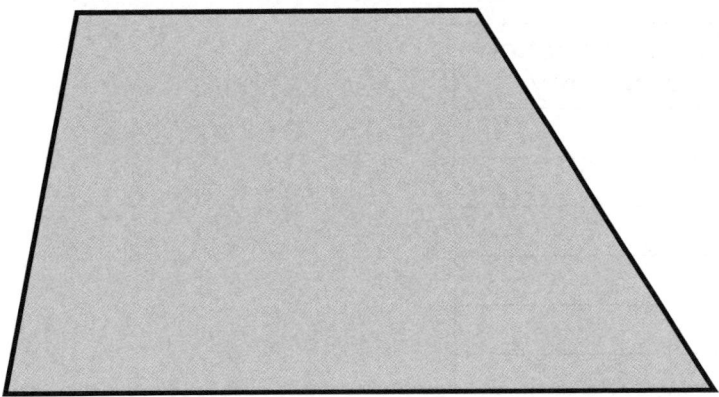

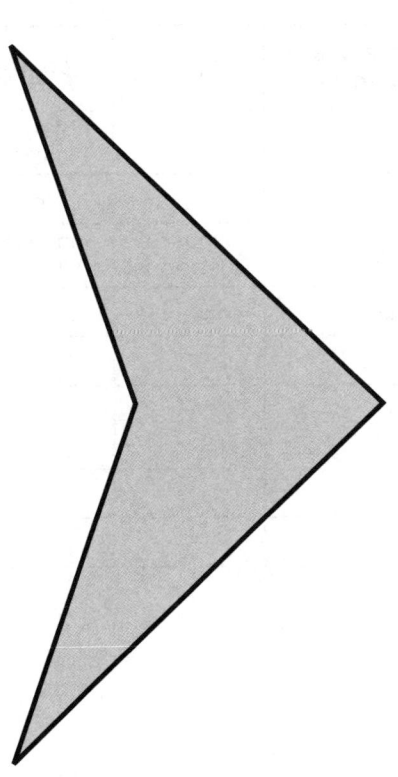

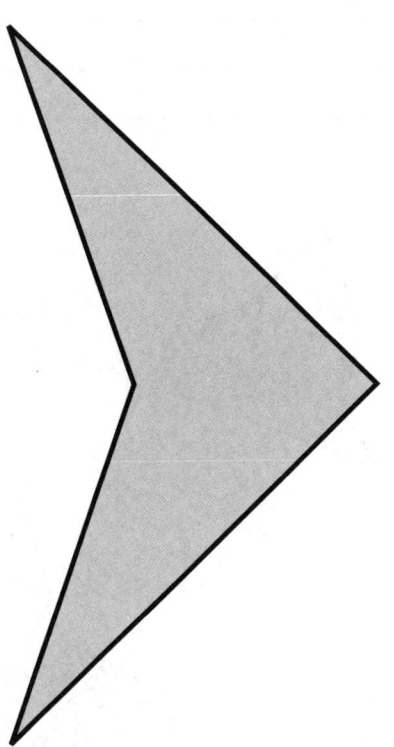

Investigation — Properties of Shapes (pages 287 and 288) continued

A

Shape	Number of lines of symmetry	Order of rotation symmetry
Equilateral △		
Isosceles △		
Square		
Rectangle		
Parallelogram		
Rhombus		
Kite		
Isosceles trapezium		
Trapezium		
Arrowhead		

Quadrilateral	Diagonals equal	Diagonals cross at right angles	Diagonals bisect each other	Diagonals bisect the angles	Sides	Angles
Square						4 right angles
Rectangle						
Parallelogram	✗	✗	✓	✗	opposite sides equal	opposite angles equal
Rhombus	✗	✓	✓	✓	4 equal	
Kite						
Isosceles trapezium						
Trapezium						
Arrowhead						

Practical (page 293)

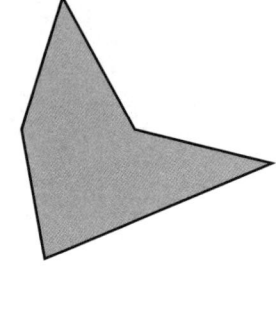
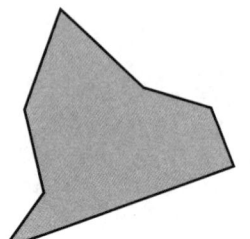

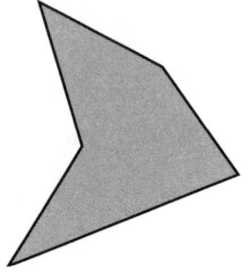

Practical (pages 295 and 296)

2

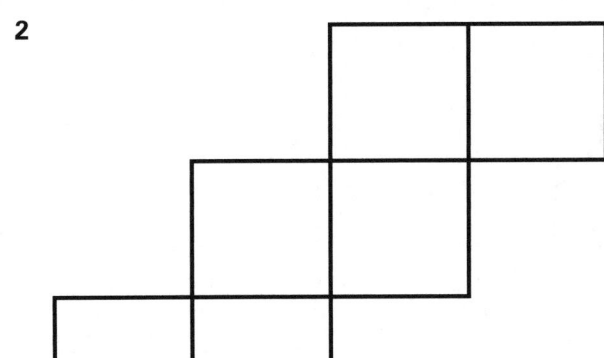
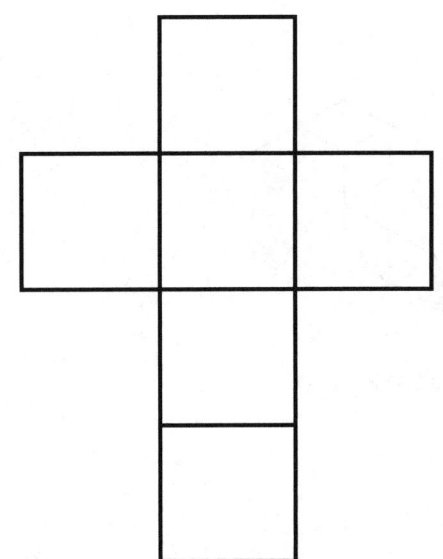

Exercise 4 (page 296)

5 a

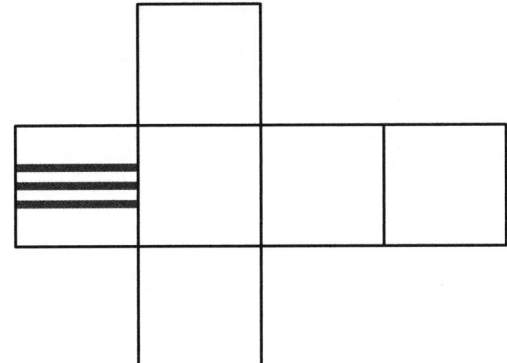

b

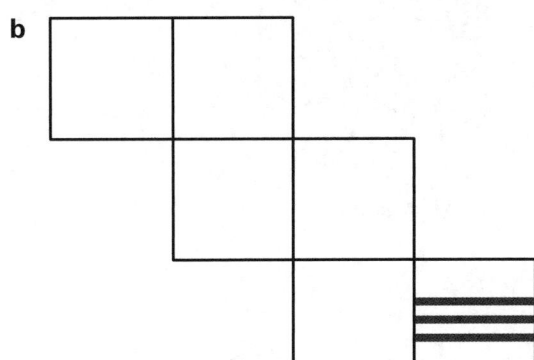

c

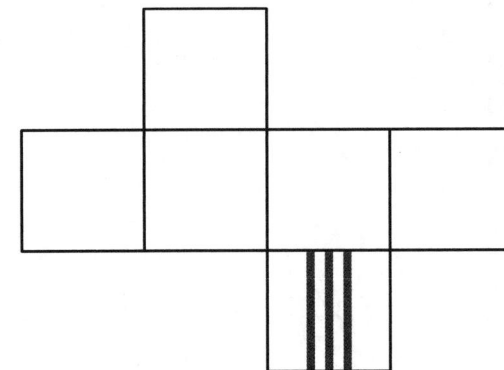

Exercise 5 (page 299)

3 a

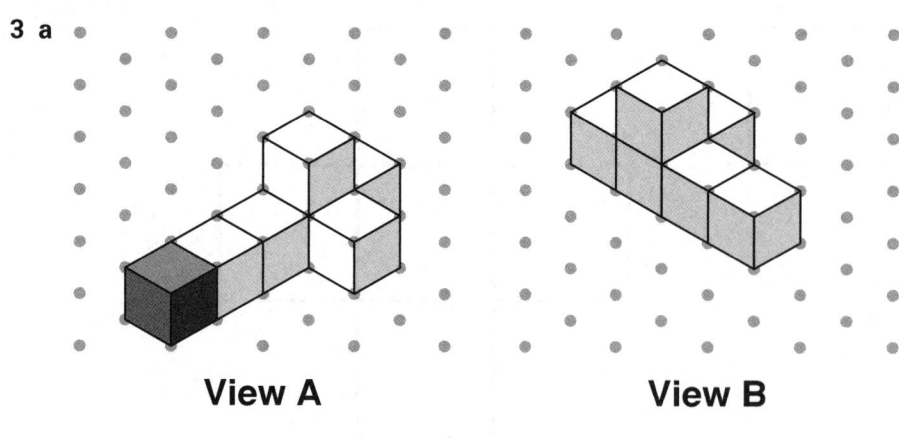

View A   View B

b

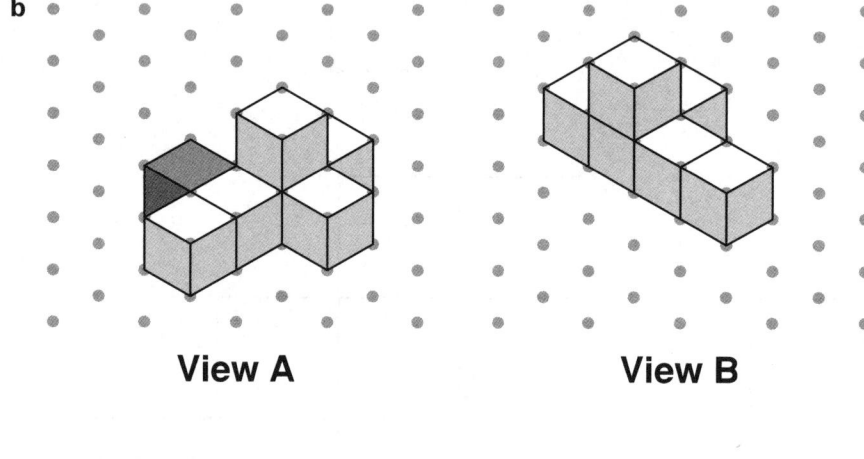

View A   View B

c

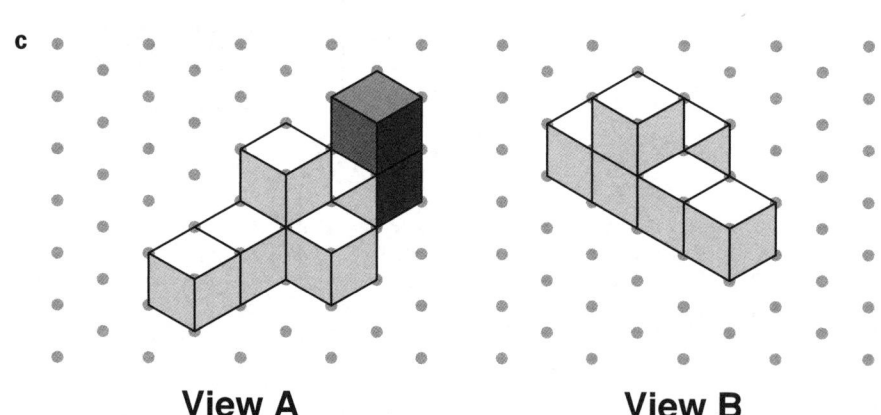

View A   View B

Exercise 6 (pages 303 and 304)

1

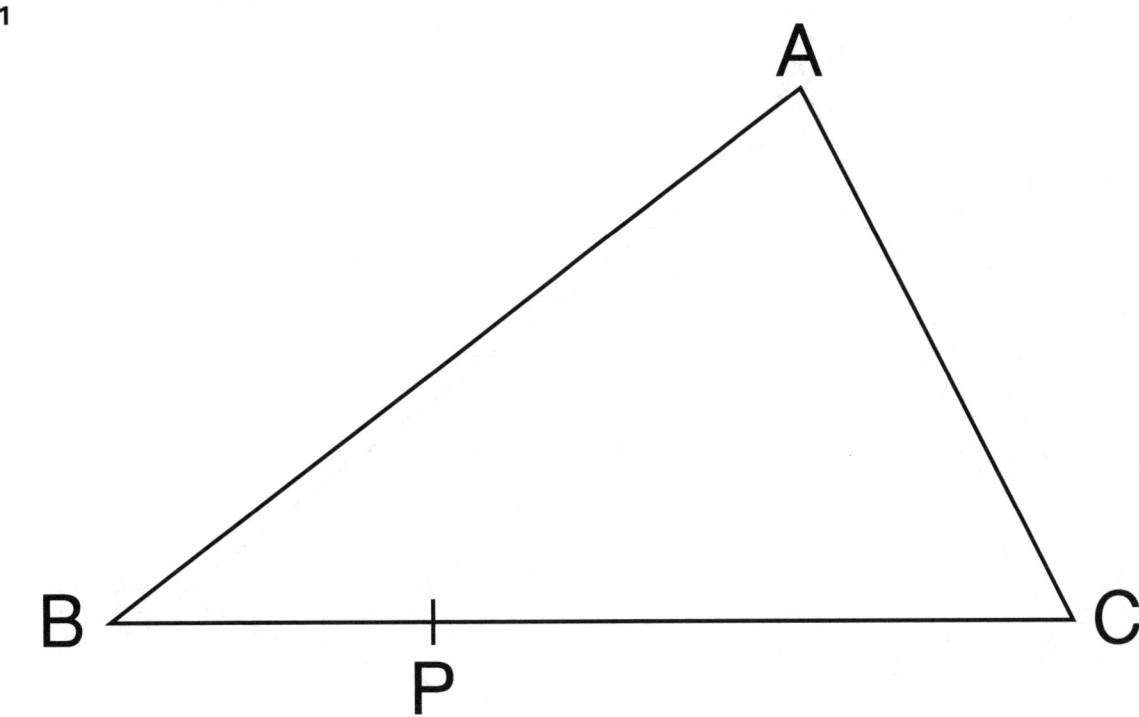

Review 1

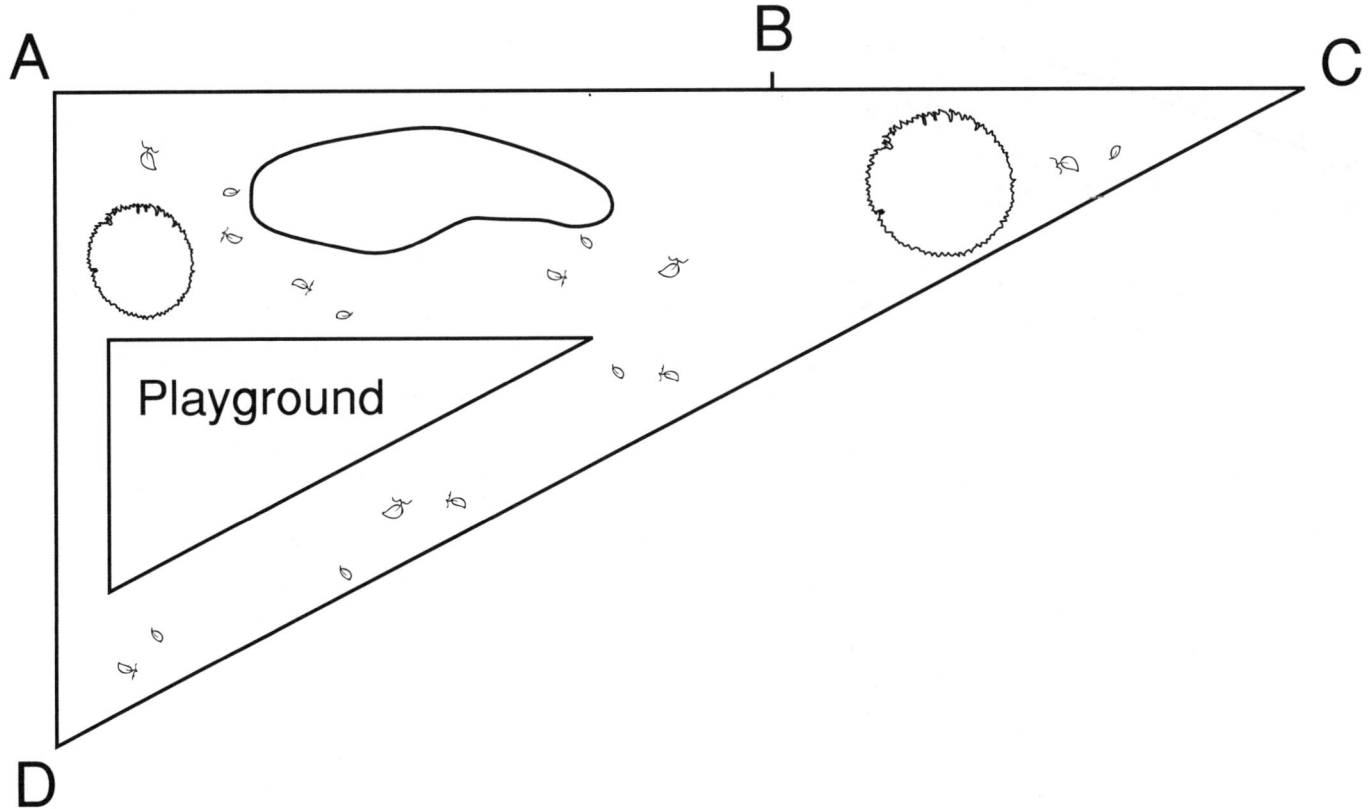

Exercise 6 (pages 303 and 304) continued

Review 2

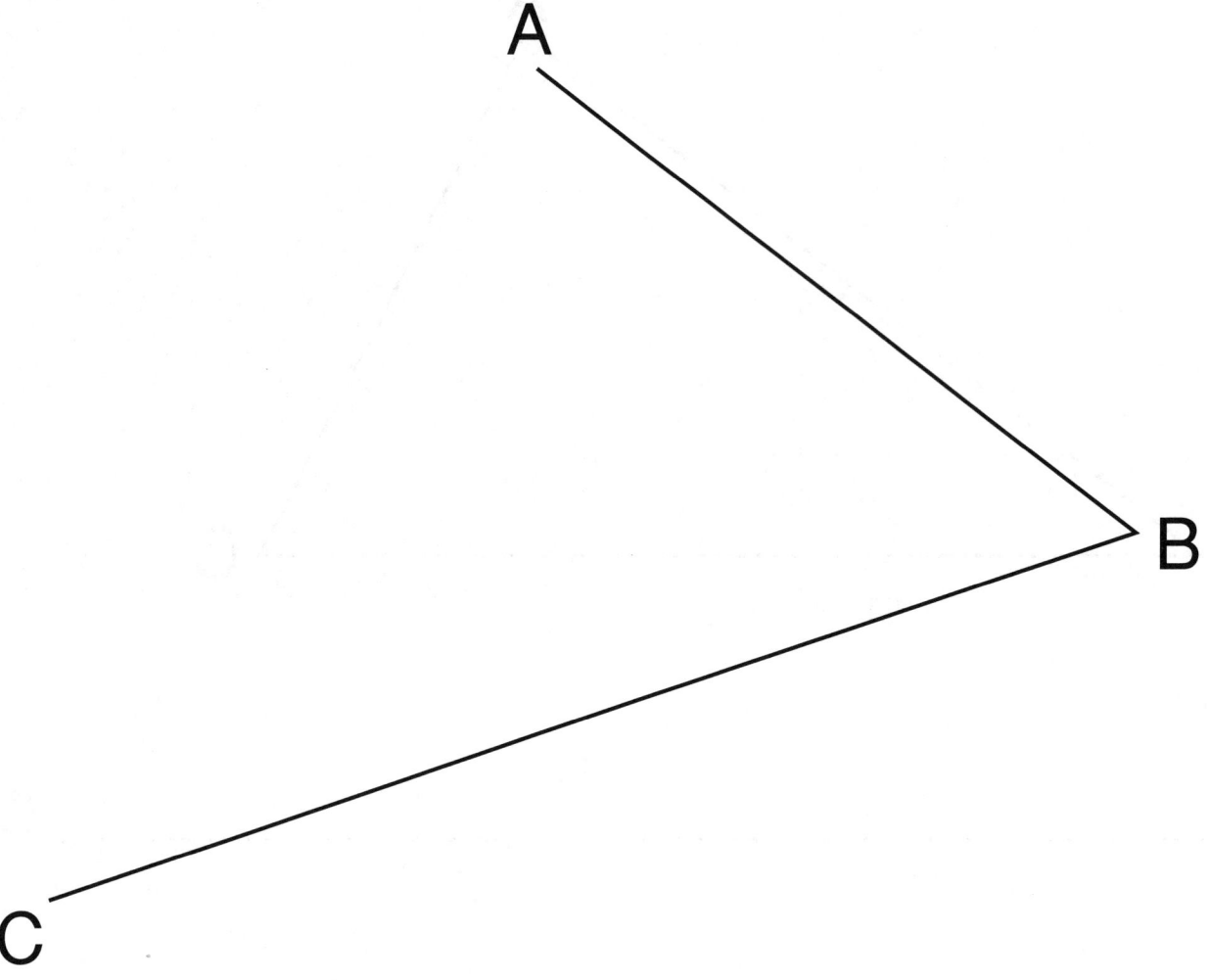

13 Coordinates and Transformations

Exercise 2 (pages 314, 315 and 316)

1 b

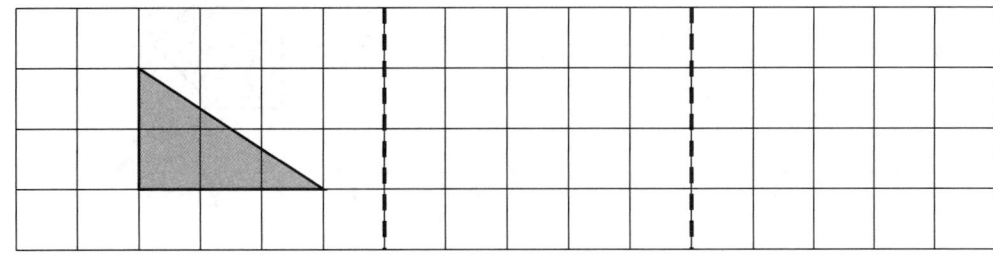

m_1 m_2

c

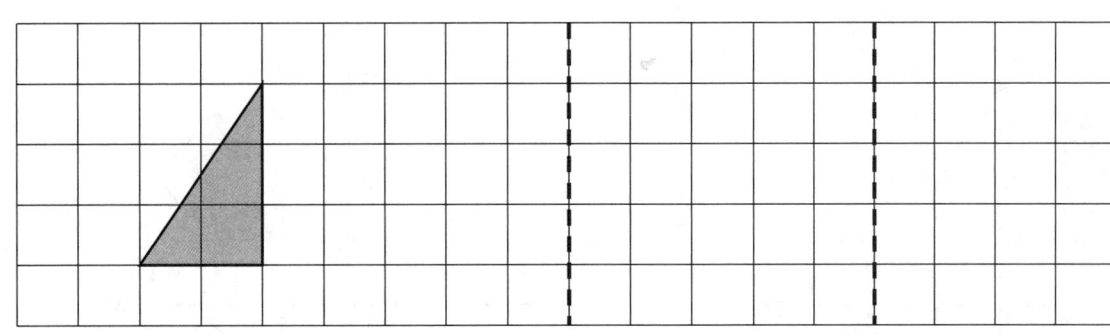

m_1 m_2

2 a

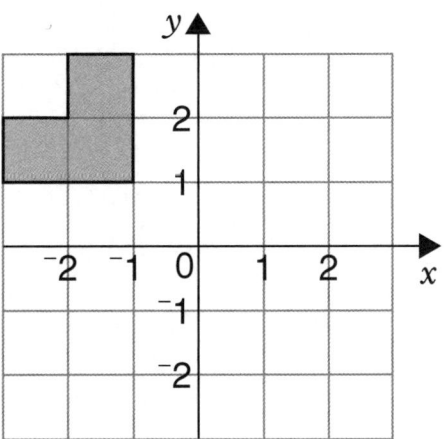

b

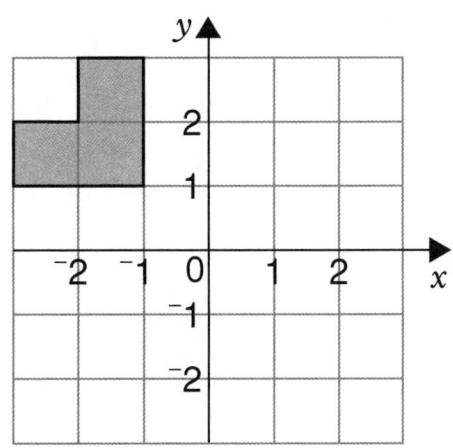

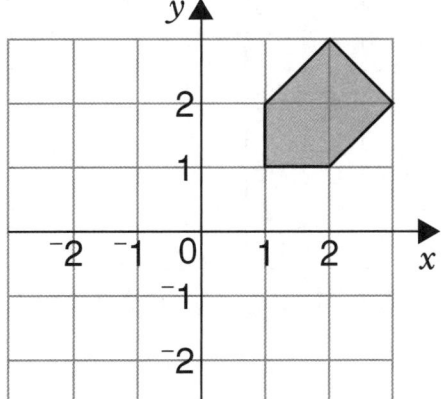

Exercise 2 (pages 314, 315 and 316) continued

3 a

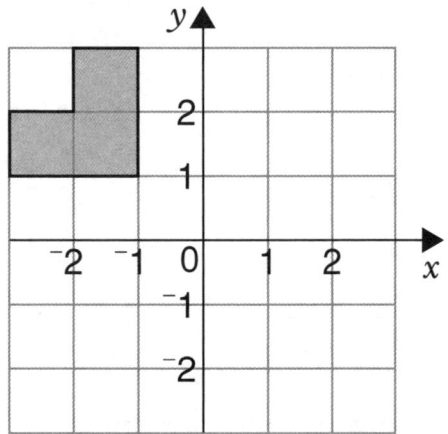

b

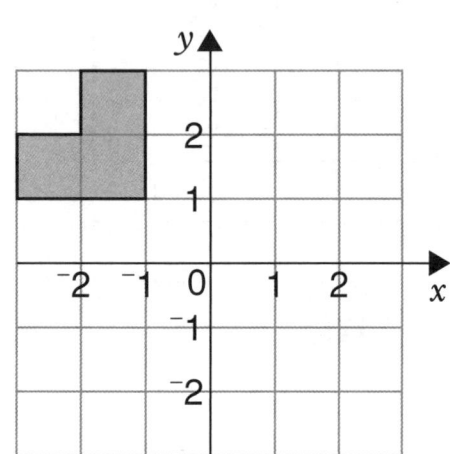

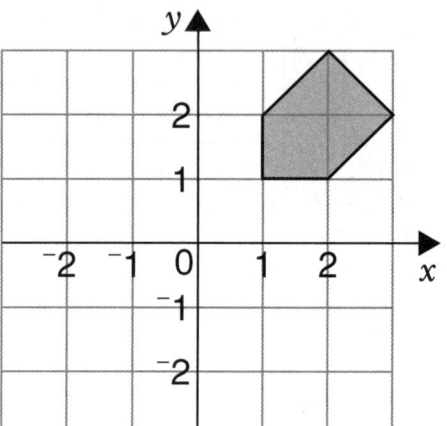

4

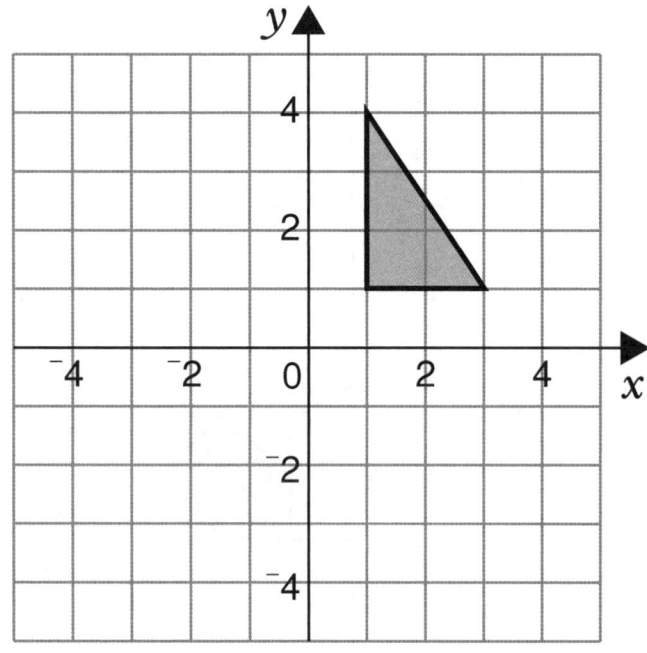

Exercise 2 (pages 314, 315 and 316) continued

Review 1

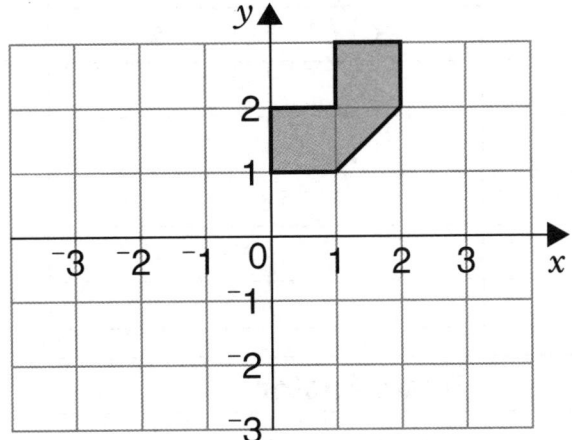

Review 2

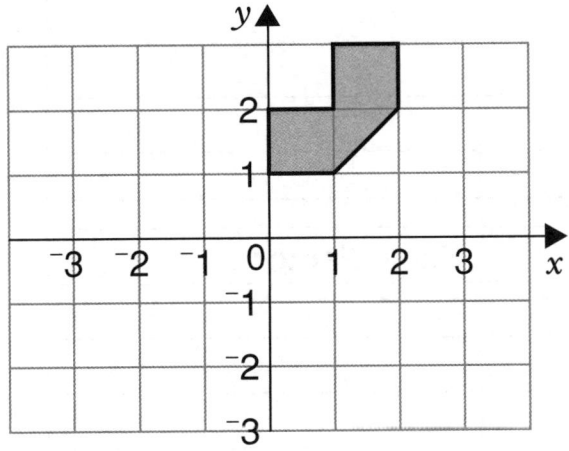

Practical (page 317)

A

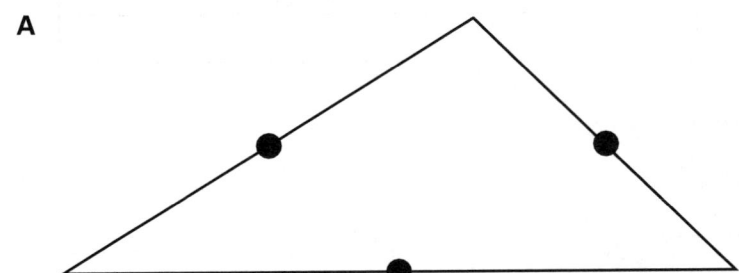

Practical (page 322)

A

Length (cm)	Length on original	Scale Factor		
		2	**3**	**4**
distance between eyes	2	4		
height of alien	6	12		
width of head	4			
length of foot	1			
length of leg	1·1			

Length (cm)	Length on original	Scale Factor		
		2	**3**	**4**
distance between eyes	2	4		
height of alien	6	12		
width of head	4			
length of foot	1			
length of leg	1·1			

Length (cm)	Length on original	Scale Factor		
		2	**3**	**4**
distance between eyes	2	4		
height of alien	6	12		
width of head	4			
length of foot	1			
length of leg	1·1			

B

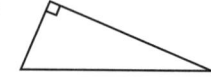  

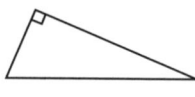  

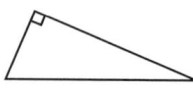 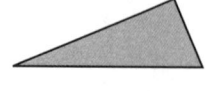 

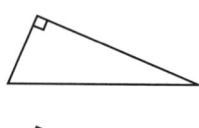 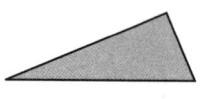 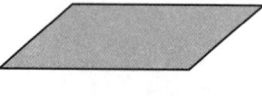

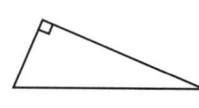 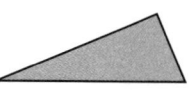 

Practical (page 323) continued

B continued

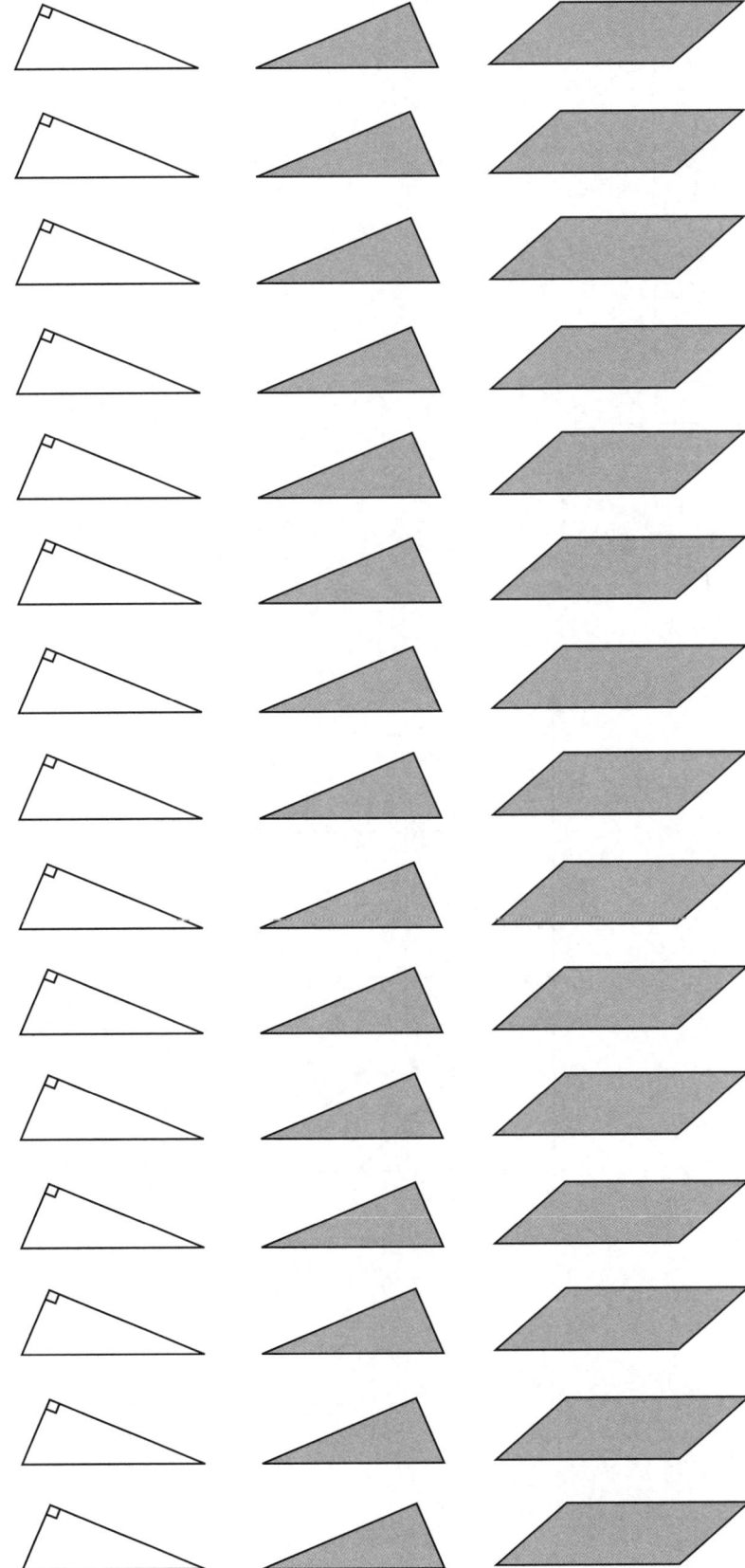

Exercise 5 (page 324)

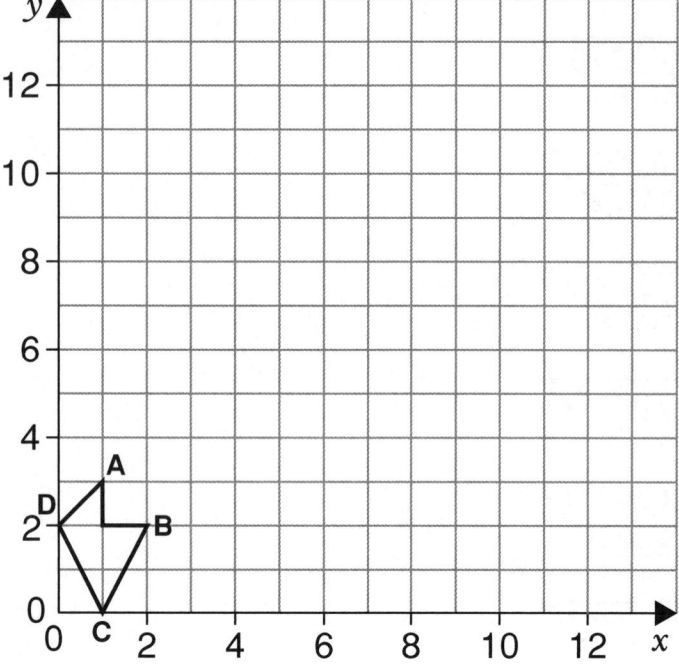

3

Test Yourself (page 333)

6 a

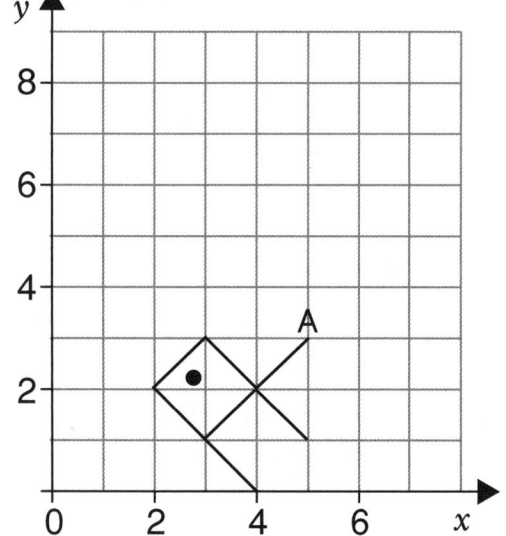

b

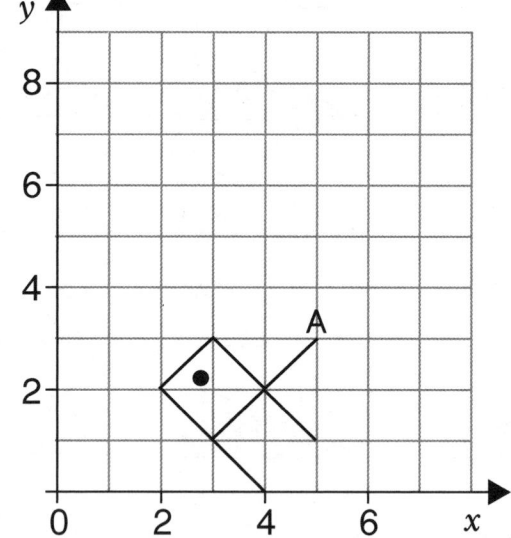

14 Measures, Perimeter, Area and Volume

Exercise 2 (page 338)

Review 1

E													E
0·42	**4000**	5·832	4	400	4200	583 200	5832	4200	4200	420	400	0·04	**4000**

				E				
5832	0·562	0·04	5832	0·562	**4000**	5·62	5·832	4

Write the letter beside each measurement above its answer in the box.

E 4 ℓ in cm^3 = **4000** **R** 4000 ℓ in m^3 **S** 4·2 ℓ in cm^3 **N** 562 ℓ in m^3

O 0·42 m^3 in ℓ **G** 5·62 cm^3 in mℓ **A** 5832 cm^3 in ℓ **I** 5832 mℓ in cm^3

V 40 ℓ in m^3 **D** 583·2 m^3 in ℓ **L** 0·4 ℓ in cm^3 **P** 420 cm^3 in ℓ

Exercise 6 (pages 347 and 348)

Aircraft	VB502	BA172		CO28	NZ2		*E1452		*C04
Bearing	080°		320°			060°		*300°	

5

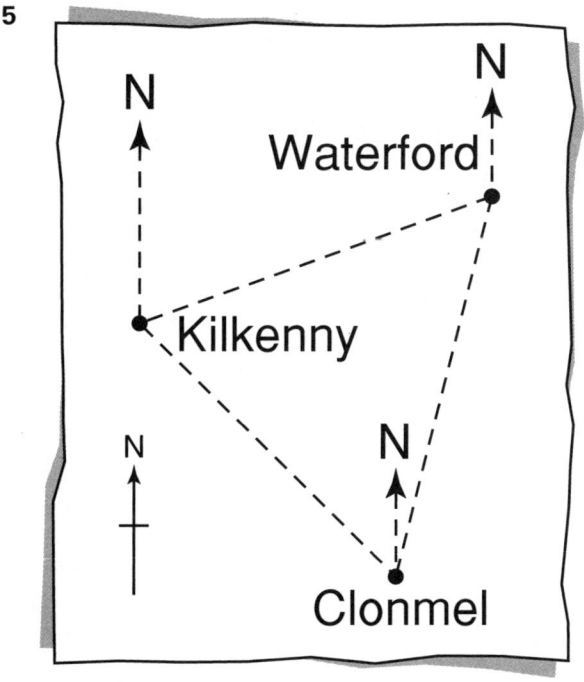

6

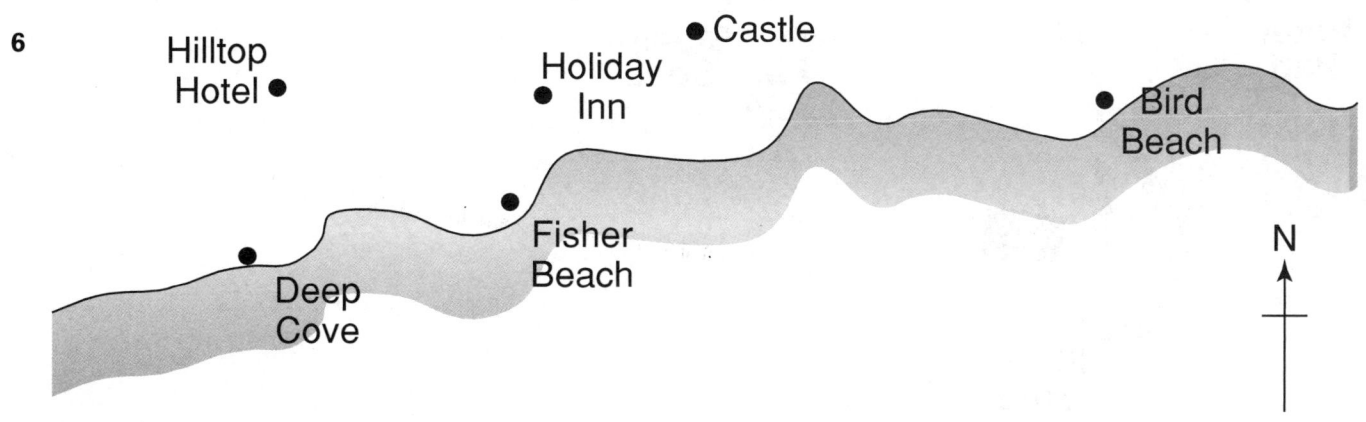

Exercise 6 (pages 347 and 348) continued

Review 3

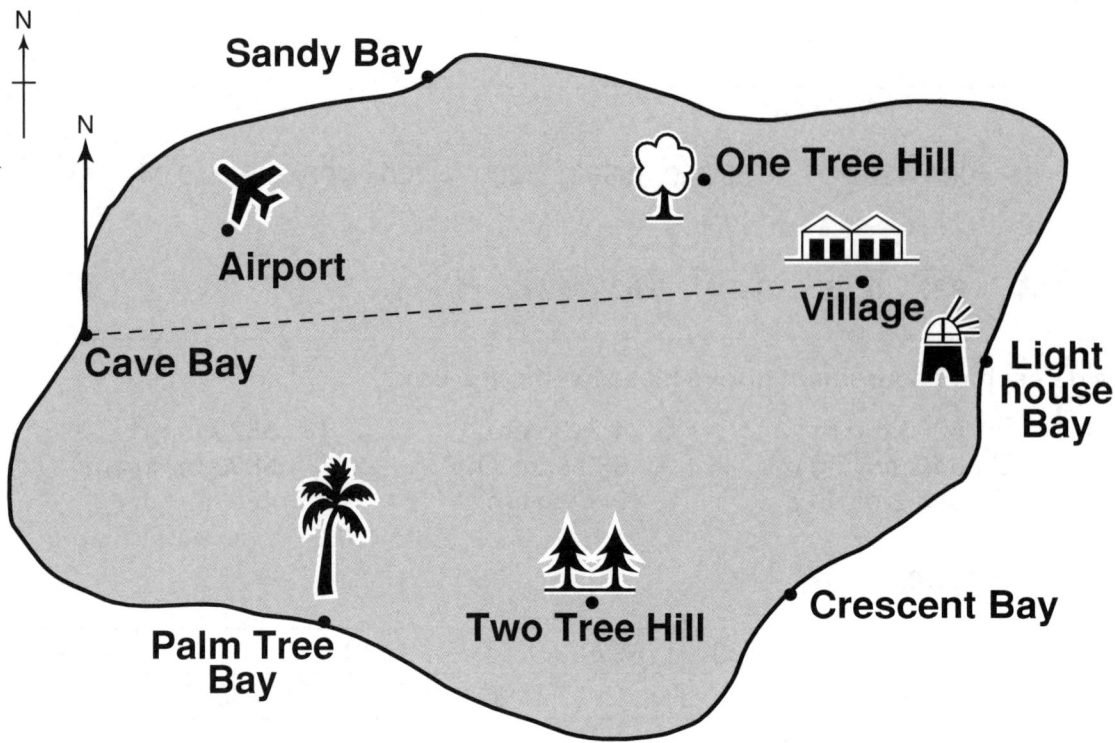

Test Yourself (page 361)

10

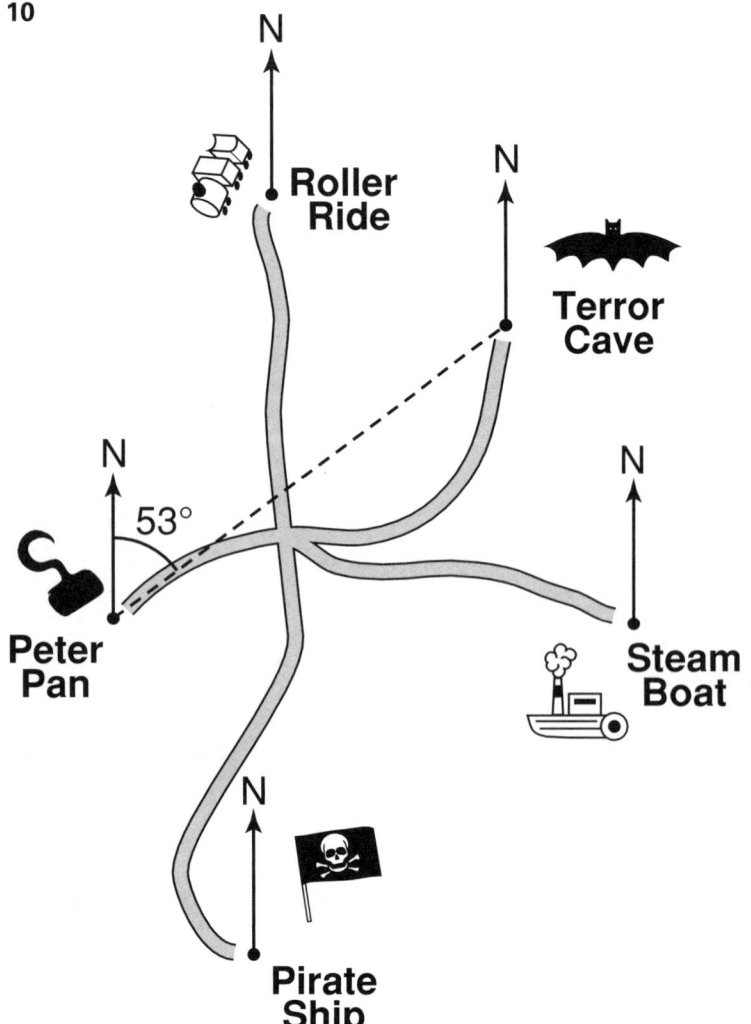

Handling Data Support

Practice Questions (page 366)

1 a

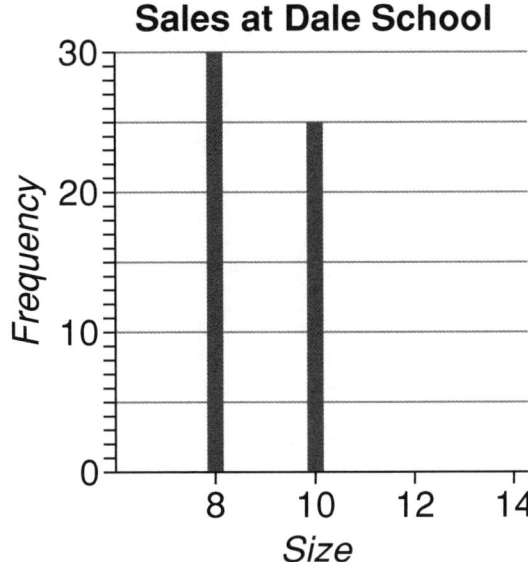

15 Data Collection

Exercise 2 (pages 373 and 374)

1

Mass of apple (g)	Tally	Frequency
$150 < m \leqslant 160$		
$160 < m \leqslant 170$		
$170 < m \leqslant 180$		
$180 < m \leqslant 190$		
$190 < m \leqslant 200$		
$200 < m \leqslant 210$		

2 a

Length of throw (x m)	Tally	Frequency
$0 \leqslant x < 4$		
$4 \leqslant x < 8$		
$8 \leqslant x < 12$		
$12 \leqslant x < 16$		
$16+$		

Exercise 2 (pages 373 and 374) continued

Review

Amount of ice cream (x mℓ)	Tally	Frequency
$1800 < x \leq 1900$		
$1900 < x \leq 2000$		
$2000 < x \leq 2100$		
$2100 < x \leq 2200$		

Test Yourself (page 380)

2 a

Time (t seconds)	Tally	Frequency
$20 < t \leq 25$		
$25 < t \leq 30$		
$30 < t \leq 35$		
$35 < t \leq 40$		

16 Analysing Data. Drawing and Interpreting Graphs

Exercise 7 (pages 392, 393 and 394)

1

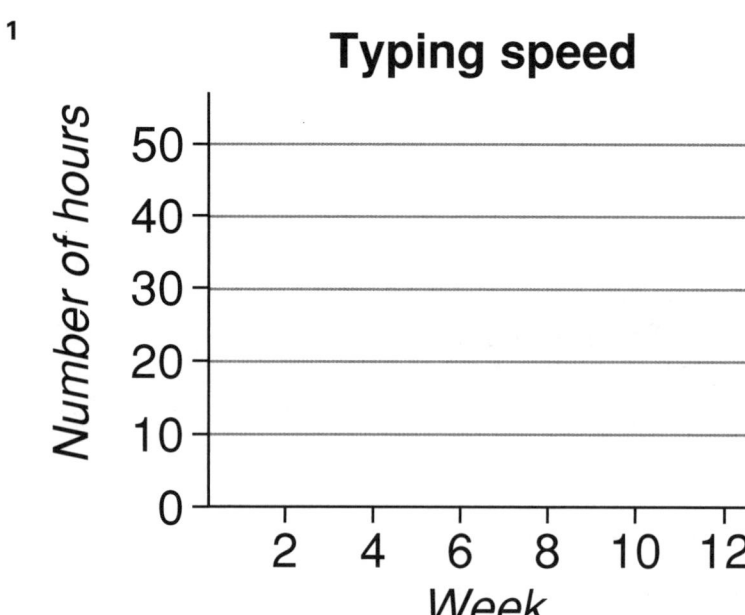

Exercise 7 (pages 392, 393 and 394) continued

2 a

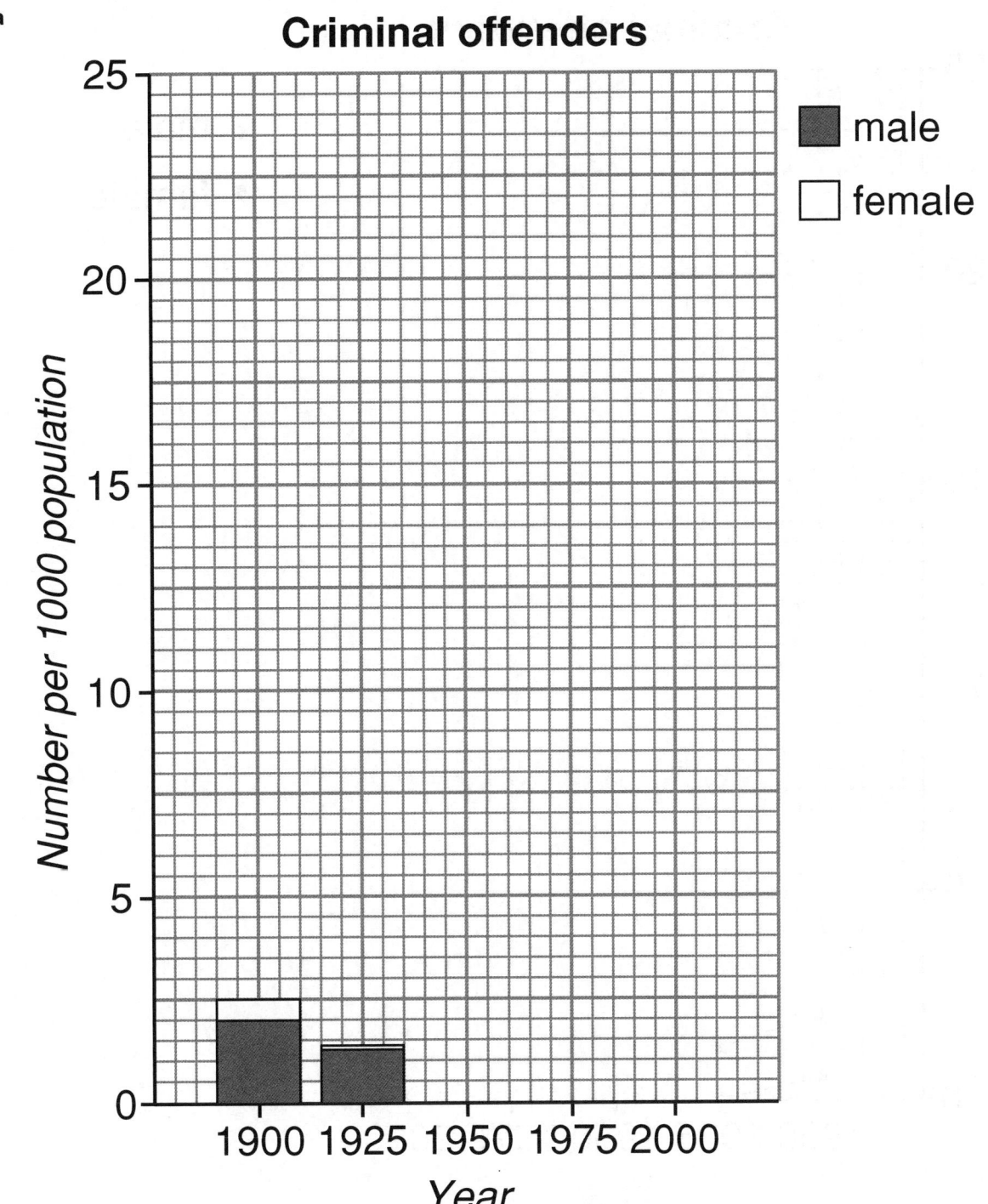

Criminal offenders

□ male
□ female

Number per 1000 population

25

20

15

10

5

0

1900 1925 1950 1975 2000

Year

Exercise 7 (pages 392, 393 and 394) continued

2 b

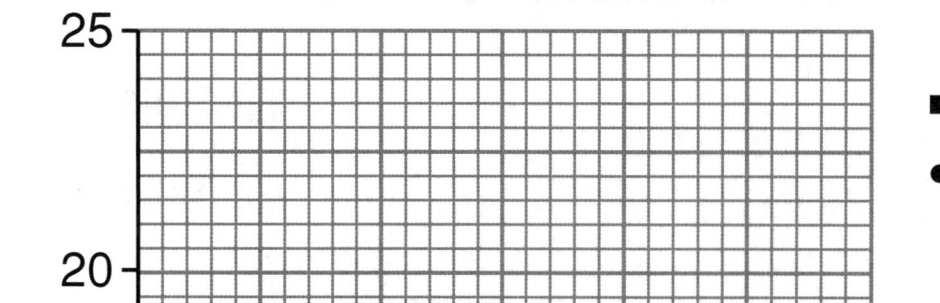

Criminal offenders

Exercise 7 continued (pages 392, 393 and 394)

3 a

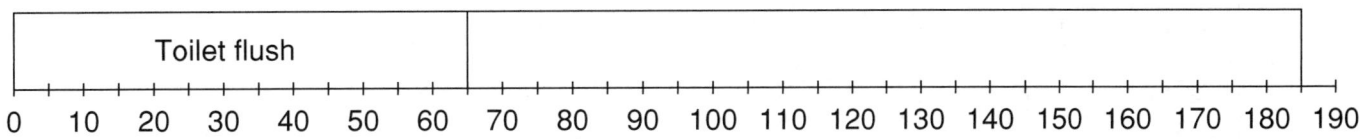

Review

a

Three top attractions

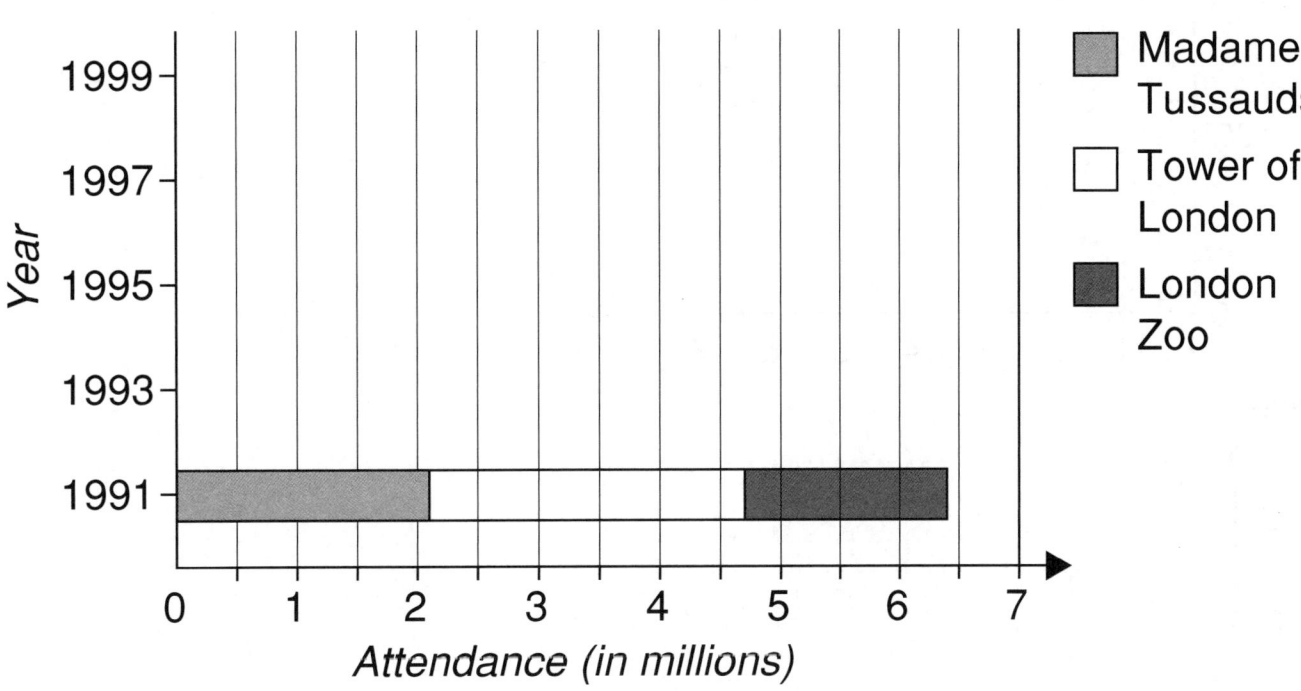

Exercise 8 (pages 395 and 396)

1

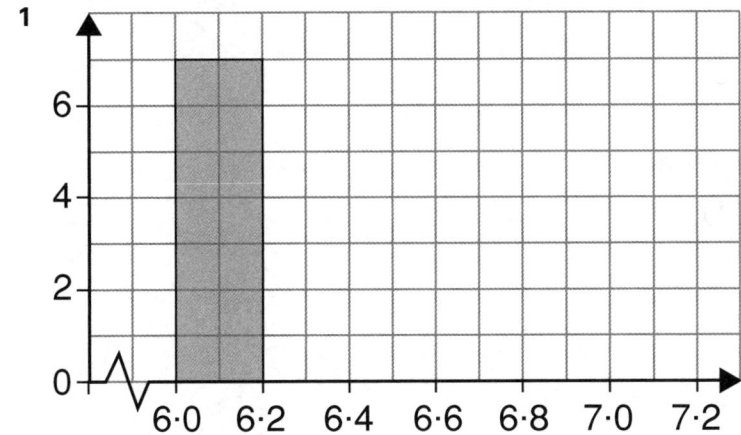

Exercise 8 (pages 395 and 396) continued

2

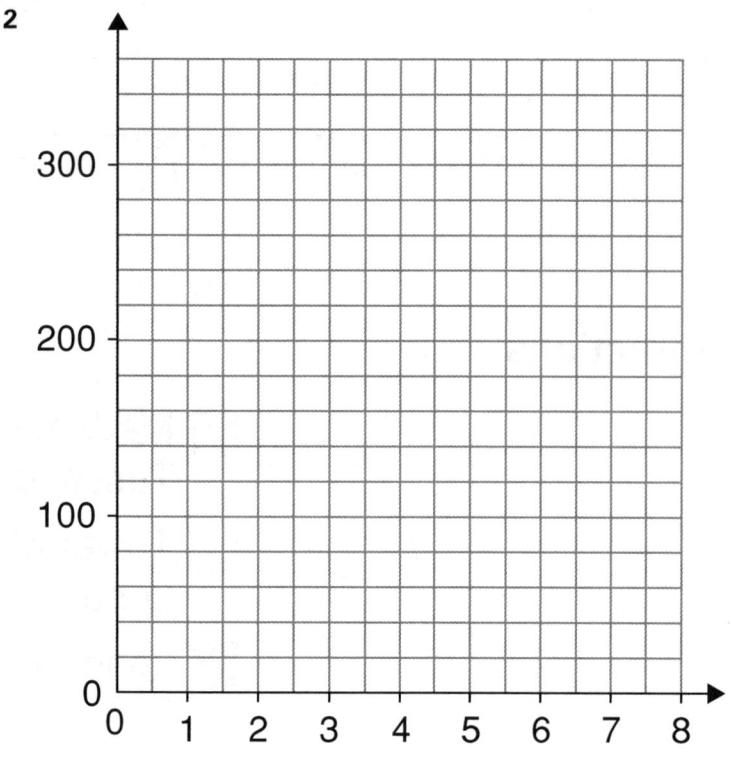

Review
a

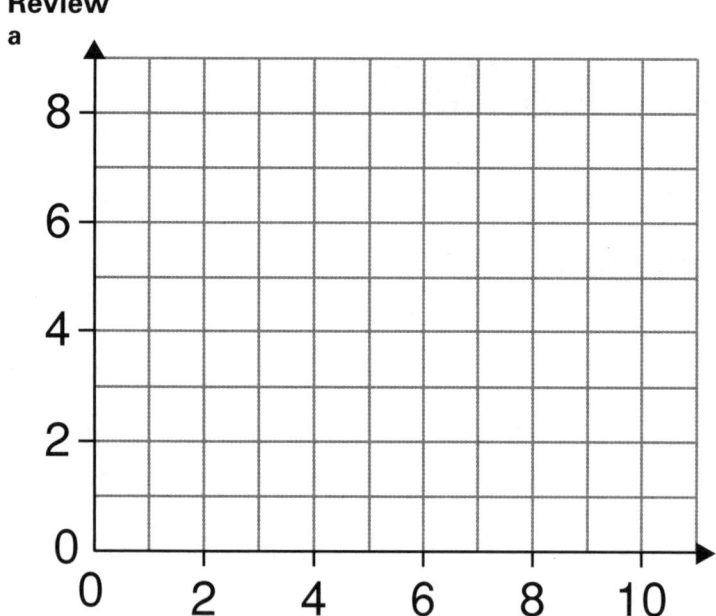

Exercise 9 (page 398)

3 a

Mobile phone colour	Number of mobile phones	Pie chart angle
black	75	
yellow	5	
grey	25	
blue	15	
Total	**120**	**360°**

Exercise 10 (pages 400 and 401)

1 a

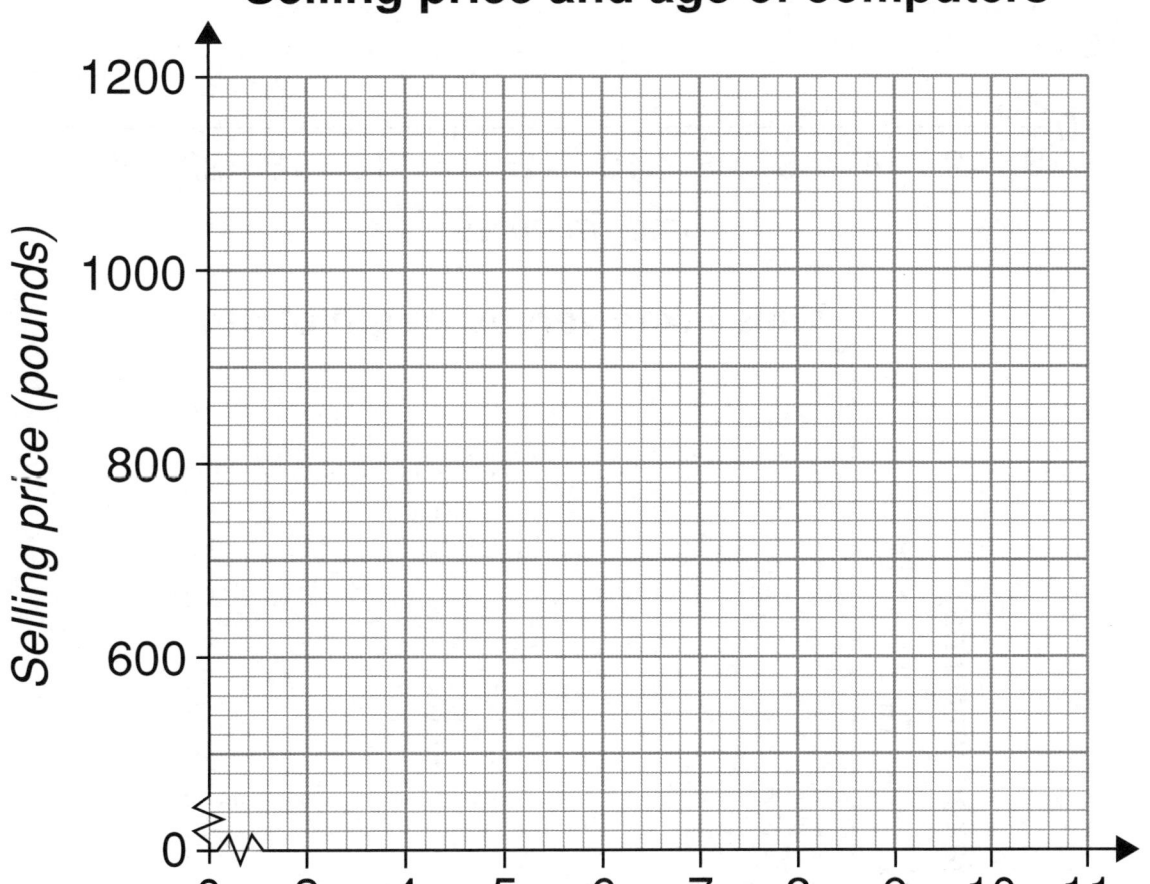

Selling price and age of computers

Exercise 10 (pages 400 and 401) continued

2 a

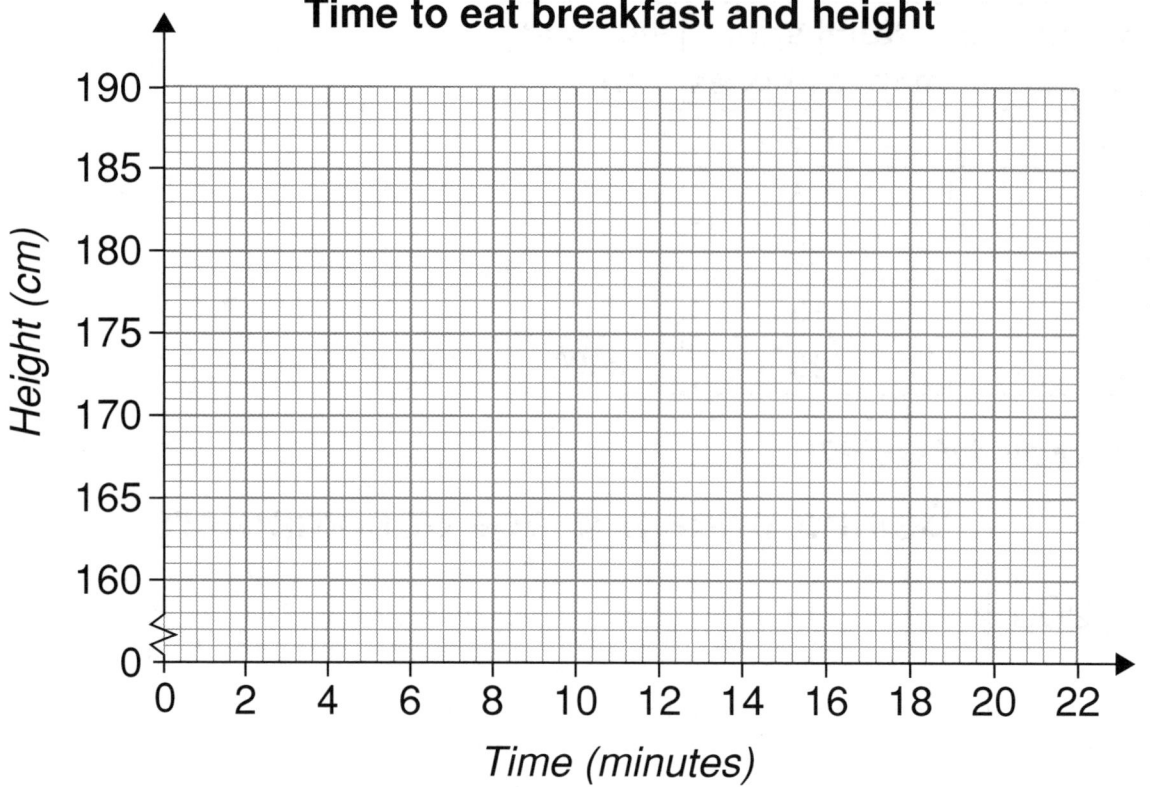

Time to eat breakfast and height

3 a

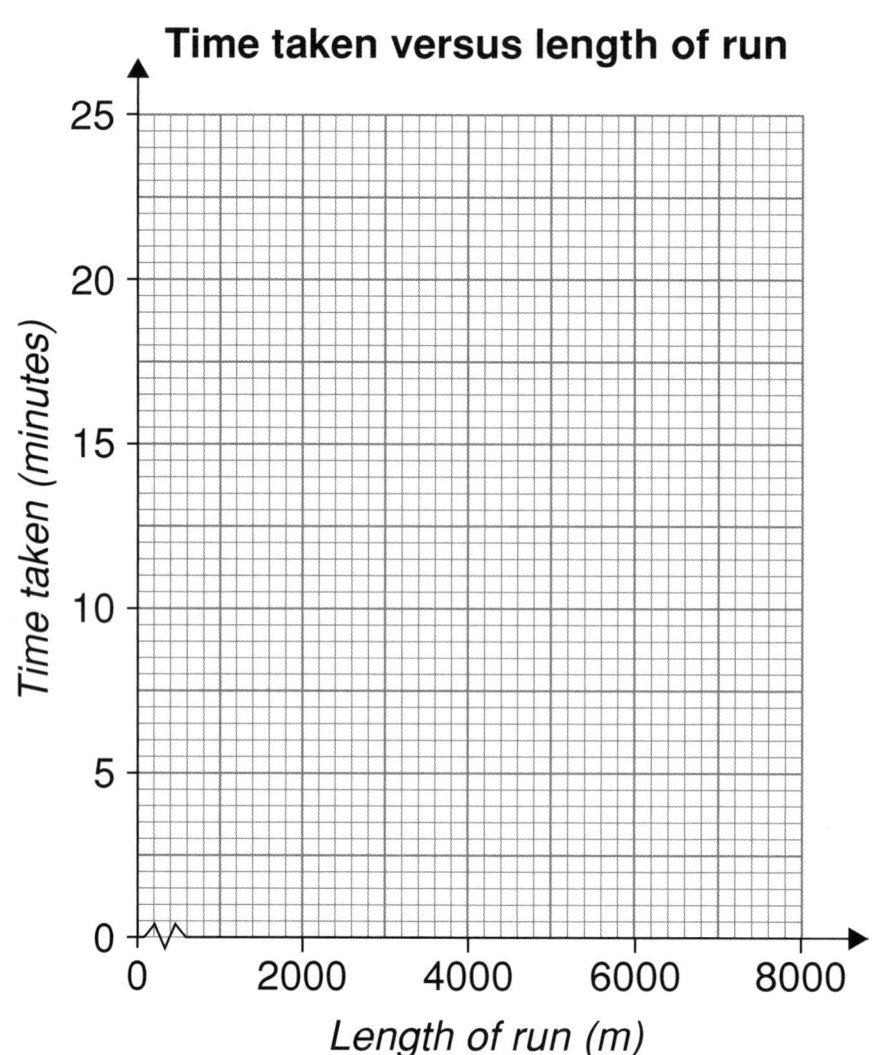

Time taken versus length of run

Exercise 11 (page 406)

9

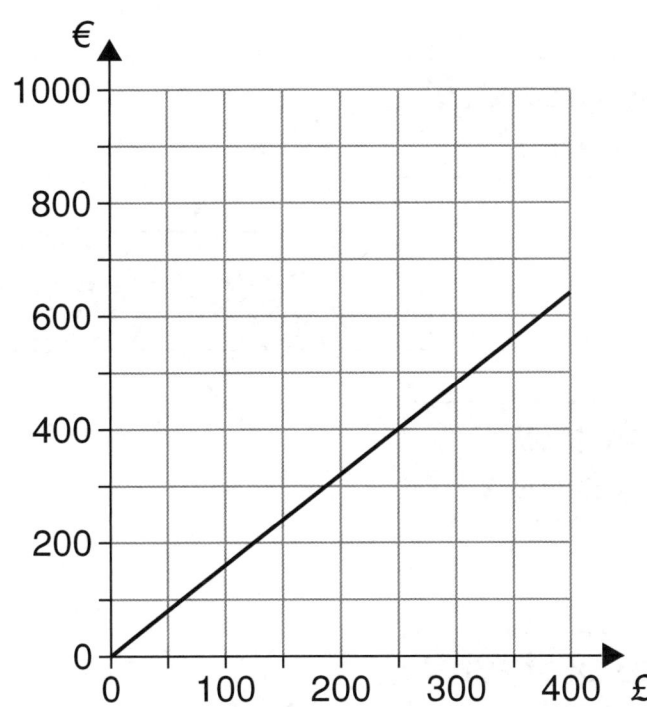

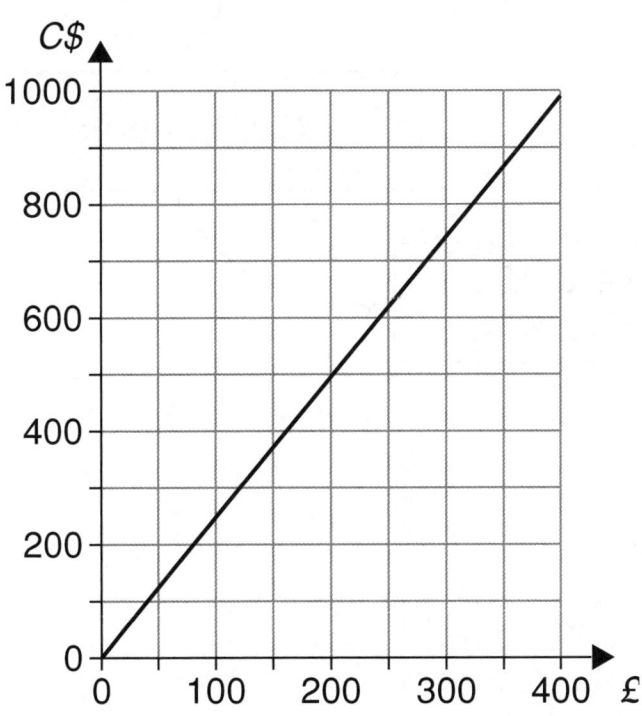

Number of £	Approximate number of €	Approximate number of C$
0		
200		
400		

Practical (page 408)

6

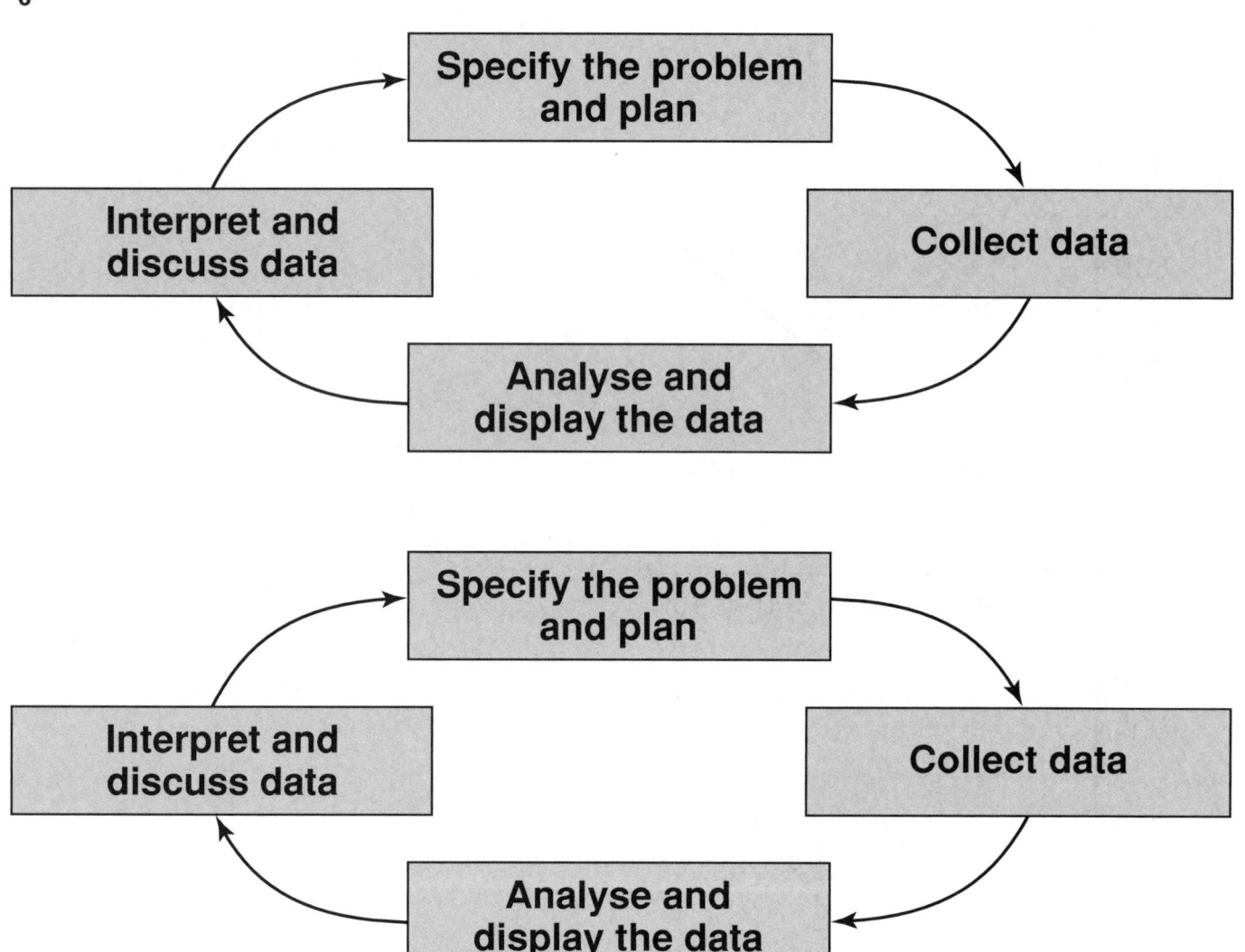

Test Yourself (page 412)

11

Handspans

New National Framework

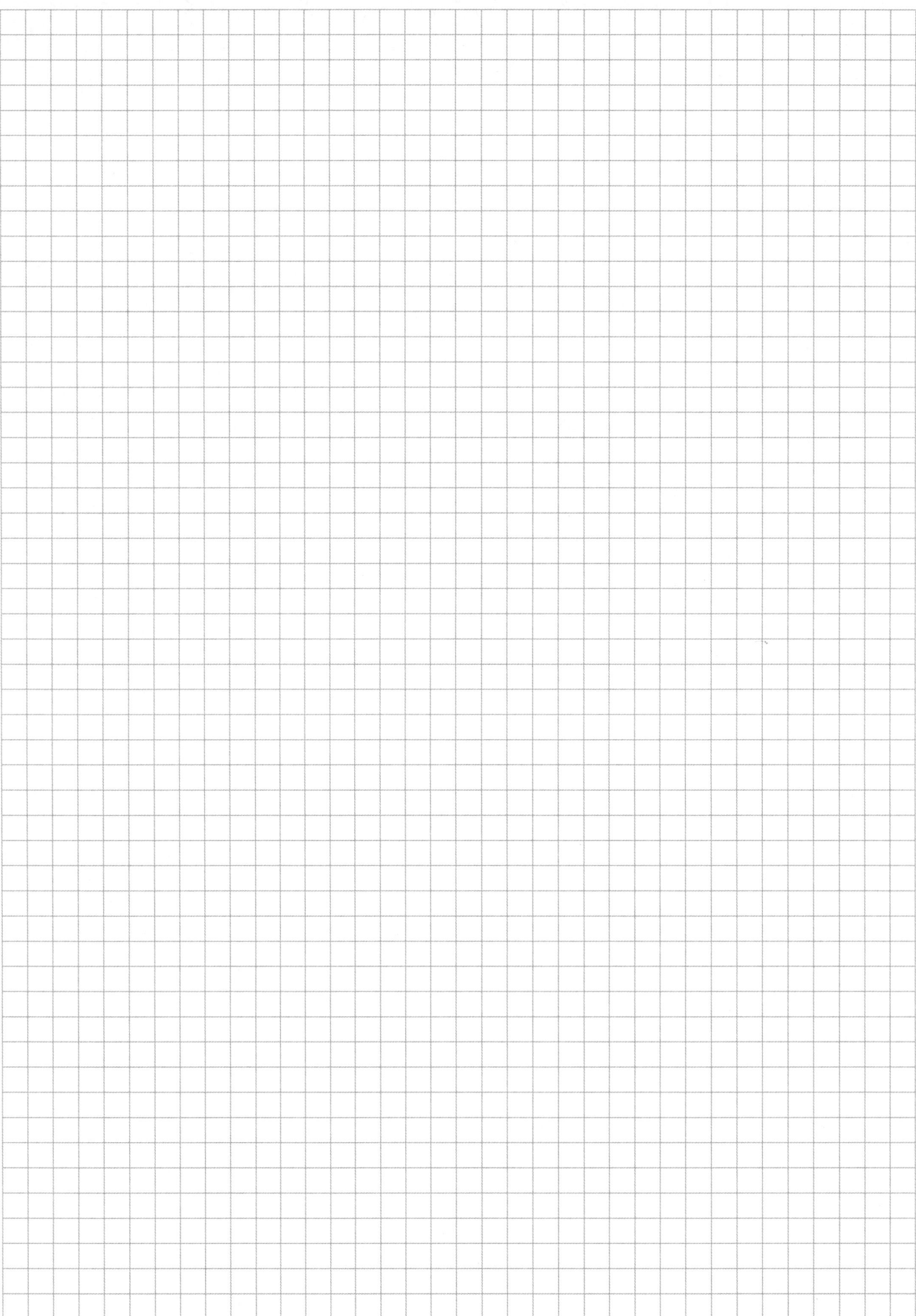

New National | Framework

New National | Framework

6

Answers

Number Support

Page 7 Practice Questions

1 a 6 hundreds or 600   **b** 6 units or 6
   c 6 hundredths or 0·06   **d** 6 tenths or 0·6
   e 6 thousandths or 0·006

2 a 5006·7   **b** 517 000·53   **c** 0·96
   d 207·084   **e** 8·60   **f** 15·69

3 a b

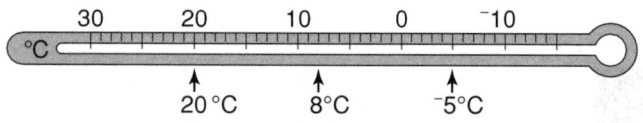

     20 °C   8°C   ⁻5°C

   c 5 °C   **d** 11 °C

4 a 24 and 1, 12 and 2, 8 and 3, 6 and 4
   b 30 and 1, 15 and 2, 10 and 3, 5 and 6
   c 48 and 1, 24 and 2, 16 and 3, 12 and 4, 6 and 8
   d 60 and 1, 30 and 2, 20 and 3, 15 and 4, 12 and 5, 10 and 6

5 a $65 \times 5 = 325$, $390 \div 65 = 6$, $12 \times 65 = 780$, $20 \times 65 = 1300$
   b One possible answer is:
     $16 \times 65 = 10 \times 65 + 6 \times 65$
              $= 650 + 390$
              $= 600 + 50 + 300 + 90$
              $= 900 + 140$
              $= 1040$

6 a $8 + 2 < 7 + 6$ **b** $6 - 3 = 1 + 2$
   c $0 > ⁻3$ **d** $⁻7 < ⁻2$
   e $3 - 2 > ⁻5$ **f** $5 - 5 > 4 - 6$

7 a $(4 + 2) \times 3 = 18$ and $4 + (2 \times 3) = 10$
   b $(2 + 4) \times (6 + 3 + 1) = 60$
   c $(4 + 5 + 1) \times 5 = 50$
   d $4 + (5 + 1) \times 5 = 34$

8 2

9 a Baffin Island 200 000, Great Britain 80 000, Greenland 840 000, Honshu 90 000, Sumatra 100 000
   b Greenland, Baffin Island, Sumatra, Honshu, Great Britain

10 a D   **b** C

11 a 7531
   b Because all of the digits are odd.
   c

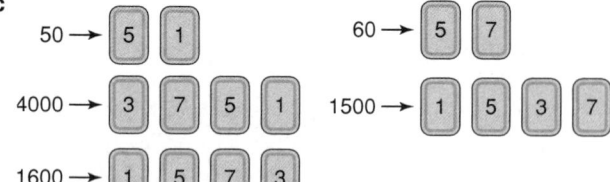

12 a ⁻9 °C, ⁻7 °C, ⁻3 °C, 0 °C, 1 °C, 4 °C,
   b ⁻5·6, ⁻5·2, ⁻5, ⁻2·5, 3·4, 4·7

13 a $1\frac{1}{2}$, 123, 54, 108
   b Possible answers are $9 \times 3 = 27$, $54 \div 2 = 27$ or $81 \div 3 = 27$ or $108 \div 4 = 27$

14 a 690   **b** 1200   **c** 39   **d** 83
   e 12·3   **f** 1190   **g** 3·6   **h** 0·058
   i 53 600   **j** 0·472   **k** 6   **l** 0·011
   m 6·32   **n** 49 630

15 a 63   **b** 44   **c** 5·9   **d** 0·4   **e** 0·16

16 a 270   **b** 226   **c** 1411   **d** 391   **e** 1079
   f 1884   **g** 349   **h** 7111   **i** 5051   **j** 8749
   k 5853   **l** 4417   **m** 7·155   **n** 3·46   **o** 4·48

17 5·72

18 a 1·83 kg, 1·8 kg, 1·491 kg, 1·38 kg, 1·34 kg, 1·3 kg, 1·287 kg
   b 2·143 kg, 2·13 kg, 2·1 kg, 2·09 kg, 2·081 kg, 2·08 kg, 2·043 kg

19 a 20   **b** 8   **c** 4   **d** 15   **e** 18   **f** 10
   g 200   **h** 56   **i** 8

20 a 1, 2, 4, 5, 10, 20   **b** 1, 2, 4, 8, 16, 32, 64
   c 1, 2, 4   **d** 4   **e** 8

21 Possible answers are:
   a 1800   **b** 600   **c** 1200   **d** 30
   e 8   **f** 11   **g** 100

22 a 1, 3, 6, 10, 15, 21
   b 1, 9, 25, 49, 81, 121

23 430 m

24 A rectangle

25 a $\frac{57}{100}$ **b** $\frac{48}{60}$ or $\frac{4}{5}$

26 a 165 mℓ **b** 5·68 mm

27 7·8 mℓ

28 a 15 °C   **b** 7 °C   **c** 25 °C   **d** 10 °C   **e** 12 °C   **f** 42 °C

29 ⁻45 m

30 a 25   **b** 20   **c** 2   **d** 2   **e** 6
   f 15   **g** 45   **h** 48   **i** 54   **j** 5
   k 121   **l** 55   **m** 128

31 a 36   **b** 8   **c** 50   **d** 90   **e** 102
   f 103   **g** 72   **h** 54   **i** 54   **j** 86
   k 64   **l** 94   **m** 48   **n** 201   **o** 153
   p 263   **q** 586   **r** 448   **s** 440   **t** 170
   u 593   **v** 5509   **w** 3128

32 0 °C

33 A rectangle

34 Possible answers are:
   a $400 + 300 = 700$ or $350 + 300 = 650$
   b $4 \times 6 = 24$ or $4·5 \times 6 = 27$
   c $1000 - 400 = 600$ or $1000 - 350 = 650$

35 a £8·30   **b** £11·70
   c Each burger is less than £2 so six will be less than £12.

36 a 1·4   **b** 1·6   **c** 1·4   **d** 0·8   **e** 0·8
   f 1·3   **g** 5·1   **h** 6·6   **i** 22·5   **j** 11·6

37 a 0·2 added   **b** 0·8 subtracted
   c 0·04 subtracted   **d** 0·07 subtracted
   e 0·5 added

38 a 51·3   **b** 88   **c** 9·73   **d** 34·08
   e 63·32   **f** 137·76   **g** 1220·8

39 a $\frac{1}{2}$ **b** $\frac{1}{3}$ **c** $\frac{3}{4}$ **d** $\frac{4}{5}$ **e** $\frac{2}{3}$
   f $\frac{4}{5}$ **g** $\frac{3}{4}$ **h** $\frac{5}{6}$ **i** $\frac{1}{9}$ **j** $\frac{3}{8}$
   k $\frac{3}{4}$ **l** $\frac{3}{5}$ **m** $\frac{3}{5}$ **n** $\frac{2}{3}$

40 a 60   **b** 360   **c** 1800   **d** 902   **e** 432
   f 3·6   **g** 18·4   **h** 6·3   **i** 215

41 a <   **b** =   **c** >   **d** <

42 a 1872   **b** 3024

43 a 130   **b** 975   **c** 0·8   **d** 0·7   **e** 0·6

44 a 13   **b** 22

45 £3·30

46 a 0·65   **b** 8·75   **c** 5·45   **d** 4·85

47 a 2   **b** 3   **c** 1   **d** 10   **e** 8   **f** 12

48 a 0·5   **b** 0·25   **c** 0·75   **d** 0·1   **e** 0·01
   f 2·5   **g** 4·75   **h** 0·3   **i** 0·42   **j** 0·39
   k 0·06   **l** 0·04   **m** 0·6   **n** 0·8   **o** 0·75
   p 0·35   **q** 1·85

49 a $\frac{3}{10}$ **b** $\frac{7}{10}$ **c** $\frac{9}{10}$ **d** $\frac{4}{5}$ **e** $\frac{1}{2}$
   f $\frac{3}{5}$ **g** $\frac{41}{100}$ **h** $\frac{77}{100}$ **i** $\frac{83}{100}$ **j** $\frac{1}{4}$
   k $\frac{7}{50}$ **l** $\frac{9}{50}$ **m** $\frac{16}{25}$ **n** $\frac{37}{50}$ **o** $\frac{9}{20}$
   p $\frac{18}{25}$ **q** $\frac{17}{20}$ **r** $\frac{47}{50}$

50 a 1 : 4   **b** 20%   **c** $\frac{4}{5}$

51 a $\frac{16}{20}$ **b** $\frac{5}{6}$

52 a 9 **b** $^-6 + {}^-5 = {}^-11$
53 5 : 1
54 a $^-7$ **b** $^-1$ **c** $^-2$ **d** $^-4{\cdot}5$ **e** $^-5{\cdot}5$
55 a 4·83 m **b** 4·94 m **c** 5·03 m **d** 4·92 m
56 a 684, 648, 468 **b** 864 **c** 864
57 a £32 **b** 5 m **c** 40 cm **d** 12 ℓ
 e £160 **f** 30 cm **g** 210 g **h** £5·50
 i 18 ℓ **j** 4·8 m **k** 35 kg
58 a $\frac{19}{25}$ **b** 76%
59 a $\frac{24}{48}$ **b** $\frac{36}{48}$ **c** $\frac{32}{48}$ **d** $\frac{40}{48}$ **e** $\frac{28}{48}$
60 a 35% **b** 65%
61 45
62 E
63 a 11 **b** 12 **c** 9 **d** 6 **e** 5
64 a 12 **b** 77 **c** 45 **d** 2·4 **e** 7·5
 f 572 **g** 1064
65 a 78 **b** 59 **c** 39 **d** 54
 e 245 **f** 569 **g** 251
 h 208 R 3 or $208\frac{3}{4}$ **i** 61 R 5 or $61\frac{5}{6}$
 j 566 R 5 or $566\frac{5}{9}$ **k** 532 R 1 or $532\frac{1}{7}$
66 a 2889 **b** 6048 **c** 3136 **d** 5274 **e** 624
 f 1148 **g** 7848 **h** 21 096 **i** 5832
67 a 974·3 **b** 973·4 **c** 3·749
68 a 25% **b** 70% **c** 60% **d** 85%
 e 400% **f** 175% **g** 73% **h** 86%
 i 4% **j** 19% **k** 114%
69 a 60 **b** 120 **c** 70 **d** 96 **e** 210
 f 153 **g** 384 **h** 212 **i** 434 **j** 920
 k 3420 **l** 3600 **m** 143 **n** 252 **o** 600
70 a 126 500 **b** 127 499
71 42 hours
72 a 300 g **b** 150 g **c** 600 g **d** 750 g
73 a 6 **b** 12 **c** 12 **d** 20 **e** $\frac{5}{3}$ or $1\frac{2}{3}$
 f $\frac{16}{3}$ or $5\frac{1}{3}$ **g** $7\frac{1}{2}$ **h** 62 **i** 36
74 Possible answers are 3 and 7, 7 and 11, 13 and 17,
 19 and 23, 37 and 41, 43 and 47.

75 $3 + {}^-5 = {}^-2$, $3 + {}^-6 = {}^-3$, $3 + {}^-7 = {}^-4$
 a $^-3$ **b** $^-5$ **c** $^-6$
76 Possible answers are:

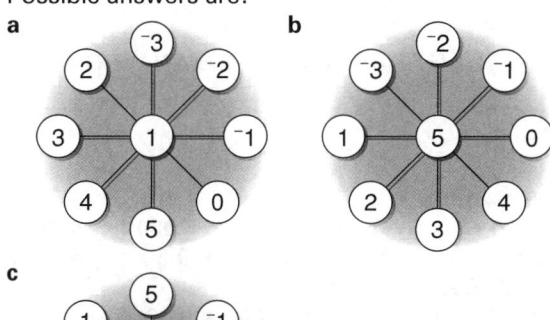

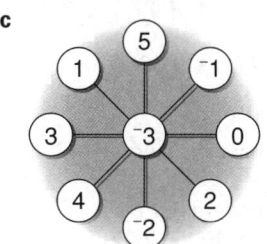

77 a Yes **b** No **c** Yes **d** No **e** Yes **f** No
78 a 169 **b** 14 **c** 1·44 **d** 1·4
79 5
80 a 4410 **b** 8·84 **c** 3 **d** 5·0
81 Possible answers are:
 a $30\ 672 \div 72 = 426$
 b $16{\cdot}3 - 7{\cdot}94 = 8{\cdot}36$
 c $(2{\cdot}645751311)^2 = 7$
82 A possible answer is:
 $400 \times 72 + 20 \times 72 + 6 \times 72 = 30\ 672$
83 a £17·55 **b** 27·88 ℓ **c** £1955·48
 d £4·73 to the nearest penny
84 a 8 days 5 hours
 b 9 minutes 32 seconds
85 Wonder Bar, 32%

Chapter 1 Place Value, Ordering and Rounding

Page 16 Exercise 1
1 a 100 **b** 10 000 **c** 10 **d** 1000 **e** 100 000 000
 f 1 000 000 000 **g** 100 000 000 000
2 a 10^3 **b** 10^6 **c** 10^2 **d** 10^5 **e** 10^7 **f** 10^9
 g 10^1 **h** 10^{10}
3 a Thirty-seven million
 b Five million, six hundred and twenty-five thousand
 c One million, forty-eight thousand, five hundred and
  seventy-six
 d Thirty million, two hundred and seventy-one
  thousand
 e One hundred and fourteen million, one hundred
  thousand
4 a Four point three times ten to the power of three
 b Two point seven times ten to the power of five
 c Five point two six two times ten to the power of
  three
 d One point zero seven times ten to the power of six
 e One point four two seven times ten to the power of
  nine
5 a 2510 **b** 4 500 003 **c** 1499
 d 5 250 007 **e** 1249 **f** 1 999 992
 g 2 249 996

Review 1
a 10^4 **b** 10^8 **c** 10^9

Review 2
a One point six times ten to the power of two
b Two point seven one times ten to the power of three
c Four point four nine seven times ten to the power of
 nine

Review 3
One hundred and eighty-six million three hundred and
twenty-four thousand

Review 4
a 3256 **b** 2 499 995

Page 17 Puzzle
426·315 or 456·312 or 476·310 or 406·317

Page 18 Exercise 2
1 a 4·68 **b** 8·425 **c** 27·19 **d** 8·22 **e** 8·1
 f 5·39 **g** 8·633 **h** 5·751 **i** 3·681 **j** 4·8
 k 4·339 **l** 9 **m** 4·9 **n** 15·99 **o** 3·999
2 a 8·82 **b** 3·897 **c** 4·13 **d** 5·371 **e** 7·657
 f 8·722 **g** 0·049 **h** 8·421 **i** 5·228 **j** 6·29
 k 3·994 **l** 15·797 **m** 54·54 **n** 7·304 **o** 5·96
 p 7·3 **q** 15·893 **r** 14·991 **s** 11·992
3 1·559 ℓ
4 The shading makes the letter T.
5 a Add 0·02 **b** Add 0·2 **c** Subtract 0·4
 d Subtract 0·5 **e** Add 0·007 **f** Subtract 0·008
 g Add 0·02 **h** Add 0·07 **i** Subtract 0·008
 j Subtract 0·009

6 0·05 g too little
7 1·42 km
8 0·02 cm

Review 1
AN ANT CAN LIFT FIFTY TIMES ITS OWN WEIGHT

Review 2
16·66 km

Page 19 Exercise 3

1 **a** 1200 **b** 3000 **c** 8000 **d** 30
e 200 **f** 180 000 **g** 0·3 **h** 0·02
i 1 400 000 **j** 0·006
2 **a** 82 **b** 159 **c** 440 **d** 0·42 **e** 0·0014
3 **a** 600 **b** 1400 **c** 3800
4 30
5 There are many possible answers. Two are 4000 × 20 and 80 × 1000.
6 There are many possible answers. One is
a 8000 ÷ 400 **b** 20 ÷ 100 **c** 400 ÷ 20 000

Review 1
a 1600 **b** 210 000 **c** 20 **d** 400 **e** 0·31

Review 2
3200

Review 3
600

Review 4
There are many possible answers. One is 800 × 20.

Review 5
There are many possible answers. One is 40 ÷ 500.

Page 20 Investigation: × and ÷ by 0·1 and 0·01

Number	× 10	÷ 0·1	× 100	÷ 0·01	÷ 10	× 0·1	÷ 100	× 0·01
4·7	47	47	470	470	0·47	0·47	0·047	0·047
0·8	8	8	80	80	0·08	0·08	0·008	0·008
0·92	9·2	9·2	92	92	0·092	0·092	0·0092	0·0092
16	160	160	1600	1600	1·6	1·6	0·16	0·16
12·3	123	123	1230	1230	1·23	1·23	0·123	0·123
572	5720	5720	57200	57200	57·2	57·2	5·72	5·72
46·23	462·3	462·3	4623	4623	4·623	4·623	0·4623	0·4623
5271	52710	52710	527100	527100	527·1	527·1	52·71	52·71

1 Multiplying by 10 and dividing by 0·1 give the same answer.
Multiplying by 100 and dividing by 0·01 give the same answer.
Dividing by 10 and multiplying by 0·1 give the same answer.
Dividing by 100 and multiplying by 0·01 give the same answer.
2 All positive numbers when multiplied by 10 get bigger.
All positive numbers when multiplied by 0·1 or 0·01 get smaller.
All positive numbers when divided by 0·1 or 0·01 get bigger.
Multiplying by 0·1 or 0·01 makes a positive number smaller.
Dividing by 0·1 or 0·01 makes a positive number bigger.

Page 21 Discussion

Dot 1 One possible answer is: $5·6 × 0·01 = 5·6 × \frac{1}{100}$
$= 5·6 ÷ 100$
$= 0·056$
So 5·6 × 0·01 is the same as 5·6 ÷ 100.
Dot 2 One possible answer is: $18 ÷ 0·01 = 18 ÷ \frac{1}{100}$.
This is read as 'how many hundredths are there in 18?'
In 1 there are 100 hundredths so in 18 there are 18 × 100 hundredths.
So 18 ÷ 0·01 = 18 × 100.
Dot 3 Yes Robert is right.
$0·6 × 0·7 = 6 × 0·1 × 7 × 0·1$
$= 6 × 7 × 0·1 × 0·1$
$= 42 ÷ 10 ÷ 10$
$= 4·2 ÷ 10$
$= \mathbf{0·42}$
Dot 4 0·2 × 0·3 = 0·06, 0·5 × 0·4 = 0·2, 0·48 ÷ 0·6 = 0·8, 0·28 ÷ 0·7 = 0·4
Dot 5 When we multiply a positive number by a number less than 1, the answer is less than the positive number.
When we divide a positive number by a number less than 1, the answer is more than the positive number.

Page 23 Exercise 4

1 **a** C **b** B **c** D **d** A **e** D **f** B **g** C
2 **a** 0·8 **b** 3·2 **c** 50 **d** 1·7 **e** 8·32
f 8·6 **g** 503 **h** 0·016 **i** 0·002 **j** 4·78
k 160 **l** 0·048 **m** 1·4 **n** 5·64 **o** 3·84
p 8·5 **q** 76·53 **r** 8200 **s** ⁻0·586 **t** ⁻0·0018
u ⁻0·14 **v** ⁻6·2 **w** ⁻8·4
3 **a** 0·1 **b** 0·1 **c** 0·01 **d** 0·01 **e** 0·01
f 0·1 **g** 0·01 **h** 4 **i** 120 **j** 32
k 500 **l** ⁻6 **m** ⁻520 **n** ⁻7·9
4 **a** 2 **b** 1·2 **c** 2·4 **d** 2·8
e 1·8 **f** 3·5 **g** 2·7 **h** 3
i ⁻3·2 **j** ⁻4·8 **k** ⁻6·3
5 0·18 ℓ or 180 mℓ
6 **a** 3 × 0·002 = 0·006
3 × 0·0002 = 0·0006
3 × 0·00002 = 0·00006
b 0·005 × 4 = 0·02
0·0005 × 4 = 0·002
0·00005 × 4 = 0·0002
c 7 × 0·003 = 0·021
7 × 0·0003 = 0·0021
7 × 0·00003 = 0·00021
d 0·5 × 0·007 = 0·0035
0·5 × 0·0007 = 0·00035
0·5 × 0·00007 = 0·000035
e 0·4 × 0·009 = 0·0036
0·4 × 0·0009 = 0·00036
0·4 × 0·00009 = 0·000036
f 0·0008 × 0·7 = 0·00056
0·00008 × 0·7 = 0·000056
0·000008 × 0·7 = 0·0000056
g 0·006 × 1·2 = 0·0072
0·0006 × 1·2 = 0·00072
0·00006 × 1·2 = 0·000072
7 **a** 0·036 **b** 0·04 **c** 0·0042 **d** 0·0012
e 0·06 **f** 0·0132
8 120 m

9 a A rectangle with sides 0·6 and 0·4 has 24 squares out of 100. 24 out of 100 is 0·24

b A rectangle with sides 0·5 and 0·3 has 15 squares out of 100. 15 out of 100 is 0·15

c A rectangle of 18 squares out of 100 has sides 0·2 and 0·9

Review 1
a D **b** B **c** C **d** A

Review 2
THE TOWER OF LONDON WAS ONCE USED AS A ZOO

Review 3
0·24 kg or 240 g

Review 4
a 0·15 **b** 0·028 **c** 0·072 **d** 0·066

Page 25 Exercise 5
1 a T   **b** F   **c** F   **d** T   **e** F   **f** T
 g F   **h** T   **i** T   **j** T   **k** F   **l** T
2 a < **b** < **c** > **d** < **e** < **f** < **g** > **h** <
3 a 0·08 kg, 0·081 kg, 0·082 kg, 0·11 kg, 0·128 kg
 b 20·563 m, 20·65 m, 21·057 m, 21·07 m, 24·792 m
4 a 7·234 **b** 8·724 **c** 0·423
5 a < **b** > **c** < **d** > **e** > **f** > **g** > **h** > **i** <
6 a 5·608 kg, 5·68 kg, 5681 g, 5·8 kg, 5860 g
 b 12·3 ℓ, 12·368 ℓ, 12·3862 ℓ, 12 600 mℓ, 123 682 mℓ
7 a T **b** T **c** F **d** F **e** T **f** T **g** T **h** F
8 a 7·25 **b** 3·27 **c** 7·25 **d** 0·085
 e 0·025 **f** 4·015 **g** 3·265 **h** 5·735
 i ⁻1·3 **j** ⁻0·35 **k** ⁻0·035
9 There are many possible answers. Three are: 0·415 and 0·485, 0·1 and 0·8, 0·38 and 0·52.
10 a There are many possible answers. Ten are: 81·43, 81·445, 81·46, 81·47, 81·48, 81·49, 81·5, 81·511, 81·52, 81·533.
 b z could have an infinite number of values.
11 a x is greater than 1·50 m but less than or equal to 1·55 m.
 b red **c** yellow
 d 1·63 – green, 1·54 – red, 1·72 – purple, 1·69 – yellow

Review 1
a F **b** T **c** T **d** T **e** F **f** F **g** F **h** F

Review 2
0·6842 kg, 0·685 kg, 0·6851 kg, 0·6858 kg, 0·69 kg

Review 3
a 7·212 < 7·234 < 7·259
b 6·7941 < 6·8032 < 6·832
c 0·30894 < 0·3096 < 0·30961
d 0·00057 < 0·0045 < 0·00507

Review 4
a 6·3421 kg, 6·324 kg, 6·32 kg, 6·3024 kg, 6·2431 kg
b 0·8927 km, 0·8641 km, 0·864 km, 0·8064 km
c 14 300 mℓ, 14·2794 ℓ, 14 279 mℓ, 14·2789 ℓ, 14·2783 ℓ

Review 5
a 0·75 **b** 0·075 **c** ⁻0·55 **d** ⁻2·05

Review 6
a There are many possible answers. Ten are: 5·87, 5·88, 5·89, 5·9, 5·915, 5·92, 5·93, 5·941, 5·95, 5·97.
b No. There is an infinite number of values q could be because you can keep adding decimal places.

Page 28 Exercise 6
1 a 320   **b** 5900   **c** 6000   **d** 1830
 e 30 000 **f** 900 000 **g** 43 300 **h** 90 000
 i 700 000 **j** 68 400 **k** 59 000 **l** 70 000
 m 100 000 **n** 6 000 000 **o** 6 000 000
2 a Nearest 1000 **b** Nearest 10 000
3 a Nearest 10 000 **b** Nearest 100 000
4 Possible answers are:
 Chinese 800 million, English 450 million,
 Hindi 300 million, Spanish 300 million,
 Russian 250 million.
5 Hunter, because 9 276 000 is closer to nine and a half million than to nine million.
6 a 12 500 **b** 13 499
7 a 175 000 **b** 184 999
8 Possible answers are:
 a £2501, £2540, £2527 **b** £2480, £2451, £2492

Review 1
a 4700   **b** 9000   **c** 40 000   **d** 4690
e 47 900 **f** 20 000 **g** 900 000 **h** 6 000 000

Review 2
Nearest hundred

Review 3
a 24 650 **b** 24 749

Page 31 Exercise 7
1 i a 3   **b** 7   **c** 1   **d** 1   **e** 2
    f 3   **g** 13   **h** 4   **i** 1   **j** 23
    k 1   **l** 1   **m** 1   **n** 325
  ii a 2·82 **b** 6·83 **c** 0·73 **d** 1·29 **e** 1·84
    f 2·90 **g** 13·00 **h** 4·00 **i** 1·00 **j** 22·88
    k 0·86 **l** 0·90 **m** 0·99 **n** 325·09
  iii a 2·8 **b** 6·8 **c** 0·7 **d** 1·3 **e** 1·8
    f 2·9 **g** 13·0 **h** 4·0 **i** 1·0 **j** 22·9
    k 0·9 **l** 0·9 **m** 1·0 **n** 325·1
2 a 3·42 **b** 24·0 **c** 0·00 **d** 14·00 **e** 65
 f 2·1 **g** 125·70 **h** 5 **i** 14·0 **j** 1·64
 k 13·65 **l** 0·1 **m** 7·0 **n** 8·00 **o** 0·10
3 a 2·8 kg **b** 1·4 kg **c** 1·7 kg
4 a 2·$\dot{6}$ or 2·67 **b** 4·$\dot{7}$ or 4·78 **c** 1·27
 d 20·53   **e** 0·05   **f** 0·24 or 0·243$\dot{6}$
 g 7·6$\dot{3}$ or 7·63 **h** 17·29 **i** 6·62 or 6·617$\dot{3}\dot{6}$
 j 37·89 or 37·89$\dot{3}$ **k** 0·01 or 0·011383$\dot{3}$
5 £17
6 1·1 m
7 2·06 m
8 Possible answers are:
 a 4 seconds to the nearest second
 b 1·9 ℓ per day to the nearest tenth of a litre
 c 47p to the nearest penny
 d 0·3 mℓ to the nearest tenth of a millilitre

Review 1
a 6·25 **b** 45·27 **c** 0·37 **d** 0·02 **e** 10·04 **f** 33·00

Review 2
a 6·3 **b** 45·3 **c** 0·4 **d** 0·0 **e** 10·0 **f** 33·0

Review 3
a 3·6 km **b** 7·3 kg **c** 8·5 m

Review 4
a 4·2 or 4·1$\dot{6}$ **b** 10·3 or 10·$\dot{3}$ **c** 0·1 or 0·13$\dot{7}$
d 1·7 or 1·$\dot{6}$ **e** 2·3 or 2·$\dot{3}$

Review 5
£53·20

Review 6
Possible answers are:
a 78 mm to the nearest millimetre
b 1·3 mm to the nearest tenth of a millimetre

Chapter 2 Integers, Power and Roots

Page 37 Exercise 1
1 a A is 3, B is 2, C is ⁻7, D is ⁻2, E is 2
  b ⁻2   c ⁻1   d ⁻1
2

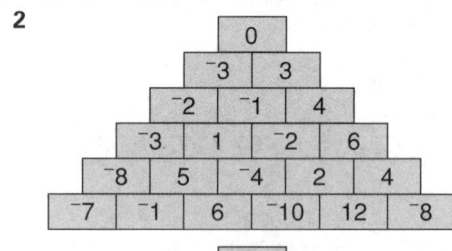

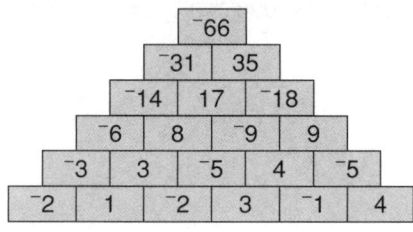

3 a ⁻5, 4, ⁻5, 6, ⁻8, 8, ⁻8, 5, ⁻5, 1
  b ⁻10
  c ⁻3
  d Subtract the answer to **c** from the answer to **b**.
    The sum is ⁻7.
4 a 0   b 0   c 0   d 0   e 3
  f 4   g 4   h ⁻6   i 4   j ⁻8
  k 11   l 1   m ⁻12   n ⁻25   o 126
  p 9   q ⁻26   r 211   s ⁻120
5 a 10   b 20   c 20   d ⁻20   e ⁻55   f ⁻13
  g ⁻18   h ⁻17   i ⁻26   j 64   k 15
6 a 8   b ⁻3   c 5   d ⁻1   e ⁻2
  f 2   g ⁻4   h ⁻4   i ⁻9   j ⁻16
  k 6   l ⁻11   m ⁻34   n ⁻36
7 a There are many possible answers.
    One is: ⁻16 and ⁻8.
  b There are many possible answers.
    One is: 3 − 24 = ⁻21.
8 13

Review 1
a 1 b ⁻3 c ⁻36

Review 2
a 23 b 19 c 10

Review 3
a ⁻28 b ⁻58 c ⁻12 d 11 e ⁻15 f ⁻10
g ⁻30 h ⁻17

Review 4
There are many possible answers. Two are: 5 − 40 = ⁻35
and ⁻70 − ⁻35 = ⁻35.

Page 39 Puzzle

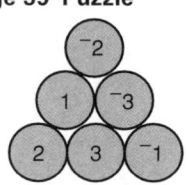

Page 40 Exercise 2
1 a ⁻3 × 4 = ⁻12   b ⁻3 × ⁻4 = 12
    ⁻4 × 4 = ⁻16   ⁻4 × ⁻4 = 16
    ⁻5 × 4 = ⁻20   ⁻5 × ⁻4 = 20
  c 3 × 4 = 12   d 3 × ⁻4 = ⁻12
    4 × 4 = 16   4 × ⁻4 = ⁻16
    5 × 4 = 20   5 × ⁻4 = ⁻20
2 a ⁻15   b ⁻24   c ⁻15   d ⁻16   e ⁻28
  f ⁻27   g ⁻18   h 12   i 20
  j 32   k ⁻30   l ⁻30   m 30
3 a ⁻4 b 6 c ⁻24 d 24
4 a

×	2	⁻5	⁻3	4
⁻1	⁻2	5	3	⁻4
3	6	⁻15	⁻9	12
⁻2	⁻4	10	6	⁻8
5	10	⁻25	⁻15	20

b

×	⁻1	2	3	⁻4
3	⁻3	6	9	⁻12
⁻1	1	⁻2	⁻3	4
⁻4	4	⁻8	⁻12	16
⁻6	6	⁻12	⁻18	24

c

×	⁻6	2	⁻1	8
⁻4	24	⁻8	4	⁻32
7	⁻42	14	⁻7	56
⁻6	36	⁻12	6	⁻48
⁻1	6	⁻2	1	⁻8

d

×	⁻3	7	⁻2	⁻8
6	⁻18	42	⁻12	⁻48
⁻1	3	⁻7	2	8
⁻7	21	⁻49	14	56
2	⁻6	14	⁻4	⁻16

5 a ⁻1 × 4 = ⁻4, ⁻4 ÷ 4 = ⁻1, ⁻4 ÷ ⁻1 = 4
  b 2 × ⁻9 = ⁻18, ⁻18 ÷ ⁻9 = 2, ⁻18 ÷ 2 = ⁻9
  c ⁻3 × ⁻7 = 21, 21 ÷ ⁻7 = ⁻3, 21 ÷ ⁻3 = ⁻7
  d ⁻48 ÷ 8 = ⁻6, ⁻6 × 8 = ⁻48, 8 × ⁻6 = ⁻48
  e ⁻60 ÷ ⁻4 = 15, 15 × ⁻4 = ⁻60, ⁻4 × 15 = ⁻60
6 a ⁻3   b ⁻3   c ⁻2   d ⁻2   e 3
  f 4   g 2   h ⁻2   i ⁻4   j 3
  k ⁻2   l 2   m ⁻4   n ⁻4
7 a ⁻8   b ⁻4   c ⁻8   d ⁻6   e 4
  f 10   g ⁻9   h ⁻4   i ⁻4   j ⁻7
  k ⁻4   l ⁻5   m 8
8 a 3 b ⁻6 c 3 d 4
9 a 3 b 4 c 6 d 12 e 9
10 There are many possible ways. One is:

×	⁻5	⁻1	3	⁻3
2	⁻10	⁻2	6	⁻6
⁻2	10	2	⁻6	6
⁻8	40	8	⁻24	24
⁻10	50	10	⁻30	30

11 There are many possible answers. Some are:
    ⁻4 × 2 = ⁻8, ⁻8 × 1 = ⁻8, ⁻24 ÷ 3 = ⁻8, 40 ÷ ⁻5 = ⁻8.
12 a £100 b £260 c £816

Review 1
a 24   b ⁻42   c ⁻40   d 21   e ⁻27
f ⁻64   g ⁻7   h 4   i ⁻6   j 6
k ⁻6   l ⁻5   m ⁻9   n 2

Review 2
a 7 b 3 × ⁻8 = ⁻24, ⁻24 ÷ 3 = ⁻8, ⁻24 ÷ ⁻8 = 3

Review 3
⁻20 and ⁻4

Page 42 Exercise 3

1	a	62	b	⁻945	c	⁻219	d	⁻3854
	e	378	f	32	g	60·5	h	36·8
	i	⁻953·8	j	⁻288·75	k	⁻18	l	68
	m	⁻4·9	n	⁻8·5	o	⁻0·41	p	⁻1245·28
	q	19·2	r	51 440	s	31		

2 a ⁻1·9   b ⁻5·0   c 10·7   d ⁻5·1   e 1·7

3 a

+	⁻84	253	⁻461	514
⁻198	⁻282	55	⁻659	316
⁻257	⁻341	⁻4	⁻718	257
346	262	599	⁻115	860
⁻98	⁻182	155	⁻559	416

b

×	47	⁻14	⁻28	8·6
⁻6·4	⁻300·8	89·6	179·2	⁻55·04
⁻72	⁻3384	1008	2016	⁻619·2
95	4465	⁻1330	⁻2660	817
⁻69	⁻3243	966	1932	⁻593·4

Review

THE STUDY OF STUPIDITY IS CALLED MONOLOGY

Page 44 Exercise 4

1	a	⁻14	b	11	c	10	d	⁻2
	e	⁻7	f	⁻7	g	⁻4	h	⁻24
	i	21	j	⁻3	k	⁻5	l	⁻20
	m	12	n	⁻9	o	⁻17	p	⁻1
	q	3	r	⁻4	s	4	t	⁻2
	u	⁻5	v	2	w	7·5	x	24

2 $2 \times {}^-5 \times ({}^-3 + {}^-2) = 50$

Review

A swan

Page 45 Discussion

● Zenta's numbers work because 4 and 5 have no common factors. 10 and 2 have common factor 2. This means if the number is divisible by 10, it will also be divisible by 2.

● Josh's numbers are right because 3 and 4 have no common factors. 2 and 6 have common factor of 2.

Page 45 Exercise 5

1 a 1278, 3600, 4536, 6804, 24 056
   b 1278, 3600, 4536, 6804, 12 645
   c 3600, 4536, 6804, 24 056
   d 3600, 12 645
   e 3600, 4536, 24 056
   f 1278, 3600, 4536, 6804, 12 645
   g 3600

2 a A because 3 × 8 = 24 and 3 and 8 have no common factors other than 1.
   b B because 5 × 8 = 40 and 5 and 8 have no common factors other than 1.

3 a 1278, 3600, 4536, 6804   b 3600, 12 645
   c 3600, 4536, 6804   d 1278, 3600, 4536, 6804
   e 3600   f 3600, 4536
   g 3600

4 a 516, 3468, 4332   b 3510, 14 715
   c 1008, 3564, 6570, 12 510 d 980, 1460
   e 2352, 23 568   f 1360, 58 360

5 a 14 586   b 40 500   c 38 808   d 10 488

6 She should continue to divide 2029 by prime numbers until the answer she gets is less than the number she is dividing by.

7 360

Review 1

B because 2 × 9 = 18 and 2 and 9 have no common factors other than 1.

Review 2

a All of them
b 1590, 2880, 6030, 41 520
c 2136, 2880, 14 076, 41 520
d 2880, 13 266, 6030, 14 076
e 2880, 41 520
f 2136, 2880, 41 520
g 1590, 2880, 6030, 41 520

Review 3

144

Page 46 Investigation: Consecutive Numbers

The sum of any five consecutive numbers is divisible by 5. If the first number is n, the second is $n + 1$, the third is $n + 2$, the fourth is $n + 3$ and the fifth is $n + 4$.

$n + n + 1 + n + 2 + n + 3 + n + 4 = 5n + 10$

$5n + 10$ is always divisible by 5 because both $5n$ and 10 are divisible by 5.

Page 46 Investigation: Divisibility by 7

1 This method works for all numbers that are divisible by 7.

2 A number is divisible by 11 if the difference between the sum of the odd-placed digits and the sum of the even-placed digits is equal to 0 or is a multiple of 11.
   Examples
   4235 is divisible by 11 because $(4 + 3) - (2 + 5) = 0$.
   64 856 is divisible by 11 because $(6 + 8 + 6) - (4 + 5) = 11$.

Page 48 Exercise 6

1 a

2	60
2	30
3	15
	5

b

2	210
3	105
5	35
	7

c

3	225
3	75
5	25
	5

d

2	560
2	280
2	140
2	70
5	35
	7

2 a

2	88
2	44
2	22
	11

$88 = 2^3 \times 11$

b

2	148
2	74
	37

$148 = 2^2 \times 37$

c

2	168
2	84
2	42
3	21
	7

$168 = 2^3 \times 3 \times 7$

d

2	300
2	150
3	75
5	25
	5

$300 = 2^2 \times 3 \times 5^2$

e

2	396
2	198
3	99
3	33
	11

$396 = 2^2 \times 3^2 \times 11$

f

2	4680
2	2340
2	1170
3	585
3	195
5	65
	13

$4680 = 2^3 \times 3^2 \times 5 \times 13$

3 a

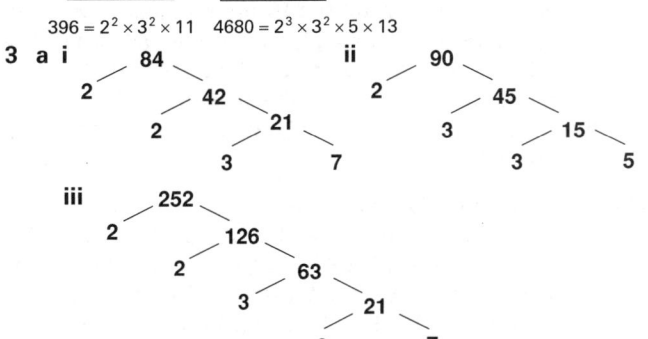

b The two possible answers are:

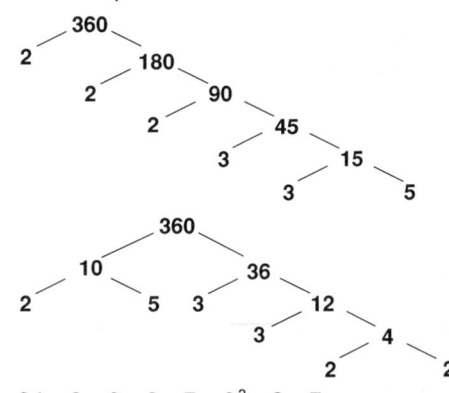

4 a $84 = 2 \times 2 \times 3 \times 7 = 2^2 \times 3 \times 7$
  b $90 = 2 \times 3 \times 3 \times 5 = 2 \times 3^2 \times 5$
  c $252 = 2 \times 2 \times 3 \times 3 \times 7 = 2^2 \times 3^2 \times 7$
5 a $110 = 2 \times 5 \times 11$ **b** $45 = 3^2 \times 5$
  c $70 = 2 \times 5 \times 7$ **d** $100 = 2^2 \times 5^2$
  e $350 = 2 \times 5^2 \times 7$ **f** $93 = 3 \times 31$
  g $1224 = 2^3 \times 3^2 \times 17$ **h** $1725 = 3 \times 5^2 \times 23$
  i $2580 = 2^2 \times 3 \times 5 \times 43$

Review 1

a

2	24
2	12
2	6
	3

$24 = 2^3 \times 3$

b

2	120
2	60
2	30
3	15
	5

$120 = 2^3 \times 3 \times 5$

c

2	198
3	99
3	33
	11

$198 = 2 \times 3^2 \times 11$

d

2	144
2	72
2	36
2	18
3	9
	3

$144 = 2^4 \times 3^2$

e

2	1836
2	918
3	459
3	153
3	51
	17

$1836 = 2^2 \times 3^3 \times 17$

Review 2

a A is 10, B is 2, C is 5, D is 11
b A is 2, B is 3, C is 3, D is 7
c A is 35, B is 5, C is 7

Review 3

a $220 = 2^2 \times 5 \times 11$ **b** $126 = 2 \times 3^2 \times 7$
c $315 = 3^2 \times 5 \times 7$ **d** $702 = 2 \times 3^3 \times 13$
e $812 = 2^2 \times 7 \times 29$ **f** $1755 = 3^3 \times 5 \times 13$

Page 49 Puzzle

1 2 and 11, 2 and 31, 2 and 41, 2 and 61, 2 and 71
2 4, 25 and 49. They are the squares of prime numbers.
3 42, 60, 66, 70, 78, 84, 90
4 If each of 17, 37, 47, 67 and 97 are multiplied with each of 13, 23, 43, 53, 73 and 83, the answers all end in 1. If two of the prime numbers 19, 29, 59, 79 and 89 are multiplied with each other, the answers end in 1.
If two of 11, 31, 41, 61 and 71 are multiplied together, the answers end in 1.

Page 49 Investigations: Factors

1 The two-digit numbers are: 10, 12, 18, 20, 21, 24, 27, 30, 36, 40, 42, 45, 48, 50, 54, 60, 63, 70, 72, 80, 81, 84, 90. There are many 3-digit numbers. The first ten are: 100, 102, 108, 110, 111, 112, 114, 117, 120, 126.
2 84, has factors: 1, 2, 3, 4, 6, 7, 12, 14, 21, 28, 42, 84. 60, 72, 90 and 96 have exactly 12 factors.
3 11, 22, 36 and 84.

4 It is true for all squares of four numbers. If the number in the top-left has factors a and b then the number in the top-right has factors a and c, the bottom-left has factors d and b and the bottom-right has factors d and c.
Multiplying opposite corners we get $ab \times dc = abcd$ and $ac \times db = abcd$.

Example

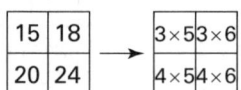

Page 50 Exercise 7

1 a 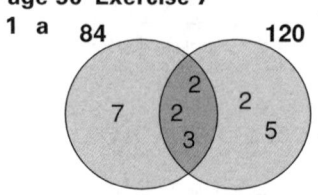   **b**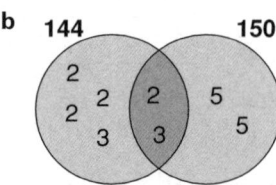

c 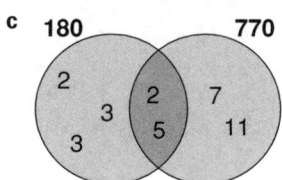   **d**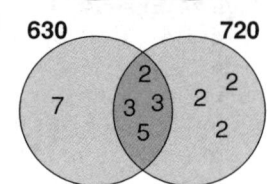

2 a 12   **b** 6   **c** 10   **d** 90
3 a 840   **b** 3600   **c** 13 860   **d** 5040
4 a 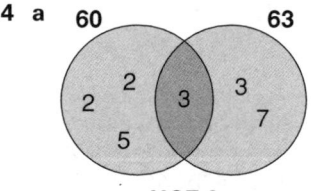   **b**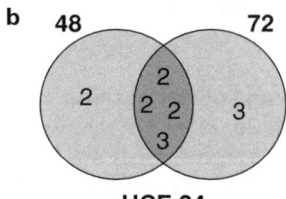

HCF 3   HCF 24

c 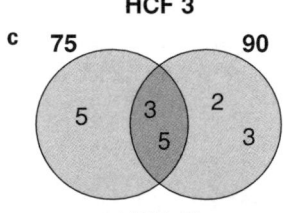   **d**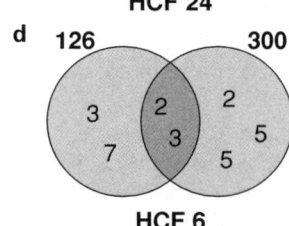

HCF 15   HCF 6

e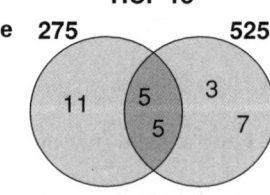

HCF 25

5 a 1260   **b** 144   **c** 450   **d** 6300   **e** 5775
6 a $\frac{8}{15}$ **b** $\frac{3}{8}$ **c** $\frac{2}{7}$ **d** $\frac{4}{9}$ **e** $\frac{5}{11}$
7 a HCF $= 2^3 \times 3^2$ LCM $= 2^4 \times 3^3$
  b HCF $= 2^2 \times 3 \times 5$ LCM $= 2^3 \times 3^2 \times 5^2$
  c HCF $= 3 \times 5$,   LCM $= 2 \times 3^2 \times 5 \times 11$
  d HCF $= 2 \times 3 \times 7$,   LCM $= 2^2 \times 3^2 \times 7^2$
  e HCF $= 2 \times 3 \times 5^2$,   LCM $= 2^2 \times 3^2 \times 5^3$
  f HCF $= 2 \times 7$,   LCM $= 2 \times 3 \times 5 \times 7^2 \times 11$
  g HCF $= 2^3 \times 3 \times 5^2$,   LCM $= 2^4 \times 3^2 \times 5^3$
  h HCF $= 5 \times 7^2$,   LCM $= 2 \times 5^2 \times 7^2 \times 11$
  i HCF $= 5 \times 29$,   LCM $= 3^2 \times 5^2 \times 7 \times 29$
  j HCF $= 11 \times 41$,   LCM $= 5 \times 7 \times 11 \times 41$

Review 1
a 36   **b** 70

Review 2
a 720   **b** 1400

Review 3

1.8	4		2.2	5	3.2
4		4.7	2		4
0		5		5.1	
	6.3	0		7.2	8
8.3		9.2			
10.6	6		11.7	9	2

Page 52 Exercise 8

1 **a** 1   **b** 64   **c** 49   **d** 64   **e** 36
   f 4   **g** 6   **h** 8   **i** 81   **j** 9
   k 7   **l** 144   **m** 8   **n** 1
2 **a** 3, ⁻3 **b** 5, ⁻5 **c** 7, ⁻7 **d** 12, ⁻12 **e** 11, ⁻11
3 Possible answers are:
   a $\sqrt{256} = \sqrt{4 \times 64} = 16$ **b** $\sqrt{400} = \sqrt{4 \times 100} = 20$
   c $\sqrt{324} = \sqrt{4 \times 81} = 18$ **d** $\sqrt{441} = \sqrt{9 \times 49} = 21$
   e $\sqrt{484} = \sqrt{4 \times 121} = 22$
4 **a** $1 < \sqrt{3} < 2$ **b** $2 < \sqrt{7} < 3$
   c $3 < \sqrt{10} < 4$ **d** $3 < \sqrt{11} < 4$
   e $2 < \sqrt{5} < 3$
5 **a** 28·09   **b** 0·3136   **c** 44·89   **d** 127·69
   e 14·39   **f** 17·66   **g** 70·73   **h** 8·43
   i 1·79   **j** 8·06   **k** 24·74   **l** 27·11
   m 1·25   **n** 1·19   **o** 2·19   **p** 5·05
   q 38·01   **r** 0·36   **s** 9·75
6 **a** About 4·4   **b** About 6·2   **c** About 8·5
   d About 9·7   **e** About 11·4
7 **a** About 22   **b** About 3·8   **c** About 26
   d About 140   **e** About 88
8 1, 36, 1225
9 6 and 4
10 **a** $60 = 2^2 \times 3 \times 5$ **b** 15 **c** 900
11 **a** 3·3 **b** 4·1 **c** 2·24

Review 1
a 36 **b** 144 **c** 25 **d** 7 **e** 144

Review 2
a 9, ⁻9 **b** 10, ⁻10

Review 3
Possible answers are:
a $\sqrt{196} = \sqrt{4 \times 49} = 14$ **b** $\sqrt{729} = \sqrt{9 \times 81} = 27$

Review 4
$\sqrt{5}$ lies between $\sqrt{4}$ and $\sqrt{9}$ so $\sqrt{5}$ lies between 2 and 3.

Review 5
a 67·2 **b** 53·3 **c** 3·6 **d** 9·4 **e** 1·3

Page 54 Puzzle

1 11, 12, 13, 14, 15, 16
2 53 and 54
3 **a** $28^2 = 784$
   b $35^2 = 1225$ or $45^2 = 2025$ or $55^2 = 3025$ or $65^2 = 4225$
      or $75^2 = 5625$ or $85^2 = 7225$ or $95^2 = 9025$
   c Any number from 39 to 99 that has a 9 or a 1 as the
      second digit.
   d 75 **e** 888 **f** 39
4 1892

Page 54 Investigations: Sums and Differences

1 The difference of the squares in the first table is equal
   to the sum of the two consecutive numbers.
   a 19 **b** 51 **c** 165 **d** 215
   The difference of the squares in the second table is
   equal to double the sum of the pair of numbers which
   differ by 2.
   a 68 **b** 84 **c** 316 **d** 544
   The difference of the squares of numbers which differ
   by 3 is three times the sum of the pair of numbers
   which differ by 3.
   a 171 **b** 561
   The difference of the squares of numbers which differ
   by n is n times the sum of the pair of numbers which
   differ by n.
   a 180 **b** 720 **c** 6560
2 **a** 2, 5, 8, 10, 13, 17, 18, 20, 25, 26, 29, 32, 34, 37, 40,
      41, 45, 50, 52, 53, 58, 61, 65, 68, 72, 73, 74, 80, 82,
      85, 89, 90, 97 and 98 can be written as the sum of
      two square numbers.
      If the numbers are written in the form of a pyramid,
      patterns emerge.

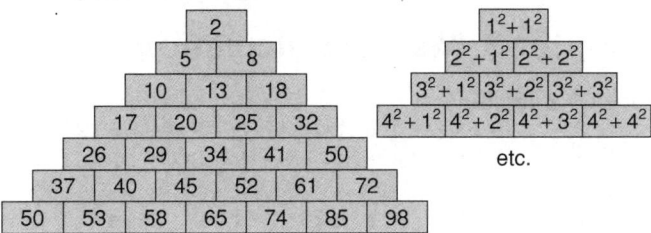

      etc.

      There are patterns in the differences between
      numbers in the same row and down the left-hand
      and right-hand sides of the pyramid.

   b 3, 5, 7, 8, 9, 11, 12, 13, 15, 16, 17, 19, 20, 21, 23, 24,
      25, 27, 28 and 29 can be written as the difference
      between two square numbers.

Page 56 Exercise 9

1 **a** 6^2 **b** 4^2 **c** 5^3 **d** 2^2 **e** 8^3 **f** 12^3
2 **a** 27   **b** 4   **c** 27   **d** 10   **e** 64
   f 5   **g** 3   **h** 4   **i** 2   **j** 3
   k 1   **l** 4   **m** 10
3 **a** 2   **b** 8   **c** 4   **d** 27   **e** 10
   f 125   **g** 1   **h** 5   **i** 1   **j** 64
   k ⁻27   **l** ⁻64   **m** ⁻1000 **n** 0·001 **o** 216
   p 0·027 **q** 0·064 **r** ⁻0·125
4 **a** 9261   **b** 3375   **c** ⁻729   **d** 4·913
   e 551·368 **f** ⁻50·653 **g** 7   **h** 6
   i 9   **j** 12   **k** 40
5 **a** 2·3 **b** 2·6 **c** 3·4 **d** 3·9 **e** 5·0 **f** 9·9
6 $4^3 = 13 + 15 + 17 + 19$, $5^3 = 21 + 23 + 25 + 27 + 29$
7 64
8 $1729 = 12^3 + 1^3$ or $9^3 + 10^3$

Review 1
a 125 **b** 3 **c** 1000 **d** 1 **e** 8 **f** 5 **g** 1000

Review 2
a 1000 **b** 3 **c** 125 **d** 4 **e** 1 **f** 5

Review 3
a ⁻512   b 4096   c 2·197   d 11   e 1·1
f 105·15   g 4·53   h 5·30   i 6·93

Review 4
$2 = 1^3 + 1^3$, $9 = 1^3 + 2^3$, $16 = 2^3 + 2^3$, $28 = 1^3 + 3^3$,
$35 = 2^3 + 3^3$, $54 = 3^3 + 3^3$, $65 = 1^3 + 4^3$, $72 = 2^3 + 4^3$,
$91 = 3^3 + 4^3$

Page 57 Puzzle

  1 13   **2** 18   **3** 15   **4** 29 or 39

In **1**, **2** and **3** there is only one answer because the last
digit of the answer tells us what the last digit of the
number must be and there is only one 2-digit number with
this last digit that gives a 4-digit answer.
e.g. in **1**, the last digit must be 3 and $3^3 = 27$, $13^3 = 2197$,
$23^3 = 12\ 167$, ... so it must be 13.
In **4**, there are two 2-digit numbers ending in 9 that give a
5-digit answer. The 2-digit number must end in 9 because
9^3 is the only cube that ends in 9.

Page 57 Investigation: End Digits

1 9^{12} will end in 1 because 9 to the power of an even
number ends in 1, 9^{15} will end in 9 because 9 to the
power of an odd number ends in 9.
2 The pattern of last digits of 4^1, 4^2, 4^3, 4^4, ... is 4, 6, 4,
6, ...
The last digit of 4^{21} is 4 and of 4^{30} is 6.
The pattern of last digits of the powers of 2 is 2, 4, 8, 6,
2, 4, 8, 6, ...
The pattern of last digits of the powers of 3 is 3, 9, 7, 1,
3, 9, 7, 1, ...
2^{11} ends in 8, 2^{12} ends in 6, 2^{13} ends in 2, 2^{14} ends in 4.
3^9 ends in 3, 3^{12} ends in 1, 3^{15} ends in 7 and 3^{22} ends
in 9.
The pattern of last digits of the power of 7 is 7, 9, 3, 1,
7, 9, 3, 1, ...
The pattern of last digits of the power of 8 is 8, 4, 2, 6,
8, 4, 2, 6.
7^{41} ends in 7 and 8^{41} ends in 8.

Chapter 3 Mental Calculation

Page 63 Exercise 1

1 a 3   b ⁻13   c ⁻10   d ⁻2   e 3
   f 16   g 0   h 1   i ⁻0·7   j ⁻1
   k 1·9
2 a 73   b 22   c 5·29   d 29   e 33
   f 548   g 4·64
3 a 75   b 88   c 107
4 a 34   b Yes
   c It has the same numbers as the Melancholia square.
5 a 780   b 60   c 1390   d 570
   e 12 100   f 1400   g 1700
6 Only one way to get the total is given.
   a 10, 25, 25, 50   b 25, 30, 30, 50
   c 25, 25, 30, 50   d No
   e 30, 30, 30, 50
7 a 5·9   b 3·7   c 1·9   d 7·5
   e 11·2   f 1·7   g 9·3   h 17·1
   i 3·5   j 3·4   k 18·1   l 1·19
   m 0·99   n 0·28   o 0·24   p 0·11
   q 0·025   r 0·036   s 0·025
8 a 200   b 150   c 1100   d 157
   e 121   f 2   g 19·4   h 6·9
   i 4·7   j ⁻2·2   k ⁻3·04
9 a 405   b 785   c 503   d 343
   e 470   f 128   g 409   h 236
   i 3290   j 4697   k 4391   l 4528
10 a 7·76   b 19·56   c 4·17   d 0·27   e 5·38
   f 8·14   g 1·23   h 7·28   i 6·56
11 a 1342   b 885   c 5·12   d 25·2
   e 6·93   f 8180   g 1·48   h 8·634 and 7·98
12 a 172   b 49   c 586   d 2508   e 1·7
   f 3·3   g 13·5   h 3·2   i 0·6
13 There are many possible answers. You can begin with
any number in one of the circles. A possible answer is:

a   b

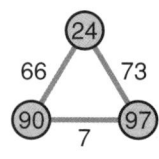 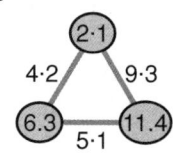

14 Possible answers are:
   a 13 + 24 + 56 + 7 = 100
   b 54 + 73 − 26 − 1 = 100
15 17 + 19 + 21 + 23 = 80
16 a 1·87 km, 1·31 km, 1·71 km, 1·62 km, 1·84 km
   b 16·7 km
17 There are many possible answers.
   One is: 2·3 + 5·78 + 1·92 = 10.

Review 1
a 5   b ⁻2   c 0·8   d 0·9   e ⁻1·6

Review 2
a 138   b 880   c 120   d 60
e 7600   f 1300   g 4·5   h 2·5
i 1   j 0·015   k 527   l 4·61
m 6·5   n ⁻0·3

Review 3
Yes, we find the second magic square by adding 7 to

3	8	13	18
17	14	7	4
16	11	10	5
6	9	12	15

13	23	24	10
16	18	21	15
19	17	14	20
22	12	11	25

every number in the first magic square and then rotating
the square 90° clockwise.

Review 4
THE MOST COMMON FORM OF CANCER IS SKIN
CANCER

Page 65 Investigation: Magic Squares

1

16	3	2	13
5	10	11	8
9	6	7	12
4	15	14	1

One possible answer with the

12
1

and other blocks in different positions is:

2	13	16	3
11	8	5	10
7	12	9	6
14	1	4	15

2 One possible way is:

0·5	1·9	0·9
1·5	1·1	0·7
1·3	0·3	1·7

Another possible way is:

0·9	0·7	1·7
1·9	1·1	0·3
0·5	1·5	1·3

3 Two possible ways are:

1·9	2·0	1·5
1·4	1·8	2·2
2·1	1·6	1·7

and

1·9	0·7	2·8
2·7	1·8	0·9
0·8	2·9	1·7

4 No. One possible way of making it a magic square by putting a decimal point in each number is:

15·5	1·50	2·50	12·5
4·50	10·5	9·5	7·50
8·50	6·5	5·50	11·5
3·50	13·5	14·5	0·5

If all of these numbers are multiplied or divided by 10, 100 etc. this will also form a magic square.

Page 66 Puzzles

1 Possible answers are:

a
```
  1513
−  731
  ────
   782
```

b
```
  7425
+ 7433
 ─────
 14 858
```

2
```
   6·55
+  5·66
  ─────
  12·21
```

Page 68 Exercise 2

1 a 1200   **b** 3000   **c** 8000   **d** 3
e 20   **f** 18 000   **g** 30   **h** 56 000
i 20   **j** 2   **k** 40   **l** 14 000
m 2   **n** 3   **o** 16 000

2 a 640   **b** 1300   **c** 1260   **d** 1320
e 1350   **f** 1840   **g** 496   **h** 1410
i 3084   **j** 400   **k** 1400   **l** 6900
m 192   **n** 625   **o** ⁻264   **p** ⁻630
q ⁻480   **r** 608   **s** ⁻406

3 a 0·38   **b** 1·6   **c** 3   **d** 2·4
e 2·46   **f** 8·63   **g** 2·28   **h** 0·35
i 0·24   **j** ⁻0·28   **k** 0·72   **l** ⁻0·05
m ⁻0·008   **n** 0·15   **o** 0·035   **p** 0·045
q 0·12   **r** 0·015   **s** 0·006

4 a 12·4   **b** 8·4   **c** 7·2   **d** 24
e 23·2   **f** 23·8   **g** 20   **h** ⁻40
i 25·6   **j** 23·4   **k** ⁻18·5   **l** 15·6
m 31·9   **n** 40·8   **o** ⁻85·5   **p** 170
q 180·6   **r** ⁻102·3   **s** 64·32   **t** 12·12
u 54·36   **v** 32·32   **w** 48·24   **x** 28·56

5 a $7 \times 3·65 = 25·55$, $25·55 \div 7 = 3·65$, $25·55 \div 3·65 = 7$
b i 42·86   **ii** 1·32   **iii** 565·752

6 a 110   **b** 35   **c** 45   **d** 45
e 120·5   **f** 22·5   **g** 211·5   **h** 64
i 0·6   **j** 7·1   **k** 0·67   **l** 4·92
m 15·5   **n** 2·3   **o** 0·85   **p** 25
q 7   **r** 51

7 a

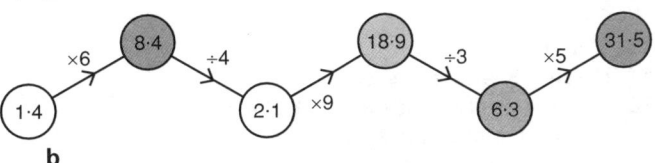

b

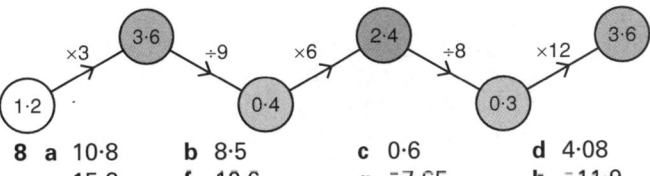

8 a 10·8   **b** 8·5   **c** 0·6   **d** 4·08
e 15·3   **f** 12·6   **g** ⁻7·65   **h** ⁻11·9
i ⁻4·5   **j** 11

9 a 720   **b** 73·5   **c** 16   **d** 9·6   **e** 9·8   **f** 396

10 Possible answers are:
a $0·02 \times 3 = 0·06$ **b** $1·2 \div 3 = 0·4$
c $0·8 \times 0·4 = 0·32$ **d** $12 \times 0·12 = 1·44$

11 a 8·228   **b** 4·84   **c** 484   **d** 0·8228   **e** 0·484

12 a $1·6 \times 2·5 + 3 = 7$
b $(1·6 + 3·4) \times 4 = 20$
c $(1·6 + 3·4) \div (8 − 3) = 1$

Review 1

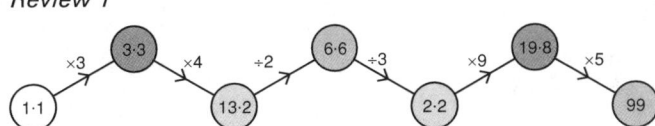

Review 2
a 6·3   **b** ⁻0·48   **c** 1·84   **d** 0·04   **e** 0·025

Review 3
a 12   **b** 2·85   **c** ⁻6

Review 4
NO TWO CORNFLAKES LOOK THE SAME

Review 5
a 1920   **b** 36·2   **c** 31   **d** 43·2   **e** 598

Review 6
15 and 30

New National | Framework

Page 70 Exercise 3

1 a 6 499 999 **b** 4500 **c** 135 **d** 2·5
e 2 hours 35 minutes **f** 56° **g** 5 m **h** 81
i 26 cm **j** 52 **k** 12 **l** 186
m £1·05 **n** 25 **o** 49 **p** 100°
q 0·05 **r** 2·5
2 52 kg
3 96
4 a 32 **b** 160 **c** 120
5 One possible answer is 5, 6, 7, 8 in one square, 10, 12, 1, 3 in another and 9, 11, 2, 4 in the other. There are other answers.
6 a £37·40 **b** £61
7 a 252 **b** 1400
8 a £15 **b** £16 **c** £45
9 57 km
10 a i 11 years **ii** 44 years **iii** 47 years
iv 34 years **v** 54 years **vi** 63 years
vii 17 years **viii** 72 years
c 125 years
11 106
12 23·5 cm
13 121 cm
14 a 7 and 8 **b** 20 and 5
c 18 and 6 **d** 0·4 and 0·8
15 6 and 10
16 11·8 g
17 He sent 3 letters overseas.
18 a One possible answer is: $(4 - 3) \times 5 \times 6$.
b Possible answers are:
$(4 \times 3) + 6 - 5 \times 3 + 1 = 4$
$4 \times (3 + 6) - 5 \times 3 + 1 = 22$
$4 \times (3 + 6 - 5) \times 3 + 1 = 49$
$4 \times (3 + 6 - 5 \times 3) + 1 = {}^-23$
$4 \times (3 + 6 - 5 \times 3 + 1) = {}^-20$
$4 \times 3 + (6 - 5) \times 3 + 1 = 16$
$4 \times 3 + 6 - (5 \times 3 + 1) = 2$
$4 \times 3 + 6 - 5 \times (3 + 1) = {}^-2$
$(4 \times 3 + 6 - 5) \times 3 + 1 = 40$
There are other ways to make some of these answers.
19 a There are two possible answers. She could have bought 4 packets of tennis balls and 5 packets of golf balls or 9 packets of tennis balls and 2 packets of golf balls.
b 2 packets of tennis balls and 3 packets of golf balls.

Review 1
600

Review 2
a £2485 **b** £576

Review 3
a £44·80 **b** 11

Review 4
a 14 **b** 13 **c** 17

Review 5
£7·10

Review 6
2 milkshakes, 3 thickshakes and 4 juices

Page 73 Puzzles

1 84 **2** 192 **3** 232 and 520

Page 74 Exercise 4

1 b and **e**

Review
Exact number

Page 75 Exercise 5

1 a A **b** C **c** C **d** B **e** B **f** C **g** C **h** C **i** B
2 Possible answers are given. Other answers are possible.
a $100 \times 80 = 8000$ **b** $300 \times 80 = 24\,000$
c $200 \times 200 = 40\,000$ **d** $600 \times 300 = 180\,000$
e $200 \div 20 = 10$ **f** $500 \div 50 = 10$
g $400 \div 40 = 10$ **h** $600 \div 30 = 20$
i $200 \times 80 = 16\,000$ **j** $800 \times 50 = 40\,000$
k $700 \div 35 = 20$ **l** $500 \div 25 = 20$
3 Possible answers are:
a 400 **b** 7 **c** 10 **d** 8 **e** 18
f 72 **g** 1 **h** 15 **i** 35 **j** 4
k 60 **l** 150 **m** 50 **n** 40 **o** 160
4 Possible answers are
a $\frac{100 \times 30}{20} = \frac{3000}{20}$
$= 150$
b $\frac{90 \times 90}{10} = \frac{8100}{10}$
$= 810$
c $\frac{50 \times 80}{20} = \frac{4000}{20}$
$= 200$
d $\frac{20 \times 20}{5} = \frac{400}{5}$
$= 80$
e $\frac{900}{3 \times 3} = \frac{900}{9}$
$= 100$
f $\frac{700}{5 \times 2} = \frac{700}{10}$
$= 70$
g $\frac{100 + 80}{80 - 20} = \frac{180}{60}$
$= 3$
h $\frac{140 + 70}{90 - 20} = \frac{210}{70}$
$= 3$
i $\frac{17 + 11}{4 - 1} = \frac{28}{3}$
$\approx \frac{27}{3}$
$= 9$
j $\frac{19 + 12}{10 - 7} = \frac{31}{3}$
$\approx \frac{30}{3}$
$= 10$
k $12 \div (4 + 4) = 12 \div 8$
$= 1·5$
l $8 \div (4 - 2) = 8 \div 2$
$= 4$
m $(18 - 5) \div (1 + 10) = 13 \div 11$
$\approx 13 \div 10$
$= 1·3$
n $(30 - 5) \div (11 - 5) = 25 \div 6$
$\approx 24 \div 6$
$= 4$
o $(8 - 5) \div (3 + 1) = 3 \div 4$
$\approx 4 \div 4$ (or 0·75)
$= 1$
5 a About 10·5 **b** About 9·5
c About 8·5 **d** About 20·5
e About 63
6 Possible answers are:
a £25 000 + £7000 **b** 12 000 − 9000
c 30×20 **d** 50×20
e $800 \div 40$ **f** $600 \div 20$
g $18 \div 18$ **h** 1×20 or $1·4 \times 20$
i 1×6 **j** $2·5 \times 1$
k 12×20 **l** 80×2
7 a 14 m **b** 20 m **c** 30 m

Review 1
a B **b** C **c** A **d** A **e** A

Review 2
Possible answers are:
a $200 \times 100 = 20\,000$ **b** $600 \times 700 = 420\,000$
c $600 \div 30 = 20$ **d** $900 \div 30 = 30$
e $200 \div 20 = 10$ **f** $400 \div 20 = 20$
g $700 \div 35 = 20$ **h** $600 \div 30 = 20$

Review 3
Possible answers are:
a 60 **b** 7 **c** 400 or 360 **d** 8
e 200 **f** 20 **g** 100 or 90 **h** 120

Review 4
a $\frac{100 \times 50}{20} = \frac{5000}{20}$
$= 250$
b $\frac{400}{6 + 2} = \frac{400}{8}$
$= 50$
c $\frac{22 + 9}{9 - 3} = \frac{31}{6}$
$\approx \frac{30}{6}$
$= 5$

d $(16 - 9) \div (4 + 4) = 7 \div 8$
$\approx 8 \div 8$
$= 1$

Chapter 4 Written and Calculator Calculation

Page 81 Exercise 1

1
a 70·86   **b** 282·09   **c** 53·84   **d** 12·45
e 37·44   **f** 48·74   **g** 43·16   **h** 7·48
i 24·01   **j** 9·43   **k** 8·48   **l** 79·4
m 33·45   **n** 63·86   **o** 50·27

2 **a** 108·43   **b** 262·26
   c There are many possible answers.
   d There are many possible answers.

3 **a** 14·34 mm   **b** 1·66 mm

4 1·02 ℓ

5 £142·44

6 125 cm

7 **a** The largest possible sum is 188·91. A possible way of getting this is 97·4 + 6·31 + 85·2.
   b The smallest possible sum is 44·19. A possible way of getting this is 23·8 + 4·79 + 15·6.

8 Possible answers are:
   a 57·3 + 12·64 = 69·94   **b** 65·27 − 14·3 = 50·97

Review 1
LIGHT CAN TRAVEL AROUND THE EARTH AT THE EQUATOR SEVEN AND A HALF TIMES PER SECOND

Review 2
15·2 g

Review 3
777 g

Page 83 Investigation: Ten pounds and eighty-nine pence
Yes you always get £10·89 unless the first and last digits are the same. *Example* £3·43 − £3·43 = 0 in step 2.

Page 84 Exercise 2

1
a 13 340   **b** 7614   **c** 8748   **d** 13 632
e 19 663   **f** 40 132   **g** 30 285   **h** 33 388
i 67 166

2 7168 km

3
a 224·8   **b** 256·2   **c** 408·1   **d** 386·1
e 154   **f** 324   **g** 97·2   **h** 7·44
i 10·01   **j** 62·79   **k** 47·34   **l** 6·352
m 4·315   **n** 3·432

4
a 105·4   **b** 41·6   **c** 128·8   **d** 131·2
e 270·2   **f** 685   **g** 4993·2   **h** 8384·8
i 2456·4   **j** 375·12   **k** 285·48   **l** 307·38
m 406·56   **n** 188·33   **o** 320·12

5
a 52·92   **b** 40·56   **c** 23·68   **d** 44·94
e 63·36   **f** 115·75   **g** 31·84   **h** 100·98
i 13·962   **j** 38·622   **k** 54·891   **l** 44·467
m 20·864   **n** 22·116   **o** 17·955

6 **a** £122·76   **b** £28·73   **c** £82·63   **d** £51·43   **e** £73·79

7 **a** £95·25   **b** £4·75

8 **a** 28, 29 and 30   **b** 89 and 71

Review 1
DREAMT IS THE ONLY ENGLISH WORD THAT ENDS IN THE LETTERS MT

Review 5
Possible answers are:
a 20×90 **b** 10×17 **c** $50 \div 25$ **d** 20×8
e $800 \div 40$

Review 2
a £27·90   **b** £101·92   **c** £28·70   **d** £20·64   **e** £24·67

Review 3
25 and 35

Review 4
US $29·16

Page 85 Puzzles
1 **a** $A = 1, B = 2, C = 5$ **b** $A = 1, B = 2, C = 4$

Page 86 Exercise 3

1 **a** 18   **b** 6   **c** 9

2 **a** 56   **b** 8

3 **a** 52   **b** 4

4 **a** 12   **b** 13

5
a 12·8   **b** 18·5   **c** 26·8   **d** 23·75
e 13·5   **f** 8·5   **g** 13·25   **h** 3·7
i 2·4   **j** 1·6   **k** 3·65   **l** 1·25
m 3·2   **n** 2·6   **o** 4·2   **p** 4·6
q 3·25   **r** 4·5

6
a 34·8   **b** 18·9   **c** 36·7   **d** 29·5
e 3·6   **f** 3·8   **g** 6·7   **h** 4·6
i 2·4   **j** 1·7   **k** 2·2   **l** 3·1
m 2·2

7 18·5

8 **a** £6·30   **b** £8·20

9 4·6 m

10 24 for 86p. Possible answers are 'this is the most popular size so it is able to be sold a little more cheaply' or 'there is most competition from other brands in this size'.

11 $92·8 \div 16 = 5·8$

Review 1
a 34   **b** 23

Review 2
a 16·8   **b** 8·25   **c** 6·4

Review 3
a 34·2   **b** 35·5   **c** 6·3   **d** 3·3

Review 4
48 visits for £140. A possible answer is 'the gym wants people to join up for the greatest number of visits'.

Page 88 Exercise 4

1 **a** D   **b** D   **c** B   **d** C   **e** B   **f** B   **g** C   **h** C   **i** A

2
a $820 \div 2$ **b** $930 \div 3$ **c** $7200 \div 4$
d $84\,600 \div 9$ **e** $3840 \div 24$ **f** $7230 \div 84$
g $9610 \div 39$ **h** $4820 \div 32$ **i** $5960 \div 16$

3
a 90   **b** 30   **c** 80   **d** 270
e 84   **f** 550   **g** 910   **h** 1205
i 450   **j** 4300   **k** 4250   **l** 1145
m 6480   **n** 153·$\dot{3}$ **o** 6087·5   **p** 12 133·3

4
a 90   **b** 65   **c** 90   **d** 50
e 250   **f** 85   **g** 102·5   **h** 41·25
i 204·8   **j** 142·5   **k** 82·5   **l** 178·75
m 273·6   **n** 60·3125

5 a 873·3   **b** 554·3   **c** 993·3   **d** 5755·6
e 131·5   **f** 154·5   **g** 171·5   **h** 195·1
i 82·8   **j** 188·5   **k** 188·4   **l** 17·8
m 32·0   **n** 264·7
6 a $850   **b** $2723·33   **c** $1573·33   **d** $298·67

Review 1
a 860 ÷ 3   **b** 91 400 ÷ 7   **c** 830 ÷ 59   **d** 4270 ÷ 61

Review 2
a 40   **b** 2400   **c** 1170   **d** 405   **e** 45

Review 3
a 432·2   **b** 6042·9   **c** 144·7   **d** 66·8   **e** 86·2

Page 90 Exercise 5
1 a No, the answer should be bigger than 125 because we are multiplying by a number bigger than 1.
   b Yes
   c No, the answer should be smaller than £3·85 because we are dividing by a number bigger than 1.
   d No, because the answer should be within the range of the data. It should be between 1·46 and 1·72.
2 a Smaller   **b** Bigger   **c** Smaller
   d Bigger   **e** Bigger   **f** Smaller
   g Bigger   **h** Bigger   **i** Smaller
3 b, d and **f**
4
5 C
6 Possible answers are:
   a 512 ÷ 2 ÷ 8 or 512 ÷ 4 ÷ 4
   b 8·3 × 1 + 8·3 × 0·1   **c** 586 × 100 − 586
7 a and **c**
8 Possible answers are:
   a 8·7 × 3·9 = 8 × 3·9 + 0·7 × 3·9
   b 19·63 + 52·71 = 20 + 50 − 0·37 + 2·71
   c 51 × 17
   d 2688 ÷ 28
   e 4·123 105 6²
9 Possible answers are:
   a B because 48 is just less than 50 and 89 is just less than 90 so this answer is less than 50 × 90 = 4500. The last digit is 2 because 8 × 9 = 72.
   b A because 53 is just more than 50 and 32 is just more than 30 so 53 × 32 is just more than 30 so 50 × 30 = 1500. The last digit is 6 because 3 × 2 = 6.
   c A because 79 is just less than 80 and so 79 × 1·5 is just less than 80 × 1·5 = 120.
   d C because 539 is more than 500 and 0·72 is just more than 0·7 so 539 × 0·72 is more than 500 × 0·7 = 350.
   e C because 488 is less than 500 and 1·47 is just less than 1·5 so 488 × 1·47 is less than 500 × 1·5 = 750.

Review 1
No, the answer should be bigger than 21·7 g because she is multiplying by a number bigger than 1.

Review 2
There are many possible answers. One is:
a 33·58 ÷ 7·3   **b** 60 × 16   **c** 80 × 39 + 7 × 39
d 106·02 ÷ 1·86

Review 3
a B because 27 is just less than 30 and 48 is just less than 50 so 27 × 48 is just less than 30 × 50 = 1500. The last digit must be 6 because 7 × 8 = 56.
b C because 5·2 is just more than 5 and 3·1 is just more than 3 so 5·2 × 3·1 is just more than 5 × 3 = 15. The last digit is 2 because 2 × 1 = 2.

Page 92 Exercise 6
1 a 8   **b** 8·03   **c** 22·88   **d** ⁻6·35   **e** 11·616
2 a He forgot to key the right bracket after the 5.
   b He forgot to key the right bracket after the 2.
   c She forgot to put brackets around 4 × 2.
3 a 1·11   **b** 2·43   **c** 24·67   **d** 1·04
   e 3·45   **f** 1·19   **g** 2·33   **h** 2·28
   i 3·09   **j** 4·23   **k** 2·66   **l** 15·20
   m 22·25   **n** 109·51   **o** 0·83
4 a 10·8   **b** 304   **c** 85   **d** 7·8   **e** 0·7
5 a 4·2 × (6 + 3 − 3) × 4 + 1·8 = 102·6
   b 4·2 × (6 + 3 − 3 × 4 + 1·8) = ⁻5·04
6 Possible answers are:
   a 1·9 + 3·6 × 0·2 = 2·62
   b 12·8 − 6·9 ÷ 3 = 10·5
   c 12·4 ÷ 4 − 1·8 = 1·3
   d 8·4 × 4 + 1·6 × 2 = 36·8
   e 4(11·6 × 0·5 − 2·2) = 14·4
   f 2·4 + 4(6·8 − 2·7) = 18·8
   g (6·9 + 2·4 ÷ 3) + 1·8 = 9·5
   h 3·4 × 2 − 6·6 ÷ 3 = 4·6
   i (2·7 + 11 ÷ 5) ÷ 7 = 0·7

Review 1
BEES HAVE FIVE EYES

Review 2
180 ÷ (6·39 + 4·3 × 2·7) − 6·8 = 3·2

Page 94 Exercise 7
1 £22·70
2 4·7 km
3 £20·26
4 a 8750   **b** 13 125   **c** 21 000   **d** 32 550   **e** 65 625
5 a £68·25   **b** £84·90   **c** £39·45   **d** £41·70

Review
£2·65

Chapter 5 Fractions

Page 97 The whole sound of music
1 c and d
2 One possible answer is:

3 a > **b** = **c** < **d** < **e** =

Page 98 Investigation: Unit Fractions
1 8 × 3 into
$\frac{1}{2}, \frac{1}{3}, \frac{1}{8}, \frac{1}{24}$

✓	✓	✓	✓	✗	✗	✗	✗
✓	✓	✓	✓	✗	✗	✗	✗
✓	✓	✓	✓	o	o	o	◇

8 × 3 into
$\frac{1}{2}, \frac{1}{4}, \frac{1}{6}, \frac{1}{12}$

✓	✓	✓	✓	✗	✗	✗	✗
✓	✓	✓	✓	✗	✗	o	o
✓	✓	✓	✓	o	o	◇	◇

An 8 × 3 rectangle can be divided into 5 parts of $\frac{1}{2}, \frac{1}{8}, \frac{1}{12}, \frac{1}{4}, \frac{1}{24}$.

An 8 × 3 rectangle can be divided into 6 parts of $\frac{1}{3}, \frac{1}{4}, \frac{1}{6}, \frac{1}{8}, \frac{1}{12}, \frac{1}{24}$.

2 3 parts: $\frac{1}{2}, \frac{1}{3}, \frac{1}{6}$ 4 parts: $\frac{1}{2}, \frac{1}{4}, \frac{1}{6}, \frac{1}{12}$
It cannot be divided into more than 4 parts, each a different unit fraction.

3 30 squares can be divided into 3 parts: $\frac{1}{2}, \frac{1}{3}, \frac{1}{6}$;
4 parts: $\frac{1}{2}, \frac{1}{3}, \frac{1}{10}, \frac{1}{15}$;   5 parts: $\frac{1}{2}, \frac{1}{5}, \frac{1}{6}, \frac{1}{10}, \frac{1}{30}$.
16 squares cannot be divided into parts, each a different unit fraction. Nor can 15 squares.
18 squares can be divided into 3 parts: $\frac{1}{2}, \frac{1}{3}, \frac{1}{6}$ or 4 parts: $\frac{1}{2}, \frac{1}{3}, \frac{1}{9}, \frac{1}{18}$.
40 squares can be divided into 4 parts: $\frac{1}{2}, \frac{1}{4}, \frac{1}{5}, \frac{1}{20}$;
5 parts: $\frac{1}{2}, \frac{1}{4}, \frac{1}{8}, \frac{1}{10}, \frac{1}{40}$;
6 parts: $\frac{1}{2}, \frac{1}{5}, \frac{1}{8}, \frac{1}{10}, \frac{1}{20}, \frac{1}{40}$.

Note The unit fractions possible are found by finding the factors of the number. For example, because 24 has factors 1, 2, 3, 4, 6, 8, 12 and 24, the unit fractions $\frac{1}{2}, \frac{1}{3}, \frac{1}{4}, \frac{1}{6}, \frac{1}{8}, \frac{1}{12}$ and $\frac{1}{24}$ are possible.

Page 99 Exercise 1
1 a $\frac{2}{3}$ b $\frac{3}{8}$ c $\frac{5}{12}$
2 a About $\frac{1}{4}$ b About $\frac{1}{6}$
3 a About $\frac{2}{3}$ b About $\frac{1}{8}$

Review 1
$\frac{2}{5}$

Review 2
a About $\frac{1}{4}$ b About $\frac{1}{3}$

Page 100 Exercise 2
1 a $\frac{1}{2}$ b $\frac{1}{3}$ c $\frac{3}{4}$ d $\frac{9}{20}$
2 a $\frac{1}{4}$ b $\frac{1}{4}$ c $\frac{3}{8}$ d $\frac{1}{20}$ e $\frac{7}{20}$
3 a $\frac{1}{4}$ b $\frac{1}{6}$ c $\frac{5}{12}$ d $\frac{3}{10}$ e $\frac{1}{10}$
4 a $\frac{2}{3}$ b $\frac{2}{3}$ c $\frac{3}{4}$ d $\frac{1}{4}$ e $\frac{2}{3}$ f $\frac{1}{6}$ g $\frac{4}{7}$
5 a 60 b $\frac{23}{60}$ c $\frac{37}{60}$
6 a 30 b $\frac{4}{15}$ c Higher
7 a $\frac{1}{50}$ b $\frac{49}{50}$
8 $\frac{7}{16}$
9 $\frac{29}{30}$
10 $\frac{21}{100}$
11 a $\frac{1}{4}$ b $\frac{3}{20}$ c $\frac{1}{10}$ d $\frac{1}{20}$
12 3

Review 1
a $\frac{1}{4}$ b $\frac{3}{4}$ c $\frac{9}{20}$ d $\frac{13}{20}$

Review 2
a $\frac{2}{3}$ b $\frac{5}{9}$ c $\frac{3}{10}$

Review 3
$\frac{3}{4}$

Review 4
a $\frac{1}{4}$ b $\frac{1}{3}$ c $\frac{1}{12}$ d $\frac{5}{12}$

Page 102 Exercise 3
1 a $\frac{3}{5}$ b $\frac{41}{50}$ c $\frac{1}{25}$ d $\frac{157}{1000}$ e $\frac{931}{1000}$
f $\frac{1}{4}$ g $\frac{4}{5}$ h $\frac{1}{8}$ i $\frac{111}{200}$ j $\frac{5}{8}$
k $\frac{17}{200}$ l $\frac{7}{125}$ m $\frac{7}{250}$ n $\frac{1}{200}$ o $\frac{1}{125}$
p $1\frac{7}{10}$ q $1\frac{2}{5}$ r $2\frac{43}{100}$ s $7\frac{7}{25}$ t $3\frac{13}{20}$
u $4\frac{4}{125}$ v $3\frac{1}{200}$

Review
CHILDREN GROW FASTER IN THE SPRING

Page 104 Investigation: Prime Factors and Recurring Decimals
Fractions with denominators that have prime factors other than 2 or 5 will recur if written in decimal form.

Page 104 Exercise 4
1 a 0·6 b 0·05 c 0·375 d 0·15 e 0·16
f 0·95 g 0·68 h 0·9 i 0·3 j 0·54
k 0·92 l 1·6 m 2·75 n 4·625 o 1·1
p 2·08 q 1·2 r 3·5
2 a 0·14 b 0·56 c 0·57 d 8·67 e 4·17
f 1·86 g 1·17 h 2·06 i 1·36
3 a D b A
4 a 0·85 b 0·82 c 0·65 d 0·61 e 0·64
f 1·28 g 1·35
5 a 0·$\dot{2}$ b 0·41$\dot{6}$ c 0·4$\dot{5}$ d 0·$\dot{6}$
6 a 0·$\dot{1}$ b 0·$\dot{6}$ c 0·4$\dot{5}$ d 0·1$\dot{6}$ e 0·8$\dot{3}$
f 0·1$\dot{8}$ g 2·8$\dot{3}$ h 5·$\dot{2}$ i 8·$\dot{6}$ j 1·$\dot{4}$
k 2·$\dot{3}$
7 a 0·9 b 0·375 c 0·65 d 0·9$\dot{1}$
8 a 1·$\dot{3}$ b 1·$\dot{6}$ c 2 d 2·$\dot{3}$
9 a 0·$\dot{0}\dot{9}$, 0·$\dot{1}\dot{8}$, 0·$\dot{2}\dot{7}$, 0·$\dot{3}\dot{6}$
b The repeating digits are the multiples of 9.
c i 0·$\dot{4}\dot{5}$ ii 0·$\dot{5}\dot{4}$ iii 0·$\dot{6}\dot{3}$ iv 0·$\dot{7}\dot{2}$

Review 1
a 0·8 b 0·35 c 0·44 d 0·65
e 0·76 f 0·62 g 1·7 h 4·875
i 1·3 j 2·16 k 5·4 l 5·75
m 2·14 n 1·71

Review 2
a 0·77 b 0·62 c 0·68 d 0·15 e 7·13

Review 3
a 0·$\dot{1}$, 0·$\dot{2}$, 0·$\dot{3}$, 0·$\dot{4}$, 0·$\dot{5}$
b The repeating digits are the whole numbers 1, 2, 3, ...
c 0·$\dot{6}$, 0·$\dot{7}$, 0·$\dot{8}$

Page 106 Exercise 5
1 a < b > c < d < e <
f > g > h < i > j >
k > l < m < n <
2 a > b < c > d < e >
f > g <
3 Merlin Street
4 Seth
5 Paul
6 Kieran
7 a 11 ÷ 6 b 29 ÷ 8 c $\frac{22}{9}$ d 56 ÷ 9
8 a $\frac{1}{4}, \frac{1}{3}, \frac{5}{12}, \frac{5}{6}$ b $\frac{1}{4}, \frac{3}{8}, \frac{7}{10}, \frac{7}{32}$ c $\frac{1}{2}, \frac{2}{5}, \frac{7}{10}, \frac{4}{5}, \frac{9}{10}$
d $\frac{1}{8}, \frac{1}{6}, \frac{7}{24}, \frac{1}{3}$ e $\frac{17}{30}, \frac{11}{15}, \frac{4}{5}, \frac{5}{6}, \frac{7}{30}$ f $\frac{5}{8}, \frac{3}{4}, \frac{19}{24}, \frac{5}{6}, \frac{7}{8}, \frac{11}{12}$
9 a $\frac{5}{8}, \frac{8}{17}, \frac{4}{9}, \frac{3}{7}$ b $\frac{7}{9}, \frac{8}{11}, \frac{5}{3}, \frac{5}{8}$ c $\frac{7}{8}, \frac{22}{7}, \frac{4}{5}, \frac{15}{19}$
10 a

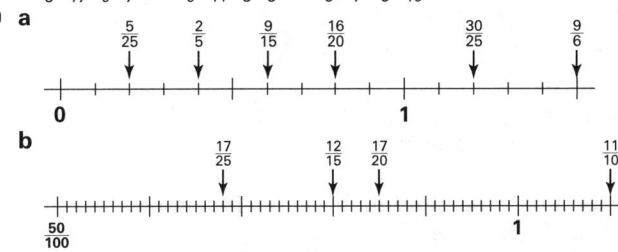

b

11 a $\frac{5}{6}$ b $\frac{4}{9}$ c $\frac{8}{15}$ d $\frac{11}{15}$ e $\frac{17}{35}$

Review 1
a < b < c > d < e < f > g >

Review 2
Runner

Review 3
a $\frac{7}{8}, \frac{13}{16}, \frac{3}{4}, \frac{1}{2}, \frac{7}{16}$ b $\frac{17}{20}, \frac{4}{5}, \frac{3}{4}, \frac{7}{10}, \frac{13}{20}, \frac{1}{2}$

Review 4

a

$$\frac{10}{25} \quad \frac{12}{20} \quad \frac{12}{15} \quad \frac{24}{20} \quad \frac{12}{8}$$

(number line from 0 to 2)

b

$$\frac{13}{25} \quad \frac{24}{30} \quad \frac{19}{20} \quad \frac{6}{5}$$

$\frac{50}{100}$ 1

(number line)

Page 108 Puzzle
$\frac{27}{36}$

Page 109 Exercise 6

1 **a** $\frac{3}{5}$ **b** $\frac{4}{7}$ **c** $\frac{1}{4}$ **d** $1\frac{2}{3}$ **e** $1\frac{1}{3}$
f $\frac{5}{8}$ **g** $\frac{11}{14}$ **h** $\frac{5}{8}$ **i** $\frac{1}{6}$ **j** $\frac{3}{4}$
k $\frac{1}{2}$ **l** $\frac{1}{2}$ **m** $\frac{3}{4}$ **n** $\frac{1}{10}$ **o** $\frac{2}{3}$
p $\frac{7}{12}$ **q** $\frac{31}{40}$ **r** $\frac{7}{24}$ **s** $\frac{14}{15}$ **t** $\frac{1}{20}$
u $\frac{1}{12}$ **v** $\frac{9}{10}$ **w** $\frac{1}{24}$ **x** $1\frac{1}{15}$ **y** $1\frac{7}{20}$
z $\frac{17}{24}$

2 **a** 1 **b** 0 **c** $1\frac{3}{20}$ **d** $\frac{5}{6}$ **e** $1\frac{13}{24}$
f $1\frac{11}{24}$ **g** $\frac{11}{12}$ **h** $1\frac{1}{4}$ **i** $1\frac{1}{6}$ **j** $\frac{29}{40}$

3 **a** $\frac{53}{56}$ **b** $\frac{11}{84}$ **c** $2\frac{79}{180}$ **d** $\frac{7}{12}$ **e** $2\frac{11}{56}$
f $\frac{7}{72}$ **g** $1\frac{2}{45}$ **h** $1\frac{103}{105}$

4 $\frac{1}{6}$

5 $\frac{3}{10}$

6 $\frac{19}{40}$

7 Glass

8 **a** $1\frac{1}{6}$ **b** $\frac{3}{4}$ **c** $\frac{1}{4}$

9 **a** One possible answer: $\frac{1}{3} + \frac{4}{6}$
b One possible answer: $\frac{5}{6} - \frac{1}{3} = \frac{1}{2}$
c One possible answer is $\frac{1}{2} + \frac{1}{3} + \frac{1}{6} = 1$

10 Three possible answers are $\frac{1}{3} + \frac{4}{6}$ and $\frac{1}{2} + \frac{4}{8}$ and $\frac{3}{4} + \frac{2}{8}$.

Review 1
a $\frac{1}{8}$ **b** $1\frac{5}{8}$ **c** $1\frac{2}{9}$ **d** $\frac{1}{6}$ **e** $1\frac{5}{12}$ **f** $\frac{4}{15}$ **g** $1\frac{9}{40}$ **h** $1\frac{11}{24}$

Review 2
$\frac{1}{30}$

Review 3
Two possible answers are $\frac{4}{8} + \frac{3}{2}$ and $\frac{2}{3} + \frac{8}{6}$.

Page 110 Puzzle
Possible answers are:
a $\frac{1}{4} + \frac{2}{4} = \frac{3}{4}, \frac{1}{4} + \frac{1}{2} = \frac{3}{4}$
b $\frac{1}{5} + \frac{5}{10} = \frac{7}{10}, \frac{2}{5} + \frac{3}{10} = \frac{7}{10}, \frac{3}{5} + \frac{1}{10} = \frac{7}{10}$
c $\frac{1}{5} + \frac{1}{10} = \frac{3}{10}$
d $\frac{1}{16} + \frac{5}{8} = \frac{11}{16}, \frac{3}{16} + \frac{4}{8} = \frac{11}{16}, \frac{5}{16} + \frac{3}{8} = \frac{11}{16}, \frac{7}{16} + \frac{2}{8} = \frac{11}{16}, \frac{9}{16} + \frac{1}{8} = \frac{11}{16}$
There are many other answers.

Page 110 Investigation: Egyptian Fractions
1 **a** $\frac{7}{10}$ **b** $\frac{11}{24}$ **c** $\frac{13}{30}$ **d** $\frac{4}{15}$
2 Possible answers are:

(Egyptian numeral diagrams a–f)

3 **a** (Egyptian numeral diagrams a, b, c)
c (Egyptian numeral diagrams, "or" alternative)

Page 112 Exercise 7

1 **a** 16 **b** 24 **c** 35 **d** 45 **e** 27
f 105 **g** 99 **h** 117 **i** 160
2 **a** T **b** T **c** T **d** F **e** F
f T **g** T
3 **a** $8\frac{1}{4}$ **b** $10\frac{5}{8}$ **c** $24\frac{2}{3}$ **d** $24\frac{1}{6}$ **e** $62\frac{2}{9}$
f $29\frac{5}{7}$ **g** $23\frac{1}{5}$ **h** $7\frac{7}{8}$ m **i** $16\frac{2}{3}$ m **j** $5\frac{1}{7}$ cm
k $9\frac{1}{3}$ mm **l** $86\frac{2}{3}$ g **m** $4\frac{14}{25}$ ℓ **n** $28\frac{7}{10}$ km **o** $42\frac{1}{2}$ m
4 **a** $13\frac{1}{3}$ m **b** $20\frac{4}{7}$ cm **c** $19\frac{4}{5}$ cm **d** $40\frac{4}{5}$ g
5 **a** **i** $\frac{1}{7} \times 5 = \frac{5}{7}$ **ii** $\frac{2}{7} \times 5 = \frac{10}{7}$ **iii** $\frac{4}{5} \times 5 = \frac{20}{5}$
  $\frac{1}{7} \times 6 = \frac{6}{7}$ $\frac{2}{7} \times 6 = \frac{12}{7}$ $\frac{4}{5} \times 6 = \frac{24}{5}$
 b **i** $\frac{1}{7} \times {}^-2 = \frac{-2}{7}$ **ii** $\frac{2}{7} \times {}^-2 = \frac{-4}{7}$ **iii** $\frac{4}{5} \times {}^-2 = \frac{-8}{5}$
  $\frac{1}{7} \times {}^-1 = \frac{-1}{7}$ $\frac{2}{7} \times {}^-1 = \frac{-2}{7}$ $\frac{4}{5} \times {}^-1 = \frac{-4}{5}$
  $\frac{1}{7} \times 0 = 0$ $\frac{2}{7} \times 0 = 0$ $\frac{4}{5} \times 0 = 0$
6 **a** 10 **b** 8 **c** $7\frac{1}{2}$ **d** 14 **e** $18\frac{2}{3}$
f 6 **g** 10 **h** $10\frac{1}{2}$ **i** 10 **j** $6\frac{2}{5}$
k $12\frac{4}{5}$ **l** $4\frac{4}{5}$ **m** 24 **n** $6\frac{3}{4}$ **o** $32\frac{1}{2}$
p $22\frac{1}{2}$
7 One possible answer is $\frac{2}{3} \times 6 = 4$.

Review 1
a $14\frac{1}{4}$ **b** $43\frac{3}{4}$ **c** $26\frac{4}{7}$

Review 2
$6\frac{1}{8}$ hours

Review 3
a $7\frac{1}{2}$ **b** $15\frac{3}{4}$ **c** $7\frac{1}{5}$ **d** $9\frac{1}{3}$ **e** $37\frac{1}{3}$ **f** $18\frac{2}{5}$ **g** $38\frac{1}{2}$

Page 114 Exercise 8

1 **a** 60 **b** 30 **c** 20
2 **a** 120 **b** 60 **c** 40 **d** 30
3 **a** $3 \div \frac{1}{4}$ **b** $7 \div \frac{1}{3}$ **c** $4 \div \frac{1}{5}$ **d** $6 \div \frac{1}{8}$
4 **a** 16 **b** 30 **c** 24 **d** 40 **e** 42
f 36 **g** 100 **h** 32 **i** 96
5 **a** 32 **b** 55 **c** 60
6 **a** 9 **b** 8 **c** 5 **d** 12
7 **a** 15 **b** 18 **c** 12 **d** 20
8 Increases

Review 1
a $6 \div \frac{1}{4}$ **b** $8 \div \frac{1}{3}$.

Review 2
a 30 **b** 32 **c** 18 **d** 63

Review 3
a 45 **b** 56

Review 4
a 12 **b** 12

Review 5
a 18 **b** 10 **c** 32

Chapter 6 Percentages, Fractions Decimals

Page 120 Exercise 1

1

Fraction	Decimal	Percentage
$\frac{2}{5}$	0·4	40%
$\frac{4}{5}$	0·8	80%
$\frac{3}{5}$	0·6	60%
$\frac{13}{20}$	0·65	65%
$\frac{11}{20}$	0·55	55%
$\frac{9}{20}$	0·45	45%
$\frac{7}{20}$	0·35	35%

Fraction	Decimal	Percentage
$\frac{4}{25}$	0·16	16%
$\frac{3}{100}$	0·03	3%
$\frac{3}{50}$	0·06	6%
$\frac{1}{8}$	0·125	12·5%
$1\frac{3}{4}$	1·75	175%
$1\frac{9}{25}$	1·36	136%
$\frac{1}{3}$	0·$\dot{3}$	$33\frac{1}{3}$%

2 a 40% **b** 28% **c** 65% **d** 54%
e 60% **f** 70% **g** $12\frac{1}{2}$% **h** $33\frac{1}{3}$%
i $66\frac{2}{3}$% **j** 146% **k** 236%

3 a 56% **b** 56·5% **c** 123% **d** 87·5%
e 568% **f** 208% **g** 145% **h** 146·5%
i 1265% **j** 0·75% **k** 0·95% **l** 400·2%
m 1508·6%

4 a 42·9% **b** 44·4% **c** 41·2% **d** 68·4%
e 52·9% **f** 111·8% **g** 86·6% **h** 11·3%
i 47·1% **j** 65·6% **k** 76·7% **l** 33·8%
m 141·7% **n** 128·6% **o** 161·9% **p** 502·7%
q 809·4%

5 a $33\frac{1}{3}$% **b** $66\frac{2}{3}$% **c** $12\frac{1}{2}$% **d** 30% **e** 40%
6 37·5%
7 10%
8 4%
9 8·3%
10 a $\frac{29}{206}$ **b** 12·6% **c** $\frac{25}{206}$ **d** 0·12
e 31·1% **f** 15·0% **g** 61·2%
11 a F **b** T **c** T **d** T **e** F **f** F **g** T **h** T
i T **j** T **k** T
12 a T **b** T **c** F **d** F **e** T **f** F **g** T
13 a 0·65 **b** $\frac{3}{8}$ **c** $\frac{32}{50}$ **d** 85%
14 a Science 80% **b** Maths 80%
15 Asha because £14 out of £150 is 9·3% (1 d.p.) and £16 out of £200 is 8%.

Review 1
a i 39% **ii** 68·5% **iii** 176·4%
b i $\frac{14}{25}$ **ii** $\frac{1}{25}$ **iii** $1\frac{11}{20}$

Review 2
Fruito, 75%

Review 3
a $\frac{1}{4}$, 0·6, 70%, $\frac{3}{4}$ **b** $\frac{1}{8}$, 0·3, $\frac{2}{5}$, 0·45 **c** 120%, $1\frac{3}{8}$, $1\frac{2}{5}$, $1\frac{1}{2}$
d 0·25, 0·5, $\frac{14}{25}$, $\frac{25}{40}$ **e** $\frac{5}{8}$, $\frac{7}{11}$, 69%, 0·78

Review 4
A STARFISH CAN TURN ITS STOMACH INSIDE OUT

Page 122 Discussion
No. Megan's results are probably correct. Because she rounded them to the nearest per cent, they probably all rounded up giving an extra 1% total.

Page 124 Exercise 2
1 a 24 **b** 18 **c** 7·5 **d** 18
e 165 m **f** 25·5 ℓ **g** 24·5 mm **h** 67·5 km
i 117 g **j** 32 kg **k** £100 **l** 264 cm
m 128 ℓ **n** 3500 mm **o** 62 m **p** 30·6 g
q 38 km **r** 31·2 sec **s** £435 **t** 36 m
u 28·8 ℓ **v** 4·68 km **w** 18·6 kg
2 £17·50
3 Yes
4 a £22·50 **b** £27 **c** £36
5 a 78 kBytes **b** 102 kBytes **c** 25·2 kBytes
6 a 0·32 m **b** 3·12 ℓ **c** 4·92 cm **d** £2·40
e 14·28 ℓ **f** 7·8 m **g** £7·55 **h** 18·6 m
i 1·625 km
7 a 63 **b** 77·5 **c** 50 **d** £8·75
8 Some possible answers are: 50% of 64, 20% of 160, 10% of 320, 5% of 640.

Review 1
a £24 **b** 52 m **c** 102 ℓ **d** 14 cm
e 300 g **f** 5775 m **g** 652·8 km **h** 4950 mℓ
i 14·7 m **j** £261

Review 2
27

Review 3
£42

Page 125 Exercise 3
1 a 3·2 m **b** 20·8 ℓ **c** 74 cm **d** £8·84
e 20·16 km **f** 22·8 g **g** 65 mm **h** 82 m
i £94·12 **j** 194·7 km **k** 982·8 cm
2 a £45·05 **b** £51·87 **c** £39·82 **d** £352·75
e £48·51 **f** £290·54 **g** £3·96 **h** £37·95
i £11·90 **j** £10·44 **k** £9·02 **l** £18
3 a 173·76 kg **b** 44·16 m **c** 24·12 cm **d** 1·596 km
e 8·25 g **f** 11·84 kg **g** 73·01 m **h** 94·29 cm
4 £11·20
5 a £79·20 **b** £11·40
6 £6365·76
7 £382·50
8 BB's bags by 68p
9 a 3·382 kg
b England 53 712 160, Wales 3 054 720, Scotland 5 027 560, Northern Ireland 1 845 560
10 5·44 m^2
11 a £7·50 **b** £22·50 **c** £108
12 a About £400 **b** About £600 **c** About £900
13 a 492 **b** 320
14 a 555 **b** 444

Review 1
a 4·5 m **b** 188 m **c** £20·14 **d** 23·12 km
e 249·9 cm **f** £290
~~23·8~~

Review 2
a 173·6 kg **b** 27·54 m **c** 505·2 mm **d** 2542 cm
e 1·206 g **f** 249·21 kg **g** 11 218·4 g **h** 12·18 m
i £14 **j** £108·75 **k** 1·925 km

Review 3
a 4465 **b** Echo Music by £3

Review 4
a 186 **b** 117

Page 127 Discussion
- They have doubled in value.
- They would be worth 6 times the original amount.
- £0
- An increase of 50% will give 150% of the original value.
- 1·25
- The answers in order are 1·25, 1·35, 1·85, 0·95.
- $0·9 \times 0·85 = 0·765$
  $= 76·5\%$
  $100 - 76·5 = 23·5$
  They decreased by 23·5%.

Page 129 Exercise 4
1 a 88   **b** 150   **c** 156   **d** 156
   e 133   **f** 54·8   **g** 5·175
2 a £189 **b** £4·80 **c** £108
3 £20·70
4 £1040
5 Julieanne £5·50, Daniel £7·70, Witek £11, Simon £3·30, Cathy £6·05
6 3680
7 £1248
8 a £345·60 **b** £14·40
9 a 18 **b** $\frac{3}{18} = \frac{1}{6}$
10 a £76·38 **b** £11·34 **c** £105·16
11 Less. The horse increased in value by 12% of £500 to £560. It then decreased in value by 12% of £560 to £492·80. So 12% of the higher value is greater than 12% of £500.
12 In the sale, items were 80% of the original price. Red dot items are 85% of 80% of the original price. $0·85 \times 0·8 = 0·68$. This is 68% of the original price. The original price has been reduced by $100\% - 68\% = 32\%$
13 a 0·7225
   b 27·75% because 85% of 85% = $0·85 \times 0·85 = 0·7225 = 72·25\%$ and $100\% - 72·25\% = 27·75\%$.

Review 1
a 269·7 **b** 4142

Review 2
144 g

Review 3
a £1·65 **b** £2·55 **c** £5·25 **d** £2·43

Page 131 Puzzle
70 km/h

Page 131 Investigation: Population increase
It will take 18 years for the rabbit population to double. If it is 2500 at the start it will take 18 years to double. No matter what the starting number it will always take 18 years to double if the rate of increase is 4%.
Other rates of increase will take different times for the population to double.
6% will take 12 years, 8% will take 10 years, 10% will take 8 years.

Page 131 Exercise 5
1 64 g
2 a 400 mℓ or 0·4 ℓ **b** 500 mℓ or 0·5 ℓ **c** 1·5 ℓ
3 0·75%
4 a £42·30 **b** £6·35
5 £452
6 £8227·56
7 30·2%
8 £12·75
9 a 1 **b** 70%
10 About 2 g
11 a 42% **b** 38% **c** 21%
   d They add to 101%. This is because we rounded the answers to the nearest % and all of them round up. If we had rounded them to 1 d.p. we would have got 41·7%, 37·5% and 20·8% which add to 100%.
12 £8
13 It doesn't matter because $17\frac{1}{2}\%$ of $12\frac{1}{2}\%$ of an amount is the same as $12\frac{1}{2}\%$ of $17\frac{1}{2}\%$ of the amount.

Review 1
6%

Review 2
a £8946 **b** £6888·42

Review 3
14·1%

Review 4
a About £4 million **b** About £9 million

7 Ratio and Proportion

Page 138 Exercise 1
1 a 10 ℓ **b** 7·5 ℓ **c** 6·25 ℓ **d** 22·5 ℓ **e** 10·25 ℓ
2 a £18   **b** £4·50   **c** £3·60   **d** £63
   e £93·75   **f** £3·85   **g** £220·80   **h** £392
   i 16   **j** 30
3 £46·75
4 £3
5 600
6 a 6 ℓ **b** $4\frac{4}{5}$ ℓ or 4·8 ℓ **c** $7\frac{1}{5}$ ℓ or 7·2 ℓ **d** 0·9 ℓ
7 a $4\frac{4}{5}$ mℓ or 4·8 mℓ **b** $14\frac{2}{5}$ mℓ or 14·4 mℓ
   c 21 mℓ **d** $1\frac{4}{5}$ mℓ or 1·8 mℓ
8 £6·20
9 a NZ$3200 **b** £125
10 a 27·2 km **b** 7·9 miles (1 d.p.)

Review 1
a £12·15 **b** £11·34

Review 2
£0·60 or 60p

Review 3
62

Page 140 Exercise 2
1 a 3 : 2 : 5   **b** 2 : 4 : 1   **c** 4 : 3 : 5   **d** 3 : 1 : 4
   e 1 : 3 : 4   **f** 4 : 7 : 12   **g** 4 : 1 : 5   **h** 8 : 20 : 3
   i 9 : 20 : 16
2 50 : 2 : 1
3 a 27 : 5 : 4 **b** 3 : 1
4 a 6 : 2 : 3 **b** 4 : 6 : 3
5 a 5 : 6   **b** 3 : 4   **c** 7 : 4   **d** 1 : 3
   e 11 : 15   **f** 31 : 50   **g** 5 : 3   **h** 3 : 1
   i 1 : 3
6 3 : 4
7 a 2 : 5   **b** 1 : 10   **c** 53 : 200   **d** 13 : 60
   e 1 : 5   **f** 3 : 2   **g** 9 : 40   **h** 50 : 17
   i 5 : 12   **j** 60 : 7

8 111 : 89
9 3 : 1
10 36 : 19
11 4 : 24 : 1
12 $3 : 5z$

Review 1
a 2 : 3 : 10 **b** 10 : 4 : 5 **c** 3 : 4 : 5

Review 2
11 : 6 : 2

Review 3
a 1 : 8 **b** 17 : 12 **c** 1 : 4

Review 4
a 15 : 2 **b** 7 : 24 **c** 41 : 200

Review 5
1 : 3

Page 142 Exercise 3
1 a 4 : 5 : 3 **b** $\frac{1}{3}$ or $33\frac{1}{3}$% or $0\cdot\dot{3}$
  c $\frac{1}{4}$ or 25% or 0·25 **d** $\frac{2}{3}$ or $66\frac{2}{3}$% or $0\cdot\dot{6}$
2 a i 2 : 4 : 5 **ii** 4 : 5 : 2
  b i $\frac{2}{11}$ **ii** $\frac{5}{11}$ **iii** $\frac{4}{11}$
  c i $\frac{2}{5}$ **ii** 2 : 5 **iii** The fraction of Maha's points that Nicola scored is the same as the ratio of Nicola's points to Maha's points.
3 a 4 : 7 **b** 5 : 7 : 4 **c** 35% **d** $\frac{1}{5}$ **e** 0·25
4 a i 4 : 15 **ii** 15 : 4 : 10 **b i** $\frac{4}{29}$ **ii** $\frac{10}{29}$ **iii** $\frac{15}{29}$

Review
a 4 : 9 **b** 3 : 9 : 4
c $\frac{3}{16}$ or 18·75% or 0·1875
d 25%

Page 143 Exercise 4
1 900 g
2 8000 cm or 80 m
3 1500 g of flour and 1000 g of sugar
4 a 1 kg **b** $1\frac{1}{2}$ kg
5 Selma is 84 and Mira is 6.
6 a 0·8 ℓ of juice and 1·4 ℓ of tea
  b 3·2 ℓ
7 980 hectares
8 a 60 **b** 10 **c** Group B
9 a i 2·25 kg **ii** 1350 mℓ **iii** 67·5 g **iv** 1125 g
  b

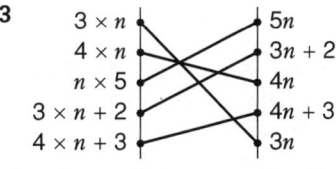

Lasagne
2.2 lb mince
1 onion
1 pint cheese sauce
1 oz tomato paste
$\frac{1}{2}$ lb lasagne sheets

10 12

11 a 0·5 km **b** 2·3 km **c** 7·65 km **d** 10·6 km
12 £2·50 and £6·25
13 40 cm

Review 1
Susan Smith got 20 and Jack Horner got 5.

Review 2
a Junior is $1\frac{1}{2}$ km, senior is $4\frac{1}{2}$ km **b** 8 km

Review 3
£50

Review 4
a 1·25 kg **b** 375 mℓ **c** 37·5 g

Page 145 Puzzle
Team B

Page 146 Exercise 5
1 a £30 and £20
  b £400 and £2000
  c £16, £48 and £64
  d £16 000, £24 000 and £32 000
  e £510, £170 and £170
  f £900, £1200 and £1500
  g £120, £240 and £360
  h £3840, £2880 and £2880
  i £8·50, £42·50 and £34
  j £49·50, £33 and £66
2 Deirdre £60 000, Siobhan £120 000
3 a 1 : 2 : 4 **b** 1 cubic metre
4 20 kg nitrogen, 20 kg potash, 30 kg lime
5 40°, 60°, 80°
6 Bob £18 000, Rick £54 000, Pete £108 000
7 Jake and Abbie should pay £47·11 each and Jenni should pay £70·67. One of them will have to pay 1p more to make up the amount to £164·90.
8 Len should get £230 000, Sharren £657 143 and Denver £262 857.

Review 1
a £16, £24 and £40 **b** £36, £72 and £84
c 325 g, 2275 g and 2600 g

Review 2
10 hectares

Review 3
7·5 kg of manure, 4·5 kg of blood and bone and 3 kg of loam.

Review 4
Margo got £225, Richard £168·75 and Lyn £56·25.

Page 147 Investigation: Triangle Paths
In all three triangles, the path that begins at A will finish at A. This will also be true if A divides the side in the ratio 1 : 3 or 2 : 3. Any ratio will give the same result.

Algebra Support

Page 153 Practice Questions
1 a £10 **b** £12
2 a 50 minutes
  b 165 minutes or 2 hours 45 minutes
  c 80 minutes or 1 hour 20 minutes

3

$3 \times n$		$5n$
$4 \times n$		$3n + 2$
$n \times 5$		$4n$
$3 \times n + 2$		$4n + 3$
$4 \times n + 3$		$3n$

4 a $3n$ **b** $4y$ **c** $4(x + 2)$ **d** $3(t - 4)$ **e** a^2
5 a 20p **b** 80p **c** 140p or £1·40
6 a 3, 4·5, 6, 7·5, 9, 10·5
  b 2, ‾1, ‾4, ‾7, ‾10, ‾13

7 a $a + b = b + a$ **b** $ab = ba$
   c $pqr = qrp$ **d** $(x + y) + z$

8 a $n - 5$ **b** $n + 6$ **c** $7n$
   d $\frac{n}{4}$ **e** $4n + 3$ **f** $5n - 2$
   g $6(n + 4)$ **h** $2(n - 3)$ **i** n^2

9 a 6 **b** 7 **c** 1 **d** 7 **e** 17 **f** 20

10 a 2, 5, 8, 11, 14
   b ⁻1, 4, 9, 14, 19
   c 20, 15, 10, 5, 0
   d 1·5, 1·8, 2·1, 2·4, 2·7

11 a $r - 4$ **b** $s + 6$ **c** $t - 3$ **d** $p - 7$

12 a

x	⁻1	0	2
y	⁻2	0	4

b

x	0	3	6
y	⁻3	0	3

c

x	5	8	10
y	5	2	0

13 a

x	⁻3	0	1
y	⁻1	2	3

   b (⁻3, ⁻1), (0, 2), (1, 3)

c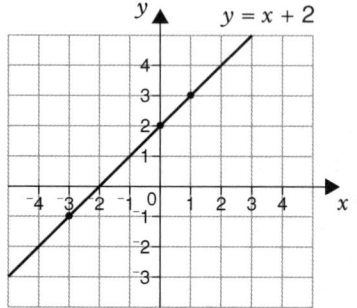

d Yes

e There are many possible answers.

14 a $2y$ **b** $3x$ **c** $7a$ **d** $10x$
   e $7y$ **f** $4p$ **g** $9b$ **h** $6n$
   i $9m$ **j** $7p$

15 a A **b** E **c** B **d** C

16 a 10, 13, 9, 11 **b** 3, 1, 11, 5

17 a 1   **b** 1   **c** n **d** x
   e 3   **f** 6   **g** $4m$ **h** $2x$
   i $4a$ **j** $5n$ **k** $6x$

18 Any number greater than 1.

19 a A   **b** $w = 10m$

20 a

Tree number	1	2	3	4	5	6
Number of strips	4	6	8	10	12	14

   b The first tree is made of 4 strips. Each time a new tree is drawn, two more strips are added.

   c The base of every tree needs 2 strips. There are $n \times 2$ branches on each tree where n is the tree number. The number of strips needed for tree n is $n \times 2 + 2$.

21 a and **c**

22 a

Hours	1	4	8	10
Cost (£)	£170	£230	£310	£350

b

Hire charges

c It tells us the scale doesn't start at 0.
d About £330
e About 6 hours

23 a 4, 5, 6, 7, 8   **b** 0, 1, 2, 3, 4
   c 3, 5, 7, 9, 11   **d** 9, 8, 7, 6, 5

24 a 23 **b** 19 **c** 41 **d** ⁻10

25 a $n + 4 = 10$ **b** $3n = 7$
   c $3n = 15$ **d** $2n + 3 = 26$

26 a

x	3	5	10	12
y	0	4	14	18

   b $y = 2(x - 3)$ or $x \rightarrow 2(x - 3)$

27 a There are many possible answers. Two are 0, 7, 14, 21, 28, ... and 3, 10, 17, 24, 31, ...
   b No

28 a 7 **b** 21 **c** 9 **d** 27 **e** 40 **f** 60 **g** 1 **h** 2 **i** 10

29 a $x \rightarrow$ [multiply by 2] $\rightarrow$ [subtract 1] $\rightarrow y$
   b $y = 2(x - 1)$

30 a $3n = 33$; $n = 11$ **b** $5x - 5 = 35$; $x = £8$
   c $9y = 180°$; $y = 20°$

Chapter 8 Expressions, Formulae and Equations

Page 157 Stand ins

1 $A = 5, B = 4, C = 6, D = 2, E = 3$
2 $P = 3, Q = 2, R = 6, S = 5, T = 4$

Page 158 Exercise 1

1 $3n - 4$, $\frac{5x}{2}$ and $3(x + 7)$ are expressions. $3n - 4 = 7$ and $x + y = 8$ are equations. Equations have an equals sign.
2 a Equation   **b** A particular value

Review
a Expression. It has no equals sign. x can have any value.
b Equation. a has a particular value.

Page 159 Exercise 2

1 No, he can't tell just by looking at it because he is not told if a and b stand for particular things or if they are just numbers.
2 Formula, because c and n stand for particular things and $c = 3n + 2$ gives the relationship between these.

Review
$D = 5p - 2$ is a formula, $5b - 2$ is an expression and $5m - 2 = 18$ is an equation.

Page 159 Exercise 3

1 a $y = 2x - 4$ is a function, $a + b = 7$ is an equation, $3x + 7$ is an expression, $C = 4b - T$ is a formula.
   b $F = ma$ is a formula, $5x + 3 = 18$ is an equation, $\frac{7n}{p}$ is an expression, $y = \frac{1}{2}x - 3$ is a function.

c $\frac{9y-2}{x}$ is an expression, $v = u^2 + 2as$ is a formula,
$y = \frac{x}{7} - 3$ is a function, $3a + 14 = {}^-7$ is an equation.

d $\frac{5a+21}{9} = 17$ is an equation, $P = VI$ is a formula, $\frac{7n-8}{2m}$
is an expression, $y = \frac{3x-2}{4}$ is a function.

2 Yes

Review

a $y = 3x + 7$ is a function, $2x + 4 = 7$ is an equation, $5n - 3$
is an expression, $v = u + at$ is a formula

b $P = nRT$ is a formula, $2p + n$ is an expression,
$2p + 7 = {}^-3$ is an equation, $y = \frac{x+3}{7}$ is a function.

Page 160 Exercise 4

1 a $4n$ **b** $7p$ **c** $5(x+4)$ **d** ab
  e st **f** $7(p+q)$ **g** abc **h** $a(b+c)$
  i $3mn$ **j** $\frac{n+m}{p}$ **k** $\frac{p+4}{q}$ **l** $\frac{a+b}{b}$
  m mnp **n** $\frac{3x+4}{y}$ **o** $\frac{5y}{7}$ **p** $\frac{10n}{13}$
  q $3xwz$ **r** $2x^2$ **s** $4y^2$

2 a $4 \times p$ **b** $4 \times (x+3)$ **c** $8 \times (n-2)$ **d** $x \times x$
  e $5 \times b \times b$ **f** $4 \times a \times b$ **g** $7 \times q \times q$ **h** $8 \times p \times q$
  i $4n \times 4n$ or $4 \times n \times 4 \times n$
  j $2 \times m \times 2 \times m$ or $2m \times 2m$
  k $3 \times p \times p$

3 a $3d$ **b** $\frac{n}{3}$ **c** mn **d** $2m+n$ **e** $2mn$
  f $\frac{a+b}{2}$ **g** $3(f+g)$

4 a $4x\,\text{m}^2$ **b** $12n\,\text{cm}^2$ **c** $12p^2\,\text{km}^2$

Review 1
a $4p$ **b** $4(b-3)$ **c** a^3 **d** $4m^2$ **e** $\frac{x+y}{a}$ **f** $35a$

Review 2
a pq **b** $3a+b$ **c** $3(x+y)$

Page 161 Exercise 5

1 The left-hand side of the equals does not equal the
right hand side. So $96 - 32 \neq 96 - 30$. The line
following this is not equal to the one above.
$96 - 30 \neq 66 - 2$.
It should be written as $96 - 32 = 96 - 30 - 2$
$= 66 - 2$.

2 a $5 \times y$ **b** $2 \times y$ **c** y^2 **d** y^2 **e** $(y-2)$
  f $(y+4)$ **g** $y+2$ **h** $8-y$ **i** y^2

3 a True **b** True **c** False **d** False

Review
a The equals sign is not true. $83 - 27 \neq 83 - 30$ and
$83 - 30 \neq 53 + 3$.

b Liam has done the operations in the wrong order. He
should have multiplied 6×4 first, then subtracted 3, to
get 21.

c $c + d + e \neq cde$, $c + d + e = d + e + c$

Page 162 Exercise 6

1 a $m + l = 7, 7 - m = l$ **b** $\frac{b}{3} = a, \frac{b}{a} = 3$
  c $3y = x, 3 = \frac{x}{y}$ **d** $\frac{b}{a} = 5, b = 5 \times a$
  e $\frac{6}{p} = q, 6 = qp$

2 a $b = p + y, r - g = b, 2p + y = r, r - y = 2p$
  b There are many possible answers. Some are:
  $b - y = p, r - b = g, r - 2p = y, 3y = b, \frac{b}{3} = y, 5y = r,$
  $\frac{r}{5} = y, \frac{g}{2} = y$

3 a Subtracting 8 **b** Dividing by 9
  c Subtracting 2 then dividing by 6
  d Adding 4 then multiplying by 3
  e Dividing by 3 then subtracting 2
  f Multiplying by 2 then adding 4

4 a C **b** B **c** C **d** B **e** C

Review 1
a $r + y = g, y = \frac{r}{3}, g = 2b$

b Possible answers are $g = r + y, g = 2b, g = 4y, g = b + 2y$.

c Possible answers are $r = g - y, r = b + y, r = 3y, r = 2b - y,$
$b = g - 2y, b = 2y, b = r - y, y = \frac{g}{4}, y = r - b, y = \frac{r}{3}, y = \frac{b}{2},$
$y = 2b - r$. There are many other answers.

Review 2
a B **b** A

Page 165 Exercise 7

1 a $6a$ **b** $8b$ **c** $15n$ **d** $12c$
  e $21m$ **f** ${}^-6x$ **g** ${}^-15p$ **h** ${}^-24m$
  i ${}^-15b$ **j** ${}^-28h$ **k** ${}^-15b$ **l** $4a$
  m $12q$ **n** $18y$ **o** ${}^-15e$ **p** ${}^-42m$
  q $12x$ **r** ${}^-35b$ **s** ${}^-48p$

2 a p^3 **b** r^4 **c** a^5 **d** q^7

3 a n^2 means 'a number multiplied by itself' and $2n$
means 'a number multiplied by 2'.
  b m^3 means 'a number multiplied by itself and then by
itself again' and $3m$ means 'a number multiplied by
3'.
  c $2b^2$ means 'a number multiplied by itself then the
answer multiplied by 2' and $(2b)^2$ means 'a number
multiplied by 2 and then the answer multiplied by
itself'.

4 a a^3 **b** p^3 **c** d^4 **d** n^4
  e m^3 **f** x^6 **g** $2p^2$ **h** $4y^3$
  i $6y^2$ **j** $6x^2$ **k** $4a^2$ **l** $15b^2$
  m $12m^2$ **n** $16b^3$

5 a b **b** $2n$ **c** $5x$ **d** 3
  e 7 **f** $\frac{3x}{2}$ **g** $\frac{5n}{3}$ **h** $\frac{7p}{4}$
  i $\frac{7y}{3}$ **j** $\frac{9a}{5}$ **k** $\frac{5b}{3}$ **l** $\frac{7y}{5}$
  m $\frac{9p}{5}$

6 a a **b** m **c** b^2 **d** a^2
  e r **f** p^3 **g** x^4 **h** n^2
  i q **j** p **k** x^2 **l** 1
  m 1 **n** a^3 **o** y **p** $\frac{1}{a}$
  q $\frac{1}{n}$

7 a n^2 **b** n^2 **c** $3m$ **d** $5b$ **e** $2a$ **f** $\frac{2x^2}{3}$

8 a $4x$ **b** $8n$ **c** $9y$

Review 1
a $20b$ **b** $21x$ **c** ${}^-10a$ **d** $24n$ **e** ${}^-56p$

Review 2
a x^4 **b** y^6 **c** r^3 **d** n^8

Review 3
IN ITS FIRST YEAR ONLY TWENTY-FIVE BOTTLES OF
COCA-COLA SOLD

Page 166 Exercise 8

1 a $4a + 8$ **b** $3x + 15$ **c** $5y + 15$ **d** $7b - 14$
  e $9a + 36$ **f** $8m - 48$ **g** $7n + 7$ **h** $4p - 8$
  i $6b - 30$ **j** $17c + 17$ **k** $14p - 28$ **l** $16m + 32$
  m $24n - 24$ **n** $2a - 26$ **o** $3n + 36$ **p** ${}^-2x - 2$
  q ${}^-3n - 9$ **r** ${}^-4a + 12$ **s** ${}^-5a + 20$

2 a $4a + 10$ **b** $15a + 18$ **c** $12x + 3$ **d** $180x - 20$
  e $30d - 174$

3 a $3x + 3y$ **b** $4a - 4b$ **c** $ab + ac$ **d** $pq - pr$
  e $2x + 6a$ **f** $3a - 6b$ **g** $6x - 15a$ **h** $8a + 4x$
  i $3x - 9a$ **j** $6y - 3x$ **k** $24a + 20b$ **l** $152a - 24b$
  m $203x - 35y$ **n** $300p - 192q$

Review
a $5x + 20$ **b** $7y + 14$ **c** $3n + 9$ **d** $8a - 16$
e $^-10p - 50$ **f** $^-3x + 12$ **g** $24x - 72$ **h** $4a + 4b$
i $6x + 3$ **j** $8a - 4$ **k** $6b + 12$ **l** $28x - 7y$
m $16n - 32m$ **n** $15x + 10y$

Page 169 Exercise 9

1 **a** $8n + 2a$ **b** $7a + 4n$ **c** $9x + 2a$ **d** $7b + 2a$
e $3a + 4x$ **f** $3n + 2a$ **g** $3x + 3b$ **h** $4x + 2n$
i $7n + 6c$ **j** $3a + 7b$ **k** $3a + b$ **l** $3n + 2a$
m $2a + 7b$ **n** $x + 2y$ **o** $7p - 8q$ **p** $6a - 16b$
q $8x - 7y$ **r** $^-n + m$
2 **a** $8a + 1$ **b** $5x + 5$ **c** $6y + 4$ **d** $9b + 5$
e $4a + 15$ **f** $5b + 4$ **g** $11y + 3$ **h** $y - 7$
i $^-12$ **j** $9y - 13$ **k** $a - 10$
3 **a** $6x^2$ **b** $12x^2$ **c** $12x^2$ **d** $3x^2$
e $6x^2$ **f** $9x^2$ **g** $10x^2$ **h** $7x^2$
i $5x^2$ **j** $6x^2$ **k** $4x^2$ **l** $9x^2$
m $8x^2 + 3x$ **n** $12x^2 + 4x$ **o** $8x^2 + 7x$
4 **a** $16x + 12$ **b** $12x + 4y - 6$ **c** $18p + 8q - 54$
5 $40x + 14y - 2$
6 **a** $4b + 5a$ **b** $15a + 7b$ **c** $4y - 2x$ **d** $x + 6y$
e $4p + 3q$ **f** $2p - 3q$ **g** $8p$ **h** $3u - 2t$
i $t - u$ **j** $2u + 3t$ **k** $^-7m + 6n$ **l** $2m - 3n$
m $^-5m - 7n$
7 Possible answers are:
a

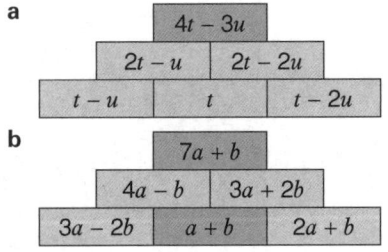

		$4t - 3u$		
$2t - u$		$2t - 2u$	
$t - u$		t		$t - 2u$

b

		$7a + b$		
$4a - b$		$3a + 2b$	
$3a - 2b$		$a + b$		$2a + b$

8 The sum of each row, column and diagonal when simplified is $3a$.

Review 1
THE AVERAGE HEIGHT OF A MAN IN THE MIDDLE AGES WAS FIVE FEET SIX INCHES

Review 2
$10x + 10y$

Review 3
a $a + 6b$ **b** $2a - 3b$ **c** $5a + 3b$

Page 171 Exercise 10

1 **a** $5x + 14$ **b** $7a + 18$ **c** $5x + 18$ **d** $5a + 17$
e $5a - 14$ **f** $10b + 23$ **g** $9x + 10$ **h** $9 + 8a$
i $19 + 10x$ **j** $7a + 6$ **k** $15y + 1$ **l** $12a + 7$
m $4 - 2y$
2 **a** $6 - 2a$ **b** $4 - 3b$ **c** $8 - 4x$ **d** $18 - 2m$
e $21 - 3b$ **f** $22 - 3x$ **g** $9 - b$ **h** $12 - n$
i $2a + 16$ **j** $n + 11$ **k** $x + 2$ **l** 3
m $7m + 2n$ **n** ^-12b **o** $7t + 3u$ **p** $26x + 22y$
q $17p - 3q$

Review
a $7x + 30$ **b** $10a + 14$ **c** $6x + 22$ **d** $12y - 15$
e $5n + 7$ **f** $7 - 3a$ **g** $7 - b$ **h** 6

Page 172 Exercise 11

1 **a** 8 **b** 6 **c** 16 **d** 6 **e** 12
f 18 **g** 3 **h** 8 **i** 28 **j** 5
k 32 **l** 8 **m** 27 **n** 64 **o** 128

2 **a** 9 **b** 8 **c** 8 **d** 12 **e** 10
f 29 **g** 64 **h** 129 **i** 6 **j** 20
k 50 **l** 25 **m** 4
3 **a** 6 **b** 5 **c** 9 **d** 25 **e** 4
4 **a** 18 **b** 15 **c** 32 **d** 6 **e** $2·4$
5 **a** $9·5$ **b** $6·5$ **c** 15 **d** $14·5$ **e** $^-4·4$
6 **a** 4 **b** $^-10$ **c** 24 **d** $^-18$ **e** 1 **f** 3
7 **a** 7 **b** $^-5·5$ **c** $22·5$ **d** $^-13·5$ **e** $0·4$ **f** 9
8 **a** 7 **b** 6 **c** $^-3$ **d** $^-7$
9 **a** 101 **b** 82 **c** 170 **d** 577
10 **a** 15 **b** 55 **c** 136
11 **a** 45 **b** 91 **c** 351

Review 1
GIRAFFES ONLY SLEEP FOR FIVE MINUTES A DAY

Review 2
a 6 **b** $2\frac{1}{2}$ or $2·5$ **c** 32 **d** 20 **e** 1 **f** 8 **g** 0 **h** 8 **i** 6

Review 3
a 9 **b** 90 **c** 170 **d** 135

Page 174 Investigation: Substituting

3 **a** It is a magic square with a row, column and diagonal total of 21.
b All values of x, y and z make this a magic square. If positive values are required, $x > y + z$.
If x is not greater than $y + z$, some of the boxes will have negative numbers. The square will still be a magic square.
c $x = 5$, $y = 3$, $z = 1$
d No. There are other magic squares with the numbers 1 to 9 which **look** different, but each row and column has the same numbers as in this one but in different places.
To get these magic squares the positions of the expressions would have to be changed.
e All decimal values of x, y and z make a magic square. For all boxes to have positive numbers, $x > y + z$.

Page 175 Exercise 12

1 **a** 10 m   **b** 12·8 m   **c** 12 m
2 **a** 30   **b** 1·8   **c** 9·6
3 **a** 277 K   **b** 283 K   **c** 323 K   **d** 302 K   **e** 272 K
f 264 K   **g** 254 K
4 **a** 80p   **b** £3·20   **c** £36·30   **d** £48·90   **e** £53
5 **a** 1280 cm^2 **b** 16 184 mm^2 **c** 338·44 cm^2
d 0·8608 m^2
6 **a** 20·5 m   **b** 22 m   **c** 28 m   **d** 20·125 m
7 **a** £5   **b** £12·80   **c** £4·50   **d** £787·50   **e** £409·06
8 **i** **a** 12 mℓ **b** 8 mℓ **c** 6 mℓ **d** 2 mℓ **e** 14 mℓ
ii **a** 6 mℓ **b** 2 mℓ **c** 3 mℓ **d** 5 mℓ
iii You still get a smaller dose than the normal adult dose. The formula can only be used for children.
9 **a** 8 m/s   **b** 10 m/s   **c** 9 m/s

Review 1
a 133 cm^2 **b** 76 cm^2 **c** 43·9 cm^2 **d** 49·75 cm^2

Review 2
a 27 mℓ **b** 41 mℓ **c** 40·5 mℓ **d** 135·5 mℓ **e** 142·25 mℓ

Review 3
a 44·1 m   **b** 490 m

Page 176 Puzzle

a
$$4531 \quad \text{or} \quad 7692$$
$$-4270 \qquad\quad -7350$$
$$\overline{261} \qquad\quad \overline{342}$$

b
$$23938 \quad \text{or} \quad 25758$$
$$-6405 \qquad\quad -6403$$
$$\overline{17533} \qquad\quad \overline{19355}$$

Page 177 Exercise 13

1 a 3 **b** 6 **c** 0·6
2 a 60 **b** 42
3 a 6 **b** 5 **c** 8 **d** 7
4 a 1 **b** 7 **c** 3
5 a £15 **b** 50 **c** £31·50 **d** 78
6 a 3 **b** 3·5

Review
a 3 **b** 0·9 **c** 3 **d** 2·25

Page 178 Exercise 14

1 a $30n$ **b** $29n + 4$
2 a $10p$ **b** $14p - 3$
3 a £$(12t)$ **b** £$(20t - 2)$
4 a $2n$ **b** $10n$
5 a $(3n + 5)$ kg
　 b Pan 1 $(2n + 2)$ kg, Pan 2 $(4n + 5)$ kg, Pan 3 $(5n + 4)$ kg
6 a $3p$ **b** $3p - 4$ **c** $\frac{3p}{4} - 2$
7 £$3(2x + 3)$
8 $\frac{t}{5} + 3$

Review 1
a $20n$ **b** $24n - 8$

Review 2
a $8x + 5$ **b** $8n - 3$ **c** $3(8x + 5)$

Page 180 Exercise 15

1 a $2a + 2b$ **b** $a + a + b + b = 2a + 2b$
2 a and b Two possible expressions are
　 $a + a + a + a + a + b + b + b + b$ and $5a + 4b$.
　 c $a + a + a + a + a + b + b + b + b = 5a + 4b$
3 a Yes. The area of the whole shape is $8 \times 6 = 48$. The area of the cupboard is $2 \times n$. The area of the new living room is $48 - 2n$.
　 b $2(8 - n) + 32 = 16 - 2n + 32$
　　　　　　　　　 $= 48 - 2n$
　 c A possible expression is $6(8 - n) + 4n$.
　 d $6(8 - n) + 4n = 48 - 6n + 4n$
　　　　　　　　　　 $= 48 - 2n$
4 a Three possible expressions are $36 - 3x$,
　 $24 + 3(4 - x)$ and $6x + 9(4 - x)$.
　 $24 + 3(4 - x) = 24 + 12 - 3x$
　　　　　　　　 $= 36 - 3x$
　 and
　 $6x + 9(4 - x) = 6x + 36 - 9x$
　　　　　　　　 $= 36 + 6x - 9x$
　　　　　　　　 $= 36 - 3x$
　 b Three possible expressions are $80 - 3p$,
　 $8(10 - p) + 5p$ and $50 + 3(10 - p)$.
　 $8(10 - p) + 5p = 80 - 8p + 5p$
　　　　　　　　 $= 80 - 3p$
　 and
　 $50 + 3(10 - p) = 50 + 30 - 3p$
　　　　　　　　 $= 80 - 3p$
　 c Three possible expressions are $5000 - 20n$,
　 $50(100 - n) + 30n$ and $2 \times 15 \times 100 + 20(100 - n)$.
　 $50(100 - n) + 30n = 5000 - 50n + 30n$
　　　　　　　　　 $= 5000 - 20n$
　 and
　 $2 \times 15 \times 100 + 20(100 - n) = 3000 + 2000 - 20n$
　　　　　　　　　　　 $= 5000 - 20n$

5 a $2(8n + 4n) = 24n$ **b** $6n$ **c** $36n^2$
　 d Area of rectangle $= 32n^2$. The square has the greater area by $4n^2$.

Review
Three possible expressions are $10b - 12$, $10(b - 4) + 28$ and $3(b - 4) + 7b$.
$10(b - 4) + 28 = 10b - 40 + 28 \qquad 3(b - 4) + 7b = 3b - 12 + 7b$
$ = 10b - 12 \qquad\qquad\qquad\quad = 3b + 7b - 12$
$ = 10b - 12$

Page 181 Puzzle

1 $n + n + 1 + n + 2 = 3n + 3$. $3n + 3$ is always divisible by 3.
　 $\frac{3n + 3}{3} = n + 1$ because $\frac{3n + 3}{3} = \frac{3(n + 1)}{3} = n + 1$.
2 $n + n + 1 + n + 2 + n + 3 + n + 4 = 5n + 10$. $5n + 10$ is always divisible by 5. $\frac{5n + 10}{5} = n + 2$ because $\frac{5n + 10}{5} = \frac{5(n + 2)}{5} = n + 2$.

Page 181 Exercise 16

1 a $f + s = 1500$ **b** $s = 2f$ **c** $s + 400 = 3f$ **d** $\frac{s}{2} = \frac{2f}{5}$
2 a $c = 4m$ **b** $c - 4 = 3m$ **c** $\frac{c}{2} = 2m$
3 a $x + y = 66$ **b** $\frac{y}{2} = \frac{3}{5}x$ **c** $y - 24 = 2(x - 24)$

Review
a $p + q = 1000$ **b** $p - 200 = q$ **c** $p + 300 = 3q$ **d** $\frac{q}{2} = \frac{p}{3}$

Page 182 Exercise 17

1 a $T = 2x$ **b** $W = 20a$ **c** $g = \frac{p}{8}$ **d** $C = 20d + 50$
　 e $A = 4s + 6$
2 $n = \frac{n_1 + n_2}{2}$

Review
a $c = 3h$ **b** $c = 20n + 500$

Page 183 Investigation: Deriving Formulae

1 $p_1 = 4t + 2$
　 If each table seated 3 along each side the formula is $p = 6t + 2$. If each table seated 3 along each side and 2 at each end the formula is $p = 6t + 4$. If each table seated a along each side and b at each end the formula is $p = 2at + 2b$.
2 $f = 4n + 2$
3 $s = (n - 2) \times 180°$
4 $G = n^2 + 12n + 28$

Page 185 Exercise 18

1 a 5·5 **b** 3·5 **c** 6·4 **d** 6 **e** 8·25
　 f 2·5625 **g** 4 **h** 2·75 **i** 2·125 **j** 10·75
　 k 12·8 **l** 4·5 **m** 7·125 **n** 10·6 **o** 21·25
　 p 2 **q** 3 **r** 3 **s** 5 **t** 7
　 u 6 **v** 2·5 **w** 3·5 **x** 5
2 a 3 **b** 6 **c** 8 **d** 10 **e** 0
　 f 2 **g** 16 **h** 3 **i** 1 **j** 8
　 k 1 **l** 5 **m** 3 **n** 2·5 **o** 2·75
　 p 1·8 **q** 1·2 **r** 2·5 **s** 3 **t** 1
3 a $\frac{n + 5}{2} = 11$; $n = 17$ **b** $2(n - 8) = 30$; $n = 23$
　 c $2n + 3 = 15$; $n = 6$ **d** $3n - 5 = 19$; $n = 8$
4 a $3(p - 5) = 60$; $p = £25$ **b** $9(n + 2) = 45$; $n = £3$
　 c $4(n + 3) = 36$; $n = 6$

Review 1
a 3·4 **b** 8 **c** 5·25 **d** 3 **e** 5

Review 2
a 0 **b** 7 **c** 1 **d** 7 **e** 4 **f** 6 **g** 2·4 **h** 1·75

Review 3
$5(n + 25) = 200$; $n = £15$

Page 186 Exercise 19

1 **a** 4 **b** 2 **c** 4 **d** 5 **e** 6 **f** 1 **g** 5 **h** 1·5
   i 3·5 **j** 11 **k** 3 **l** 10
2 **a** $4w + 8$ **b** $4w + 8 = 56$ **c** 12
3 $x + x + 2 + 2x = 18$, $x = £4$. Marjorie spent £4,
   Alan spent £6 and Olivia spent £8.
4 $n + n + 12 + n + 8 = 80$, $n = 20$. The first box had 20
   pieces, the second 32 and the third 28.
5 $n + n + 2 + n + 4 + n + 6 = 96$ or $4n + 12 = 96$, $n = 21$
6 **a** $n = 14$
   b i $24 + n + n + 16 = 76$ or $2n + 40 = 76$, $n = 18$
      ii 20 **iii** 8
7 **a** $l + 2l = 204$, $l = 68$ km. The first leg is 68 km.
   b $e + e - 17 = 151$, $e = 84$. Ana got 84 marks in her
      Maths exam.
   c $5 + 3s = 16$, $s = 4m^3$. The smaller lorry carried 4 m^3
      of shingle.
8 $(30 - x) + (23 - x) = 25$
   $\qquad 53 - 2x = 25$
   $\qquad\quad 2x = 28$
   $\qquad\quad\ x = 14$

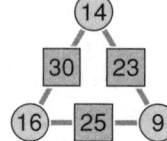

Review 1
a 4 **b** 7 **c** 8 **d** 4 **e** 3·5 **f** $x = 3$

Review 2
$4w + 20 = 80$; $w = 15$ m

Review 3
a $2l - 3 = 13$; $l = 8$ cm. Each piece was 8 cm.
b $\frac{5n - 6}{2} = 7$; $n = 4$ **c** $l + 2l - 2 = 16$; $l = 6$ cm.
   She kept 6 cm.

Review 4
$(22 - n) + (20 - n) = 26$
$\qquad 42 - 2n = 26$
$\qquad\quad 2n = 16$
$\qquad\quad\ \mathbf{n = 8}$

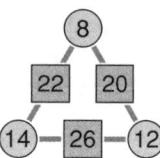

Page 189 Exercise 20

1 **a** 2 **b** 9 **c** 11 **d** 15 **e** 6 **f** 10 **g** 11 **h** 12
   i 4 **j** 12 **k** 2
2 **a** 3 **b** 4 **c** 5 **d** 3 **e** 6 **f** 3 **g** 5 **h** 6
   i 4 **j** 8 **k** 3
3 **a** 6 **b** 6 **c** 3 **d** 4 **e** 2·5 **f** 7·5 **g** 3·75
4 **a** 8 **b** 12 **c** 10 **d** 2 **e** 8 **f** 48 **g** 128
5 **a** 2 **b** 2 **c** 2 **d** 1 **e** 7

Review

¹1	1		²2	³1
2		⁴4	⁵3	⁶2
	⁷1	⁸1		4
⁹1	3	¹⁰4	¹¹2	
	¹²1	¹³2	¹⁴1	
¹⁵3		0		1

Page 190 Exercise 21

1 **a** 10   **b** 2   **c** 14   **d** 2   **e** 4
   f 5   **g** 5   **h** 4   **i** 2   **j** 3
   k 5   **l** 2·5   **m** $^-$2   **n** $^-$6   **o** 4
   p 3   **q** $^-$8   **r** $^-$1
2 **a** $n = 2·5°$ **b** $n = 34°$ **c** $n = 22°$
3 $2n + 13 = 3n + 8$; $n = 5$. There are 5 discs in each box.
4 $6(p - 3) = 4p - 4$; $p = 7$

5 **a** $3p - 1 = 4p - 9$; $p = 8$ **b** $2(p + 6) = 4p - 2$; $p = 7$
   c $4p + 6 = 5p + 2·5$; $p = 3·5$

Review 1
a 2 **b** 12 **c** 1 **d** 2 **e** $^-$0·5 **f** 9

Review 2
a $4x + 6 = 3x + 16$; $x = 10$ There are 10 chocolates in each
   box.
b $5x - 3 = 4x + 12$; $x = 15$. There are 15 chocolates in each
   box.

Page 191 Puzzle
The answer is 25.

Page 192 Exercise 22

1 **a**

Number of pizzas	1	2	3	4
Cost (£)	4	8	12	16

   b The ratio is 4 : 1 and is constant.
   c Yes. As the number of pizzas increases, the cost
      increases. The ratio of *number : cost* is always the
      same.
   d Yes

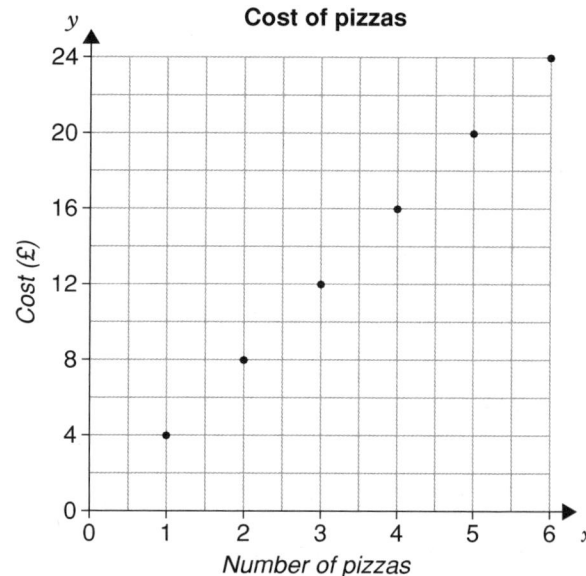

   e $y = 4x$
   f £32. One possible answer is 'I substituted 8 into the
      equation $y = 4x$ to get 32'.

2 **a**

Number	1	2	3	4	5	...
Original price	£4	£8	£12	£16	£20	...
Sale price	£3	£6	£9	£12	£15	...

   b The ratio is 3 : 4 and is constant. **c** Yes

d The points lie in a straight line. This is what we would expect because the ratio is constant.

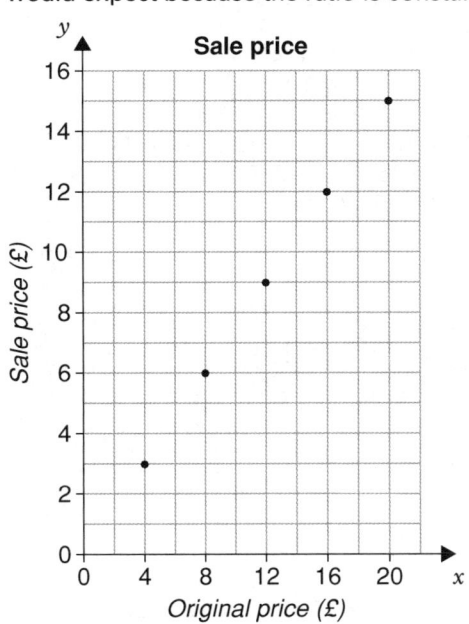

e $y = \frac{3}{4}x$ **f** £36

3 **a** Possible answers are (2, 250), (4, 500), (6, 750), (8, 1000), (10, 1250).

b Yes, it is 125 : 1.

c

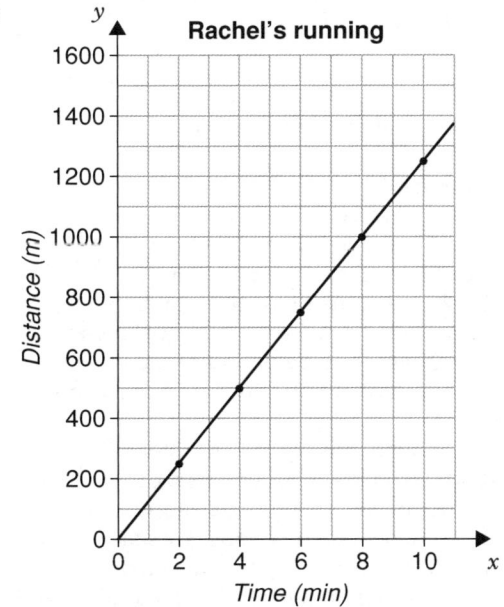

d $y = 125x$ **e i** 2250 m   **ii** 1125 m

Review

a

Miles	5	10	15	20	25
Kilometres	8	16	24	32	40

b 8 : 5. The ratio is constant, the same, for every pair.

c Yes. The number of kilometres increases constantly as the number of miles increases constantly.

d The points lie in a straight line.

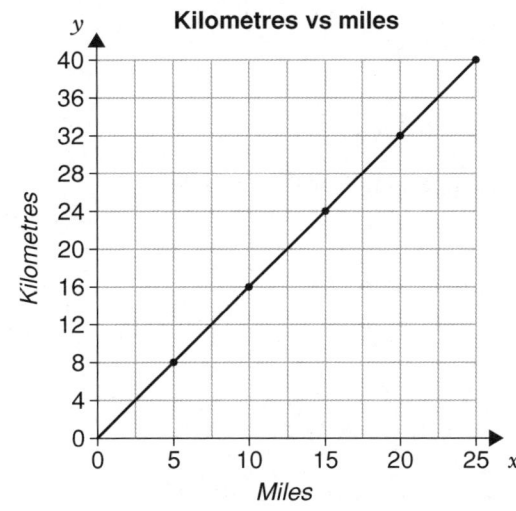

e $y = \frac{8}{5}x$ **f i** 72 km   **ii** 168 km

---

9 Sequences and Functions

Page 200 Exercise 1

1 **a** 1, 5, 9, 13, 17, 21
   b 20, 17, 14, 11, 8, 5
   c 1, 1·4, 1·8, 2·2, 2·6, 3, 3·4
   d 1, 2, 4, 8, 16, 32, 64
   e 2, 3, 5, 8, 13, 21
   f 3, 5, 8, 12, 17, 23

Review

a 50, 60, 70, 80, 90, 100
b 3, 9, 27, 81, 243

Page 202 Exercise 2

1 **a** 2, 6, 18, 54, 162
   b 1000, 100, 10, 1, $\frac{1}{10}$ (or 0·1)
   c 1, 0·5, 0·25, 0·125, 0·0625
   d 1, ⁻3, 9, ⁻27, 81
   e ⁻2, 1, ⁻0·5, 0·25, ⁻0·125
   f $\frac{1}{32}, \frac{1}{16}, \frac{1}{8}, \frac{1}{4}, \frac{1}{2}$

2 **a** 16, 32   **b** 81, 243   **c** 48, 96
   d 8, 4   **e** 0·1, 0·01 or $\frac{1}{10}, \frac{1}{100}$ **f** 0·08, 0·016
   g $\frac{1}{32}, \frac{1}{64}$

Review

a 3, 6, 12, 24, 48, 96
b 5, ⁻25, 125, ⁻625, 3125, ⁻15 625
c 100, 20, 4, 0·8, 0·16, 0·032
d ⁻3, 3, ⁻3, 3, ⁻3, 3

Page 203 Exercise 3

1 **a** 0, 1, 3, 6, 10, 15   **b** 5, 7, 10, 14, 19, 25
  c 100, 99, 96, 91, 84, 75   **d** 2, 4, 8, 14, 22, 32
  e 10, 8, 5, 1, ⁻4, ⁻10
2 **a** 5, 9, 13, 17, 21, ...   **b** 21 cm
3 **a** 1, 2, 4, 7, 11, 16   **b** 8th
4 **a** 7, 12, 17, 22, 27   **b** 7 km

Review 1
a 2, 3, 5, 8, 12, 17   **b** 1, 4, 9, 16, 25, 36
c 60, 58, 54, 48, 40, 30

Review 2
a 12, 17, 22, 27, 32, 37, ...   **b** 37

Page 205 Investigation: Pascal's Triangle

																															Sum of numbers
row 0															1																$1 = 2^0$
row 1														1		1															$2 = 2^1$
row 2													1		2		1														$4 = 2^2$
row 3												1		3		3		1													$8 = 2^3$
row 4											1		4		6		4		1												$16 = 2^4$

1 5 10 10 5 1 $32 = 2^5$
1 6 15 20 15 6 1 $64 = 2^6$
1 7 21 35 35 21 7 1 $128 = 2^7$
1 8 28 56 70 56 28 8 1 $256 = 2^8$
1 9 36 84 126 126 84 36 9 1 $512 = 2^9$
1 10 45 120 210 252 210 120 45 10 1 $1024 = 2^{10}$
1 11 55 165 330 462 462 330 165 55 11 1 $2048 = 2^{11}$
1 12 66 220 495 792 924 792 495 220 66 12 1 $4096 = 2^{12}$
1 13 78 286 715 1287 1716 1716 1287 715 286 78 13 1 $8192 = 2^{13}$
1 14 91 364 1001 2002 3003 3432 3003 2002 1001 364 91 14 1 $16384 = 2^{14}$
1 15 105 455 1365 3003 5005 6435 6435 5005 3003 1365 455 105 15 1 $32768 = 2^{15}$

The second numbers in each row, except row 0, give the row number. They are the counting numbers.
The third number in each row (from row 3 onwards) are the triangular numbers.
The sum of the second and third numbers in each row is always a triangular number.
The sums of the rows are all powers of 2.
Rows 0, 1, 3, 7, 15 contain only odd numbers.
Other facts about Pascal's Triangle:

In every odd-numbered row, the middle numbers are the same.
Patterns are formed if you write down more rows and then colour the multiples of 2 *or* 3 *or* 5 *or* 7.
This is the pattern formed for the odd numbers.

Page 205 Exercise 4

1 **a, b** and **d**
2 **a** 6   **b** 2   **c** 72   **d** ⁻12   **e** 4
  f 0·3   **g** 1⅗   **h** 1·25
3 **a** 5   **b** 3   **c** ⁻9   **d** 3   **e** ⁻2
  f 0·3   **g** ⁻⅖   **h** ⁻0·25
4 **a** 1, 4, 7, 10, 13, 16
  b 10, 15, 20, 25, 30, 35
  c 7, 14, 21, 28, 35, 42
  d 100, 95, 90, 85, 80, 75
  e 63, 54, 45, 36, 27, 18
  f 4, 3, 2, 1, 0, ⁻1
  g 0, ⁻3, ⁻6, ⁻9, ⁻12, ⁻15
  h ⁻20, ⁻17, ⁻14, ⁻11, ⁻8, ⁻5
  i 0·5, 0·75, 1, 1·25, 1·5, 1·75
5 **a** 5, 5·5, 6, 6·5, 7, 7·5
  b 2, 1·8, 1·6, 1·4, 1·2, 1
  c 5·5, 7, 8·5, 10, 11·5, 13
  d ⁻2, ⁻4, ⁻6, ⁻8, ⁻10, ⁻12

e 2, ⁻3, ⁻8, ⁻13, ⁻18, ⁻23
f 0·3, 0·9, 1·5, 2·1, 2·7, 3·3
g ⁻3, ⁻7, ⁻11, ⁻15, ⁻19, ⁻23
h 102, 110, 118, 126, 134, 142
i 1·5, 2·2, 2·9, 3·6, 4·3, 5
j 98, 96·1, 94·2, 92·3, 90·4, 88·5
6 **a** 9 and 13, 15   **b** 12 and 20, 24
  c 22, 20 and 16   **d** 87, 80, 73 and 59
7 One possible answer is:

8 36
9 **a** 1 or 2 or 3 or 5 or 6 or 10 or 15 or 30
  b 1 or 2 or 3 or 4 or 6 or 8 or 12 or 16 or 24 or 48

Review 1
a 7, 17, 27, 37, 47   **b** 3, 7, 11, 15, 19
c 3, ⁻2, ⁻7, ⁻12, ⁻17   **d** ⁻10, ⁻7, ⁻4, ⁻1, 2
e ⁻11, ⁻14, ⁻17, ⁻20, ⁻23

Review 2
a $a = 11, d = 6$ **b** $a = 14, d = ⁻5$
c $a = ⁻6, d = 3$ **d** $a = 0·4, d = ⁻0·4$

Page 207 Exercise 5

1 Possible answers are:
  a 60, 72, 84; ascends by equal steps
  b 20, 23, 26; ascends by equal steps
  c 16, 22, 29; ascends by unequal steps
  d 25, 12·5, 6·25; descends by unequal steps
  e 15, 21, 28; ascends by unequal steps
  f 10 000, 100 000, 1 000 000; ascends by unequal steps
  g 20, 15, 9; descends by unequal steps
  h 16, 8, 4; descends by unequal steps
  i 96, 91, 86; descends by equal steps
  j 20, 8, ⁻6; descends by unequal steps
  k 64, 49, 36; descends by unequal steps
2 **a** Possible answers are 0,1,3, 7, 15, 31, 63, ... *rule* double and add 1 or 0, 1, 3, 6, 10, 15, 21, ... *rule* count on in steps of 1, 2, 3, 4, 5, ...
  b Possible answers are 2, 4, 6, 8, 10, 12, 14, ... *rule* add 2 or 2, 4, 6, 10, 16, 26, 42, ... *rule* add the two previous terms together.
  c Possible answers are 2, 5, 9, 14, 20, 27, 35, ... *rule* count on in steps of 3, 4, 5, 6, 7, ... or 2, 5, 9, 16, 27, 45, ... *rule* add the two previous terms then add 2.

Review 1
a 15, 18, 21; **A**   **b** 96, 90, 84; **C**
c 100, 121, 144; **B**   **d** 5, 0·5, 0·05; **D**
e 16, 22, 29; **B**   **f** 44, 58, 74; **B**

Review 2
Possible answers are:
2, 4, 8, 16, 32, ... *rule* double
2, 4, 8, 14, 22, 32, ... *rule* count on in steps of 2, 4, 6, 8, 10, ...

Page 209 Exercise 6

1 a 6, 10, 14, 18, 22, 26
  b 2, ⁻1, ⁻4, ⁻7, ⁻10, ⁻13
  c 1, 2, 4, 8, 16, 32
  d 4, ⁻4, 4, ⁻4, 4, ⁻4
  e 0, 1, 3, 6, 10, 15
  f 100 000, 10 000, 1000, 100, 10, 1
  g 1, ⁻1, 1, ⁻1, 1, ⁻1
  h 3, 4, 7, 12, 19, 28
  i 2, 4, 8, 14, 22, 32
  j 100, 99, 97, 94, 90, 85
  k 1, 3, 4, 7, 11, 18
  l 2, 3, 5, 8, 13, 21
  m 1, 5, 13, 29, 61, 125
  n 1, ⁻5, ⁻4, ⁻9, ⁻13, ⁻22
  o ⁻3, 4·5, ⁻3, 4·5, ⁻3, 4·5

2 a 5, 9, 13, 17, 21   b 0, 3, 6, 9, 12
  c ⁻4, ⁻2, 0, 2, 4   d 9, 8, 7, 6, 5
  e 48, 46, 44, 42, 40   f 1·5, 2, 2·5, 3, 3·5
  g 60, 54, 48, 42, 36   h $1\frac{1}{2}, 2\frac{1}{2}, 3\frac{1}{2}, 4\frac{1}{2}, 5\frac{1}{2}$
  i 0·2, 0·4, 0·6 0·8, 1

3 i a 81   b 57   c 34   d ⁻10   e 10
     f 11   g ⁻54   h $20\frac{1}{2}$ i 4
  ii a 101   b 72   c 44   d ⁻15   e 0
     f 13·5 g ⁻84   h $25\frac{1}{2}$ i 5
  iii a 69   b 48   c 28   d ⁻7   e 16
     f 9·5   g ⁻36   h $17\frac{1}{2}$ i 3·4
  iv a 153   b 111   c 70   d ⁻28   e ⁻26
     f 20   g ⁻162 h $38\frac{1}{2}$ i 7·6

4 a *first term* 5 *rule* add 4
  b *first term* 0 *rule* add 3
  c *first term* ⁻4 *rule* add 2
  d *first term* 9 *rule* subtract 1
  e *first term* 48 *rule* subtract 2
  f *first term* 1·5 *rule* add 0·5
  g *first term* 60 *rule* subtract 6
  h *first term* $1\frac{1}{2}$ *rule* add 1
  i *first term* 0·2 *rule* add 0·2

5 a 2, 5, 8, 11, 14, 17   b 10, 20, 40, 80, 160, 320
  c 7, 8, 11, 16, 23, 32

6 Possible answers are:
  a *first term* 4 *rule* add 2
  b *first term* 3 *rule* add 2
  c *first term* 4 *rule* add 4
  d *first term* 10 *rule* multiply by 10
  e *first term* 6 *rule* add 10

7 ⁻3, 2, ⁻1

Review 1
a 400 000, 80 000, 16 000, 3200, 640
b ⁻1, ⁻2, ⁻4, ⁻8, ⁻16
c 1, 2, 5, 10, 17, ...
d 1, 1, 2, 3, 5
e ⁻4, 10·5, ⁻18·5, 39·5, ⁻76·5

Review 2
a 9, 14, 19, 24, 29   b 90, 81, 72, 63, 54
c 1·8, 3·8, 5·8, 7·8, 9·8 d 105, 100, 95, 90, 85
e 1·5, 2·5, 3·5, 4·5, 5·5 f 0·1, 0·2, 0·3, 0·4, 0·5

Review 3
a 109 b ⁻90 c 41·8 d 5 e 21·5 f 2·1

Review 4
a *first term* 9 *rule* add 5
b *first term* 90 *rule* subtract 9
c *first term* 1·8 *rule* add 2
d *first term* 105 *rule* subtract 5
e *first term* 1·5 *rule* add 1
f *first term* 0·1 *rule* add 0·1

Review 5
Possible answers are:
a i 6 or 12 or any multiple of 6   ii Any odd number
  is correct.   iii Any number less than ⁻12 is correct.
b i 0 or any positive or negative multiple of 5 is correct.
  ii Any number that ends in 5 is correct.

Page 210 Investigation: Fibonacci sequence

1 The sum of the first six terms is 20, which is one less
  than the eighth term.   The sum of the first seven
  terms is 33, which is one less than the ninth term.
  The sum of the first nine terms is 88.
2 Richie is right.
3 The sequence of rabbits at the beginning of each
  month is 2, 2, 4, 6, 10, 16, 26, 42, 68, ...
  This sequence is made by starting with 2, 2 and
  adding the two previous terms to get the next term.
4 The sequence given by the number of cows at the
  beginning of each year is 1, 1, 2, 3, 5, 8, ... which is the
  Fibonacci sequence.

Page 212 Exercise 7

1 Possible answers are:
  a i Multiples of 5 starting at 10
    ii Multiples of 5 starting at 25
    iii Multiples of 5 starting at 0
  b i Numbers 1 more than multiples of 5 starting at 6
    or numbers with a difference of 5 between
    consecutive terms, starting at 6.
    ii Numbers with a difference of 5 between
    consecutive terms, starting at 7.
    iii Numbers 2 less than multiples of 5 starting at 3 **or**
    numbers with a difference of 5 between
    consecutive terms, starting at 3.

2 a 6   b 3   c 7   d 11   e 15   f 9   g 9

3 0 and any positive or negative multiple of 3

4 a Ascending multiples of 8 between consecutive
    terms starting at 16
  b Numbers with a difference of 8 starting at 10
  c Numbers with a difference of 8 between
    consecutive terms starting at 0.
    All terms except the first, 0, are multiples of 8.

5 a 8 times table backwards, starting at 80
  b $T(n) = b - 7n$ where b is any number

6 Any number between 0 and 9

7 a Ascending multiples of 5 starting at 5
  b Ascending numbers with a difference of 5 between
    consecutive terms starting at 3 or ascending
    numbers two less than a multiple of 5 starting at 3
  c Ascending multiples of 5 starting at 10
  d Ascending numbers with a difference of 5 between
    consecutive terms, starting at 0, all except the first
    term 0 are multiples of 5
  e Ascending numbers with a difference of 10
    between consecutive terms, starting at 12, **or**
    ascending numbers which are two more than a
    multiple of 10 starting at 12
  f Ascending numbers with a difference of 7 between
    consecutive terms, starting at 4, **or** all numbers 3 less
    than a multiple of 7 starting at 4
  g Descending numbers, with a difference of 2
    between consecutive terms, starting at 4
  h 9 times table descending, starting at 90

8 a 9n   b 5n − 3   c 44 − 4n

Review 1
a Any value of b that is a multiple of 6.
b Numbers 1 less than the multiples of 6, starting at 5, **or**
  numbers with a difference of 6 between consecutive
  terms, starting at 5

Review 2
a Multiples of 4, starting at 4
b Multiples of 4, starting at 8
c Multiples of 4, starting at 20
d Numbers 1 less than multiples of 4 with a difference of 4 between consecutive terms, starting at 3
e Numbers with a difference of 4 between consecutive terms, starting at 6

Review 3
a The 6 times table backwards
b $T(n) = 6n + b$ where b is any number
c $T(n) = a - 7n$ where a is any number

Page 214 Exercise 8
1 a

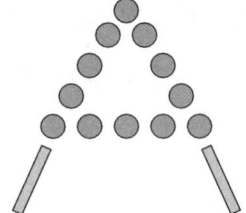

b 3, 6, 9, 12, 15, 18, 21, ...
We add 3 each time because we add one more circle to each side of the triangle to make the next triangle.
c $3n$
Each term is 3 lots of n circles where n is the logo size. So in logo size 1 there are 3×1 circles, in size 2, 3×2 circles, in size 3, 3×3 circles and so on up to size n when there are $3 \times n$ circles.

2 a 1, 4, 9, 16, ... The next stack has another layer. Each added layer increases by the next odd number 3, 5, 7, 9, ...
b The nth stack will have $n \times n$ or n^2 boxes.
The sequence gives the square numbers. Each stack can be arranged to give a square.

3 a

Alien number	1	2	3	4	5
Number of red squares	2	4	6	8	10
Number of green squares	4	8	12	16	20
Total number of squares	6	12	18	24	30

b The number red squares is twice the alien number. The expression is $2n$.
c There are n squares at each corner of the alien. There are $4n$ squares in total.
d Add the expressions in **b** and **c**. The expression for the total number of squares is $6n$.

4 a

1	2	3	4	...	10	...	20	...	30
6	10	14	18	...	42	...	82	...	122

b No

5 a

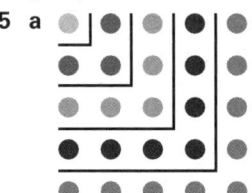

b i 4   **ii** 9   **iii** 16   **iv** 25
c i 36   **ii** 400   **iii** n^2
d i The sum of the dots in the diagonals is
$1 + 2 + 3 + 4 + 3 + 2 + 1$
The total number of dots is 16 or 4^2.
So $1 + 2 + 3 + 4 + 3 + 2 + 1 = 4^2$.
ii The sum of the dots in the diagonals of diagram 4 is $1 + 2 + 3 + 4 + 5 + 4 + 3 + 2 + 1$.
The total number of dots is 25 or 5^2.
So $1 + 2 + 3 + 4 + 5 + 4 + 3 + 2 + 1 = 5^2$.
iii The sum of the dots in the diagonals of diagram n is $(n + 1)^2$.

6 a The results will depend on what you drew.
b The number of slabs needed to surround a pond of length l and width w is $2l + 2w + 4$.
There are 4 slabs, one at each corner in every pond.
There are l slabs along each length of the pond, so $2l$ altogether.
There are w slabs along each width of the pond, so $2w$ altogether.
The total number of slabs is $2l + 2w + 4$.

7 Each vertex has $n - 3$ diagonals from it because it doesn't have a diagonal to the two adjacent vertices or to itself. In each shape there are n vertices, but each diagonal joins two vertices so in total there are $\frac{1}{2}n \times (n - 3)$ diagonals. This gives the formula $d = \frac{1}{2}n(n - 3)$ for the number of diagonals in an n-sided polygon.

Review 1
a 8, 16, 24, 32, 40, 48, ... We add 8 each time because to make the next number of stars we need 8 extra dots.
b $8n$
There are 8 dots in each star so each term is $8 \times n$ where n is the number of stars.

Review 2
a 1, 4, 9, 16, 25, 36, ...
b 64
The squares in the middle are increasing by the next odd number 3, 5, 7, 9, 11, ... **or** the sequence made is the sequence of square numbers and the 8th square number is 64.
c n^2
d $n^2 + 4$
In each shape there are n^2 green squares and in each shape there are 4 yellow circles. The total number of pieces of felt is $n^2 + 4$.

Page 218 Exercise 9
1 a $T(n) = 2n$ **b** $T(n) = 3n$
   c $T(n) = 6n$ **d** $T(n) = 5n + 1$
   e $T(n) = 5n - 1$ **f** $T(n) = 4n - 1$
   g $T(n) = 0.3n$ **h** $T(n) = 0.4n + 0.1$
   i $T(n) = 10n + 100$ **j** $T(n) = 15n + 45$
   k $T(n) = 2n - 0.5$ **l** $T(n) = 110 - 10n$
   m $T(n) = 35 - 5n$ **n** $T(n) = 48 - 8n$
   o $T(n) = 5n - 18$ **p** $T(n) = 10 - 6n$
2 a 39, 44, 49
   b We add 5 to get the next term.
   c 104
   d $5n + 4$
3 i a 44, 51, 58
     b We add 7 to get the next term.
     c 142
     d $7n + 2$

ii a 85, 98, 111
   b We add 13 to get the next term.
   c 254
d $13n - 6$
iii a 68, 80, 92
   b We add 12 to get the next term.
   c 236
d $12n - 4$
4 The formulae needed are:
   a $4N - 1$ **b** $6N + 3$
   c $90 - 5N$ **d** $0·2N + 0·5$

Review
a $T(n) = 2n - 1$ **b** $T(n) = 6n - 2$
c $T(n) = 70 - 10n$ **d** $T(n) = 11 - 6n$

Page 219 Exercise 10

1 a

Input	Output
5	17
2	8
0	2
12	38

b

Input	Output
4	1
10	4
24	11
68	33

c

Input	Output
11	42
21	72
‾2	3
‾5	‾6

d

Input	Output
7	27
1·5	5
0·5	1
‾1	‾5

2 a

Input	Output
4	6
7	15
0	‾6
3·5	4·5

b

Input	Output
25	7
0·5	2·1
‾5	1
‾15	‾1

c

Input	Output
0	8
$\frac{1}{2}$	9
6·5	21
‾3	2

d

Input	Output
7	2
$5\frac{1}{2}$	$1\frac{1}{4}$
9·6	3·3
‾5	‾4

3 a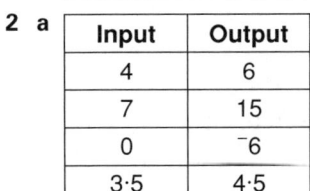

b

4 a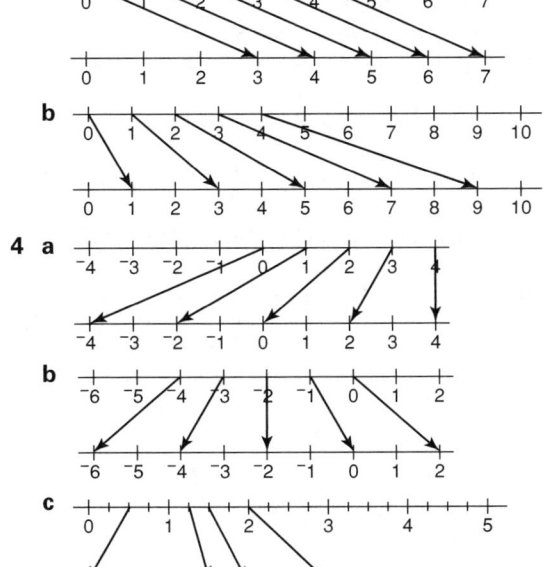

b

c

5 a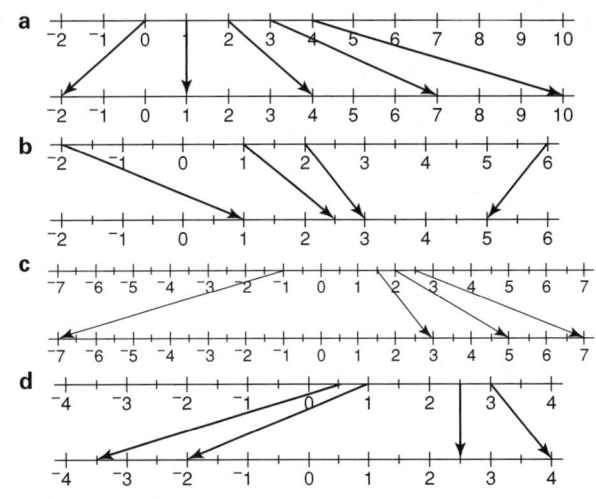

b

c

d

6 a i $x \rightarrow x + 2$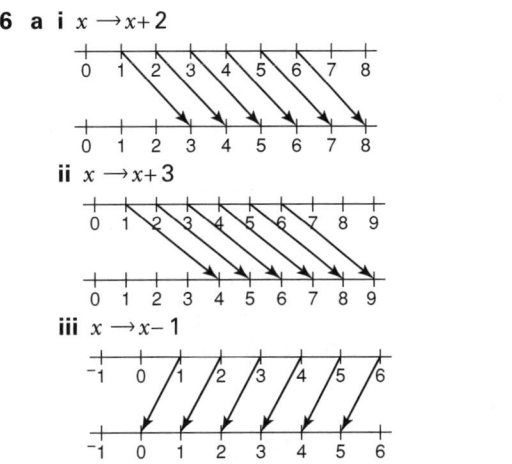

ii $x \rightarrow x + 3$

iii $x \rightarrow x - 1$

b The lines are parallel.
c Mapping arrows on a mapping diagram for a function of the form $x \rightarrow x + c$ will be parallel.

7 a and b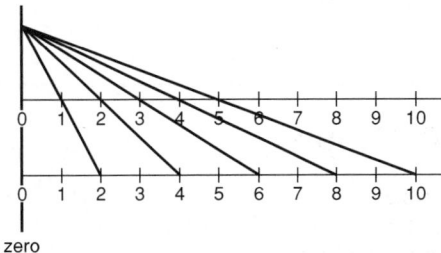

zero
line

The lines all meet at one point on the zero line.
c i and **ii** both give lines which, when extended backwards, meet at a point on the zero line.
d Mapping arrows for multiples, if extended backwards, meet at a point on the zero line.

Review 1

a

Input	Output
2	6
15	58
2·5	8
$\frac{1}{2}$	0

b

Input	Output
3	1
8	3·5
2·4	0·7
$2\frac{1}{2}$	$\frac{3}{4}$

Review 2

Input	Output
8	8
$6\frac{1}{2}$	5
0	‾8
‾1	‾10

Review 3

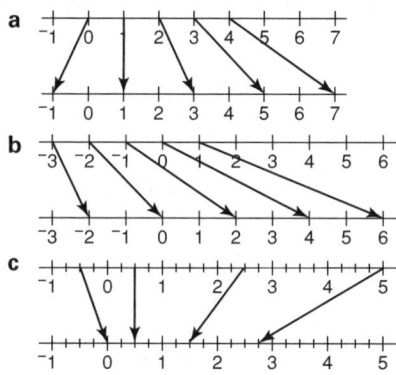

Review 4

a B   b A   c B   d A   e A   f B

Page 223 Exercise 11

1 a $x \rightarrow 2x + 1$ b $x \rightarrow 2x - 1$
   c $x \rightarrow 3x + 2$ d $x \rightarrow 2(x + 3)$
   e $x \rightarrow 3(x - 1)$ f $x \rightarrow \frac{x}{2} + 2$
   g $x \rightarrow \frac{x}{2} - 1$

2 a $x \rightarrow 2x - 1$ b $x \rightarrow 2x + 5$
   c $x \rightarrow 3x - 2$ d $x \rightarrow 2x - 4$
   e $x \rightarrow 3x + 4$

Review

a $x \rightarrow 3x + 6$ b $x \rightarrow 2x - 5$
c $x \rightarrow 4x + 3$ d $x \rightarrow 3x - 4$

Page 225 Exercise 12

1 a add 8   b subtract 6   c subtract 4
   d multiply by 6   e divide by 6   f multiply by 5

2 a $x \rightarrow 3x + 3$ or $x \rightarrow 3(x + 1)$
   b $x \rightarrow 4x + 4$ or $x \rightarrow 4(x + 1)$
   c $x \rightarrow 5x + 5$ or $x \rightarrow 5(x + 1)$

Chapter 10 Graphs

Page 232 It's a puzzle
You can join the shapes you cut out together to make the picture.

Page 234 Discussion
The points will lie on an 'imagined straight line'.
It is not possible to have a term number of $1\frac{1}{2}$, $2\frac{1}{4}$, 3·7 or any other non-whole number.
Isaac should not draw a straight line through the points because the intermediate values do not exist.

The difference between the graph of a function and the graph of a sequence with an equivalent equation is that the graph of the function is a straight line and every point on that line (an infinite number) satisfies the equation of the line, whereas the graph of a sequence is a set of discrete points which lie on an imagined straight line. The graph of a sequence will only have points in the first and/or fourth quadrant. This is because n only has positive values.

3 a Subtract 2, divide by 3
   b Add 4, multiply by 3
   c Multiply by 2, subtract 3
   d Divided by 3, add 6
4 a No. Reversing the operations gives a different function. The function for the first machine is $x \rightarrow 4(x + 3)$ and for the second machine is $x \rightarrow 4x + 3$.
   b Yes. For example, multiplying by 4 then dividing by 2 will give the same output as dividing by 2 then multiplying by 4.
5 a 10, 1, 4   b 40, 12, 24   c 4, 12, 7
   d 3, 4, 10   e 8, 20, 100   f 5, 7, 12
6 a $x \rightarrow \frac{x - 2}{3}$ b $x \rightarrow 3(x + 3)$
   c $x \rightarrow \frac{x}{4} - 2$
7 a $x \rightarrow \frac{x - 1}{2}$ b $x \rightarrow \frac{x - 2}{3}$
   c $x \rightarrow \frac{x + 1}{3}$ d $x \rightarrow 2(x - 2)$
   e $x \rightarrow \frac{x}{2} - 3$ f $x \rightarrow \frac{x}{2} + 1$
   g $x \rightarrow \frac{x}{4} + 3$ h $x \rightarrow 2x + 3$
   i $x \rightarrow 3x + 1$ j $x \rightarrow \sqrt{x - 2}$
   k $x \rightarrow \sqrt{x + 3}$

Review 1

a add 11   b multiply by 24   c add 2   d multiply by 3

Review 2
$x \rightarrow 2x + 2$ or $x \rightarrow 2(x + 1)$

Review 3
a Subtract 2, divide by 3
b Subtract 1, multiply by 4
c Divide by 3, add 4

Review 4
a 6   b 39   c 13, 15, 18   d 16, 100, 72

Review 5
a $x \rightarrow \frac{x + 3}{2}$ b $x \rightarrow 4(x - 1)$
c $x \rightarrow \frac{x}{2} + 4$ d $x \rightarrow 3x + 2$

Page 234 Exercise 1
1 a (2, 4), (1, 2), (0, 0), (⁻1, ⁻2), (⁻2, ⁻4)
   b (2, 9), (1, 7), (0, 5), (⁻1, 3), (⁻2, 1)
   c (2, 3), (1, 0), (0, ⁻3), (⁻1, ⁻6), (⁻2, ⁻9)
   d (3, 4), (2, 6), (1, 8), (0, 10), (⁻1, 12)
   e (3, ⁻8), (2, ⁻4), (1, 0), (0, 4), (⁻1, 8), (⁻2, 12)
2 i a It will cross the x- and y-axes at (0, 0).
      b It will cross the x-axis at $(⁻2\frac{1}{2}, 0)$ and y-axis at (0, 5).
      c It will cross the x-axis at (1, 0) and the y-axis at (0, ⁻3).
      d It will cross the x-axis at (5, 0) and the y-axis at (0, 10).
      e It will cross the x-axis at (1, 0) and the y-axis at (0, 4).
   ii a 1   b 6   c ⁻$1\frac{1}{2}$ d 9   e 2
3 a, c, e, f
4 a, d, e
5 a

x	⁻3	⁻2	⁻1	1	2
y	⁻8	⁻6	⁻4	0	2

   b (⁻3, ⁻8), (⁻2, ⁻6), (⁻1, ⁻4), (1, 0), (2, 2)
   c Your line should cut the y-axis at (0, ⁻2) and the x-axis at (1, 0).

d No, because $(^-\frac{1}{2}, 2)$ does not satisfy the equation
   $y = 2x - 2,$ $2 \neq 2 \times ^-\frac{1}{2} - 2$
e Yes, because $(24, 46)$ does satisfy the equation
   $y = 2x - 2,$ $46 = 2 \times 24 - 2$
f $(^-4, ^-10), (^-3, ^-8), (^-0.5, ^-3), (^-1.25, ^-4.5)$

6 a

x	0	1	2	3
y	$^-4$	$^-1$	2	5

b $(0, ^-4), (1, ^-1), (2, 2), (3, 5)$

c

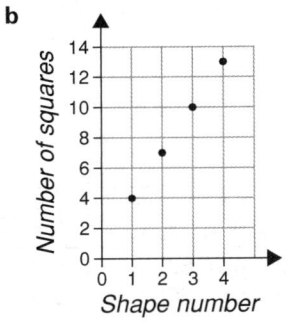

d Yes

7 a

Shape number	1	2	3	4
Number of squares	4	7	10	13

b

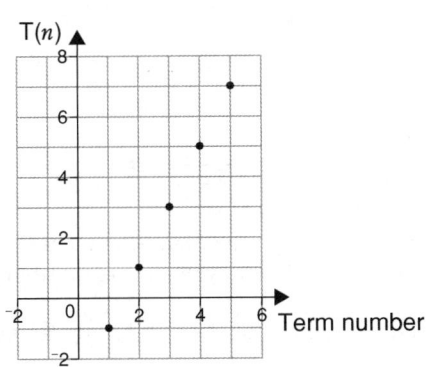

Yes, the points do lie in a straight line.

c 19

8 a

Term number	1	2	3	4	5
$T(n)$	$^-1$	1	3	5	7

b

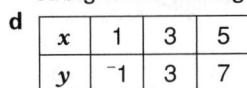

c Yes, because $T(n) = 9$ will lie on the imagined straight line through the points.

d

x	1	3	5
y	$^-1$	3	7

e

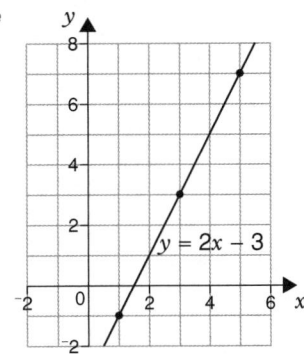

f Because **b** is the graph of the sequence $T(n) = 2n - 3$ and n can only take whole number values. The points in between these are meaningless. **e** is the graph of $y = 2x - 3$ and x can take any value. The points in between the plotted ones are meaningful.

9 a i The line should cross the x-axis at $(2, 0)$ and the y-axis at $(0, ^-6)$.
   ii The line should cross the x-axis at $(3, 0)$ and the y-axis at $(0, 12)$.
   iii The line should cross the x-axis at $(^-2, 0)$ and the y-axis at $(0, 1)$.

b i, ii, iii The graphs will be a set of points which lie on an imagined straight line. The imagined straight lines are the same as those drawn in **a i, ii, iii**

c All of the intermediate points of a function have meaning but the intermediate points of a sequence are meaningless because we can't have fractional term numbers.

Review 1
b, c, d, e

Review 2
a

x	$^-1$	0	2
y	$^-4$	$^-2$	2

b $(^-1, ^-4), (0, ^-2), (2, 2)$

c

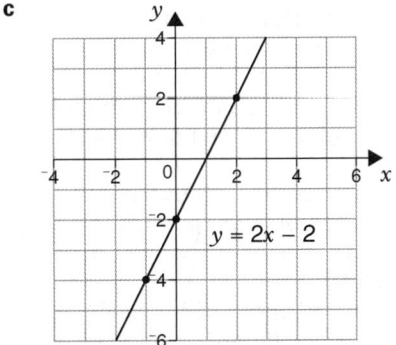

d There are many possible answers.
   One is: $(1, 0)$ and $(5, 8)$.
   Both satisfy the equation $y = 2x - 2$.
e $(3, 4)$ is a point on the line.
f No
g Yes, because $(4\frac{1}{2}, 7)$ satisfies the equation $y = 2x - 2$.

Review 3

a

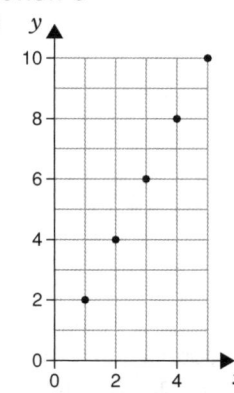

b Yes, because 22 is an even number which would lie on the imagined straight line through the other points.

c No, because the term numbers are all whole numbers and so the points in between are meaningless.

Page 236 Investigation: Features of graphs

A 1 The graphs all have the same slope but each cuts the y-axis in a different place.

   a All of the equations of the graphs have the same number, 2, multiplying x, and so all have the same slope.

   b The number on the end of each equation tells you where the line cuts the y-axis.

2 m represents the gradient.

c represents where the line crosses the y-axis.

$y = x$ is a straight line through the origin which is at a 45° angle to the x-axis.

$y = 3x - 2$ is a straight line steeper than $y = x$. It cuts the y-axis at $^-2$.

$y = 4x + 2$ is a straight line steeper than $y = 3x - 2$. It cuts the y-axis at 2.

$y = x - 4$ is a straight line with the same gradient as $y = x$. It cuts the y-axis at $^-4$.

$y = 3x + 3$ is a straight line with the same gradient as $y = 3x - 2$. It cuts the y-axis at 3.

$y = 2x - 1$ is a straight line steeper than $y = x$ but less steep than $y = 3x - 2$. It cuts the y-axis at $^-1$.

B 2 i All have the same gradient (╱) but each crosses the y-axis at a different point. The point where it crosses is given by the number at the end of the equation.

   ii All have the same gradient (╱) but each crosses the y-axis at a different point. The point where it crosses is given by the number at the end of the equation.

   iii All have the same gradient (╲) but each crosses the y-axis at a different point. The point where it crosses is given by the number at the beginning of the equation.

   iv All cross the y-axis at (0, 2). Each has a different gradient.

C There are many possible answers.

   a $y = mx + 3$ where m is any value

   b $y = mx - 2$ where m is any value

   c $y = mx - 4$ where m is any value

Page 237 Exercise 2

1 a $y = 3x - 5$

   b $y = 2x + 4$, $y = 3x - 5$, $y = \frac{1}{2}x - 2$, $y = x + 2$, $y = x - 5$

   c $y = x + 2$

   d $y = 3x - 5$, $y = x - 5$

   e $y = 8 - x$ and $y = ^-x$ would be parallel and $y = x + 2$ and $y = x - 5$ would be parallel.

---

2 a 2 **b** 3 **c** 3 **d** $\frac{1}{2}$ **e** $\frac{1}{2}$ **f** $\frac{1}{3}$ **g** $^-2$
   h $^-5$ **i** 2 **j** $^-3$ **k** $\frac{1}{2}$

3 a (0, 0) **b** (0, 0) **c** (0, 2) **d** (0, $^-4$) **e** (0, 0)
   f (0, 2) **g** (0, 4) **h** (0, 3) **i** (0, $^-1\cdot5$) **j** (0, 0)
   k (0, $^-2$)

4 $y = 2x + 3$ and $y = 2x - 3$ or $y = 3$ and $y = 2$

5 There are many possible answers. For **a** any correct answer will be of the form $y = 2x + c$ where c is any value. For **b** any correct answer will be of the form $y = ^-3x + c$ or $y = c - 3x$ where c is any value.

6 There are many possible answers. For **a** any correct answer will be of the form $y = mx + 4$ where m is any value. For **b** any correct answer will be of the form $y = mx - 2$ where m is any value.

7 a D **b** C **c** A **d** B **e** E **f** F

8 a $x = 9$ **b** $y = x + 3$ **c** $y = x - 1$ **d** $y = x - 3$

9 a 2 **b** 3 **c** $^-1$

Review 1

a $y = \frac{1}{2}x - 2$ **b** $y = 3 - x$, $y = ^-x$ **c** $y = x - 4$

d $y = 2$ **e** $y = 3 - x$, $y = ^-x$

Review 2

a F, $x = ^-3$ **b** C, $y = ^-2x + 4$ **c** A, $y = x$

d D, $y = x + 4$ **e** E, $y = ^-3$ **f** B, $y = ^-2x$

Review 3

5

Page 241 Exercise 3

1 a Yes

   b Racehorse about 65 km/h, antelope about 56 km/h, deer about 45 km/h

2 a

No. of slices sold	1	2	3	4	5	6	7	8	9	10
Money earned (£)	0·20	0·40	0·60	0·80	1·00	1·20	1·40	1·60	1·80	2·00

   b (1, 0·20), (2, 0·40), (3, 0·60), (4, 0·80), (5, 1·00), (6, 1·20), (7, 1·40), (8, 1·60), (9, 1·80), (10, 2·00)

   c

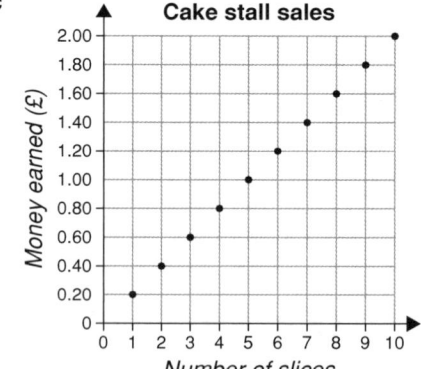

   d No. The intermediate points have no meaning because the cake stall only sold whole numbers of slices.

3 a About 90 cm **b** About 96 cm **c** About $3\frac{1}{2}$ years

   d

Age of girl (in years)	Height in cm at start of year (approximate)	Height in cm at end of year (approximate)	Approximate growth in cm
1 to 2	74	86	12
2 to 3	86	94	8
3 to 4	94	101·5	7·5

   e About 2 cm

   f The heights of both boys and girls increase with age. The increase is not constant. It lessens as time increases.

4 a

Number of classes	0	10	20	30
Total cost (£)	0	26	52	78

b and **d**

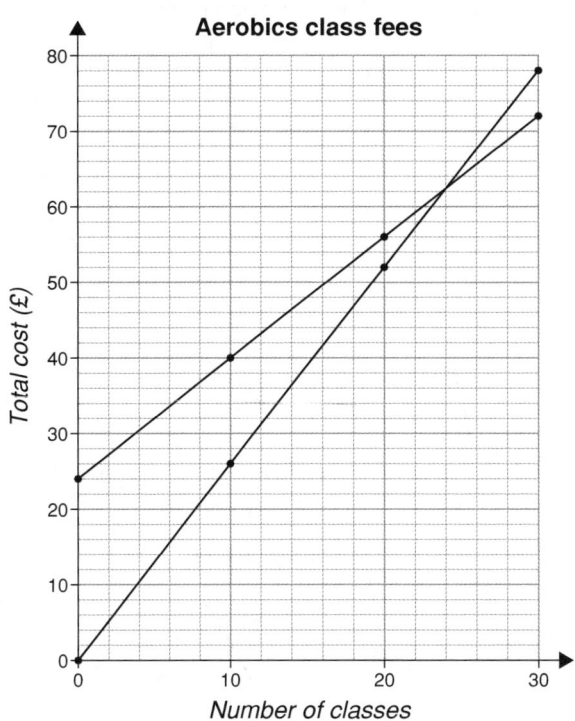

Aerobics class fees

c

Number of classes	0	10	20	30
Total cost (£)	24	40	56	72

e 24

f Paying the yearly fee and £1·60 per class is cheaper by about £2.

5 a

s	10	50	100
d	9	29	54

b (10, 9), (50, 29), (100, 54)

c

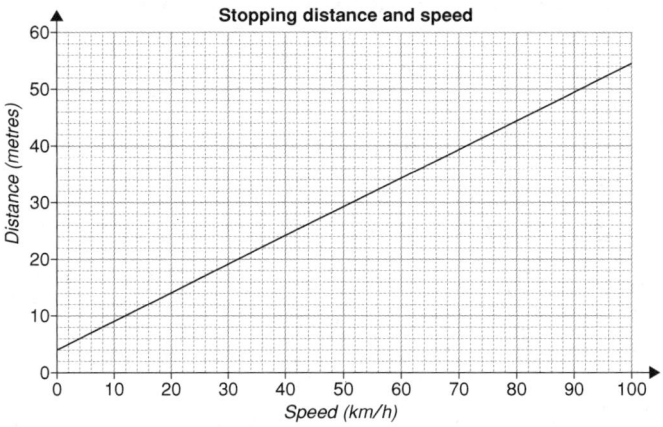

Stopping distance and speed

d About 39 metres.

d About 72 km/h.

6 a $d = 2 \cdot 50p$

b

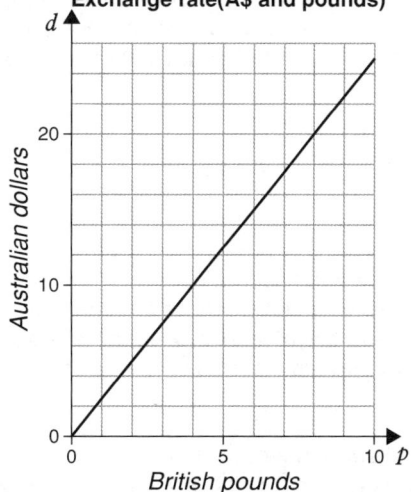

Exchange rate(A$ and pounds)

c About $15 **d** About £8·80 **e** About £150

7 a

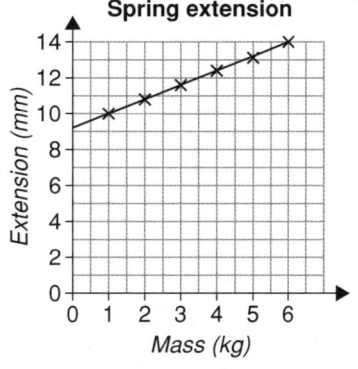

Spring extension

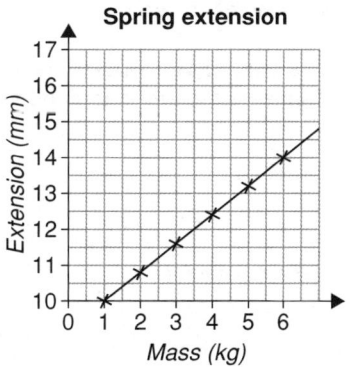

Spring extension

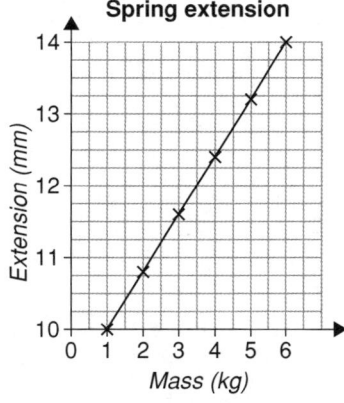

Spring extension

b The look of the graph changes when the scale is changed. The gradient looks different.

c Probably the third set.

d His conclusion that the spring extends more each time a kilogram is added is correct. Whether it extends 'lots' each time depends on an individuals perception of 'lots'

8 The graph is misleading because the vertical scale does not begin at zero. It appears that the number of male arthritis sufferers has increased a lot. In fact the increase is only about 15 more, out of about 530 or about 3%.

Review 1

a About 48 °C

b About 4 minutes

c 70 °C

d About 49 °C

e About 31 °C

f About 0·9 minutes or 54 seconds

g The liquid decreases in temperature as time increases but not at a constant rate. It decreases more rapidly at first and this decrease lessens as time goes on.

Review 2

a

No. of times lawns mown	1	2	3	4	5
Money earned (£)	1·50	3·00	4·50	6·00	7·50

b

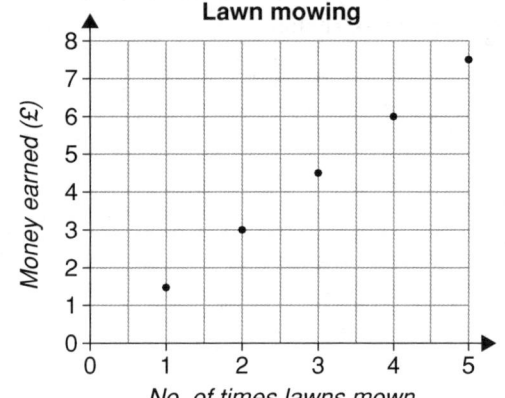

Lawn mowing

c Yes

d No, because the lawn is not mowed a fractional number of times. The intermediate points are meaningless.

Review 3

a (10, 50), (40, 104), (100, 212)

b

c About 167 °F

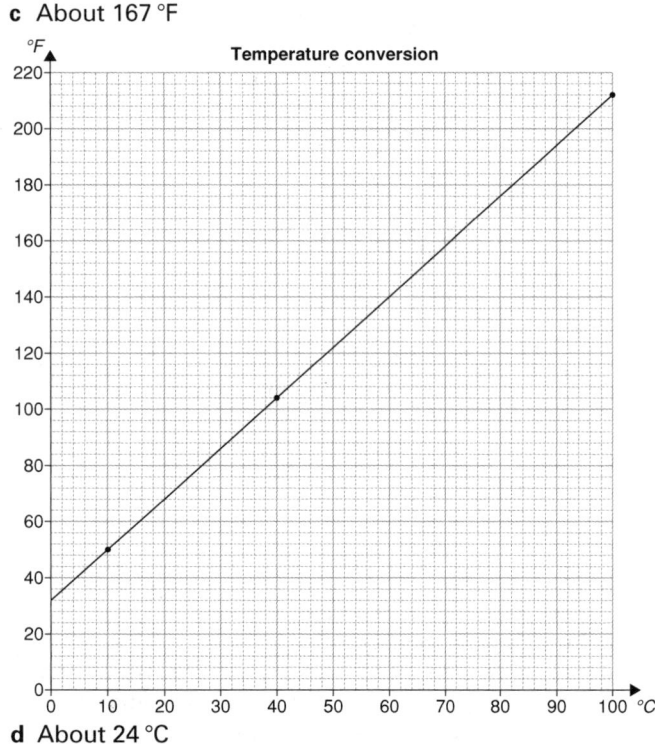

Temperature conversion

d About 24 °C

Page 245 Investigation: Collision Course

They will collide at (⁻4, ⁻5).

Page 246 Exercise 4

1

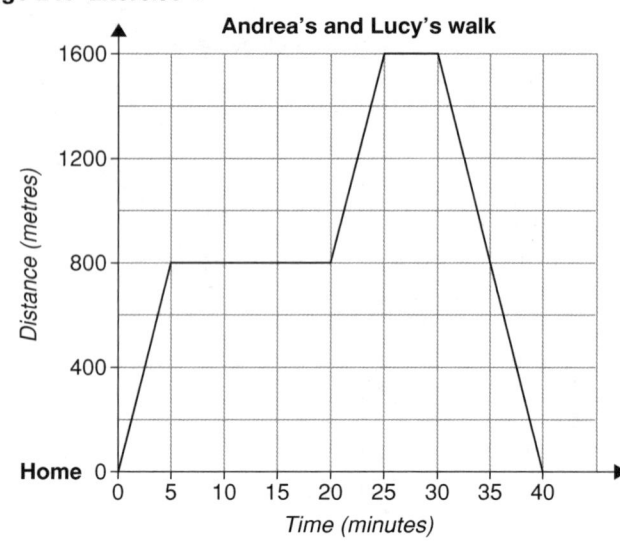

Andrea's and Lucy's walk

2

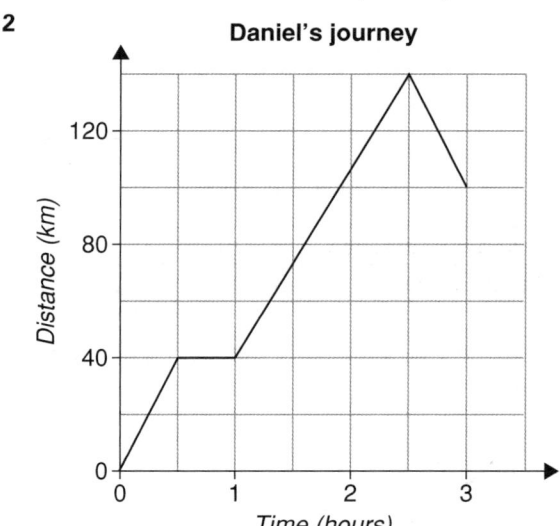

Daniel's journey

3 a 40 km   **b** 3 hours   **c** ½ hour

4 a 80 km   **b** ½ hour   **c** ½ hour

  d 40 km   **e** 1 hour

Review

Training run

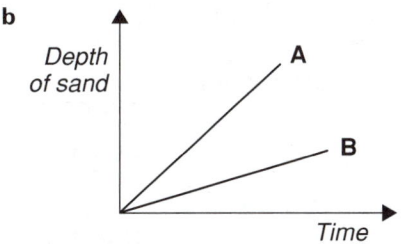
Page 248 Discussion

Dot 1

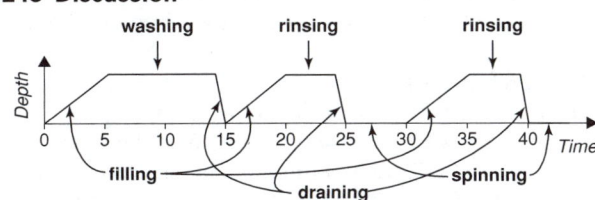

Dot 2 A possible answer is 'Anna knitted for a while and then stopped. Then the cat played with her knitting and unravelled it all. Anna knitted again but didn't knit quite as much as she had before. She stopped for a while then knitted some more. She had to unpick a little of it because she had made a mistake. Then she finished the scarf.'

Dot 3 A possible answer is 'This could be the graph of the temperature in a room. The heater was turned on and the room heated up. The thermostat kept it at the same temperature for a while. Someone turned the heater off and the room cooled down again. The heater was turned on again and the thermostat was set to a higher temperature. The room heated up to this higher temperature and then stayed at that for a while. The heater was then turned off.'

Dot 5 Graph B.

Page 249 Exercise 5

1 D

1 a A balloon is being blown up. The line slopes up, showing that volume is increasing with time, when air is being blown into the balloon. The line is horizontal when the person is taking a breath. The volume does not increase during this time.

b Each time Sam spends money the line becomes vertical indicating that the amount of money has instantly decreased. When he is not spending money the line is horizontal because the amount of money stays the same.

c The pulse rate increases with time, then remains steady, then decreases with time, probably after the person has stopped running.

d When the pasta is first put into the pot, the temperature decreases because the pasta is frozen. The pasta then slowly heats up as time increases and the temperature of the water in the pot rises. The temperature reaches a point where it stays the same with time (probably boiling point).

3 a 30 ℓ   **b** 1200 km   **c** Twice   **d** 60 ℓ   **e** $\frac{1}{2}$ **f** 80 ℓ
g Yes   **h** 15 km/ℓ

4

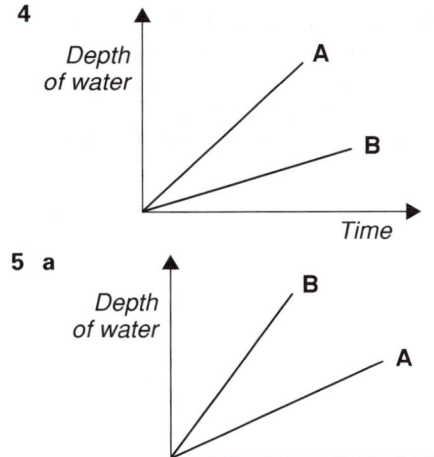

5 a

b

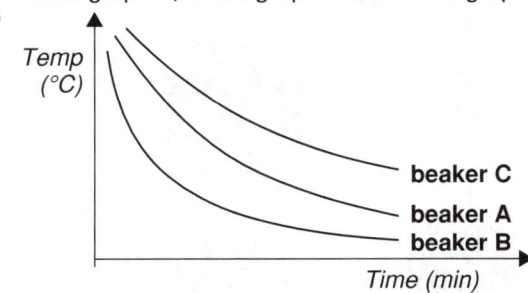

6 a Sandy, 3, Suzie, 1200   **b** Sandy
c $6\frac{1}{2}$ minutes   **d** $\frac{1}{2}$ minute
e Sandy's speed was fast at first, then decreased a lot to a very slow speed. It increased again at the end. Suzie's speed was slower at first then she sped up. She then decreased to a slow speed and then at the end she sped up, but later than Sandy.

7 a Omar
b Logan built up speed very quickly then settled at a medium to fast pace. At about 10 seconds he built up speed and ran very fast to come second. Omar built up speed quite slowly but then ran fast all the way to come first. Tom built up to a medium to fast pace quite quickly but after about 6 seconds he gradually slowed to a little less than a medium pace and he came third.

8 A has graph **P**, **B** has graph **R** and **C** has graph **Q**.

9

10 a No, the manager should not be worried. Last year's graph shows that the trend is for sales to drop between August and November, and then sales sharply increase in December.
b Sales might increase in December because of Christmas.

Review 1
a Whenever some sweets are eaten, the number of sweets left decreases suddenly. The graph is then vertical. There are times when no sweets are being eaten and the number of sweets left remains the same. The graph is then horizontal.
b A possible answer is 'over time, the skier skied down the mountain, then stopped, then skied down to the ski lift.'
c Whenever people get into the lift, the mass increases sharply and whenever people get out of the lift, the mass decreases sharply. Between floors, no one gets in or out and so the mass remains the same (the line is horizontal).

Review 2
Friday and Sunday

Review 3
a Jess   **b** Jasper   **c** About 1 minute

Review 4
a Possible answers are:

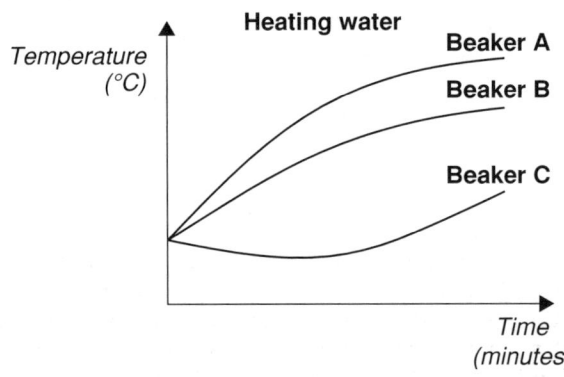

Heating water

b The temperature of the water increases over time but not constantly. It increases less as time goes on.

Page 253 Discussion

Dot 1 It costs £1·50 to laminate a 12 cm² poster.
It costs £1·50 to laminate a 10 cm² poster.
It costs £2·50 to laminate a 60 cm² poster.
It is called a step graph because the cost increases in steps.

Shape Space and Measures Support

Page 264 Practice Questions

1 a AB 5·7 cm, CD 3·8 cm, EF 10·1cm, GH 13·4 cm
   b AB and EF   **c** AB and CD and CD and EF
2 a 62 **b** 27
3 a Blue **b** Red **c** Purple **d** Green
4 $x = 37°$, $y = 145°$, $z = 325°$
5 a No **b** Yes **c** Yes
6 a

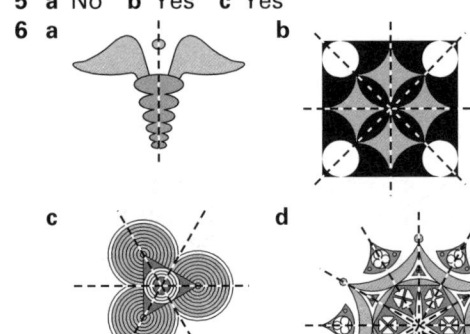

7 a B **b** E **c** A **d** G **e** D **f** C **g** H **h** J **i** F
8 a None **b** 1 **c** None **d** None **e** 1 **f** 2 **g** 1
   h None **i** 3
9 a MAKE UP ONE LIKE THIS YOURSELF
10 a ∠T or ∠PTR or ∠RTP or PT̂R or RT̂P
   b ∠NPM or ∠MPN or NP̂M or MP̂N
   c ∠ADB or ∠BDA or AD̂B or BD̂A
11 Possible answers are:
   a STR **b** XYZ
12 a t **b** z
13 a 3 **b** 6 **c** 3 **d** 4
14 a 40   **b** 3000   **c** 5000   **d** 6   **e** 42
   f 3800   **g** 52   **h** 8   **i** 4300   **j** 4·2
   k 9   **l** 8200   **m** 9650   **n** 4360   **o** 8·4
   p 5·86   **q** 0·3   **r** 7·52   **s** 60   **t** 5·2
   u 6   **v** 420   **w** 0·032
16 COWS HAVE FOUR STOMACHS
17 a 7 cm by 10 cm or 5 cm by 14 cm **b** 3 times
18 b 4, it is a square
   c There are many possible answers. Two are: (⁻2, 2)
      and (1, 0) or (⁻5, ⁻3) and (⁻2, ⁻5). It is possible to
      make a square. The coordinates of the other two
      vertices could be (⁻3, 0) and (0, ⁻2)
   d second quadrant

19 a

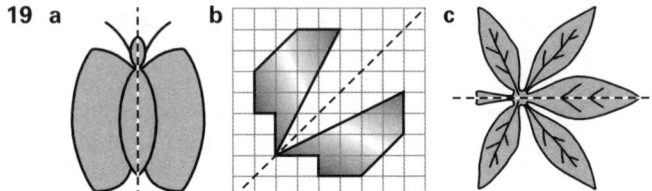

b   **c**

20 a 118°   **b** 57°   **c** 88°   **d** 75°   **e** 60°
   f $f = 38°$, $g = 31°$ **g** 65°   **h** 62°
   i $j = 67°$, $k = 61°$
21 a 10 **b** 4
22

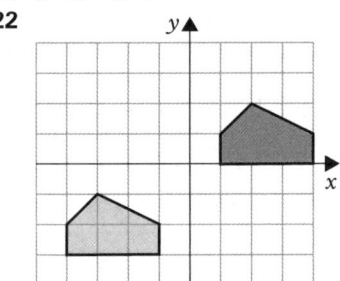

23 5 units left and 3 units down
24 a C **b** B **c** A
25 a 10 m **b** 28 cm **c** 30 mm **d** 12 km
26 a 6 m² **b** 48 cm² **c** 30 mm² **d** 6 km²
27 a 24·8 **b** 757 **c** 1·854
28 a All of them
   b There are eight other possible ways.
      Two are:

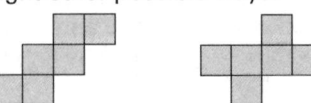

29 3
30 a $a = 132°$, $b = 48°$ **b** $c = 67·5°$, $d = 67·5°$
31 a A is at 2·2 cm, B is at 2·6 cm, C is at 1·8 cm
   b A is at 0·34 volts, B is at 0·23 volts, C is at 0·27 volts
   c A is at 500 g, B is at 850 g, C is at 100 g
   d A is at about 0·56 kg, B is at about 0·43 kg, C is at about 0·17 kg
32 a B **b** A

33 a (0, 0) 90°

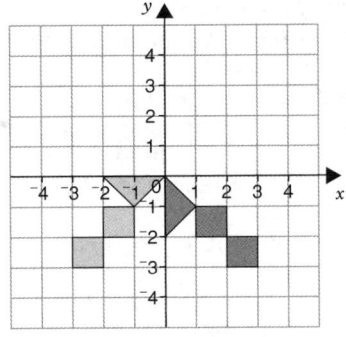

b (⁻1, 1) 180°

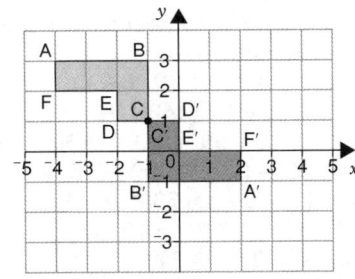

34 a i 8 km **ii** 1 kg **iii** 4·5 ℓ **iv** 600 mℓ or 0·6 ℓ
   b i 4·4 lb **ii** 1 gallon **iii** 1·75 pints **iv** 10 miles
35 South
36 a About 7 oz   **b** About 430 g
   c 200 g is about 7 oz. 1 kg is 5 times 200 g so 1 kg is
    about 5×7 oz = 35 oz.
37 80°
38 a 4 rolls **b** 40 **c** 80 km **d** 22
39 It will have a right angle and two
   angles of 45°. The line of
   symmetry cuts the base in half at
   right-angles giving a 90° angle
   and it cuts the 90° angle in half
   giving a 45° angle. The other angle is 45:

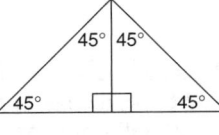

40 A square
41 Possible answers are given. The nets are not full size.
   a

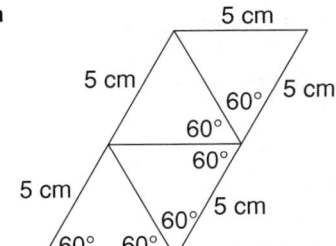

b

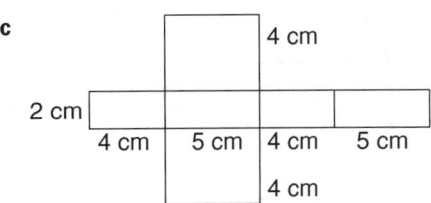

c

d

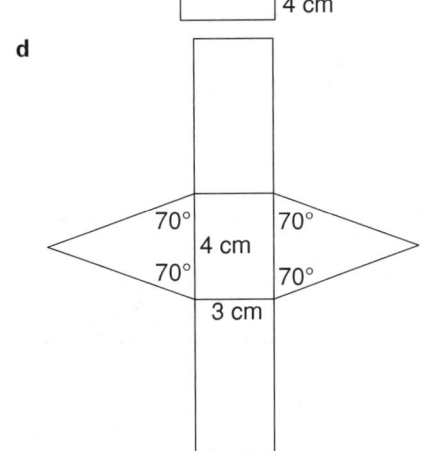

42 76 cm²
43 C and D
44 a 38 m² **b** 78 m²
45 348 cm²

---

11 Lines and Angles

Page 273 Exercise 1
1 i a a **b** d **c** b **d** f **ii a** b **b** f **c** d **d** d
2 Any three of e and f, f and g, a and d, a and b, b and c,
   c and d.
3 a i and **ii** **b iii** and **v**
4 a a and b, c and d, c and e, a and f
   b n and m, k and t, o and p, x and t, h and r, o and m,
    o and n, p and m, p and n, x and y, x and z, y and z
5 a 50° (corresponding angles)
   b 40° (alternate angles)
   c 60° (corresponding angles)
   d 121° (alternate angles)
   e 71° (alternate angles)
6 a g, i, c **b** j, l **c** m **d** f, h

Review 1
a 78° (alternate angles)
b 82° (corresponding angles)
c 108° (alternate angles)

Review 2
a e, c, h **b** d, f, h **c** c, k, g

Page 275 Exercise 2
Possible answers are given. Often there are other ways to
prove it.
   1 a   Shaded $\angle$ = 68° (alternate angles)
      $68° + 87° + x = 180°$ (angles in a triangle add to 180°)
         $x = 180° - 87° - 68°$
         $x = 25°$

b Shaded ∠ = 72° (alternate angles)
 72° + x = 180° (angles in a straight
  line add to 180°)
  x = 180° − 72°
  x = 108°

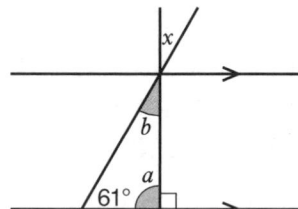

c ∠a = 90° (angles in a straight line add to
   180°)
 90° + 61° + b = 180° (angles in a triangle add to 180°)
   b = 29°
   x = 29° (vertically opposite angles equal)

2 One possible answer is:
Label the unknown angles that you need x and y.
(You can give them other labels.)

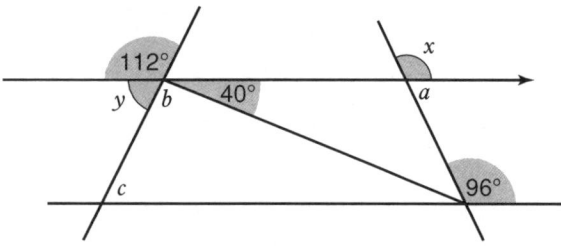

   x = 96° (corresponding angles)
 96° + a = 180° (angles on a straight line add to 180°)
   a = 84°
 b + 40° = 112° (vertically opposite angles equal)
   b = 72°
 y + 112° = 180° (angles on a straight line add to 180°)
   y = 180° − 112°
   y = 68°
   c = 68° (alternate angles)

3 ∠BDF = 70° (alternate angles on parallel lines CA and
   DF)
  x = 70° (alternate angles on parallel lines BD and
   FE)
 ∠FBD + 55° + 70° = 180° (angles on a straight line add
   to 180°)
  ∠FBD = 180° − 55° − 70°
  ∠FBD = 55°
  ∠BFA = 55° (alternate angles on parallel
   lines AE and BD)
 y + 55° + 55° = 180° (angles in a triangle add to 180°)
  y = 180° − 55° − 55°
  y = 70°

4 The angles with a dot are
equal (alternate angles).
The angles with a cross
are equal (alternate
angles).
So the opposite angles
of the parallelogram are
equal as each is the sum of an angle with a dot and an
angle with a cross.

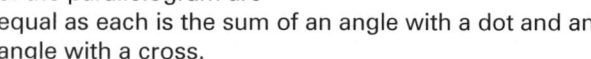

5 a = 59° (alternate angles)
 b + 59° + 62° = 180° (angles in a triangle add to 180°)
   b = 59°
 c = 62° (alternate angles)
 d + 62° + 59° = 180° (angles on a straight line add to
   180°)
   d = 180° − 62° − 59°
   d = 59°
 e + 62° + 59° = 180° (angles on a straight line add to
   180°)
   e = 180° − 62° − 59°
   e = 59°

6 Only answers are given to this question. Working
should be shown.
 a x + 20° + 68° = 180°; x = 92°
 b x + 20° = 68°; x = 48°
 c x + 20° + x + 50° + 168° = 360°; x = 61°
 d 3x + 60° + 5x = 180°; x = 15°
 e 4x + 2x + 5x + x = 360°; x = 30°
 f 4x + 60° = 120°; x = 15°
 g x + 34° = 2x; x = 34°
 h 2x + 56° = 5x − 10°; x = 22°
 i 3x + 25° + 68° = 180°; x = 29° (the adjacent angle to
   3x + 25° is the corresponding angle to 68°)
 j 3x + 38° + 5x − 10° = 180°; x = 19° (the angle adjacent
   to 5x − 10° is the alternate angle to 3x + 38°)

7 a The corresponding angles, 67° and 68°, are not
   equal.
 b GH and IJ. The alternate angles 111° and 111° are
   equal.

Review 1
Possible reasons are given
a = 56° (vertically opposite angles equal)
b = 56° (alternate angles)
c = 56° (vertically opposite angles equal)
d = 66° (corresponding angles)
e = 66° (alternate angles)
f = 114° (angles on a straight line add to 180°)
g = 62° (alternate angles)
h = 118° (angles on a straight line add to 180°)
i = 62° (vertically opposite angles equal)

Review 2
Call the unknown angle that is needed x.
x = 64° (corresponding angles)
a + 64° + 50° = 180° (angles on a straight line add to 180°)
a = 180° − 64° − 50°
a = 66°
b = a + 64° (corresponding angles)
b = 66° + 64°
b = 130°

Review 3
a 30° **b** 18°

Review 4
The alternate angles 39° and 39° are equal.

Page 277 Investigation: Angles of a triangle
 1 x = a (alternate angles)
   z = b (alternate angles)
   x + c + z = 180° (angles on a straight line add to 180°)
   We can substitute a for x and b for z in this equation.
   a + c + b = 180°
   so **a + b + c = 180°**

2 $x + y = d$
$x = b$ (alternate angles)
$y = a$ (corresponding angles)
$a + b = x + y$
so $a + b = d$
Another way of doing this is:
$x + y + c = 180°$ (angles in a straight line add to 180°)
$a + b + c = 180°$ (angles in a triangle add to 180°)
so $x + y = a + b$
and $x + y = d$
so $a + b = d$

Page 279 Exercise 3

1 **a** 61° **b** 53° **c** 120° **d** 140° **e** 52° **f** 67°
2 **a** $x = 40°$ (exterior angle of a triangle equals sum of
two opposite interior angles)
   $y = 80°$ (angles on a straight line add to 180°)
b $x = 47°$ (angles in a triangle add to 180°)
   $y = 108°$ (angles on a straight line add to 180° **or**
exterior angle of a triangle equals sum of
two opposite interior angles)
c Shaded angle $= x$ because base angles
of an isosceles triangle are equal.
   $2x = 104°$ (exterior angle of a triangle
equals sum of two opposite
interior angles)
   $x = 52°$
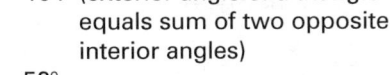
d $x = 65°$ (base angles of isosceles triangle are equal)
   $y = 115°$ (exterior angle of a triangle equals sum of
two opposite interior angles)
3 **a**
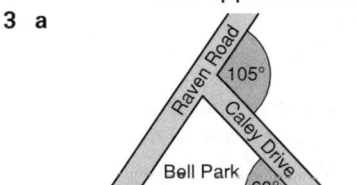
b The three angles are 60°, 75° and 45°.
4 **a** $x = 45°$ because angles at a point equal 360° and
there are eight equal angles at the centre.
   $360° ÷ 8 = 45°$
b The triangle is isosceles so
   $x + y + y = 180°$ (angles in a triangle add to 180°)
   $45 + 2y = 180°$
   $2y = 180° - 45°$
   $2y = 135°$
   $y = 67·5°$
5 A reflex angle is greater than 180°. The sum of all three
angles in a triangle is 180° so one of the angles cannot
be greater than 180°.
6 Only answers are given. Pupils should show working.
a $x = 82°$, $y = 16°$
b $a = 48°$, $b = 114°$
c $m = 135°$, $n = 88°$
d $p = 47·3°$, $q = 85·4°$
e $h = 55·7°$, $i = 58·3°$, $j = 66·9°$
f $r = 90°$, $s = 103°$
7 **a** 60° **b** 10°
8 $a = 34° + 32°$ (exterior angle of a triangle equals sum of
two opposite interior angles)
   $= 66°$
   $b + 24° + 90° = 180°$ (angle sum of triangle)
   $b = 180° - 24° - 90°$
   $= 66°$
so angle a = angle b

9 $p + q + y = 180°$ (angles on a straight line add to 180°)
   $p = x$ (alternate angles)
   $q = z$ (alternate angles)
   $p + q + y = 180°$
so $x + z + y = 180°$
10 A possible answer is:
   $a = 35°$ (alternate angles)
   $a + b = 75°$ (base angles of isosceles triangle are
equal)
   $35° + b = 75°$
   $b = 40°$
   $c + a + b + 75° = 180°$ (angles in a triangle add to 180°)
   $c + 75° + 75° = 180°$
   $c = 180° - 75° - 75°$
   $c = 30°$
11 **a**
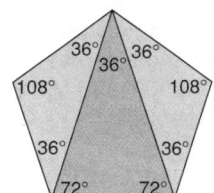
b They fit together to make a new triangle because
two of the angles, 108° and 72°, (one from each
triangle) fit together to make a straight line when
the matching edges are placed together. The
matching edges are equal because they were the
sides of a regular pentagon. The angles of the new
triangle are 36°, 108° and 36°.

Review 1

¹6	4		²5	³2
5		⁴1		5
	⁵1	2	2	
⁶6		9		⁷7
⁸5	8		⁹7	5

Review 2
a Name the unknown angles needed x and y.
   $x = 104°$ (corresponding angles)
   $x + c + 36° = 180°$ (angles on a
straight line
add to 180°)
   $104° + 36° + c = 180°$
   $c = 180° - 104° - 36°$
   $c = 40°$
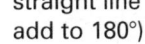
   $y = 36°$ (alternate angles)
   $b = 36°$ (vertically opposite angles equal)
b $b + 40° + 90° = 180°$ (angles in a triangle add to 180°)
   $b = 50°$
   $a = 50°$ (alternate angles)
   $c = 50°$ (vertically opposite angles)
c Call the unknown angles
needed x and y.
   $x = 37°$ (alternate angles)
   $x + c + 68° = 180°$ (angles on a
straight line
add to 180°)

   $c = 75°$
   $g = 75°$ (corresponding angles)
   $e = 75°$ (alternate angles)
   $d = 37°$ (corresponding angles)
   $h = \frac{(180 - 68)}{2}$ (isosceles triangle)
   $h = 56°$

Review 3

$x + y + z = 180°$ (angles in a triangle add to 180°)

$a + z = 180°$ (angles on a straight line add to 180°)

so $a = x + y$

Page 281 Puzzle

1 One possible answer is:

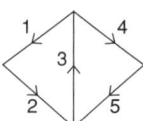

2 One possible answer is:

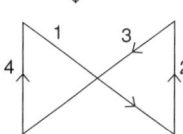

Page 281 Investigation: Finding the angle sum

$p + q + r = 180°$ (interior angles of a triangle add to 180°)

$s + t + u = 180°$ (interior angles of a triangle add to 180°)

The sum of the angles of the quadrilateral

$= p + q + r + s + t + u$

$= (p + q + r) + (s + t + u)$

$= 180° + 180°$

$= 360°$

Page 282 Exercise 4

1 a 103°   **b** 137°   **c** 93°   **d** 108°   **e** 48°

  f 222°   **g** 122°

2 a 360° **b** 360° **c** 360° **d** 360°

3 Possible answers are:

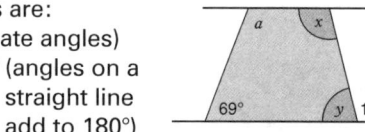

  a $x = 124°$ (alternate angles)

  $y + 124° = 180°$ (angles on a
          straight line
          add to 180°)

  $y = \mathbf{56°}$

  $a + x + y + 69° = 360°$ (angles in a quadrilateral
          add to 360°)

  $a + 124° + 56° + 69° = 360°$

  $a = \mathbf{111°}$

  b $x = 45°$ (base angles of isosceles triangle are equal)

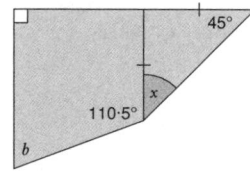

  $b + 90° + 45° + x + 110·5° = 360°$ (angles in a
          quadrilateral add to
          360°)

  $b = 360° - 90° - 45° - 45° - 110·5°$

  $b = \mathbf{69·5°}$

  c $x = 45°$ (angles on a straight
          line add to 180°)

  $y + 40° + 45° = 180°$
  (angles in a triangle add to
  180°)

  $y = 95°$

  $z = 95°$ (vertically opposite
          angles equal)

  $w + 126° = 180°$ (angles on a
  straight line add to 180°)

  $w = \mathbf{54°}$

  $90° + z + w + c = 360°$ (angles in a quadrilateral add to
          360°)

  $90° + 95° + 54° + c = 360°$

  $c = 360° - 95° - 90° - 54°$

  $c = \mathbf{121°}$

4 a $a + 90° + 90° + 68·5° = 360°$ (angles in a quadrilateral
                  add to 360°)

  $a = 111·5°$

  $b = 111·5°$ (vertically opposite angles)

  $x + b + 29° = 180°$ (angles in a triangle add to
          180°)

  $x + 111·5 + 29° = 180°$

  $x = \mathbf{39·5°}$

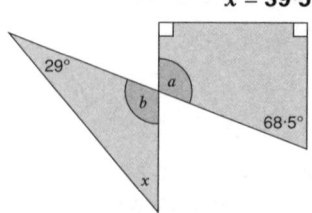

  b a, b and c all equal 60°
  because the angles of an
  equilateral triangle are all 60°.

  $x + 100° + 25° = 180°$ (angles
  in a triangle add to 180°)

  $x = 180° - 100° - 25°$

  $x = \mathbf{55°}$

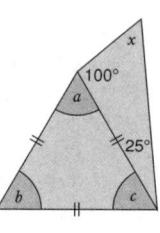

5 a $x = 90°$; $y = 90°$ (corners of a square)

Not to scale

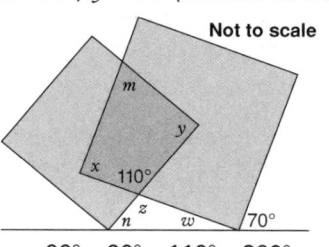

  $m + 90° + 90° + 110° = 360°$

  $m = \mathbf{70°}$

  $z = 110°$ (vertically opposite angles equal)

  $w + 90° + 70° = 180°$ (angles on a straight line add to
          180°)

  $w = 20°$

  $n + z + w = 180°$ (angles in a triangle add to 180°)

  $n = 180° - 20° - 110°$

  $n = \mathbf{50°}$

  b $x = 90°$, $y = 90°$ (corners of a rectangle)

  $p + 90° + 90° + 41° = 360°$
  (angles in a quadrilateral
  add to 360°)

  $p = \mathbf{139°}$

  $a = 139°$ (vertically opposite
  angles)

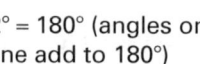

  $b + 90° + 62° = 180°$ (angles on
  a straight line add to 180°)

  $b = \mathbf{28°}$

  $139° + 28° + c = 180°$ (angles in a triangle add to 180°)

  $c = \mathbf{13°}$

  $q + 90° + 13° = 180°$ (angles on a straight line add to
          180°)

  $q = \mathbf{77°}$

Review

a $x = \mathbf{54°}$ (angles in a quadrilateral add to 360°)

b $x = \mathbf{102°}$ (angles in a quadrilateral add to 360° and
          opposite angles of a rhombus are equal)

c $x = \mathbf{154°}$ (angles in a quadrilateral add to 360° and a kite
          has one pair of equal angles)

Chapter 12 Shape, Construction and Loci

Page 285 Piecing it Together

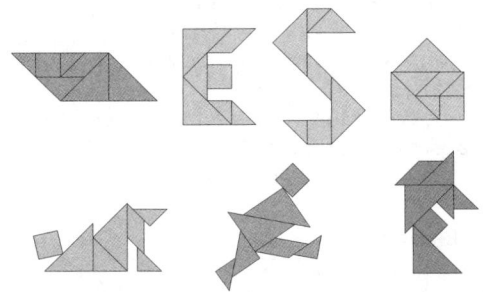

Page 286 Discussion
Dylan might answer 'a quadrilateral' or he might answer 'a kite'.

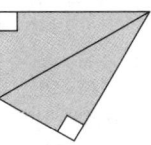

Page 286 Exercise 1
1 Another parallelogram or a hexagon. If one side of the parallelogram is twice the other, it is possible to get a rhombus.

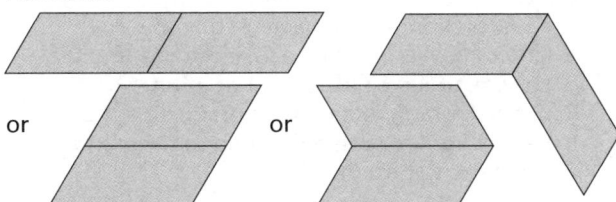

or   or

2 A parallelogram or a hexagon.

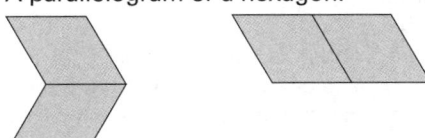

3 A rhombus, kite, square, triangle, parallelogram

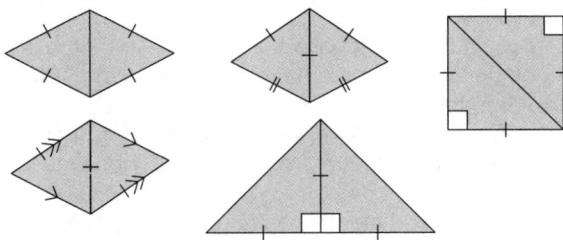

4 Possible answers are:
  a A rectangle, a square, a triangle, a trapezium, a pentagon, a parallelogram, a hexagon. There are many other possible shapes if the quarters are arranged so that sides do not match.

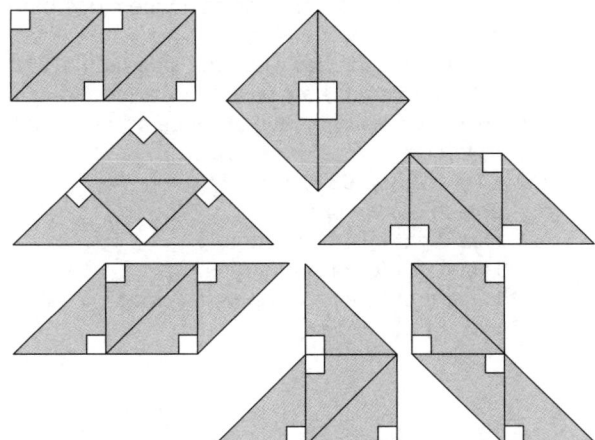

There are other ways of making a hexagon.

b A parallelogram:

or

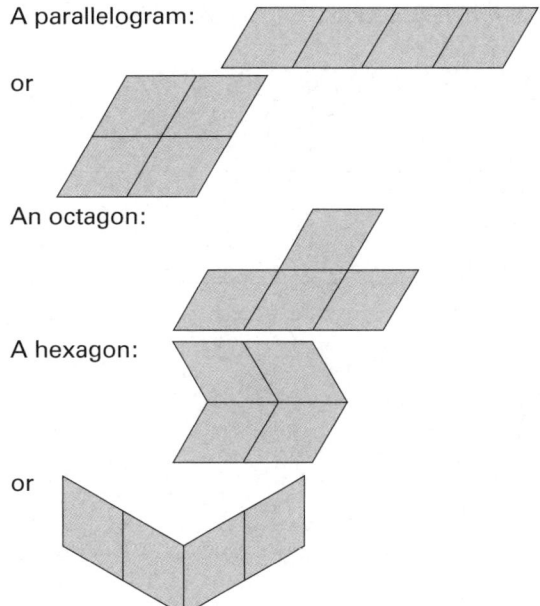

An octagon:

A hexagon:

or

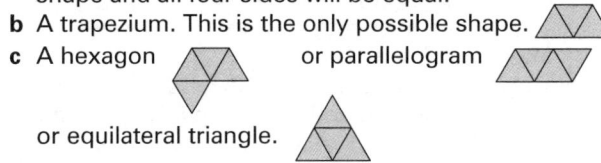
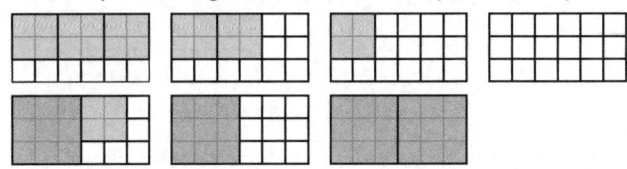
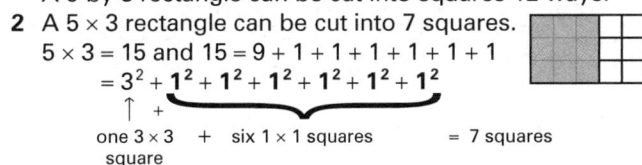
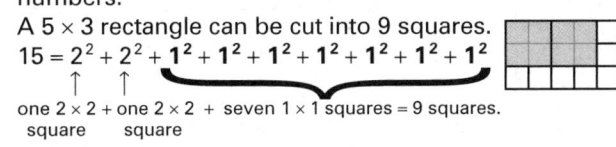

There are other shapes.
5 a A rhombus, because when two equilateral triangles are put together we will always get a four-sided shape and all four sides will be equal.
  b A trapezium. This is the only possible shape.
  c A hexagon   or parallelogram

   or equilateral triangle.

Review
An arrowhead (delta), kite, parallelogram, rectangle, triangle

Page 287 Investigation: Rectangles and Squares
1 A 6 by 3 rectangle can be cut into squares 7 ways.

A 7 by 3 rectangle can be cut into squares 8 ways.
An 8 by 3 rectangle can be cut into squares 10 ways.
A 9 by 3 rectangle can be cut into squares 12 ways.
2 A 5×3 rectangle can be cut into 7 squares.
   $5 \times 3 = 15$ and $15 = 9 + 1 + 1 + 1 + 1 + 1 + 1$
   $= 3^2 + \mathbf{1^2 + 1^2 + 1^2 + 1^2 + 1^2 + 1^2}$

   one 3×3 +   six 1×1 squares   = 7 squares
   square

A 5×3 rectangle cannot be cut into 8 squares because there is no way we can write 15 as the sum of 8 square numbers.
A 5×3 rectangle can be cut into 9 squares.
   $15 = 2^2 + 2^2 + \mathbf{1^2 + 1^2 + 1^2 + 1^2 + 1^2 + 1^2 + 1^2}$

   one 2×2 + one 2×2 + seven 1×1 squares = 9 squares.
   square   square

Note: $15 = 2^2 + 2^2 + 2^2 + 1^2 + 1^2 + 1^2$ **but** a 5×3 rectangle can't be cut into three 2×2 squares and 3 single squares.

Page 287 Investigation: Properties
See ICT Support section.

Page 287 Investigation: Properties of Shapes
A 1 and 2

Shape	Number of lines of symmetry	Order of rotation symmetry
Equilateral △	3	3
Isosceles △	1	1
Square	4	4
Rectangle	2	2
Parallelogram	0	2
Rhombus	2	2
Kite	1	1
Isosceles trapezium	1	1
Trapezium	0	1
Arrowhead	1	1

3 ● Sides are equal in length because a rhombus has two lines of symmetry. When it is folded along these lines the sides fit exactly on top of each other.

● The diagonals cross at right angles because all four angles at the point where the diagonals intersect are equal. $360° ÷ 4 = 90°$. This can be shown by folding along the diagonals.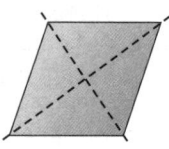

● The diagonals bisect each other because when the rhombus is folded along one line of symmetry, one half of a diagonal fits exactly on top of the other.

● The diagonals bisect the angles because when the rhombus is folded along its line of symmetry, one half of the angle sits exactly on top of the other.

Quadrilateral	Diagonals equal	Diagonals cross at right angles	Diagonals bisect each other	Diagonals bisect the angles	Sides	Angles
Square	✓	✓	✓	✓	4 equal	4 right angles
Rectangle	✓	X	✓	X	opposite sides equal	4 right angles
Parallelogram	X	X	✓	X	opposite sides equal	opposite angles equal
Rhombus	X	✓	✓	✓	4 equal	opposite angles equal
Kite	X	✓	X	X	2 pairs of adjacent sides equal	1 pair of equal angles
Isosceles trapezium	✓	X	X	X	1 pair equal 1 pair parallel	2 pairs of equal angles
Trapezium	X	X	X	X	X	no equal angles
Arrowhead	X	✓ outside the shape	X	X	2 pairs of adjacent sides equal	1 pair of equal angles 1 reflex angle

B 1

[grid diagrams of quadrilaterals and triangles on dot grids]

There are many different groups these quadrilaterals could be put into.
It is not possible to construct a triangle with a reflex angle because a reflex angle is greater than 180°. The sum of the angles of a triangle is to 180°.

2 ● It is possible to construct a trapezium with one line of symmetry.

● It is not possible to make a rhombus that is not a square, because the only three different rhombuses that can be made are all squares.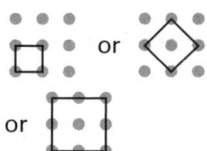

● It is not possible to make an equilateral triangle because it is not possible to find 3 dots on a $3 × 3$ grid that are equal distances apart.

Page 289 Exercise 2
1 a True b True c False d False
   e True f False g False
2 a $∠A = 64°$, AB = 6 cm   b $∠A = 121°$, AB = 8 m
   c $∠A = 129°$, AB = 7 cm   d $∠A = 68°$, AB = 12 mm
   e $∠A = 29°$, AB = 10 m   f $∠A = 108°$, AB = 40 mm
3 a rectangle, rhombus
   b square
   c parallelogram, trapezium (not isosceles), scalene triangle
   d rectangle, square, equilateral triangle
   e square, rhombus, equilateral triangle
   f kite, trapezium, arrowhead, isosceles triangle, scalene triangle, isosceles trapezium
   g parallelogram, rectangle, rhombus
   h parallelogram, rectangle, square, rhombus, isosceles trapezium
   i square, rhombus
   j square, rhombus
4 a C b B c D d B
5 a rectangle or rhombus   b parallelogram
6 a True because isosceles triangles must have two equal sides and all equilateral triangles have two equal sides.
   b False because only rectangles with equal sides are squares.
   c True because a square is a rectangle which has equal sides. Squares have all the properties that rectangles have.
   d True, a square is a rhombus with four right angles. Squares have all the properties that rhombuses have.
   e True. A quadrilateral is a shape with four sides and a kite has four sides.
   f False. All four-sided shapes are not parallelograms. Only four-sided shapes that have opposite sides equal and parallel are parallelograms.
   g True. Those rectangles with equal sides are squares.
   h False. A parallelogram has opposite sides parallel and a kite doesn't.
   i True. A rhombus has opposite sides equal and parallel and so it is a parallelogram. A parallelogram doesn't necessarily have all sides equal so not all parallelograms are rhombuses.
7 a $x = 70°$ b $x = 63°$ c $x = 50°$
8 a A possible answer is:
      A Has it got four right angles?
      B Has it got four right angles?
      C Has it got one pair of parallel sides?
      D Has it got a reflex angle?

b A possible answer is:

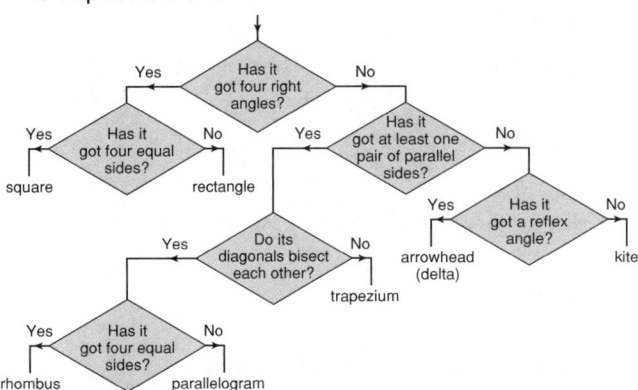

There are other possible answers.

c A possible answer is:

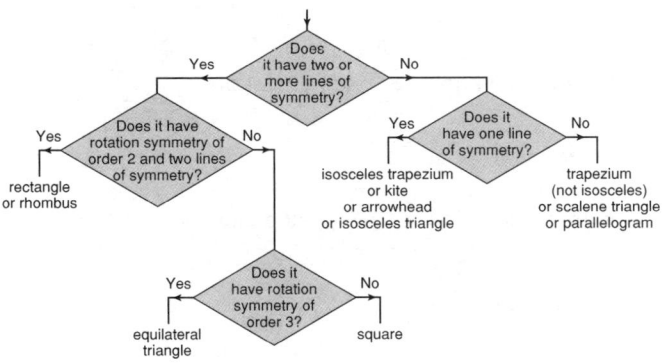

Review 1

a isosceles triangle, kite, arrowhead, isosceles trapezium

b square

c rectangle, square, isosceles trapezium

Review 2

a rhombus

b parallelogram, rectangle, isosceles trapezium

Review 3

a True. Quadrilaterals have four sides and some four-sided shapes are trapeziums (have one pair of parallel sides).

b True. A rhombus has four equal sides and some rhombuses have four right angles which means they are squares.

c False. An equilateral triangle must have three equal sides and all isosceles triangles only have two equal sides.

Review 4

List A Four equal sides and diagonals bisect the angles and the diagonals bisect at right angles. **List B** Four right angles, diagonals equal. **List C** Diagonals which cross at right angles. **List D** Diagonals which bisect each other and opposite sides parallel.

Page 291 Investigation: Polygons and right angles

It is not possible to have two right angles in a triangle.
It is possible to have two right angles in a quadrilateral.
It is not possible to have just three right angles in a quadrilateral.
It is possible to have four right angles in a quadrilateral, for example in a square or rectangle.

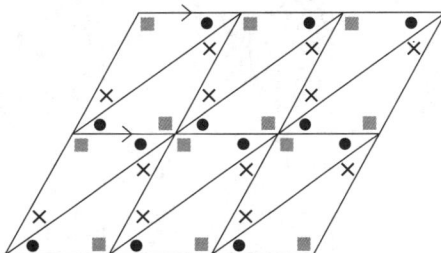

The greatest number of right angles it is possible to have in

a 5-sided polygon is 3
a 6-sided polygon is 5
a 7-sided polygon is 5
an 8-sided polygon is 6
a 9-sided polygon is 7
a 10-sided polygon is 7.

If n is odd then the number of right-angles possible is $n - 2$ and $n > 2$.
If n is even, the number of right-angles possible is $\frac{n}{2} + 2$.

Page 292 Investigation: Tessellations

1 All triangles tessellate.
   When a triangle is rotated about the mid-point of one of its sides, a parallelogram is formed. This diagram shows how we can use alternate and corresponding angles and angles on a straight line adding to 180° to show that all triangles tessellate.

□, ● and × add to 180° (angles in a triangle add to 180°) so when placed together they make a straight line.

2 b All polyiamonds tessellate.

Page 293 Exercise 3

1 A and V, P and S, R and K, Q and G, T and B, U and C, O and I, D and J, N and L and H, M and F

2 They are not the same size. Corresponding sides are not equal in length.

3 a JKL and MNO

   b In JKL   **i** angle L   **ii** angle K   **iii** angle J.
       In MNO   **i** angle M   **ii** angle N   **iii** angle O.
   c In JKL   **i** LK   **ii** KJ   **iii** LJ.
       In MNO   **i** MN   **ii** NO   **iii** MO.

4 a parallelogram, arrowhead, kite

   b rectangle, kite, parallelogram, triangle

   c kite, rhombus, parallelogram

   d rhombus

An example of a reason for the parallelogram in **a** is 'If we rotate one of the triangles 180° and then match the sides, the resulting shape will always be a parallelogram because by rotating the △ 180° the corresponding sides become parallel to each other. There are then two pairs of equal parallel sides.'

Review

a All except JKL, PNT and HBJ.

b For DEF, angle F, for GHI, angle G, for MNO, angle M, for PQR, angle R, for STU, angle T, for VWX, angle W.

c For DEF, FE, for GHI, GI, for MNO, MN, for PQR, RQ, for STU, TS, for VWX, WX.

Page 296 Exercise 4

1 a 8 **b** 4

2 A possible answer is:
This net is not full size.

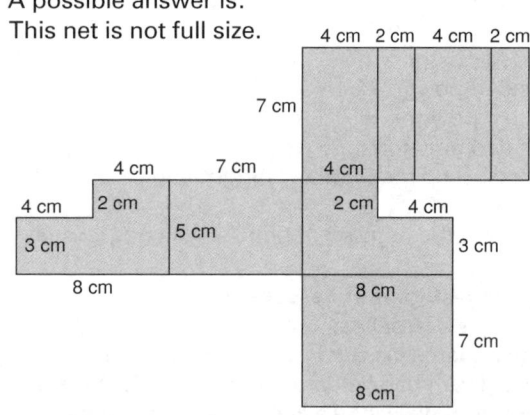

3 a   **b**   **c**

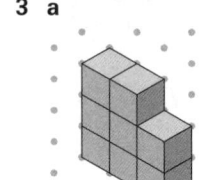 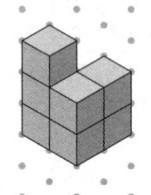 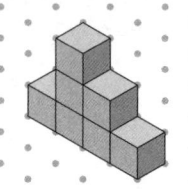

4 a   **b**   **c**

  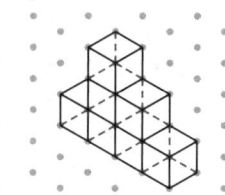

5 Possible answers are:

a   **b**

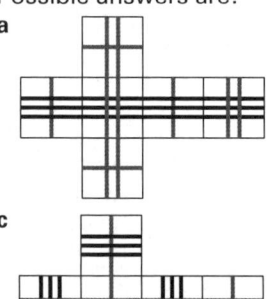 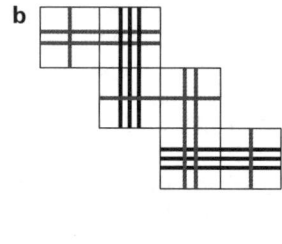

c

Review 1

a i 12 **ii** 6 **iii** 8 **iv** 2   **b i** 12 **ii** 6 **iii** 8 **iv** 2

Review 2 and Review 3

a   **b**   **c**

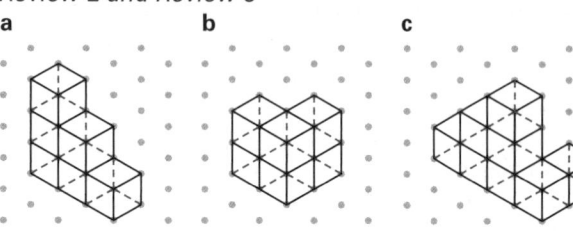

Page 297 Puzzle

   and △ ,   and ☆ ,   and

Page 298 Exercise 5

1 a B **b** D **c** A **d** C

2 a front elevation   side elevation

plan view

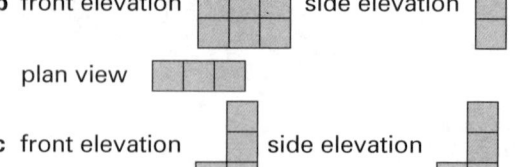

b front elevation   side elevation

plan view

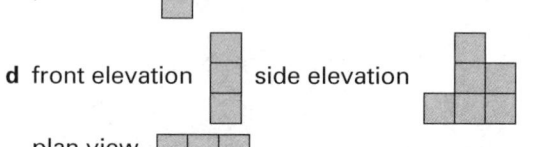

c front elevation   side elevation

plan view

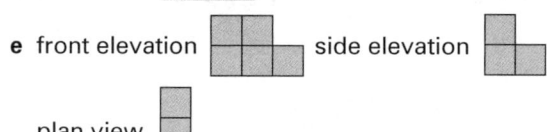

d front elevation   side elevation

plan view

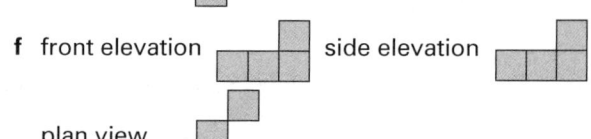

e front elevation   side elevation

plan view

f front elevation   side elevation

plan view

3 a   **b**

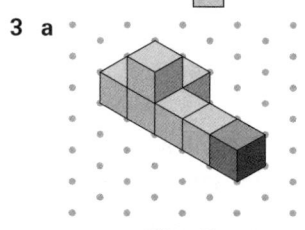 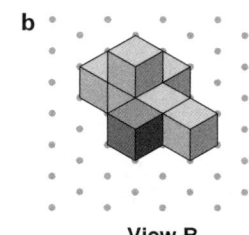

View B   **View B**

c

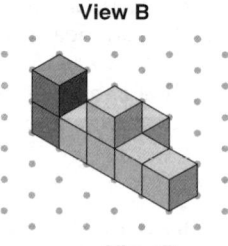

View B

4 a   **b**   **c**

d   **e**

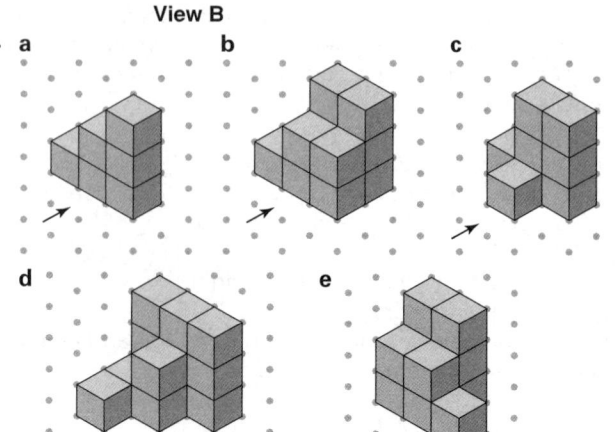

5 a A, B and C **b** A and B **c** A and D
6 Possible answers are:

a

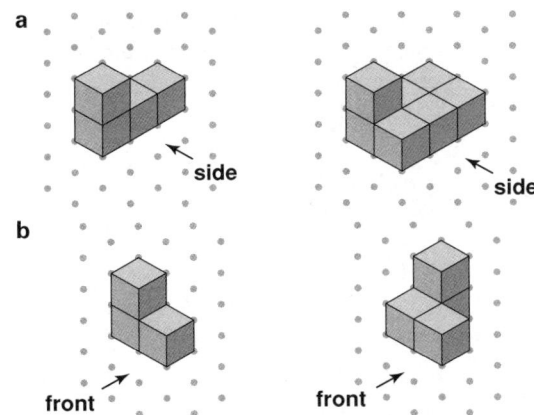

b

There are other answers.

Review 1

a

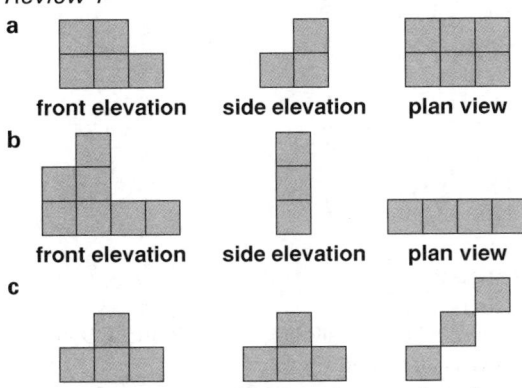

front elevation   side elevation   plan view

b

front elevation   side elevation   plan view

c

front elevation   side elevation   plan view

Review 2

a   **b**   **c**

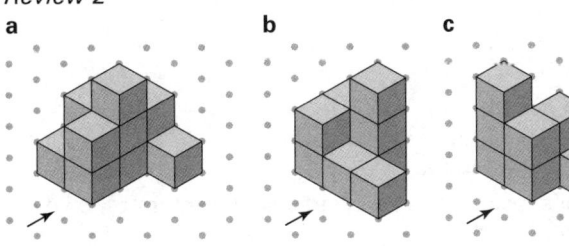

Page 300 Practical

2 Shape 1   **Shape 2**

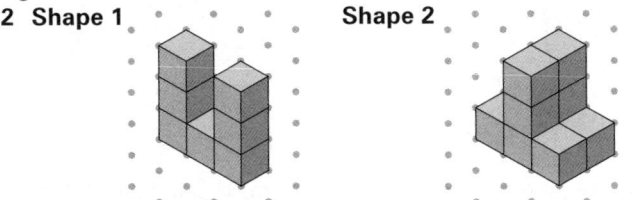

3 It is possible to build shape 1 and shape 3.
It is possible to build shape 1 in more than one way.

Page 302 Discussion

Perpendicular bisector of the line BC
BPCQ is a rhombus because all the sides have been made equal during construction.
R is the mid-point of BC because the diagonals of a rhombus bisect each other.
PQ is the perpendicular bisector of BC because the diagonals of a rhombus bisect at right angles, and PQ and BC are the diagonals of the rhombus.

Bisector of angle P
PACB is a rhombus because all the sides have been made equal during construction.
PC is a diagonal of the rhombus and the diagonals of a rhombus bisect the angles.
Line from A perpendicular to BC
AQRP is a rhombus because all the sides have been made equal during construction.
AR and PQ are the diagonals of a rhombus and the diagonals of a rhombus cross at right angles.
Perpendicular from a point P on a line segment BC
BQCR is a rhombus because the sides have been made equal during construction.
RQ and BC are diagonals of the rhombus and the diagonals of a rhombus cross at right angles.

Page 303 Exercise 6

1 c 19 mm
2 Pupil's accurate drawing
3 Draw the hexagon by drawing equilateral triangles. The triangle shown is right-angled because one diagonal creates an isosceles triangle, shown in grey.
The interior angle of a hexagon is 120°, so the two base angles of the grey triangle are 30°. One of the diagonals is a line of symmetry so makes the angles shown.

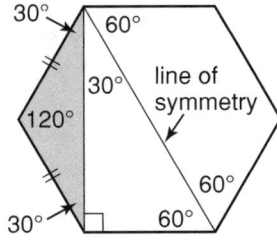

Page 305 Discussion

To construct ABCD use a protractor and ruler.
1 Draw a line 4·2 cm long and label it DC.
2 At D, construct an 87° angle with your protractor.
3 Draw the other arm of the angle 2·8 cm long. Label this line DA.
4 At A, construct a 111° angle with your protractor.
5 Draw the other arm of the angle 3 cm long. Label this line AB.
6 Join C to B to complete the quadrilateral.
To construct PQRS use a ruler and compasses.
1 Draw a line 4·5 cm long. Label it SR.
2 Open the compasses out to 5·6 cm. With the point on S, draw an arc.
3 Open the compasses out to 3 cm. With the point on R, draw an arc.
4 Label the point where the arcs drawn in **2** and **3** meet Q.
5 Join S to Q and R to Q.
6 Open the compasses out to 4·6 cm. With the point on S, draw an arc.
7 Open the compasses out to 6·3 cm. With the point on Q, draw an arc.
8 Label the point where these arcs meet P.
9 Join S to P and Q to P to complete the quadrilateral.

Page 305 Exercise 7

1 a 90° **b** 47° (to the nearest degree)
  c 55° (to the nearest degree)
  d 51° (to the nearest degree)
2 a Length of dashed line = 5·5 cm to the nearest mm
  b Size of red angle = 64° to the nearest degree
  c Length of dashed line = 5·7 cm to the nearest mm

5 Possible answers are:

a

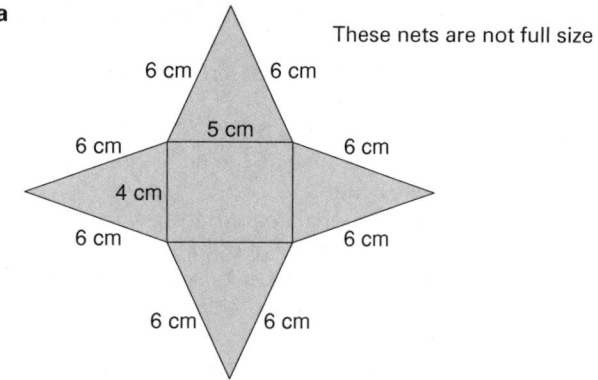

These nets are not full size

b

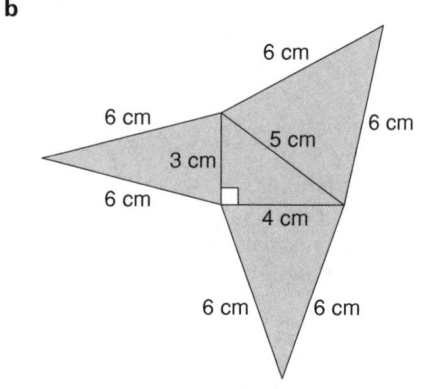

c

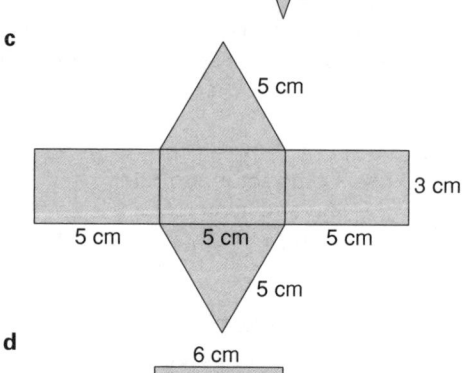

d

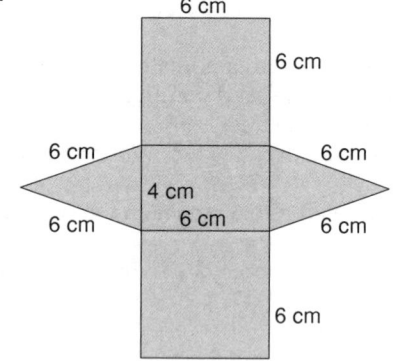

Review 1
Length of dashed line = 5·1 cm to the nearest mm

Review 4

a

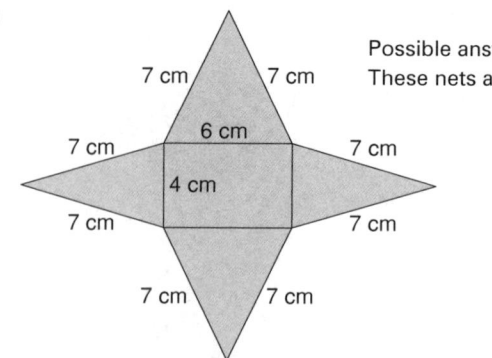

Possible answers are given.
These nets are not full size.

b

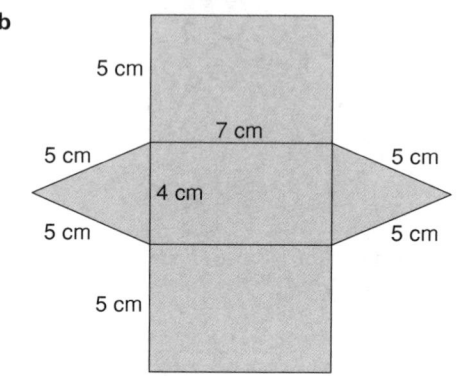

Page 308 Exercise 8
1 Jenni's locus is a circle.
2 Jenni's locus is a line parallel to both AB and CD and exactly half way between them.

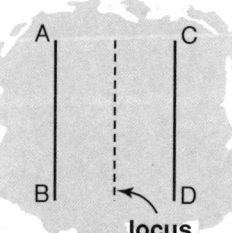

3 Jenni's locus is the perpendicular bisector of the line joining R and Q.

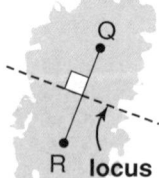

4 Jenni's locus is a line parallel to ST and exactly 5 m from ST.

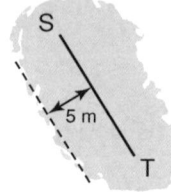

5 Jenni's locus is the bisector of the angle between DE and HG.

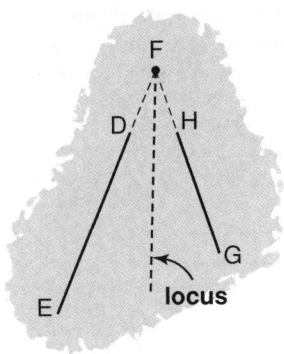

locus

Review
a The yellow counters lie on a circle.

b The yellow counters lie on the perpendicular bisector of the line joining the red and blue counters.

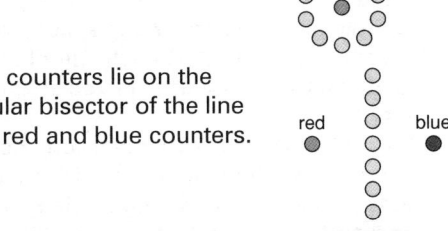
red   blue

c The red counters lie on the bisector of the angle between the adjacent edges of the table.

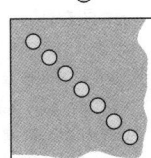

13 Coordinates and Transformations

Page 313 Exercise 1
1 a A(2, 3), B(4, 3), C(4, 1), D(1, 1)
  b i A'(2, ⁻3), B'(4, ⁻3), C'(4, ⁻1), D'(1, ⁻1)
   ii A'(⁻2, 3), B'(⁻4, 3), C'(⁻4, 1), D'(⁻1, 1)
2 a (⁻2, 1)
  b i (5, 4), (7, 0), (3, ⁻2), (1, 2)
   ii (⁻2, ⁻3), (⁻4, 1), (0, 3), (2, ⁻1)
   iii (⁻2, 3), (⁻4, ⁻1), (0, ⁻3), (2, 1)

Review
a (1, 1), (1, 2), (2, 4), (2, 3), (3, 3), (2, 1)
b i (⁻4, ⁻2), (⁻4, ⁻1), (⁻3, 1), (⁻3, 0), (⁻2, 0), (⁻3, ⁻2)
  ii (1, ⁻1), (1, ⁻2), (2, ⁻4), (2, ⁻3), (3, ⁻3), (2, ⁻1)
  iii (⁻1, 1), (⁻1, 2), (⁻2, 4), (⁻2, 3), (⁻3, 3), (⁻2, 1)
  iv (⁻1, ⁻1), (⁻1, ⁻2), (⁻2, ⁻4), (⁻2, ⁻3), (⁻3, ⁻3), (⁻2, ⁻1)

Page 314 Exercise 2
1 d A
2 a B **b** B
  c Rotation of 180° about the intersection of the two perpendicular lines
3 a 1 unit right and 2 units up
  b 3 units left and 2 units down
4 a A **b** B **c** Rotation 360° about (0, 0) **d** B
5 a Translation **b** Half-turn rotation
  c Rotation **d** Translation
6 a B **b** D **c** A **d** C
7 a C **b** E **c** B **d** F **e** D **f** A
8 There are many possible answers. One is:
  a Rotation 180° about (1, 2) followed by translation 2 squares left and 4 squares down
  b Reflection in the *x*-axis followed by translation 1 square down
  c Rotation 180° about (1, 1) followed by translation 6 squares right and 2 squares down.
9 There are many possible answers. One is:
  a i Rotation of 180° about X followed by translation 1 unit right
   ii Rotation of 90° about Y followed by translation 1 unit right

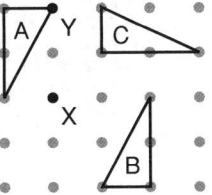

Review 1
c B

Review 2
c 270° about (0, 0), B

Review 3
a D **b** B **c** A **d** C

Review 4
a C **b** D **c** B **d** A

Review 5
There are many possible answers. One is:
a Reflection in the *y*-axis followed by reflection in the *x*-axis
b Rotation 180° about (⁻2, 0) followed by reflection in the line *x* = 2

Page 317 Practical
A Rotating a triangle about the mid-point of any of its sides will always give a parallelogram because the alternate angles formed are equal.
All parallelograms tessellate.

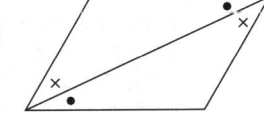

Page 318 Practical
A The maximum number of sides a polygon can have on a 3 × 3 pinboard is 7.

Page 318 Exercise 3
1 a 1 line of symmetry, no rotation symmetry
  b No lines of symmetry, rotation symmetry of order 2
  c No lines of symmetry, rotation symmetry of order 3
  d 4 lines of symmetry, rotation symmetry of order 4
  e 1 line of symmetry, no rotation symmetry
2 a 3 lines of symmetry, rotation symmetry of order 3
  b 6 lines of symmetry, rotation symmetry of order 6
  c 4 lines of symmetry, rotation symmetry of order 4
  d 1 line of symmetry, no rotation symmetry
  e No lines of symmetry, rotation symmetry of order 2
  f 2 lines of symmetry, rotation symmetry of order 2
  g 2 lines of symmetry, rotation symmetry of order 2
  h No lines of symmetry, no rotation symmetry
  i 1 line of symmetry, no rotation symmetry
  j 1 line of symmetry, no rotation symmetry

3 a If we fold an isosceles triangle along its line of symmetry, the base angles fit exactly on top of one another.

b If we fold an equilateral triangle along each of its three lines of symmetry, each pair of angles fit exactly on top of one another, so all three must be equal.

c If we fold an isosceles triangle along its line of symmetry, we get a right-angled triangle on top of another congruent right-anged triangle. The line of symmetry must therefore go through the vertex and be the perpendicular bisector of the third side.

d If we rotate a parallelogram about its centre, it looks exactly the same after a half-turn. Therefore opposite angles and opposite sides must be equal.

e If we fold a rhombus along its lines of symmetry (the diagonals) one half of the diagonal fits exactly on top of the other half so the diagonals must bisect each other.

f A parallelogram has rotation symmetry of order 2. If we rotate a parallelogram a half-turn about its centre, it looks exactly the same. Therefore, the diagonals must bisect each other.

g A parallelogram has rotation symmetry of order 2. If we rotate a parallelogram a half-turn about its centre, it looks exactly the same. Therefore the angle between a pair of opposite sides and a diagonal will fit exactly on top of each other, so they must be equal.

4

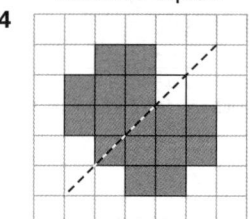

5 Possible answers are:

a

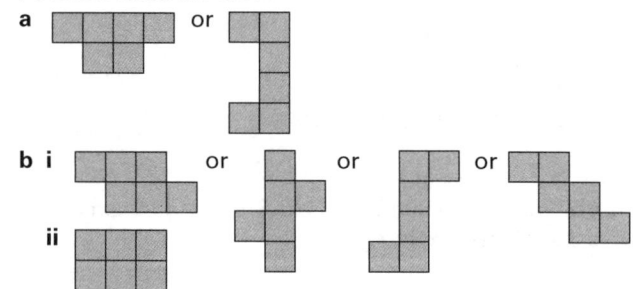

Review 1

a 1 line of symmetry, no rotation symmetry

b 3 lines of symmetry, rotation symmetry of order 3

c No lines of symmetry, rotation symmetry of order 2

Review 2

a If we fold a rhombus along its lines of symmetry (the diagonals) opposite angles and opposite sides fit exactly on top of each other.

b If we fold a rhombus along its lines of symmetry (the diagonals) half of the angle at each vertex fits exactly on top of the other half.

Review 3

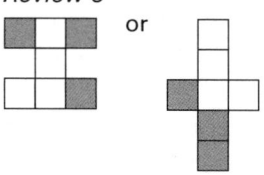

Page 319 Investigation: Pentominoes

The 12 pentominoes are shown.

The ones with line symmetry have the lines drawn on them. The ones which have rotation symmetry have a tick beside them. The order of rotation symmetry is written underneath them.

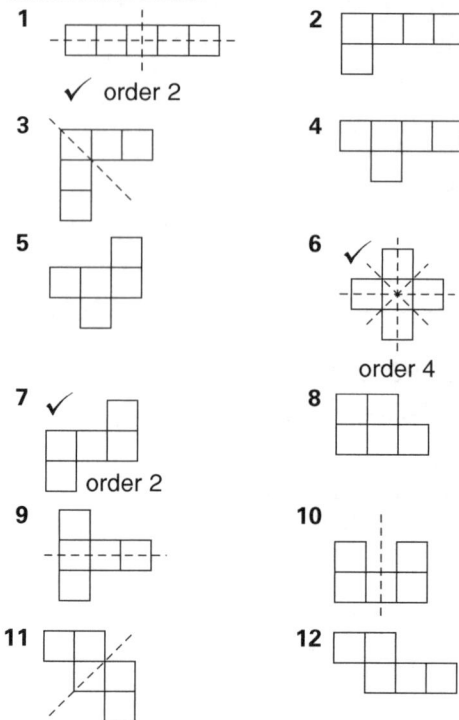

All 12 pentominoes will tessellate.

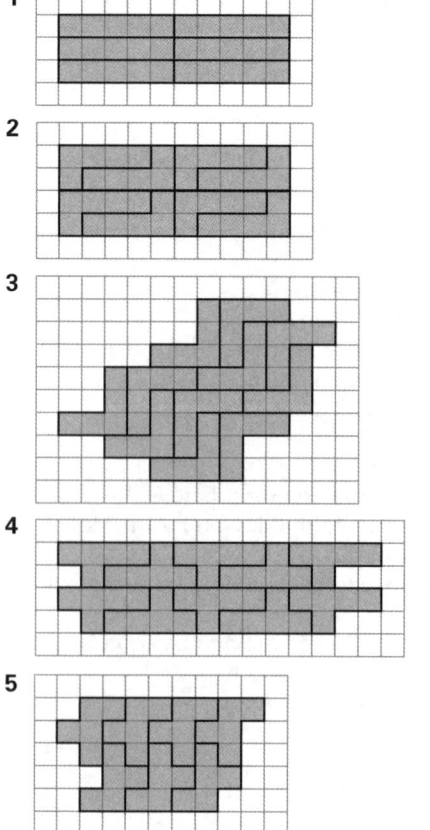

6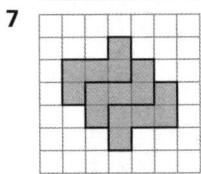

7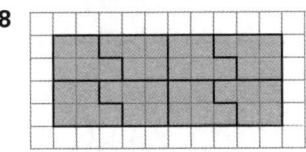

8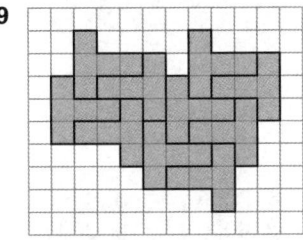

9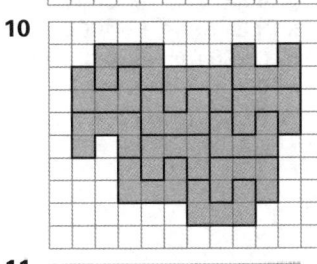

10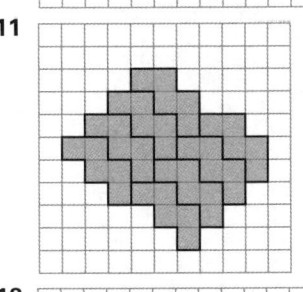

11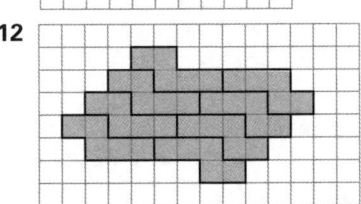

12

Page 320 Exercise 4
1 a 2   **b** 2   **c** 3
2 a 2   **b** 2   **c** 2

Review
a 4   **b** 2   **c** 3

Page 321 Discussion
The centre of enlargement is in a different position for each.

Page 322 Practical
A

Length (cm)	Length on original	Scale Factor		
		2	3	4
distance between eyes	2	4	6	8
height of alien	6	12	18	24
width of head	4	8	12	16
length of foot	1	2	3	4
length of leg	1·1	2·2	3·3	4·4

For the enlargement, scale factor 2, the ratio of corresponding lengths is always 2 : 1.
For scale factor 3, the ratio of corresponding lengths is always 3 : 1.
For scale factor 4, the ratio of corresponding lengths is always 4 : 1.
The angles of a shape are the same on an enlargement as on the original shape.

Page 323 Exercise 5
1 a A'(6, 6)   **b** A'(12, 0)   **c** A'(10, 6)
   d A'(2, 6)   **e** A'(9, 3)
2 a The purple lines have doubled in length.
      The shaded angles have stayed the same size.
   b The purple lines have become three times as long.
      The shaded angles have stayed the same size.
   c angles, lengths
3 a A(1, 3), B(2, 2), C(1, 0), D(0, 2)
   b A'(2, 6), B'(4, 4), C'(2, 0), D'(0, 4)
   c A'(3, 9), B'(6, 6), C'(3, 0), D'(0, 6)
   d A'(4, 12), B'(8, 8), C'(4, 0), D'(0, 8)
   e The coordinates are multiplied by the scale factor.
   f No
4 a P'(⁻2, 6), Q'(8, 6), R'(6, 0)
   b P'(⁻4, 4), Q'(6, 4), R'(4, ⁻2)

Review 1
a A'(8, 12)   **b** A'(5, 8)

Page 325 Exercise 6
1 a About 0·9 m   **b** About 2·7 m
   c About 5·4 m   **d** About 3·6 m
2 About 13 m
3 800 cm × 1200 cm or 8 m by 12 m
4 a 16 800   **b** 16·8 m
5 37·5 km
6 a 12 m   **b** 10·5 m   **c** 3·6 m   **d** 5·4 m
7 a 125 km   **b** 12·5 km   **c** 375 km
8 1 cm represents 1 m or 1 cm represents 100 cm
9 1 cm represents 50 cm
10 1 inch represents 1800 inches
11 a 1 cm represents 1 m or 1 cm represents 100 cm
    b 1 cm represents $\frac{1}{2}$ km or 1 cm represents 500 m or
       1 cm represents 50 000 cm
    c 1 cm represents 125 m or 1 cm represents
       12 500 cm

Review 1
About 18 m

Review 2
a 30 m   **b** 64 m

Review 3
1 cm represents 2 m or 1 cm represents 200 cm

Page 328 Exercise 7

1 On the drawing 0·5 m will be 1 cm, 2 m will be 4 cm, 4 m will be 8 cm.

2 On the drawing, 12 km will be 6 cm, 3·2 km will be 1·6 cm, 5·4 km will be 2·7 cm, 8 km will be 4 cm, 5 km will be 2·5 cm.

Review

On the drawing, 60 m will be 12 cm, 1·5 m will be 0·3 cm, 20 m will be 4 cm, 15 m will be 3 cm, 30 m will be 6 cm, 3·5 m will be 0·7 cm.

Page 329 Exercise 8

1 a (5, 4) b (5, 8) c (0, 3) d (0·5, 1)
2 a (5, 7) b (⁻3, 5) c (4, 2) d (2, ⁻3·5)

Review

a (7, 4) b (6, 4) c (⁻5, 6·5) d (⁻4·5, ⁻2)

Page 330 Exercise 9

1 a (3, 7)   b (2, 5)   c (2·5, 6)   d (5·5, 4·5)
  e (2, 0·5) f (0, ⁻1·5) g (⁻3·5, ⁻6·5)
2 a (3, 4)   b (4, 8)   c (⁻4·5, 3)   d (4, ⁻4)
  e (0·5, ⁻0·5) f (⁻2·5, ⁻4) g (⁻7, ⁻5·5) h (⁻3·5, ⁻2)
  i (2, ⁻5)   j (9, ⁻2)

Review

a (4, 6) b (⁻6, 5) c (⁻8, ⁻6·5) d (0·5, 2)

14 Measures, Perimeter, Area and Volume

Page 335 Exercise 1

1 a 0·7 m   b 3·7 cm   c 2·8 m
  d 14·5 m   e 0·8 ℓ   f 1·564 kg
  g 0·36 m   h 0·036 m   i 0·86 m
  j 0·05 ℓ   k 0·003 m   l 0·009 kg
2 a 7 tonnes   b 5 tonnes   c 8·7 tonnes
  d 5400 kg   e 8700 kg   f 9830 kg
  g 4·75 tonnes h 0·42 tonnes   i 0·325 tonnes
  j 0·079 tonnes k 0·096 tonnes l 6 kg
  m 0·005 tonnes n 30 kg   o 0·003 tonnes
3 a 40 000 m²   b 36 000 m²   c 830 000 m²
  d 57 200 m²   e 5 ha   f 7 ha
  g 12 ha   h 0·5 ha   i 8·6423 ha
  j 0·7925 ha   k 0·008 ha   l 0·0654 ha
4 a 240 minutes   b 330 minutes
  c 195 minutes   d 395 minutes
5 a 3 hours 0 minutes
  b 1 hour 35 minutes
  c 8 hours 20 minutes
  d 7 hours 6 minutes
6 a 96 hours   b 162 hours   c 240 hours
  d 543 hours   e 7·7 hours
7 a 110 years 4 months
  b 4 decades 8 years 11 months
  c 16 years 30 weeks
  d 2 days 18 hours 40 minutes
8 Hyde Park 2 550 000 m², Kensington Gardens 1 110 000 m², Regents Park 1 970 000 m², Kew Gardens 1 200 000 m²
9 7 ha
10 200 m
11 530 kg
12 20 February 1962
13 a 2·25 m² b 1·8 kg
14 a 100 cm b 250 cm c 2·5 m² d 25 000 cm²
15 B
16 a 0200 on 3 January 2005
   b 1600 on 11 February 2010
18 a 136·6 km b 73 c 6
19 244·755 m²

Review 1

11 weeks 3 days

Review 2

a 5 hours 50 minutes   b 2 years 112 days
c 145 years 1 month   d 5 days 21 hours

Review 3

a 8 ha b 12 000 m²

Review 4

a 2320 kg b 2·32 tonnes

Review 5

1922 hours on 9 January 2005

Page 337 Exercise 2

1 a 3000 cm³   b 5000 cm³   c 5 cm³   d 9 cm³
  e 5 m³   f 7 m³   g 4000 ℓ   h 3000 ℓ
  i 4 ℓ   j 8·6 cm³   k 9·2 cm³   l 4600 cm³
  m 8300 cm³   n 0·82 cm³   o 0·896 m³   p 0·437 m³
  q 4600 ℓ   r 820 ℓ   s 43 mℓ   t 8·42 ℓ
  u 0·523 ℓ   v 0·084 ℓ   w 0·043 m³
2 a 0·26 ℓ b 260 cm³
3 a 1·2 ℓ b 1200 cm³
4 a 5·15 ℓ b 5150 cm³ c 34
5 a No b 450 cm³
6 a 30 ℓ b 30 000 cm³ c 0·6 m³

Review 1

PEARLS DISSOLVE IN VINEGAR

Review 2

a 150 cℓ b 375 cm³

Page 338 Puzzle

1 8:21
2 1 p.m. on 10 June
3 1948

Page 339 Exercise 3

All of these answers are approximate.

1 a 17½ pints b 3½ pints c 8·75 pints d 0·875 pints
  e 35 pints
2 a 11 lb b 4·4 lb c 22 lb d 6·6 lb e 8·8 lb
3 a 25 miles   b 50 miles   c 30 miles
  d 22·5 miles   e 57·5 miles   f 71·9 or 72 miles
4 a 2 kg b 600 g c 150 g d 6 kg e 9 kg f 12·3 kg
5 2·7 kg
6 a 9 litres b 270 litres c 10 litres d 4 litres
  e 12·6 litres or 13·2 litres depending on how you work it out.
7 a 32 km b 320 km c 80 km d 19·2 km e 83·2 km
8 a 15 cm b 5 cm c 45 cm d 90 cm e 270 cm
9 a 5 m b 4
10 b 10 inches by 7·2 inches
11 a 15·4 lb b £7·83 c About 8·3 oz
12 6 gallons
13 5·25 pints

Review
30 miles ≃ 48 km, 4 gallons ≃ 18 litres, 4 metres ≃ 12 feet,
11 lb ≃ 5 kg, 15 inches ≃ 37·5 cm

Page 341 Exercise 4
1 a Nearest mm   b Nearest g
   c Nearest m   d Nearest km
   e Nearest g   f Nearest cm
   g Nearest 100 g   h Nearest mℓ
   i Nearest tonne   j Nearest 100 mℓ
   k Nearest minute   l Nearest second
2 a B   b C   c A
3 a C   b B   c D
4 a A   b C   c D
5 a C   b A   c D

Review 1
a Nearest cm   b Nearest tonne   c Nearest minute
d Nearest 100 mℓ e Nearest mm

Review 2
a D   b B

Page 343 Exercise 5
Answers may be different from those given.
1 a 15 cm ⩽ depth of step ⩽ 20 cm
   b 1 ℓ ⩽ water in jug ⩽ 3 ℓ
   c 1 mm ⩽ width of key ⩽ 3 mm
   d 2000 cm^2 ⩽ area of desk ⩽ 4000 cm^2
   e 150 mℓ ⩽ volume of mug ⩽ 300 mℓ
   f 1 m ⩽ height of window ⩽ 2·5 m but depends on
      individual
   g Depends on weather
   h 80 cm^2 ⩽ area of piece of toast ⩽ 140 cm^2
   i 200 cm^2 ⩽ area of computer screen ⩽ 1000 cm^2
   j 10 cm ⩽ width of face ⩽ 20 cm
   k 8 kg ⩽ mass of bike ⩽ 15 kg
   l 3 kg ⩽ mass of desk ⩽ 20 kg

Review
Possible answers are:
a 30 g < mass of lightbulb < 200 g
b 1·8 m < length of floor-length curtain < 4 m
c 10 ℓ < capacity of kitchen sink < 40 ℓ
d 1·5 tonnes < mass of car < 2·5 tonnes
e 3 m < length of long jump pit < 10 m
f 300 mℓ < capacity of a thermos flask < 2000 mℓ

Page 345 Discussion
We can use the fact that the
North lines are parallel to find
the required angle.
∠x = 360° − 224° (angles at a
      point add to 360°)
   = 136°
∠y = 180 − 136°
   = 44°
Shaded angle = 44°
(corresponding angles)

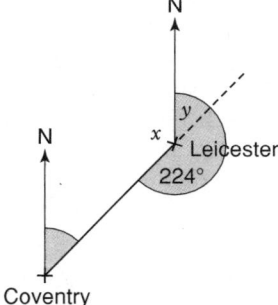
Coventry

Page 346 Exercise 6
1 a C b A c D d D e B f C
2

Aircraft	VB502	BA172	EI456	CO28	NZ2	CX201	*E1452	BA36	*CO4
Bearing	080°	030°	320°	170°	220°	060°	110°	300°	070°

3 a i 058° ii 154° iii 224° iv 286° v 218°
   b i 238° ii 334° iii 044° iv 106° v 038°

4 a 255° b 340° c 110°
5 a 070° b 135° c 194° d 250° e 315°
6 a 060° b 115° c 260° d 045° e 294° f 99°
7 a 245° b 095°
8 022°

Review 1
a D b B

Review 2
a 090° b 135° c 315°

Review 3
a Cave Bay b 085° c 340° d 160°

Page 350 Exercise 7
1 a i 10 m^2 ii 7·5 cm^2 iii 160 mm^2 iv 31·5 cm^2
       v 32 cm^2
   b i 15·4 m ii 13·8 cm
2 1 029 600 km^2
3 a 28 cm^2 b 6·46 m^2 c 10 mm^2 d 0·96 m^2
   e 6·16 m^2
4 1·25 cm
5 Because the base and perpendicular height of each is
   the same and they are all either rectangles or
   parallelograms.
6 a i 28 cm^2 ii 81 m^2 iii 22·58 m^2 iv 7·455 cm^2
   b 37 m
7 26·25 m^2
8 3·5 m
9 100 m
10 There are many possible answers. One is: a = 5 cm,
   b = 3 cm, h = 2·5 cm.
11 146 cm^2
12 a 1·2 m b 0·8 m
13 48
14 44 cm
15 42 cm

Review 1
a 50 cm^2 b 82·5 cm^2

Review 2
a i 28 cm^2 ii 250 m^2 iii 6·72 m^2
b i 24 cm ii 70 m iii 11 m

Review 3
12 m

Review 4
300 cm^2

Page 353 Investigations
Moving vertices
1 The area of the parallelograms is always the same
   because the height and the base are always the same.
2 The area of the triangles is always the same because
   the height and the base are always the same.
3 The area of the triangles are increasing each time the
   vertex A is translated. Each time the vertex is
   translated one square up, the area increases by 2
   units.

Vertex at	A	A$_1$	A$_2$	A$_3$	A$_4$	...
Area of triangle (square units)	8	10	12	14	16	...

Two possible explanations are:

- This is because if we put a rectanlge around the triangle, it would increase in area 4 units each time the vertex moved up one unit. The area of a triangle is half the area of a rectangle, so its area increases by half of $4 = 2$.
- The base of the triangle is always 4. Each time the vertex moves up a unit, the area increases by $\frac{1}{2} \times 1 \times 4 = 2$.

        increase in height

Hexagons

Hetty's shape has a perimeter of $22n$.
The perimeter of the smallest shape made with five regular hexagons of side n is $16n$.

Number of regular hexagons	5	6	7	8	9	10	11	12
Smallest perimeter	$16n$	$18n$	$18n$	$20n$	$22n$	$22n$	$24n$	$26n$

Triangles in cubes

There are four different triangles.
The equilateral triangle has the greatest area because it has the biggest base and the biggest height.

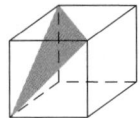 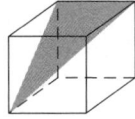 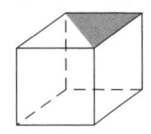 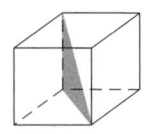

| equilateral | right-angled scalene | right-angled isosceles | right-angled scalene |

Page 354 Exercise 8

1 a $4 \cdot 5$ units^2 b $4 \cdot 5$ units^2 c $9 \cdot 5$ units^2
    d $13 \cdot 5$ units^2 e $27 \cdot 5$ units^2
2 a $6\frac{1}{2}$ units^2 b 20 units^2

Review
a $10 \cdot 5$ units^2 b 21 units^2

Page 356 Exercise 9

1 a 30 cm^3 b 64 m^3 c 880 cm^3
    d $1347 \cdot 84$ cm^3 e $163 \cdot 8$ cm^3
2 a 62 cm^2 b 96 m^2 c 622 cm^2
    d $733 \cdot 44$ cm^2 e $182 \cdot 6$ cm^2
3 a 6   b 168   c 4464 cm^2
4 60
5 8

6 a $55\ 200$ cm^3 b $1\ 152\ 000$ cm^3 or $1 \cdot 152$ m^3
    c $371\ 200$ cm^3 or $0 \cdot 3712$ m^3
7 $11\ 504$ cm^2
8 10 will fit in the 6 m long container, 20 will fit in the 12 m long container and 20 will fit in the 14 m long container.
9 a 392 cm^2
    b The lengths are 5 cm, 8 cm and 12 cm.
10 64
11 A square prism with a 2 cm by 2 cm end.
12 Volume of shaded piece = $62 \cdot 5$ cm^3, surface area of shaded piece = $106 \cdot 5$ cm^2
13 128

Review 1
a Volume = 27 cm^3, surface area = 54 cm^2
b Volume = 162 cm^3, surface area = 198 cm^2
c Volume = $815 \cdot 625$ cm^3, surface area = $535 \cdot 5$ cm^2

Review 2
972 cm^3

Review 3
16 if they are allowed to be above the top of the tray, 10 if they aren't.

Review 4
a 3 m^3
b $182\ 000$ cm^2 or $18 \cdot 2$ m^2

Review 5
33 cm $\times 33$ cm $\times 34$ cm

Page 358 Investigation: Making boxes

The dimensions could be any three lengths with a product of 24 cm^3. Examples are 24 cm $\times$ 1 cm $\times$ 1 cm, 8 cm $\times$ 2 cm $\times$ 1·5 cm, 6 cm $\times$ 2 cm $\times$ 2 cm, 2·5 cm $\times$ 3·2 cm $\times$ 3 cm, ...
The net which uses the smallest area of cardboard is always the one with all lengths equal, i.e. the net of a cube.
The lengths of each side will be closest to
$^3\sqrt{24} = 2 \cdot 88$ cm (not exact)
        $= 2 \cdot 9$ cm
For 48 cm^3, this is 3·63 cm. (This is not exact.)
For 30 cm^3, this is 3·11 cm. (This is not exact.)
For 36 cm^3, this is 3·30 cm. (This is not exact.)

Handling Data Support

Page 366 Practice Questions

1 a

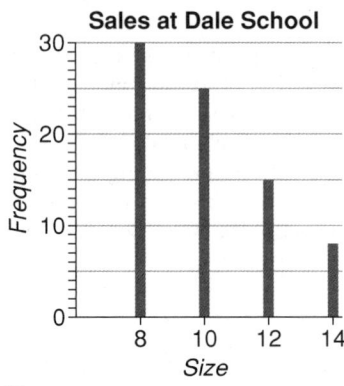

Sales at Dale School

b 25
c Some pupils may have bought more than one pair.
d Dale school, because there were more smaller sizes sold at Dale School.

2 Possible answers are:
    a Very likely   b Very likely   c Very unlikely
    d Impossible   e Certain
3 Possible answers are: e, a, b, c, d or e, b, a, c, d.
4 a 5   b 4   c So he can order most of this size.
5 a 2   b 12   c A
6 a 20–29   b

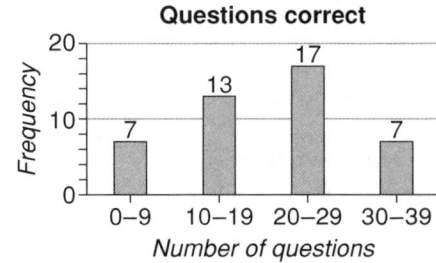

Questions correct

7 a 40 **b** 35

c It was probably hot on Saturday and cooler on Sunday because lots of ice-cream and cold drinks were sold on Saturday and very few of these were sold on Sunday. On Sunday more hot drinks were sold.

8 a mean = 19, median = 16, range = 35

b mean = £5693·38, median = £5460·50, range = £4780

c mean = 11·36 g, median = 11 g, range = 8 g

9 a 97·2 g **b** 25 g **c** 97 g

10 10 and 15

11 a week B **b** week A **c** week C

12 a 357 **b** 198 **c** 94

13 a Geography **b** French **c** A Yes, B No, C Yes D Yes

d Last week may not have been typical of the amount of homework time Rick usually spends on maths each week.

14 a Even chance **b** Less than even chance

c Impossible **d** Better than even chance

e Less than even chance **f** Impossible

g Certain

15 a a, b, c, d, e, f, g, h, i, j, k, l, m, n, o, p, q, r, s, t, u, v, w, x, y, z

b Yes

c No, because there are more consonants than vowels.

16 a C **b** A **c** B

17 2·4

18 a Green, because there are fewer green sections than any other colour.

b

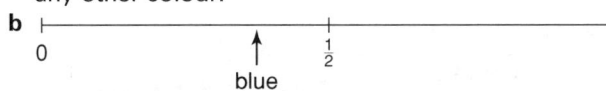

0   blue $\frac{1}{2}$ 1

c $\frac{1}{4}$ or 0·25 or 25%

19 a Nick, Jackie, Caroline **b** Rishi, Caroline

20 a Thursday **b** £7000 **c** No

21 a $\frac{1}{2}$ **b** $\frac{1}{3}$ **c** $\frac{2}{3}$ **d** 0 **e** 1

22 a

Questions correct	Tally	Frequency
0–4	I	1
5–9	II	2
10–14	IIII	4
15–19	ЖИ	5
20–24	IIII	4
25–29	ЖИ III	8
30–34	III	3
35–39	III	3

b

Questions correct	Tally	Frequency
0–9	III	3
10–19	ЖИ IIII	9
20–29	ЖИ ЖИ II	12
30–39	ЖИ I	6

c The class intervals given in **b**, because they show the spread of data better.

d

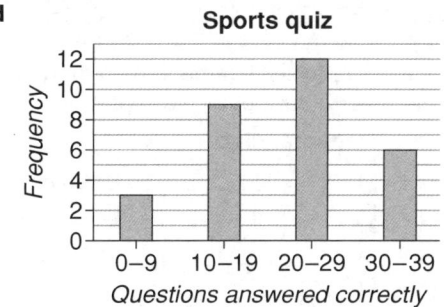

Sports quiz

Frequency (vertical axis: 0, 2, 4, 6, 8, 10, 12)

Questions answered correctly (horizontal axis: 0–9, 10–19, 20–29, 30–39)

e B

23 a There are some blue cubes in the bag but because Tina puts the cube back each time, some of the blue cubes may have been taken out more than once or some may not have been taken out at all.

b There may be white cubes in the bag but Tina may not have taken one out.

24 Possible answers are:

a Bene might find out that girls skip faster than boys and that there is a greater range of skipping speeds for boys.

b From each Year 8 pupil, Bene needs to collect whether they are a boy or a girl and the time to skip a particular distance.

c One possible answer is:

> Are you a boy or a girl?   Boy ☐   Girl ☐
>
> How long does it take you to skip 80 m? _____

or Bene could choose a sample of Year 8 boys and girls and time them to skip 80 m. The data could be collected on a sheet like this.

Boys		
Time (seconds)	Tally	Frequency
$12 \leqslant t < 15$		
$15 \leqslant t < 18$		
$18 \leqslant t < 21$		
$21 \leqslant t < 24$		

Girls		
Time (seconds)	Tally	Frequency
$12 \leqslant t < 15$		
$15 \leqslant t < 18$		
$18 \leqslant t < 21$		
$21 \leqslant t < 24$		

d As a survey from a sample of people. About 20 or 30 boys and 20 or 30 girls would be a good sample size.

25 a Plastic drink bottles **b** B

c Overall the Parker family recycled more items than the Banks family. The only thing the Banks family recycled more of than the Parker family was plastic milk bottles.

26 The pie charts show the proportion that like comedy best not the actual number of people. There were more British people surveyed. The smaller proportion of 800 British people will still be a greater number than the larger proportion of the 250 American people.

27 a 11–20, 78; 21–30, 42; 31–40, 34; 41–50, 26;
51–60, 16; 60+, 4

b Squash is a very physical game and so there are
likely to be fewer older people who play.

28 a

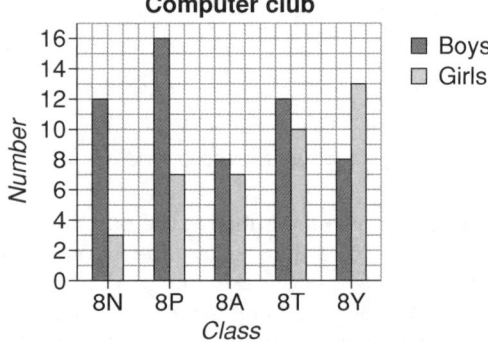

Computer club

■ Boys
□ Girls

b In all except one class, more boys than girls belong
to the computer club.

29 Both classes have the same mean number of pupils
per week away. 9L have a much more variable number
of pupils away. This is shown by the larger range.

Chapter 15 Data Collection

Page 372 Exercise 1

1 a Discrete **b** Continuous **c** Discrete
 d Discrete **e** Continuous **f** Continuous
 g Discrete **h** Continuous **i** Discrete
 j Discrete **k** Discrete **l** Discrete
 m Continuous **n** Continuous

Review
a Continuous **b** Discrete **c** Continuous
d Continuous **e** Discrete

Page 373 Exercise 2

1 a

Mass of apple (g)	Tally	Frequency
150 < m ≤ 160	III	3
160 < m ≤ 170	ЖH	5
170 < m ≤ 180	III	3
180 < m ≤ 190	III	3
190 < m ≤ 200	II	2
200 < m ≤ 210	II	2

b 5 **c** 3 **d** 7

2 a

Length of throw (x m)	Tally	Frequency
0 ≤ x < 4	I	1
4 ≤ x < 8	III	3
8 ≤ x < 12	ЖH I	6
12 ≤ x < 16	ЖH I	6
16+	IIII	4

b 3 **c** 10
d The class intervals $0 \le x < 1$, $1 \le x < 2$, would give
too many class intervals and the data would be too
widely distributed.

3

Time (minutes)	Tally	Frequency
14–	ЖH I	6
18–	ЖH I	6
22–	ЖH III	8
26–	IIII	4
30–34	ЖH I	6

4

Handspan (mm)	Tally	Frequency
160 < h ≤ 170	ЖH	5
170 < h ≤ 180	ЖH I	6
180 < h ≤ 190	ЖH II	7
190 < h ≤ 200	ЖH IIII	9
200 < h ≤ 210	ЖH III	8
210 < h ≤ 220	I	1
220 < h ≤ 230	III	3

Review

a

Amount of ice cream (x mℓ)	Tally	Frequency
1800 < x ≤ 1900	IIII	4
1900 < x ≤ 2000	ЖH II	7
2000 < x ≤ 2100	ЖH I	6
2100 < x ≤ 2200	III	3

b 7 **c** 9 **d** 17
e These would give too few class intervals and the first
class interval would have no tallies.

Page 375 Exercise 3

1 a 8 **b** 16 **c** 12
2 a 22 **b** 16 **c** 44
d A lot of the teenagers liked romance best and some
liked mystery or science fiction. None of the
children liked romance best. A lot of them liked
mystery best and some liked animal or science
fiction.
3 a 4 **b** 8 **c** 16 **d** 0
e Vegetarian, and the next best choice would be pasta.
4 Possible answers are

	Car	Bus
Cathedral		
Castle		

5

Time (min)	Car	Bus	Train	Walk	Coach	Other
1–5						
6–10						
11–15						
16–20						
20+						

Review 1
a 12 **b** 28 **c** 20
d There are more male than female cyclists and runners but more female than male swimmers. There are more males than females in the triathlon.

Review 2
A possible answer is

Age	M	F
60–		
65–		
70–		
75–		
80–		
85–		
90–		
95+		

Page 376 Discussion
Some possible related questions might be:

1 Are there variations at different times of the day or week or year?
Is the transport chosen related to cost?
Is the transport chosen related to any particular events like festivals, school holidays, ...

2 Does the depth of the water in the different parts of the river vary? If so, is the light intensity at different depths different?
Does the number in invertebrate communities change depending on the time of year?

3 Does the light intensity differ at different times of the day or year?
Does the weather affect the light intensity?

4 Are there differences between teams at different levels?
What is the best time to leave the match to buy food, if you don't want to queue at half-time?

5 What factors affect viewing habits? Time spent travelling to work or school, hours of work or homework, amount of sleep needed, ...

Page 377 Discussion
Doing a survey would give more accurate information about which pie is most popular because the sales records might be distorted by the availability of certain pies. Using the sales records would be quicker and use fewer resources. The survey method would give greater insight and allow further questions such as whether there is any difference in the popularity of the pies in winter or summer or between different year groups.

Page 377 Discussion
If a sample is too small it is not representative of the whole group.
If a sample is too big it costs too much money and takes too much time to collect.

Page 378 Exercise 4
Possible answers are:

A a Is there a difference for different year groups?
Is there a difference for different days of the week?

b The average time pupils spend moving between classes in a day is 15 minutes.
There is very little difference between year groups.

c Data needed for each class: year group, total time, in minutes, spent moving between classes today, day of week.

d
> Have one person in your class fill this in.
>
> class ☐   year group ☐
>
> total time spent moving (minutes) between classes ☐

e About 20 to 30 classes chosen randomly from the whole school (in some schools it would be sensible to ask every class).

f Primary

B a What factors affect viewing habits? Hours of work? Hours of homework? Amount of time spent travelling to work or school?

b Adults watch more sport on TV than teenagers.
Adults who work full time watch less sport than adults who work part-time.
Adults and teenagers who travel more watch less sport.

c Data needed from each person: age, number of hours of sport watched on TV each week, hours of work, hours of homework, time spent travelling.

d A possible questionnaire could be:

> Age ☐
> Number of hours spent watching sport on TV ☐
> Hours of work each week ☐
> Hours of homework each week ☐
> Hours of time spent travelling each week ☐

Note: You could give a selection of answers for each question.

e About 30 to 50 adults and 30 to 50 teenagers.

f Primary

C a Does water-cress grow best in sand or soil? Does temperature affect growth? Does amount of water affect growth? Does amount of light affect growth?

b Water-cress needs light to grow. It needs to be damp to grow. Water-cress grows best when planted in warm conditions.

c Data needed: sand or soil, light or dark, wet or dry, height of plants to the nearest mm.

d One possible answer is:

Height (mm) after	Full daylight					Dark cupboard				
	Plant 1	Plant 2	Plant 3	Plant 4	Plant 5	Plant 1	Plant 2	Plant 3	Plant 4	Plant 5
2 days										
4 days										
6 days										
8 days										
10 days										

A similar collection sheet would be needed for wet soil and dry soil.
A similar collection sheet would be needed for sand or soil.
When testing one set of conditions, the other conditions must be kept the same for all plants.

e At least 5 plants in each set of conditions.

f Primary.

D a Does the number of each type of invertebrate differ at different depths or parts of the river?
Is the number of each type of invertebrate affected by plant growth?
Does the swiftness of the current affect the number of invertebrates?

  b There are fewer invertebrates where the current is swifter.
There are more invertebrates in places where plants grow.
There are fewer invertebrates in deep water.

  c Number of invertebrates of each type, depth of water, swiftness of water, plant growth.

  d One possible answer is:

Invertebrate indicator animals	Swift water	Still water
Bloodworm		
Caddis fly larva		
Freshwater shrimp		
Mayfly nymph		
Rat tailed maggot		
Sludge worm		
Stonefly nymph		
Water louse		

  A similar collection sheet is needed for no plant growth, plant growth and deep water, shallow water.

  e Choose sites as follows: fast flowing water, still water, no plant growth, plant growth, deep water, shallow water.

  f Primary.

E a Is the light intensity at some sites affected by the time of day more than at others?

  b The light intensity is less in shady areas of the school.
Some parts of the school have a high variation of light intensity depending on the time of day.

  c Light intensity at several different sites around the school at several different times of the day.

	Light Intensity			
Time of day	site A	site B	site C	site D
9 a.m.				
12 p.m.				
3 p.m.				

  d One possible answer is:

  e A sample size of four sites and three times a day is reasonable.

  f Primary.

F a Is this affected by the temperature in the room?
Is this affected by the volume of water used?

  b The uninsulated test tubes cooled more quickly. The ones in a cold room cooled more quickly than those in a warm room.

  c The temperature of each test tube every 30 seconds using a data logger or graphical calculator. The temperature in the room.

  d A data logger (or graphical calculator) could be used to collect the data.

  e Three insulated and three uninsulated test tubes would be a good sample size.
To test if the temperature of the room made a difference, three insulated and three uninsulated test tubes would need to be tested at about three different temperatures, for example, in a fridge, at room temperature, and in an oven at about 50° C.

  f Primary.

G a Does this change from year to year?
Does this depend on the weather?

  b Teams in higher divisions score, on average, more points per game than teams in lower divisions.

  or First division soccer teams score, on average, more points per game than second division soccer teams.

  c Data could be collected from the Internet.

  d Once collected from the Internet the data needs to

First Division team	Total points for season	No. of games	Average
⋮	⋮	⋮	⋮
Grand totals			

  be put onto a collection sheet. A possible collection sheet is:
A similar sheet would be needed for second division teams.

  e To get an accurate picture, survey as many teams as possible in each division.
To check if the result is the same each year, this would need to be done for at least the last 5–10 years.

  f Secondary.

Chapter 16 Analysing Data. Drawing and Interpreting Graphs

Page 382 Exercise 1

1 a $1 \cdot 0 \leqslant d < 1 \cdot 2$ and $1 \cdot 2 \leqslant d < 1 \cdot 4$ **b** 0·81 m

2 a 50 minutes **b** 45 minutes

  c So they know roughly when to expect the first runner, perhaps for photos or timing, and how long they will have to wait for the last finisher.
It might also be useful to compare the ranges between years.

3 a Toronto range 27·1 °C, Manchester range 12·7 °C

  b Toronto has a greater range of average temperatures than Manchester. The temperature is less consistent in Toronto.

Review
a 2– and 6– **b** 6 minutes 50 seconds

Page 383 Exercise 2

1 3·49 (2 d.p.)

2 3·37 visitors per patient (2 d.p.)

3 4·65 (2 d.p.)

Review
4·49

Page 384 Exercise 3

1 a 8   b 7·75   c 3·6   d 4·6
  e 28·5   f 17   g 19·5 (1 d.p.)   h 123

2 a £2·95   b £2·45
  c £2·70; Yes, because there are the same numbers of values in each data set.

Review
3 g

Page 385 Discussion

dot 1 There are many possible answers.
dot 2 The range gives a simple measure of how spread out the data is. It tells us how consistent the data is.
dot 3 **Stan**: mean 5·6 (1 d.p.), median 4, mode 4
Bob: mean 4·6 (1 d.p.), median 5, mode 5
Stan is the better player, if we exclude the 9th hole. Bob has the lower mean overall however because the extreme 9th hole value for Stan makes his mean higher. It could be argued either way.
Using the median and mode, Stan is the better player.
Using the mean, Bob is the better player.
It is probably acceptable to use any of these measures but this would depend on what the club wanted in a player.

Page 386 Exercise 4

1 a 32·5 seconds   b 32·5 seconds
  c 34·88 seconds   d 31 seconds

2 a i 1·6 ii 2·8 b i 2 ii 4
  c His right foot, because the mean and mode are both higher for this foot.

3 a £40 394   b £1442·64 (to the nearest penny)

4 a

	Summit	Clogwyn	Llanberis
Mean	56·3	89	82·15
Median	57·5	88	81·5
Range	57	18	11

  b Summit because it has the largest range.

5 a £72 b £67 c The median is reduced by x.

6 70 and 130

7 36 and 46

8 a It was higher than 73%.   b 80%

Review 1
a i 19·5 ii 21   b 21   c No

Review 2
70 and 82

Page 387 Puzzle

1 Possible answers are:
  a 10, 12, 14, 15, 16, 18, 20
  b 10, 11, 12, 14, 16, 17, 20
  c 10, 12, 12, 12, 14, 16, 20
  d 10, 12, 13, 14, 15, 15, 20

2 a 8 b 8
  c 7 if the median and the mode are both 7.
  d 14 if the mean and the median are both 7.

Page 388 Exercise 5

1 a median 36, range 46, mode 37
  b median 4·8 hours, range 7·5 hours, mode 6·9 hours

2 a

```
5 | 0
4 | 0 0 1 1 2 2 3 3 4 6 8
3 | 0 0 1 2 3 3 4 5 5 5 6 6 6 6 6 7 7 8 9 9
2 | 8 9 9
```

  b 20
  c 11
  d median 36, range 22, mode 36
  e median 41, range 28, modes 45 and 36
  f The fathers are generally older than the mothers, shown by the higher median. There is a greater range in the age of the fathers.

Review
median 1·6 cm, range 2·8 cm, modes 1·4 cm and 2·0 cm

Page 389 Exercise 6

1 Possible answers are: Lufta because her mean is slightly higher and her range larger. This means she could score a very high score **or** Sophie because her mean is only slightly lower than Lufta's and her range is smaller. This means her scores are more consistent and reliable.

2 a **Ben**: mean 13, median 12·1, modes 12·0 and 12·1, range 4·8
    Josh: mean 12·4, median 12·4, mode 12·4, range 0·2
  b Possible answers are: Ben because he has a faster median time and faster mode times **or** Josh because he has a faster mean time and a smaller range so his times are more consistent.

3 a **Brand A**: mean = 161·25, median = 183·5, range = 205
    Brand B: mean = 176·35, median = 176, range = 24
    The median tells us that half of the Brand A light bulbs will last longer than 183·5 hours, whereas half of the Brand B light bulbs will last longer than 176 hours. The range of hours is much greater for Brand A, indicating the number of hours lasted is much less consistent. The mean for Brand A is lower than for Brand B. On 'average', Brand B light bulbs last longer.
  b If you chose Brand A your reasons could be: Because the median is higher, which means that over half of the time the light bulb will last longer than Brand B. The range is high, which indicates some light bulbs last a very long time.
    If you chose Brand B your reasons could be: Brand B has a higher mean and a much lower range. This indicates the number of hours lasted is much more consistent. A light bulb is much more likely to last a length of time fairly close to the mean.

4 All answers are rounded to 1 d.p.
London average temperature: mean = 10·9 °C,
median = 10·5 °C, range = 13·8 °C
average rainfall: mean = 49·2 mm,
median = 45·4 mm, range = 36·4 m
Bangkok average temperature: mean = 28·0 °C,
median = 28·3 °C (1 d.p.), range = 4·6 °C
average rainfall: mean = 120·4 mm,
median = 111·8 mm (1 d.p.),
range = 310·5 mm
a The temperatures in London have a lower mean and
median and a much greater range than the
temperatures in Bangkok.
The temperatures in Bangkok have a very small
range, indicating the temperature each month is
fairly consistent.
The rainfall in Bangkok has a very high range, which
indicates it varies a lot.
The mean and median rainfalls in Bangkok are much
higher than in London.
The weather in London is colder and drier than in
Bangkok.
b The mean describes the average temperatures well
because they are fairly evenly spread but is not so
good for the rainfall because the data is not evenly
spread. It is impossible to tell from the means alone
that Bangkok has 6 dry months and 6 wet months.
c Reasons for London could be:
The mean temperature is lower and the range is
greater. I like to experience a range of temperatures.
Reasons for Bangkok could be:
The mean temperature is higher and the range is
lower. This means the temperature is about the
same all the time. I like hot climates.
5 a Teenage
mean 25·3   median 26·5   mode 29   range 40
Adult
mean 14·9   median 14   mode 14   range 21
b Teenagers watch more TV than adults, shown by
the higher mean, median and mode. Adults have
more consistent viewing hours, shown by the
smaller range.

Review
a Mean = 82·4 g, median = 84 g, range = 84 g
b Mean = 82·9 g, median = 87 g, range = 38 g
c The mean mass of both breeds is very similar. The
median mass of breed B is a little higher than breed A.
The range of breed A is much higher than that of breed
B. This indicates that the masses of breed A mice are
much more widely spread.
d It could be argued that breed A or breed B is better.
Possible answers are:
breed A because it has a lower median, indicating that
the mice are smaller on average. The median describes
breed A better because the data is not evenly spread
or breed B is better because it has a smaller range,
indicating that the masses of the mice are more
consistent and clustered around the mean.

Page 392 Exercise 7
1 a

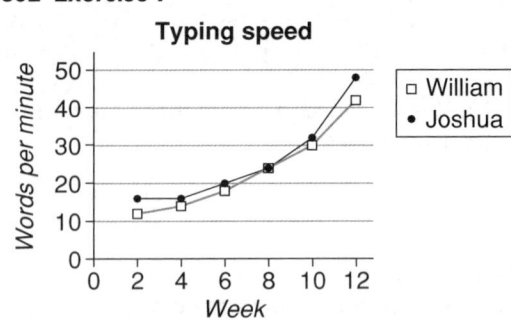

b Joshua made slightly better progress than William.
c Yes, the intermediate points could be used to
estimate typing speed.
2 a

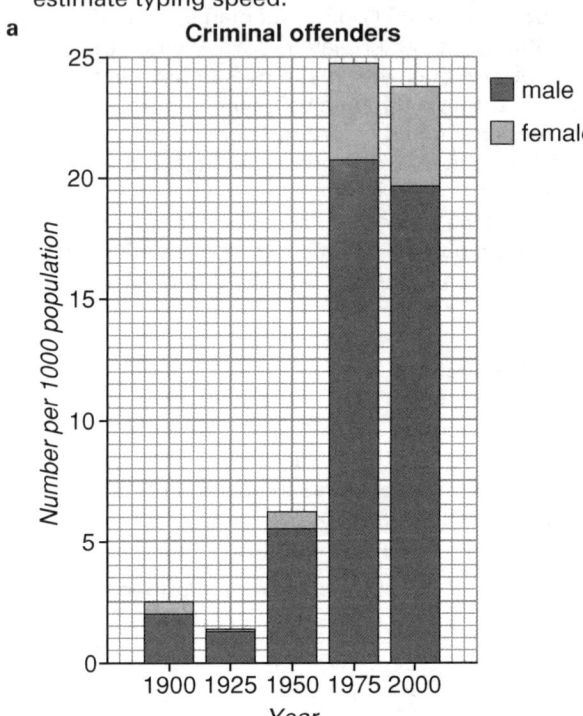

b

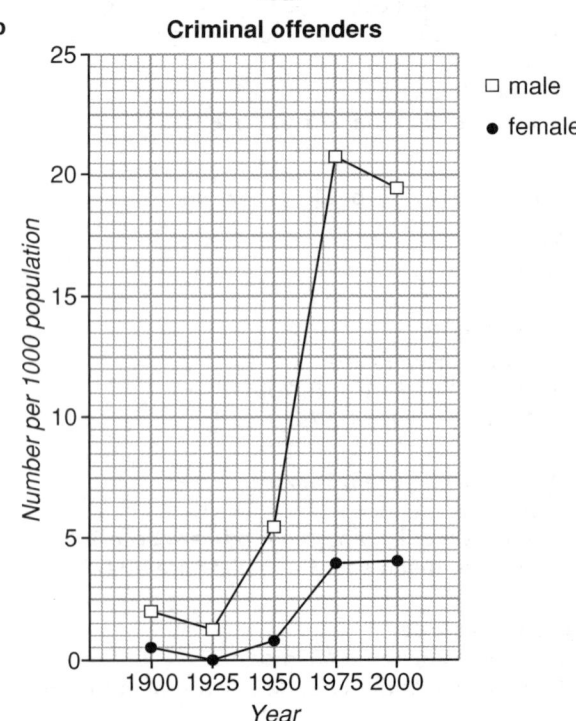

c No, there is 25 years between each point and there
is likely to be a lot of fluctuation in this time.

d You could choose either as long as you give a reason. One possible answer could be:
The line graph because it is easier to see the differences between male and female trends.
Another possible answer is:
The compound bar graph because it is easier to see the trend in overall offending.

e The number of both male and female offenders increased a lot between 1950 and 1975. There has been a slight drop in male offenders and a very slight increase in female offenders since 1975.

3 a

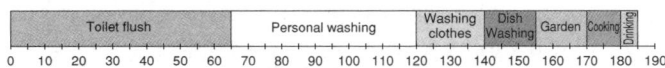

b Personal washing, because Julia's family used much more than the average in this area.

4 a

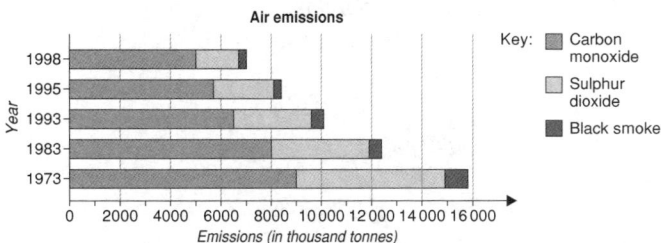

b Emissions of carbon monoxide, sulphur dioxide and black smoke all decreased between 1973 and 1998. A possible reason is that engines and factories emitting these gases have become more efficient and have been designed so they don't emit so much of these gases into the atmosphere. Another possible reason could be that anti-pollution laws have been introduced.

5 a

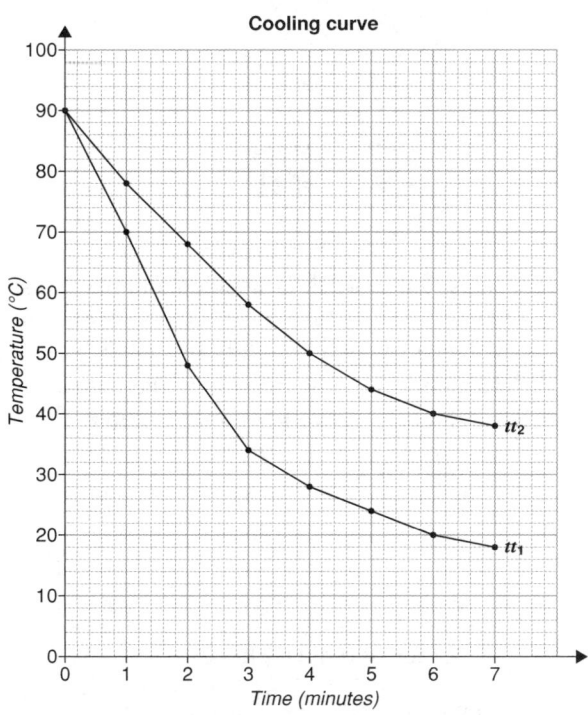

b The intermediate points do have meaning. It is possible to estimate the temperature for times in between the ones plotted.

c One possible answer is: A lone test tube cools at a faster rate than a test tube surrounded by other hot test tubes.

d One possible answer is: Yes, because when animals huddle they lose heat less rapidly.

Review

a

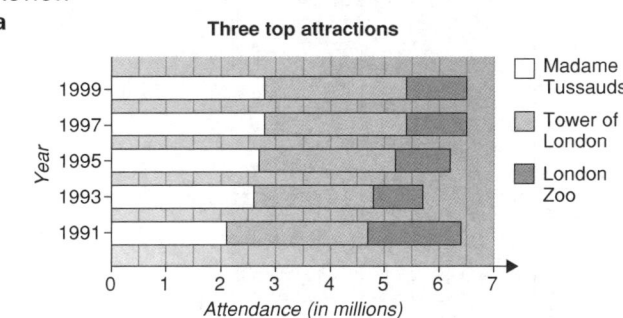

b

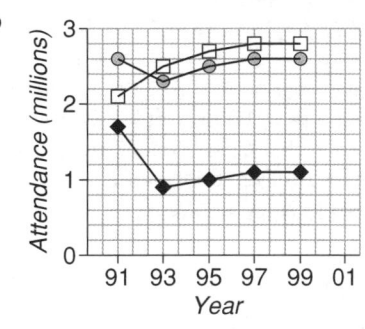

c No, because we cannot estimate the attendance between ends of years.

Page 395 Exercise 8

1

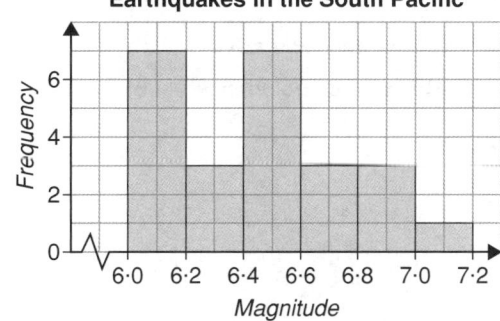

2 a

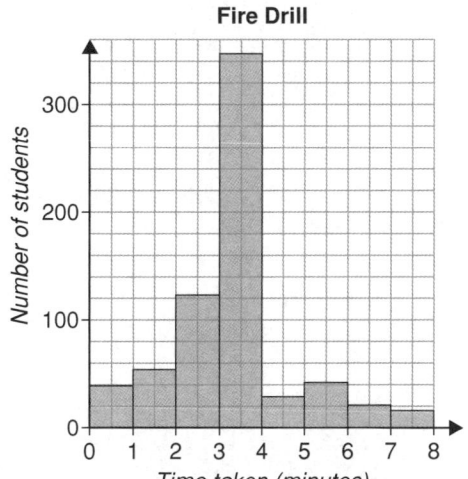

b 216   **c** 79   **d** 671   **e** 4th minute (3–4)

f A possible answer is: Most pupils had evacuated in less than 4 minutes.
There were still 108 pupils not evacuated. In a real fire this could be lethal, especially for the last 16 pupils, who took between 7 and 8 minutes to evacuate.

3 a

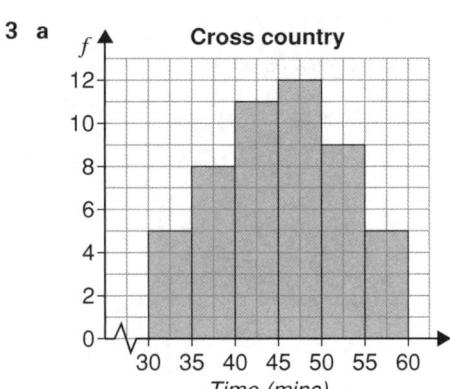

Cross country

b 45 **c** 37

d No, because the class interval 55 ⩽ *t* < 60 includes those that took 55 minutes. We do not know how many of the 5 people in this class took 55 minutes and how many took more than 55 minutes.

4 a

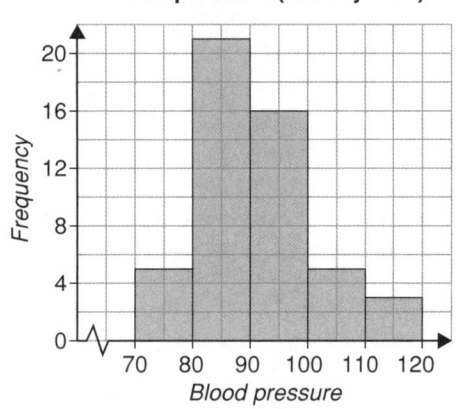

Blood pressure (healthy men)

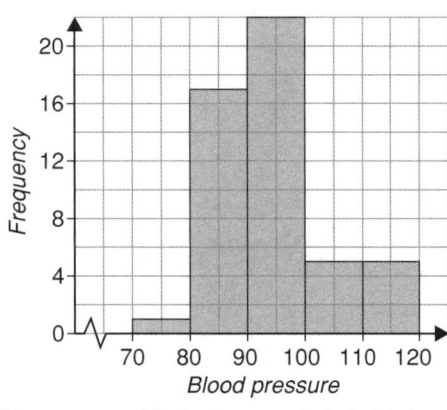

Blood pressure (men with flu)

b More men with flu have a slightly higher blood pressure than healthy men.

Review

a

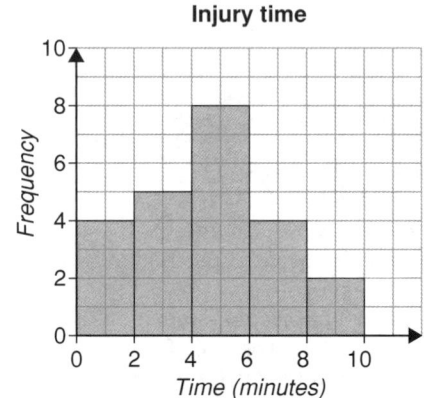

Injury time

b Rory cannot tell this from the graph because in the class interval 4 ⩽ *t* < 6 minutes, we do not know how many were 5 minutes.

c Possible answers are:
The most common injury time in a game is 4 ⩽ *t* < 6 minutes. The range of injury time in the games studied was 10 minutes.

Page 398 Exercise 9

The pie charts are smaller than the ones requested in the exercise.

1 b and **e**

2 a sleeping $\frac{9}{24} \times 360° = 135°$
playing sport $\frac{5}{24} \times 360° = 75°$
doing homework $\frac{2}{24} \times 360° = 30°$
eating $\frac{1}{24} \times 360° = 15°$
friends $\frac{7}{24} \times 360° = 105°$

b

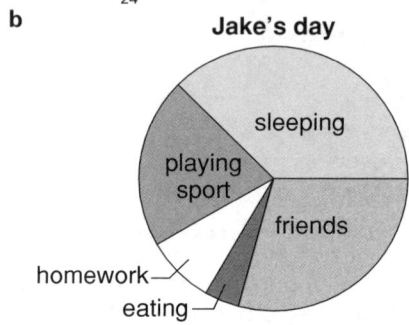

Jake's day

Activity	Sleeping	Playing sport	Doing homework	Eating	Friends
Hours	7	3	3	2	9
Pie chart angle	105°	45°	45°	30°	135°

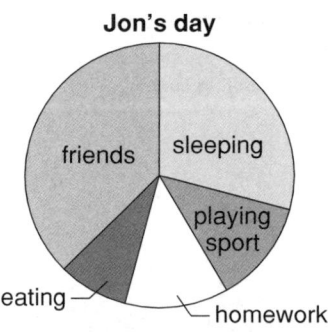

Jon's day

c Jake spent more time sleeping and playing sport than Jon while Jon spent more time eating and with friends. Jake did less homework than Jon.

3 a

Mobile phone colour	Number of mobile phones	Pie chart angle
black	75	225°
yellow	5	15°
grey	25	75°
blue	15	45°
Total	**120**	**360°**

b **Colour of mobile phones**

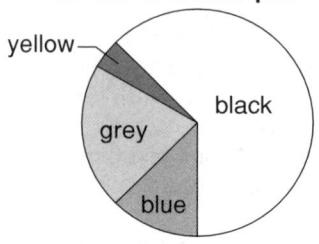

c A possible answer is:
The information will help the shop owner decide how many mobile phones of each colour she should stock.

4 a Angles of sectors are rounded to the nearest degree.
State nursery 5°, state primary 191°, state secondary 139°, non-maintained schools 22°, special schools 4°
The angles add to 361° because more of them round up than down and so the rounding errors do not offset each other.
When this happens we must either use halves of a degree or round one of the angles down, usually the largest angle.
We will round state primary down to 190° in this case.

b

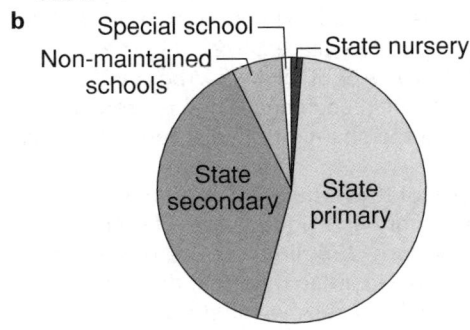

Review 1

a grade A $\frac{24}{120} \times 360° = 72°$
grade B $\frac{58}{120} \times 360° = 174°$
grade C $\frac{33}{120} \times 360° = 99°$
grade D $\frac{5}{120} \times 360° = 15°$

b

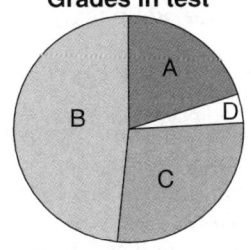

Grades in test

Review 2

a **Age of Beatle's fans**

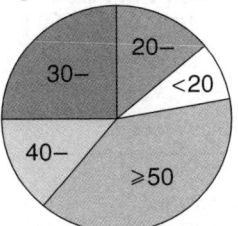

Angles are:

<20	30°
20–	50°
30–	90°
40–	50°
≥50	140°

b

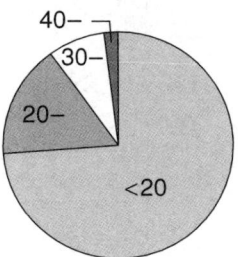

Angles are:

<20	266°
20–	58°
30–	29°
40–	7°
≥50	0°

c The Beatles fans are much older in general. Most are 40 and over. Most fans of the other group are under 20.

e It is useful to know the ages of fans so that food, seating, security and items for sale are appropriate for the age group likely to attend the concerts.

Page 400 Exercise 10

1 a

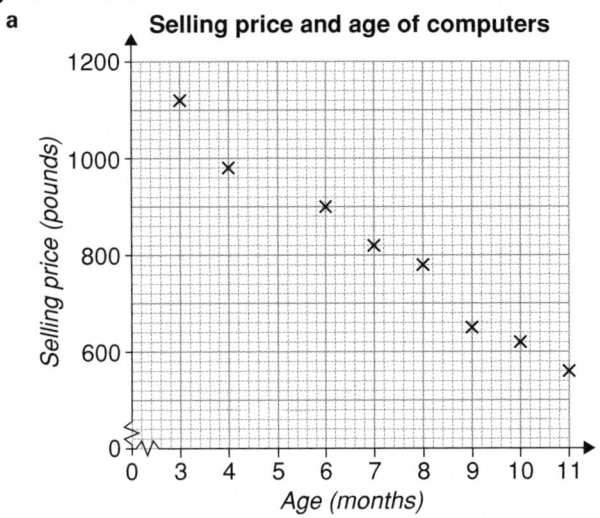

b It decreases.

2 a

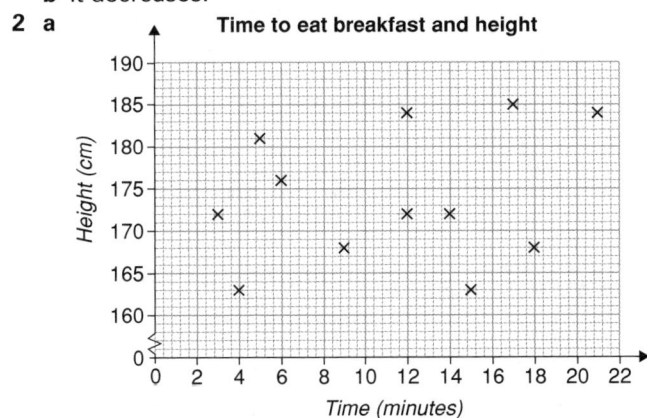

b No

3 a

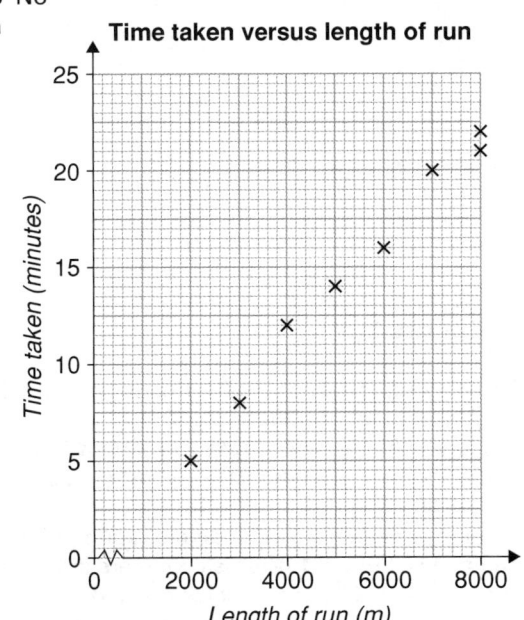

b It increases.

4 a

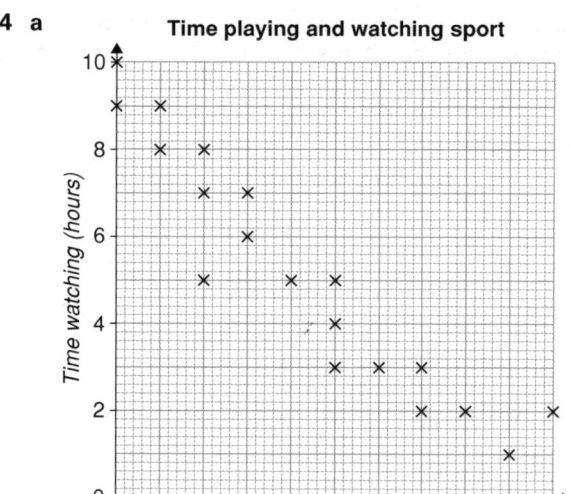

Time playing and watching sport

b Yes

5 a

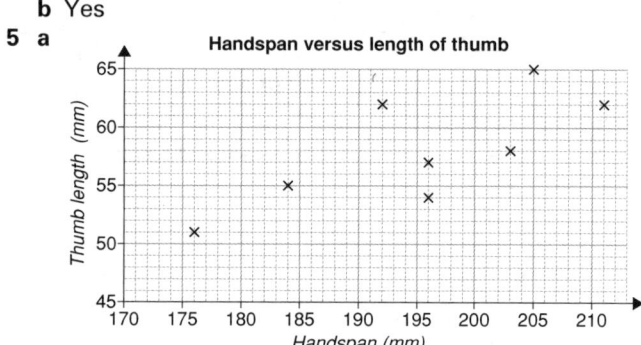

Handspan versus length of thumb

b Yes

Review

a

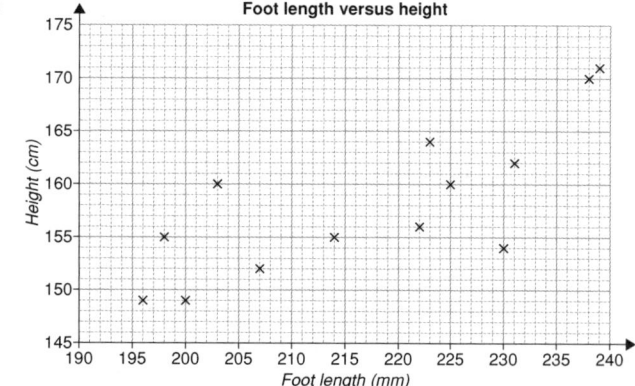

Foot length versus height

b Yes

Page 403 Exercise 11

1 a £2

  b Yes. Although most of Julia's friends got from £2 to £4, only one got more than £2·50.

2 a i About 30%   **ii** About 20%   **iii** About 20%

  b i About 40%   **ii** About 15%   **iii** About 20%

  c About 400

  d B. There is a greater percentage of younger females than males. 50% of the female workers are under 40 whereas only 40% of the males are under 40. Since there is about the same number of males as females, the number of younger (under 40) females is greater.

3 September would be a good choice because the average temperature is only a little below that of August but it is much less windy.

4 a About £350   **b** About 40 seconds

5 a The sales of CDs has increased greatly since 1983. The sales of cassettes rose between 1983 and 1988 and since then they have fallen.
The sales of LPs fell slowly overall from 1973 to when they fell sharply in 1988. From 1993 sales have been almost nil.
The sales of singles has risen and fallen within a fairly small range from 1973 to 2002

  b Discuss the answer to this question. It will change as time goes on.

6 The two biggest changes are coal and gas.
Coal usage has decreased by 20% and gas has increased by $23\frac{1}{2}$%.
Oil has dropped from 4% to $1\frac{1}{2}$% and imports have dropped by 1%. Other fuels have increased by 1%.
Hydro and Nuclear have both dropped by $\frac{1}{2}$%.

7 We can only get an approximate answer for **a**, **b** and **c**. They are given to the nearest 5%.

  a About 10%   **b** About 15%   **c** About 80%

  d The population in each age range is predicted to increase but the shape of the distribution remains the same.

  e The shapes of the population pyramids for the United Kingdom and Uganda are different. In the United Kingdom pyramid there is a bulge in the middle. In the Ugandan pyramid, the number of people in each successive older age group decreases.
From 2000 to 2025, in the United Kingdom the number of people in every age range up to 54 decreases. In every age range over 54 the number of people increases. This indicates an ageing population boom.
In Uganda from 2000 to 2025, the population is predicted to increase in every age group fairly evenly.
There is a much lower life expectancy in Uganda than in the United Kingdom.

8 Smokers are more likely to believe the more positive statements about smoking such as 'smoking can help you calm down' and 'smoking can put you in a better mood'. Non-smokers are more likely to believe negative statements about smoking such as 'smokers are more boring than people who don't smoke'.

9

Number of £	Approximate number of €	Approximate number of C$
0	0	0
200	320	490
400	640	990

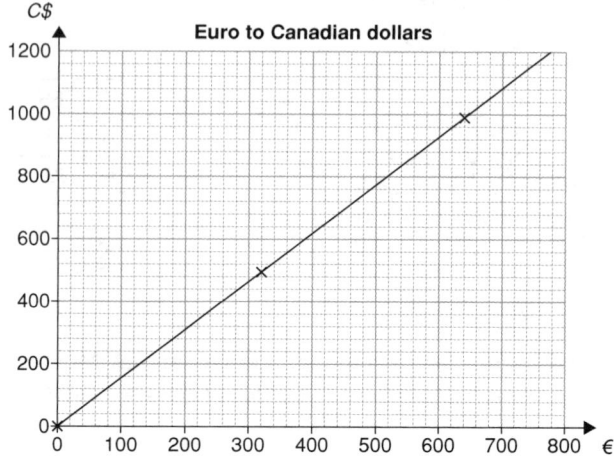

Euro to Canadian dollars

Review 1

a The percentage of households with only one car and two or more cars has increased overall from 1961 to 1971. From 1971 to 2001 the percentage of households with two or more cars increased but the percentage with one car only stayed about the same. This indicates that some of the households who had only one car have now got more cars and those who didn't have a car at all, now have one.

b Possible answers are: petrol prices and availability, other fuel developments, price of cars, British economy, other transport developments.

Review 2

a 30–34   **b** About 22%

c In each age range from 60 onwards, there are more females than males. Below this age, the numbers are approximately the same.

Review 3

It is most likely to measure 125° because 14 out of 20 pupils measured it as 124°, 125° or 126° and the majority of these 14 measured it as 125°. The 6 who measured it as 55° probably measured the wrong angle.

17 Probability

Page 414 Exercise 1

1 All

2 Red, because there were more red hats sold than any other colour.

3 Card B, because this has the greatest proportion of 'win' squares.

4 Pack 1, because this has the greater proportion of chocolate-coated bars. $\frac{8}{20} > \frac{6}{20}$

5 Box 1, as this has a greater proportion of oranges. $\frac{12}{30} > \frac{10}{26}$ [0·4 > 0·38 (2 d.p.)]

6 a Other **b** Other **c** Europe **d** 12

Review 1

Land, because the area of land is greater than the area of water.

Review 2

a Celia, because she has a greater proportion of red balloons. $\frac{10}{22} > \frac{12}{36}$.

b Samantha, because she has the greatest proportion of red balloons. $\frac{6}{11} > \frac{10}{22} > \frac{12}{36}$. (0·55 > 0·45 > 0·33)

Page 416 Exercise 2

1 $\frac{3}{5}$

2 0·7

3 35%

4 a $\frac{5}{18}$ **b** $\frac{13}{18}$

5 a $\frac{7}{15}$ **b** $\frac{8}{15}$ **c** $\frac{6}{15}$ or $\frac{2}{5}$ **d** $\frac{9}{15}$ or $\frac{3}{5}$ **e** $\frac{13}{15}$ **f** $\frac{2}{15}$

6 a $\frac{1}{2}$ **b** $\frac{1}{4}$ **c** $\frac{1}{13}$ **d** $\frac{12}{13}$ **e** $\frac{10}{13}$ **f** $\frac{1}{2}$ **g** $\frac{2}{13}$
   h $\frac{10}{52}$ or $\frac{5}{26}$ **i** $\frac{42}{52}$ or $\frac{21}{26}$

7

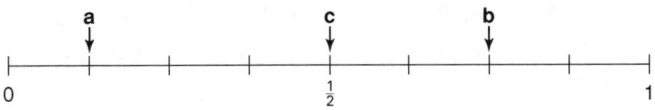

8 a $\frac{1}{200}$
   b They both have the same chance because they both have 4 tickets.
   c 10
   d $\frac{3}{200}$

9 The outcomes are not equally likely because there are not the same numbers of pupils, staff and parents.

10 a i 6 **ii** 4 **iii** 0 **iv** 2 **b i** 1 **ii** $\frac{5}{6}$ **iii** $\frac{1}{2}$

11 a i 6 **ii** 8 **iii** 2 **b** $\frac{1}{8}$

12 a 1 **b** 2

Review 1

88%

Review 2

a $\frac{2}{5}$ **b** $\frac{3}{5}$ **c** $\frac{7}{25}$ **d** $\frac{17}{25}$

Review 3

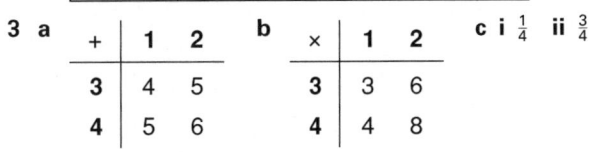

Page 418 Discussion

● It is not a fair game if each group gets 1 point.
  It is a fair game if group A gets 2 points and group B gets 3 points.
  This is because the probability of it stopping on an odd number is $\frac{3}{5}$ and on an even number is $\frac{2}{5}$.
  $2 \times \frac{3}{5} = 3 \times \frac{2}{5}$.

● A biased spinner is one that is not equally likely to stop on each of the sections.
  There are many possible examples. One is:

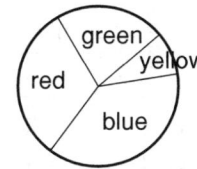

Page 418 Discussion

Craig is right because the outcomes when a coin is tossed twice are HH, HT, TH, TT. Each outcome is equally likely so there is a one in four chance of getting two heads.

Page 419 Exercise 3

1

1st	Jan	Jan	Laura	Laura	Caryl	Caryl
2nd	Laura	Caryl	Jan	Caryl	Jan	Laura

2 a

Josie	vanilla	vanilla	vanilla	chocolate	chocolate
Charlotte	chocolate	raspberry	vanilla	chocolate	raspberry

Josie	chocolate	raspberry	raspberry	raspberry
Charlotte	vanilla	chocolate	raspberry	vanilla

b

Morning tea	banana	banana	banana	apple	apple
Lunch	banana	apple	orange	banana	apple

Morning tea	apple	orange	orange	orange
Lunch	orange	banana	apple	orange

c boy girl, boy boy, girl girl, girl boy

d

round 1	Jack	Jack	Jack	Queen	Queen
round 2	King	Queen	Jack	King	Queen

round 1	Queen	King	King	King
round 2	Jack	King	Queen	Jack

3 a

+	1	2
3	4	5
4	5	6

b

×	1	2
3	3	6
4	4	8

c i $\frac{1}{4}$ **ii** $\frac{3}{4}$

4 a $\frac{1}{12}$ **b** $\frac{1}{4}$ **c** $\frac{1}{3}$ **d** $\frac{1}{4}$ **e** $\frac{1}{2}$ **f** 0

5 a

DICE 2

+	1	2	3	4	5	6
1	2	3	4	5	6	7
2	3	4	5	6	7	8
3	4	5	6	7	8	9
4	5	6	7	8	9	10
5	6	7	8	9	10	11
6	7	8	9	10	11	12

(left label: D I C E 1)

b i $\frac{1}{36}$ **ii** $\frac{4}{36}$ or $\frac{1}{9}$ **iii** $\frac{4}{36}$ or $\frac{1}{9}$ **iv** $\frac{1}{36}$
v 0 **vi** $\frac{15}{36}$ or $\frac{5}{12}$ **vii** $\frac{21}{36}$ or $\frac{7}{12}$ **viii** $\frac{15}{36}$ or $\frac{5}{12}$
ix $\frac{5}{36}$ **x** $\frac{15}{36}$ or $\frac{5}{12}$ **xi** $\frac{21}{36}$ or $\frac{7}{12}$ **xii** $\frac{10}{36}$ or $\frac{5}{18}$

c 7 **d** 2 or 12 **e** $\frac{6}{36}$ or $\frac{1}{6}$ **f** $\frac{6}{36}$ or $\frac{1}{6}$

Review 1
a $\frac{1}{4}$ **b** $\frac{1}{2}$

Review 2
a

	2nd spin		
	2	**4**	**7**
2	4	6	9
4	6	8	11
7	9	11	14

(left label: 1st spin)

b i $\frac{2}{9}$ **ii** $\frac{1}{9}$ **iii** 0 **iv** $\frac{6}{9}$ or $\frac{2}{3}$ **v** $\frac{6}{9}$ or $\frac{2}{3}$
vi $\frac{3}{9}$ or $\frac{1}{3}$ **vii** $\frac{5}{9}$ **viii** $\frac{2}{9}$ **ix** $\frac{3}{9}$ or $\frac{1}{3}$ **x** $\frac{5}{9}$

Page 421 Discussion
● It is very unlikely the outcomes will be exactly the same. It is most likely they will be different each time.
● Melanie, because she tossed the dice more times.

Page 421 Exercise 4
1 $\frac{33}{40}$
2 $\frac{100}{500}$ or $\frac{1}{5}$
3 $\frac{150}{200}$ or $\frac{3}{4}$
4 $\frac{17}{20}$
5 $\frac{29}{50}$
6 a i $\frac{19}{50}$ **ii** $\frac{16}{50}$ or $\frac{8}{25}$ **iii** $\frac{6}{50}$ or $\frac{3}{25}$
   b i $\frac{40}{100}$ or $\frac{2}{5}$ **ii** $\frac{33}{100}$ **iii** $\frac{11}{100}$
   c b, because the number of people surveyed is greater
7 a $\frac{54}{100}$ or $\frac{27}{50}$ or about $\frac{1}{2}$ **b** $\frac{5}{100}$ or $\frac{1}{20}$
8 a $\frac{7}{50}$
   b $\frac{43}{50}$ because he has to score double forty (80) or less than double forty (80) so the probabilities must add to 1.

Review 1
$\frac{17}{35}$

Review 2
i a $\frac{27}{50}$ **b** $\frac{1}{50}$
ii Yes. The accuracy increases if the number surveyed increases.

New National Framework

MATHEMATICS 8

Teacher Resource Pack

New National Framework Mathematics Teacher Resource Pack 8 Core provides you with a comprehensive range of resources to support the 8 Core pupil book.

☑ **Worksheets** providing comprehensive additional practice and consolidation to support work in the four support chapter sections.

☑ **Chapter Reviews** providing a set of photocopiable write-on question sheets for additional practice, extension and consolidation work. They can also be used for homework.

☑ **ICT Support** with full, comprehensive teachers' notes on the use of spreadsheets, dynamic geometry packages (Geometer's Sketchpad) LOGO (MSW LOGO) and a graphing package (Omnigraph).

☑ **Resource sheets** of pupil book diagrams, graphs and tables that can be photocopied.

☑ **Graphic calculator support** for use with the three most popular graphic calculators from Casio, Sharp and Texas. Worksheets are also provided.

☑ **Pupil Book answers**.

Comprehensive web support is available at www.nelsonthornes.com. This includes full mappings to the framework for teaching mathematics, the sample medium term plans and a range of additional resources.

Each Teacher Resource Pack is supported by a Teacher Planning Pack. Assessment Resource Banks and Starter Support Packs are also available for years 7, 8 and 9.

nelson thornes

www.nelsonthornes.com

ISBN 0-7487-7886-1

9 780748 778867